SI PREFIXES

Fraction	Prefix	Symbol	Multiple	Prefix	Symbol
10^{-1}	deci	d	10	deka	da
10^{-2}	centi	c	10^2	hecto	h
10^{-3}	milli	m	10^3	kilo	k
10^{-6}	micro	μ	10^6	mega	M
10^{-9}	nano	n	10^9	giga	G
10^{-12}	pico	p	10^{12}	tera	T
10^{-15}	femto	f	10^{15}	peta	P
10^{-18}	atto	a	10^{18}	exa	E

GREEK ALPHABET

Alpha	A	α	Iota	I	ι	Rho	P	ρ
Beta	B	β	Kappa	K	κ	Sigma	Σ	σ
Gamma	Γ	γ	Lambda	Λ	λ	Tau	T	τ
Delta	Δ	δ	Mu	M	μ	Upsilon	Υ	υ
Epsilon	E	ϵ, ε	Nu	N	ν	Phi	Φ	ϕ
Zeta	Z	ζ	Xi	Ξ	ξ	Chi	X	χ
Eta	H	η	Omicron	O	o	Psi	Ψ	ψ
Theta	Θ	θ	Pi	$\prod$	π	Omega	Ω	ω

SOME COMMONLY USED NON-SI UNITS

Unit	Quantity	Symbol	SI value
Angstrom	length	Å	10^{-10} m = 100 pm
Calorie	energy	cal	4.184 J (defined)
Debye	dipole moment	D	3.3356×10^{-30} C·m
			0.39345 au
Gauss	magnetic field strength	G	10^{-4} T

E_h	cm^{-1}	Hz
$2.293\ 710 \times 10^{17}$	$5.034\ 11 \times 10^{22}$	$1.509\ 189 \times 10^{33}$
$3.808\ 798 \times 10^{-4}$	83.5935	$2.506\ 069 \times 10^{12}$
$3.674\ 931 \times 10^{-2}$	8065.54	$2.417\ 988 \times 10^{14}$
1	$2.194\ 7463 \times 10^5$	$6.579\ 684 \times 10^{15}$
$4.556\ 355 \times 10^{-6}$	1	$2.997\ 925 \times 10^{10}$
$1.519\ 830 \times 10^{-16}$	$3.335\ 64 \times 10^{-11}$	1

Problems and Solutions

to accompany

DONALD A. McQUARRIE'S

QUANTUM CHEMISTRY
Second Edition

Helen O. Leung
AMHERST COLLEGE

Mark D. Marshall

AMHERST COLLEGE

University Science Books
Sausalito, California

University Science Books
www.uscibooks.com

Text Design: Paul Anagnostopoulos
Cover Design: George Kelvin
Composition: Windfall Software, using ZzTEX
Printer & Binder: Victor Graphics

This book is printed on acid-free paper.

ISBN-13: 978-1-891389-52-8
ISBN-10: 1-891389-52-1

Library of Congress Control Number 2007931880

Printed in the United States of America
10 9 8 7 6 5 4 3 2 1

Contents

Preface   vii

CHAPTER 1   **The Dawn of the Quantum Theory**   1

MATHCHAPTER A   **Complex Numbers**   31

CHAPTER 2   **The Classical Wave Equation**   43

MATHCHAPTER B   **Probability and Statistics**   87

CHAPTER 3   **The Schrödinger Equation and a Particle in a Box**   99

MATHCHAPTER C   **Vectors**   133

CHAPTER 4   **The Postulates and General Principles of Quantum Mechanics**   143

MATHCHAPTER D   **Series and Limits**   211

CHAPTER 5   **The Harmonic Oscillator and Vibrational Spectroscopy**   225

MATHCHAPTER E   **Spherical Coordinates**   265

CHAPTER 6   **The Rigid Rotator and Rotational Spectroscopy**   279

MATHCHAPTER F   **Determinants**   331

CHAPTER 7   **The Hydrogen Atom**   337

MATHCHAPTER G   **Matrices**   373

CHAPTER 8   **Approximation Methods**   387

MATHCHAPTER H   **Matrix Eigenvalue Problems**   459

CHAPTER 9   **Many-Electron Atoms**   469

CHAPTER 10   **The Chemical Bond: One- and Two-Electron Molecules**   513

CHAPTER 11   **Qualitative Theory of Chemical Bonding**   567

CHAPTER 12   **The Hartree–Fock–Roothaan Method**   613

Preface

For as long as we have been teaching quantum chemistry, there has been Donald McQuarrie's *Quantum Chemistry,* and we have both used the first edition in our classes for many years. This solutions manual has been prepared for the second edition of *Quantum Chemistry,* and as Professor McQuarrie notes in his preface to the new edition, the ideas of basic quantum mechanics are timeless, but the practice of quantum mechanics has changed dramatically. The problems and solutions contained in this manual reflect both principles. Some of them could have been written for the very first textbooks of quantum chemistry, such as Pauling and Wilson's 1935 text, yet are still as relevant to understanding and mastering the subject today as they were decades ago. Other problems would have been beyond the reach of undergraduates just a few years ago, but are now routinely done using widely available software packages or the internet.

Working through problems of both types is essential to learning the ideas as well as the practice of quantum chemistry, and we hope that this solutions manual will encourage students to do many problems—perhaps even more than they are assigned! Our own understanding of quantum mechanics continues to be enhanced as we work through problems whether they are in a textbook or come from our research laboratories. Several times during this project Professor McQuarrie remarked to us, "I hope that you guys like doing problems. I happen to love it." Perhaps it is unrealistic to expect students to share this sentiment, but we hope you will at least share the satisfaction of seeing how well the basic principles of quantum mechanics work, and how they can be applied to explain so much of chemistry.

In working numerical problems, we have used the values of fundamental constants and conversion factors that appear inside the front cover of the textbook to the full precision given there. Extra significant figures are retained in intermediate calculations with the final answer rounded to the appropriate precision. Since this is not an introductory textbook, we occasionally depart from strict adherence to this policy if doing so helps convey an idea the problem is addressing. Likewise, we did not think it necessary to write out explicitly the steps in every unit conversion or detail every algebraic manipulation. We would echo Professor McQuarrie's suggestions to learn *Mathematica, MathCad,* or some similar program. Many problems that we have worked out in detail could be done straightforwardly and much faster with one or the other, and in fact, we used both to check our work.

We are grateful for the opportunity to work with Donald McQuarrie in preparing this solutions manual. His insight, advice, and enthusiasm have been most valuable to us. Many others helped us bring this book to print, and we wish to acknowledge their contributions. Jane Ellis of University Science Books organized all the details, smoothed out wrinkles as they occurred, and would keep us on track if we began to drift. Carole McQuarrie provided us with her LaTeX source for the problems. Mervin Hanson took our graphics files and transformed them into illustrations suitable for publication. Paul Anagnostopoulos and Jacqui Scarlott at Windfall Software endured our bad habits in LaTeX, merged the questions from the textbook into our source for the solutions, and composed the book. Finally, they say you can best judge executives by the quality of the team they assemble, and by this measure Bruce and Kathy Armbruster are outstanding. They are supportive and encouraging, and it is a true pleasure having them as our publishers.

And yes, Don, we love working problems, too.

Helen O. Leung
Mark D. Marshall
Amherst, Massachusetts

The Dawn of the Quantum Theory

PROBLEMS AND SOLUTIONS

1–1. Radiation in the ultraviolet region of the electromagnetic spectrum is usually described in terms of wavelength, λ, and is given in nanometers (10^{-9} m). Calculate the values of ν, $\tilde{\nu}$, and E for ultraviolet radiation with $\lambda = 200$ nm and compare your results with those in Figure 1.19.

$$\nu = \frac{c}{\lambda} = \frac{2.997\,924\,58 \times 10^8\,\text{m·s}^{-1}}{200. \times 10^{-9}\,\text{m}} = 1.50 \times 10^{15}\,\text{Hz}$$

$$\tilde{\nu} = \frac{1}{\lambda} = \frac{1}{200. \times 10^{-9}\,\text{m}} = 5.00 \times 10^6\,\text{m}^{-1} = 5.00 \times 10^4\,\text{cm}^{-1}$$

$$E = h\nu = \frac{hc}{\lambda} = \frac{\left(6.626\,0755 \times 10^{-34}\,\text{J·s}\right)\left(2.997\,924\,58 \times 10^8\,\text{m·s}^{-1}\right)}{200. \times 10^{-9}\,\text{m}}$$

$$= 1.50 \times 10^{15}\,\text{Hz} = 9.93 \times 10^{-19}\,\text{J}$$

The results are as expected from an inspection of Figure 1.19.

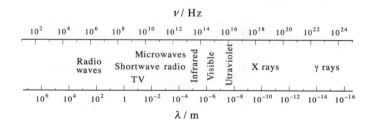

FIGURE 1.19
The regions of electromagnetic radiation.

1–2. Radiation in the infrared region is often expressed in terms of wave numbers, $\tilde{\nu} = 1/\lambda$. A typical value of $\tilde{\nu}$ in this region is 10^3 cm^{-1}. Calculate the values of ν, λ, and E for radiation with $\tilde{\nu} = 10^3$ cm^{-1} and compare your results with those in Figure 1.19.

$$v = \frac{c}{\lambda} = c\tilde{v} = \left(2.997\,924\,58 \times 10^8 \text{ m·s}^{-1}\right) \left(10^3 \text{ cm}^{-1}\right) \left(\frac{100 \text{ cm}}{1 \text{ m}}\right) = 3 \times 10^{13} \text{ Hz}$$

$$\lambda = \frac{1}{\tilde{v}} = 10^{-3} \text{ cm} = 10^{-5} \text{ m}$$

$$E = hv = hc\tilde{v} = 2 \times 10^{-20} \text{ J}$$

The results are as expected from an inspection of Figure 1.19.

1–3. Past the infrared region, in the direction of lower energies, is the microwave region. In this region, radiation is usually characterized by its frequency, v, expressed in units of megahertz (MHz), where the unit hertz (Hz) is a cycle per second. A typical microwave frequency is 2.0×10^4 MHz. Calculate the values of $\tilde{v}$, λ, and E for this radiation and compare your results with those in Figure 1.19.

$$\tilde{v} = \frac{1}{\lambda} = \frac{v}{c} = \left(\frac{2.0 \times 10^4 \text{ MHz}}{2.997\,924\,58 \times 10^8 \text{ m·s}^{-1}}\right) \left(\frac{10^6 \text{ Hz}}{1 \text{ MHz}}\right) = 67 \text{ m}^{-1} = 0.67 \text{ cm}^{-1}$$

$$\lambda = \frac{c}{v} = \frac{2.997\,924\,58 \times 10^8 \text{ m·s}^{-1}}{2.0 \times 10^{10} \text{ Hz}} = 1.5 \times 10^{-2} \text{ m} = 1.5 \text{ cm}$$

$$E = hv = \left(6.626\,0755 \times 10^{-34} \text{ J·s}\right) \left(2.0 \times 10^{10} \text{ Hz}\right) = 1.3 \times 10^{-23} \text{ J}$$

1–4. Planck's principal assumption was that the energies of the electronic oscillators can have only the values $E = nhv$ and that $\Delta E = hv$. As $v \to 0$, then $\Delta E \to 0$ and E is essentially continuous. Thus, we should expect the nonclassical Planck distribution to go over to the classical Rayleigh–Jeans distribution at low frequencies, where $\Delta E \to 0$. Show that Equation 1.2 reduces to Equation 1.1 as $v \to 0$. (Recall that $e^x = 1 + x + (x^2/2!) + \cdots$, or, in other words, that $e^x \approx 1 + x$ when x is small.)

As $v \to 0$, $\frac{hv}{k_B T}$ becomes small, and $e^{hv/k_B T} \approx 1 + \frac{hv}{k_B T}$. Using this approximation in Equation 1.2 gives

$$\rho_v(T)\,dv = \frac{8\pi h}{c^3} \frac{v^3\,dv}{e^{hv/k_B T} - 1}$$

$$\approx \frac{8\pi h}{c^3} \frac{v^3\,dv}{1 + \frac{hv}{k_B T} - 1}$$

$$= \frac{8\pi k_B T}{c^3} v^2\,dv$$

which is equivalent to Equation 1.1

1–5. Before Planck's theoretical work on blackbody radiation, Wien showed empirically that (Equation 1.4)

$$\lambda_{\max} T = 2.90 \times 10^{-3} \, \text{m} \cdot \text{K}$$

where $\lambda_{\max}$ is the wavelength at which the blackbody spectrum has its maximum value at a temperature T. This expression is called the Wien displacement law; derive it from Planck's theoretical expression for the blackbody distribution by differentiating Equation 1.3 with respect to λ. *Hint:* Set $hc/\lambda_{\max} k_B T = x$ and derive the intermediate result $e^{-x} + (x/5) = 1$. This equation cannot be solved analytically but must be solved numerically. Solve it by iteration on a hand calculator, and show that $x = 4.965$ is the solution.

The maximum in the wavelength version of the Planck radiation law is found where the derivative with respect to λ is equal to zero. Start with Equation 1.3 and apply both the product and quotient rules for derivatives.

$$\frac{d\rho_\lambda(T)}{d\lambda} = \frac{d}{d\lambda}\left(\frac{8\pi hc}{\lambda^5} \frac{1}{e^{hc/\lambda k_B T} - 1}\right)$$

$$= -5\frac{8\pi hc}{\lambda^6}\frac{1}{e^{hc/\lambda k_B T} - 1} + \frac{8\pi hc}{\lambda^5}\frac{e^{hc/\lambda k_B T}}{\left(e^{hc/\lambda k_B T} - 1\right)^2}\frac{hc}{\lambda^2 k_B T}$$

At $\lambda = \lambda_{\max}$, this equals zero.

$$-5\frac{8\pi hc}{\lambda_{\max}^6}\frac{1}{e^{hc/\lambda_{\max} k_B T} - 1} + \frac{8\pi hc}{\lambda_{\max}^5}\frac{e^{hc/\lambda_{\max} k_B T}}{\left(e^{hc/\lambda_{\max} k_B T} - 1\right)^2}\frac{hc}{\lambda_{\max}^2 k_B T} = 0$$

$$-e^{hc/\lambda_{\max} k_B T} + 1 + e^{hc/\lambda_{\max} k_B T}\frac{hc}{5\lambda_{\max} k_B T} = 0$$

$$-1 + e^{-hc/\lambda_{\max} k_B T} + \frac{hc}{5\lambda_{\max} k_B T} = 0$$

Now make use of the suggested substitution, $hc/\left(\lambda_{\max} k_B T\right) = x$, to get $e^{-x} + \frac{x}{5} = 1$, or $x = 5 - 5e^{-x}$. Using a guess of $x = 5$ gives $x = 5 - 5e^{-5} = 4.966$, and the process converges in the fourth iteration at $x = 4.965$. Recalling the definition of x,

$$\frac{hc}{\lambda_{\max} k_B T} = 4.965$$

$$\lambda_{\max} T = \frac{hc}{4.965 k_B}$$

$$= \frac{\left(6.626\,0755 \times 10^{-34} \, \text{J} \cdot \text{s}\right)\left(2.997\,924\,58 \times 10^8 \, \text{m} \cdot \text{s}^{-1}\right)}{4.965\left(1.380\,658 \times 10^{-23} \, \text{J} \cdot \text{K}^{-1}\right)}$$

$$= 2.90 \times 10^{-3} \, \text{m} \cdot \text{K}$$

which is the Wien displacement law.

1–6. At what wavelength does the maximum in the energy-density distribution function for a blackbody occur if (a) $T = 300$ K, (b) $T = 3000$ K, and (c) $T = 10\,000$ K?

Use the Wien displacement law for the three different temperatures.

(a)

$$\lambda_{max} = \frac{2.90 \times 10^{-3}\,\text{m}\cdot\text{K}}{T} = \frac{2.90 \times 10^{-3}\,\text{m}\cdot\text{K}}{300\,\text{K}} = 9.67 \times 10^{-6}\,\text{m}$$

(b)

$$\lambda_{max} = \frac{2.90 \times 10^{-3}\,\text{m}\cdot\text{K}}{3000\,\text{K}} = 9.67 \times 10^{-7}\,\text{m}$$

(c)

$$\lambda_{max} = \frac{2.90 \times 10^{-3}\,\text{m}\cdot\text{K}}{10\,000\,\text{K}} = 2.90 \times 10^{-7}\,\text{m}$$

1–7. Sirius, one of the hottest known stars, has approximately a blackbody spectrum with $\lambda_{max} = 260$ nm. Estimate the surface temperature of Sirius.

The Wien displacement law may be used to estimate the surface temperature of the star.

$$T = \frac{2.90 \times 10^{-3}\,\text{m}\cdot\text{K}}{\lambda_{max}} = \frac{2.90 \times 10^{-3}\,\text{m}\cdot\text{K}}{260 \times 10^{-9}\,\text{m}} = 11\,000\,\text{K}$$

1–8. The temperature of the fireball in a thermonuclear explosion can reach temperatures of approximately 10^7 K. What value of λ_{max} does this correspond to? In what region of the spectrum is this wavelength found (cf. Figure 1.19)?

Treating the fireball as the blackbody, the Wien displacement law gives

$$\lambda_{max} = \frac{2.90 \times 10^{-3}\,\text{m}\cdot\text{K}}{T} = \frac{2.90 \times 10^{-3}\,\text{m}\cdot\text{K}}{10^7\,\text{K}} = 3 \times 10^{-10}\,\text{m}$$

which Figure 1.19 indicates is in the X-ray region of the electromagnetic spectrum.

1–9. We can use the Planck distribution to derive the *Stefan–Boltzmann law*, which gives the total energy density emitted by a blackbody as a function of temperature. Derive the Stefan–Boltzmann law by integrating the Planck distribution over all frequencies. *Hint:* You'll need to use the integral $\int_0^{\infty} dx\, x^3/(e^x - 1) = \pi^4/15$.

The total energy density emitted by a blackbody is found by integrating Equation 1.2 over all frequencies, making the substitutions $x = h\nu/k_B T$ and $dx = h\,d\nu/k_B T$.

$$E_V = \int_0^\infty \rho_\nu(T)\,d\nu$$

$$= \frac{8\pi h}{c^3} \int_0^\infty \frac{\nu^3\,d\nu}{e^{h\nu/k_B T} - 1}$$

$$= \frac{8\pi k_B^4 T^4}{h^3 c^3} \int_0^\infty \frac{x^3\,dx}{e^x - 1}$$

$$= \frac{8\pi k_B^4 T^4}{h^3 c^3} \frac{\pi^4}{15}$$

$$= \frac{8\pi^5 k_B^4 T^4}{15 h^3 c^3}$$

$$= 4\frac{\sigma}{c} T^4$$

This gives a theoretical value for the Stefan-Boltzmann constant, σ, of

$$\sigma = \frac{2\pi^5 k_B^4}{15 h^3 c^2}$$

$$= \frac{2\pi^5 \left(1.380\,658 \times 10^{-23}\text{ J·K}^{-1}\right)^4}{15 \left(6.626\,0755 \times 10^{-34}\text{ J·s}\right)^3 \left(2.997\,924\,58 \times 10^8\text{ m·s}^{-1}\right)^2}$$

$$= 5.671 \times 10^{-8}\text{ J·m}^{-2}\text{·K}^{-4}\text{·s}^{-1}$$

which agrees nicely with the experimental value, $\sigma = 5.6697 \times 10^{-8}$ J·m^{-2}·K^{-4}·s^{-1}.

1–10. Can you derive the temperature dependence of the result in Problem 1–9 without evaluating the integral?

Yes, since $x = h\nu/k_B T$ is unitless, the integral $\int_0^\infty \frac{x^3\,dx}{e^x - 1}$ in the previous problem is also unitless. Therefore, the temperature dependence is seen to be T^4 without the need for explicit evaluation of the integral.

1–11. Calculate the energy of a photon for a wavelength of 100 pm (about one atomic diameter).

$$E = \frac{hc}{\lambda} = \frac{\left(6.626\,0755 \times 10^{-34}\text{ J·s}\right)\left(2.997\,924\,58 \times 10^8\text{ m·s}^{-1}\right)}{100 \times 10^{-12}\text{ m}} = 2 \times 10^{-15}\text{ J}$$

1–12. Express Planck's radiation law in terms of λ (and $d\lambda$) by using the relationship $\lambda\nu = c$.

Planck's radiation law in terms of ν is given by Equation 1.2.

$$\rho_\nu(T)\, d\nu = \frac{8\pi h}{c^3}\, \frac{\nu^3 \, d\nu}{e^{h\nu/k_B T} - 1}$$

Make the substitutions $\nu = c/\lambda$ and $d\nu = -c\, d\lambda/\lambda^2$ to give the law in terms of λ.

$$\rho_\lambda(T)\, d\lambda = \frac{8\pi h}{c^3}\, \frac{(c/\lambda)^3 \left(-c/\lambda^2\right)\, d\lambda}{e^{hc/\lambda k_B T} - 1}$$

$$= -\frac{8\pi hc}{\lambda^5}\, \frac{d\lambda}{e^{hc/\lambda k_B T} - 1}$$

The negative sign occurs because the intervals $d\nu$ and $d\lambda$ have opposite signs (an increase in frequency leads to a decrease in wavelength). Regardless of the form chosen, it is conventional in applying the radiation law to consider the interval as increasing. This is formally accomplished by substituting $-d\lambda$ for $d\lambda$ in the last line of the equation to give

$$\rho_\lambda(T)\, d\lambda = \frac{8\pi hc}{\lambda^5}\, \frac{d\lambda}{e^{hc/\lambda k_B T} - 1}$$

1–13.   Calculate the number of photons in a 2.00 mJ light pulse at (a) 1.06 μm, (b) 537 nm, and (c) 266 nm.

The total energy of the light pulse is equal to the energy of each photon times the number of photons, $E_{\text{pulse}} = nh\nu = nhc/\lambda$.

(a)

$$n = \frac{\left(2.00 \times 10^{-3}\ \text{J}\right)\left(1.06 \times 10^{-6}\ \text{m}\right)}{\left(6.626\,0755 \times 10^{-34}\ \text{J·s}\right)\left(2.997\,924\,58 \times 10^8\ \text{m·s}^{-1}\right)} = 1.07 \times 10^{16}$$

(b)

$$n = \frac{\left(2.00 \times 10^{-3}\ \text{J}\right)\left(537 \times 10^{-9}\ \text{m}\right)}{\left(6.626\,0755 \times 10^{-34}\ \text{J·s}\right)\left(2.997\,924\,58 \times 10^8\ \text{m·s}^{-1}\right)} = 5.41 \times 10^{15}$$

(c)

$$n = \frac{\left(2.00 \times 10^{-3}\ \text{J}\right)\left(266 \times 10^{-9}\ \text{m}\right)}{\left(6.626\,0755 \times 10^{-34}\ \text{J·s}\right)\left(2.997\,924\,58 \times 10^8\ \text{m·s}^{-1}\right)} = 2.68 \times 10^{15}$$

1–14.   The mean temperature of the earth's surface is 288 K. What is the maximum wavelength of the earth's blackbody radiation? What part of the spectrum does this wavelength correspond to?

The desired wavelength is found using the Wien displacement law, Equation 1.4.

$$\lambda_{\text{max}} = \frac{2.90 \times 10^{-3}\ \text{m·K}}{T} = \frac{2.90 \times 10^{-3}\ \text{m·K}}{288\ \text{K}} = 1.01 \times 10^{-5}\ \text{m}$$

This wavelength corresponds to the infrared region of the spectrum.

1–15. A helium-neon laser (used in supermarket scanners) emits light at 632.8 nm. Calculate the frequency of this light. What is the energy of the photon generated by this laser?

$$\nu = \frac{c}{\lambda} = \frac{2.997\,924\,58 \times 10^{8}\ \text{m·s}^{-1}}{632.8 \times 10^{-9}\ \text{m}} = 4.738 \times 10^{14}\ \text{Hz}$$

$$E = \frac{hc}{\lambda} = \frac{\left(6.626\,0755 \times 10^{-34}\ \text{J·s}\right)\left(2.997\,924\,58 \times 10^{8}\ \text{m·s}^{-1}\right)}{632.8 \times 10^{-9}\ \text{m}} = 3.139 \times 10^{-19}\ \text{J}$$

1–16. The power output of a laser is measured in units of watts (W), where one watt is equal to one joule per second. ($1\ \text{W} = 1\ \text{J·s}^{-1}$.) What is the number of photons emitted per second by a 1.00 mW nitrogen laser? The wavelength emitted by a nitrogen laser is 337 nm.

A 1.00 mW laser delivers 1.00×10^{-3} J in a second. This amount of energy is equal to the energy of a single laser photon times the number of photons, n, emitted each second.

$$n = \frac{1.00 \times 10^{-3}\ \text{J}}{hc/\lambda}$$

$$= \left(1.00 \times 10^{-3}\ \text{J}\right) \frac{337 \times 10^{-9}\ \text{m}}{\left(6.626\,0755 \times 10^{-34}\ \text{J·s}\right)\left(2.997\,924\,58 \times 10^{8}\ \text{m·s}^{-1}\right)}$$

$$= 1.70 \times 10^{15}$$

1–17. A household lightbulb is a blackbody radiator. Many lightbulbs use tungsten filaments that are heated by an electric current. What temperature is needed so that $\lambda_{\text{max}} = 550$ nm?

The Wien displacement law, Equation 1.4, provides the connection between the temperature and wavelength at which $\rho_{\lambda}(T)$ is a maximum for a blackbody radiator.

$$T = \frac{2.90 \times 10^{-3}\ \text{m·K}}{\lambda_{\text{max}}} = \frac{2.90 \times 10^{-3}\ \text{m·K}}{550 \times 10^{-9}\ \text{m}} = 5300\ \text{K}$$

1–18. The threshold wavelength for potassium metal is 564 nm. What is its work function? What is the kinetic energy of electrons ejected if radiation of wavelength 410 nm is used?

Equation 1.11, which provides the relationship between the work function, ϕ, and the threshold frequency, ν_0, may be rewritten in terms of the threshold wavelength: $hc/\lambda_0 = \phi$. For potassium,

$$\phi = \frac{\left(6.626\,0755 \times 10^{-34}\,\text{J·s}\right)\left(2.997\,924\,58 \times 10^{8}\,\text{m·s}^{-1}\right)}{564 \times 10^{-9}\,\text{m}}$$

$$= 3.52 \times 10^{-19}\,\text{J}$$

$$= 2.20\,\text{eV}$$

Similarly, Equation 1.12 may be modified to give the kinetic energy of the photoelectron in terms of the wavelength of the incident photon.

$$T = hc\left(\frac{1}{\lambda} - \frac{1}{\lambda_0}\right)$$

$$= \left(6.626\,0755 \times 10^{-34}\,\text{J·s}\right)\left(2.997\,924\,58 \times 10^{8}\,\text{m·s}^{-1}\right)$$

$$\times \left(\frac{1}{410 \times 10^{-9}\,\text{m}} - \frac{1}{564 \times 10^{-9}\,\text{m}}\right)$$

$$= 1.32 \times 10^{-19}\,\text{J}$$

1–19.   Given that the work function of chromium is 4.40 eV, calculate the kinetic energy of electrons emitted from a chromium surface that is irradiated with ultraviolet radiation of wavelength 200 nm.

The work function must first be converted to joules.

$$\phi = 4.40\,\text{eV}\left(\frac{1.602\,177\,33 \times 10^{-19}\,\text{J}}{1\,\text{eV}}\right) = 7.05 \times 10^{-19}\,\text{J}$$

This must be compared with the energy of the photon to ensure the radiation is above the threshold frequency.

$$E = \frac{hc}{\lambda} = \frac{\left(6.626\,0755 \times 10^{-34}\,\text{J·s}\right)\left(2.997\,924\,58 \times 10^{8}\,\text{m·s}^{-1}\right)}{200 \times 10^{-9}\,\text{m}} = 9.93 \times 10^{-19}\,\text{J}$$

Indeed, the photon has sufficient energy to eject electrons from the chromium surface, and the kinetic energy of these photoelectrons will be the difference between the photon energy and the work function.

$$T = 9.93 \times 10^{-19}\,\text{J} - 7.05 \times 10^{-19}\,\text{J} = 2.88 \times 10^{-19}\,\text{J}$$

1–20.   When a clean surface of silver is irradiated with light of wavelength 230 nm, the kinetic energy of the ejected electrons is found to be 0.805 eV. Calculate the work function and the threshold frequency of silver.

The kinetic energy of the photoelectrons is given by

$$T = \frac{hc}{\lambda} - \phi$$

so that

$$\phi = \frac{\left(6.626\,0755 \times 10^{-34}\,\text{J·s}\right)\left(2.997\,924\,58 \times 10^{8}\,\text{m·s}^{-1}\right)}{230 \times 10^{-9}\,\text{m}}$$

$$- 0.805\,\text{eV}\left(\frac{1.602\,177\,33 \times 10^{-19}\,\text{J}}{1\,\text{eV}}\right)$$

$$= 7.35 \times 10^{-19}\,\text{J}$$

$$= 4.59\,\text{eV}$$

The threshold frequency is related to the work function via $\phi = h\nu_0$, so that

$$\nu_0 = \frac{7.35 \times 10^{-19}\,\text{J}}{6.626\,0755 \times 10^{-34}\,\text{J·s}} = 1.11 \times 10^{15}\,\text{Hz}$$

1–21. Some data for the kinetic energy of ejected electrons as a function of the wavelength of the incident radiation for the photoelectron effect for sodium metal are shown below.

λ/nm	100	200	300	400	500
KE/eV	10.1	3.94	1.88	0.842	0.222

Plot these data to obtain a straight line, and calculate h from the slope of the line and the work function ϕ from its intercept with the horizontal axis.

From $T = h\nu - \phi$, a graph of photoelectron kinetic energy versus photon frequency has the form of a straight line graph with slope h and intercept $-\phi$. First, the wavelengths are converted to frequency using $\nu\lambda = c$, and the kinetic energies given in the problem are converted to joules.

$\nu/10^{15}$ Hz	3.00	1.50	1.00	0.749	0.600
$T/10^{-19}$ J	16.2	6.31	3.01	1.35	0.356

The data do indeed make a straight line graph.

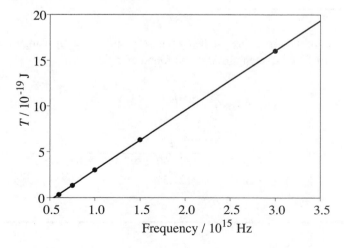

A least-squares fit to the data results in a slope of $h = 6.60 \times 10^{-34}$ J·s, and an intercept equal to -3.59×10^{-19} J so that $\phi = 3.59 \times 10^{-19}$ J $= 2.24$ eV.

1–22. Show that Planck's constant has dimensions of angular momentum.

Angular momentum, $l = mvr$, has the units $(\text{kg})\left(\text{m}\cdot\text{s}^{-1}\right)(\text{m}) = \text{kg}\cdot\text{m}^2\cdot\text{s}^{-1}$, which is the same as Planck's constant, $\text{J}\cdot\text{s} = \text{kg}\cdot\text{m}^2\cdot\text{s}^{-1}$.

1–23. What is the wavelength and frequency of electromagnetic radiation having an energy of 1 rydberg (a rydberg is equal to 109 680 cm^{-1})? Convert 1 rydberg to electron volts and to kilojoules/mole (kJ$\cdot$mol^{-1}).

$$\lambda = \frac{1}{\tilde{\nu}} = \frac{1}{109\ 680\ \text{cm}^{-1}} = 9.174 \times 10^{-6}\ \text{cm} = 91.174\ \text{nm}$$

$$\nu = c\tilde{\nu} = \left(2.997\ 924\ 58 \times 10^8\ \text{m}\cdot\text{s}^{-1}\right)\left(109\ 680\ \text{cm}^{-1}\right) = 3.2861 \times 10^{15}\ \text{Hz}$$

$$E = hc\tilde{\nu} = \left(6.626\ 0755 \times 10^{-34}\ \text{J}\cdot\text{s}\right)\left(2.997\ 924\ 58 \times 10^8\ \text{m}\cdot\text{s}^{-1}\right)\left(109\ 680\ \text{cm}^{-1}\right)\left(\frac{100\ \text{cm}}{1\ \text{m}}\right)$$

$$= 2.17874 \times 10^{-18}\ \text{J}$$

$$= 13.599\ \text{eV}$$

This is the energy for one photon. For 1 mole of photons, the energy is

$$E = 2.17874 \times 10^{-18}\ \text{J}\left(6.022\ 1367 \times 10^{23}\ \text{mol}^{-1}\right)$$

$$= 1312.1\ \text{kJ}\cdot\text{mol}^{-1}$$

1–24. Use the Rydberg formula (Equation 1.14) to calculate the wavelengths of the first three lines of the Lyman series.

The frequencies of the first 3 lines in the Lyman series are found by using $n_1 = 1$ and $n_2 = 2, 3, 4$ in Equation 1.14. The wavelengths are then found using $c = \nu\lambda$.

$$\tilde{\nu} = 109\ 680\ \text{cm}^{-1}\left(\frac{1}{1^2} - \frac{1}{n_2^2}\right) \qquad n_2 = 2, 3, 4$$

n_2	2	3	4
$\tilde{\nu}/\text{cm}^{-1}$	82 260	97 493	102 825
λ/nm	121.57	102.57	97.253

1–25. A line in the Lyman series of hydrogen has a wavelength of 1.03×10^{-7} m. Find the original energy level of the electron.

The wavelength corresponds to a frequency of $\tilde{\nu} = 1/\lambda = 9.709 \times 10^4$ cm^{-1}, and for the Lyman series,

$$\tilde{\nu} = 109\,680 \text{ cm}^{-1} \left(\frac{1}{1^2} - \frac{1}{n^2} \right).$$

This can be used to find n, the quantum number of the original energy level.

$$9.709 \times 10^4 \text{ cm}^{-1} = 109\,680 \text{ cm}^{-1} \left(\frac{1}{1^2} - \frac{1}{n^2} \right)$$

$$0.885\,19 = 1 - \frac{1}{n^2}$$

$$\frac{1}{n^2} = 0.114\,81$$

$$n = 2.9513$$

Since the quantum number must be an integer, the original energy level has $n = 3$.

1–26. A ground-state hydrogen atom absorbs a photon of light that has a wavelength of 97.2 nm. It then gives off a photon that has a wavelength of 486 nm. What is the final state of the hydrogen atom?

A ground state hydrogen atom has $n_1 = 1$, and the frequency of the absorbed photon is $\tilde{\nu} = 1/\lambda = 1.029 \times 10^5$ cm^{-1}. The Rydberg formula can be used to find the quantum number, n, of the intermediate energy level.

$$1.029 \times 10^5 \text{ cm}^{-1} = 109\,680 \text{ cm}^{-1} \left(\frac{1}{1^2} - \frac{1}{n^2} \right)$$

$$0.9380 = \frac{1}{1^2} - \frac{1}{n^2}$$

$$\frac{1}{n^2} = 0.061\,99$$

$$n = 4.016$$

Thus, the 486 nm photon is emitted from the energy level with $n = 4$. The frequency of this photon is $\tilde{\nu} = 1/\lambda = 2.058 \times 10^4$ cm^{-1}, and the Rydberg formula is employed again. Since the photon is being emitted, take $n_2 = 4$ and solve for the quantum number of the final energy level.

$$2.058 \times 10^4 \text{ cm}^{-1} = 109\,680 \text{ cm}^{-1} \left(\frac{1}{n^2} - \frac{1}{4^2} \right)$$

$$0.1876 = \frac{1}{n^2} - \frac{1}{16}$$

$$\frac{1}{n^2} = 0.2501$$

$$n = 2.00$$

Again, since the quantum number is by definition an integer, the final energy level has $n = 2$.

1–27. Show that the Lyman series occurs between 91.2 nm and 121.6 nm, that the Balmer series occurs between 364.7 nm and 656.3 nm, and that the Paschen series occurs between 821.0 nm and 1876 nm. Identify the spectral regions to which these wavelengths correspond.

The short wavelength (high frequency) limit of a spectroscopic series in the hydrogen atom may be found by taking the limit of the Rydberg formula as $n_2 \to \infty$.

$$\tilde{\nu}_{max} = \lim_{n_2 \to \infty} 109\,680 \text{ cm}^{-1} \left(\frac{1}{n_1^2} - \frac{1}{n_2^2} \right)$$

$$= \frac{109\,680 \text{ cm}^{-1}}{n_1^2}$$

The long wavelength (low frequency) limit is given by the transition between the energy levels with quantum numbers n_1 and $n_2 = n_1 + 1$, where n_1 is the quantum number of the lowest energy level of the series.

$$\tilde{\nu}_{min} = 109\,680 \text{ cm}^{-1} \left[\frac{1}{n_1^2} - \frac{1}{(n_1 + 1)^2} \right]$$

Once the limiting frequencies are determined, the corresponding wavelength limits are found using $\lambda = 1/\tilde{\nu}$.

Series	Lyman	Balmer	Paschen
n_1	1	2	3
$\tilde{\nu}_{max}/10^3 \text{ cm}^{-1}$	109.68	27.420	12.186
$\tilde{\nu}_{min}/10^3 \text{ cm}^{-1}$	82.260	15.233	5.3317
$\lambda_{short}/\text{nm}$	91.174	364.70	820.57
λ_{long}/nm	121.57	656.46	1875.6
Spectral Region	IR	Visible	UV

1–28. Calculate the wavelength and the energy of a photon associated with the series limit of the Lyman series.

As in Problem 1–27, the series limit in the hydrogen atom emission spectrum is found by taking the limit of the Rydberg formula as $n_2 \to \infty$. For the Lyman series, the limit is simply $\tilde{\nu} = 109\,680$ cm^{-1}. The wavelength and energy of the photon associated with this limit are

$$\lambda = \frac{1}{\tilde{\nu}} = 91.174 \text{ nm}$$

$$E = hc\tilde{\nu} = 2.1787 \times 10^{-18} \text{ J} = 13.599 \text{ eV}.$$

1–29. Show that the value of the Rydberg constant is $109\ 676\ \text{cm}^{-1}$ if it is expressed in terms of the reduced mass of a hydrogen atom.

The reduced mass of a hydrogen atom is

$$\mu = \frac{m_{\text{p}} m_{\text{e}}}{m_{\text{p}} + m_{\text{e}}}$$

$$= \frac{\left(1.672\ 6231 \times 10^{-27}\ \text{kg}\right)\left(9.109\ 3897 \times 10^{-31}\ \text{kg}\right)}{1.672\ 6231 \times 10^{-27}\ \text{kg} + 9.109\ 3897 \times 10^{-31}\ \text{kg}}$$

$$= 9.10\ 443 \times 10^{-31}\ \text{kg}.$$

This is used in the formula for the Rydberg constant.

$$R_{\text{H}} = \frac{\mu e^4}{8\epsilon_0^2 h^2}$$

$$= \frac{\left(9.104\ 43 \times 10^{-31}\ \text{kg}\right)\left(1.602\ 177\ 33 \times 10^{-19}\ \text{C}\right)^4}{8\left(8.854\ 187\ 816 \times 10^{-12}\ \text{C}^2 \cdot \text{J}^{-1} \cdot \text{m}^{-1}\right)^2 \left(6.626\ 0755 \times 10^{-34}\ \text{J} \cdot \text{s}\right)^2}$$

$$= 2.178\ 69 \times 10^{-18}\ \text{J}$$

Conversion to cm^{-1} gives

$$R_{\text{H}} = 2.178\ 69 \times 10^{-18}\ \text{J} \left(\frac{5.034\ 11 \times 10^{22}\ \text{cm}^{-1}}{1\ \text{J}}\right) = 109\ 678\ \text{cm}^{-1}$$

1–30. Calculate the reduced mass of a nitrogen molecule in which both nitrogen atoms have an atomic mass of 14.00. Do the same for a hydrogen chloride molecule in which the chlorine atom has an atomic mass of 34.97. Hydrogen has an atomic mass of 1.008.

The reduced mass is $\mu = \dfrac{m_1 m_2}{m_1 + m_2}$.

For N_2, $m_1 = 14.00$ amu $= m_2$.

$$\mu = \frac{(14.00\ \text{amu})\ (14.00\ \text{amu})}{14.00\ \text{amu} + 14.00\ \text{amu}} = 7.00\ \text{amu}$$

For $H^{35}Cl$,

$$\mu = \frac{(1.008\ \text{amu})\ (34.97\ \text{amu})}{1.008\ \text{amu} + 34.97\ \text{amu}} = 0.9798\ \text{amu}$$

1–31. We shall learn when we study molecular spectroscopy that heteronuclear diatomic molecules absorb radiation in the microwave region and that one determines directly the so-called rotational constant of the molecule defined by

$$B = \frac{h}{8\pi^2 I}$$

where I is the moment of inertia of the molecule. Given that $B = 3.13 \times 10^5$ MHz for $H^{35}Cl$, calculate the internuclear separation for $H^{35}Cl$. The atomic mass of H and ^{35}Cl are 1.008 and 34.97, respectively.

The moment of inertia for a diatomic molecule is $I = \mu r^2$, where μ is the reduced mass and r is the internuclear separation. Thus,

$$r = \sqrt{\frac{h}{8\pi^2 \mu B}}$$

The reduced mass for $H^{35}Cl$ was found in Problem 1–30, $\mu = 0.9798$ amu $= 1.6270 \times 10^{-27}$ kg.

$$r = \sqrt{\frac{6.626\ 0755 \times 10^{-34}\ \text{J·s}}{8\pi^2\ (1.6270 \times 10^{-27}\ \text{kg})\ (3.13 \times 10^{11}\ \text{Hz})}}$$

$$= 1.28 \times 10^{-10}\ \text{m}$$

$$= 1.28\ \text{Å}$$

1–32. Derive the Bohr formula for $\tilde{\nu}$ for a nucleus of atomic number Z.

For a nucleus with atomic number Z, the Coulomb force reflects the charge of $+Ze$ on the nucleus, and Equation 1.21 becomes

$$\frac{(Ze)\,e}{4\pi\epsilon_0 r^2} = \frac{m_e v^2}{r}$$

Similarly, in all subsequent equations, e^2 is replaced by Ze^2 to arrive at

$$\tilde{\nu} = \frac{m_e Z^2 e^4}{8\epsilon_0^2 c h^3}\left(\frac{1}{n_1^2} - \frac{1}{n_2^2}\right) = Z^2 R_\text{H}\left(\frac{1}{n_1^2} - \frac{1}{n_2^2}\right)$$

which is the Bohr formula for $\tilde{\nu}$ for a nucleus of atomic number Z.

1–33. The series in the He^+ spectrum that corresponds to the set of transitions where the electron falls from a higher level into the $n = 4$ state is called the Pickering series, an important series in solar astronomy. Derive the formula for the wavelengths of the observed lines in this series. In what region of the spectrum do they occur?

The Bohr formula for a nucleus of atomic number Z was found in Problem 1–32.

$$\tilde{\nu} = Z^2 R_H \left(\frac{1}{n_1^2} - \frac{1}{n_2^2} \right)$$

For the Pickering series in He$^+$, $Z = 2$ and $n_1 = 4$,

$$\tilde{\nu} = 4 \left(109\,680 \text{ cm}^{-1} \right) \left(\frac{1}{16} - \frac{1}{n_2^2} \right) \qquad n_2 = 5, 6, 7, \ldots$$

The longest wavelength transition occurs for $n_2 = 5$, for which $\tilde{\nu} = 9871.2$ cm^{-1}, and $\lambda_{\text{long}} = 1013.0$ nm. The short wavelength limit is found as $n_2 \to \infty$, in which case $\tilde{\nu} = 27\,420$ cm^{-1} and $\lambda_{\text{short}} = 364.70$ nm. These wavelengths run from the near-infrared to the visible regions of the spectrum.

1–34. Using the Bohr theory, calculate the ionization energy (in electron volts and in kJ·mol^{-1}) of singly ionized helium.

The ionization energy is equal to the ground state series limit of the Bohr formula for the He$^+$ ion.

$$IE = \frac{4 \left(109\,680 \text{ cm}^{-1} \right)}{1^2}$$

$$= 438\,720 \text{ cm}^{-1}$$

$$= 8.7149 \times 10^{-18} \text{ J}$$

$$= 54.394 \text{ eV}$$

$$= 5248.2 \text{ kJ·mol}^{-1}$$

1–35. Show that the speed of an electron in the nth Bohr orbit is $v = e^2/2\epsilon_0 nh$. Calculate the values of v for the first few Bohr orbits.

From Equation 1.22,

$$v = \frac{nh}{2\pi m_e r}$$

and Equation 1.23 provides

$$r = \frac{\epsilon_0 h^2 n^2}{\pi m_e e^2}$$

Substituting this value for r in the equation for v gives

$$v = \frac{nh}{2\pi m_e}\left(\frac{\pi m_e e^2}{\epsilon_0 h^2 n^2}\right)$$

$$= \frac{e^2}{2\epsilon_0 n h}$$

$$= \frac{\left(1.602\,177\,33 \times 10^{-19}\,\text{C}\right)^2}{2\left(8.854\,187\,816 \times 10^{-12}\,\text{C}^2{\cdot}\text{J}^{-1}{\cdot}\text{m}^{-1}\right)\left(6.626\,0755 \times 10^{-34}\,\text{J}{\cdot}\text{s}\right)n}$$

$$= \frac{2.1877 \times 10^6\,\text{m}{\cdot}\text{s}^{-1}}{n}$$

For the first few values of n this gives

n	1	2	3
v	$2.1877 \times 10^6\,\text{m}{\cdot}\text{s}^{-1}$	$1.0938 \times 10^6\,\text{m}{\cdot}\text{s}^{-1}$	$7.2923 \times 10^5\,\text{m}{\cdot}\text{s}^{-1}$

1–36. Ionizing a hydrogen atom in its electronic ground state requires 2.179×10^{-18} J of energy. The sun's surface has a temperature of ≈ 6000 K and is composed, in part, of atomic hydrogen. Is the hydrogen present as H(g) or H^+(g)? What is the temperature required so that the maximum wavelength of the emission of a blackbody ionizes atomic hydrogen? In what region of the electromagnetic spectrum is this wavelength found?

From Equation 1.4, the wavelength at the maximum in the blackbody radiation spectrum of a 6000 K object is

$$\lambda_{\max} = \frac{2.90 \times 10^{-3}\,\text{m}{\cdot}\text{K}}{6000\,\text{K}} = 4.8 \times 10^{-7}\,\text{m}$$

Photons with this wavelength have an energy of

$$E = \frac{hc}{\lambda_{\max}}$$

$$= \frac{\left(6.626\,0755 \times 10^{-34}\,\text{J}{\cdot}\text{s}\right)\left(2.997\,924\,58 \times 10^8\,\text{m}{\cdot}\text{s}^{-1}\right)}{4.8 \times 10^{-7}\,\text{m}}$$

$$= 4.1 \times 10^{-19}\,\text{J}$$

This energy is insufficient to ionize a ground state hydrogen atom, so the atomic hydrogen at the sun's surface is present as H(g). Ionization of atomic hydrogen requires photons with an energy of 2.179×10^{-18} J or greater, or since the wavelength of a photon is $\lambda = hc/E$, a wavelength shorter than

$$\lambda = \frac{\left(6.626\,0755 \times 10^{-34}\,\text{J}{\cdot}\text{s}\right)\left(2.997\,924\,58 \times 10^8\,\text{m}{\cdot}\text{s}^{-1}\right)}{2.179 \times 10^{-18}\,\text{J}}$$

$$= 9.116 \times 10^{-8}\,\text{m}$$

which is in the UV region of the electromagnetic spectrum.

Equation 1.4 shows that a temperature of at least

$$T = \frac{2.90 \times 10^{-3} \, \text{m} \cdot \text{K}}{9.116 \times 10^{-8} \, \text{m}} = 31\,800 \, \text{K}$$

is required for photons at the maximum of a blackbody spectrum to have sufficient energy to ionize atomic hydrogen.

1–37. A proton and a negatively charged μ meson (called a *muon*) can form a short-lived species called a *mesonic atom*. The charge of a muon is the same as that on an electron and the mass of a muon is 207 m_e. Assume that the Bohr theory can be applied to such a mesonic atom and calculate the ground-state energy, the radius of the first Bohr orbit, and the energy and frequency associated with the $n = 1$ to $n = 2$ transition in a mesonic atom.

The muon is sufficiently heavy that the approximation $\mu \approx m_e$ cannot be used for the reduced mass of the mesonic atom, but rather must be calculated using Equation 1.39. (Although both the muon and the reduced mass are represented by μ, the meaning should be clear from context.)

$$\mu_\mu = \frac{m_p m_\mu}{m_p + m_\mu}$$

$$= \frac{m_p \left(207 m_e\right)}{m_p + 207 m_e}$$

$$= \frac{207 \left(1.672\,6231 \times 10^{-27} \, \text{kg}\right) \left(9.109\,3897 \times 10^{-31} \, \text{kg}\right)}{\left(1.672\,6231 \times 10^{-27} \, \text{kg}\right) + 207 \left(9.109\,3897 \times 10^{-31} \, \text{kg}\right)}$$

$$= 1.695 \times 10^{-28} \, \text{kg}$$

This reduced mass is used in place of m_e in Equations 1.27, 1.23, and 1.28 to give the ground-state energy, the radius of the first Bohr orbit, and the energy associated with the $n = 1$ to $n = 2$ transition in a mesonic atom, respectively.

$$E_1 = -\frac{\mu e^4}{8 \epsilon_0^2 h^2 \left(1^2\right)}$$

$$= -\frac{\left(1.695 \times 10^{-28} \, \text{kg}\right) \left(1.602\,177\,33 \times 10^{-19} \, \text{C}\right)^4}{8 \left(8.854\,187\,816 \times 10^{-12} \, \text{C}^2 \cdot \text{J}^{-1} \cdot \text{m}^{-1}\right)^2 \left(6.626\,0755 \times 10^{-34} \, \text{J} \cdot \text{s}\right)^2}$$

$$= -4.06 \times 10^{-16} \, \text{J}$$

$$r = \frac{\epsilon_0 h^2}{\pi \mu e^2}$$

$$= \frac{\left(8.854\,187\,816 \times 10^{-12} \, \text{C}^2 \cdot \text{J}^{-1} \cdot \text{m}^{-1}\right) \left(6.626\,0755 \times 10^{-34} \, \text{J} \cdot \text{s}\right)^2}{\pi \left(1.695 \times 10^{-28} \, \text{kg}\right) \left(1.602\,177\,33 \times 10^{-19} \, \text{C}\right)^2}$$

$$= 2.84 \times 10^{-13} \, \text{m}$$

$$\Delta E_{1\rightarrow 2} = \frac{\mu e^4}{8\epsilon_0^2 h^2}\left(\frac{1}{1^2} - \frac{1}{2^2}\right)$$

$$= \frac{\left(1.695 \times 10^{-28}\text{ kg}\right)\left(1.602\,177\,33 \times 10^{-19}\text{ C}\right)^4}{8\left(8.854\,187\,816 \times 10^{-12}\text{ C}^2\cdot\text{J}^{-1}\cdot\text{m}^{-1}\right)^2\left(6.626\,0755 \times 10^{-34}\text{ J}\cdot\text{s}\right)^2}\left(\frac{3}{4}\right)$$

$$= 3.04 \times 10^{-16}\text{ J}$$

The frequency associated with the transition is that of a photon with an energy of 3.04×10^{-16} J, namely

$$\nu = \Delta E_{1\rightarrow 2}/h = 4.59 \times 10^{17}\text{ Hz}$$

1–38. Calculate the de Broglie wavelength for (a) an electron with a kinetic energy of 100 eV, (b) a proton with a kinetic energy of 100 eV, and (c) an electron in the first Bohr orbit of a hydrogen atom.

The de Broglie wavelength of a moving particle is given by Equation 1.48:

$$\lambda = \frac{h}{p} = \frac{h}{mv} = \frac{h}{\sqrt{2mE}}$$

(a) For the 100 eV electron,

$$\lambda = \frac{6.626\,0755 \times 10^{-34}\text{ J}\cdot\text{s}}{\sqrt{2\left(9.109\,3897 \times 10^{-31}\text{ kg}\right)\left(100\text{ eV}\right)\left(1.602\,177 \times 10^{-19}\text{ J}\cdot\text{eV}^{-1}\right)}}$$

$$= 1.23 \times 10^{-10}\text{ m}$$
$$= 123\text{ pm}$$

(b) For the 100 eV proton,

$$\lambda = \frac{6.626\,0755 \times 10^{-34}\text{ J}\cdot\text{s}}{\sqrt{2\left(1.672\,6231 \times 10^{-27}\text{ kg}\right)\left(100\text{ eV}\right)\left(1.602\,177 \times 10^{-19}\text{ J}\cdot\text{eV}^{-1}\right)}}$$

$$= 2.86 \times 10^{-12}\text{ m}$$
$$= 2.86\text{ pm}$$

(c) The velocity of an electron in the first Bohr orbit of a hydrogen atom was calculated in Problem 1–35, $v = 2.1877 \times 10^6$ m·s^{-1}.

$$\lambda = \frac{6.626\,0755 \times 10^{-34}\text{ J}\cdot\text{s}}{\left(9.109\,3897 \times 10^{-31}\text{ kg}\right)\left(2.1877 \times 10^6\text{ m}\cdot\text{s}^{-1}\right)}$$

$$= 3.32 \times 10^{-10}\text{ m}$$
$$= 3.32\text{ pm}$$

1–39.   Calculate (a) the wavelength and kinetic energy of an electron in a beam of electrons accelerated by a voltage increment of 100 V and (b) the kinetic energy of an electron that has a de Broglie wavelength of 200 pm (1 picometer $= 10^{-12}$ m).

(a) An electron accelerated by a voltage increment of 100 V has $100\,\text{eV} = 1.60 \times 10^{-17}\,\text{J}$ of kinetic energy. The wavelength for such an electron was found in Problem 1–38, $\lambda = 123$ pm.

(b) The de Broglie relation $p = h/\lambda$ allows the kinetic energy of a particle to be expressed in terms of its de Broglie wavelength.

$$
\begin{aligned}
E &= \frac{p^2}{2m} \\[6pt]
&= \frac{h^2}{2m\lambda^2} \\[6pt]
&= \frac{\left(6.626\,0755 \times 10^{-34}\,\text{J·s}\right)^2}{2\left(9.109\,3897 \times 10^{-31}\,\text{kg}\right)\left(200 \times 10^{-12}\,\text{m}\right)^2} \\[6pt]
&= 6.02 \times 10^{-18}\,\text{J} \\[6pt]
&= 37.6\,\text{eV}
\end{aligned}
$$

1–40.   Through what potential must a proton initially at rest fall so that its de Broglie wavelength is 1.0×10^{-10} m?

The relationship between kinetic energy and de Broglie wavelength developed in Problem 1–39 is used to find

$$
\begin{aligned}
E &= \frac{p^2}{2m} \\[6pt]
&= \frac{h^2}{2m\lambda^2} \\[6pt]
&= \frac{\left(6.626\,0755 \times 10^{-34}\,\text{J·s}\right)^2}{2\left(1.672\,6231 \times 10^{-27}\,\text{kg}\right)\left(1.00 \times 10^{-10}\,\text{m}\right)^2} \\[6pt]
&= 1.3 \times 10^{-20}\,\text{J} \\[6pt]
&= 0.082\,\text{eV}
\end{aligned}
$$

This requires a potential difference of 0.082 V.

1–41.   Calculate the energy and wavelength associated with an α particle that has fallen through a potential difference of 4.0 V. Take the mass of an α particle to be 6.64×10^{-27} kg.

The energy of the α particle is $E = (2e)(4.0\text{ V}) = 8.0\text{ eV}$, or $E = 1.3 \times 10^{-18}$ J. The wavelength is found as in Problem 1–38.

$$\lambda = \frac{6.626\,0755 \times 10^{-34}\text{ J·s}}{\sqrt{2\left(6.64 \times 10^{-27}\text{ kg}\right)\left(1.3 \times 10^{-18}\text{ J}\right)}}$$

$$= 5.1 \times 10^{-12}\text{ m}$$

$$= 5.1\,\text{pm}$$

1–42. One of the most powerful modern techniques for studying structure is neutron diffraction. This technique involves generating a collimated beam of neutrons at a particular temperature from a high-energy neutron source and is accomplished at several accelerator facilities around the world. If the speed of a neutron is given by $v_n = (3k_BT/m)^{1/2}$, where m is the mass of a neutron, then what temperature is needed so that the neutrons have a de Broglie wavelength of 50 pm?

Using the formula given for the speed of the neutron, it is possible to determine a relationship between the temperature of the beam and the de Broglie wavelength of neutrons it contains.

$$\lambda = \frac{h}{p}$$

$$= \frac{h}{mv}$$

$$= \frac{h}{m\sqrt{3k_BT/m}}$$

$$= \frac{h}{\sqrt{3mk_BT}}$$

The mass of a neutron is 1.675×10^{-27} kg, so

$$T = \frac{h^2}{3mk_B\lambda^2}$$

$$= \frac{\left(6.626\,0755 \times 10^{-34}\text{ J·s}\right)^2}{3\left(1.675 \times 10^{-27}\text{ kg}\right)\left(1.380\,658 \times 10^{-23}\text{ J·K}^{-1}\right)\left(50 \times 10^{-12}\text{ m}\right)^2}$$

$$= 2500\text{ K}$$

1–43. Show that the speed of an electron in the nth Bohr orbit of a hydrogen-like ion with a nuclear charge Z is $v = Ze^2/2\epsilon_0nh$. Calculate the values of v for the first Bohr orbit for $Z = 10$ and $Z = 50$.

As in Problem 1–32, the increased nuclear charge for a hydrogen-like ion with a nuclear charge Z affects the Coulomb force between the nucleus and the electron so that Equation 1.22 becomes

$$\frac{Ze^2}{4\pi\epsilon_0 r^2} = \frac{m_e v^2}{r}$$

There is no effect on Planck's angular momentum quantization (Equation 1.22), but the modified Coulomb law causes changes in both Equation 1.23 and consequently the expression for the velocity of the electron derived in Problem 1–35. In each, e^2 is replaced by Ze^2 to give

$$v = \frac{nh}{2\pi m_e}\left(\frac{\pi m_e Ze^2}{\epsilon_0 h^2 n^2}\right)$$

$$= \frac{Ze^2}{2\epsilon_0 nh}$$

$$= \frac{Z\left(1.602\,177\,33\times10^{-19}\,\text{C}\right)^2}{2\left(8.854\,187\,816\times10^{-12}\,\text{C}^2\cdot\text{J}^{-1}\cdot\text{m}^{-1}\right)\left(6.626\,0755\times10^{-34}\,\text{J}\cdot\text{s}\right)n}$$

$$= 2.1877\times10^6\,\text{m}\cdot\text{s}^{-1}\frac{Z}{n}$$

For $n = 1$ this gives

Z	10	50
v	$2.1877\times10^7\,\text{m}\cdot\text{s}^{-1}$	$1.0938\times10^8\,\text{m}\cdot\text{s}^{-1}$

1–44. Two narrow slits separated by 0.10 mm are illuminated by light of wavelength 600 nm. What is the angular position of the first maximum in the interference pattern? If a detector is located 2.00 m beyond the slits, what is the distance between the central maximum and the first maximum?

The first maximum occurs when $d\sin\theta = \lambda$.

$$\sin\theta = \frac{600\times10^{-9}\,\text{m}}{0.10\times10^{-3}\,\text{m}} = 6.00\times10^{-3}$$

$$\theta = 0.34°$$

Figure 1.15 shows that the distance from the central maximum, y, is given by

$$y = l\tan\theta$$

$$= (2.00\text{ m})\tan0.34°$$

$$= 1.2\times10^{-2}\,\text{m}$$

1–45. Two narrow slits are illuminated with red light of wavelength 694.3 nm from a laser, producing a set of evenly placed bright bands on a screen located 3.00 m beyond the slits. If the distance between the bands is 1.50 cm, then what is the distance between the slits?

If the spacing between bright bands is 1.50 cm, then in particular, the distance between the central maximum and the first maximum has this value. From Figure 1.15, with y denoting the distance between these two maxima on the second screen,

$$\tan \theta = \frac{y}{l}$$

$$= \frac{1.50 \times 10^{-2}\,\text{m}}{3.00\,\text{m}}$$

$$= 5.00 \times 10^{-3}$$

The first maximum occurs when $d \sin \theta = \lambda$, and for such small angles $\sin \theta \approx \tan \theta$. (Indeed, the two values differ only in the 4th decimal place for this problem.) Thus,

$$d = \frac{\lambda}{\sin \theta}$$

$$\approx \frac{694.3 \times 10^{-9}\,\text{m}}{5.00 \times 10^{-3}}$$

$$= 1.39 \times 10^{-4}\,\text{m}$$

$$= 0.139\,\text{mm}$$

1–46. If we locate an electron to within 20 pm, then what is the uncertainty in its speed?

From $\Delta x\,\Delta p > h$, and $p = mv$,

$$\Delta v > \frac{h}{m\,\Delta x}$$

$$= \frac{6.626\,0755 \times 10^{-34}\,\text{J·s}}{(9.109\,3897 \times 10^{-31}\,\text{kg})\,(20 \times 10^{-12}\,\text{m})}$$

$$= 3.6 \times 10^{7}\,\text{m·s}^{-1}$$

1–47. What is the uncertainty of the momentum of an electron if we know its position is somewhere in a 10 pm interval? How does the value compare to momentum of an electron in the first Bohr orbit?

From $\Delta x\,\Delta p > h$,

$$\Delta p > \frac{h}{\Delta x}$$

$$= \frac{6.626\,0755 \times 10^{-34}\,\text{J·s}}{10 \times 10^{-12}\,\text{m}}$$

$$= 6.6 \times 10^{-23}\,\text{kg·m·s}^{-1}$$

This can be compared to the momentum of an electron in the first Bohr orbit, which can be found using the velocity found in Problem 1–35,

$$p = mv$$

$$= \left(9.109\,3897 \times 10^{-31}\,\text{kg}\right)\left(2.1877 \times 10^{6}\,\text{m}\cdot\text{s}^{-1}\right)$$

$$= 1.9929 \times 10^{-24}\,\text{kg}\cdot\text{m}\cdot\text{s}^{-1}$$

The uncertainty of the momentum for an electron with position known to $\Delta x = 10\,\text{pm}$ is approximately 33 times the momentum of an electron in the first Bohr orbit.

1–48.   There is also an uncertainty principle for energy and time:

$$\Delta E\,\Delta t \geq h$$

Show that both sides of this expression have the same units.

The units of $\Delta E\,\Delta t$ are those of energy $\times$ time, or J·s, which are the same as the units for h.

1–49.   The relationship introduced in Problem 1–48 has been interpreted to mean that a particle of mass m $(E = mc^2)$ can materialize from nothing provided that it returns to nothing within a time $\Delta t \leq h/mc^2$. Particles that last for time Δt or more are called *real particles*; particles that last less than time Δt are called *virtual particles*. The mass of the charged pion, a subatomic particle, is 2.5×10^{-28} kg. What is the minimum lifetime if the pion is to be considered a real particle?

The minimum lifetime for a real particle as defined in the problem is

$$\Delta t = \frac{h}{mc^2}$$

$$= \frac{6.626\,0755 \times 10^{-34}\,\text{J}\cdot\text{s}}{\left(2.5 \times 10^{-28}\,\text{kg}\right)\left(2.997\,924\,58 \times 10^{8}\,\text{m}\cdot\text{s}^{-1}\right)^2}$$

$$= 2.9 \times 10^{-23}\,\text{s}$$

1–50.   Another application of the relationship given in Problem 1–48 has to do with the excited-state energies and lifetimes of atoms and molecules. If we know that the lifetime of an exicted state is 10^{-9} s, then what is the uncertainty in the energy of this state?

The uncertainty in the energy of the state is limited by

$$\Delta E > \frac{h}{\Delta t}$$

$$= \frac{6.626\,0755 \times 10^{-34}\,\text{J}\cdot\text{s}}{10^{-9}\,\text{s}}$$

$$= 7 \times 10^{-25}\,\text{J}$$

1–51. When an excited nucleus decays, it emits a γ ray. The lifetime of an excited state of a nucleus is of the order of 10^{-12} s. What is the uncertainty in the energy of the γ ray produced? (See Problem 1–48.)

As in Problem 1–50,

$$\Delta E > \frac{h}{\Delta t}$$

$$= \frac{6.626\,0755 \times 10^{-34}\ \text{J·s}}{10^{-12}\ \text{s}}$$

$$= 7 \times 10^{-22}\ \text{J}$$

1–52. In this problem, we will prove that the inward force required to keep a mass revolving around a fixed center is $f = mv^2/r$. To prove this, let us look at the velocity and the acceleration of a revolving mass. Referring to Figure 1.20, we see that

$$|\Delta \mathbf{r}| \approx \Delta s = r\Delta\theta \tag{1.52}$$

if $\Delta\theta$ is small enough that the arc length Δs and the vector difference $|\Delta \mathbf{r}| = |\mathbf{r}_1 - \mathbf{r}_2|$ are essentially the same. In this case, then

$$v = \lim_{\Delta t \to 0} \frac{\Delta s}{\Delta t} = r \lim_{\Delta t \to 0} \frac{\Delta\theta}{\Delta t} = r\omega \tag{1.53}$$

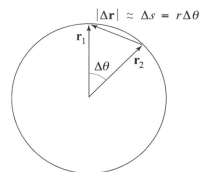

FIGURE 1.20
Diagram for defining angular speed.

If ω and r are constant, then $v = r\omega$ is constant, and because acceleration is $\lim_{t\to 0}(\Delta v/\Delta t)$, we might wonder if there is any acceleration. The answer is most definitely *yes* because velocity is a vector quantity and the direction of $\mathbf{v}$, which is the same as $\Delta \mathbf{r}$, is constantly changing even though its magnitude is not. To calculate this acceleration, draw a figure like Figure 1.20 but expressed in terms of v instead of r. From your figure, show that

$$\Delta v = |\Delta \mathbf{v}| = v\Delta\theta \tag{1.54}$$

is in direct analogy with Equation 1.52, and show that the particle experiences an acceleration given by

$$a = \lim_{\Delta t \to 0} \frac{\Delta v}{\Delta t} = v \lim_{\Delta t \to 0} \frac{\Delta \theta}{\Delta t} = v\omega \tag{1.55}$$

Thus, we see that the particle experiences an acceleration and requires an inward force equal to $ma = mv\omega = mv^2/r$ to keep it moving in its circular orbit.

The figure is quite similar to Figure 1.20, but with v in place of r.

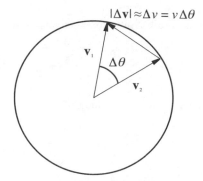

Again, if $\Delta\theta$ is small enough, the (velocity) arc length, Δv and the vector difference $|\Delta \mathbf{v}| = |\mathbf{v}_1 - \mathbf{v}_2|$ are essentially the same, and the acceleration is given by the limit in Equation 1.55. Since $a = v\omega$ and $\omega = v/r$, $ma = mv\omega = mv^2/r$.

1–53. In this problem, we shall derive the equation for the momentum of a photon, $p = h/\lambda$. You might have expected the momentum of a photon to be zero because the mass of a photon is zero, but you must remember that a photon travels at the speed of light and that we must consider its relativistic mass. For this problem, we must accept one formula from the special theory of relativity. The mass of a particle varies with its velocity according to

$$m = \frac{m_0}{[1 - (v^2/c^2)]^{1/2}} \tag{1.56}$$

where m_0, the mass as $v \to 0$, is called the *rest mass* of the particle. If we assume that the motion takes place in the x direction, the momentum is mv_x, and so

$$p_x = \frac{m_0 v_x}{[1 - (v_x^2/c^2)]^{1/2}} \tag{1.57}$$

Note that even though $m_0 = 0$ for a photon, p_x is not necessarily zero because $v_x = c$ and so $p_x \approx 0/0$, an indeterminate form. Recall that force F is equal to the rate of change of momentum or that

$$F = \frac{dp}{dt} \tag{1.58}$$

This is just Newton's second law. Kinetic energy can be defined as the work that is required to accelerate a particle from rest to some final velocity v. Because work is the integral of force times distance, the kinetic energy can be expressed as

$$T = \int_{v=0}^{v=v} F \, dx \tag{1.59}$$

where we have dropped the x subscripts for convenience. Equation 1.59 can be manipulated as

$$T = \int_{v=0}^{v=v} F\,dx = \int_{v=0}^{v=v} \frac{dp}{dt}\frac{dx}{dt}\,dt = \int_{v=0}^{v=v} v\frac{dp}{dt}\,dt$$

$$= \int_{v=0}^{v=v} v\frac{d(mv)}{dt}\,dt = \int_{v=0}^{v=v} v\,d(mv) \tag{1.60}$$

Remember now that m in these last two integrals is *not* constant but is a function of v through Equation 1.53. Note that if m were a constant, as it is in nonrelativistic (or classical) mechanics, then $T = \frac{1}{2}mv^2$, the classical result. Now substitute Equation 1.56 into Equation 1.60 to obtain

$$T = m_0 c^2 \left\{ \frac{1}{[1-(v^2/c^2)]^{1/2}} - 1 \right\} \tag{1.61}$$

To obtain this result, you need the standard integral

$$\int \frac{x\,dx}{(ax^2+b)^{3/2}} = -\frac{1}{a(ax^2+b)^{1/2}}$$

Show that Equation 1.61 reduces to the classical result as $v/c \to 0$. By combining Equations 1.56 and 1.61, show that T can be written as

$$T = (m - m_0)c^2 \tag{1.62}$$

This equation is interpreted by considering mc^2 to be the total energy E of the particle and $m_0 c^2$ to be the rest energy of the particle, so that

$$E = T + m_0 c^2$$

Lastly now, eliminate v in favor of p by using Equations 1.57 and 1.61, and show that

$$(T + m_0 c^2)^2 = (pc)^2 + (m_0 c^2)^2$$

and using Equation 1.62, write this as

$$E = [(pc)^2 + (m_0 c^2)^2]^{1/2} \tag{1.63}$$

which is our desired equation and a fundamental equation of the special theory of relativity. In the case of a photon, $m_0 = 0$ and $E = pc$. But E also equals $h\nu$ according to the quantum theory, and so we have $pc = h\nu$, which yields $p = h/\lambda$ because $c = \lambda\nu$.

Substituting Equation 1.56 into Equation 1.60 requires first finding $d(mv)$, which is

$$d(mv) = d\left\{ \frac{m_0 v}{[1-(v^2/c^2)]^{1/2}} \right\}$$

$$= \frac{m_0}{[1-(v^2/c^2)]^{1/2}}\,dv + \frac{m_0\,(v^2/c^2)}{[1-(v^2/c^2)]^{3/2}}\,dv$$

$$= \frac{m_0}{[1-(v^2/c^2)]^{3/2}}\,dv$$

Thus,

$$T = \int_{v=0}^{v=v} v \, d(mv)$$

$$= m_0 \int_{v=0}^{v=v} \frac{v}{\left[1 - (v^2/c^2)\right]^{3/2}} \, dv$$

$$= m_0 c^2 \frac{1}{\left[1 - (v^2/c^2)\right]^{1/2}} \Bigg|_{v=0}^{v=v}$$

$$= m_0 c^2 \left\{ \frac{1}{\left[1 - (v^2/c^2)\right]^{1/2}} - 1 \right\}$$

When v^2/c^2 is very small,

$$\frac{1}{\left[1 - (v^2/c^2)\right]^{1/2}} \approx 1 + \frac{v^2}{2c^2}$$

and

$$T = m_0 c^2 \left\{ \frac{1}{\left[1 - (v^2/c^2)\right]^{1/2}} - 1 \right\}$$

$$\approx m_0 c^2 \left[1 + \left(\frac{v^2}{2c^2} \right) - 1 \right]$$

$$= \frac{m_0 v^2}{2}$$

which is the classical result for kinetic energy.

Equation 1.56 gives $m_0 = m \left[1 - (v^2/c^2)\right]^{1/2}$, so

$$T = m_0 c^2 \left\{ \frac{1}{\left[1 - (v^2/c^2)\right]^{1/2}} - 1 \right\}$$

$$= \left\{ \frac{m_0 c^2}{\left[1 - (v^2/c^2)\right]^{1/2}} - m_0 c^2 \right\}$$

$$= \left\{ \frac{m \left[1 - (v^2/c^2)\right]^{1/2} c^2}{\left[1 - (v^2/c^2)\right]^{1/2}} - m_0 c^2 \right\}$$

$$= mc^2 - m_0 c^2$$

$$= \left(m - m_0\right) c^2$$

$$= E - m_0 c^2$$

Squaring Equation 1.57 gives

$$p^2 = \frac{m_0^2 v^2}{1 - (v^2/c^2)}$$

$$c^2 p^2 - p^2 v^2 = m_0^2 v^2 c^2$$

$$c^2 p^2 = v^2 \left(m_0^2 c^2 + p^2 \right)$$

$$\frac{v^2}{c^2} = \frac{p^2}{p^2 + m_0^2 c^2}$$

Then, starting with $E^2 = (T + m_0 c^2)^2$ and substituting using Equation 1.61,

$$\left(T + m_0 c^2\right)^2 = \left\{ \frac{m_0 c^2}{[1 - (v^2/c^2)]^{1/2}} \right\}^2$$

$$= \frac{m_0^2 c^4}{1 - \frac{p^2}{p^2 + m_0^2 c^2}}$$

$$= \left(p^2 + m_0^2 c^2 \right) \frac{m_0^2 c^4}{p^2 + m_0^2 c^2 - p^2}$$

$$= c^2 \left(p^2 + m_0^2 c^2 \right)$$

$$= (pc)^2 + \left(m_0 c^2 \right)^2$$

Taking the square root gives $E = \left[(pc)^2 + (m_0 c^2)^2 \right]^{1/2}$ as desired.

1–54.   In this problem we shall derive an expression for the interference pattern for a two-slit experiment. First show that $y(z) = A \cos[2\pi z/\lambda]$ represents a wave with amplitude A and wavelength λ. Now argue that $y(z, t) = A \cos[2\pi (z - vt)/\lambda]$ represents a similar wave that has been moved (translated) to the right by a distance vt. We say that $y(z, t) = A \cos[2\pi (z - vt)/\lambda]$ represents a wave form that is traveling to the right (a traveling wave) with a velocity v. If, for example, the wave is an electromagnetic wave, then $y(z, t)$ represents the electric field at a point (z, t) and we write $E(z, t) = E_0 \cos[2\pi (z - vt)/\lambda]$. If we let z_0 be the distance $S_1 P$ in Figure 1.15, then we can write the electric field at the point P as a superposition of the waves coming from the two slits, or

$$E(\theta) = E_0 \cos \left[\frac{2\pi}{\lambda} (z_0 - vt) \right] + E_0 \cos \left[\frac{2\pi}{\lambda} (z_0 + d \sin \theta - vt) \right]$$

where we write $E(\theta)$ to emphasize its dependence on the angle θ. Now use the trigonometric identity

$$\cos \alpha + \cos \beta = 2 \cos \left(\frac{\alpha + \beta}{2} \right) \cos \left(\frac{\alpha - \beta}{2} \right)$$

to write $E(\theta)$ as

$$E(\theta) = 2E_0 \cos \left(\frac{\pi d \sin \theta}{\lambda} \right) \cos \left[\frac{2\pi}{\lambda} \left(z_0 - vt + \frac{d \sin \theta}{2} \right) \right]$$

The intensity of a wave is given by the square of its amplitude (this is proven in Problem 2–18), and so

$$I(\theta) = 4E_0^2 \cos^2\left(\frac{\pi d \sin\theta}{\lambda}\right) \cos^2\left[\frac{2\pi}{\lambda}\left(z_0 - vt + \frac{d\sin\theta}{2}\right)\right]$$

The recording on the screen in Figure 1.15 is an average of $I(\theta)$ over a period of time that amounts to many cycles of the wave. Using the relation $\cos^2\alpha = \frac{1}{2}(1 + \cos 2\alpha)$, show that the average of the term $\cos^2[2\pi(z_0 - vt + d\sin\theta/2)/\lambda]$ in $I(\theta)$ is equal to 1/2, giving

$$I(\theta) = 2E_0^2 \cos^2\left(\frac{\pi d \sin\theta}{\lambda}\right)$$

as our desired result. (We will derive this same result more easily in Section 2.6 using complex numbers.) Plot $I(\theta)/E_0^2$ against θ for typical values of a two-slit interference experiment, $d = 0.010$ mm and $\lambda = 6000$ Å and compare your result to Figure 2.10.

As z varies, the term $\cos(2\pi z/\lambda)$ oscillates between -1 and $+1$, so that $y(z) = A\cos(2\pi z/\lambda)$ has amplitude A as it varies from $-A$ to A. A wave repeats in space every wavelength, so find Δz such that $y(z + \Delta z) = y(z)$.

$$\cos\frac{2\pi(z + \Delta z)}{\lambda} = \cos\frac{2\pi z}{\lambda}\cos\frac{2\pi\Delta z}{\lambda} - \sin\frac{2\pi z}{\lambda}\sin\frac{2\pi\Delta z}{\lambda} = \cos\frac{2\pi z}{\lambda}$$

This requires

$$\cos\frac{2\pi\Delta z}{\lambda} = 1$$

$$\sin\frac{2\pi\Delta z}{\lambda} = 0$$

which occurs when

$$\frac{2\pi\Delta z}{\lambda} = 2n\pi \qquad n = 0, \pm 1, \pm 2, \ldots$$

$$\Delta z = n\lambda$$

Thus, the wave repeats itself at intervals of λ, which means the wavelength is λ.

Similarly, the wave $y(z) = A\cos\left[\frac{2\pi}{\lambda}(z - vt)\right]$ has the same value at $z' = z + vt$ as the original $y(z) = A\cos\frac{2\pi z}{\lambda}$ has at z. Thus, it is a copy of the original, just shifted to the right (toward positive z) by a distance vt.

Now, writing the expression for the electric field as the superposition of two waves as suggested, and using the trigonometric identity

$$\cos\alpha + \cos\beta = 2\cos\left(\frac{\alpha + \beta}{2}\right)\cos\left(\frac{\alpha - \beta}{2}\right)$$

with $\alpha = \frac{2\pi}{\lambda}(z_0 + d\sin\theta - vt)$ and $\beta = \frac{2\pi}{\lambda}(z_0 - vt)$,

$$E(\theta) = E_0 \cos\left[\frac{2\pi}{\lambda}\left(z_0 - vt\right)\right] + E_0 \cos\left[\frac{2\pi}{\lambda}\left(z_0 + d\sin\theta - vt\right)\right]$$

$$= 2E_0 \cos\left(\frac{\pi d \sin\theta}{\lambda}\right)\cos\left[\frac{2\pi}{\lambda}\left(z_0 - vt + \frac{d\sin\theta}{2}\right)\right]$$

Squaring this last result gives the intensity, $I(\theta)$, as noted in the problem. The observed recording on the screen corresponds to an average of $I(\theta)$ over time. Only the second $\cos^2$ term in $I(\theta)$ has any time dependence, and it has the form $\cos^2(\phi - vt)$. One cycle of the wave occurs for $t = 2\pi/v$, so taking the average over N complete cycles is

$$\frac{1}{2N\pi/v}\int_{t=0}^{t=2N\pi/v}\cos^2(\phi - vt)\,dt = \frac{1}{2N\pi/v}\int_{t=0}^{t=2N\pi/v}\frac{1 + \cos(2\phi - 2vt)}{2}\,dt$$

$$= \frac{1}{2N\pi/v}\left[\frac{t}{2} - \frac{1}{4v}\sin(2\phi - 2vt)\right]\Bigg|_{t=0}^{t=2N\pi/v}$$

$$= \frac{1}{2}$$

Thus,

$$I(\theta) = 2E_0^2 \cos^2\left(\frac{\pi d \sin\theta}{\lambda}\right)$$

as desired. (Strictly speaking, this is not an expression for the intensity of the wave, which is simply the square of the amplitude, but rather, as stated, the time average of the intensity.)

A plot of $I(\theta)/E_0^2$ against θ for the slit separation and wavelength specified reproduces Figure 2.10.

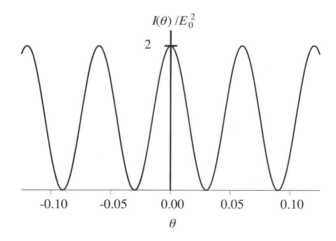

Complex Numbers

PROBLEMS AND SOLUTIONS

A–1. Find the real and imaginary parts of the following quantities:

(a) $(2-i)^3$ (b) $e^{\pi i/2}$

(c) $e^{-2+i\pi/2}$ (d) $(\sqrt{2}+2i)e^{-i\pi/2}$

In each case express the given complex number in the form $z = x + iy$ and identify $\mathrm{Re}(z) = x$, $\mathrm{Im}(z) = y$.

(a)

$$(2-i)^3 = \left[2^3 + 3(2)^2(-i) + 3(2)(-i)^2 + (-i)^3\right]$$
$$= 8 - 12i - 6 + i$$
$$= 2 - 11i$$

(b)

$$e^{i\pi/2} = \cos\left(\frac{\pi}{2}\right) + i\sin\left(\frac{\pi}{2}\right)$$
$$= 0 + i$$

(c)

$$e^{-2+i\pi/2} = e^{-2}e^{i\pi/2}$$
$$= 0 + e^{-2}i$$

(d)

$$(\sqrt{2}+2i)e^{-i\pi/2} = (\sqrt{2}+2i)\left[\cos\left(-\frac{\pi}{2}\right) + i\sin\left(-\frac{\pi}{2}\right)\right]$$
$$= (\sqrt{2}+2i)(-i)$$
$$= 2 - i\sqrt{2}$$

A–2.   If $z = x + 2iy$, then find

(a) $\text{Re}(z^*)$ (b) $\text{Re}(z^2)$

(c) $\text{Im}(z^2)$ (d) $\text{Re}(zz^*)$

(e) $\text{Im}(zz^*)$

(a) $z^* = x - 2iy$, $\text{Re}(z^*) = x$
(b) $z^2 = x^2 - 4y^2 + 4ixy$, $\text{Re}(z^2) = x^2 - 4y^2$
(c) $\text{Im}(z^2) = 4xy$
(d) $zz^* = x^2 + 4y^2$, $\text{Re}(zz^*) = x^2 + 4y^2$. More generally, since zz^* is purely real, $\text{Re}(zz^*) = zz^*$ for any complex number z.
(e) $\text{Im}(zz^*) = 0$. This can be seen in particular from the value for zz^* in the previous part, but since zz^* is always purely real, $\text{Im}(zz^*) = 0$ for any complex number z.

A–3.   Express the following complex numbers in the form $re^{i\theta}$:

(a) $6i$ (b) $4 - \sqrt{2}i$

(c) $-1 - 2i$ (d) $\pi + ei$

If $z = x + iy$, then $z = re^{i\theta}$ with $r = \sqrt{x^2 + y^2}$ and $\tan\theta = y/x$, although care should be taken to place θ in the correct quadrant through consideration of the signs of x and y individually.

(a) The point is on the y-axis in the complex plane, 6 units distant from the origin, $z = 6e^{i\pi/2}$.

(b) $r = \sqrt{16 + 2} = 3\sqrt{2}$, $\tan\theta = \dfrac{-\sqrt{2}}{4}$, and θ is in the 4th quadrant. Thus, $\theta = -0.340$, and $z = 3\sqrt{2}e^{-0.340i}$.

(c) $r = \sqrt{1 + 4} = \sqrt{5}$, $\tan\theta = \dfrac{-2}{-1}$, and θ is in the 3rd quadrant. Thus, $\theta = 1.107 + \pi$, and $z = \sqrt{5}e^{(1.107 + \pi)i}$.

(d) $r = \sqrt{\pi^2 + e^2}$, $\tan\theta = \dfrac{e}{\pi}$, and θ is in the 1st quadrant. Thus, $\theta = 0.713$, and $z = \sqrt{\pi^2 + e^2}e^{0.713i}$.

A–4.   Express the following complex numbers in the form $x + iy$:

(a) $e^{\pi/4i}$ (b) $6e^{2\pi i/3}$

(c) $e^{-(\pi/4)i + \ln 2}$ (d) $e^{-2\pi i} + e^{4\pi i}$

If $z = re^{i\theta}$, then $z = r\cos\theta + ir\sin\theta$.

(a) $z = (1)\cos\dfrac{\pi}{4} + (1)i\sin\dfrac{\pi}{4} = \dfrac{\sqrt{2}}{2} + i\dfrac{\sqrt{2}}{2}$

(b) $z = 6\cos\dfrac{2\pi}{3} + 6i\sin\dfrac{2\pi}{3} = -3 + i3\sqrt{3}$

(c) $z = e^{-(\pi/4)i + \ln 2} = e^{\ln 2}e^{-(\pi/4)i} = 2e^{-(\pi/4)i} = 2 \cos\left(-\dfrac{\pi}{4}\right) + 2i \sin\left(-\dfrac{\pi}{4}\right) = \sqrt{2} - i\sqrt{2}$

(d) $z = [\cos(-2\pi) + i \sin(-2\pi) + \cos(4\pi) + i \sin(4\pi)] = 2$

A–5. Prove that $e^{i\pi} = -1$. Comment on the nature of the numbers in this relation.

As in Problem A–4, $e^{i\pi} = \cos\pi + i \sin\pi = -1$. This is quite remarkable in that the result of raising e, a transcendental number, to the power of another transcendental number (π) times the purely imaginary number, i, is a simple integer.

A–6. Show that

$$\cos\theta = \frac{e^{i\theta} + e^{-i\theta}}{2}$$

and that

$$\sin\theta = \frac{e^{i\theta} - e^{-i\theta}}{2i}$$

Using Equation A.6,

$$e^{i\theta} = \cos\theta + i \sin\theta$$
$$e^{-i\theta} = \cos\theta - i \sin\theta$$

Then, adding these two equations gives

$$e^{i\theta} + e^{-i\theta} = 2\cos\theta$$

or

$$\cos\theta = \frac{e^{i\theta} + e^{-i\theta}}{2}$$

On the other hand, subtracting the two equations gives

$$e^{i\theta} - e^{-i\theta} = 2i \sin\theta$$

or

$$\sin\theta = \frac{e^{i\theta} + e^{-i\theta}}{2i}$$

A–7. Use Equation A.6 to derive

$$z^n = r^n(\cos\theta + i \sin\theta)^n = r^n(\cos n\theta + i \sin n\theta)$$

and from this, the formula of de Moivre:

$$(\cos \theta + i \sin \theta)^n = \cos n\theta + i \sin n\theta$$

On the one hand for $z = re^{i\theta}$,

$$z^n = r^n \left(e^{i\theta}\right)^n$$
$$= r^n \left(\cos \theta + i \sin \theta\right)^n$$

but also

$$z^n = r^n \left(e^{i\theta}\right)^n$$
$$= r^n e^{in\theta}$$
$$= r^n \left(\cos n\theta + i \sin n\theta\right)$$

Thus,

$$z^n = r^n \left(\cos \theta + i \sin \theta\right)^n = r^n \left(\cos n\theta + i \sin n\theta\right)$$

Since this last equation is true for arbitrary r, it follows that

$$(\cos \theta + i \sin \theta)^n = \cos n\theta + i \sin n\theta$$

A–8. Use the formula of de Moivre, which is given in Problem A–7, to derive the trigonometric identities

$$\cos 2\theta = \cos^2 \theta - \sin^2 \theta$$
$$\sin 2\theta = 2 \sin \theta \cos \theta$$
$$\cos 3\theta = \cos^3 \theta - 3 \cos \theta \sin^2 \theta$$
$$= 4 \cos^3 \theta - 3 \cos \theta$$
$$\sin 3\theta = 3 \cos^2 \theta \sin \theta - \sin^3 \theta$$
$$= 3 \sin \theta - 4 \sin^3 \theta$$

Starting with the formula of De Moirve for $n = 2$,

$$(\cos \theta + i \sin \theta)^2 = \cos^2 \theta - \sin^2 \theta + 2i \sin \theta \cos \theta$$
$$= \cos 2\theta + i \sin 2\theta$$

Equating the real and imaginary parts (which must separately equal each other) of this equation

$$\cos 2\theta = \cos^2 \theta - \sin^2 \theta$$
$$\sin 2\theta = 2 \sin \theta \cos \theta$$

Similarly for $n = 3$,

$$(\cos\theta + i\sin\theta)^3 = \cos^3\theta - 3\cos\theta\sin^2\theta + 3i\sin\theta\cos^2\theta - i\sin^3\theta$$

$$= \cos^3\theta - 3\cos\theta\left(1 - \cos^2\theta\right) + 3i\sin\theta\left(1 - \sin^2\theta\right) - i\sin^3\theta$$

$$= 4\cos^3\theta - 3\cos\theta + 3i\sin\theta - 4i\sin^3\theta$$

$$= \cos 3\theta + i\sin 3\theta$$

and again equating real and imaginary parts,

$$\cos 3\theta = \cos^3\theta - 3\cos\theta\sin^2\theta$$

$$= 4\cos^3\theta - 3\cos\theta$$

$$\sin 3\theta = 3\cos^2\theta\sin\theta - \sin^3\theta$$

$$= 3\sin\theta - 4\sin^3\theta$$

A–9.   Consider the set of functions

$$\Phi_m(\phi) = \frac{1}{\sqrt{2\pi}}e^{im\phi} \qquad \begin{cases} m = 0, \pm 1, \pm 2, \ldots \\ 0 \leq \phi \leq 2\pi \end{cases}$$

First show that

$$\int_0^{2\pi} d\phi\,\Phi_m(\phi) = \begin{cases} 0 & \text{for all values of } m \neq 0 \\ \sqrt{2\pi} & m = 0 \end{cases}$$

Now show that

$$\int_0^{2\pi} d\phi\,\Phi_m^*(\phi)\Phi_n(\phi) = \begin{cases} 0 & m \neq n \\ 1 & m = n \end{cases}$$

If $m = 0$, then $\Phi_0(\phi) = 1/\sqrt{2\pi}$, and

$$\int_0^{2\pi} d\phi\,\Phi_0(\phi) = \frac{1}{\sqrt{2\pi}}\int_0^{2\pi} d\phi$$

$$= \frac{\phi}{\sqrt{2\pi}}\Big|_0^{2\pi}$$

$$= \sqrt{2\pi}$$

Otherwise, if $m \neq 0$,

$$\int_0^{2\pi} d\phi\,\Phi_m(\phi) = \frac{1}{\sqrt{2\pi}}\int_0^{2\pi} d\phi\,e^{im\phi}$$

$$= \frac{e^{im\phi}}{im\sqrt{2\pi}}\Big|_0^{2\pi}$$

$$= 0$$

Considering the integral of the product of two of these functions when $m = n$,

$$\int_0^{2\pi} d\phi \, \Phi_m^*(\phi)\Phi_m(\phi) = \frac{1}{2\pi} \int_0^{2\pi} d\phi \, e^{-im\phi}e^{im\phi}$$

$$= \frac{1}{2\pi} \int_0^{2\pi} d\phi$$

$$= \frac{\phi}{2\pi}\bigg|_0^{2\pi}$$

$$= 1$$

Otherwise, for $m \neq n$,

$$\int_0^{2\pi} d\phi \, \Phi_m^*(\phi)\Phi_n(\phi) = \frac{1}{2\pi} \int_0^{2\pi} d\phi \, e^{-im\phi}e^{in\phi}$$

$$= \frac{1}{2\pi} \int_0^{2\pi} d\phi e^{i(n-m)\phi}$$

$$= \frac{e^{i(n-m)\phi}}{2i(n-m)\pi}\bigg|_0^{2\pi}$$

$$= \frac{e^{2(n-m)\pi i} - 1}{2\pi}$$

$$= 0$$

where the last result follows because for m and n each equal to an integer, $m - n =$ an integer.

A–10. This problem offers a derivation of Euler's formula. Start with

$$f(\theta) = \ln(\cos\theta + i\sin\theta) \tag{1}$$

Show that

$$\frac{df}{d\theta} = i \tag{2}$$

Now integrate both sides of equation 2 to obtain

$$f(\theta) = \ln(\cos\theta + i\sin\theta) = i\theta + c \tag{3}$$

where c is a constant of integration. Show that $c = 0$ and then exponentiate equation 3 to obtain Euler's formula.

The chain rule provides the result for the derivative with respect to θ of the logarithm of a function of θ, $\dfrac{d\ln g(\theta)}{d\theta} = \dfrac{dg(\theta)/d\theta}{g(\theta)}$. Thus,

$$\frac{df(\theta)}{d\theta} = \frac{d}{d\theta}\ln(\cos\theta + i\sin\theta)$$

$$= \frac{-\sin\theta + i\cos\theta}{\cos\theta + i\sin\theta}$$

$$= \frac{i^2\sin\theta + i\cos\theta}{\cos\theta + i\sin\theta}$$

$$= i\frac{\cos\theta + i\sin\theta}{\cos\theta + i\sin\theta}$$

$$= i$$

Integrating both sides of the equation gives

$$\int \frac{df(\theta)}{d\theta}\,d\theta = \int i\,d\theta$$

$$f(\theta) = i\theta + c$$

where c is a constant of integration. At $\theta = 0$, $f(\theta) = \ln(\cos 0 + i\sin 0) = \ln 1 = 0$, but also $f(\theta) = i(0) + c$, which implies $c = 0$, and therefore

$$\ln(\cos\theta + i\sin\theta) = i\theta$$

Exponentiation of this last result then gives,

$$e^{i\theta} = e^{\ln(\cos\theta + i\sin\theta)}$$

$$= \cos\theta + i\sin\theta$$

A–11. Using Euler's formula and assuming that x represents a real number, show that $\cos ix$ and $-i\sin ix$ are equivalent to real functions of the real variable x. These functions are defined as the hyperbolic cosine and hyperbolic sine functions, $\cosh x$ and $\sinh x$, respectively. Sketch these functions. Do they oscillate like $\sin x$ and $\cos x$?

Euler's formula gives

$$\cos ix = \frac{e^{i^2 x} + e^{-i^2 x}}{2}$$

$$= \frac{e^{-x} + e^{x}}{2} = \frac{e^{x} + e^{-x}}{2}$$

$$= \cosh x$$

and

$$-i\sin ix = -i\frac{e^{i^2 x} - e^{-i^2 x}}{2i}$$

$$= -i\frac{e^{-x} - e^{x}}{2i} = \frac{e^{x} - e^{-x}}{2}$$

$$= \sinh x$$

Both functions are seen to be real functions of the real variable x. They are sketched below. Although neither function oscillates like $\sin x$ and $\cos x$, $\sinh x$ is seen to be an odd function of x, while $\cosh x$ is an even function.

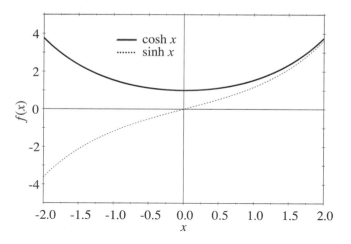

A–12. Show that $\sinh ix = i \sin x$ and that $\cosh ix = \cos x$. (See the previous problem.)

Using the result of the previous problem,

$$\sinh ix = \frac{e^{ix} - e^{-ix}}{2}$$

$$= i\frac{e^{ix} - e^{-ix}}{2i}$$

$$= i \sin x$$

and

$$\cosh ix = \frac{e^{ix} + e^{-ix}}{2}$$

$$= \cos x$$

A–13. Evaluate i^i.

$$i^i = \left(e^{i\pi/2}\right)^i$$

$$= e^{i^2\pi/2}$$

$$= e^{-\pi/2}$$

$$\approx 0.2079$$

A–14. The equation $x^2 = 1$ has two distinct roots, $x = \pm 1$. The equation $x^N = 1$ has N distinct roots, called the N roots of unity. This problem shows how to find the N roots of unity. We shall see that some of the roots turn out to be complex, so let's write the equation as $z^N = 1$. Now let $z = e^{i\theta}$ and obtain $e^{iN\theta} = 1$. Show that this must be equivalent to $e^{iN\theta} = 1$, or

$$\cos N\theta + i \sin N\theta = 1$$

Now argue that $N\theta = 2\pi n$, where n has the N distinct values $0, 1, 2, \ldots, N-1$ or that the N roots of unity are given by

$$z = e^{2\pi i n/N} \qquad n = 0, 1, 2, \ldots, N-1$$

Show that we obtain $z = 1$ and $z = \pm 1$, for $N = 1$ and $N = 2$, respectively. Now show that

$$z = 1, \ -\frac{1}{2} + i\frac{\sqrt{3}}{2}, \ \text{and} -\frac{1}{2} - i\frac{\sqrt{3}}{2}$$

for $N = 3$. Show that each of these roots is of unit magnitude. Plot these three roots in the complex plane. Now show that $z = 1, i, -1$, and $-i$ for $N = 4$ and that

$$z = 1, \ -1, \ \frac{1}{2} \pm i\frac{\sqrt{3}}{2}, \ \text{and} -\frac{1}{2} \pm i\frac{\sqrt{3}}{2}$$

for $N = 6$. Plot the four roots for $N = 4$ and the six roots for $N = 6$ in the complex plane. Compare the plots for $N = 3$, $N = 4$, and $N = 6$. Do you see a pattern?

Letting $z = re^{i\theta}$, and if $z^N = 1$, then

$$z^N = \left(re^{i\theta}\right)^N = r^N e^{iN\theta} = 1$$

Furthermore, $\left|z^N\right| = |1| = 1$, so

$$\left|z^N\right| = \left|r^N e^{iN\theta}\right|$$

$$= \sqrt{r^N e^{iN\theta} r^N e^{-iN\theta}}$$

$$= r^N$$

$$= 1$$

Thus, $r = 1$ and $z^N = e^{iN\theta} = 1$, or using Euler's formula,

$$\cos N\theta + i \sin N\theta = 1$$

or

$$\cos N\theta = 1$$

$$\sin N\theta = 0$$

Consequently, $N\theta = 2\pi n$, or $\theta = (2\pi n)/N$. Since the trigonometric functions are periodic with period 2π, $\theta + 2\pi$ is not distinct from θ, and only the N values given by $n = 0, 1, 2, \ldots, N-1$ are distinct. As a result,

$$z = e^{2\pi i n/N} \qquad n = 0, 1, 2, \ldots, N-1$$

For $N = 1$, only $n = 0$ occurs, and $z = e^0 = 1$. For $N = 2$, there are two well-known roots corresponding to $n = 0, 1$: $z = e^0 = 1$ and $z = e^{\pi i} = -1$.

For $N = 3$,

$$z = e^0 = 1 \qquad\qquad n = 0$$

$$z = e^{2\pi i/3} = \cos\frac{2\pi}{3} + i\sin\frac{2\pi}{3} = -\frac{1}{2} + i\frac{\sqrt{3}}{2} \qquad n = 1$$

$$z = e^{4\pi i/3} = \cos\frac{4\pi}{3} + i\sin\frac{4\pi}{3} = -\frac{1}{2} - i\frac{\sqrt{3}}{2} \qquad n = 2$$

The root $z = 1$ is easily seen to be of unit magnitude. For the other 2 roots,

$$|z| = \sqrt{\left(\frac{1}{2}\right)^2 + \left(\frac{\sqrt{3}}{2}\right)^2} = 1$$

Of course, this result is expected, since $\left|e^{2\pi i n/N}\right| = 1$ in general.

For $N = 4$,

$$z = e^0 = 1 \qquad\qquad n = 0$$

$$z = e^{2\pi i/4} = \cos\frac{\pi}{2} + i\sin\frac{\pi}{2} = 0 + i \qquad n = 1$$

$$z = e^{4\pi i/4} = \cos\pi + i\sin\pi = -1 + 0i \qquad n = 2$$

$$z = e^{6\pi i/4} = \cos\frac{3\pi}{2} + i\sin\frac{3\pi}{2} = 0 - i \qquad n = 3$$

Finally, for $N = 6$,

$$z = e^0 = 1 \qquad\qquad n = 0$$

$$z = e^{2\pi i/6} = \cos\frac{\pi}{3} + i\sin\frac{\pi}{3} = \frac{1}{2} + i\frac{\sqrt{3}}{2} \qquad n = 1$$

$$z = e^{4\pi i/6} = \cos\frac{2\pi}{3} + i\sin\frac{2\pi}{3} = -\frac{1}{2} + i\frac{\sqrt{3}}{2} \qquad n = 2$$

$$z = e^{6\pi i/6} = \cos\pi + i\sin\pi = -1 + 0i \qquad n = 3$$

$$z = e^{8\pi i/6} = \cos\frac{4\pi}{3} + i\sin\frac{4\pi}{3} = -\frac{1}{2} - i\frac{\sqrt{3}}{2} \qquad n = 4$$

$$z = e^{10\pi i/6} = \cos\frac{5\pi}{3} + i\sin\frac{5\pi}{3} = \frac{1}{2} - i\frac{\sqrt{3}}{2} \qquad n = 5$$

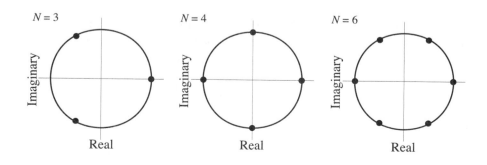

Since $|z| = 1$ for any N^{th} root of 1, all the roots lie on the circumference of a unit circle in the complex plane. One root is always $z = 1$ and lies on the x-axis at $x = 1$, and the remaining $N - 2$ roots are symmetrically distributed on the unit circle, so that all the angles between the roots are equal.

A–15. Using the results of Problem A–14, find the three distinct roots of $x^3 = 8$.

To find the desired roots, start by letting $z = x/2$. Then,

$$z^3 = \frac{x^3}{8} = \frac{8}{8} = 1$$

Problem A–14 can then be used to determine

$$z = 1, \quad -\frac{1}{2} + i\frac{\sqrt{3}}{2}, \quad \text{and} \quad -\frac{1}{2} - i\frac{\sqrt{3}}{2}$$

and $x = 2z$, or

$$x = 2, \quad -1 + i\sqrt{3}, \quad \text{and} \quad -1 - i\sqrt{3}$$

A–16. The *Schwartz inequality* says that if $z_1 = x_1 + iy_1$ and $z_2 = x_2 + iy_2$, then $x_1x_2 + y_1y_2 \leq |z_1| \cdot |z_2|$. To prove this inequality, start with its square

$$(x_1x_2 + y_1y_2)^2 \leq |z_1|^2 |z_2|^2 = (x_1^2 + y_1^2)(x_2^2 + y_2^2)$$

Now use the fact that $(x_1y_2 - x_2y_1)^2 \geq 0$ to prove the inequality.

As suggested, start with the square of the left hand side of the inequality. Since adding a positive quantity to one number results in a larger number, it must be true that

$$\left(x_1x_2 + y_1y_2\right)^2 \leq \left(x_1x_2 + y_1y_2\right)^2 + (x_1y_2 - x_2y_1)^2$$

Now continue by expanding the squared binomials,

$$\left(x_1x_2 + y_1y_2\right)^2 + (x_1y_2 - x_2y_1)^2 = x_1^2x_2^2 + 2x_1x_2y_1y_2 + y_1^2y_2^2 + x_1^2y_2^2 - 2x_1x_2y_1y_2 + x_2^2y_1^2$$

$$= x_1^2x_2^2 + y_1^2y_2^2 + x_1^2y_2^2 + x_2^2y_1^2$$

$$= \left(x_1^2 + y_1^2\right)\left(x_2^2 + y_2^2\right)$$

$$= |z_1|^2 |z_2|^2$$

Thus,

$$\left(x_1x_2 + y_1y_2\right)^2 \leq |z_1|^2 |z_2|^2$$

Finally, take the (positive) square root of both sides of the inequality to give the Schwartz inequality.

$$(x_1x_2 + y_1y_2) \leq |z_1| \cdot |z_2|$$

A–17.   The *triangle inequality* says that if z_1 and z_2 are complex numbers, then $|z_1 + z_2| \leq |z_1| + |z_2|$. To prove this inequality, start with

$$|z_1 + z_2|^2 = (x_1 + x_2)^2 + (y_1 + y_2)^2 = x_1^2 + y_1^2 + x_2^2 + y_2^2 + 2x_1x_2 + 2y_1y_2$$

Now use the Schwartz inequality (previous problem) to prove the inequality. Why do you think this is called the triangle inequality?

The information given in the problem gets us to the point

$$|z_1 + z_2|^2 = x_1^2 + y_1^2 + x_2^2 + y_2^2 + 2x_1x_2 + 2y_1y_2$$

$$= |z_1|^2 + |z_2|^2 + 2\left(x_1x_2 + y_1y_2\right)$$

Now apply the Schwartz inequality

$$|z_1 + z_2|^2 \leq |z_1|^2 + |z_2|^2 + 2|z_1| \cdot |z_2|$$

$$= \left(|z_1| + |z_2|\right)^2$$

Taking the (positive) square root of both sides completes the proof.

$$|z_1 + z_2| \leq |z_1| + |z_2|$$

If z_1 and z_2 are considered vectors in the complex plane, then by vector addition, they along with $z_1 + z_2$ form a triangle in the complex plane. The triangle inequality may then be interpreted as saying that the sum of the lengths of two sides of a triangle is greater than or equal to the length of the third.

CHAPTER 2

The Classical Wave Equation

PROBLEMS AND SOLUTIONS

2–1. Find the general solutions to the following differential equations.

(a) $\dfrac{d^2y}{dx^2} - 4\dfrac{dy}{dx} + 3y = 0$

(b) $\dfrac{d^2y}{dx^2} + 6\dfrac{dy}{dx} = 0$

(c) $\dfrac{dy}{dx} + 3y = 0$

(d) $\dfrac{d^2y}{dx^2} + 2\dfrac{dy}{dx} - y = 0$

(e) $\dfrac{d^2y}{dx^2} - 3\dfrac{dy}{dx} + 2y = 0$

The differential equations in this problem are ordinary linear differential equations. Their solutions are of the form $y(x) = e^{\alpha x}$. To evaluate α, substitute $y(x)$ into each equation. Note that $dy/dx = \alpha e^{\alpha x}$ and $d^2y/dx^2 = \alpha^2 e^{\alpha x}$. Additionally, because $e^{\alpha x}$ is nonzero, the entire differential equation can be divided through by this term. The general solution is a linear combination of the individual solutions. In the following, c_1 and c_2 represent arbitrary constants.

(a)

$$\frac{d^2y}{dx^2} - 4\frac{dy}{dx} + 3y = 0$$

$$\alpha^2 e^{\alpha x} - 4\alpha e^{\alpha x} + 3e^{\alpha x} = 0$$

$$\alpha^2 - 4\alpha + 3 = 0$$

$$(\alpha - 3)(\alpha - 1) = 0$$

$$\alpha = 3 \text{ or } \alpha = 1$$

The two solutions are $y(x) = e^{3x}$ and $y(x) = e^x$. The general solution is

$$y(x) = c_1 e^{3x} + c_2 e^x$$

(b)

$$\frac{d^2y}{dx^2} + 6\frac{dy}{dx} = 0$$

$$\alpha^2 e^{\alpha x} + 6\alpha e^{\alpha x} = 0$$

$$\alpha^2 + 6\alpha = 0$$

$$\alpha\,(\alpha + 6) = 0$$

$$\alpha = 0 \text{ or } \alpha = -6$$

The two solutions are $y(x) = 1$ and $y(x) = e^{-6x}$. The general solution is

$$y(x) = c_1 + c_2 e^{-6x}$$

(c)

$$\frac{dy}{dx} + 3y = 0$$

$$\alpha e^{\alpha x} + 3e^{\alpha x} = 0$$

$$\alpha + 3 = 0$$

$$\alpha = -3$$

The solution is $y(x) = e^{-3x}$. The general solution is

$$y(x) = c_1 e^{-3x}$$

(d)

$$\frac{d^2y}{dx^2} + 2\frac{dy}{dx} - y = 0$$

$$\alpha^2 e^{\alpha x} + 2\alpha e^{\alpha x} - e^{\alpha x} = 0$$

$$\alpha^2 + 2\alpha - 1 = 0$$

$$\alpha = \frac{1}{2}\left(-2 \pm \sqrt{4 + 4}\right) = -1 \pm \sqrt{2}$$

The two solutions are $y(x) = e^{\left(-1+\sqrt{2}\right)x}$ and $y(x) = e^{\left(-1-\sqrt{2}\right)x}$. The general solution is

$$y(x) = c_1 e^{\left(-1+\sqrt{2}\right)x} + c_2 e^{\left(-1-\sqrt{2}\right)x}$$

(e)

$$\frac{d^2y}{dx^2} - 3\frac{dy}{dx} + 2y = 0$$

$$\alpha^2 e^{\alpha x} - 3\alpha e^{\alpha x} + 2e^{\alpha x} = 0$$

$$\alpha^2 - 3\alpha + 2 = 0$$

$$(\alpha - 2)\,(\alpha - 1) = 0$$

$$\alpha = 2 \text{ or } \alpha = 1$$

The two solutions are $y(x) = e^{2x}$ and $y(x) = e^x$. The general solution is

$$y(x) = c_1 e^{2x} + c_2 e^x$$

2–2. Solve the following differential equations:

(a) $\dfrac{d^2y}{dx^2} - 4y = 0 \qquad y(0) = 2 \qquad \dfrac{dy}{dx}(\text{at } x = 0) = 4$

(b) $\dfrac{d^2y}{dx^2} - 5\dfrac{dy}{dx} + 6y = 0 \qquad y(0) = -1 \qquad \dfrac{dy}{dx}(\text{at } x = 0) = 0$

(c) $\dfrac{dy}{dx} - 2y = 0 \qquad y(0) = 2$

After obtaining the general solution to each of the differential equations using the method described in Problem 2–1, the arbitrary constants are determined by applying the boundary conditions.

(a)

$$\frac{d^2y}{dx^2} - 4y = 0$$

$$\alpha^2 e^{\alpha x} - 4e^{\alpha x} = 0$$

$$\alpha^2 - 4 = 0$$

$$\alpha = \pm 2$$

The general solution is

$$y(x) = c_1 e^{2x} + c_2 e^{-2x}$$

and the derivative of $y(x)$ with respect to x is

$$\frac{dy}{dx} = 2c_1 e^{2x} - 2c_2 e^{-2x}$$

The boundary conditions on $y(x)$ and dy/dx at $x = 0$ give

$$c_1 + c_2 = 2$$

$$2c_1 - 2c_2 = 4$$

Solving these simultaneous equations gives $c_1 = 2$ and $c_2 = 0$. The solution for the differential equation with the specified boundary conditions is

$$y(x) = 2e^{2x}$$

(b)

$$\frac{d^2y}{dx^2} - 5\frac{dy}{dx} + 6y = 0$$

$$\alpha^2 e^{\alpha x} - 5\alpha e^{\alpha x} + 6e^{\alpha x} = 0$$

$$\alpha^2 - 5\alpha + 6 = 0$$

$$(\alpha - 3)(\alpha - 2) = 0$$

$$\alpha = 3 \text{ or } \alpha = 2$$

The general solution is

$$y(x) = c_1 e^{3x} + c_2 e^{2x}$$

and the derivative of $y(x)$ with respect to x is

$$\frac{dy}{dx} = 3c_1 e^{3x} + 2c_2 e^{2x}$$

The boundary conditions on $y(x)$ and dy/dx at $x = 0$ give

$$c_1 + c_2 = -1$$

$$3c_1 + 2c_2 = 0$$

Solving these simultaneous equations gives $c_1 = 2$ and $c_2 = -3$. The solution for the differential equation with the specified boundary conditions is

$$y(x) = 2e^{3x} - 3e^{2x}$$

(c)

$$\frac{dy}{dx} - 2y = 0$$

$$\alpha e^{\alpha x} - 2e^{\alpha x} = 0$$

$$\alpha - 2 = 0$$

$$\alpha = 2$$

The general solution is

$$y(x) = c_1 e^{2x}$$

The boundary condition on $y(x)$ at $x = 0$ gives

$$c_1 = 2$$

Thus, the solution for the differential equation with the specified boundary condition is

$$y(x) = 2e^{2x}$$

2–3. Prove that $x(t) = \cos \omega t$ oscillates with a frequency $\nu = \omega/2\pi$. Prove that $x(t) = A \cos \omega t + B \sin \omega t$ oscillates with the same frequency, $\omega/2\pi$.

If τ is the period of oscillation and n is an integer, then

$$x(t + n\tau) = x(t)$$

For $x(t) = \cos \omega t$,

$$x(t + n\tau) = x(t)$$

$$\cos(\omega t + n\omega\tau) = \cos \omega t$$

For this relation to be satisfied, $\omega\tau = 2\pi$, or

$$\tau = \frac{2\pi}{\omega}$$

Thus, the frequency of oscillation is

$$\nu = \frac{1}{\tau} = \frac{\omega}{2\pi}$$

Similarly, for $x(t) = A \cos \omega t + B \sin \omega t$,

$$x(t + n\tau) = x(t)$$

$$A \cos(\omega t + n\omega\tau) + B \sin(\omega t + n\omega\tau) = A \cos \omega t + B \sin \omega t$$

For this relation to be satisfied, $\omega\tau = 2\pi$. In other words, both $\cos \omega t$ and $\sin \omega t$ oscillate with the same period,

$$\tau = \frac{2\pi}{\omega}$$

Thus, the frequency of oscillation is

$$\nu = \frac{1}{\tau} = \frac{\omega}{2\pi}$$

2–4. Solve the following differential equations:

(a) $\dfrac{d^2x}{dt^2} + \omega^2 x(t) = 0$ $x(0) = 0$ $\dfrac{dx}{dt}$ (at $t = 0$) $= v_0$

(b) $\dfrac{d^2x}{dt^2} + \omega^2 x(t) = 0$ $x(0) = A$ $\dfrac{dx}{dt}$ (at $t = 0$) $= v_0$

Prove in both cases that $x(t)$ oscillates with frequency $\omega/2\pi$.

As illustrated in Example 2–3, the general solution to $\dfrac{d^2x}{dt^2} + \omega^2 x(t) = 0$ is

$$x(t) = c_3 \cos \omega t + c_4 \sin \omega t$$

The derivative of $x(t)$ with respect to t is

$$\frac{dx}{dt} = -c_3\omega \sin \omega t + c_4\omega \cos \omega t$$

Apply the initial conditions to the general solution and to its derivative for each part of the problem.

(a) The initial conditions on $x(t)$ and dx/dt at $t = 0$ give

$$c_3 = 0$$

$$c_4 \omega = v_0$$

The last equation gives $c_4 = v_0/\omega$. Thus, the solution for the differential equation with the specified boundary conditions is

$$x(t) = \frac{v_0}{\omega} \sin \omega t$$

As shown in Problem 2–3, $\sin \omega t$ oscillates with a frequency of $\omega/2\pi$. Thus, $x(t)$ also oscillates with this frequency.

(b) The initial conditions on $x(t)$ and dx/dt at $t = 0$ give

$$c_3 = A$$

$$c_4 \omega = v_0$$

The last equation gives $c_4 = v_0/\omega$. Thus, the solution for the differential equation with the specified boundary conditions is

$$x(t) = A \cos \omega t + \frac{v_0}{\omega} \sin \omega t$$

As shown in Problem 2–3, both $\cos \omega t$ and $\sin \omega t$ oscillate with a frequency of $\omega/2\pi$. Thus, $x(t)$ also oscillates with this frequency.

2–5. The general solution to the differential equation

$$\frac{d^2 x}{dt^2} + \omega^2 x(t) = 0$$

is

$$x(t) = c_1 \cos \omega t + c_2 \sin \omega t$$

For convenience, we often write this solution in the equivalent forms

$$x(t) = A \sin(\omega t + \phi) \qquad \text{or} \qquad x(t) = B \cos(\omega t + \psi)$$

Show that all three of these expressions for $x(t)$ are equivalent. Derive equations for A and ϕ in terms of c_1 and c_2, and for B and ψ in terms of c_1 and c_2. Show that all three forms of $x(t)$ oscillate with frequency $\omega/2\pi$. *Hint:* Use the trigonometric identities

$$\sin(\alpha + \beta) = \sin \alpha \cos \beta + \cos \alpha \sin \beta \qquad \text{and} \qquad \cos(\alpha + \beta) = \cos \alpha \cos \beta - \sin \alpha \sin \beta$$

First examine the function $x(t) = A \sin(\omega t + \phi)$. Expanding this function, we have

$$x(t) = A \sin \omega t \cos \phi + A \cos \omega t \sin \phi$$

For this expression to be equivalent to $x(t) = c_1 \cos \omega t + c_2 \sin \omega t$, the following conditions must apply:

$$A \sin \phi = c_1$$

$$A \cos \phi = c_2$$

To write ϕ in terms of c_1 and c_2, divide $A \sin \phi$ by $A \cos \phi$ to give

$$\frac{A \sin \phi}{A \cos \phi} = \frac{c_1}{c_2}$$

$$\tan \phi = \frac{c_1}{c_2}$$

$$\phi = \tan^{-1} \frac{c_1}{c_2}$$

To write A in terms of c_1 and c_2, square $A \sin \phi$ and $A \cos \phi$, respectively, and sum.

$$A^2 \sin^2 \phi + A^2 \cos^2 \phi = c_1^2 + c_2^2$$

$$A^2 \left(\sin^2 \phi + \cos^2 \phi \right) = c_1^2 + c_2^2$$

$$A^2 = c_1^2 + c_2^2$$

$$A = \left(c_1^2 + c_2^2 \right)^{1/2}$$

Now consider the function $x(t) = B \cos(\omega t + \psi)$. Expanding this function gives

$$x(t) = B \cos \omega t \cos \psi - B \sin \omega t \sin \psi$$

For this expression to be equivalent to $x(t) = c_1 \cos \omega t + c_2 \sin \omega t$, the following conditions must apply:

$$B \cos \psi = c_1$$

$$B \sin \psi = -c_2$$

To write ψ in terms of c_1 and c_2, divide $B \sin \psi$ by $B \cos \psi$ to give

$$\frac{B \sin \psi}{B \cos \psi} = -\frac{c_2}{c_1}$$

$$\tan \psi = -\frac{c_2}{c_1}$$

$$\psi = \tan^{-1} \left(-\frac{c_1}{c_2} \right)$$

To write B in terms of c_1 and c_2, square $B \cos \psi$ and $B \sin \psi$, respectively, and sum.

$$B^2 \cos^2 \psi + B^2 \sin^2 \psi = c_1^2 + c_2^2$$

$$B^2 \left(\cos^2 \psi + \sin^2 \psi \right) = c_1^2 + c_2^2$$

$$B^2 = c_1^2 + c_2^2$$

$$B = \left(c_1^2 + c_2^2 \right)^{1/2}$$

Problem 2–3 shows that $x(t) = c_1 \cos \omega t + c_2 \sin \omega t$ oscillates with frequency $\omega/2\pi$. For $x(t) = A \sin(\omega t + \phi)$, if τ is the period of oscillation and n is an integer, then $x(t + n\tau) = x(t)$, that is,

$$A \sin [\omega (t + n\tau) + \phi] = A \sin(\omega t + \phi)$$

$$\sin (\omega t + n\omega\tau + \phi) = \sin(\omega t + \phi)$$

For this relation to be satisfied, $\omega\tau = 2\pi$, or

$$\tau = \frac{2\pi}{\omega}$$

Thus, the frequency of oscillation is

$$\nu = \frac{1}{\tau} = \frac{\omega}{2\pi}$$

Similarly, for $x(t) = B \cos(\omega t + \psi)$, if τ is the period of oscillation and n is an integer, then $x(t + n\tau) = x(t)$, that is,

$$B \cos [\omega (t + n\tau) + \psi] = B \cos(\omega t + \psi)$$

$$\cos (\omega t + n\omega\tau + \psi) = \cos(\omega t + \psi)$$

For this relation to be satisfied, $\omega\tau = 2\pi$, or

$$\tau = \frac{2\pi}{\omega}$$

Thus, the frequency of oscillation is

$$\nu = \frac{1}{\tau} = \frac{\omega}{2\pi}$$

All three forms are seen to have the same oscillation frequency, $\omega/2\pi$. Note that both ϕ and ψ affect only the phase of the oscillation, not the frequency.

2–6. In all the differential equations that we have discussed so far, the values of the exponents α that we have found have been either real or purely imaginary. Let us consider a case in which α turns out to be complex. Consider the equation

$$\frac{d^2 y}{dx^2} + 2\frac{dy}{dx} + 10y = 0$$

If we substitute $y(x) = e^{\alpha x}$ into this equation, we find that $\alpha^2 + 2\alpha + 10 = 0$ or that $\alpha = -1 \pm 3i$. The general solution is

$$y(x) = c_1 e^{(-1+3i)x} + c_2 e^{(-1-3i)x}$$

$$= c_1 e^{-x} e^{3ix} + c_2 e^{-x} e^{-3ix}$$

Show that $y(x)$ can be written in the equivalent form

$$y(x) = e^{-x}(c_3 \cos 3x + c_4 \sin 3x)$$

Thus we see that complex values of the α's lead to trigonometric solutions modulated by an exponential factor. Solve the following equations:

(a) $\dfrac{d^2 y}{dx^2} + 2\dfrac{dy}{dx} + 2y = 0$

(b) $\dfrac{d^2 y}{dx^2} - 6\dfrac{dy}{dx} + 25y = 0$

(c) $\dfrac{d^2 y}{dx^2} + 2\beta\dfrac{dy}{dx} + (\beta^2 + \omega^2)y = 0$

(d) $\dfrac{d^2 y}{dx^2} + 4\dfrac{dy}{dx} + 5y = 0 \qquad y(0) = 1 \qquad \dfrac{dy}{dx}(\text{at } x = 0) = -3$

Using $e^{\pm ix} = \cos x \pm i \sin x$, $y(x)$ can be written in terms of trigonometric functions:

$$y(x) = c_1 e^{-x} e^{3ix} + c_2 e^{-x} e^{-3ix}$$

$$= e^{-x}\left(c_1 e^{3ix} + c_2 e^{-3ix}\right)$$

$$= e^{-x}\left(c_1 \cos 3x + c_1 i \sin 3x + c_2 \cos 3x - c_2 i \sin 3x\right)$$

$$= e^{-x}\left[(c_1 + c_2)\cos 3x + (c_1 i - c_2 i)\sin 3x\right]$$

$$= e^{-x}(c_3 \cos 3x + c_4 \sin 3x)$$

whereupon we set $c_3 = c_1 + c_2$ and $c_4 = i(c_1 - c_2)$.

To solve the differential equations in the rest of the problems, substitute $y(x) = e^{\alpha x}$ into each equation and solve for α.

(a)

$$\frac{d^2 y}{dx^2} + 2\frac{dy}{dx} + 2y = 0$$

$$\alpha^2 + 2\alpha + 2 = 0$$

$$\alpha = \frac{-2 \pm \sqrt{4-8}}{2} = -1 \pm i$$

The general solution is $y(x) = c_1 e^{(-1+i)x} + c_2 e^{(-1-i)x}$, or, written in terms of trigonometric functions, $y(x) = e^{-x}(c_3 \cos x + c_4 \sin x)$.

(b)

$$\frac{d^2 y}{dx^2} - 6\frac{dy}{dx} + 25y = 0$$

$$\alpha^2 - 6\alpha + 25 = 0$$

$$\alpha = \frac{6 \pm \sqrt{36 - 100}}{2} = 3 \pm 4i$$

The general solution is $y(x) = c_1 e^{(3+4i)x} + c_2 e^{(3-4i)x}$, or, written in terms of trigonometric functions, $y(x) = e^{3x}(c_3 \cos 4x + c_4 \sin 4x)$.

(c)

$$\frac{d^2 y}{dx^2} + 2\beta \frac{dy}{dx} + (\beta^2 + \omega^2)y = 0$$

$$\alpha^2 + 2\beta\alpha + (\beta^2 + \omega^2) = 0$$

$$\alpha = \frac{-2\beta \pm \sqrt{4\beta^2 - 4(\beta^2 + \omega^2)}}{2} = -\beta \pm i\omega$$

The general solution is $y(x) = c_1 e^{(-\beta+i\omega)x} + c_2 e^{(-\beta-i\omega)x}$, or, written in terms of trigonometric functions, $y(x) = e^{-\beta x}(c_3 \cos \omega x + c_4 \sin \omega x)$.

(d)

$$\frac{d^2 y}{dx^2} + 4\frac{dy}{dx} + 5y = 0$$

$$\alpha^2 + 4\alpha + 5 = 0$$

$$\alpha = \frac{-4 \pm \sqrt{16 - 20}}{2} = -2 \pm i$$

The general solution is $y(x) = c_1 e^{(-2+i)x} + c_2 e^{(-2-i)x}$, or, written in terms of trigonometric functions, $y(x) = e^{-2x}(c_3 \cos x + c_4 \sin x)$.

The derivative of $y(x)$ with respect to x is

$$\frac{dy}{dx} = -2e^{-2x}(c_3 \cos x + c_4 \sin x) + e^{-2x}(-c_3 \sin x + c_4 \cos x)$$

The boundary conditions on $y(x)$ and dy/dx at $x = 0$ give

$$c_3 = 1$$

$$-2c_3 + c_4 = -3$$

The last equation gives $c_4 = -1$. Therefore, the solution for the differential equation with the specified boundary conditions is

$$y(x) = e^{-2x}(\cos x - \sin x)$$

2–7. This problem develops the idea of a classical harmonic oscillator. Consider a mass m attached to a spring, as shown in Figure 2.11. Suppose there is no gravitational force acting on m so that the only force is from the spring. Let the relaxed or undistorted length of the spring be x_0. Hooke's law says that the force acting on the mass m is $f = -k(x - x_0)$, where k is a constant characteristic of the spring and is called the force constant of the spring. Note that the minus sign indicates the direction of the force: to the left if $x > x_0$ (extended) and to the right if $x < x_0$ (compressed). The momentum of the mass is

$$p = m\frac{dx}{dt} = m\frac{d(x - x_0)}{dt}$$

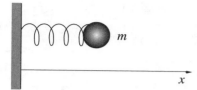

FIGURE 2.11
A body of mass m connected to a wall by a spring.

Newton's second law says that the rate of change of momentum is equal to a force

$$\frac{dp}{dt} = f$$

Replacing $f(x)$ by Hooke's law, show that

$$m\frac{d^2x}{dt^2} = -k(x - x_0)$$

Upon letting $\xi = x - x_0$ be the displacement of the spring from its undistorted length, then

$$m\frac{d^2\xi}{dt^2} + k\xi = 0$$

Given that the mass starts at $\xi = 0$ with an intial velocity v_0, show that the displacement is given by

$$\xi(t) = v_0 \left(\frac{m}{k}\right)^{1/2} \sin\left[\left(\frac{k}{m}\right)^{1/2} t\right]$$

Interpret and discuss this solution. What does the motion look like? What is the frequency? What is the amplitude?

Replacing $f(x)$ by Hooke's law in $\frac{dp}{dt} = f$, we have

$$\frac{dp}{dt} = -k\left(x - x_0\right)$$

The momentum of the mass is $p = m\frac{dx}{dt} = m\frac{d(x - x_0)}{dt}$. Thus,

$$\frac{d}{dt}\left(m\frac{dx}{dt}\right) = \frac{d}{dt}\left[m\frac{d(x - x_0)}{dt}\right] = -k\left(x - x_0\right)$$

$$m\frac{d^2x}{dt^2} = m\frac{d^2(x - x_0)}{dt^2} = -k\left(x - x_0\right)$$

Letting $\xi = x - x_0$, the above equation becomes

$$m\frac{d^2\xi}{dt^2} = -k\xi$$

$$\frac{d^2\xi}{dt^2} + \frac{k}{m}\xi = 0$$

The general solution to this differential equation is analogous to that given in Example 2-3:

$$\xi(t) = c_3 \cos\left(\sqrt{\frac{k}{m}}\,t\right) + c_4 \sin\left(\sqrt{\frac{k}{m}}\,t\right)$$

The derivative of $\xi(t)$ with respect to t is

$$\frac{d\xi}{dt} = -c_3\sqrt{\frac{k}{m}}\sin\left(\sqrt{\frac{k}{m}}\,t\right) + c_4\sqrt{\frac{k}{m}}\cos\left(\sqrt{\frac{k}{m}}\,t\right)$$

The initial conditions on $\xi(t)$ and $d\xi/dt$ (that is, the velocity of the oscillator) at $x = 0$ give

$$c_3 = 0$$

$$c_4\sqrt{\frac{k}{m}} = v_0$$

The last equation gives $c_4 = v_0\sqrt{m/k}$. Therefore, the solution for the differential equation with the specified initial conditions is

$$\xi(t) = v_0\sqrt{\frac{m}{k}}\sin\left(\sqrt{\frac{k}{m}}\,t\right)$$

According to this solution, the motion of the classical harmonic oscillator is sinusoidal. The frequency is $\dfrac{1}{2\pi}\sqrt{\dfrac{k}{m}}$ (See Problem 2–3). The amplitude of the oscillator is $v_0\sqrt{\dfrac{m}{k}}$.

2–8. Modify Problem 2–7 to the case where the mass is moving through a viscous medium with a viscous force proportional to but opposite the velocity. Show that the equation of motion is

$$m\frac{d^2\xi}{dt^2} + \gamma\frac{d\xi}{dt} + k\xi = 0$$

where γ is the viscous drag coefficient. Solve this equation and discuss the behavior of $\xi(t)$ for various values of m, γ, and k. This system is called a *damped harmonic oscillator*.

There are two forces acting on the mass. One is due to the spring, and as shown in Problem 2–7, the restoring force is $-k\xi$. The other is a viscous force proportional but opposite to the velocity, that is, $-\gamma\,(dx/dt)$ where γ is the proportionality constant, the viscous drag coefficient. Since $dx/dt = d\xi/dt$, the force due to the viscous medium is $-\gamma\,(d\xi/dt)$. Putting these together, the equation of motion is

$$m\frac{d^2\xi}{dt^2} = f = -k\xi - \gamma\frac{d\xi}{dt}$$

$$m\frac{d^2\xi}{dt^2} + \gamma\frac{d\xi}{dt} + k\xi = 0$$

$$\frac{d^2\xi}{dt^2} + \frac{\gamma}{m}\frac{d\xi}{dt} + \frac{k}{m}\xi = 0$$

To solve this equation, substitute $\xi = e^{\alpha t}$ into the above equation and divide through by $e^{\alpha t}$ to give

$$\alpha^2 + \frac{\gamma}{m}\alpha + \frac{k}{m} = 0$$

$$\alpha = \frac{1}{2}\left(-\frac{\gamma}{m} \pm \sqrt{\frac{\gamma^2}{m^2} - \frac{4k}{m}}\right)$$

Therefore, the general solution to the equation of motion is

$$\xi = c_1 \exp\left[\frac{1}{2}\left(-\frac{\gamma}{m} + \sqrt{\frac{\gamma^2}{m^2} - \frac{4k}{m}}\right)t\right] + c_2 \exp\left[\frac{1}{2}\left(-\frac{\gamma}{m} - \sqrt{\frac{\gamma^2}{m^2} - \frac{4k}{m}}\right)t\right]$$

$$= \exp\left(-\frac{\gamma t}{2m}\right)\left[c_1 \exp\left(\sqrt{\frac{\gamma^2}{4m^2} - \frac{k}{m}}\,t\right) + c_2 \exp\left(-\sqrt{\frac{\gamma^2}{4m^2} - \frac{k}{m}}\,t\right)\right]$$

If $\dfrac{\gamma^2}{4m^2} < \dfrac{k}{m}$, then the displacement is

$$\xi = \exp\left(-\frac{\gamma t}{2m}\right)\left[c_3 \cos\left(\sqrt{\frac{k}{m} - \frac{\gamma^2}{4m^2}}\,t\right) + c_4 \sin\left(\sqrt{\frac{k}{m} - \frac{\gamma^2}{4m^2}}\,t\right)\right]$$

The mass oscillates with a frequency of $\nu = \dfrac{1}{2\pi}\sqrt{\dfrac{k}{m} - \dfrac{\gamma^2}{4m^2}}$. The frequency of oscillation is decreased by the viscous medium to an extent depending on the values of k, m, and γ. The greater the value of k, that is, the stronger the spring, the less the effect of the viscous medium on the frequency. On the other hand, a smaller mass or a greater viscosity leads to a greater effect.

Furthermore, if $\dfrac{\gamma^2}{4m^2} > \dfrac{k}{m}$, the mass no longer exhibits harmonic motion.

2–9. Consider the linear second-order differential equation

$$\frac{d^2y}{dx^2} + a_1(x)\frac{dy}{dx} + a_0(x)y(x) = 0$$

Note that this equation is linear because $y(x)$ and its derivatives appear only to the first power and there are no cross terms. It does not have constant coefficients, however, and there is no general, simple method for solving it like there is if the coefficients were constants. In fact, each equation of this type must be treated more or less individually. Nevertheless, because it is linear, we must have that, if $y_1(x)$ and $y_2(x)$ are any two solutions, then a linear combination,

$$y(x) = c_1 y_1(x) + c_2 y_2(x)$$

where c_1 and c_2 are constants, is also a solution. Prove that $y(x)$ is a solution.

To prove that $y(x)$ is a solution, show that $\dfrac{d^2y}{dx^2} + a_1(x)\dfrac{dy}{dx} + a_0(x)y(x) = 0$.

Given that $y(x) = c_1 y_1(x) + c_2 y_2(x)$, write $\dfrac{d^2 y}{dx^2} + a_1(x)\dfrac{dy}{dx} + a_0(x)y(x)$ in terms of y_1 and y_2:

$$\dfrac{d^2 y}{dx^2} + a_1(x)\dfrac{dy}{dx} + a_0(x)y(x)$$

$$= \dfrac{d^2}{dx^2}\left(c_1 y_1 + c_2 y_2\right) + a_1(x)\dfrac{d}{dx}\left(c_1 y_1 + c_2 y_2\right) + a_0(x)\left(c_1 y_1 + c_2 y_2\right)$$

$$= c_1\left[\dfrac{d^2 y_1}{dx^2} + a_1(x)\dfrac{dy_1}{dx} + a_0(x)y_1(x)\right] + c_2\left[\dfrac{d^2 y_2}{dx^2} + a_1(x)\dfrac{dy_2}{dx} + a_0(x)y_2(x)\right]$$

Since $y_1(x)$ and $y_2(x)$ are solutions to the differential equation, the following relations hold:

$$\dfrac{d^2 y_1}{dx^2} + a_1(x)\dfrac{dy_1}{dx} + a_0(x)y_1(x) = 0$$

$$\dfrac{d^2 y_2}{dx^2} + a_1(x)\dfrac{dy_2}{dx} + a_0(x)y_2(x) = 0$$

Substituting these relations into the expression for $\dfrac{d^2 y}{dx^2} + a_1(x)\dfrac{dy}{dx} + a_0(x)y(x)$ derived earlier gives

$$\dfrac{d^2 y}{dx^2} + a_1(x)\dfrac{dy}{dx} + a_0(x)y(x) = c_1\,(0) + c_2\,(0) = 0$$

Thus, $y(x)$, a linear combination of the solutions $y_1(x)$ and $y_2(x)$, is indeed a solution to the differential equation.

2–10. We will see in Chapter 3 that the Schrödinger equation for a particle of mass m that is constrained to move freely along a line between 0 and a is

$$\dfrac{d^2 \psi}{dx^2} + \left(\dfrac{8\pi^2 m E}{h^2}\right)\psi(x) = 0$$

with the boundary condition

$$\psi(0) = \psi(a) = 0$$

In this equation, E is the energy of the particle and $\psi(x)$ is its wave function. Solve this differential equation for $\psi(x)$, apply the boundary conditions, and show that the energy can have only the values

$$E_n = \dfrac{n^2 h^2}{8ma^2} \qquad n = 1, 2, 3, \ldots$$

or that the energy is quantized.

By analogy to Example 2–3, the solution to the Schrödinger equation for this particle is

$$\psi(x) = c_3 \cos\left(\sqrt{\dfrac{8\pi^2 m E}{h^2}}\,x\right) + c_4 \sin\left(\sqrt{\dfrac{8\pi^2 m E}{h^2}}\,x\right)$$

The boundary conditions on $\psi(x)$ at $x = 0$ and $x = a$ give

$$c_3 = 0$$

$$c_3 \cos\left(\sqrt{\frac{8\pi^2 m E}{h^2}}\, a\right) + c_4 \sin\left(\sqrt{\frac{8\pi^2 m E}{h^2}}\, a\right) = 0$$

With the value for c_3 determined, the last equation becomes

$$c_4 \sin\left(\sqrt{\frac{8\pi^2 m E}{h^2}}\, a\right) = 0$$

This relation is satisfied when $c_4 = 0$ (which, when combined with $c_3 = 0$, gives the trivial solution, $\psi = 0$ at all values of x) or

$$\sin\left(\sqrt{\frac{8\pi^2 m E}{h^2}}\, a\right) = 0$$

$$\sqrt{\frac{8\pi^2 m E}{h^2}}\, a = n\pi$$

where n is a nonzero integer. ($n = 0$ implies either $a = 0$, that is, the particle moves along a line with no length or $E = 0$. Either case gives a trivial solution.)

Rearranging the last equation to solve for E gives

$$E_n = \frac{n^2 h^2}{8 m a^2}$$

The subscript n for E denotes that energy is dependent on n. Since n can take on only nonzero integral values (1, 2, 3, ...), the energy of the particle is not continuous but quantized. The wave function for the particle is $\psi(x) = c_4 \sin\left(\sqrt{\frac{8\pi^2 m E}{h^2}}\, x\right) = c_4 \sin\dfrac{n\pi x}{a}$.

2–11. Prove that the number of nodes for a vibrating string clamped at both ends is $n - 1$ for the nth harmonic.

The motion along the vibrating string clamped at both ends is described by Equation 2.20:

$$X(x) = B \sin\frac{n\pi x}{l}$$

To find the number of nodes, first determine x where there is no displacement, that is, $X(x) = 0$. This condition is satisfied when

$$\sin\frac{n\pi x}{l} = 0$$

or when $n\pi x/l$ is an integral multiple of π. Since the end points of the string are $x = 0$ and $x = l$, the values of $n\pi x/l$ range from 0 to $n\pi$, that is,

$$\frac{n\pi x}{l} = 0, \pi, 2\pi, \ldots, n\pi$$

$$x = 0, \frac{l}{n}, \frac{2l}{n}, \ldots, l$$

According to the last equation, the number of points along the string where there is no displacement is $n + 1$. Since the two end points of the string are clamped and are not counted as nodes, the number of nodes for a vibrating string is $n - 1$.

2–12.   Prove that

$$y(x, t) = A \sin\left[\frac{2\pi}{\lambda}(x - vt)\right]$$

is a wave of wavelength λ and frequency $\nu = v/\lambda$ traveling to the right with a velocity v.

If λ is the wavelength, then $y(x + \lambda, t)$ should equal $y(x, t)$.

$$y(x + \lambda, t) = A \sin\left[\frac{2\pi}{\lambda}(x + \lambda - vt)\right]$$

$$= A \sin\left[\frac{2\pi}{\lambda}(x - vt) + 2\pi\right]$$

$$= A \sin\left[\frac{2\pi}{\lambda}(x - vt)\right]$$

$$= y(x, t)$$

Indeed, λ is the wavelength of $y(x, t)$.

If v/λ is the frequency, or equivalently, λ/v is the period, then $y(x, t + \lambda/v)$ should equal $y(x, t)$.

$$y(x, t + \lambda/v) = A \sin\left[\frac{2\pi}{\lambda}(x - vt + \lambda)\right]$$

$$= A \sin\left[\frac{2\pi}{\lambda}(x - vt) + 2\pi\right]$$

$$= A \sin\left[\frac{2\pi}{\lambda}(x - vt)\right]$$

$$= y(x, t)$$

Therefore, λ/v is the period, and v/λ is the frequency of $y(x, t)$.

At $t = 0$, the wave is represented by

$$y = A \sin\left[\frac{2\pi}{\lambda}x\right]$$

If the wave travels to the right with a velocity v, then at time t, the wave has moved a distance vt to the right. That is, the wave at $x' = x + vt$ at time t is the same as the wave at x at time $t = 0$.

The equation of the wave then becomes

$$y = A \sin\left[\frac{2\pi}{\lambda}(x - vt)\right]$$

This is the same as the equation given in the question. Thus, the wave travels to the right with a velocity v.

2–13. Sketch the normal modes of a vibrating rectangular membrane and show that they look like those shown in Figure 2.6.

The plots are done by setting $t = 0$ and $\phi_{nm} = 0$ in Equation 2.49. The equation for each mode is

$$u_{nm} = K_{nm} \sin\frac{n\pi x}{a} \sin\frac{m\pi x}{b}$$

where K_{nm} is the amplitude of the mode.

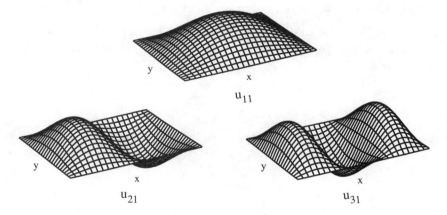

u_{11}

u_{21} u_{31}

2–14. This problem is the extension of Problem 2–10 to two dimensions. In this case, the particle is constrained to move freely over the surface of a rectangle of sides a and b. The Schrödinger equation for this problem is

$$\frac{\partial^2\psi}{\partial x^2} + \frac{\partial^2\psi}{\partial y^2} + \left(\frac{8\pi^2 mE}{h^2}\right)\psi(x, y) = 0$$

with the boundary conditions

$$\psi(0, y) = \psi(a, y) = 0 \qquad \text{for all } y, \qquad 0 \le y \le b$$

$$\psi(x, 0) = \psi(x, b) = 0 \qquad \text{for all } x, \qquad 0 \le x \le a$$

Solve this equation for $\psi(x, y)$, apply the boundary conditions, and show that the energy is quantized according to

$$E_{n_x, n_y} = \frac{n_x^2 h^2}{8ma^2} + \frac{n_y^2 h^2}{8mb^2} \qquad \begin{cases} n_x = 1, 2, 3, \ldots \\ n_y = 1, 2, 3, \ldots \end{cases}$$

The Schrödinger equation for this problem is a partial differential equation. It can be simplified using the separation of variables method. Assume

$$\psi(x, y) = X(x)Y(y)$$

and substitute this expression into the Schrödinger equation, giving

$$Y\frac{d^2X}{dx^2} + X\frac{d^2Y}{dy^2} + \left(\frac{8\pi^2 mE}{h^2}\right)XY = 0$$

Dividing both sides of this equation by $\psi(x, y) = X(x)Y(y)$ gives

$$\frac{1}{X}\frac{d^2X}{dx^2} + \frac{1}{Y}\frac{d^2Y}{dy^2} + \frac{8\pi^2 mE}{h^2} = 0$$

$$\frac{1}{X}\frac{d^2X}{dx^2} + \frac{1}{Y}\frac{d^2Y}{dy^2} = -\frac{8\pi^2 mE}{h^2}$$

Since the first term $\dfrac{1}{X}\dfrac{d^2X}{dx^2}$ depends only on x and the second term $\dfrac{1}{Y}\dfrac{d^2Y}{dy^2}$ depends only on y, each of them can be varied independently. Thus, each term must be a negative constant (for nontrivial solutions), and the two terms sum together to give $-8\pi^2 mE/h^2$. That is, letting p and q be real numbers,

$$\frac{1}{X}\frac{d^2X}{dx^2} = -p^2$$

$$\frac{1}{Y}\frac{d^2Y}{dy^2} = -q^2$$

and

$$-p^2 - q^2 = -\frac{8\pi^2 mE}{h^2}$$

$$p^2 + q^2 = \frac{8\pi^2 mE}{h^2}$$

The general solutions for the two ordinary differential equations are

$$X(x) = c_1 \cos px + c_2 \sin px$$

$$Y(y) = c_3 \cos qy + c_4 \sin qy$$

To obtain specific solutions, apply the boundary conditions. Writing these conditions in terms of X and Y gives

$$\psi(0, y) = X(0)Y(y) = 0$$

$$\psi(a, y) = X(a)Y(y) = 0$$

$$\psi(x, 0) = X(x)Y(0) = 0$$

$$\psi(x, b) = X(x)Y(b) = 0$$

which can be simplified to

$$X(0) = 0$$

$$X(a) = 0$$

$$Y(0) = 0$$

$$Y(b) = 0$$

The boundary conditions on $X(x)$ at $x = 0$ and $x = a$ give

$$c_1 = 0$$

$$c_1 \cos pa + c_2 \sin pa = 0$$

These equation gives either $c_2 = 0$ (trivial solution) or $pa = n_x \pi$ where $n_x = 1, 2, 3, ... (n_x = 0$ gives a trivial solution. See Problem 2–10.)

Similarly, applying the boundary conditions on $Y(y)$ at $y = 0$ and $y = b$ gives $c_3 = 0$ and $qb = n_y \pi$ where $n_y = 1, 2, 3, ...$ Thus,

$$p = \frac{n_x \pi}{a}$$

$$q = \frac{n_y \pi}{b}$$

and

$$X(x) = c_2 \sin \left(\frac{n_x \pi x}{a} \right)$$

$$Y(y) = c_4 \sin \left(\frac{n_y \pi y}{b} \right)$$

Combining these two functions gives

$$\psi(x, y) = X(x)Y(y) = A \sin \left(\frac{n_x \pi x}{a} \right) \sin \left(\frac{n_y \pi y}{b} \right)$$

where $A = c_2 c_4$. In fact, a linear combination of ψ's with different combinations of n_x and n_y is also a solution to the Schrödinger equation.

Recalling that $p^2 + q^2 = 8\pi^2 m E / h^2$,

$$p^2 + q^2 = \left(\frac{n_x \pi}{a} \right)^2 + \left(\frac{n_y \pi}{b} \right)^2 = \frac{8\pi^2 m E}{h^2}$$

$$E_{n_x, n_y} = \frac{n_x^2 h^2}{8ma^2} + \frac{n_y^2 h^2}{8mb^2}$$

The subscripts for E emphasize that energy depends on the quantum numbers n_x and n_y.

2–15. Extend Problems 2–10 and 2–14 to three dimensions, where a particle is constrained to move freely throughout a rectangular box of sides a, b, and c. The Schrödinger equation for this system is

$$\frac{\partial^2 \psi}{\partial x^2} + \frac{\partial^2 \psi}{\partial y^2} + \frac{\partial^2 \psi}{\partial z^2} + \left(\frac{8\pi^2 m E}{h^2} \right) \psi(x, y, z) = 0$$

and the boundary conditions are that $\psi(x, y, z)$ vanishes over all the surfaces of the box.

As in Problem 2–14, the Schrödinger equation for this problem can be simplified using the separation of variables method. Assume

$$\psi(x, y, z) = X(x)Y(y)Z(z)$$

and substitute this expression into the Schrödinger equation to give

$$YZ\frac{\partial^2 X}{\partial x^2} + XZ\frac{\partial^2 Y}{\partial y^2} + XY\frac{\partial^2 Z}{\partial z^2} + \left(\frac{8\pi^2 m E}{h^2}\right)XYZ = 0$$

Divide both sides of this equation by $\psi(x, y) = X(x)Y(y)Z(z)$,

$$\frac{1}{X}\frac{d^2 X}{dx^2} + \frac{1}{Y}\frac{d^2 Y}{dy^2} + \frac{1}{Z}\frac{d^2 Z}{dz^2} + \frac{8\pi^2 m E}{h^2} = 0$$

$$\frac{1}{X}\frac{d^2 X}{dx^2} + \frac{1}{Y}\frac{d^2 Y}{dy^2} + \frac{1}{Z}\frac{d^2 Z}{dz^2} = -\frac{8\pi^2 m E}{h^2}$$

Since each of the terms on the left side of the equality sign depends on a unique variable, it can be varied independently. Thus, each term must be a negative constant (for nontrivial solutions), and the three terms sum together to give $-8\pi^2 m E/h^2$. That is, letting p, q, r be real numbers,

$$\frac{1}{X}\frac{d^2 X}{dx^2} = -p^2$$

$$\frac{1}{Y}\frac{d^2 Y}{dy^2} = -q^2$$

$$\frac{1}{Z}\frac{d^2 Z}{dz^2} = -r^2$$

and

$$p^2 + q^2 + r^2 = \frac{8\pi^2 m E}{h^2}$$

The general solutions for the three ordinary differential equations are

$$X(x) = c_1 \cos px + c_2 \sin px$$

$$Y(y) = c_3 \cos qy + c_4 \sin qy$$

$$Z(z) = c_5 \cos rz + c_6 \sin rz$$

To obtain specific solutions, apply the boundary conditions

$$X(0) = X(a) = 0$$

$$Y(0) = Y(b) = 0$$

$$Z(0) = Z(c) = 0$$

to give

$$p = \frac{n_x \pi}{a}$$

$$q = \frac{n_y \pi}{b}$$

$$r = \frac{n_z \pi}{c}$$

where $n_x, n_y, n_z = 1, 2, 3, \ldots$ (see Problem 2–10 for details.)

The functions are

$$X(x) = c_2 \sin\left(\frac{n_x \pi x}{a}\right)$$

$$Y(y) = c_4 \sin\left(\frac{n_y \pi y}{b}\right)$$

$$Z(z) = c_6 \sin\left(\frac{n_z \pi y}{c}\right)$$

and they combine to give

$$\psi(x, y, z) = X(x)Y(y)Z(z) = A \sin\left(\frac{n_x \pi x}{a}\right) \sin\left(\frac{n_y \pi y}{b}\right) \sin\left(\frac{n_z \pi z}{c}\right)$$

where $A = c_2 c_4 c_6$. In fact, a linear combination of ψ's with different combinations of n_x, n_y, and n_z is also a solution to the Schrödinger equation.

Recalling that $p^2 + q^2 + r^2 = 8\pi^2 m E / h^2$,

$$p^2 + q^2 + r^2 = \left(\frac{n_x \pi}{a}\right)^2 + \left(\frac{n_y \pi}{b}\right)^2 + \left(\frac{n_z \pi}{c}\right)^2 = \frac{8\pi^2 m E}{h^2}$$

$$E_{n_x, n_y, n_z} = \frac{n_x^2 h^2}{8ma^2} + \frac{n_y^2 h^2}{8mb^2} + \frac{n_z^2 h^2}{8mc^2}$$

The subscripts for E emphasize that energy depends on the quantum numbers n_x, n_y, and n_z.

2–16. Show that Equations 2.46 and 2.48 are equivalent. How are G_{nm} and ϕ_{nm} in Equation 2.48 related to the quantities in Equation 2.46?

Expanding Equation 2.48 gives

$$T_{nm}(t) = G_{nm} \cos\left(\omega_{nm} t + \phi_{nm}\right) = G_{nm} \cos\left(\omega_{nm} t\right) \cos \phi_{nm} - G_{nm} \sin\left(\omega_{nm} t\right) \sin \phi_{nm}$$

For this expression to be equivalent to $T_{nm} = E_{nm} \cos\left(\omega_{nm} t\right) + F_{nm} \sin\left(\omega_{nm} t\right)$ (Equation 2.46), the following conditions must apply:

$$G_{nm} \cos \phi_{nm} = E_{nm}$$

$$G_{nm} \sin \phi_{nm} = -F_{nm}$$

2–17.   Prove that $u_n(x, t)$, the nth normal mode of a vibrating string (Equation 2.23), can be written as the superposition of two similar traveling waves moving in opposite directions. Let $\phi_n = 0$ in Equation 2.25.

Setting $\phi_n = 0$, the nth normal mode of a vibrating string is

$$u_n(x, t) = A_n \cos\left(\omega_n t\right) \sin \frac{n\pi x}{l}$$

Since $\sin \alpha \cos \beta = \dfrac{1}{2} \sin(\alpha + \beta) + \dfrac{1}{2} \sin(\alpha - \beta)$, the above equation becomes

$$u_n(x, t) = \frac{A_n}{2}\left[\sin\left(\frac{n\pi x}{l} + \omega_n t\right) + \sin\left(\frac{n\pi x}{l} - \omega_n t\right)\right]$$

Substitute $\omega_n = n\pi v/l$ into the above equation to give

$$u_n(x, t) = \frac{A_n}{2}\left\{ \sin\left[\frac{n\pi}{l}(x + vt)\right] + \sin\left[\frac{n\pi}{l}(x - vt)\right]\right\}$$

The last equation is a sum of a wave moving to the left and one moving to the right, respectively.

2–18.   This problem shows that the intensity of a wave is proportional to the square of its amplitude. Figure 2.12 illustrates the geometry of a vibrating string. Because the velocity at any point of the string is $\partial u / \partial t$, the kinetic energy, T, of the entire string is

$$T = \int_0^l \frac{1}{2}\rho \left(\frac{\partial u}{\partial t}\right)^2 dx$$

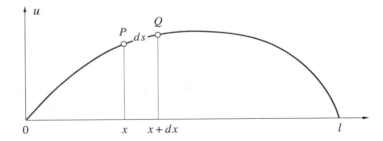

FIGURE 2.12
The geometry of a vibrating string.

where ρ is the linear mass density of the string. The potential energy is found by considering the increase in length of the small arc PQ of length ds in Figure 2.12. The segment of the string along that arc has increased its length from dx to ds. Therefore, the potential energy associated with this increase is

$$V = \int_0^l \tau(ds - dx)$$

where τ is the tension in the string. Using the fact that $(ds)^2 = (dx)^2 + (du)^2$, show that

$$V = \int_0^l \tau \left\{ \left[1 + \left(\frac{\partial u}{\partial x} \right)^2 \right]^{1/2} - 1 \right\} dx$$

Using the fact that $(1 + x)^{1/2} \approx 1 + (x/2)$ for small x, show that

$$V = \frac{1}{2} \tau \int_0^l \left(\frac{\partial u}{\partial x} \right)^2 dx$$

for small displacements.

The total energy of the vibrating string is the sum of T and V and so

$$E = \frac{\rho}{2} \int_0^l \left(\frac{\partial u}{\partial t} \right)^2 dx + \frac{\tau}{2} \int_0^l \left(\frac{\partial u}{\partial x} \right)^2 dx$$

Equation 2.25 shows that the nth normal mode can be written in the form

$$u_n(x, t) = D_n \cos(\omega_n t + \phi_n) \sin \frac{n \pi x}{l}$$

where $\omega_n = v n \pi / l$. Using this equation, show that

$$T_n = \frac{\pi^2 v^2 n^2 \rho}{4l} D_n^2 \sin^2(\omega_n t + \phi_n)$$

and

$$V_n = \frac{\pi^2 n^2 \tau}{4l} D_n^2 \cos^2(\omega_n t + \phi_n)$$

Using the fact that $v = (\tau/\rho)^{1/2}$, show that

$$E_n = \frac{\pi^2 v^2 n^2 \rho}{4l} D_n^2$$

Note that the total energy, or intensity, is proportional to the square of the amplitude. Although we have shown this proportionality only for the case of a vibrating string, it is a general result and shows that the intensity of a wave is proportional to the square of the amplitude. If we had carried everything through in complex notation instead of sines and cosines, then we would have found that E_n is proportional to $|D_n|^2$ instead of just D_n^2.

Generally, there are many normal modes present at the same time, and the complete solution is

$$u(x, t) = \sum_{n=1}^{\infty} D_n \cos(\omega_n t + \phi_n) \sin \frac{n \pi x}{l}$$

Using the fact that (see Problem 3–23)

$$\int_0^l \sin \frac{n \pi x}{l} \sin \frac{m \pi x}{l} dx = 0 \qquad \text{if } m \neq n$$

show that

$$E_n = \frac{\pi^2 v^2 \rho}{4l} \sum_{n=1}^{\infty} n^2 D_n^2$$

The relation $(ds)^2 = (dx)^2 + (du)^2$ gives

$$(ds)^2 = (dx)^2 \left[1 + \left(\frac{\partial u}{\partial x} \right)^2 \right]$$

$$ds = dx \left[1 + \left(\frac{\partial u}{\partial x} \right)^2 \right]^{1/2}$$

so that upon substitution for ds, $V = \int_0^l \tau (ds - dx)$ becomes

$$V = \int_0^l \tau \left\{ dx \left[1 + \left(\frac{\partial u}{\partial x} \right)^2 \right]^{1/2} - dx \right\}$$

$$= \int_0^l \tau \left\{ \left[1 + \left(\frac{\partial u}{\partial x} \right)^2 \right]^{1/2} - 1 \right\} dx$$

For small displacements of the string, $\partial u / \partial x$, and thus $(\partial u / \partial x)^2$, is small. Using the relation $(1 + x)^{1/2} \approx 1 + (x/2)$, the expression for V simplifies:

$$V = \int_0^l \tau \left\{ \left[1 + \frac{1}{2} \left(\frac{\partial u}{\partial x} \right)^2 \right] - 1 \right\} dx$$

$$= \frac{1}{2} \tau \int_0^l \left(\frac{\partial u}{\partial x} \right)^2 dx$$

To determine the kinetic and potential energies, take the derivatives of the nth normal mode, $u_n(x, l) = D_n \cos(\omega_n t + \phi_n) \sin(n\pi x / l)$, with respect to t and x, respectively:

$$\frac{\partial u}{\partial t} = -\omega_n D_n \sin(\omega_n t + \phi_n) \sin \frac{n\pi x}{l}$$

$$\frac{\partial u}{\partial x} = \frac{n\pi}{l} D_n \cos(\omega_n t + \phi_n) \cos \frac{n\pi x}{l}$$

Using $\partial u / \partial t$, the kinetic energy of the string is:

$$T = \int_0^l \frac{1}{2} \rho \left(\frac{\partial u}{\partial t} \right)^2 dx$$

$$= \frac{\rho}{2} \omega_n^2 D_n^2 \sin^2(\omega_n t + \phi_n) \int_0^l \sin^2 \frac{n\pi x}{l} dx$$

The integral has the form

$$\int \sin^2 ax \, dx = \frac{x}{2} - \frac{1}{4a} \sin 2ax$$

Using this form and evaluating at the limits, the kinetic energy expression becomes

$$T = \frac{\rho}{2}\omega_n^2 D_n^2 \sin^2\left(\omega_n t + \phi_n\right)\frac{l}{2}$$

Rewriting $\omega_n = vn\pi/l$,

$$T_n = \frac{\pi^2 v^2 n^2 \rho}{4l} D_n^2 \sin^2\left(\omega_n t + \phi_n\right)$$

The subscript n for T emphasizes that the kinetic energy depends on n.

Using $\partial u/\partial x$, the potential energy is

$$V = \frac{1}{2}\tau \int_0^l \left(\frac{\partial u}{\partial x}\right)^2 dx$$

$$= \frac{1}{2}\tau \left(\frac{n\pi}{l}\right)^2 D_n^2 \cos^2\left(\omega_n t + \phi_n\right)\int_0^l \cos^2\frac{n\pi x}{l}dx$$

The integral has the form

$$\int \cos^2 ax\, dx = \frac{x}{2} + \frac{1}{4a}\sin 2ax$$

Using this form and evaluating at the limits, the potential energy expression becomes

$$V_n = \frac{1}{2}\tau \left(\frac{n\pi}{l}\right)^2 D_n^2 \cos^2\left(\omega_n t + \phi_n\right)\frac{l}{2}$$

$$= \frac{n^2\pi^2\tau}{4l} D_n^2 \cos^2\left(\omega_n t + \phi_n\right)$$

The subscript n for V emphasizes that the potential energy depends on n.

The total energy is the sum of kinetic and potential energies. The expression can be simplified using $v = (\tau/\rho)^{1/2}$, or $\tau = v^2\rho$:

$$E_n = T_n + V_n$$

$$= \frac{\pi^2 v^2 n^2 \rho}{4l} D_n^2 \sin^2\left(\omega_n t + \phi_n\right) + \frac{n^2\pi^2 v^2\rho}{4l} D_n^2 \cos^2\left(\omega_n t + \phi_n\right)$$

$$= \frac{\pi^2 v^2 n^2 \rho}{4l} D_n^2$$

Using the complete solution $u(x, t) = \sum_{n=1}^{\infty} D_n \cos(\omega_n t + \phi_n)\sin\frac{n\pi x}{l}$,

$$\left(\frac{\partial u}{\partial t}\right)^2 = \sum_{m=1}^{\infty}\sum_{n=1}^{\infty}\left[-\omega_m D_m \sin\left(\omega_m t + \phi_m\right)\sin\frac{m\pi x}{l}\right]\left[-\omega_n D_n \sin\left(\omega_n t + \phi_n\right)\sin\frac{n\pi x}{l}\right]$$

$$= \sum_{m=1}^{\infty}\sum_{n=1}^{\infty}\omega_m \omega_n D_m D_n \sin\left(\omega_m t + \phi_m\right)\sin\left(\omega_n t + \phi_n\right)\sin\frac{m\pi x}{l}\sin\frac{n\pi x}{l}$$

$$\left(\frac{\partial u}{\partial x}\right)^2 = \sum_{m=1}^{\infty}\sum_{n=1}^{\infty}\left[\frac{m\pi}{l}D_m \cos\left(\omega_m t + \phi_m\right)\cos\frac{m\pi x}{l}\right]\left[\frac{n\pi}{l}D_n \cos\left(\omega_n t + \phi_n\right)\cos\frac{n\pi x}{l}\right]$$

$$= \sum_{m=1}^{\infty}\sum_{n=1}^{\infty}\frac{m\pi}{l}\frac{n\pi}{l}D_m D_n \cos\left(\omega_m t + \phi_m\right)\cos\left(\omega_n t + \phi_n\right)\cos\frac{m\pi x}{l}\cos\frac{n\pi x}{l}$$

These expressions can be used to evaluate the kinetic and potential energies. Given that $\int_0^l \sin\left(n\pi x/l\right)\sin\left(m\pi x/l\right)dx = 0$ for $m \neq n$, the only integrals that are nonzero are those with $n = m$:

$$T_n = \int_0^l \frac{1}{2}\rho\left(\frac{\partial u}{\partial t}\right)^2 dx$$

$$= \frac{\rho}{2}\sum_{n=1}^{\infty}\omega_n^2 D_n^2 \sin^2\left(\omega_n t + \phi_n\right)\int_0^l \sin^2\frac{n\pi x}{l}dx$$

$$= \frac{\pi^2 v^2 \rho}{4l}\sum_{n=1}^{\infty}n^2 D_n^2 \sin^2\left(\omega_n t + \phi_n\right)$$

$$V_n = \frac{1}{2}\tau\int_0^l \left(\frac{\partial u}{\partial x}\right)^2 dx$$

$$= \frac{1}{2}\tau\sum_{n=1}^{\infty}\left(\frac{n\pi}{l}\right)^2 D_n^2 \cos^2\left(\omega_n t + \phi_n\right)\int_0^l \cos^2\frac{n\pi x}{l}dx$$

$$= \frac{\pi^2 \tau}{4l}\sum_{n=1}^{\infty}n^2 D_n^2 \cos^2\left(\omega_n t + \phi_n\right)$$

$$= \frac{\pi^2 v^2 \rho}{4l}\sum_{n=1}^{\infty}n^2 D_n^2 \cos^2\left(\omega_n t + \phi_n\right)$$

The relations $\omega_n = v n\pi/l$ and $\tau = v^2\rho$ are used in obtaining the expressions above. The total energy is

$$E_n = T_n + V_n$$

$$= \frac{\pi^2 v^2 \rho}{4l}\sum_{n=1}^{\infty}n^2 D_n^2 \left[\sin^2\left(\omega_n t + \phi_n\right) + \cos^2\left(\omega_n t + \phi_n\right)\right]$$

$$= \frac{\pi^2 v^2 \rho}{4l}\sum_{n=1}^{\infty}n^2 D_n^2$$

2–19. In this problem, we'll generalize Equation 2.58 to the case of N slits. Figure 2.13 summarizes the geometrical setup. Realize that all the rays impinge on one point P; they are approximately parallel, as in Figure 2.9. The electric field at the point P is given by

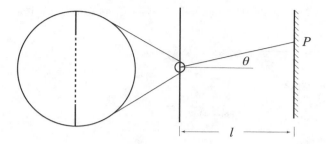

FIGURE 2.13
The geometry used in Problem 2–19 for the derivation of the interference pattern for N slits.

$$E(\theta) = E_0 e^{i(k\rho_1 - \omega t)} + E_0 e^{i(k\rho_2 - \omega t)} + \cdots + E_0 e^{i(k\rho_N - \omega t)}$$

$$= E_0 e^{-i\omega t} \sum_{n=1}^{N} e^{ik\rho_n}$$

which is the generalization of Equation 2.56. Now argue that the path difference between any two adjacent sources is

$$\delta = \rho_{n+1} - \rho_n \qquad n = 1, 2, \ldots, N-1$$

and therefore that $\rho_{n+1} - \rho_1 = n\delta$, where $\delta = d \sin \theta$. Now show that

$$E(\theta) = E_0 e^{i(k\rho_1 - \omega t)} \sum_{n=0}^{N-1} e^{ikn\delta}$$

Now use the fact that

$$\sum_{n=0}^{N-1} x^n = \frac{1 - x^N}{1 - x}$$

(see MathChapter D for a simple proof of this formula) to write $E(\theta)$ as

$$E(\theta) = E_0 e^{i(k\rho_1 - \omega t)} \left(\frac{1 - e^{ikN\delta}}{1 - e^{ik\delta}} \right)$$

Now factor $e^{ikN\delta/2}$ from the numerator and $e^{ik\delta/2}$ from the denominator to get

$$E(\theta) = E_0 e^{i(k\rho_1 - \omega t + kN\delta/2 - k\delta/2)} \left(\frac{e^{ikN\delta/2} + e^{-ikN\delta/2}}{e^{ik\delta/2} - e^{-ik\delta/2}} \right)$$

Finally, use the fact that $(e^{ix} - e^{-ix})/2i = \sin x$ and that $k = 2\pi/\lambda$ to show that the intensity of radiation at the point P is

$$I(\theta) = A^*(\theta) A(\theta) = E_0^2 \left(\frac{\sin[(\pi N d/\lambda) \sin \theta]}{\sin[(\pi d/\lambda) \sin \theta]} \right)^2 \tag{2.59}$$

Plot $I(\theta)$ against θ for $d = 0.010$ mm and $\lambda = 6000$ Å and show that it is the same result as in Figure 2.10.

According to the expression

$$E(\theta) = E_0 e^{i(k\rho_1 - \omega t)} + E_0 e^{i(k\rho_2 - \omega t)} + \cdots + E_0 e^{i(k\rho_N - \omega t)}$$

ρ_n represents the distance between slit n and P. Thus, the path difference between two sources, n and $n + 1$, adjacent to each other, δ, is $\rho_{n+1} - \rho_n$. This difference is $d \sin \theta$, as illustrated in Figure 2.9, where d is the distance between two adjacent slits. The path difference between slits 1 and $n + 1$ is just the sum of the path differences between n sets of adjacent slits, that is,

$$\rho_{n+1} - \rho_1 = n\delta$$

Using this relation, the electric field expression can be rewritten as

$$E(\theta) = E_0 e^{i(k\rho_1 - \omega t)} \left[1 + e^{ik(\rho_2 - \rho_1)} + \cdots + e^{ik(\rho_N - \rho_1)} \right]$$

$$= E_0 e^{i(k\rho_1 - \omega t)} \left[1 + e^{ik\delta} + \cdots + e^{ik(N-1)\delta} \right]$$

$$= E_0 e^{i(k\rho_1 - \omega t)} \left[1 + \sum_{n=1}^{N-1} e^{ikn\delta} \right]$$

This last expression can be simplified by noting that $e^{ikn\delta} = 1$ when $n = 0$. Thus,

$$E(\theta) = E_0 e^{i(k\rho_1 - \omega t)} \sum_{n=0}^{N-1} e^{ikn\delta}$$

Given that $\displaystyle\sum_{n=0}^{N-1} x^n = \frac{1 - x^N}{1 - x}$, the electric field expression becomes

$$E(\theta) = E_0 e^{i(k\rho_1 - \omega t)} \left(\frac{1 - e^{ikN\delta}}{1 - e^{ik\delta}} \right)$$

Factoring $e^{ikN\delta/2}$ from the numerator and $e^{ik\delta/2}$ from the denominator gives

$$E(\theta) = E_0 e^{i(k\rho_1 - \omega t)} \left[\frac{e^{ikN\delta/2} \left(e^{-ikN\delta/2} - e^{ikN\delta/2} \right)}{e^{ik\delta/2} \left(e^{-ik\delta/2} - e^{ik\delta/2} \right)} \right]$$

$$= E_0 e^{i(k\rho_1 - \omega t + kN\delta/2 - k\delta/2)} \left(\frac{e^{ikN\delta/2} - e^{-ikN\delta/2}}{e^{ik\delta/2} - e^{-ik\delta/2}} \right)$$

Using the relation $(e^{ix} - e^{-ix})/2i = \sin x$, $\delta = d \sin \theta$, and $k = 2\pi/\lambda$, the above expression becomes

$$E(\theta) = E_0 e^{i(k\rho_1 - \omega t + kN\delta/2 - k\delta/2)} \left[\frac{\sin (kN\delta/2)}{\sin (k\delta/2)} \right]$$

$$E(\theta) = E_0 e^{i(k\rho_1 - \omega t + kN\delta/2 - k\delta/2)} \left\{ \frac{\sin \left[(\pi Nd/\lambda) \sin \theta \right]}{\sin \left[(\pi d/\lambda) \sin \theta \right]} \right\}$$

The amplitude of the electric field is $A(\theta) = E_0 \left\{ \dfrac{\sin\left[(\pi N d/\lambda)\sin\theta\right]}{\sin\left[(\pi d/\lambda)\sin\theta\right\}} \right\}$. The intensity of radiation at point P is

$$I(\theta) = A^*(\theta)A(\theta) = E_0^2 \left\{ \frac{\sin[(\pi N d/\lambda)\sin\theta\,]}{\sin[(\pi d/\lambda)\sin\theta]} \right\}^2$$

To reproduce Figure 2.10, use the above equation and set $N = 2$, $d = 0.010$ mm, and $\lambda = 6000$ Å. As shown in the following figure, it is the same as Figure 2.10.

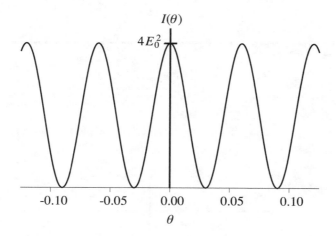

2–20. Show that Equation 2.59 from the previous problem reduces to Equation 2.58 when $N = 2$.

For $N = 2$, Equation 2.59 has the form $(\sin 2z/\sin z)^2$ where $z = (\pi d/\lambda)\sin\theta$. Since $\sin 2z = 2\sin z\cos z$, then $(\sin 2z/\sin z)^2 = 4\cos^2 z$. Thus, the intensity is

$$I(\theta) = 4E_0^2 \cos^2 \frac{\pi d\,\sin\theta}{\lambda}$$

2–21. Plot Equation 2.59 for $N = 3$.

For comparison with Figure 2.10, the following plot for $N = 3$ is constructed with $d = 0.010$ mm and $\lambda = 6000$ Å.

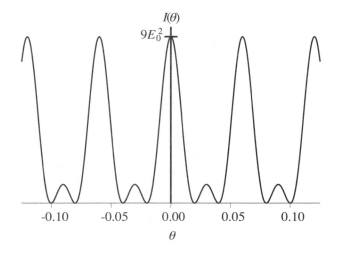

2–22.   Plot Equation 2.59 for $N = 1$. What's going on here?

For $N = 1$, there is only one slit and thus no interference pattern can be observed. As a result, the intensity is constant.

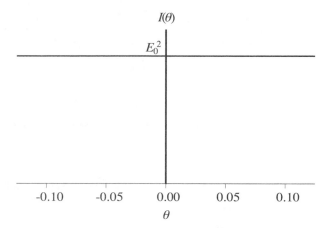

2–23.   If you plot Equation 2.59 for a number of values of N, you'll see that the number of minima between the large maxima (called the principal maxima) is equal to $N - 1$. Can you prove this?

Letting $x = (d/\lambda) \sin \theta$, Equation 2.59 becomes

$$I(\theta) = E_0^2 \frac{\sin^2 (\pi N x)}{\sin^2 \pi x}$$

The function represented by the fraction is periodic. The denominator has a period of π while the numerator has a period of π/N. Taking the numerator and denominator together, the fraction has a period of π. The maxima of $I(\theta)$ occur when both the numerator and denominator are zero, that is, within a period of the fraction, when $x = p$ and $x = p + 1$, where p is an integer [according to L'Hôpital's rule, at these points, $I(\theta) = N^2 E_0^2$]. The minima ($I = 0$) within the same period occur when the numerator is zero but the denominator is nonzero. Specifically, $I = 0$ occurs within one period when $x = 1/N, 2/N, 3/N, \ldots, (N-1)/N$. Thus, there are $N-1$ minima between the principal maxima.

Problems 2–24 through 2–29 illustrate some other applications of differential equations to classical mechanics.

Many problems in classical mechanics can be reduced to the problem of solving a differential equation with constant coefficients (cf. Problem 2–7). The basic starting point is Newton's second law, which says that the rate of change of momentum is equal to the force acting on a body. Momentum p equals mv, and so if the mass is constant, then in one dimension we have

$$\frac{dp}{dt} = m\frac{dv}{dt} = m\frac{d^2x}{dt^2} = f$$

If we are given the force as a function of x, then this equation is a differential equation for x(t), which is called the trajectory of the particle. Going back to the simple harmonic oscillator discussed in Problem 2–7, if we let x be the displacement of the mass from its equilibrium position, then Hooke's law says that f(x) = −kx, and the differential equation corresponding to Newton's second law is

$$\frac{d^2x}{dt^2} + kx(t) = 0$$

a differential equation that we have seen several times.

2–24. Consider a body falling freely from a height x_0 according to Figure 2.14. If we neglect air resistance or viscous drag, the only force acting upon the body is the gravitational force mg. Using the coordinates in Figure 2.14a, mg acts in the same direction as x and so the differential equation corresponding to Newton's second law is

$$m\frac{d^2x}{dt^2} = mg$$

Show that

$$x(t) = \frac{1}{2}gt^2 + v_0 t + x_0$$

where x_0 and v_0 are the initial values of x and v. According to Figure 2.14a, $x_0 = 0$ and so

$$x(t) = \frac{1}{2}gt^2 + v_0 t$$

If the particle is just dropped, then $v_0 = 0$ and so

$$x(t) = \frac{1}{2}gt^2$$

Discuss this solution.

Now do the same problem using Figure 2.14b as the definition of the various quantities involved, and show that although the equations may look different from those above, they say exactly the

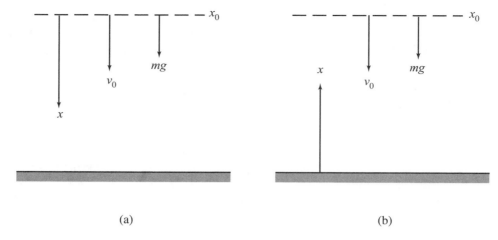

FIGURE 2.14
(a) A coordinate system for a body falling from a height x_0, and (b) a different coordinate system for a body falling from a height x_0.

same thing because the diagram we draw to define the direction of x, v_0, and mg does not affect the falling body.

Using the coordinates in Figure 2.14a and Newton's second law $m\, d^2x/dt^2 = mg$, dx/dt, the velocity of the falling body is obtained by integrating the second law once with respect to t. A second integration gives x:

$$\frac{dx}{dt} = v = gt + c_1$$

$$x = \frac{1}{2}gt^2 + c_1 t + c_2$$

The initial conditions on $x(t)$ and dx/dt at $t = 0$ give

$$c_2 = x_0$$
$$c_1 = v_0$$

Therefore, the equation of motion is

$$x = \frac{1}{2}gt^2 + v_0 t + x_0$$

If the initial position, x_0, is 0, the equation becomes $x = \frac{1}{2}gt^2 + v_0 t$.

If $v_0 = 0$, then $x = \frac{1}{2}gt^2$. The falling body has acceleration g and the distance fallen increases quadratically with time.

Using the coordinate system in Figure 2.14b, Newton's second law becomes $m\, d^2x/dt^2 = -mg$. The negative sign arises from the fact that the direction of the distance traveled, x, is defined to be opposite to the direction of the force. Integrating the second law once gives the velocity of the body, and twice gives the position of the body:

$$\frac{dx}{dt} = v = -gt + c_1$$

$$x = -\frac{1}{2}gt^2 + c_1 t + c_2$$

The initial conditions are $x(0) = x_0$ and $v(0) = -v_0$. These give

$$c_2 = x_0$$

$$c_1 = -v_0$$

The equation of motion becomes

$$x = -\frac{1}{2}gt^2 - v_0 t + x_0$$

This equation says exactly the same thing as that derived from Figure 2.14a. The difference in form arises because the directions of x in the two cases are opposite. In the coordinate system of Figure 2.14a, x increases with time whereas in the coordinate system of Figure 2.14b, x decreases with time, indicating that the body is falling.

2–25. Derive an equation for the maximum height a body will reach if it is shot straight upward with a velocity v_0. Refer to Figure 2.14b but realize that in this case v_0 points upward. How long will it take for the body to return to earth?

Referring to the coordinate system in Figure 2.14b but with v_0 pointing upward [that is, $v(0) = v_0$], the equation of motion (see Problem 2–24) is

$$x = -\frac{1}{2}gt^2 + v_0 t + x_0$$

Since the body is shot upward, $x(0) = 0$, and the above equation becomes

$$x = -\frac{1}{2}gt^2 + v_0 t$$

The velocity of the body is

$$\frac{dx}{dt} = -gt + v_0$$

The body reaches a maximum height (x_{max}) when the velocity of the body is zero, that is, $dx/dt = 0$. Call the time it takes to reach that height t_{max}.

$$-gt_{max} + v_0 = 0$$

$$t_{max} = \frac{v_0}{g}$$

The maximum height is

$$x_{max} = -\frac{1}{2}gt_{max}^2 + v_0 t_{max} = \frac{v_0^2}{2g}$$

The time it takes the body to return to Earth can be determined by setting $x = 0$:

$$x = -\frac{1}{2}gt^2 + v_0 t = 0$$

$$t\left(-\frac{1}{2}gt + v_0\right) = 0$$

$$t = 0 \text{ or } t = \frac{2v_0}{g}$$

At $t = 0$, the body is shot upward. It returns to Earth at $t = 2v_0/g$. Note that this time is twice the amount of time it takes for the body to reach a maximum height. This is because it takes the body the same amount of time to reach the maximum height from Earth and to reach the Earth from maximum height.

2–26. Consider a simple pendulum as shown in Figure 2.15. We let the length of the pendulum be l and assume that all the mass of the pendulum is concentrated at its end, as shown in Figure 2.15. A physical example of this case might be a mass suspended by a string. We assume that the motion of the pendulum is set up such that it oscillates within a plane so that we have a problem in plane polar coordinates. Let the distance along the arc in the figure describe the motion of the pendulum, so that its momentum is $mds/dt = mld\theta/dt$ and its rate of change of momentum is $mld^2\theta/dt^2$. Show that the component of force in the direction of motion is $-mg\sin\theta$, where the minus sign occurs because the direction of this force is opposite that of the angle θ. Show that the equation of motion is

$$ml\frac{d^2\theta}{dt^2} = -mg\sin\theta$$

Now assume that the motion takes place only through very small angles and show that the motion becomes that of a simple harmonic oscillator. What is the natural frequency of this harmonic oscillator? *Hint:* Use the fact that $\sin\theta \approx \theta$ for small values of θ.

The force mg acts on the mass and points downward. Its component in the direction of the motion of the mass is $-mg\sin\theta$. The negative sign indicates that the force is in the direction opposite to the motion.

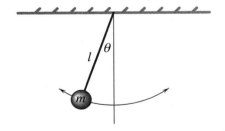

FIGURE 2.15
The coordinate system describing an oscillating pendulum.

The equation of motion is

$$\frac{dp}{dt} = f$$

$$ml\frac{d^2\theta}{dt^2} = -mg\sin\theta$$

If the motion takes place only through very small angles, then $\sin\theta \approx \theta$ and

$$ml\frac{d^2\theta}{dt^2} = -mg\theta$$

$$\frac{d^2\theta}{dt^2} = -\frac{g}{l}\theta$$

The general solution to this equation is (see Example 2–3)

$$x(t) = c_1 \cos\left(\sqrt{\frac{g}{l}}\,t\right) + c_2 \sin\left(\sqrt{\frac{g}{l}}\,t\right)$$

The natural frequency of this harmonic oscillator is $v = \dfrac{1}{2\pi}\sqrt{\dfrac{g}{l}}$. Note that v is independent of the mass of the harmonic oscillator.

2–27. Consider the motion of a pendulum like that in Problem 2–26 but swinging in a viscous medium. Suppose that the viscous force is proportional to but oppositely directed to its velocity; that is,

$$f_{\text{viscous}} = -\lambda\frac{ds}{dt} = -\lambda l\frac{d\theta}{dt}$$

where λ is a viscous drag coefficient. Show that for small angles, Newton's equation is

$$ml\frac{d^2\theta}{dt^2} + \lambda l\frac{d\theta}{dt} + mg\theta = 0$$

Show that there is no harmonic motion if

$$\lambda^2 > \frac{4m^2g}{l}$$

Does it make physical sense that the medium can be so viscous that the pendulum undergoes no harmonic motion?

The equation of motion is

$$\frac{dp}{dt} = f = -mg\sin\theta + f_{\text{viscous}}$$

$$ml\frac{d^2\theta}{dt^2} = -mg\sin\theta - \lambda l\frac{d\theta}{dt}$$

For small angles, $\sin\theta \approx \theta$ and the above equation becomes

$$ml\frac{d^2\theta}{dt^2} + \lambda l\frac{d\theta}{dt} + mg\theta = 0$$

$$\frac{d^2\theta}{dt^2} + \frac{\lambda}{m}\frac{d\theta}{dt} + \frac{g}{l}\theta = 0$$

This differential equation can be solved by letting $\theta = e^{\alpha t}$. Substituting this expression into the differential equation and dividing through by $e^{\alpha t}$ gives

$$\alpha^2 + \frac{\lambda}{m}\alpha + \frac{g}{l} = 0$$

$$\alpha = \frac{1}{2}\left(-\frac{\lambda}{m} \pm \sqrt{\frac{\lambda^2}{m^2} - \frac{4g}{l}}\right)$$

Thus, the general solution to the equation of motion is

$$\theta(t) = c_1 \exp\left[\frac{1}{2}\left(-\frac{\lambda}{m} + \sqrt{\frac{\lambda^2}{m^2} - \frac{4g}{l}}\,t\right)\right] + c_2 \exp\left[\frac{1}{2}\left(-\frac{\lambda}{m} - \sqrt{\frac{\lambda^2}{m^2} - \frac{4g}{l}}\,t\right)\right]$$

$$= \exp\left(-\frac{\lambda t}{2m}\right)\left[c_1 \exp\left(\sqrt{\frac{\lambda^2}{4m^2} - \frac{g}{l}}\,t\right) + c_2 \exp\left(-\sqrt{\frac{\lambda^2}{4m^2} - \frac{g}{l}}\,t\right)\right]$$

If $\lambda^2 < 4m^2 g/l$, then $\theta(t)$ contains the form $c_1 e^{i\beta t} + c_2 e^{-i\beta t}$ where $\beta = \sqrt{\frac{g}{l} - \frac{\lambda^2}{4m^2}}$ is a non-negative real number. This expression gives harmonic motion with a frequency of $\nu = \beta/(2\pi)$.

On the other hand, if $\lambda^2 > 4m^2 g/l$, then $\theta(t)$ contains the form $c_1 e^{\gamma t} + c_2 e^{-\gamma t}$ where $\gamma = \sqrt{\frac{\lambda^2}{4m^2} - \frac{g}{l}}$ is a non-negative real number. There is no harmonic motion.

2–28. Consider two pendulums of equal lengths and masses that are connected by a spring that obeys Hooke's law (Problem 2–7). This system is shown in Figure 2.16. Assuming that the motion takes place in a plane and that the angular displacement of each pendulum from the horizontal is small, show that the equations of motion for this system are

$$m\frac{d^2x}{dt^2} = -m\omega_0^2 x - k(x - y)$$

$$m\frac{d^2y}{dt^2} = -m\omega_0^2 y - k(y - x)$$

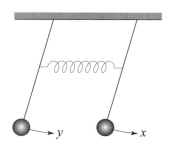

FIGURE 2.16
Two pendulums coupled by a spring that obeys Hooke's law.

where ω_0 is the natural vibrational frequency of each isolated pendulum [i.e., $\omega_0 = (g/l)^{1/2}$] and k is the force constant of the connecting spring. In order to solve these two simultaneous differential equations, assume that the two pendulums swing harmonically and so try

$$x(t) = Ae^{i\omega t} \qquad y(t) = Be^{i\omega t}$$

Substitute these expressions into the two differential equations and obtain

$$\left(\omega^2 - \omega_0^2 - \frac{k}{m}\right) A = -\frac{k}{m} B$$

$$\left(\omega^2 - \omega_0^2 - \frac{k}{m}\right) B = -\frac{k}{m} A$$

Now we have two simultaneous linear homogeneous algebraic equations for the two amplitudes A and B. We shall learn in MathChapter E that the determinant of the coefficients must vanish in order for there to be a nontrivial solution. Show that this condition gives

$$\left(\omega^2 - \omega_0^2 - \frac{k}{m}\right)^2 = \left(\frac{k}{m}\right)^2$$

Now show that there are two natural frequencies for this system—namely,

$$\omega_1^2 = \omega_0^2 \qquad \text{and} \qquad \omega_2^2 = \omega_0^2 + \frac{2k}{m}$$

Interpret the motion associated with these frequencies by substituting ω_1^2 and ω_2^2 back into the two equations for A and B. The motion associated with these values of A and B are called *normal modes*, and any complicated, general motion of this system can be written as a linear combination of these normal modes. Notice that there are two coordinates (x and y) in this problem and two normal modes. We shall see in Chapter 5 that the complicated vibrational motion of molecules can be resolved into a linear combination of natural, or normal, modes.

Two different forces operate on each pendulum, one due to gravity and the other due to the spring. The former is $-mg \sin \theta$ (see Problem 2–26), and the latter is $-k\xi$ where ξ is the displacement of the spring from its undistorted length (see Problem 2–7). Since the angular displacement of each pendulum from the horizontal is small, $\sin \theta \approx \theta$, where θ is the angle each pendulum makes with it position at rest. Additionally, x and y point in approximately the same direction. Consequently, the displacement for the pendulum whose coordinate is described by x is $x - y$ whereas the displacement for the pendulum whose coordinate is described by y is $y - x$. Applying Newton's second law to each pendulum gives

$$m\frac{d^2x}{dt^2} = -mg\theta_x - k(x - y)$$

$$m\frac{d^2y}{dt^2} = -mg\theta_y - k(y - x)$$

Since $\theta_x = x/l$ and $\theta_y = y/l$, the two equations of motion become

$$m\frac{d^2x}{dt^2} = -\frac{mgx}{l} - k(x - y)$$

$$m\frac{d^2y}{dt^2} = -\frac{mgy}{l} - k(y - x)$$

Letting $\omega_0 = (g/l)^{1/2}$, the above equations become

$$m\frac{d^2x}{dt^2} = -m\omega_0^2 x - k(x - y)$$

$$m\frac{d^2y}{dt^2} = -m\omega_0^2 y - k(y - x)$$

Assuming that the two pendulums swing harmonically, then $x(t) = Ae^{i\omega t}$ and $y(t) = Be^{i\omega t}$. Substituting these expressions into the equations of motion, and dividing throughout by $e^{i\omega t}$ gives

$$mi^2\omega^2 A = -m\omega_0^2 A - k(A - B)$$

$$mi^2\omega^2 B = -m\omega_0^2 B - k(B - A)$$

or after rearrangement

$$\left(\omega^2 - \omega_0^2 - \frac{k}{m}\right) A = -\frac{k}{m}B$$

$$\left(\omega^2 - \omega_0^2 - \frac{k}{m}\right) B = -\frac{k}{m}A$$

To solve these simultaneous equations, set the determinant of the coefficients of A and B to zero:

$$\begin{vmatrix} \omega^2 - \omega_0^2 - \frac{k}{m} & \frac{k}{m} \\ \frac{k}{m} & \omega^2 - \omega_0^2 - \frac{k}{m} \end{vmatrix} = 0$$

$$\left(\omega^2 - \omega_0^2 - \frac{k}{m}\right)^2 - \left(\frac{k}{m}\right)^2 = 0$$

$$\omega^2 - \omega_0^2 - \frac{k}{m} = \pm\frac{k}{m}$$

$$\omega_1^2 = \omega_0^2 \quad \text{or} \quad \omega_2^2 = \omega_0^2 + \frac{2k}{m}$$

The two solutions of ω are labeled with the subscripts 1 and 2, respectively.

Substituting ω_1 into the two equations of motion, $\left(\omega^2 - \omega_0^2 - \frac{k}{m}\right) A = -\frac{k}{m}B$ and $\left(\omega^2 - \omega_0^2 - \frac{k}{m}\right) B = -\frac{k}{m}A$ results in identical equations

$$-\frac{k}{m}A = -\frac{k}{m}B$$

$$A = B$$

This condition indicates that the two pendulums swing in phase.

Substituting ω_2 into the two equations of motion gives

$$\frac{k}{m} A = -\frac{k}{m} B$$

$$A = -B$$

This condition indicates the two pendulums swing against each other, 180° out of phase.

2–29. Problem 2–28 can be solved by introducing center-of-mass and relative coordinates (cf. Section 5.3). Add and subtract the differential equations for $x(t)$ and $y(t)$ and then introduce the new variables

$$\eta = x + y \qquad \text{and} \qquad \xi = x - y$$

Show that the differential equations for η and ξ are independent. Solve each one and compare your results to those of Problem 2–28.

The two equations of motions are

$$m\frac{d^2 x}{dt^2} = -m\omega_0^2 x - k(x - y)$$

$$m\frac{d^2 y}{dt^2} = -m\omega_0^2 y - k(y - x)$$

Adding and subtracting them, respectively, gives

$$m\frac{d^2}{dt^2}(x + y) = -m\omega_0^2 (x + y)$$

$$m\frac{d^2}{dt^2}(x - y) = -m\omega_0^2 (x - y) - 2k(x - y)$$

Substituting $\eta = x + y$ and $\xi = x - y$ into the above equations and dividing through by m gives

$$\frac{d^2 \eta}{dt^2} = -\omega_0^2 \eta$$

$$\frac{d^2 \xi}{dt^2} = -\omega_0^2 \xi - \frac{2k}{m} \xi$$

These two differential equations are independent.

The solution for the first equation is

$$\eta = c_1 \cos \omega_0 t + c_2 \sin \omega_0 t = C \sin(\omega_0 t + \phi)$$

The angular frequency of the center of mass motion is ω_0, which is identical to ω_1 in Problem 2–28.

The solution for the second equation is

$$\xi = c_3 \cos\left(\sqrt{\omega_0^2 + \frac{2k}{m}}\, t\right) + c_4 \sin\left(\sqrt{\omega_0^2 + \frac{2k}{m}}\, t\right) = D \sin\left(\sqrt{\omega_0^2 + \frac{2k}{m}}\, t + \psi\right)$$

The angular frequency of the relative motion is $\sqrt{\omega_0^2 + 2k/m}$, which is identical to ω_1 in Problem 2–28.

2–30. Equation 2.51 suggests that $\sin[2\pi(x + vt)/\lambda]$ and $\sin[2\pi(x - vt)/\lambda]$ are solutions to the wave equation. Now prove that

$$u(x, t) = \phi(x + vt) + \psi(x - vt)$$

where ϕ and ψ are suitably well-behaved but otherwise arbitrary functions, is also a solution. This solution is known as d'Alembert's solution of the wave equation. Give a physical interpretation of $\phi(x + vt)$ and $\psi(x - vt)$. D'Alembert's solution can be obtained by transforming the independent variables in the wave equation. Introduce the new variables

$$\eta = x + vt \qquad \text{and} \qquad \xi = x - vt$$

into Equation 2.1 and show that the wave equation becomes

$$\frac{\partial^2 u}{\partial \eta \partial \xi} = 0$$

This form of the wave equation can be solved by two successive integrations. The first integration gives

$$\frac{\partial u}{\partial \eta} = f(\eta)$$

where $f(\eta)$ is an arbitrary function of η only. Show that a second integration gives

$$u(\eta, \xi) = \int f(\eta)d\eta + \psi(\xi) = \phi(\eta) + \psi(\xi)$$

or

$$u(\eta, \xi) = \phi(x + vt) + \psi(x - vt)$$

If $\phi(x + vt)$ and $\psi(x - vt)$ are solutions to the wave equation, then a linear combination of them, $u(x, t) = \phi(x + vt) + \psi(x - vt)$ is also a solution. Now show that $\phi(x + vt)$ and $\psi(x - vt)$ satisfy the wave equation.

The derivatives of ϕ with respect to x are

$$\frac{\partial \phi}{\partial x} = \frac{\partial \phi}{\partial (x + vt)} \frac{\partial (x + vt)}{\partial x}$$

$$= \frac{\partial \phi}{\partial (x + vt)}$$

$$\frac{\partial^2 \phi}{\partial x^2} = \frac{\partial}{\partial x} \left[\frac{\partial \phi}{\partial (x + vt)} \right]$$

$$= \frac{\partial}{\partial (x + vt)} \frac{\partial (x + vt)}{\partial x} \left[\frac{\partial \phi}{\partial (x + vt)} \right]$$

$$= \frac{\partial^2 \phi}{\partial (x + vt)^2}$$

The derivatives of ϕ with respect to t are

$$\frac{\partial \phi}{\partial t} = \frac{\partial \phi}{\partial (x + vt)} \frac{\partial (x + vt)}{\partial t}$$

$$= v \frac{\partial \phi}{\partial (x + vt)}$$

$$\frac{\partial^2 \phi}{\partial t^2} = \frac{\partial}{\partial t} \left[v \frac{\partial \phi}{\partial (x + vt)} \right]$$

$$= \frac{\partial}{\partial (x + vt)} \frac{\partial (x + vt)}{\partial t} \left[v \frac{\partial \phi}{\partial (x + vt)} \right]$$

$$= v^2 \frac{\partial^2 \phi}{\partial (x + vt)^2}$$

Thus, $\dfrac{\partial^2 \phi}{\partial x^2} = \dfrac{1}{v^2} \dfrac{\partial^2 \phi}{\partial t^2}$. That is, ϕ satisfies the wave equation.

Similarly, the second derivatives of ψ with respect to x and t are

$$\frac{\partial^2 \psi}{\partial x^2} = \frac{\partial^2 \psi}{\partial (x + vt)^2}$$

$$\frac{\partial^2 \psi}{\partial t^2} = v^2 \frac{\partial^2 \psi}{\partial (x + vt)^2}$$

Thus, ψ satisfies the wave equation.

Summing the second derivatives of ϕ and ψ with respect to x, and also the second derivatives of ϕ and ψ with respect to t gives

$$\frac{\partial^2}{\partial x^2} (\phi + \psi) = \frac{\partial^2}{\partial (x + vt)^2} (\phi + \psi)$$

$$\frac{\partial^2}{\partial t^2} (\phi + \psi) = v^2 \frac{\partial^2}{\partial (x + vt)^2} (\phi + \psi)$$

Thus, for $u(x, t) = \phi(x + vt) + \psi(x - vt)$, $\dfrac{\partial^2 u}{\partial x^2} = \dfrac{1}{v^2}\dfrac{\partial^2 u}{\partial t^2}$, that is, $u(x, t)$ satisfies the wave equation. $\phi(x + vt)$ and $\psi(x - vt)$ represent waves traveling to the negative and positive direction, respectively.

Writing $\eta = x + vt$ and $\xi = x - vt$, the derivatives of u with respect to x are

$$\frac{\partial u}{\partial x} = \frac{\partial u}{\partial \eta}\frac{\partial \eta}{\partial x} + \frac{\partial u}{\partial \xi}\frac{\partial \xi}{\partial x}$$

$$= \frac{\partial u}{\partial \eta} + \frac{\partial u}{\partial \xi}$$

$$\frac{\partial^2 u}{\partial x^2} = \frac{\partial}{\partial x}\left(\frac{\partial u}{\partial \eta} + \frac{\partial u}{\partial \xi}\right)$$

$$= \left(\frac{\partial}{\partial \eta}\frac{\partial \eta}{\partial x} + \frac{\partial}{\partial \xi}\frac{\partial \xi}{\partial x}\right)\left(\frac{\partial u}{\partial \eta} + \frac{\partial u}{\partial \xi}\right)$$

$$= \frac{\partial^2 u}{\partial \eta^2} + 2\frac{\partial^2 u}{\partial \eta \partial \xi} + \frac{\partial^2 u}{\partial \xi^2}$$

The derivatives of u with respect to t are

$$\frac{\partial u}{\partial t} = \frac{\partial u}{\partial \eta}\frac{\partial \eta}{\partial t} + \frac{\partial u}{\partial \xi}\frac{\partial \xi}{\partial t}$$

$$= v\frac{\partial u}{\partial \eta} - v\frac{\partial u}{\partial \xi}$$

$$\frac{\partial^2 u}{\partial t^2} = \frac{\partial}{\partial t}\left(v\frac{\partial u}{\partial \eta} - v\frac{\partial u}{\partial \xi}\right)$$

$$= \left(\frac{\partial}{\partial \eta}\frac{\partial \eta}{\partial t} + \frac{\partial}{\partial \xi}\frac{\partial \xi}{\partial t}\right)\left(v\frac{\partial u}{\partial \eta} - v\frac{\partial u}{\partial \xi}\right)$$

$$= v^2\left(\frac{\partial^2 u}{\partial \eta^2} - 2\frac{\partial^2 u}{\partial \eta \partial \xi} + \frac{\partial^2 u}{\partial \xi^2}\right)$$

The wave equation becomes

$$\frac{\partial^2 u}{\partial x^2} = \frac{1}{v^2}\frac{\partial^2 u}{\partial t^2}$$

$$\frac{\partial^2 u}{\partial \eta^2} + 2\frac{\partial^2 u}{\partial \eta \partial \xi} + \frac{\partial^2 u}{\partial \xi^2} = \frac{1}{v^2}\left[v^2\left(\frac{\partial^2 u}{\partial \eta^2} - 2\frac{\partial^2 u}{\partial \eta \partial \xi} + \frac{\partial^2 u}{\partial \xi^2}\right)\right]$$

$$4\frac{\partial^2 u}{\partial \eta \partial \xi} = 0$$

$$\frac{\partial^2 u}{\partial \eta \partial \xi} = 0$$

Integrating the last equation over ξ gives a function independent of ξ:

$$\frac{\partial u}{\partial \eta} = f(\eta)$$

Another integration over η gives

$$u(\eta, \xi) = \int f(\eta)d\eta + \text{constant}$$

Let the integral be $\phi(\eta)$. The constant is independent of η. It is an arbitrary function of ξ which can be called $\psi(\xi)$. Thus,

$$u(\eta, \xi) = \phi(\eta) + \psi(\xi) = \phi(x + vt) + \psi(x - vt)$$

Probability and Statistics

PROBLEMS AND SOLUTIONS

B–1. Using the following table

x	$f(x)$
-6	0.05
-2	0.15
0	0.50
1	0.10
3	0.05
4	0.10
5	0.05

calculate $\langle x \rangle$ and $\langle x^2 \rangle$ and show that $\sigma_x^2 > 0$.

$$\langle x \rangle = \sum_{j=1}^{n} x_j f(x_j)$$

$$= (-6)(0.05) + (-2)(0.15) + (0)(0.50) + (1)(0.10)$$
$$+ (3)(0.05) + (4)(0.10) + (5)(0.05)$$

$$= 0.30$$

$$\langle x^2 \rangle = \sum_{j=1}^{n} x_j^2 f(x_j)$$

$$= (-6)^2(0.05) + (-2)^2(0.15) + (0)^2(0.50) + (1)^2(0.10)$$
$$+ (3)^2(0.05) + (4)^2(0.10) + (5)^2(0.05)$$

$$= 5.80$$

$$\sigma_x^2 = \langle x^2 \rangle - \langle x \rangle^2$$

$$= 5.80 - 0.09$$

$$= 5.71 > 0$$

B–2. A discrete probability distribution that is commonly used in statistics is the Poisson distribution

$$f_n = \frac{\lambda^n}{n!} e^{-\lambda} \qquad n = 0, 1, 2, \ldots$$

where λ is a positive constant. Prove that f_n is normalized. Evaluate $\langle n \rangle$ and $\langle n^2 \rangle$ and show that $\sigma^2 > 0$. Recall that

$$e^x = \sum_{n=0}^{\infty} \frac{x^n}{n!}$$

———————————————

If f_n is normalized, then $\displaystyle\sum_{n=0}^{\infty} f_n = 1$.

$$\sum_{n=0}^{\infty} f_n = \sum_{n=0}^{\infty} \frac{\lambda^n}{n!} e^{-\lambda}$$

$$= e^{-\lambda} \sum_{n=0}^{\infty} \frac{\lambda^n}{n!}$$

$$= e^{-\lambda} e^{+\lambda}$$

$$= 1$$

Thus, the Poisson distribution is normalized. To find $\langle n \rangle$ use Equation B.4.

$$\langle n \rangle = \sum_{n=0}^{\infty} n f_n$$

$$= \sum_{n=0}^{\infty} n \frac{\lambda^n}{n!} e^{-\lambda}$$

$$= e^{-\lambda} \sum_{n=0}^{\infty} n \frac{\lambda^n}{n!}$$

Since the term with $n = 0$ is zero,

$$\langle n \rangle = e^{-\lambda} \sum_{n=1}^{\infty} n \frac{\lambda^n}{n!}$$

$$= e^{-\lambda} \lambda \sum_{n=1}^{\infty} \frac{\lambda^{n-1}}{(n-1)!}$$

Now let $m = n - 1$,

$$\langle n \rangle = e^{-\lambda} \lambda \sum_{m=0}^{\infty} \frac{\lambda^m}{m!}$$

$$= e^{-\lambda} \lambda e^{+\lambda}$$

$$= \lambda$$

To find $\langle n^2 \rangle$, use Equation B.5.

$$\langle n^2 \rangle = \sum_{n=0}^{\infty} n^2 f_n$$

$$= \sum_{n=0}^{\infty} n^2 \frac{\lambda^n}{n!} e^{-\lambda}$$

Once again, the term with $n = 0$ does not contribute to the sum.

$$\langle n^2 \rangle = e^{-\lambda} \lambda \sum_{n=1}^{\infty} n \frac{\lambda^{n-1}}{(n-1)!}$$

$$= e^{-\lambda} \lambda \sum_{m=0}^{\infty} (m+1) \frac{\lambda^m}{m!}$$

$$= e^{-\lambda} \lambda \left(\sum_{m=0}^{\infty} m \frac{\lambda^m}{m!} + \sum_{m=0}^{\infty} \frac{\lambda^m}{m!} \right)$$

These two sums were evaluated earlier

$$\langle n^2 \rangle = e^{-\lambda} \lambda \left(\lambda e^{+\lambda} + e^{+\lambda} \right)$$

$$= \lambda (\lambda + 1)$$

Finally,

$$\sigma^2 = \langle n^2 \rangle - \langle n \rangle^2$$

$$= \lambda (\lambda + 1) - \lambda^2$$

$$= \lambda > 0$$

B–3. An important continuous distribution is the exponential distribution

$$p(x)dx = ce^{-\lambda x}dx \qquad 0 \le x < \infty$$

Evaluate c, $\langle x \rangle$, and σ^2, and the probability that $x \ge a$.

If $p(x)$ is normalized, $\int_0^\infty p(x)\, dx = 1$.

$$\int_0^\infty p(x)\, dx = c \int_0^\infty e^{-\lambda x}\, dx$$

$$= -\frac{c}{\lambda} e^{-\lambda x} \Big|_0^\infty$$

$$= \frac{c}{\lambda}$$

Thus, to normalize $p(x)$, $c = \lambda$. Use this value for c when finding $\langle x \rangle$.

$$\langle x \rangle = \int_0^\infty x p(x)\, dx$$

$$= \lambda \int_0^\infty x e^{-\lambda x}\, dx$$

This integral is found inside the back cover of the textbook.

$$\langle x \rangle = \lambda \frac{1!}{\lambda^2}$$

$$= \frac{1}{\lambda}$$

Finding the variance requires knowing $\langle x^2 \rangle$.

$$\langle x^2 \rangle = \int_0^\infty x^2 p(x)\, dx$$

$$= \lambda \int_0^\infty x^2 e^{-\lambda x}\, dx$$

Again, this integral is found inside the back cover of the textbook.

$$\langle x^2 \rangle = \lambda \frac{2!}{\lambda^3}$$

$$= \frac{2}{\lambda^2}$$

Thus,

$$\sigma^2 = \langle x^2 \rangle - \langle x \rangle^2$$

$$= \frac{2}{\lambda^2} - \frac{1}{\lambda^2}$$

$$= \frac{1}{\lambda^2} > 0$$

The probability that $x \geq a$ is $\int_a^\infty p(x)\, dx$.

$$\int_a^\infty p(x)\, dx = \lambda \int_a^\infty e^{-\lambda x}\, dx$$

$$= -e^{-\lambda x} \Big|_a^\infty$$

$$= e^{-\lambda a}$$

B–4.   Prove explicitly that

$$\int_{-\infty}^\infty e^{-\alpha x^2}\, dx = 2 \int_0^\infty e^{-\alpha x^2}\, dx$$

by breaking the integral from $-\infty$ to ∞ into one from $-\infty$ to 0 and another from 0 to ∞. Let $z = -x$ in the first integral and $z = x$ in the second to prove the above relation.

$$\int_{-\infty}^{\infty} e^{-\alpha x^2}\, dx = \int_{-\infty}^{0} e^{-\alpha x^2}\, dx + \int_{0}^{\infty} e^{-\alpha x^2}\, dx$$

Now, let $z = -x$ in the first integral and $z = x$ in the second.

$$\int_{-\infty}^{\infty} e^{-\alpha x^2}\, dx = -\int_{\infty}^{0} e^{-\alpha z^2}\, dz + \int_{0}^{\infty} e^{-\alpha z^2}\, dz$$

$$= \int_{0}^{\infty} e^{-\alpha z^2}\, dz + \int_{0}^{\infty} e^{-\alpha z^2}\, dz$$

$$= 2\int_{0}^{\infty} e^{-\alpha z^2}\, dz$$

$$= 2\int_{0}^{\infty} e^{-\alpha x^2}\, dx$$

B–5. By using the procedure in Problem B–4, show explicitly that

$$\int_{-\infty}^{\infty} xe^{-\alpha x^2}\, dx = 0$$

$$\int_{-\infty}^{\infty} xe^{-\alpha x^2}\, dx = \int_{-\infty}^{0} xe^{-\alpha x^2}\, dx + \int_{0}^{\infty} xe^{-\alpha x^2}\, dx$$

Now, let $z = -x$ in the first integral,

$$\int_{-\infty}^{\infty} e^{-\alpha x^2}\, dx = -\int_{0}^{\infty} ze^{-\alpha z^2}\, dz + \int_{0}^{\infty} xe^{-\alpha x^2}\, dx$$

$$= -\int_{0}^{\infty} xe^{-\alpha x^2}\, dx + \int_{0}^{\infty} xe^{-\alpha x^2}\, dx$$

$$= 0$$

B–6. Integrals of the type

$$I_n(\alpha) = \int_{-\infty}^{\infty} x^{2n} e^{-\alpha x^2}\, dx \qquad n = 0, 1, 2, \ldots$$

occur frequently in a number of applications. We can simply either look them up in a table of integrals or continue this problem. First, show that

$$I_n(\alpha) = 2\int_{0}^{\infty} x^{2n} e^{-\alpha x^2}\, dx$$

The case $n = 0$ can be handled by the following trick. Show that the square of $I_0(\alpha)$ can be written in the form

$$I_0^2(\alpha) = 4\int_{0}^{\infty}\int_{0}^{\infty} dx\, dy\, e^{-\alpha(x^2+y^2)}$$

Now convert to plane polar coordinates, letting

$$r^2 = x^2 + y^2 \quad \text{and} \quad dxdy = rdrd\theta$$

Show that the appropriate limits of integration are $0 \le r < \infty$ and $0 \le \theta \le \pi/2$ and that

$$I_0^2(\alpha) = 4 \int_0^{\pi/2} d\theta \int_0^\infty drr e^{-\alpha r^2}$$

which is elementary and gives

$$I_0^2(\alpha) = 4 \cdot \frac{\pi}{2} \cdot \frac{1}{2\alpha} = \frac{\pi}{\alpha}$$

or that

$$I_0(\alpha) = \left(\frac{\pi}{\alpha}\right)^{1/2}$$

Now prove that the $I_n(\alpha)$ may be obtained by repeated differentiation of $I_0(\alpha)$ with respect to α and, in particular, that

$$\frac{d^n I_0(\alpha)}{d\alpha^n} = (-1)^n I_n(\alpha)$$

Use this result and the fact that $I_0(\alpha) = (\pi/\alpha)^{1/2}$ to generate $I_1(\alpha)$, $I_2(\alpha)$, and so forth.

The procedure of Problem B–4 can be used, since the integrand is an even function of x.

$$I_n(\alpha) = \int_{-\infty}^\infty x^{2n} e^{-\alpha x^2} dx$$

$$= \int_{-\infty}^0 x^{2n} e^{-\alpha x^2} dx + \int_0^\infty x^{2n} e^{-\alpha x^2} dx$$

Let $z = -x$ in the first integral and $z = x$ in the second.

$$I_n(\alpha) = -\int_\infty^0 z^{2n} e^{-\alpha z^2} dz + \int_0^\infty z^{2n} e^{-\alpha z^2} dz$$

$$= 2 \int_0^\infty z^{2n} e^{-\alpha z^2} dz$$

$$= 2 \int_0^\infty x^{2n} e^{-\alpha x^2} dx$$

Since $I_n(\alpha)$ depends only on α,

$$I_0^2(\alpha) = 4 \left(\int_0^\infty e^{-\alpha x^2} dx\right) \left(\int_0^\infty e^{-\alpha y^2} dy\right) = 4 \int_0^\infty \int_0^\infty dx\, dy\, e^{-\alpha(x^2+y^2)}$$

The integration is over the entire first quadrant, $0 \le x \le \infty$ and $0 \le y \le \infty$ so that in polar coordinates, the limits of integration are $0 \le r < \infty$ and $0 \le \theta \le \pi/2$. Using the substitutions

$r^2 = x^2 + y^2$ and $dx\,dy = r\,dr\,d\theta$, the integral becomes

$$I_0^2(\alpha) = 4\int_0^{\pi/2} d\theta \int_0^\infty dr\, r e^{-\alpha r^2}$$

$$= 4\left(\frac{\pi}{2}\right)\frac{1}{2\alpha} = \frac{\pi}{\alpha}$$

Taking the square root of both sides of the equation leads directly to the desired result, $I_0(\alpha) = (\pi/\alpha)^{1/2}$.

Now, differentiate $I_0(\alpha)$ with respect to α:

$$\frac{dI_0(\alpha)}{d\alpha} = -\int_{-\infty}^\infty x^2 e^{-\alpha x^2}\,dx = -I_1(\alpha)$$

$$\frac{d^2 I_0(\alpha)}{d\alpha^2} = \int_{-\infty}^\infty x^4 e^{-\alpha x^2}\,dx = I_2(\alpha)$$

This can be repeated, and the pattern is obvious, although the general result could be proved via induction if desired.

$$\frac{d^n I_0(\alpha)}{d\alpha^n} = (-1)^n \int_{-\infty}^\infty x^{2n} e^{-\alpha x^2}\,dx = (-1)^n I_n(\alpha)$$

Starting with $I_0(\alpha) = (\pi/\alpha)^{1/2}$, application of the formula gives $I_1(\alpha) = (\pi/\alpha)^{1/2}/2\alpha$, $I_2(\alpha) = 3\,(\pi/\alpha)^{1/2}/4\alpha^2$, and so forth.

B–7.   Without using a table of integrals, show that all of the odd moments of a Gaussian distribution are zero. Using the results derived in Problem B–6, calculate $\langle x^4 \rangle$ for a Gaussian distribution.

Since the odd moments of a Gaussian distribution are found by integration over an odd integrand, the procedure of Problem B–5 is used.

$$\left(2\pi a^2\right)^{-1/2}\int_{-\infty}^\infty x^{2n+1} e^{-x^2/2a^2}\,dx = \left(2\pi a^2\right)^{-1/2}\left[\int_{-\infty}^0 x^{2n+1} e^{-x^2/2a^2}\,dx + \int_0^\infty x^{2n+1} e^{-x^2/2a^2}\,dx\right]$$

$$= \left(2\pi a^2\right)^{-1/2}\left[-\int_0^\infty z^{2n+1} e^{-z^2/2a^2}\,dz + \int_0^\infty x^{2n+1} e^{-x^2/2a^2}\,dx\right]$$

$$= \left(2\pi a^2\right)^{-1/2}\left[-\int_0^\infty x^{2n+1} e^{-x^2/2a^2}\,dx + \int_0^\infty x^{2n+1} e^{-x^2/2a^2}\,dx\right]$$

$$= 0$$

For the Gaussian distribution, $\langle x^4 \rangle$ can be expressed in terms of the integral $I_2(\alpha)$ of Problem B–6.

$$\langle x^4 \rangle = \left(2\pi a^2\right)^{-1/2}\int_{-\infty}^\infty x^4 e^{-x^2/2a^2}\,dx = \left(2\pi a^2\right)^{-1/2} I_2(1/2a^2) = 3a^4$$

B–8. Consider a particle to be constrained to lie along a one-dimensional segment 0 to a. We will learn in the next chapter that the probability that the particle is found to lie between x and $x + dx$ is given by

$$p(x)dx = \frac{2}{a} \sin^2 \frac{n\pi x}{a} dx$$

where $n = 1, 2, 3, \ldots$. First show that $p(x)$ is normalized. Now show that the average position of the particle along the line segment is $a/2$. Is this result physically reasonable? The integrals that you need are (*The CRC Handbook of Chemistry and Physics* or *The CRC Standard Mathematical Tables*; CRC Press: Boca Raton, FL)

$$\int \sin^2 \alpha x dx = \frac{x}{2} - \frac{\sin 2\alpha x}{4\alpha}$$

and

$$\int x \sin^2 \alpha x dx = \frac{x^2}{4} - \frac{x \sin 2\alpha x}{4\alpha} - \frac{\cos 2\alpha x}{8\alpha^2}$$

To show that $p(x)$ is normalized, evaluate the integral $\int_0^a p(x)\, dx$.

$$\int_0^a p(x)\, dx = \frac{2}{a} \int_0^a \sin^2 \frac{n\pi x}{a}\, dx$$

$$= \frac{2}{a} \left[\frac{x}{2} - \frac{\sin(2n\pi x/a)}{4(n\pi/a)} \right]\Bigg|_0^a$$

$$= \frac{2}{a} \left(\frac{a}{2} \right) = 1$$

The average position of the particle is given by $\langle x \rangle$.

$$\langle x \rangle = \int_0^a xp(x)\, dx$$

$$= \frac{2}{a} \int_0^a x \sin^2 \frac{n\pi x}{a}\, dx$$

$$= \frac{2}{a} \left[\frac{x^2}{4} - \frac{x \sin(2n\pi x/a)}{4(n\pi/a)} - \frac{\cos(2n\pi x/a)}{8(n\pi/a)^2} \right]\Bigg|_0^a$$

$$= \frac{2}{a} \left(\frac{a^2}{4} \right) = \frac{a}{2}$$

This result makes physical sense because the probability distribution is symmetric about the center of the box.

B–9. Show that $\langle x \rangle^2 = a^2/4$ and that the variance associated with the probability distribution given in Problem B–8 is given by $\left(\dfrac{a}{2\pi n} \right)^2 \left(\dfrac{\pi^2 n^2}{3} - 2 \right)$. The necessary integral is (CRC tables)

$$\int x^2 \sin^2 \alpha x \, dx = \frac{x^3}{6} - \left(\frac{x^2}{4\alpha} - \frac{1}{8\alpha^3}\right) \sin 2\alpha x - \frac{x \cos 2\alpha x}{4\alpha^2}$$

In Problem B–8 it was shown that $\langle x \rangle = a/2$, so that $\langle x \rangle^2 = a^2/4$. To find the variance, it is necessary to determine $\langle x^2 \rangle$.

$$\langle x^2 \rangle = \int_0^a x^2 p(x) \, dx$$

$$= \frac{2}{a} \int_0^a x^2 \sin^2 \frac{n\pi x}{a} \, dx$$

$$= \frac{2}{a} \left[\frac{x^3}{6} - \left(\frac{ax^2}{4n\pi} - \frac{a^3}{8n^3\pi^3}\right) \sin \frac{2n\pi x}{a} - \frac{a^2 x \cos^2(2n\pi x/a)}{4n^2\pi^2}\right]_0^a$$

$$= \frac{2}{a} \left(\frac{a^3}{6} - \frac{a^3}{4n^2\pi^2}\right)$$

$$= \frac{a^2}{3} - \frac{a^2}{2n^2\pi^2}$$

The variance is $\sigma_x^2 = \langle x^2 \rangle - \langle x \rangle^2$.

$$\sigma_x^2 = \frac{a^2}{3} - \frac{a^2}{2n^2\pi^2} - \frac{a^2}{4} = \frac{a^2}{12} - \frac{a^2}{2n^2\pi^2} = \left(\frac{a}{2\pi n}\right)^2 \left(\frac{\pi^2 n^2}{3} - 2\right)$$

B–10.   Show that

$$\sigma_x = (\langle x^2 \rangle - \langle x \rangle^2)^{1/2}$$

for a particle in a box is less than a, the width of the box, for any value of n. If σ_x is the uncertainty in the position of the particle, could σ_x ever be larger than a?

From Problem B–9 and for any value of n,

$$\sigma_x = \frac{a}{2\pi n} \sqrt{\frac{\pi^2 n^2}{3} - 2}$$

$$= a \sqrt{\frac{1}{12} - \frac{1}{2\pi^2 n^2}}$$

$$< a \sqrt{\frac{1}{12}} = \frac{a}{2\sqrt{3}} < a$$

Thus, σ_x is smaller than a for any value of n. This makes sense because if the particle is known to be somewhere inside the interval $0 \le x \le a$ and never outside, then the uncertainty in its position cannot be larger than the length of interval.

B–11.   All the definite integrals used in Problems B–8 and B–9 can be evaluated from

$$I(\beta) = \int_0^a e^{\beta x} \sin^2 \frac{n \pi x}{a} \, dx$$

Show that the above integrals are given by $I(0)$, $I'(0)$, and $I''(0)$, respectively, where the primes denote differentiation with respect to β. Using a table of integrals, evaluate $I(\beta)$ and then the above three integrals by differentiation.

Taking the first and second derivatives of $I(\beta)$ with respect to β gives

$$I'(\beta) = \int_0^a x e^{\beta x} \sin^2 \frac{n \pi x}{a} \, dx$$

$$I''(\beta) = \int_0^a x^2 e^{\beta x} \sin^2 \frac{n \pi x}{a} \, dx$$

For $\beta = 0$, the corresponding equations for $I(0)$, $I'(0)$, and $I''(0)$ are

$$I(0) = \int_0^a \sin^2 \frac{n \pi x}{a} \, dx$$

$$I'(0) = \int_0^a x \sin^2 \frac{n \pi x}{a} \, dx$$

$$I''(0) = \int_0^a x^2 \sin^2 \frac{n \pi x}{a} \, dx$$

A table of integrals provides the general form (perhaps after some assistance from trigonometric identities)

$$\int e^{\beta x} \sin^2 bx \, dx = \frac{e^{\beta x}}{2\beta} - \frac{e^{\beta x}}{\beta^2 + 4b^2} \left(\frac{\beta \cos 2bx}{2} + b \sin 2bx \right)$$

and so

$$I(\beta) = \frac{e^{\beta a} - 1}{2\beta} - \frac{\beta}{2} \left(\frac{e^{\beta a} - 1}{\beta^2 - 4n^2 \pi^2 a^{-2}} \right)$$

Now, knowing that the behavior of the integral is desired as $\beta \to 0$, expand $I(\beta)$ in a Maclaurin series, and on the right hand side of the last equation, use the series expansion for $e^{\beta a}$.

$$I(\beta) = I(0) + \beta I'(0) + \frac{\beta^2}{2} I''(0) + O(\beta^3)$$

$$I(\beta) = \frac{a}{2} + \frac{a^2}{4}\beta + \frac{a^3}{12}\beta^2 - \frac{1}{4n^2\pi^2 a^{-2}} \left(\frac{\beta^2 a}{2} + \frac{\beta^3 a^2}{4} \right) + O(\beta^3)$$

$$= \frac{a}{2} + \frac{a^2}{4}\beta + \left(\frac{a^3}{6} - \frac{a^3}{4\pi^2 n^2} \right) \frac{\beta^2}{2} + O(\beta^3)$$

Comparing the coefficients of the powers of β gives

$$I(0) = \frac{a}{2} \qquad I'(0) = \frac{a^2}{4} \qquad I''(0) = \frac{a^3}{6} - \frac{a^3}{4\pi^2 n^2}$$

B–12. Using the probability distribution given in Problem B–8, calculate the probability that the particle will be found between 0 and $a/2$. The necessary integral is given in Problem B–8.

The probability that the particle will be found between 0 and $a/2$ is $\int_0^{a/2} p(x)\,dx$.

$$\int_0^{a/2} p(x)\,dx = \frac{2}{a} \int_0^{a/2} \sin^2 \frac{n\pi x}{a}\,dx$$

$$= \frac{2}{a} \left[\frac{x}{2} - \frac{\sin(2n\pi x/a)}{4(n\pi/a)} \right]_0^{a/2}$$

$$= \frac{2}{a} \left(\frac{a}{4} \right) = \frac{1}{2}$$

This makes sense; the probability that the particle will be found in half the box is 0.5.

CHAPTER 3

The Schrödinger Equation and a Particle in a Box

PROBLEMS AND SOLUTIONS

3–1. Evaluate $g = \hat{A}f$, where $\hat{A}$ and f are given below:

$\hat{A}$	f
(a) SQRT	x^4
(b) $\dfrac{d^3}{dx^3} + x^3$	e^{-ax}
(c) $\displaystyle\int_0^1 dx$	$x^3 - 2x + 3$
(d) $\dfrac{\partial^2}{\partial x^2} + \dfrac{\partial^2}{\partial y^2} + \dfrac{\partial^2}{\partial z^2}$	$x^3y^2z^4$

(a) $\text{SQRT}(x^4) = \pm x^2$

(b) $\dfrac{d^3 e^{-ax}}{dx^3} + x^3 e^{-ax} = -a^3 e^{-ax} + x^3 e^{-ax} = e^{-ax}\left(x^3 - a^3\right)$

(c) $\displaystyle\int_0^1 \left(x^3 - 2x + 3\right)\,dx = \left.\dfrac{x^4}{4} - x^2 + 3x\right|_0^1 = \dfrac{9}{4}$

(d) $\dfrac{\partial^2(x^3y^2z^4)}{\partial x^2} + \dfrac{\partial^2(x^3y^2z^4)}{\partial y^2} + \dfrac{\partial^2(x^3y^2z^4)}{\partial z^2} = 6xy^2z^4 + 2x^3z^4 + 12x^3y^2z^2$

3–2. Determine whether the following operators are linear or nonlinear:

(a) $\hat{A}f(x) = \text{SQR}\,f(x)$ [square $f(x)$]

(b) $\hat{A}f(x) = f^*(x)$ [form the complex conjugate of $f(x)$]

(c) $\hat{A}f(x) = 0$ [multiply $f(x)$ by zero]

(d) $\hat{A}f(x) = [f(x)]^{-1}$ [take the reciprocal of $f(x)$]

(e) $\hat{A}f(x) = f(0)$ [evaluate $f(x)$ at $x = 0$]

(f) $\hat{A}f(x) = \ln f(x)$ [take the logarithm of $f(x)$]

An operator, $\hat{A}$, is linear if $\hat{A}[c_1 f_1(x) + c_2 f_2(x)] = c_1 \hat{A} f_1(x) + c_2 \hat{A} f_2(x)$.

(a)

$$\hat{A}\left[c_1 f_1(x) + c_2 f_2(x)\right] = \left[c_1 f_1(x) + c_2 f_2(x)\right]^2$$

$$= c_1^2 f_1^2(x) + 2c_1 f_1(x) c_2 f_2(x) + c_2^2 f_2^2(x)$$

$$c_1 \hat{A} f_1(x) + c_2 \hat{A} f_2(x) = c_1 f_1^2(x) + c_2 f_2^2(x)$$

$$\neq \hat{A}\left[c_1 f_1(x) + c_2 f_2(x)\right]$$

Nonlinear

(b)

$$\hat{A}\left[c_1 f_1(x) + c_2 f_2(x)\right] = c_1^* f_1^*(x) + c_2^* f_2^*(x)$$

$$c_1 \hat{A} f_1(x) + c_2 \hat{A} f_2(x) = c_1 f_1^*(x) + c_2 f_2^*(x)$$

$$\neq \hat{A}\left[c_1 f_1(x) + c_2 f_2(x)\right]$$

Nonlinear

(c)

$$\hat{A}\left[c_1 f_1(x) + c_2 f_2(x)\right] = 0$$

$$c_1 \hat{A} f_1(x) + c_2 \hat{A} f_2(x) = c_1(0) + c_2(0) = 0$$

$$= \hat{A}\left[c_1 f_1(x) + c_2 f_2(x)\right]$$

Linear

(d)

$$\hat{A}\left[c_1 f_1(x) + c_2 f_2(x)\right] = \left[c_1 f_1(x) + c_2 f_2(x)\right]^{-1}$$

$$c_1 \hat{A} f_1(x) + c_2 \hat{A} f_2(x) = \frac{c_1}{f_1(x)} + \frac{c_2}{f_2(x)}$$

$$\neq \hat{A}\left[c_1 f_1(x) + c_2 f_2(x)\right]$$

Nonlinear

(e)

$$\hat{A}\left[c_1 f_1(x) + c_2 f_2(x)\right] = \left[c_1 f_1(0) + c_2 f_2(0)\right]$$

$$c_1 \hat{A} f_1(x) + c_2 \hat{A} f_2(x) = c_1 f_1(0) + c_2 f_2(0)$$

$$= \hat{A}\left[c_1 f_1(x) + c_2 f_2(x)\right]$$

Linear

(f)

$$\hat{A}\left[c_1 f_1(x) + c_2 f_2(x)\right] = \ln\left[c_1 f_1(x) + c_2 f_2(x)\right]$$

$$c_1 \hat{A} f_1(x) + c_2 \hat{A} f_2(x) = c_1 \ln f_1(x) + c_2 \ln f_2(x)$$

$$\neq \hat{A}\left[c_1 f_1(x) + c_2 f_2(x)\right]$$

Nonlinear

3–3. In each case, show that $f(x)$ is an eigenfunction of the operator given. Find the eigenvalue.

$\hat{A}$	$f(x)$
(a) $\dfrac{d^2}{dx^2}$	$\cos \omega x$
(b) $\dfrac{d}{dt}$	$e^{i\omega t}$
(c) $\dfrac{d^2}{dx^2} + 2\dfrac{d}{dx} + 3$	$e^{\alpha x}$
(d) $\dfrac{\partial}{\partial y}$	$x^2 e^{6y}$

(a) $\hat{A}f(x) = \dfrac{d^2 \cos \omega x}{dx^2} = -\omega^2 \cos \omega x$; eigenvalue $= -\omega^2$

(b) $\hat{A}f(x) = \dfrac{de^{i\omega t}}{dt} = i\omega e^{i\omega t}$; eigenvalue $= i\omega$

(c) $\hat{A}f(x) = \dfrac{d^2 \left(e^{\alpha x}\right)}{dx^2} + 2\dfrac{d \left(e^{\alpha x}\right)}{dx} + 3\left(e^{\alpha x}\right) = \left(\alpha^2 + 2\alpha + 3\right) e^{\alpha x}$; eigenvalue $= \alpha^2 + 2\alpha + 3$

(d) $\hat{A}f(x) = \dfrac{\partial \left(x^2 e^{6y}\right)}{\partial y} = 6x^2 e^{6y}$; eigenvalue $= 6$

3–4. Show that $(\cos ax)(\cos by)(\cos cz)$ is an eigenfunction of the operator,

$$\nabla^2 = \frac{\partial^2}{\partial x^2} + \frac{\partial^2}{\partial y^2} + \frac{\partial^2}{\partial z^2}$$

which is called the Laplacian operator.

$$\nabla^2 (\cos ax)(\cos by)(\cos cz) = \frac{\partial^2 (\cos ax)(\cos by)(\cos cz)}{\partial x^2} + \frac{\partial^2 (\cos ax)(\cos by)(\cos cz)}{\partial y^2}$$

$$+ \frac{\partial^2 (\cos ax)(\cos by)(\cos cz)}{\partial z^2}$$

$$= -a^2 (\cos ax)(\cos by)(\cos cz) - b^2 (\cos ax)(\cos by)(\cos cz)$$

$$- c^2 (\cos ax)(\cos by)(\cos cz)$$

$$= -\left(a^2 + b^2 + c^2\right)(\cos ax)(\cos by)(\cos cz)$$

The eigenvalue of the Laplacian operator for the eigenfunction $(\cos ax)(\cos by)(\cos cz)$ is $-\left(a^2 + b^2 + c^2\right)$.

3–5. Write out the operator $\hat{A}^2$ for $\hat{A} =$

(a) $\dfrac{d^2}{dx^2}$ (b) $\dfrac{d}{dx} + x$ (c) $\dfrac{d^2}{dx^2} - 2x\dfrac{d}{dx} + 1$

Hint: Be sure to include $f(x)$ before carrying out the operations.

In each case, the operator $\hat{A}^2$ is found by determining an expression for

$$\hat{A}^2 f(x) = \hat{A}\left[\hat{A}f(x)\right]$$

(a) $\hat{A}\left[\hat{A}f(x)\right] = \dfrac{d^2}{dx^2}\left[\dfrac{d^2 f(x)}{dx^2}\right] = \dfrac{d^4 f(x)}{dx^4}$

The operator $\hat{A}^2$ is

$$\hat{A}^2 = \dfrac{d^4}{dx^4}$$

(b)

$$\hat{A}\left[\hat{A}f(x)\right] = \left(\dfrac{d}{dx} + x\right)\left[\dfrac{df(x)}{dx} + xf(x)\right]$$

$$= \dfrac{d^2 f(x)}{dx^2} + x\dfrac{df(x)}{dx} + f(x)\dfrac{dx}{dx} + x\dfrac{df(x)}{dx} + x^2 f(x)$$

$$= \dfrac{d^2 f(x)}{dx^2} + 2x\dfrac{df(x)}{dx} + f(x) + x^2 f(x)$$

Thus, the operator $\hat{A}^2$ is

$$\hat{A}^2 = \dfrac{d^2}{dx^2} + 2x\dfrac{d}{dx} + 1 + x^2$$

(c) In determining the square of this operator it will be necessary to evaluate $\dfrac{d^2}{dx^2}\left[2x\dfrac{df(x)}{dx}\right]$, which requires some care.

$$\dfrac{d^2}{dx^2}\left[2x\dfrac{df(x)}{dx}\right] = \dfrac{d}{dx}\left\{\dfrac{d}{dx}\left[2x\dfrac{df(x)}{dx}\right]\right\}$$

$$= \dfrac{d}{dx}\left[2\dfrac{df(x)}{dx} + 2x\dfrac{d^2 f(x)}{dx^2}\right]$$

$$= 2\dfrac{d^2 f(x)}{dx^2} + 2\dfrac{d^2 f(x)}{dx^2} + 2x\dfrac{d^3 f(x)}{dx^3}$$

$$= 4\dfrac{d^2 f(x)}{dx^2} + 2x\dfrac{d^3 f(x)}{dx^3}$$

Continuing on to determine $\hat{A}^2$,

$$\hat{A}\left[\hat{A}f(x)\right] = \left(\frac{d^2}{dx^2} - 2x\frac{d}{dx} + 1\right)\left[\frac{d^2 f(x)}{dx^2} - 2x\frac{df(x)}{dx} + f(x)\right]$$

$$= \frac{d^4 f(x)}{dx^4} - 4\frac{d^2 f(x)}{dx^2} - 2x\frac{d^3 f(x)}{dx^3} + \frac{d^2 f(x)}{dx^2}$$

$$- 2x\frac{d^3 f(x)}{dx^3} + 4x\frac{df(x)}{dx} + 4x^2\frac{d^2 f(x)}{dx^2} - 2x\frac{df(x)}{dx}$$

$$+ \frac{d^2 f(x)}{dx^2} - 2x\frac{df(x)}{dx} + f(x)$$

$$= \frac{d^4 f(x)}{dx^4} - 4x\frac{d^3 f(x)}{dx^3} + \left(4x^2 - 2\right)\frac{d^2 f(x)}{dx^2} + f(x)$$

Thus, the operator $\hat{A}^2$ is

$$\hat{A}^2 = \frac{d^4}{dx^4} - 4x\frac{d^3}{dx^3} + \left(4x^2 - 2\right)\frac{d^2}{dx^2} + 1$$

3–6. Determine whether or not the following pairs of operators commute.

	$\hat{A}$	$\hat{B}$
(a)	$\dfrac{d}{dx}$	$\dfrac{d^2}{dx^2} + 2\dfrac{d}{dx}$
(b)	x	$\dfrac{d}{dx}$
(c)	SQR	SQRT
(d)	$\dfrac{\partial}{\partial x}$	$\dfrac{\partial}{\partial y}$

In each case determine $\hat{A}\hat{B}f(x)$ and $\hat{B}\hat{A}f(x)$ and compare the two results for equality.

(a)

$$\hat{A}\hat{B}f(x) = \frac{d}{dx}\left[\frac{d^2 f(x)}{dx^2} + 2\frac{df(x)}{dx}\right]$$

$$= \frac{d^3 f(x)}{dx^3} + 2\frac{d^2 f(x)}{dx^2}$$

$$\hat{B}\hat{A}f(x) = \left(\frac{d^2}{dx^2} + 2\frac{d}{dx}\right)\frac{df(x)}{dx}$$

$$= \frac{d^3 f(x)}{dx^3} + 2\frac{d^2 f(x)}{dx^2}$$

$$= \hat{A}\hat{B}f(x)$$

The two operators commute.

(b)

$$\hat{A}\hat{B}f(x) = x\frac{df(x)}{dx}$$

$$\hat{B}\hat{A}f(x) = \left(\frac{d}{dx}\right)xf(x)$$

$$= f(x) + x\frac{df(x)}{dx}$$

$$\neq \hat{A}\hat{B}f(x)$$

The two operators do not commute.

(c)

$$\hat{A}\hat{B}f(x) = \text{SQR}\left[\text{SQRT}\,f(x)\right]$$

$$= f(x)$$

$$\hat{B}\hat{A}f(x) = \text{SQRT}\left[\text{SQR}\,f(x)\right]$$

$$= \text{SQRT}\left[f(x)\right]^2$$

$$= \pm f(x)$$

$$\neq \hat{A}\hat{B}f(x)$$

The two operators do not commute.

(d) For these two operators, use a function of two (or more) variables, $f(x, y)$.

$$\hat{A}\hat{B}f(x) = \frac{\partial}{\partial x}\left[\frac{\partial f(x, y)}{\partial y}\right] = \frac{\partial^2 f(x, y)}{\partial x\partial y}$$

$$\hat{B}\hat{A}f(x) = \frac{\partial}{\partial y}\left[\frac{\partial f(x, y)}{\partial x}\right] = \frac{\partial^2 f(x, y)}{\partial y\partial x}$$

For well behaved functions, $\dfrac{\partial^2 f(x, y)}{\partial x\partial y} = \dfrac{\partial^2 f(x, y)}{\partial y\partial x}$, and the operators commute.

3–7. In ordinary algebra, $(P + Q)(P - Q) = P^2 - Q^2$. Expand $(\hat{P} + \hat{Q})(\hat{P} - \hat{Q})$. Under what conditions do we find the same result as in the case of ordinary algebra?

Remembering that operators always require a function upon which to operate,

$$\left(\hat{P} + \hat{Q}\right)\left(\hat{P} - \hat{Q}\right)f(x) = \left(\hat{P} + \hat{Q}\right)\left[\hat{P}f(x) - \hat{Q}f(x)\right]$$

$$= \left(\hat{P} + \hat{Q}\right)\hat{P}f(x) - \left(\hat{P} + \hat{Q}\right)\hat{Q}f(x)$$

$$= \hat{P}^2 f(x) + \hat{Q}\hat{P}f(x) - \hat{P}\hat{Q}f(x) + \hat{Q}^2 f(x)$$

Thus, $(\hat{P} + \hat{Q})(\hat{P} - \hat{Q}) = \hat{P}^2 + \hat{Q}\hat{P} - \hat{P}\hat{Q} + \hat{Q}^2$. The two "cross terms" do not cancel when $\hat{P}$ and $\hat{Q}$ do not commute, since $\hat{Q}\hat{P} \neq \hat{P}\hat{Q}$. The same result as in the case of ordinary algebra is only obtained when the two operators do commute.

3–8. If we operate on the particle-in-a-box wave functions (Equations 3.27) with the momentum operator (Equation 3.11), we find

$$\hat{P} B \sin \frac{n\pi x}{a} = -i\hbar B \frac{\partial}{\partial x} \left(\sin \frac{n\pi x}{a} \right)$$

$$= -\frac{i\hbar n\pi}{a} B \cos \frac{n\pi x}{a}$$

Note that this is *not* an eigenvalue equation, and so we say that the momentum of a particle in a box does not have a fixed, definite value. Although the particle does not have a definite momentum, we can use the classical equation $E = p^2/2m$ to define formally some sort of effective momentum. Using Equation 3.21 for E, show that $p = nh/2a$ and that the de Broglie wavelengths associated with these momenta are $\lambda = h/p = 2a/n$. Show that this last equation says that an integral number of half-wavelengths fit into the box or that Figure 3.2 corresponds to standing de Broglie waves or matter waves.

Equation 3.21 gives the energy for the particle in a box as

$$E_n = \frac{h^2 n^2}{8ma^2} \qquad n = 1, 2, \ldots$$

The energy of the particle is purely kinetic energy, so that the classical equation $E = p^2/2m$ suggests

$$\frac{p^2}{2m} = \frac{h^2 n^2}{8ma^2}$$

$$p^2 = \frac{h^2 n^2}{4a^2}$$

$$p = \frac{nh}{2a}$$

The de Broglie relation gives the wavelength associated with the particle as

$$\lambda = \frac{h}{p}$$

$$= \frac{2ah}{nh}$$

$$= \frac{2a}{n}$$

This last equation can be rewritten as

$$n\frac{\lambda}{2} = a$$

or that an integral number of half-wavelengths is equal to the length of the box. The wave has a constant value equal to zero at each end of the box, and the matter waves in Figure 3.2 are standing waves.

3–9. In Section 3.5, we applied the equations for a particle in a box to the π electrons in butadiene. This simple model is called the free-electron model. Using the same argument, show that the length of hexatriene can be estimated to be 867 pm. Show that the first electronic transition is predicted to occur at 2.8×10^4 cm^{-1}. (Remember that hexatriene has six π electrons.)

In applying the free-electron model to hexatriene, the Pauli exclusion principle requires that the six π electrons fill the first three particle-in-a-box levels. The length of the box can be estimated as equal to three C=C bond lengths plus two C–C bond lengths, plus the distance of a carbon atom radius at each end, or

$$3 \times 135 \text{ pm} + 2 \times 154 \text{ pm} + 2 \times 77.0 \text{ pm} = 867 \text{ pm}$$

The first excited state is that which has one electron elevated from the $n = 3$ state to the $n = 4$ state. The energy required to make this transition is

$$\Delta E = \frac{h^2}{8ma^2}\left(4^2 - 3^2\right)$$

$$= \frac{\left(6.626\,0755 \times 10^{-34} \text{ J·s}\right)^2 7}{8\left(9.109\,3897 \times 10^{-31}\text{ kg}\right)\left(867 \times 10^{-12}\text{ m}\right)^2}$$

$$= 5.610 \times 10^{-19} \text{ J}$$

$$= hc\tilde{\nu}$$

Thus,

$$\tilde{\nu} = \frac{5.610 \times 10^{-19} \text{ J}}{\left(6.626\,0755 \times 10^{-34}\text{ J·s}\right)\left(2.997\,924\,58 \times 10^{10}\text{ cm·s}^{-1}\right)}$$

$$= 2.82 \times 10^4 \text{ cm}^{-1}$$

3–10. Prove that if $\psi(x)$ is a solution to the Schrödinger equation, then any constant times $\psi(x)$ is also a solution.

If $\psi(x)$ is a solution of the Schrödinger equation, then $\hat{H}\psi(x) = E\psi(x)$. The Hamiltonian operator, $\hat{H}$, is a linear operator, so if c is a constant,

$$\hat{H}c\psi(x) = c\hat{H}\psi(x)$$

$$= cE\psi(x)$$

$$= E\left[c\psi(x)\right]$$

where the last step follows because both c and E are numbers whose multiplication commutes. This shows that $c\psi(x)$ is an eigenfunction of $\hat{H}$ with eigenvalue equal to E, and it follows that $c\psi(x)$ is also a solution to the Schrödinger equation.

3–11. In this problem, we will prove that the form of the Schrödinger equation imposes the condition that the first derivative of a wave function be continuous. The Schrödinger equation is

$$\frac{d^2\psi}{dx^2} + \frac{2m}{\hbar^2}[E - V(x)]\psi(x) = 0$$

If we integrate both sides from $a - \epsilon$ to $a + \epsilon$, where a is an arbitrary value of x and ϵ is infinitesimally small, then we have

$$\left.\frac{d\psi}{dx}\right|_{x=a+\epsilon} - \left.\frac{d\psi}{dx}\right|_{x=a-\epsilon} = \frac{2m}{\hbar^2}\int_{a-\epsilon}^{a+\epsilon}[V(x) - E]\psi(x)dx$$

Now show that $d\psi/dx$ is continuous if $V(x)$ is continuous.

Suppose now that $V(x)$ is *not* continuous at $x = a$, as in

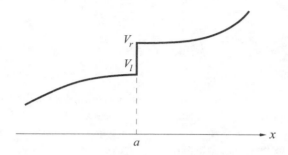

Show that

$$\left.\frac{d\psi}{dx}\right|_{x=a+\epsilon} - \left.\frac{d\psi}{dx}\right|_{x=a-\epsilon} = \frac{2m}{\hbar^2}(V_l + V_r - 2E)\psi(a)\epsilon$$

so that $d\psi/dx$ is continuous even if $V(x)$ has a *finite* discontinuity. What if $V(x)$ has an infinite discontinuity, as in the problem of a particle in a box? Are the first derivatives of the wave functions continuous at the boundaries of the box?

The first derivative of a wave function is continuous if

$$\lim_{\epsilon\to 0}\left(\left.\frac{d\psi}{dx}\right|_{x=a+\epsilon} - \left.\frac{d\psi}{dx}\right|_{x=a-\epsilon}\right) = 0$$

Start with

$$\left.\frac{d\psi}{dx}\right|_{x=a+\epsilon} - \left.\frac{d\psi}{dx}\right|_{x=a-\epsilon} = \frac{2m}{\hbar^2}\int_{a-\epsilon}^{a+\epsilon}[V(x) - E]\,\psi(x)\,dx \qquad (1)$$

If $V(x)$ is continuous, then $\lim_{\epsilon\to 0} V(a \pm \epsilon) = V(a)$. A similar result holds for $\psi(x)$, which is already known to be continuous. Thus,

$$\lim_{\epsilon\to 0}\frac{2m}{\hbar^2}\int_{a-\epsilon}^{a+\epsilon}[V(x) - E]\,\psi(x)\,dx = \frac{2m}{\hbar^2}[V(a) - E]\,\psi(a)\lim_{\epsilon\to 0}\int_{a-\epsilon}^{a+\epsilon}dx = 0 \qquad (2)$$

Combining Equations 1 and 2 shows that

$$\lim_{\epsilon\to 0}\left(\left.\frac{d\psi}{dx}\right|_{x=a+\epsilon} - \left.\frac{d\psi}{dx}\right|_{x=a-\epsilon}\right) = 0$$

and therefore $d\psi/dx$ is continuous.

Now suppose that $V(x)$ has a finite discontinuity at $x = a$. Then divide the integral into two parts, $a - \epsilon$ to a, and a to $a + \epsilon$. In the limit as $\epsilon \to 0$,

$$\frac{2m}{\hbar^2} \int_{a-\epsilon}^{a+\epsilon} [V(x) - E]\psi(x)\,dx = \frac{2m}{\hbar^2} \left\{ \int_{a-\epsilon}^{a} [V(x) - E]\psi(x)\,dx + \int_{a}^{a+\epsilon} [V(x) - E]\psi(x)\,dx \right\}$$

$$= \frac{2m}{\hbar^2} \left[(V_l - E)\,\psi(a) \int_{a-\epsilon}^{a} dx + (V_r - E)\,\psi(a) \int_{a}^{a+\epsilon} dx \right]$$

$$= \frac{2m}{\hbar^2} (V_l + V_r - 2E)\,\psi(a)\epsilon$$

As $\epsilon \to 0$, $\lim_{\epsilon \to 0} (V_l + V_r - 2E)\,\psi(a)\epsilon = 0$, and $d\psi/dx$ remains continuous even though $V(x)$ has a finite discontinuity.

If, however, the discontinuity in $V(x)$ at the point a is infinite, then there is no finite limiting value for $V(x)$ upon approaching from one side, and the expression $\frac{2m}{\hbar^2} \int_{a-\epsilon}^{a+\epsilon} [V(x) - E]\psi(x)\,dx$ cannot be integrated and will not equal zero in the limit $\epsilon \to 0$. Thus, $d\psi/dx$ is not a continuous function in this case. This can be seen for the particle in a box, which has infinite discontinuities in the potential at the boundaries of the box. The first derivatives of the particle-in-a-box wave functions are not continuous at the boundaries of the box. Outside the box, $d\psi/dx = 0$, while inside the box at $x = 0$, $d\psi/dx = (\sqrt{2/a})n\pi/a$, and at $x = a$, $d\psi/dx = (-1)^n(\sqrt{2/a})n\pi/a$.

3–12. Show that the probability associated with the state ψ_n for a particle in a one-dimensional box of length a obeys the following relationships:

$$\text{Prob}(0 \le x \le a/4) = \text{Prob}(3a/4 \le x \le a) = \begin{cases} \dfrac{1}{4} & n \text{ even} \\[2mm] \dfrac{1}{4} - \dfrac{(-1)^{\frac{n-1}{2}}}{2\pi n} & n \text{ odd} \end{cases}$$

and

$$\text{Prob}(a/4 \le x \le a/2) = \text{Prob}(a/2 \le x \le 3a/4) = \begin{cases} \dfrac{1}{4} & n \text{ even} \\[2mm] \dfrac{1}{4} + \dfrac{(-1)^{\frac{n-1}{2}}}{2\pi n} & n \text{ odd} \end{cases}$$

For a particle in a one-dimensional box of length a, $\psi(x) = (2/a)^{1/2} \sin(n\pi x/a)$, and

$$\text{Prob}\,(c \leq x \leq d) = \int_c^d \psi^*(x)\psi(x)\,dx$$

$$= \frac{2}{a} \int_c^d \sin^2\left(\frac{n\pi x}{a}\right) dx$$

$$= \frac{2}{a} \left[\frac{x}{2} - \frac{\sin(2n\pi x/a)}{4(n\pi/a)}\right]_{x=c}^{x=d}$$

$$= \left[\frac{x}{a} - \frac{\sin(2n\pi x/a)}{2n\pi}\right]_{x=c}^{x=d}$$

$$= \frac{d}{a} - \frac{\sin(2n\pi d/a)}{2n\pi} - \frac{c}{a} + \frac{\sin(2n\pi c/a)}{2n\pi}$$

$$= \frac{d-c}{a} - \left(\frac{1}{2n\pi}\right)\left[\sin\left(\frac{2n\pi d}{a}\right) - \sin\left(\frac{2n\pi c}{a}\right)\right]$$

For each of the regions under consideration, $d - c = a/4$, and only the sin terms need be considered individually. (Recall that $\sin n\pi = \sin 2n\pi = 0$ for integer n.)

c	d	Prob $(c \leq x \leq d)$
0	$\dfrac{a}{4}$	$\dfrac{1}{4} - \dfrac{1}{2n\pi}\sin\left(\dfrac{n\pi}{2}\right) - 0$
$\dfrac{a}{4}$	$\dfrac{a}{2}$	$\dfrac{1}{4} - 0 + \dfrac{1}{2n\pi}\sin\left(\dfrac{n\pi}{2}\right)$
$\dfrac{a}{2}$	$\dfrac{3a}{4}$	$\dfrac{1}{4} - \dfrac{1}{2n\pi}\sin\left(\dfrac{3n\pi}{2}\right) - 0$
$\dfrac{3a}{4}$	a	$\dfrac{1}{4} - 0 + \dfrac{1}{2n\pi}\sin\left(\dfrac{3n\pi}{2}\right)$

For n even, $\sin\dfrac{n\pi}{2} = 0$ and $\sin\dfrac{3n\pi}{2} = 0$; for n odd, $\sin\dfrac{n\pi}{2} = (-1)^{(n-1)/2}$ and $\sin\dfrac{3n\pi}{2} = -(-1)^{(n-1)/2}$. Therefore,

$$\text{Prob}\,(0 \leq x \leq a/4) = \text{Prob}\,(3a/4 \leq x \leq a) = \begin{cases} \dfrac{1}{4} & n \text{ even} \\[2mm] \dfrac{1}{4} - \dfrac{(-1)^{\frac{n-1}{2}}}{2\pi n} & n \text{ odd} \end{cases}$$

and

$$\text{Prob}\,(a/4 \leq x \leq a/2) = \text{Prob}\,(a/2 \leq x \leq 3a/4) = \begin{cases} \dfrac{1}{4} & n \text{ even} \\[2mm] \dfrac{1}{4} + \dfrac{(-1)^{\frac{n-1}{2}}}{2\pi n} & n \text{ odd} \end{cases}$$

3–13. What are the units, if any, for the wave function of a particle in a one-dimensional box?

Inspection of the wave function for a particle in a one-dimensional box, $\psi(x) = (2/a)^{1/2} \sin(n\pi x/a)$, shows that $\psi(x)$ has units $m^{-1/2}$. This makes sense because the probability, $\psi^*(x)\psi(x)\,dx$, must be unitless.

3–14.   Using a table of integrals, show that

$$\int_0^a \sin^2 \frac{n\pi x}{a} dx = \frac{a}{2}$$

$$\int_0^a x \sin^2 \frac{n\pi x}{a} dx = \frac{a^2}{4}$$

and

$$\int_0^a x^2 \sin^2 \frac{n\pi x}{a} dx = \left(\frac{a}{2\pi n}\right)^3 \left(\frac{4\pi^3 n^3}{3} - 2n\pi\right)$$

All of these integrals can be evaluated from

$$I(\beta) = \int_0^a e^{\beta x} \sin^2 \frac{n\pi x}{a} dx$$

Show that the above integrals are given by $I(0)$, $I'(0)$, and $I''(0)$, respectively, where the primes denote differentiation with respect to β. Using a table of integrals, evaluate $I(\beta)$ and then the above three integrals by differentiation.

_____

The table of integrals in *The CRC Handbook of Chemistry and Physics* has the integral

$$\int \sin^2 cx \, dx = \frac{x}{2} - \frac{1}{4c} \sin 2cx$$

Using $c = n\pi/a$ and adding the desired limits of integration gives

$$\int_0^a \sin^2 \frac{n\pi x}{a} dx = \left(\frac{x}{2} - \frac{a}{4n\pi} \sin \frac{2n\pi x}{a}\right)\Big|_0^a$$

$$= \frac{a}{2}$$

The second integral is also found in the tables.

$$\int x \sin^2 cx \, dx = \frac{x^2}{4} - \frac{x \sin 2cx}{4c} - \frac{\cos 2cx}{8c^2}$$

Using $c = n\pi/a$ and adding the desired limits of integration gives

$$\int_0^a x \sin^2 \frac{n\pi x}{a} dx = \left[\frac{x^2}{4} - \frac{x \sin(2n\pi x/a)}{4(n\pi/a)} - \frac{\cos(2n\pi x/a)}{8(n\pi/a)^2}\right]\Big|_0^a$$

$$= \frac{a^2}{4}$$

Finally, from the tables,

$$\int x^2 \sin^2 cx\, dx = \frac{x^3}{6} - \left(\frac{x^2}{4c} - \frac{1}{8c^3}\right)\sin 2cx - \frac{x\cos 2cx}{4c^2}$$

Once again using $c = n\pi/a$ and adding the desired limits of integration turns this into

$$\int_0^a x^2 \sin^2 \frac{n\pi x}{a}\, dx = \left\{ \frac{x^3}{6} - \left[\frac{x^2}{4(n\pi/a)} - \frac{1}{8(n\pi/a)^3}\right]\sin 2(n\pi x/a) - \frac{x\cos(2n\pi x/a)}{4(n\pi/a)^2} \right\}\Big|_0^a$$

$$= \frac{a^3}{6} - \frac{a^3}{4n^2\pi^2} = \left(\frac{a}{2\pi n}\right)^3 \left(\frac{4\pi^3 n^3}{3} - 2n\pi\right)$$

The remainder of the answer to this problem may be found in the solution to Problem B–11.

3–15.   Show that

$$\langle x \rangle = \frac{a}{2}$$

for all the states of a particle in a box. Is this result physically reasonable?

$$\langle x \rangle = \int_0^a x\psi^*(x)\psi(x)\, dx$$

$$= \frac{2}{a}\int_0^a x\sin^2\frac{n\pi x}{a}\, dx$$

$$= \frac{2}{a}\cdot\frac{a^2}{4} = \frac{a}{2}$$

Thus, for any n, $\langle x \rangle = a/2$. The result is physically reasonable, since the probability of finding the particle at any position x inside the box, $\psi^*(x)\psi(x)$, is symmetric about the center of the box for any value of n, and the particle is equally likely to be found at $a/2 + \delta$ as at $a/2 - \delta$. Thus, the average of a series of measurements will give the center of the box, or $a/2$.

3–16.   Show that $\langle p \rangle = 0$ for all states of a one-dimensional box of length a.

Using Equation 3.37,

$$\langle p \rangle = \int_0^a \psi^*(x)\hat{P}\psi(x)\, dx$$

$$= \frac{2}{a}\int_0^a \sin\frac{n\pi x}{a}\left(-i\hbar\frac{d}{dx}\right)\sin\frac{n\pi x}{a}\, dx$$

$$= -\frac{2i\hbar n\pi}{a^2}\int_0^a \sin\frac{n\pi x}{a}\cos\frac{n\pi x}{a}\, dx$$

$$= 0$$

The result holds for any integer value of n.

3–17.   Show that

$$\sigma_x = (\langle x^2 \rangle - \langle x \rangle^2)^{1/2}$$

for a particle in a box is less than a, the width of the box, for any value of n. If σ_x is the uncertainty in the position of the particle, could σ_x ever be larger than a?

From the solutions to Problems B–8 and B–9 (or Equations 3.31 and 3.32 in the text),

$$\langle x \rangle = \frac{a}{2}$$

$$\langle x^2 \rangle = \frac{a^2}{3} - \frac{a^2}{2n^2\pi^2}$$

Then, as detailed in Problems B–9 and B–10,

$$\sigma_x = \left(\langle x^2 \rangle - \langle x \rangle^2 \right)^{1/2}$$

$$= \left(\frac{a^2}{3} - \frac{a^2}{2n^2\pi^2} - \frac{a^2}{4} \right)^{1/2}$$

$$= a \left(\frac{1}{12} - \frac{1}{2n^2\pi^2} \right)^{1/2}$$

$$< a \sqrt{\frac{1}{12}} = \frac{a}{2\sqrt{3}} < a$$

Thus, σ_x is smaller than a for any value of n. This makes sense because if the particle is known to be somewhere inside the interval $0 \le x \le a$ and never outside, then the uncertainty in its position cannot be larger than the length of interval.

3–18. A classical particle in a box has an equi-likelihood of being found anywhere within the region $0 \le x \le a$. Consequently, its probability distribution is

$$p(x)dx = \frac{dx}{a} \qquad 0 \le x \le a$$

Show that $\langle x \rangle = a/2$ and $\langle x^2 \rangle = a^2/3$ for this system. Now show that $\langle x^2 \rangle$ (Equation 3.32) and σ_x (Equation 3.33) for a quantum-mechanical particle in a box take on the classical values as $n \to \infty$. This result is an example of the *correspondence principle*.

Using the probability distribution given for the classical particle,

$$\langle x \rangle = \int_0^a x p(x)\, dx = \frac{1}{a} \int_0^a x\, dx = \frac{x^2}{2a} \Big|_0^a = \frac{a}{2}$$

$$\langle x^2 \rangle = \int_0^a x^2 p(x)\, dx = \frac{1}{a} \int_0^a x^2\, dx = \frac{x^3}{3a} \Big|_0^a = \frac{a^2}{3}$$

Thus, for the classical particle,

$$\sigma_x = \left(\langle x^2 \rangle - \langle x \rangle^2 \right)^{1/2} = \left(\frac{a^2}{3} - \frac{a^2}{4} \right)^{1/2} = \frac{a}{2\sqrt{3}}$$

The limiting values of $\langle x^2 \rangle$ and σ_x for the quantum mechanical particle as $n \to \infty$ are

$$\lim_{n\to\infty} \langle x^2 \rangle = \lim_{n\to\infty} \left(\frac{a^2}{3} - \frac{a^2}{2n^2\pi^2} \right) = \frac{a^2}{3}$$

$$\lim_{n\to\infty} \sigma_x = \lim_{n\to\infty} \left[a \left(\frac{1}{12} - \frac{1}{2\pi^2 n^2} \right)^{1/2} \right] = \frac{a}{2\sqrt{3}}$$

In the limit as $n \to \infty$, the quantum mechanical results are seen to take on the classical values

---

3–19. Using the trigonometric identity

$$\sin 2\theta = 2 \sin\theta \cos\theta$$

show that

$$\int_0^a \sin\frac{n\pi x}{a} \cos\frac{n\pi x}{a} dx = 0$$

The trigonometric identity can be used to simplify the integrand.

$$\int_0^a \sin\frac{n\pi x}{a} \cos\frac{n\pi x}{a} \, dx = \frac{1}{2} \int_0^a \sin\frac{2n\pi x}{a}$$

$$= -\frac{a}{4n\pi} \cos\frac{2n\pi x}{a} \Big|_0^a$$

$$= 0$$

3–20. Prove that

$$\int_0^a e^{\pm i2\pi nx/a} dx = 0 \quad n \neq 0$$

$$\int_0^a e^{\pm i2\pi nx/a} \, dx = \pm\frac{a}{2\pi ni} e^{\pm i2\pi nx/a} \Big|_0^a$$

$$= \pm\frac{a}{2\pi ni} \left(e^{\pm i2\pi n} - 1 \right)$$

$$= 0$$

since $e^{\pm i2\pi n} = 1$ for integer n. Note that if $n = 0$ the integral becomes

$$\int_0^a dx = a$$

3–21. Using the trigonometric identity

$$\sin \alpha \sin \beta = \frac{1}{2} \cos(\alpha - \beta) - \frac{1}{2} \cos(\alpha + \beta)$$

show that the particle-in-a-box wave functions (Equations 3.27) satisfy the relation

$$\int_0^a \psi_n^*(x)\psi_m(x)\, dx = 0 \qquad m \neq n$$

(The asterisk in this case is superfluous because the functions are real.) If a set of functions satisfies the above integral condition, we say that the set is *orthogonal* and, in particular, that $\psi_m(x)$ is orthogonal to $\psi_n(x)$. If, in addition, the functions are normalized, then we say that the set is *orthonormal*.

If $m \neq n$,

$$
\begin{aligned}
\int_0^a \psi_n^*(x)\psi_m(x)\, dx &= \frac{2}{a} \int_0^a \sin\frac{n\pi x}{a} \sin\frac{m\pi x}{a}\, dx \\
&= \frac{1}{a} \int_0^a \cos\frac{(n-m)\pi x}{a}\, dx + \frac{1}{a} \int_0^a \cos\frac{(n+m)\pi x}{a}\, dx \\
&= \frac{1}{(n-m)\pi} \sin\frac{(n-m)\pi x}{a}\Big|_0^a + \frac{1}{(n+m)\pi} \sin\frac{(n+m)\pi x}{a}\Big|_0^a \\
&= 0
\end{aligned}
$$

since $\sin N\pi = 0$ for any integer, N, and since the sum or difference of two integers is also an integer.

3–22. Prove that the set of functions

$$\psi_n(x) = a^{-1/2} e^{i\pi nx/a} \qquad n = 0, \pm 1, \pm 2, \ldots$$

is orthonormal (cf. Problem 3–21) over the interval $-a \leq x \leq a$. A compact way to express orthonormality in the ψ_n is to write

$$\int_{-a}^a \psi_m^*(x)\psi_n\, dx = \delta_{mn}$$

The symbol δ_{mn} is called a Kronecker delta and is defined by

$$\delta_{mn} = \begin{cases} 1 & \text{if } m = n \\ 0 & \text{if } m \neq n \end{cases}$$

If $m \neq n$,

$$\int_{-a}^{a} \psi_m^*(x)\psi_n(x)\, dx = \frac{1}{2a} \int_{-a}^{a} e^{-i\pi mx/a} e^{i\pi nx/a}\, dx$$

$$= \frac{1}{2a} \int_{-a}^{a} e^{i\pi(n-m)x/a}\, dx$$

$$= \frac{1}{2(n-m)i\pi} \left. e^{i\pi(n-m)x/a} \right|_{-a}^{a}$$

$$= \frac{1}{(n-m)\pi} \left[\frac{e^{i\pi(n-m)} - e^{-i\pi(n-m)}}{2i} \right]$$

$$= \frac{1}{(n-m)\pi} \sin\left[(n-m)\pi\right]$$

$$= 0$$

On the other hand, if $m = n$,

$$\int_{-a}^{a} \psi_n^*(x)\psi_n(x)\, dx = \frac{1}{2a} \int_{-a}^{a} e^{-i\pi nx/a} e^{i\pi nx/a}\, dx$$

$$= \frac{1}{2a} \int_{-a}^{a} dx$$

$$= \frac{1}{2a} x \Big|_{-a}^{a}$$

$$= 1$$

Thus,

$$\int_{-a}^{a} \psi_m^*(x)\psi_n(x)\, dx = \delta_{mn}$$

3–23. In problems dealing with a particle in a box, we often need to evaluate integrals of the type

$$\int_0^a \sin\frac{n\pi x}{a} \sin\frac{m\pi x}{a}\, dx \qquad \text{and} \qquad \int_0^a \cos\frac{n\pi x}{a} \cos\frac{m\pi x}{a}\, dx$$

Integrals such as these are easy to evaluate if you convert the trigonometric functions to complex exponentials by using the identities (see MathChapter A)

$$\cos\theta = \frac{e^{i\theta} + e^{-i\theta}}{2} \qquad \text{and} \qquad \sin\theta = \frac{e^{i\theta} - e^{-i\theta}}{2i}$$

and then realize that the set of functions

$$\psi_n(x) = a^{-1/2} e^{in\pi x/a} \qquad n = 0, \pm 1, \pm 2, \ldots$$

is orthonormal on the interval $-a \leq x \leq a$ (Problem 3–22). Show that

$$\int_0^a \sin\frac{n\pi x}{a} \sin\frac{m\pi x}{a}\, dx = \int_0^a \cos\frac{n\pi x}{a} \cos\frac{m\pi x}{a}\, dx = \frac{a}{2}\delta_{nm}$$

where δ_{nm} is the Kronecker delta (defined in Problem 3–22). Also show that

$$\int_0^a \cos \frac{n\pi x}{a} \sin \frac{m\pi x}{a} dx = 0$$

—————————————

Because the integrands are even functions of x, it follows that

$$\int_0^a \sin \frac{n\pi x}{a} \sin \frac{m\pi x}{a} dx = \frac{1}{2} \int_{-a}^a \sin \frac{n\pi x}{a} \sin \frac{m\pi x}{a} dx$$

and

$$\int_0^a \cos \frac{n\pi x}{a} \cos \frac{m\pi x}{a} dx = \frac{1}{2} \int_{-a}^a \cos \frac{n\pi x}{a} \cos \frac{m\pi x}{a} dx$$

Since these are particle-in-a-box wave functions, n, $m > 0$. Now use the Euler relation in each integral. First,

$$\frac{1}{2} \int_{-a}^a \sin \frac{n\pi x}{a} \sin \frac{m\pi x}{a} dx = -\frac{1}{8} \int_{-a}^a \left(e^{in\pi x/a} - e^{-in\pi x/a} \right) \left(e^{im\pi x/a} - e^{-im\pi x/a} \right) dx$$

$$= -\frac{a}{4} \int_{-a}^a \left(\frac{1}{2a} e^{in\pi x/a} e^{im\pi x/a} - \frac{1}{2a} e^{in\pi x/a} e^{-im\pi x/a} \right.$$

$$\left. - \frac{1}{2a} e^{-in\pi x/a} e^{im\pi x/a} + \frac{1}{2a} e^{-in\pi x/a} e^{-im\pi x/a} \right) dx$$

$$= -\frac{a}{4} \int_{-a}^a \left[\psi_{-n}(x)^* \psi_m(x) - \psi_{-n}(x)^* \psi_{-m}(x) \right.$$

$$\left. - \psi_n(x)^* \psi_m(x) + \psi_n(x)^* \psi_{-m}(x) \right] dx$$

$$= -\frac{a}{4} \left(\delta_{-nm} - \delta_{-n,-m} - \delta_{nm} + \delta_{n,-m} \right)$$

$$= \frac{a}{2} \delta_{nm}$$

because if $n = m$, then also $-n = -m$. Additionally, for positive values of n and m, n can never equal $-m$.

The argument is very similar for the second integral,

$$\frac{1}{2}\int_{-a}^{a}\cos\frac{n\pi x}{a}\cos\frac{m\pi x}{a}\,dx = \frac{1}{8}\int_{-a}^{a}\left(e^{in\pi x/a}+e^{-in\pi x/a}\right)\left(e^{im\pi x/a}+e^{-im\pi x/a}\right)\,dx$$

$$= \frac{a}{4}\int_{-a}^{a}\left(\frac{1}{2a}e^{in\pi x/a}e^{im\pi x/a}+\frac{1}{2a}e^{in\pi x/a}e^{-im\pi x/a}\right.$$

$$\left.+\frac{1}{2a}e^{-in\pi x/a}e^{im\pi x/a}+\frac{1}{2a}e^{-in\pi x/a}e^{-im\pi x/a}\right)\,dx$$

$$= \frac{a}{4}\int_{-a}^{a}\left[\psi_{-n}(x)^{*}\psi_{m}(x)+\psi_{-n}(x)^{*}\psi_{-m}(x)\right.$$

$$\left.+\psi_{n}(x)^{*}\psi_{m}(x)+\psi_{n}(x)^{*}\psi_{-m}(x)\right]\,dx$$

$$= \frac{a}{4}\left(\delta_{-nm}+\delta_{-n,-m}+\delta_{nm}+\delta_{n,-m}\right)$$

$$= \frac{a}{2}\delta_{nm}$$

Thus,

$$\int_{0}^{a}\sin\frac{n\pi x}{a}\sin\frac{m\pi x}{a}\,dx = \int_{0}^{a}\cos\frac{n\pi x}{a}\cos\frac{m\pi x}{a}\,dx = \frac{a}{2}\delta_{nm}$$

The same method, utilizing the orthogonality of the set of functions $\psi_n(x)$ over the interval $-a \leq x \leq a$, cannot be used for the final integral, since the integrand is not an even function of x. In fact, it is an odd function of x. The integral can be evaluated by using the Euler identities to substitute for $\sin(n\pi x/a)$ and $\cos(n\pi x/a)$, but it is simpler to use the trigonometric identity

$$\cos\beta\sin\alpha = \frac{1}{2}\sin\left(\alpha+\beta\right)+\frac{1}{2}\sin\left(\alpha-\beta\right)$$

For $m = n$, the identity is simply

$$\cos\alpha\sin\alpha = \frac{1}{2}\sin 2\alpha$$

and the integral is

$$\int_{0}^{a}\cos\frac{n\pi x}{a}\sin\frac{n\pi x}{a}\,dx = \frac{1}{2}\int_{0}^{a}\sin\frac{2n\pi x}{a}\,dx$$

$$= -\frac{\cos 2n\pi x/a}{2n\pi/a}\bigg|_{0}^{a}$$

$$= -\frac{\cos 2n\pi - 1}{2n\pi/a}$$

$$= 0$$

But for $n \neq m$,

$$\int_0^a \cos\frac{n\pi x}{a}\sin\frac{m\pi x}{a}\,dx = \frac{1}{2}\int_0^a \sin\frac{(m+n)\pi x}{a}\,dx + \frac{1}{2}\int_0^a \sin\frac{(m-n)\pi x}{a}\,dx$$

$$= -\frac{\cos(m+n)\pi x/a}{2(m+n)\pi/a}\bigg|_0^a - \frac{\cos(m-n)\pi x/a}{2(m-n)\pi/a}\bigg|_0^a$$

$$= -\frac{\cos(m+n)\pi - 1}{2(m+n)\pi/a} - \frac{\cos(m-n)\pi - 1}{2(m-n)\pi/a}$$

$$= -\frac{(-1)^{m+n} - 1}{2(m+n)\pi/a} - \frac{(-1)^{m-n} - 1}{2(m-n)\pi/a}$$

$$= \begin{cases} 0 & m+n \text{ even} \\ \dfrac{2m}{m^2-n^2}\left(\dfrac{a}{\pi}\right) & m+n \text{ odd} \end{cases}$$

since for integer m and n, if $m+n$ is even, so too is $m-n$. Since $2n$ is also an even integer, the last equation can represent the result for $m=n$, too. Thus, in general for m and n integers,

$$\int_0^a \cos\frac{n\pi x}{a}\sin\frac{m\pi x}{a}\,dx = \begin{cases} 0 & m+n \text{ even} \\ \dfrac{2m}{m^2-n^2} & m+n \text{ odd} \end{cases}$$

and the integral is not always equal to zero.

3–24. Show that the set of functions

$$\phi_n(\theta) = (2\pi)^{-1/2}e^{in\theta} \qquad 0 \le \theta \le 2\pi$$

is orthonormal (Problem 3–21).

If $m \ne n$,

$$\int_0^{2\pi} \phi_m^*(\theta)\phi_n(\theta)\,d\theta = \frac{1}{2\pi}\int_0^{2\pi} e^{-im\theta}e^{in\theta}\,d\theta$$

$$= \frac{1}{2\pi}\int_0^{2\pi} e^{i(n-m)\theta}\,d\theta$$

$$= \frac{1}{2(n-m)i\pi} e^{i(n-m)\theta}\bigg|_0^{2\pi}$$

$$= \frac{1}{2(n-m)i\pi}\left[e^{2(n-m)\pi i} - 1\right]$$

$$= 0$$

since $e^{2(n-m)\pi i} = 1$ for m and n integers. On the other hand, if $m = n$,

$$\int_0^{2\pi} \phi_n^*(\theta)\phi_n(\theta)\, d\theta = \frac{1}{2\pi}\int_0^{2\pi} e^{-in\theta} e^{in\theta}\, d\theta$$

$$= \frac{1}{2\pi}\int_0^{2\pi} d\theta$$

$$= \frac{\theta}{2\pi}\Big|_0^{2\pi}$$

$$= 1$$

Thus,

$$\int_0^{2\pi} \phi_m^*(\theta)\phi_n(\theta)\, d\theta = \delta_{mn}$$

and the set of functions is orthonormal.

3–25. In going from Equation 3.34 to 3.35, we multiplied Equation 3.34 from the left by $\psi^*(x)$ and then integrated over all values of x to obtain Equation 3.35. Does it make any difference whether we multiplied from the left or the right?

Multiplication of two function is commutative. As long as it is realized that $\hat{H}$ operates only on $\psi_n(x)$, and that $\hat{H}\psi_n(x)$ is itself just a function, it makes no differernce whether $\psi_n^*(x)$ (or any other function) is multiplied from the left or from the right.

3–26. Calculate $\langle x \rangle$ and $\langle x^2 \rangle$ for the $n = 2$ state of a particle in a one-dimensional box of length a. Show that

$$\sigma_x = \frac{a}{4\pi}\left(\frac{4\pi^2}{3} - 2\right)^{1/2}$$

The general expressions for $\langle x \rangle$, $\langle x^2 \rangle$, and σ_x for the particle in a box are found in Equations 3.31 through 3.33.

$$\langle x \rangle = \frac{a}{2}$$

$$\langle x^2 \rangle = \frac{a^2}{3} - \frac{a^2}{2n^2\pi^2}$$

$$\sigma_x = \frac{a}{2\pi n}\left(\frac{\pi^2 n^2}{3} - 2\right)^{1/2}$$

For $n = 2$,

$$\langle x \rangle = \frac{a}{2}$$

$$\langle x^2 \rangle = \frac{a^2}{3} - \frac{a^2}{8\pi^2}$$

$$\sigma_x = \frac{a}{4\pi} \left(\frac{4\pi^2}{3} - 2 \right)^{1/2}$$

3–27. Calculate $\langle p \rangle$ and $\langle p^2 \rangle$ for the $n = 2$ state of a particle in a one-dimensional box of length a. Show that

$$\sigma_p = \frac{h}{a}$$

The results for $\langle p \rangle$, $\langle p^2 \rangle$, and σ_p for arbitrary n of the particle in a box are found in Equations 3.38 through 3.41.

$$\langle p \rangle = 0$$

$$\langle p^2 \rangle = \frac{n^2 \pi^2 \hbar^2}{a^2}$$

$$\sigma_p = \frac{n \pi \hbar}{a}$$

For $n = 2$, $\langle p \rangle = 0$, $\langle p^2 \rangle = h^2/a^2$, and $\sigma_p = h/a$.

3–28. Consider a particle of mass m in a one-dimensional box of length a. Its average energy is given by

$$\langle E \rangle = \frac{1}{2m} \langle p^2 \rangle$$

Because $\langle p \rangle = 0$, $\langle p^2 \rangle = \sigma_p^2$, where σ_p can be called the uncertainty in p. Using the uncertainty principle, show that the energy must be at least as large as $\hbar^2/8ma^2$ because σ_x, the uncertainty in x, cannot be larger than a.

Starting with Equation 3.43 and the inequality $\sigma_x \le a$,

$$\frac{\hbar}{2\sigma_p} < \sigma_x \leq a$$

gives

$$\frac{\hbar}{2a} \leq \sigma_p$$

Both sides are positive, so squaring preserves the inequality

$$\frac{\hbar^2}{4a^2} \leq \sigma_p^2$$

Since $\langle p^2 \rangle = \sigma_p^2$ and $\langle E \rangle = \langle p^2 \rangle / 2m$, then

$$\frac{\hbar^2}{8ma^2} \leq \langle E \rangle$$

3–29. Discuss the degeneracies of the first few energy levels of a particle in a three-dimensional box when $a \neq b \neq c$.

In general, when $a \neq b \neq c$, the first few energy levels of a particle in a three-dimensional box will not be degenerate.

3–30. Show that the normalized wave function for a particle in a three-dimensional box with sides of length a, b, and c is

$$\psi(x, y, z) = \left(\frac{8}{abc}\right)^{1/2} \sin \frac{n_x \pi x}{a} \sin \frac{n_y \pi y}{b} \sin \frac{n_z \pi z}{c}$$

Section 3.9 of the text shows that

$$\psi(x, y, z) = A_x A_y A_z \sin \frac{n_x \pi x}{a} \sin \frac{n_y \pi y}{b} \sin \frac{n_z \pi z}{c}$$

is a solution to the Schrödinger equation for a particle in a three-dimensional box, and that it has the proper boundary conditions. Direct substitution back into Equation 3.44 gives the energy eigenvalues of Equation 3.57. All that remains is to show that the normalization condition requires $A_x A_y A_z = (8/abc)^{1/2}$.

Normalization requires that

$$1 = \int_0^a dx \int_0^b dy \int_0^c dz \, \psi^*(x, y, z)\psi(x, y, z)$$

$$= \left(A_x A_y A_z\right)^2 \int_0^c \sin^2 \frac{n_z \pi z}{c} \, dz \int_0^b \sin^2 \frac{n_y \pi y}{b} \, dy \int_0^a \sin^2 \frac{n_x \pi x}{a} \, dx$$

In Problem 3–14, it was shown that $\displaystyle\int_0^a \sin^2 \frac{n_x \pi x}{a} \, dx = \frac{a}{2}$. Thus,

$$\int_0^a \int_0^b \int_0^c dz\, dy\, dx\; \psi^*(x,y,z)\psi(x,y,z) = 1 = \left(A_x A_y A_z\right)^2 \left(\frac{c}{2}\right)\left(\frac{b}{2}\right)\left(\frac{a}{2}\right)$$

or

$$A_x A_y A_z = \left(\frac{8}{abc}\right)^{1/2}$$

3–31. Show that $\langle \mathbf{p}\rangle = 0$ for the ground state of a particle in a three-dimensional box with sides of length a, b, and c.

From Equations 3.58 and 3.59,

$$\hat{\mathbf{P}} = -i\hbar\left(\mathbf{i}\frac{\partial}{\partial x} + \mathbf{j}\frac{\partial}{\partial y} + \mathbf{k}\frac{\partial}{\partial z}\right)$$

and,

$$\langle \mathbf{p}\rangle = \int_0^a dx \int_0^b dy \int_0^c dz\; \psi^*(x,y,z)\hat{\mathbf{P}}\psi(x,y,z)$$

$$= -i\hbar\mathbf{i}\int_0^a \sin\frac{n_x\pi x}{a}\frac{\partial}{\partial x}\left(\sin\frac{n_x\pi x}{a}\right)dx \int_0^b \sin^2\frac{n_y\pi y}{b}\,dy \int_0^c \sin^2\frac{n_z\pi z}{c}\,dz$$

$$-\,i\hbar\mathbf{j}\int_0^a \sin^2\frac{n_x\pi x}{a}\,dx \int_0^b \sin\frac{n_y\pi y}{b}\frac{\partial}{\partial y}\left(\sin\frac{n_y\pi y}{b}\right)dy \int_0^c \sin^2\frac{n_z\pi z}{c}\,dz$$

$$-\,i\hbar\mathbf{k}\int_0^a \sin^2\frac{n_x\pi x}{a}\,dx \int_0^b \sin^2\frac{n_y\pi y}{b}\,dy \int_0^c \sin\frac{n_z\pi z}{c}\frac{\partial}{\partial z}\left(\sin\frac{n_z\pi z}{c}\right)dz$$

$$= -i\hbar\mathbf{i}\frac{n_x\pi}{a}\int_0^a \sin\frac{n_x\pi x}{a}\cos\frac{n_x\pi x}{a}\,dx \int_0^b \sin^2\frac{n_y\pi y}{b}\,dy \int_0^c \sin^2\frac{n_z\pi z}{c}\,dz$$

$$-\,i\hbar\mathbf{j}\frac{n_y\pi}{b}\int_0^a \sin^2\frac{n_x\pi x}{a}\,dx \int_0^b \sin\frac{n_y\pi y}{b}\cos\frac{n_y\pi y}{b}\,dy \int_0^c \sin^2\frac{n_z\pi z}{c}\,dz$$

$$-\,i\hbar\mathbf{k}\frac{n_z\pi}{c}\int_0^a \sin^2\frac{n_x\pi x}{a}\,dx \int_0^b \sin^2\frac{n_y\pi y}{b}\,dy \int_0^c \sin\frac{n_z\pi z}{c}\cos\frac{n_z\pi z}{c}\,dz$$

Each of the three sets of integrals contains a multiplicative factor like $\displaystyle\int_0^a \sin\frac{n_x\pi x}{a}\cos\frac{n_x\pi x}{a}\,dx$ that was shown to be zero in Problem 3–19. Thus, each of the three terms equals zero, and $\langle \mathbf{p}\rangle = 0$ for the particle in the three-dimensional box.

3–32. What are the degeneracies of the first four energy levels for a particle in a three-dimensional box with $a = b = 1.5c$?

Substituting $b = a$ and $c = a/1.5$ into Equation 3.57 gives

$$E = \frac{h^2}{8m} \left(\frac{n_x^2 + n_y^2 + 2.25n_z^2}{a^2} \right)$$

The first four energy levels are

Energy level	(n_x, n_y, n_z)	Degeneracy	$E / (h^2/8ma^2)$
E_{111}	(1, 1, 1)	1	4.25
E_{211}	(2, 1, 1) , (1, 2, 1)	2	7.25
E_{221}	(2, 2, 1)	1	10.25
E_{112}	(1, 1, 2)	1	11.00

3–33. The Schrödinger equation for a particle of mass m constrained to move on a circle of radius a is

$$-\frac{\hbar^2}{2I} \frac{d^2\psi}{d\theta^2} = E\psi(\theta) \qquad 0 \leq \theta \leq 2\pi$$

where $I = ma^2$ is the moment of inertia and θ is the angle that describes the position of the particle around the ring. Show by direct substitution that the solutions to this equation are

$$\psi(\theta) = Ae^{in\theta}$$

where $n = \pm(2IE)^{1/2}/\hbar$. Argue that the appropriate boundary condition is $\psi(\theta) = \psi(\theta + 2\pi)$ and use this condition to show that

$$E = \frac{n^2\hbar^2}{2I} \qquad n = 0, \pm1, \pm2, \ldots$$

Show that the normalization constant A is $(2\pi)^{-1/2}$. Discuss how you might use these results for a free-electron model of benzene.

The differential equation can be written as

$$\frac{d^2\psi}{d\theta^2} + \frac{2IE}{\hbar^2}\psi(\theta) = 0$$

Substituting $\psi(\theta) = Ae^{in\theta}$ gives

$$-n^2 Ae^{in\theta} + \frac{2IE}{\hbar^2}Ae^{in\theta} = 0$$

or $n = \pm (2IE)^{1/2}/\hbar$. Since $\psi(\theta)$ represents the amplitude of the matter wave at the point θ, and because θ and $\theta + 2\pi$ describe the same point in space, the wave function must repeat its value every 2π radians, or $\psi(\theta) = \psi(\theta + 2\pi)$. Then,

$$Ae^{in\theta} = Ae^{in(\theta+2\pi)}$$

$$1 = e^{i2\pi n}$$

$$1 = \cos 2\pi n + i \sin 2\pi n$$

This is only true if n is an integer, in which case,

$$E = \frac{n^2 \hbar^2}{2I} \qquad n = 0, \pm 1, \pm 2, \ldots$$

The lowest energy level is singly degenerate, while all the other energy levels are double degenerate, since the states with quantum numbers n and $-n$ have the same energy. The normalization condition requires that

$$\int_0^{2\pi} \psi^*(\theta)\psi(\theta)\,d\theta = 1$$

$$A^2 \int_0^{2\pi} e^{-in\theta} e^{in\theta}\,d\theta = 1$$

$$A^2 \int_0^{2\pi} d\theta = 1$$

$$A = \frac{1}{\sqrt{2\pi}}$$

Using the particle in a box as the basis for the free-electron model for linear conjugated hydrocarbons in Section 3.5 suggests a similar model based on this "particle on a ring" for cyclic conjugated hydrocarbons. There are six π electrons in benzene, and the results above indicate that there will be two electrons in each of the three energy levels $n = 0$ and ± 1. The first electronic transition would be a $n = \pm 1 \rightarrow n = \pm 2$ transition, and the frequency associated with this transition would be

$$\tilde{\nu} = \frac{\hbar}{4\pi c I}\left(2^2 - 1^2\right)$$

3–34. Set up the problem of a particle in a box with its walls located at $-a$ and $+a$. Show that the energies are equal to those of a box with walls located at 0 and $2a$. (These energies may be obtained from the results that we derived in the chapter simply by replacing a by $2a$.) Show, however, that the wave functions are not the same and are given by

$$\psi_n(x) = \begin{cases} \dfrac{1}{a^{1/2}} \sin \dfrac{n\pi x}{2a} & n \text{ even} \\[2mm] \dfrac{1}{a^{1/2}} \cos \dfrac{n\pi x}{2a} & n \text{ odd} \end{cases}$$

Does it bother you that the wave functions seem to depend upon whether the walls are located at $\pm a$ or 0 and $2a$? Surely the particle "knows" only that it has a region of length $2a$ in which to move and cannot be affected by where you place the origin for the two sets of wave functions. What does this tell you? Do you think that any experimentally observable properties depend upon where you choose to place the origin of the x axis? Show that $\sigma_x \sigma_p > \hbar/2$, exactly as we obtained in Section 3.8.

The general solution to the Schrödinger equation for a particle in a one-dimensional box is (Section 3.5)

$$\psi(x) = A \cos kx + B \sin kx \qquad k = \frac{(2mE)^{1/2}}{\hbar}$$

For this problem, however, the boundary conditions are $\psi(-a) = 0$ and $\psi(a) = 0$, so

$$\psi(-a) = A\cos(-ka) + B\sin(-ka) = A\cos ka - B\sin ka = 0$$

and

$$\psi(a) = A\cos(ka) + B\sin(ka) = A\cos ka + B\sin ka = 0$$

Adding and subtracting these two equations gives

$$A\cos ka = 0 \qquad \text{and} \qquad B\sin ka = 0$$

The boundary conditions are satisfied by taking

$$k = \frac{n\pi}{2a}$$

where $n = 1, 2, \ldots$ and setting $B = 0$ when n is odd or $A = 0$ when n is even. (The trivial solution with both A and B equal to zero is excluded.) Thus,

$$\psi_n(x) = \begin{cases} B\sin\dfrac{n\pi x}{2a} & n \text{ even} \\[2mm] A\cos\dfrac{n\pi x}{2a} & n \text{ odd} \end{cases}$$

The normalization constants A and B are both equal to $a^{-1/2}$. (See Problem 3–23 for the necessary integrals.) The energy, E, is found using the two expressions for the variable k:

$$\frac{(2mE)^{1/2}}{\hbar} = k = \frac{n\pi}{2a}$$

$$E = \frac{h^2 n^2}{32ma^2}$$

This is the same result that would be obtained for a box of length $2a$ with walls located at $x = 0$ and $x = 2a$ (Equation 3.21):

$$E_n = \frac{h^2 n^2}{8m(2a)^2} \qquad n = 1, 2, \ldots$$

In fact, the wave functions have precisely the same form in the box regardless of whether the walls are at $x = 0$ and $x = 2a$ or at $x = \pm a$. All that has changed is the mathematical expression of that form due to a shift in the coordinate system. (It is a simple application of trigonometric identities to show that the wave functions of Section 3.5 are recovered if $x' = x + a$ is substituted into the wave functions obtained in this problem.) No experimentally observable property can depend on the choice of coordinate system. Since σ_x and σ_p are observable properties, $\sigma_x \sigma_p \geq \hbar/2$ as in Section 3.8.

3–35. The quantized energies of a particle in a box result from the boundary conditions, or from the fact that the particle is restricted to a finite region. In this problem, we investigate the quantum-mechanical problem of a free particle, one that is not restricted to a finite region. The potential energy $V(x)$ is equal to zero and the Schrödinger equation is

$$\frac{d^2\psi}{dx^2} + \frac{2mE}{\hbar^2}\psi(x) = 0 \qquad -\infty < x < \infty$$

Note that the particle can lie anywhere along the x axis in this problem. Show that the two solutions of this Schrödinger equation are

$$\psi_1(x) = A_1 e^{i(2mE)^{1/2}x/\hbar} = A_1 e^{ikx}$$

and

$$\psi_2(x) = A_2 e^{-i(2mE)^{1/2}x/\hbar} = A_2 e^{-ikx}$$

where

$$k = \frac{(2mE)^{1/2}}{\hbar}$$

Show that if E is allowed to take on negative values, then the wave functions become unbounded for large x. Therefore, we will require that the energy, E, be a positive quantity.

To get a physical interpretation of the states that $\psi_1(x)$ and $\psi_2(x)$ describe, operate on $\psi_1(x)$ and $\psi_2(x)$ with the momentum operator $\hat{P}$ (Equation 3.11), and show that

$$\hat{P}\psi_1 = -i\hbar\frac{d\psi_1}{dx} = \hbar k \psi_1$$

and

$$\hat{P}\psi_2 = -i\hbar\frac{d\psi_2}{dx} = -\hbar k \psi_2$$

Notice that these are eigenvalue equations. Our interpretation of these two equations is that ψ_1 describes a free particle with fixed momentum $\hbar k$ and that ψ_2 describes a particle with fixed momentum $-\hbar k$. Thus, ψ_1 describes a particle moving to the right and ψ_2 describes a particle moving to the left, both with a fixed momentum. Notice also that there are no restrictions on k, and so the particle can have any value of momentum. Now show that

$$E = \frac{\hbar^2 k^2}{2m}$$

Notice that the energy is not quantized; the energy of the particle can have any positive value in this case because no boundaries are associated with this problem.

Last, show that $\psi_1^*(x)\psi_1(x) = A_1^* A_1 = |A_1|^2 = \text{constant}$, and that $\psi_2^*(x)\psi_2(x) = A_2^* A_2 = |A_2|^2 = \text{constant}$. Discuss this result in terms of the probabilistic interpretation of $\psi^*\psi$. Also discuss the application of the uncertainty principle to this problem. What are σ_p and σ_x?

Example 2–4 gives the solutions to this Schrödinger equation:

$$\psi_1(x) = A_1 e^{ikx} \qquad \psi_2(x) = A_2 e^{-ikx}$$

with

$$k = \frac{(2mE)^{1/2}}{\hbar}$$

If E is less than zero, then $k = i(-2mE)^{1/2}/\hbar$, and

$$\lim_{x\to-\infty}\psi_1(x) = \lim_{x\to-\infty} A_1 e^{-x(-2mE)^{1/2}/\hbar}$$

$$\lim_{x\to\infty}\psi_2(x) = \lim_{x\to\infty} A_2 e^{x(-2mE)^{1/2}/\hbar}$$

both diverge. Therefore, E must be positive. Using Equation 3.11 for the momentum operator gives

$$\hat{P}\psi_1 = -i\hbar\frac{d\psi_1}{dx} = -i\hbar\frac{d}{dx}\left(A_1 e^{ikx}\right)$$

$$= -i^2\hbar k A_1 e^{ikx} = \hbar k A_1 e^{ikx} = \hbar k\psi_1(x)$$

$$\hat{P}\psi_2 = -i\hbar\frac{d\psi_2}{dx} = -i\hbar\frac{d}{dx}\left(A_2 e^{-ikx}\right)$$

$$= i^2\hbar k A_2 e^{ikx} = -\hbar k A_2 e^{ikx} = -\hbar k\psi_2(x)$$

Thus, for the free particle, there are two possible values for the momentum, $\hbar k$ and $-\hbar k$. For the free particle, all energy is kinetic energy. Since the kinetic energy is $p^2/2m$, it follows that for the two possible values of the momentum

$$E = \frac{p^2}{2m} = \frac{(\pm\hbar k)^2}{2m} = \frac{\hbar^2 k^2}{2m}$$

Finally,

$$\psi_1^*(x)\psi_1(x) = \left(A_1 e^{ikx}\right)^*\left(A_1 e^{ikx}\right)$$

$$= A_1^* A_1 = |A_1|^2 = \text{constant}$$

$$\psi_2^*(x)\psi_2(x) = \left(A_2 e^{-ikx}\right)^*\left(A_2 e^{-ikx}\right)$$

$$= A_2^* A_2 = |A_2|^2 = \text{constant}$$

Since $\psi^*(x)\psi(x)$ is a constant, the particle is equally likely to be found anywhere along the x axis. Thus, there is an infinite uncertainty in the position of the particle. This is completely consistent with the uncertainty principle because the momentum of the particle is known exactly ($\sigma_p = 0$).

3–36. Derive the equation for the allowed energies of a particle in a one-dimensional box by assuming that the particle is described by standing de Broglie waves within the box.

A standing wave requires that an integral number of half-wavelengths fits exactly into the box, or

$$\frac{n\lambda}{2} = a \quad \text{or} \quad \lambda = \frac{2a}{n}$$

The de Broglie relationship is

$$\lambda = \frac{h}{p}$$

Combining the two results for λ and solving for p gives

$$p = \frac{nh}{2a}$$

The corresponding energy (purely kinetic) is

$$E = \frac{p^2}{2m} = \frac{n^2h^2}{8ma^2}$$

3–37. In Chapter 4, we will encounter the time-dependent Schrödinger equation

$$\hat{H}\Psi(x, t) = i\hbar \frac{\partial \Psi(x, t)}{\partial t}$$

where Ψ is now a function of both position and time. Show that if the Hamiltonian operator does not contain time explicitly [$\hat{H} = \hat{H}(x)$], then this partial differential equation can be separated into two ordinary differential equations by setting $\Psi(x, t) = \psi(x)f(t)$. What is the separation constant in this problem? Generalize this result to three dimensions. What is the function $f(t)$?

If the Hamiltonian operator does not contain time explicitly, then it has no effect on $f(t)$. That is, $\hat{H}\psi(x)f(t) = f(t)\hat{H}\psi(x)$. (Recall, too, that multiplication of functions is commutative.) Thus, making the substitution $\Psi(x, t) = \psi(x)f(x)$:

$$\hat{H}\Psi(x, t) = i\hbar \frac{\partial \Psi(x, t)}{\partial t}$$

$$\hat{H}\psi(x)f(t) = i\hbar \frac{\partial \psi(x)f(t)}{\partial t}$$

$$f(t)\hat{H}\psi(x) = i\hbar\psi(x)\frac{df(t)}{dt}$$

Dividing by $\psi(x)f(t)$,

$$\frac{\hat{H}\psi(x)}{\psi(x)} = i\hbar\frac{df(t)/dt}{f(t)} \tag{1}$$

The left hand side of Equation 1 depends only on x, and the right hand side depends only on t. Since these are two independent variables, the only way for the equation to be valid for all values of x and t is for each side to be equal to the same constant, E. Therefore,

$$\frac{\hat{H}\psi(x)}{\psi(x)} = E$$

$$i\hbar\frac{df(t)/dt}{f(t)} = E$$

A completely analogous result would be obtained in three dimensions with $\Psi(x, y, z, t) = \psi(x, y, z)f(t)$, as the Hamiltonian operator, $\hat{H}(x, y, z)$ would still have no effect on $f(t)$, and the partial derivative with respect to time would have no effect on $\psi(x, y, z)$. The time-dependent Schrödinger equation would separate into a part that depends only on the spatial coordinates

x, y, z and a part depending only on t, giving

$$\frac{\hat{H}\psi(x, y, z)}{\psi(x, y, z)} = E$$

$$i\hbar\frac{df(t)/dt}{f(t)} = E$$

In either case, the first of these two equations is just the time independent Schrödinger equation, which allows the identification of separation constant E as the energy of the system. The second equation is the simple differential equation

$$\frac{df(t)}{dt} - \frac{E}{i\hbar}f(t) = 0$$

that is easily solved to give

$$f(t) = e^{-iEt/\hbar}$$

3–38. Using the result of Problem 3–37, what is the time-dependent wave function for the ground state of a particle in a one-dimensional box of length a? Use this wave function to evaluate the average value of x. Are you surprised?

Problem 3–37 shows that the time-dependent wave function for a system is $\Psi(x, t) = \psi(x)e^{-iEt/\hbar}$, where $\psi(x)$ is the solution of the time-independent Schrödinger equation $\hat{H}\psi(x) = E\psi(x)$ with eigenvalue E. Thus, for the one-dimensional box of length a,

$$\Psi_n(x, t) = \left(\frac{2}{a}\right)^{1/2} \sin\left(\frac{n\pi x}{a}\right) e^{-iE_n t/\hbar}$$

with

$$E_n = \frac{n^2 h^2}{8ma^2}$$

The average value of x is found as usual,

$$\langle x \rangle = \int_0^a \Psi_n^*(x) x \Psi_n(x)\, dx$$

$$= \frac{2}{a}\int_0^a \sin\left(\frac{n\pi x}{a}\right) e^{iE_n t/\hbar} x \sin\left(\frac{n\pi x}{a}\right) e^{-iE_n t/\hbar}\, dx$$

$$= \frac{2}{a}\int_0^a \sin\left(\frac{n\pi x}{a}\right) x \sin\left(\frac{n\pi x}{a}\right)\, dx$$

$$= \frac{a}{2}$$

That precisely the same result is obtained for the time-dependent wave function as for the time-independent wave function should not be surprising, since the physical system remains perfectly symmetric about the center of the box.

3–39.   We can use the wave functions of Problem 3–34 to illustrate some fundamental symmetry properties of wave functions. Show that the wave functions are alternately symmetric and antisymmetric or even and odd with respect to the operation $x \to -x$, which is a reflection through the $x = 0$ line. This symmetry property of the wave function is a consequence of the symmetry of the Hamiltonian operator, as we shall now show. The Schrödinger equation may be written as

$$\hat{H}(x)\psi_n(x) = E_n\psi_n(x)$$

Reflection through the $x = 0$ line gives $x \to -x$, and so

$$\hat{H}(-x)\psi_n(-x) = E_n\psi_n(-x)$$

Now show that $\hat{H}(x) = \hat{H}(-x)$ (i.e., that $\hat{H}$ is symmetric), and so show that

$$\hat{H}(x)\psi_n(-x) = E_n\psi_n(-x)$$

Thus, we see that $\psi_n(-x)$ is also an eigenfunction of $\hat{H}$ belonging to the same eigenvalue E_n. Now, if there is only one eigenfunction associated with each eigenvalue (we call this a *nondegenerate case*), then argue that $\psi_n(x)$ and $\psi_n(-x)$ must differ by a multiplicative constant [i.e., that $\psi_n(-x) = c\psi_n(x)$]. By applying the inversion operation again to this equation, show that $c = \pm 1$ and that all the wave functions must be either even or odd with respect to reflection through the $x = 0$ line because the Hamiltonian operator is symmetric. Thus, we see that the symmetry of the Hamiltonian operator influences the symmetry of the wave functions.

For the wave functions of Problem 3–34 with odd n

$$\psi_n(-x) = \frac{1}{a^{1/2}} \cos\left(-\frac{n\pi x}{2a}\right) = \frac{1}{a^{1/2}} \cos\left(\frac{n\pi x}{2a}\right) = \psi_n(x)$$

and the wave functions for odd n are symmetric. For even n,

$$\psi_n(-x) = \frac{1}{a^{1/2}} \sin\left(-\frac{n\pi x}{2a}\right) = -\frac{1}{a^{1/2}} \sin\left(\frac{n\pi x}{2a}\right) = -\psi_n(x)$$

and the wave functions for even n are antisymmetric. The Hamiltonian operator is seen to be symmetric,

$$\hat{H}(x) = -\frac{\hbar^2}{2m}\frac{d^2}{dx^2}$$

$$\hat{H}(-x) = -\frac{\hbar^2}{2m}\frac{d^2}{d\,(-x)^2} = -\frac{\hbar^2}{2m}\frac{d^2}{dx^2} = \hat{H}(x)$$

and from $\hat{H}(-x)\psi_n(-x) = E_n\psi_n(-x)$ it follows that $\hat{H}(x)\psi_n(-x) = E_n\psi_n(-x)$, which is to say that $\psi_n(-x)$ is also an eigenfunction of $\hat{H}$ with the same eigenvalue E_n as $\psi_n(x)$. For the nondegenerate case, however, there is only one unique eigenfunction with the eigenvalue E_n. Therefore, the two functions $\psi_n(x)$ and $\psi_n(-x)$ can only differ by a constant, or $\psi_n(-x) = c\psi_n(x)$. Repeating the reflection operation,

$$\psi_n(x) = \psi_n\left[-(-x)\right] = c\psi_n(-x) = c^2\psi_n(x)$$

which leads to the conclusion that $c = \pm 1$ and consequently to the conclusion (assuming the nondegenerate case) that if the Hamiltonian is symmetric with respect to reflection through the $x = 0$ line, then all the $\psi_n(x)$ are either even or odd with respect to that same reflection. The Hamiltonian and wave functions from Problem 3–34 are an illustration of this principle.

Vectors

PROBLEMS AND SOLUTIONS

C–1. Find the length of the vector $\mathbf{v} = 2\,\mathbf{i} - \mathbf{j} + 3\,\mathbf{k}$.

Use Equation C.5:

$$v = |\mathbf{v}| = \left(v_x^2 + v_y^2 + v_z^2\right)^{1/2}$$
$$= \left[2^2 + (-1)^2 + 3^2\right]^{1/2}$$
$$= \sqrt{14}$$

C–2. Find the length of the vector $\mathbf{r} = x\,\mathbf{i} + y\,\mathbf{j}$ and of the vector $\mathbf{r} = x\,\mathbf{i} + y\,\mathbf{j} + z\,\mathbf{k}$.

Equation C.5 provides the length of a vector. For the vector $\mathbf{r} = x\,\mathbf{i} + y\,\mathbf{j}$,

$$r = |\mathbf{r}| = \left(r_x^2 + r_y^2 + r_z^2\right)^{1/2}$$
$$= \left(x^2 + y^2 + 0^2\right)^{1/2}$$
$$= \left(x^2 + y^2\right)^{1/2}$$

and for the vector $\mathbf{r} = x\,\mathbf{i} + y\,\mathbf{j} + z\,\mathbf{k}$,

$$r = |\mathbf{r}| = \left(r_x^2 + r_y^2 + r_z^2\right)^{1/2}$$
$$= \left(x^2 + y^2 + z^2\right)^{1/2}$$

C–3. Prove that $\mathbf{u} \cdot \mathbf{v} = 0$ if $\mathbf{u}$ and $\mathbf{v}$ are perpendicular to each other. Two vectors that are perpendicular to each other are said to be orthogonal.

If **u** and **v** are perpendicular to each other, then the angle θ between them is 90°. Then from the definition of the scalar product (Equation C.6),

$$\mathbf{u} \cdot \mathbf{v} = |\mathbf{u}||\mathbf{v}| \cos \theta = |\mathbf{u}||\mathbf{v}| \cos 90° = 0$$

C–4. Show that the vectors $\mathbf{u} = 2\,\mathbf{i} - 4\,\mathbf{j} - 2\,\mathbf{k}$ and $\mathbf{v} = 3\,\mathbf{i} + 4\,\mathbf{j} - 5\,\mathbf{k}$ are orthogonal.

Find the scalar product of the two vectors using Equation C.9,

$$\mathbf{u} \cdot \mathbf{v} = u_x v_x + u_y v_y + u_z v_z$$
$$= (2)(3) + (-4)(4) + (-2)(-5)$$
$$= 0$$

Since $\mathbf{u} \cdot \mathbf{v} = 0$, the two vectors are orthogonal. (See Problem C–3.)

C–5. Show that the vector $\mathbf{r} = 2\,\mathbf{i} - 3\,\mathbf{k}$ lies entirely in a plane perpendicular to the y axis.

The y axis can be represented by the vector $\mathbf{u} = y\mathbf{j}$, where y can take on any real value. Now take the scalar product of this vector with $\mathbf{r}$.

$$\mathbf{u} \cdot \mathbf{r} = u_x r_x + u_y r_y + u_z r_z$$
$$= (2)(0) + (0)(y) + (-3)(0)$$
$$= 0$$

The scalar product is zero, and therefore, the vector $\mathbf{r}$ is perpendicular to the y axis.

C–6. Find the angle between the two vectors $\mathbf{u} = -\mathbf{i} + 2\,\mathbf{j} + \mathbf{k}$ and $\mathbf{v} = 3\,\mathbf{i} - \mathbf{j} + 2\,\mathbf{k}$.

First find the scalar product using Equation C.9:

$$\mathbf{u} \cdot \mathbf{v} = u_x v_x + u_y v_y + u_z v_z$$
$$= (-1)(3) + (2)(-1) + (1)(2) = -3$$

Then use Equation C.6,

$$|\mathbf{u}||\mathbf{v}| \cos \theta = -3$$

$$\cos \theta = \frac{-3}{(1 + 4 + 1)^{1/2} (9 + 1 + 4)^{1/2}} = -0.327$$

$$\theta = 109°$$

C–7. Show that the set of vectors $(1/\sqrt{3}, 1/\sqrt{3}, 0, 1/\sqrt{3})$, $(1/\sqrt{3}, -1/\sqrt{3}, 1/\sqrt{3}, 0)$, $(0, 1/\sqrt{3}, 1/\sqrt{3}, -1/\sqrt{3})$, and $(1/\sqrt{3}, 0, -1/\sqrt{3}, -1/\sqrt{3})$ is orthonormal.

The four vectors are

$$\mathbf{v}_1 = (1/\sqrt{3}, 1/\sqrt{3}, 0, 1/\sqrt{3})$$

$$\mathbf{v}_2 = (1/\sqrt{3}, -1/\sqrt{3}, 1/\sqrt{3}, 0)$$

$$\mathbf{v}_3 = (0, 1/\sqrt{3}, 1/\sqrt{3}, -1/\sqrt{3})$$

$$\mathbf{v}_4 = (1/\sqrt{3}, 0, -1/\sqrt{3}, -1/\sqrt{3})$$

Their lengths are given by

$$\left(\mathbf{v}_1 \cdot \mathbf{v}_1\right)^{1/2} = \left(\frac{1}{3} + \frac{1}{3} + 0 + \frac{1}{3}\right)^{1/2} = 1$$

$$\left(\mathbf{v}_2 \cdot \mathbf{v}_2\right)^{1/2} = \left(\frac{1}{3} + \frac{1}{3} + \frac{1}{3} + 0\right)^{1/2} = 1$$

$$\left(\mathbf{v}_3 \cdot \mathbf{v}_3\right)^{1/2} = \left(0 + \frac{1}{3} + \frac{1}{3} + \frac{1}{3}\right)^{1/2} = 1$$

$$\left(\mathbf{v}_4 \cdot \mathbf{v}_4\right)^{1/2} = \left(\frac{1}{3} + 0 + \frac{1}{3} + \frac{1}{3}\right)^{1/2} = 1$$

The dot products between the different vectors are

$$\left(\mathbf{v}_1 \cdot \mathbf{v}_2\right) = \left(\frac{1}{3} - \frac{1}{3} + 0 + 0\right) = 0$$

$$\left(\mathbf{v}_1 \cdot \mathbf{v}_3\right) = \left(0 + \frac{1}{3} + 0 - \frac{1}{3}\right) = 0$$

$$\left(\mathbf{v}_1 \cdot \mathbf{v}_4\right) = \left(\frac{1}{3} + 0 + 0 - \frac{1}{3}\right) = 0$$

$$\left(\mathbf{v}_2 \cdot \mathbf{v}_3\right) = \left(0 - \frac{1}{3} + \frac{1}{3} + 0\right) = 0$$

$$\left(\mathbf{v}_2 \cdot \mathbf{v}_4\right) = \left(\frac{1}{3} + 0 - \frac{1}{3} + 0\right) = 0$$

$$\left(\mathbf{v}_3 \cdot \mathbf{v}_4\right) = \left(0 + 0 - \frac{1}{3} + \frac{1}{3}\right) = 0$$

Each vector has a length equal to 1, and they are each orthogonal to all of the others. Therefore the set of vectors is orthonormal.

C–8. Determine $\mathbf{w} = \mathbf{u} \times \mathbf{v}$ given that $\mathbf{u} = -\mathbf{i} + 2\mathbf{j} + \mathbf{k}$ and $\mathbf{v} = 3\mathbf{i} - \mathbf{j} + 2\mathbf{k}$. What is $\mathbf{v} \times \mathbf{u}$ equal to?

The cross product between the two vectors is found using Equation C.17:

$$\mathbf{u} \times \mathbf{v} = \left(u_y v_z - u_z v_y\right) \mathbf{i} + \left(u_z v_x - u_x v_z\right) \mathbf{j} + \left(u_x v_y - u_y v_x\right) \mathbf{k}$$

$$= (4 + 1) \mathbf{i} + (3 + 2) \mathbf{j} + (1 - 6) \mathbf{k}$$

$$= 5\,\mathbf{i} + 5\,\mathbf{j} - 5\,\mathbf{k}$$

From Equation C.15, $\mathbf{u} \times \mathbf{v} = -\mathbf{v} \times \mathbf{u}$, so

$$\mathbf{v} \times \mathbf{u} = -5\,\mathbf{i} - 5\,\mathbf{j} + 5\,\mathbf{k}$$

C–9. Show that $\mathbf{u} \times \mathbf{u} = 0$.

Again, the cross product is found using Equation C.17:

$$\mathbf{u} \times \mathbf{u} = \left(u_y u_z - u_z u_y\right) \mathbf{i} + \left(u_z u_x - u_x u_z\right) \mathbf{j} + \left(u_x u_y - u_y u_x\right) \mathbf{k}$$

$$= 0$$

C–10. Using Equation C.16, prove that $\mathbf{u} \times \mathbf{v}$ is given by Equation C.17.

$$\mathbf{u} = u_x \mathbf{i} + u_y \mathbf{j} + u_z \mathbf{k}$$

$$\mathbf{v} = v_x \mathbf{i} + v_y \mathbf{j} + v_z \mathbf{k}$$

$$\mathbf{u} \times \mathbf{v} = \left(u_x \mathbf{i} + u_y \mathbf{j} + u_z \mathbf{k}\right) \times \left(v_x \mathbf{i} + v_y \mathbf{j} + v_z \mathbf{k}\right) \tag{1}$$

The components of the two vectors, u_x, u_y, u_z, v_x, v_y, and v_z, are all scalars, so that Equation 1 can be written as

$$\mathbf{u} \times \mathbf{v} = u_x v_x \, (\mathbf{i} \times \mathbf{i}) + u_x v_y \, (\mathbf{i} \times \mathbf{j}) + u_x v_z \, (\mathbf{i} \times \mathbf{k})$$

$$+ u_y v_x \, (\mathbf{j} \times \mathbf{i}) + u_y v_y \, (\mathbf{j} \times \mathbf{j}) + u_y v_z \, (\mathbf{j} \times \mathbf{k})$$

$$+ u_z v_x \, (\mathbf{k} \times \mathbf{i}) + u_z v_y \, (\mathbf{k} \times \mathbf{j}) + u_z v_z \, (\mathbf{k} \times \mathbf{k})$$

Now use Equation C.16 to simplify the vector products of the unit vectors.

$$\mathbf{u} \times \mathbf{v} = 0 + u_x v_y \, \mathbf{k} - u_x v_z \, \mathbf{j} - u_y v_x \, \mathbf{k} + 0 + u_y v_z \, \mathbf{i} + u_z v_x \, \mathbf{j} - u_z v_y \, \mathbf{i} + 0$$

$$= \left(u_y v_z - u_z v_y\right) \mathbf{i} + \left(u_z v_x - u_x v_z\right) \mathbf{j} + \left(u_x v_y - u_y v_x\right) \mathbf{k}$$

which is Equation C.17.

C–11. Show that $l = |\mathbf{l}| = m v r$ for circular motion.

Equation C.19 defines the angular momentum, $\mathbf{l} = \mathbf{r} \times \mathbf{p}$. Since $\mathbf{p} = m\mathbf{v}$, $\mathbf{l} = m\mathbf{r} \times \mathbf{v}$. For circular motion, the angle between the direction of travel and the velocity vector is always $\theta = 90°$.

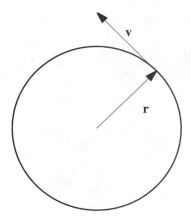

Equation C.14 then gives

$$m\mathbf{r} \times \mathbf{v} = m\,|\,\mathbf{r}\,||\,\mathbf{v}\,|\mathbf{c}\sin\theta$$

$$= mrv\mathbf{c}\sin 90°$$

$$= mvr\mathbf{c}$$

The vector $\mathbf{c}$ is a unit vector of length one, and so $|\,\mathbf{l}\,| = mvr$.

C–12.   Show that

$$\frac{d}{dt}(\mathbf{u}\cdot\mathbf{v}) = \frac{d\mathbf{u}}{dt}\cdot\mathbf{v} + \mathbf{u}\cdot\frac{d\mathbf{v}}{dt}$$

and

$$\frac{d}{dt}(\mathbf{u}\times\mathbf{v}) = \frac{d\mathbf{u}}{dt}\times\mathbf{v} + \mathbf{u}\times\frac{d\mathbf{v}}{dt}$$

Take

$$\mathbf{u} = u_x(t)\,\mathbf{i} + u_y(t)\,\mathbf{j} + u_z(t)\,\mathbf{k}$$

$$\mathbf{v} = v_x(t)\,\mathbf{i} + v_y(t)\,\mathbf{j} + v_z(t)\,\mathbf{k}$$

Then the first derivatives of these two vectors with respect to time are

$$\frac{d\mathbf{u}}{dt} = \frac{du_x(t)}{dt}\,\mathbf{i} + \frac{du_y(t)}{dt}\,\mathbf{j} + \frac{du_z(t)}{dt}\,\mathbf{k}$$

$$\frac{d\mathbf{v}}{dt} = \frac{dv_x(t)}{dt}\,\mathbf{i} + \frac{dv_y(t)}{dt}\,\mathbf{j} + \frac{dv_z(t)}{dt}\,\mathbf{k}$$

From Equation C.9,

$$\mathbf{u}\cdot\mathbf{v} = u_x(t)v_x(t) + u_y(t)v_y(t) + u_z(t)v_z(t)$$

So

$$\frac{d}{dt}(\mathbf{u} \cdot \mathbf{v}) = \frac{d}{dt}\left[u_x(t)v_x(t)\right] + \frac{d}{dt}\left[u_y(t)v_y(t)\right] + \frac{d}{dt}\left[u_z(t)v_z(t)\right]$$

$$= \frac{du_x(t)}{dt}v_x(t) + \frac{du_y(t)}{dt}v_y(t) + \frac{du_z(t)}{dt}v_z(t) + u_x(t)\frac{dv_x(t)}{dt} + u_y(t)\frac{dv_y(t)}{dt} + u_z(t)\frac{dv_z(t)}{dt}$$

$$= \frac{d\mathbf{u}}{dt}\cdot\mathbf{v} + \mathbf{u}\cdot\frac{d\mathbf{v}}{dt}$$

Equation C.17 gives

$$\mathbf{u}\times\mathbf{v} = \left(u_yv_z - u_zv_y\right)\mathbf{i} + \left(u_zv_x - u_xv_z\right)\mathbf{j} + \left(u_xv_y - u_yv_x\right)\mathbf{k}$$

So that

$$\frac{d}{dt}(\mathbf{u}\times\mathbf{v}) = \frac{d}{dt}\left(u_yv_z - u_zv_y\right)\mathbf{i} + \frac{d}{dt}\left(u_zv_x - u_xv_z\right)\mathbf{j} + \frac{d}{dt}\left(u_xv_y - u_yv_x\right)\mathbf{k}$$

$$= \left[\frac{du_y(t)}{dt}v_z(t) + u_y(t)\frac{dv_z(t)}{dt} - \frac{du_z(t)}{dt}v_y(t) - u_z(t)\frac{dv_y(t)}{dt}\right]\mathbf{i}$$

$$+ \left[\frac{du_z(t)}{dt}v_x(t) + u_z(t)\frac{dv_x(t)}{dt} - \frac{du_x(t)}{dt}v_z(t) - u_x(t)\frac{dv_z(t)}{dt}\right]\mathbf{j}$$

$$+ \left[\frac{du_x(t)}{dt}v_y(t) + u_x(t)\frac{dv_y(t)}{dt} - \frac{du_y(t)}{dt}v_x(t) - u_y(t)\frac{dv_x(t)}{dt}\right]\mathbf{k}$$

$$= \left[\frac{du_y(t)}{dt}v_z(t) - \frac{du_z(t)}{dt}v_y(t)\right]\mathbf{i} + \left[\frac{du_z(t)}{dt}v_x(t) - \frac{du_x(t)}{dt}v_z(t)\right]\mathbf{j}$$

$$+ \left[\frac{du_x(t)}{dt}v_y(t) - \frac{du_y(t)}{dt}v_x(t)\right]\mathbf{k} + \left[u_y(t)\frac{dv_z(t)}{dt} - u_z(t)\frac{dv_y(t)}{dt}\right]\mathbf{i}$$

$$+ \left[u_z(t)\frac{dv_x(t)}{dt} - u_x(t)\frac{dv_z(t)}{dt}\right]\mathbf{j} + \left[u_x(t)\frac{dv_y(t)}{dt} - u_y(t)\frac{dv_x(t)}{dt}\right]\mathbf{k}$$

$$= \frac{d\mathbf{u}}{dt}\times\mathbf{v} + \mathbf{u}\times\frac{d\mathbf{v}}{dt}$$

C–13.   Using the results of Problem C–12, prove that

$$\mathbf{u}\times\frac{d^2\mathbf{u}}{dt^2} = \frac{d}{dt}\left(\mathbf{u}\times\frac{d\mathbf{u}}{dt}\right)$$

From Problem C–12,

$$\frac{d}{dt}\left(\mathbf{u}\times\frac{d\mathbf{u}}{dt}\right) = \frac{d\mathbf{u}}{dt}\times\frac{d\mathbf{u}}{dt} + \mathbf{u}\times\frac{d}{dt}\left(\frac{d\mathbf{u}}{dt}\right)$$

The first term on the right of this expression is zero (Problem C–9). Thus,

$$\frac{d}{dt}\left(\mathbf{u} \times \frac{d\mathbf{u}}{dt}\right) = \mathbf{u} \times \frac{d}{dt}\left(\frac{d\mathbf{u}}{dt}\right)$$

$$= \mathbf{u} \times \frac{d^2\mathbf{u}}{dt^2}$$

C–14. In vector notation, Newton's equations for a single particle are

$$m\frac{d^2\mathbf{r}}{dt^2} = \mathbf{F}(x, y, z)$$

By operating on this equation from the left by $\mathbf{r} \times$ and using the result of Problem C–13, show that

$$m\frac{d}{dt}\left(\mathbf{r} \times \frac{d\mathbf{r}}{dt}\right) = \mathbf{r} \times \mathbf{F}$$

Because momentum is defined as $\mathbf{p} = m\mathbf{v} = m\dfrac{d\mathbf{r}}{dt}$, the above expression reads

$$\frac{d}{dt}(\mathbf{r} \times \mathbf{p}) = \mathbf{r} \times \mathbf{F}$$

But $\mathbf{r} \times \mathbf{p} = \mathbf{l}$, the angular momentum, and so we have

$$\frac{d\mathbf{l}}{dt} = \mathbf{r} \times \mathbf{F}$$

This is the form of Newton's equation for a rotating system. Notice that $d\mathbf{l}/dt = 0$, or that angular momentum is conserved if $\mathbf{r} \times \mathbf{F} = 0$. Can you identify $\mathbf{r} \times \mathbf{F}$?

Start with Newton's equations of motion in vector form

$$m\frac{d^2\mathbf{r}}{dt^2} = \mathbf{F}(x, y, z)$$

Then operate from the left by $\mathbf{r} \times$ to give

$$\mathbf{r} \times m\frac{d^2\mathbf{r}}{dt^2} = \mathbf{r} \times \mathbf{F}$$

This last equation can be rewritten using the results of Problem C–13,

$$m\frac{d}{dt}\left(\mathbf{r} \times \frac{d\mathbf{r}}{dt}\right) = \mathbf{r} \times \mathbf{F}$$

$$\frac{d}{dt}\left(\mathbf{r} \times \mathbf{p}\right) = \mathbf{r} \times \mathbf{F}$$

$$\frac{d\mathbf{l}}{dt} = \mathbf{r} \times \mathbf{F}$$

The quantity $\mathbf{r} \times \mathbf{F}$ is called torque.

C–15. Find the gradient of $f(x, y, z) = x^2 - yz + xz^2$ at the point $(1, 1, 1)$.

The gradient of a function is defined by Equation C.24,

$$\nabla f(x, y, z) = \mathbf{i}\frac{\partial f}{\partial x} + \mathbf{j}\frac{\partial f}{\partial y} + \mathbf{k}\frac{\partial f}{\partial z}$$

$$= \mathbf{i}\left(2x + z^2\right) + \mathbf{j}(-z) + \mathbf{k}(-y + 2xz)$$

Evaluating at the point $(1, 1, 1)$ gives $3\,\mathbf{i} - \mathbf{j} + \mathbf{k}$.

C–16. The electrostatic potential produced by a dipole moment μ located at the origin and directed along the x axis is given by

$$\phi(x, y, z) = \frac{\mu x}{(x^2 + y^2 + z^2)^{3/2}} \qquad (x, y, z \neq 0)$$

Derive an expression for the electric field associated with this potential.

The electric field associated with a potential is given by Equation C.26,

$$\mathbf{E} = -\nabla\phi(x, y, z)$$

$$= -\mathbf{i}\frac{\partial\phi}{\partial x} - \mathbf{j}\frac{\partial\phi}{\partial y} - \mathbf{k}\frac{\partial\phi}{\partial z}$$

$$= \mathbf{i}\left[\frac{3\mu x^2}{\left(x^2 + y^2 + z^2\right)^{5/2}} - \frac{\mu}{\left(x^2 + y^2 + z^2\right)^{3/2}}\right]$$

$$+ \mathbf{j}\left[\frac{3\mu xy}{\left(x^2 + y^2 + z^2\right)^{5/2}}\right] + \mathbf{k}\left[\frac{3\mu xz}{\left(x^2 + y^2 + z^2\right)^{5/2}}\right]$$

C–17. We proved the *Schwartz inequality* for complex numbers in Problem A–16. For vectors, the Schwartz inequality takes the form

$$(\mathbf{u} \cdot \mathbf{v})^2 \leq |\mathbf{u}|^2|\mathbf{v}|^2$$

Why do you think that this is so? Do you see a parallel between this result for two-dimensional vectors and the complex number version?

Take the definition of the scalar product between two vectors (Equation C.6) and square both sides,

$$\mathbf{u} \cdot \mathbf{v} = |\,\mathbf{u}\,||\,\mathbf{v}\,|\cos\theta$$

$$(\mathbf{u} \cdot \mathbf{v})^2 = |\,\mathbf{u}\,|^2|\,\mathbf{v}\,|^2 \cos^2\theta$$

Since $\cos^2\theta \leq 1$ for any value of θ, the angle between $\mathbf{u}$ and $\mathbf{v}$, it follows immediately that

$$(\mathbf{u} \cdot \mathbf{v})^2 \leq |\,\mathbf{u}\,|^2|\,\mathbf{v}\,|^2 \tag{1}$$

If the Schwartz inequality is written out in terms of components for two-dimensional vectors,

$$\left(u_x v_x + u_y v_y\right)^2 \le \left(u_x^2 + u_y^2\right)\left(v_x^2 + v_y^2\right)$$

it is seen to be directly analogous to the complex number version of Problem A–16. The similarity is made even stronger by taking (positive) square root of both sides of Equation 1:

$$|(\mathbf{u} \cdot \mathbf{v})| \le |\mathbf{u}|\,|\mathbf{v}|$$

Since the right hand side of the inequality is necessarily positive, it is possible to remove the absolute value from the left hand side, and write

$$(\mathbf{u} \cdot \mathbf{v}) \le |\mathbf{u}|\,|\mathbf{v}|$$

C–18.　We proved the *triangle inequality* for complex numbers in Problem A–17. For vectors, the triangle inequality takes the form

$$|\mathbf{u} + \mathbf{v}| \le |\mathbf{u}| + |\mathbf{v}|$$

Prove this inequality by starting with

$$|\mathbf{u} + \mathbf{v}|^2 = |\mathbf{u}|^2 + |\mathbf{v}|^2 + 2\mathbf{u} \cdot \mathbf{v}$$

and then using the Schwartz inequality (previous problem). Why do you think this is called the triangle inequality?

From the definition of length of a vector given in Equation C.11,

$$| \mathbf{u} + \mathbf{v} |^2 = (\mathbf{u} + \mathbf{v}) \cdot (\mathbf{u} + \mathbf{v})$$

$$= \mathbf{u} \cdot \mathbf{u} + \mathbf{v} \cdot \mathbf{v} + 2\,\mathbf{u} \cdot \mathbf{v}$$

$$= | \mathbf{u} |^2 + | \mathbf{v} |^2 + 2\,\mathbf{u} \cdot \mathbf{v}$$

Using the Schwartz inequality in the form at the end of Problem C–17, $(\mathbf{u} \cdot \mathbf{v}) \le |\mathbf{u}|\,|\mathbf{v}|$, gives,

$$| \mathbf{u} + \mathbf{v} |^2 \le | \mathbf{u} |^2 + | \mathbf{v} |^2 + 2| \mathbf{u} |\,|\mathbf{v}| = (| \mathbf{u} | + | \mathbf{v} |)^2$$

Taking the (positive) square root of both sides of the equation leads to the triangle inequality for vectors

$$| \mathbf{u} + \mathbf{v} | \le | \mathbf{u} | + | \mathbf{v} |$$

Reference to Figure C.1 shows that the three vectors $\mathbf{u}$, $\mathbf{v}$, and $\mathbf{u} + \mathbf{v}$ form a triangle. The inequality relating the lengths of these three vectors is then seen to be equivalent to a statement regarding the lengths of the three sides of a triangle, which gives the inequality its name.

The Postulates and General Principles
of Quantum Mechanics

PROBLEMS AND SOLUTIONS

4–1. Which of the following candidates for wave functions are normalizable over the indicated intervals?

(a) $e^{-x^2/2}$ $(-\infty, \infty)$

(b) e^{-x} $(-\infty, \infty)$

(c) $e^{i\theta}$ $(0, 2\pi)$

(d) $\cosh x$ $(0, \infty)$

(e) xe^{-x} $(0, \infty)$

Normalize those that can be normalized. Are the others suitable wave functions?

Because the function in (b) diverges as $x \to -\infty$ and that in (d) diverges as $x \to \infty$, they are not normalizable. A wave function has a probabilistic interpretation; thus, it must be normalizable. Consequently, the functions in (b) and (d) are not suitable wave functions.

To normalize the functions in (a), (c), and (e), determine the integral $c = \int_{-\infty}^{\infty} f^*(\tau) f(\tau) d\tau$. The normalized function is $c^{-1/2} f$.

(a)

$$\int_{-\infty}^{\infty} e^{-x^2/2} e^{-x^2/2} dx = 2 \int_0^{\infty} e^{-x^2} dx$$

$$= \sqrt{\pi}$$

The normalized function is $\pi^{-1/4} e^{-x^2/2}$.

(b)

$$\int_0^{2\pi} e^{-i\theta} e^{i\theta} d\theta = \int_0^{2\pi} d\theta$$

$$= 2\pi$$

The normalized function is $(2\pi)^{-1/2} e^{i\theta}$.

(c)

$$\int_0^{\infty} xe^{-x} xe^{-x} dx = 2 \int_0^{\infty} x^2 e^{-2x} dx$$

$$= \frac{1}{4}$$

143

The normalized function is $2xe^{-x}$.

4–2. Which of the following wave functions are normalized over the indicated two-dimensional intervals?

(a) $e^{-(x^2+y^2)/2}$ $0 \leq x < \infty, 0 \leq y < \infty$

(b) $e^{-(x+y)/2}$ $0 \leq x < \infty, 0 \leq y < \infty$

(c) $\left(\dfrac{4}{ab}\right)^{1/2} \sin\dfrac{\pi x}{a} \sin\dfrac{\pi y}{b}$ $0 \leq x \leq a, 0 \leq y \leq b$

Normalize those that aren't.

(a)

$$\int_0^\infty dy \int_0^\infty dx\, e^{-(x^2+y^2)} = \int_0^\infty e^{-y^2}\, dy \int_0^\infty e^{-x^2}\, dx$$

$$= \left(\frac{\sqrt{\pi}}{2}\right)\left(\frac{\sqrt{\pi}}{2}\right)$$

$$= \frac{\pi}{4}$$

The normalized function is $(2/\sqrt{\pi})e^{-(x^2+y^2)/2}$.

(b)

$$\int_0^\infty dy \int_0^\infty dx\, e^{-(x+y)} = \int_0^\infty e^{-y}\, dy \int_0^\infty e^{-x}\, dx$$

$$= \left(-e^{-y}\Big|_0^\infty\right)\left(-e^{-x}\Big|_0^\infty\right)$$

$$= 1$$

The function is normalized.

(c)

$$\int_0^b dy \int_0^a dx\, \frac{4}{ab} \sin^2\frac{\pi x}{a} \sin^2\frac{\pi y}{b} = \frac{4}{ab} \int_0^a \sin^2\frac{\pi y}{b}\, dy \int_0^b \sin^2\frac{\pi x}{a}\, dx$$

$$= \frac{4}{ab}\left(\frac{y}{2} - \frac{b}{4\pi}\sin\frac{2\pi y}{b}\Big|_0^b\right)\left(\frac{x}{2} - \frac{a}{4\pi}\sin\frac{2\pi x}{a}\Big|_0^a\right)$$

$$= 1$$

The function is normalized.

4–3. Why does $\psi^*\psi$ have to be everywhere real, nonnegative, finite, and of definite value?

$\psi^*\psi$ is the probability density of a system at the point (x, y, z). For this probabilistic interpretation to be valid, $\psi^*\psi$ must be everywhere real, nonnegative, finite, and of definite value.

4–4. In this problem, we will prove that the form of the Schrödinger equation imposes the condition that the first derivative of a wave function be continuous. The Schrödinger equation is

$$\frac{d^2\psi}{dx^2} + \frac{2m}{\hbar^2}[E - V(x)]\psi(x) = 0$$

If we integrate both sides from $a - \epsilon$ to $a + \epsilon$, where a is an arbitrary value of x and ϵ is infinitesimally small, then we have

$$\frac{d\psi}{dx}\Big|_{x=a+\epsilon} - \frac{d\psi}{dx}\Big|_{x=a-\epsilon} = \frac{2m}{\hbar^2}\int_{a-\epsilon}^{a+\epsilon}[V(x) - E]\psi(x)dx$$

Now show that $d\psi/dx$ is continuous if $V(x)$ is continuous.

Suppose now that $V(x)$ is *not* continuous at $x = a$, as in

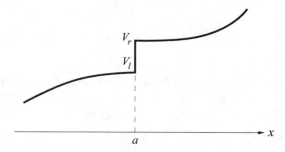

Show that

$$\frac{d\psi}{dx}\Big|_{x=a+\epsilon} - \frac{d\psi}{dx}\Big|_{x=a-\epsilon} = \frac{2m}{\hbar^2}(V_l + V_r - 2E)\psi(a)\epsilon$$

so that $d\psi/dx$ is continuous even if $V(x)$ has a *finite* discontinuity. What if $V(x)$ has an infinite discontinuity, as in the problem of a particle in a box? Are the first derivatives of the wave functions continuous at the boundaries of the box?

The first derivative of a wave function is continuous if

$$\lim_{\epsilon \to 0}\left(\frac{d\psi}{dx}\Big|_{x=a+\epsilon} - \frac{d\psi}{dx}\Big|_{x=a-\epsilon}\right) = 0$$

Start with

$$\frac{d\psi}{dx}\Big|_{x=a+\epsilon} - \frac{d\psi}{dx}\Big|_{x=a-\epsilon} = \frac{2m}{\hbar^2}\int_{a-\epsilon}^{a+\epsilon}[V(x) - E]\,\psi(x)\,dx \tag{1}$$

If $V(x)$ is continuous, then $\lim_{\epsilon \to 0} V(a \pm \epsilon) = V(a)$. A similar result holds for $\psi(x)$, which is already known to be continuous. Thus,

$$\lim_{\epsilon \to 0}\frac{2m}{\hbar^2}\int_{a-\epsilon}^{a+\epsilon}[V(x) - E]\,\psi(x)\,dx = \frac{2m}{\hbar^2}[V(a) - E]\,\psi(a)\lim_{\epsilon \to 0}\int_{a-\epsilon}^{a+\epsilon}dx = 0 \tag{2}$$

Combining Equations 1 and 2 shows that

$$\lim_{\epsilon \to 0} \left(\frac{d\psi}{dx}\bigg|_{x=a+\epsilon} - \frac{d\psi}{dx}\bigg|_{x=a-\epsilon} \right) = 0$$

and therefore $d\psi/dx$ is continuous.

Now suppose that $V(x)$ has a finite discontinuity at $x = a$. Then divide the integral into two parts, $a - \epsilon$ to a, and a to $a + \epsilon$. In the limit as $\epsilon \to 0$,

$$\frac{2m}{\hbar^2} \int_{a-\epsilon}^{a+\epsilon} [V(x) - E]\psi(x)\,dx = \frac{2m}{\hbar^2} \left\{ \int_{a-\epsilon}^{a} [V(x) - E]\psi(x)\,dx + \int_{a}^{a+\epsilon} [V(x) - E]\psi(x)\,dx \right\}$$

$$= \frac{2m}{\hbar^2} \left[(V_l - E)\psi(a) \int_{a-\epsilon}^{a} dx + (V_r - E)\psi(a) \int_{a}^{a+\epsilon} dx \right]$$

$$= \frac{2m}{\hbar^2} (V_l + V_r - 2E)\psi(a)\epsilon$$

As $\epsilon \to 0$, $\lim_{\epsilon \to 0} (V_l + V_r - 2E)\psi(a)\epsilon = 0$, and $d\psi/dx$ remains continuous even though $V(x)$ has a finite discontinuity.

If, however, the discontinuity in $V(x)$ at the point a is infinite, then there is no finite limiting value for $V(x)$ upon approaching from one side, and the expression $\frac{2m}{\hbar^2} \int_{a-\epsilon}^{a+\epsilon} [V(x) - E]\psi(x)\,dx$ cannot be integrated and will not equal zero in the limit $\epsilon \to 0$. Thus, $d\psi/dx$ is not a continuous function in this case. This can be seen for the particle in a box, which has infinite discontinuities in the potential at the boundaries of the box. The first derivatives of the particle-in-a-box wave functions are not continuous at the boundaries of the box. Outside the box, $d\psi/dx = 0$, while inside the box at $x = 0$, $d\psi/dx = (\sqrt{2/a})n\pi/a$, and at $x = a$, $d\psi/dx = (-1)^n(\sqrt{2/a})n\pi/a$.

4–5. Determine whether the following functions are acceptable or not as state functions over the indicated intervals.

(a) $\dfrac{1}{x}$ $(0, \infty)$ (b) $e^{-2x}\sinh x$ $(0, \infty)$

(c) $e^{-x}\cos x$ $(0, \infty)$ (d) e^x $(-\infty, \infty)$

(e) $e^{-x}\sinh x$ $(0, \infty)$

(a) Not acceptable. The function is not defined at $x = 0$. More importantly, the function cannot be normalized.
(b) Acceptable.
(c) Acceptable.
(d) Not acceptable. The function diverges as $x \to \infty$ and thus is not normalizable.
(e) Not acceptable. The function is not normalizable.

4–6. Consider the linear differential equation

$$a(x)y''(x) + b(x)y'(x) + c(x)y(x) = 0$$

where $y''(x)$ and $y'(x)$ are standard notation for d^2y/dx^2 and dy/dx, respectively. Show that if $y_1(x)$ and $y_2(x)$ are each solutions to the above differential equation, then so is $y(x) = c_1y_1(x) + c_2y_2(x)$, where c_1 and c_2 are constants.

To show that $y(x)$ is a solution, show that $a(x)y''(x) + b(x)y'(x) + c(x)y(x) = 0$.

Given that $y(x) = c_1y_1(x) + c_2y_2(x)$, write $a(x)y''(x) + b(x)y'(x) + c(x)y(x)$ in terms of y_1 and y_2:

$$a(x)y''(x) + b(x)y'(x) + c(x)y(x)$$
$$= a(x)\left[c_1y_1''(x) + c_2y_2''(x)\right] + b(x)\left[c_1y_1'(x) + c_2y_2'(x)\right] + c(x)\left[c_1y_1(x) + c_2y_2(x)\right]$$
$$= c_1\left[a(x)y_1''(x) + b(x)y_1'(x) + c(x)y_1(x)\right] + c_2\left[a(x)y_2''(x) + b(x)y_2'(x) + c(x)y_2(x)\right]$$

Since $y_1(x)$ and $y_2(x)$ are solutions to the differential equation, the following relations hold:

$$a(x)y_1''(x) + b(x)y_1'(x) + c(x)y_1(x) = 0$$
$$a(x)y_2''(x) + b(x)y_2'(x) + c(x)y_2(x) = 0$$

Substituting these relations into the expression for $a(x)y''(x) + b(x)y'(x) + c(x)y(x)$ derived earlier gives

$$a(x)y''(x) + b(x)y'(x) + c(x)y(x) = c_1(0) + c_2(0) = 0$$

Thus, $y(x)$ is indeed a solution to the differential equation.

4–7. Calculate the values of $\sigma_E^2 = \langle E^2 \rangle - \langle E \rangle^2$ for a particle in a box in the state described by

$$\psi(x) = \left(\frac{630}{a^9}\right)^{1/2} x^2(a-x)^2 \qquad 0 \leq x \leq a$$

Since $\langle E \rangle = \int_0^a \psi^* \hat{H} \psi \, dx$ and $\langle E^2 \rangle = \int_0^a \psi^* \hat{H}^2 \psi \, dx$, first evaluate $\hat{H}\psi$ and $\hat{H}^2\psi$.

$$\hat{H}\psi = -\frac{\hbar^2}{2m}\frac{d^2}{dx^2}\left[\left(\frac{630}{a^9}\right)^{1/2}x^2(a-x)^2\right]$$

$$= -\frac{\hbar^2}{2m}\left(\frac{630}{a^9}\right)^{1/2}\frac{d^2}{dx^2}\left(a^2x^2 - 2ax^3 + x^4\right)$$

$$= -\frac{\hbar^2}{2m}\left(\frac{630}{a^9}\right)^{1/2}\frac{d}{dx}\left(2a^2x - 6ax^2 + 4x^3\right)$$

$$= -\frac{\hbar^2}{2m}\left(\frac{630}{a^9}\right)^{1/2}\left(2a^2 - 12ax + 12x^2\right)$$

$$\hat{H}^2\psi = -\frac{\hbar^2}{2m}\frac{d^2}{dx^2}\hat{H}\psi$$

$$= \frac{\hbar^4}{4m^2}\left(\frac{630}{a^9}\right)^{1/2}\frac{d^2}{dx^2}\left(2a^2 - 12ax + 12x^2\right)$$

$$= \frac{\hbar^4}{4m^2}\left(\frac{630}{a^9}\right)^{1/2}\frac{d}{dx}\left(-12a + 24x\right)$$

$$= \frac{\hbar^4}{4m^2}\left(\frac{630}{a^9}\right)^{1/2}(24)$$

Now evaluate $\langle E \rangle$ and $\langle E^2 \rangle$:

$$\langle E \rangle = \int_0^a \psi^*\hat{H}\psi\, dx$$

$$= -\frac{\hbar^2}{2m}\left(\frac{630}{a^9}\right)\int_0^a x^2(a-x)^2\left(2a^2 - 12ax + 12x^2\right)\, dx$$

$$= -\frac{\hbar^2}{2m}\left(\frac{630}{a^9}\right)\int_0^a \left(2a^4x^2 - 16a^3x^3 + 38a^2x^4 - 36ax^5 + 12x^6\right)\, dx$$

$$= -\frac{\hbar^2}{2m}\left(\frac{630}{a^9}\right)\left(\frac{2a^4x^3}{3} - \frac{16a^3x^4}{4} + \frac{38a^2x^5}{5} - \frac{36ax^6}{6} + \frac{12x^7}{7}\right)\Bigg|_0^a$$

$$= \frac{6\hbar^2}{ma^2}$$

$$\langle E^2 \rangle = \int_0^a \psi^*\hat{H}^2\psi\, dx$$

$$= \frac{\hbar^4}{4m^2}\left(\frac{630}{a^9}\right)(24)\int_0^a x^2(a-x)^2\, dx$$

$$= \frac{\hbar^4}{4m^2}\left(\frac{630}{a^9}\right)(24)\int_0^a \left(a^2x^2 - 2ax^3 + x^4\right)\, dx$$

$$= \frac{\hbar^4}{4m^2}\left(\frac{630}{a^9}\right)(24)\left(\frac{a^2x^3}{3} - \frac{2ax^4}{4} + \frac{x^5}{5}\right)\Bigg|_0^a$$

$$= \frac{126\hbar^4}{m^2a^4}$$

Thus, the energy variance is

$$\sigma_E^2 = \langle E^2 \rangle - \langle E \rangle^2$$

$$= \frac{126\hbar^2}{m^2 a^4} - \left(\frac{6\hbar^2}{ma^2} \right)^2$$

$$= \frac{90\hbar^4}{m^2 a^4}$$

4–8. Consider a free particle constrained to move over the rectangular region $0 \le x \le a$, $0 \le y \le b$. The energy eigenfunctions of this system are

$$\psi_{n_x,n_y}(x, y) = \left(\frac{4}{ab} \right)^{1/2} \sin \frac{n_x \pi x}{a} \sin \frac{n_y \pi y}{b} \qquad \begin{cases} n_x = 1, 2, 3, \dots \\ n_y = 1, 2, 3, \dots \end{cases}$$

The Hamiltonian operator for this system is

$$\hat{H} = -\frac{\hbar^2}{2m} \left(\frac{\partial^2}{\partial x^2} + \frac{\partial^2}{\partial y^2} \right)$$

Show that if the system is one of its eigenstates, then

$$\sigma_E^2 = \langle E^2 \rangle - \langle E \rangle^2 = 0$$

If the system is in one of its eigenstates, then

$$\hat{H} \psi_{n_x,n_y} = E_{n_x,n_y} \psi_{n_x,n_y}$$

$$\hat{H}^2 \psi_{n_x,n_y} = E_{n_x,n_y}^2 \psi_{n_x,n_y}$$

Therefore,

$$\langle E \rangle = \int_0^b dy \int_0^a dx \, \psi_{n_x,n_y}^* \hat{H} \psi_{n_x,n_y}$$

$$= E_{n_x,n_y} \int_0^b dy \int_0^a dx \, \psi_{n_x,n_y}^* \psi_{n_x,n_y}$$

$$= E_{n_x,n_y}$$

$$\langle E^2 \rangle = \int_0^b dy \int_0^a dx \, \psi_{n_x,n_y}^* \hat{H}^2 \psi_{n_x,n_y}$$

$$= E_{n_x,n_y}^2 \int_0^b dy \int_0^a dx \, \psi_{n_x,n_y}^* \psi_{n_x,n_y}$$

$$= E_{n_x,n_y}^2$$

giving the energy variance as

$$\sigma_E^2 = \langle E^2 \rangle - \langle E \rangle^2 = E_{n_x,n_y}^2 - E_{n_x,n_y}^2 = 0$$

4–9. The momentum operator in two dimensions is

$$\hat{P} = -i\hbar \left(\mathbf{i}\, \frac{\partial}{\partial x} + \mathbf{j}\, \frac{\partial}{\partial y} \right)$$

Using the wave function given in Problem 4–8, calculate the value of $\langle\, p\, \rangle$ and then

$$\sigma_p^2 = \langle\, p^2\, \rangle - \langle\, p\, \rangle^2$$

Compare your result with σ_p^2 in the one-dimensional case.

First evaluate $\hat{P}\psi_{n_x,n_y}$ and $\hat{P}^2\psi_{n_x,n_y}$:

$$\hat{P}\psi_{n_x,n_y} = -i\hbar \left(\mathbf{i}\, \frac{\partial}{\partial x} + \mathbf{j}\, \frac{\partial}{\partial y} \right) \left[\left(\frac{4}{ab} \right)^{1/2} \sin \frac{n_x \pi x}{a} \sin \frac{n_y \pi y}{b} \right]$$

$$= -i\hbar \left(\frac{4}{ab} \right)^{1/2} \left(\mathbf{i}\, \frac{n_x \pi}{a} \cos \frac{n_x \pi x}{a} \sin \frac{n_y \pi y}{b} + \mathbf{j}\, \frac{n_y \pi}{b} \sin \frac{n_x \pi x}{a} \cos \frac{n_y \pi y}{b} \right)$$

$$\hat{P}^2\psi_{n_x,n_y} = \left[-i\hbar \left(\mathbf{i}\, \frac{\partial}{\partial x} + \mathbf{j}\, \frac{\partial}{\partial y} \right) \right] \left[-i\hbar \left(\mathbf{i}\, \frac{\partial}{\partial x} + \mathbf{j}\, \frac{\partial}{\partial y} \right) \right] \psi_{n_x,n_y}$$

$$= -\hbar^2 \left(\frac{\partial^2}{\partial x^2} + \frac{\partial^2}{\partial y^2} \right) \psi_{n_x,n_y}$$

$$= 2m\hat{H}\psi_{n_x,n_y}$$

$$= 2m \left(\frac{n_x h^2}{8ma^2} + \frac{n_y h^2}{8mb^2} \right) \psi_{n_x,n_y}$$

$$= \left(\frac{n_x^2 h^2}{4a^2} + \frac{n_y^2 h^2}{4b^2} \right) \psi_{n_x,n_y}$$

In determining the operator $\hat{P}^2 = \hat{P} \cdot \hat{P}$, note that $\mathbf{i} \cdot \mathbf{j}$ and $\mathbf{j} \cdot \mathbf{i}$ equal 0. In addition, the ψ_{n_x,n_y} are the eigenfunctions of $\hat{H}$.

Now evaluate $\langle p \rangle$:

$$\langle p \rangle = \int_0^b dy \int_0^a dx \, \psi^*_{n_x,n_y} \hat{P} \psi_{n_x,n_y}$$

$$= -i\hbar \left(\frac{4}{ab} \right) \int_0^b dy \int_0^a dx \, \sin \frac{n_x \pi x}{a} \sin \frac{n_y \pi y}{b} \left(\mathbf{i} \frac{n_x \pi}{a} \cos \frac{n_x \pi x}{a} \sin \frac{n_y \pi y}{b} + \mathbf{j} \frac{n_y \pi}{b} \sin \frac{n_x \pi x}{a} \cos \frac{n_y \pi y}{b} \right)$$

$$= -i\hbar \left(\frac{4}{ab} \right) \left(\mathbf{i} \frac{n_x \pi}{a} \int_0^b \sin^2 \frac{n_y \pi y}{b} \, dy \int_0^a \sin \frac{n_x \pi x}{a} \cos \frac{n_x \pi x}{a} \, dx \right.$$

$$\left. + \mathbf{j} \frac{n_y \pi}{b} \int_0^b \sin \frac{n_y \pi y}{b} \cos \frac{n_y \pi y}{b} \, dy \int_0^a \sin^2 \frac{n_x \pi x}{a} \, dx \right)$$

$$= 0$$

The last result follows because

$$\int_0^a \sin \left(n_x \pi x/a \right) \cos \left(n_x \pi x/a \right) \, dx = \int_0^b \sin \left(n_y \pi y/b \right) \cos \left(n_y \pi y/b \right) \, dy = 0.$$

The expression for $\langle p^2 \rangle$ is

$$\langle p^2 \rangle = \int_0^b dy \int_0^a dx \, \psi^*_{n_x,n_y} \hat{P}^2 \psi_{n_x,n_y}$$

$$= \left(\frac{n_x^2 h^2}{4a^2} + \frac{n_y^2 h^2}{4b^2} \right) \int_0^b dy \int_0^a dx \, \psi^*_{n_x,n_y} \psi_{n_x,n_y}$$

$$= \frac{n_x^2 h^2}{4a^2} + \frac{n_y^2 h^2}{4b^2}$$

The momentum variance is

$$\sigma_p^2 = \langle p^2 \rangle - \langle p \rangle^2$$

$$= \langle p^2 \rangle$$

$$= \frac{h^2}{4} \left(\frac{n_x^2}{a^2} + \frac{n_y^2}{b^2} \right)$$

This result is an extension of σ_p^2 in the one-dimensional case where only the first term need be included.

4–10. Suppose that a particle in a two-dimensional box (cf. Problem 4–8) is in the state

$$\psi(x, y) = \frac{30}{(a^5 b^5)^{1/2}} x(a - x) y(b - y)$$

Show that $\psi(x, y)$ is normalized, and then calculate the value of $\langle E \rangle$ associated with the state described by $\psi(x, y)$.

Show that $\psi(x, y)$ is normalized:

$$\int_0^b dy \int_0^a dx \, \psi^*(x, y)\psi(x, y) = \frac{900}{a^5 b^5} \int_0^b y^2 (b - y)^2 \, dy \int_0^a x^2 (a - x)^2 \, dx$$

$$= \frac{900}{a^5 b^5} \int_0^b \left(b^2 y^2 - 2by^3 + y^4\right) dy \int_0^a \left(a^2 x^2 - 2ax^3 + x^4\right) dx$$

$$= \frac{900}{a^5 b^5} \left.\left(\frac{b^2 y^3}{3} - \frac{2by^4}{4} + \frac{y^5}{5}\right)\right|_0^b \left.\left(\frac{a^2 x^3}{3} - \frac{2ax^4}{4} + \frac{x^5}{5}\right)\right|_0^a$$

$$= \frac{900}{a^5 b^5} \left(\frac{b^5}{30}\right)\left(\frac{a^5}{30}\right)$$

$$= 1$$

To find the value of $\langle E \rangle$, first determine $\hat{H}\psi(x, y)$:

$$\hat{H}\psi(x, y) = -\frac{\hbar^2}{2m}\left(\frac{\partial^2}{\partial x^2} + \frac{\partial^2}{\partial y^2}\right)\left[\frac{30}{(a^5 b^5)^{1/2}}x(a - x)y(b - y)\right]$$

$$= -\frac{\hbar^2}{2m}\left[\frac{30}{(a^5 b^5)^{1/2}}\right]\left\{y(b - y)\frac{\partial^2}{\partial x^2}[x(a - x)] + x(a - x)\frac{\partial^2}{\partial y^2}[y(a - y)]\right\}$$

$$= \frac{\hbar^2}{m}\left[\frac{30}{(a^5 b^5)^{1/2}}\right][y(b - y) + x(a - x)]$$

Now use the last expression to evaluate $\langle E \rangle$:

$$\langle E \rangle = \int_0^b dy \int_0^a dx \, \psi^*(x, y)\hat{H}\psi(x, y)$$

$$= \frac{\hbar^2}{m}\left(\frac{900}{a^5 b^5}\right)\int_0^b dy \int_0^a dx \, x(a - x)y(b - y)[y(b - y) + x(a - x)]$$

$$= \frac{\hbar^2}{m}\left(\frac{900}{a^5 b^5}\right)\left[\int_0^b y^2 (b - y)^2 \, dy \int_0^a x (a - x) \, dx\right.$$

$$\left. + \int_0^b y (b - y) \, dy \int_0^a x^2 (a - x)^2 \, dx\right]$$

Two of these integrals were evaluated earlier in this problem:

$$\int_0^a x^2 (a - x)^2 \, dx = \frac{a^5}{30}$$

$$\int_0^b y^2 (b - y)^2 \, dy = \frac{b^5}{30}$$

The other two integrals in the $\langle E \rangle$ expression are

$$\int_0^a x (a - x) \, dx = \frac{a^3}{6}$$

$$\int_0^b y (b - y) \, dy = \frac{b^3}{6}$$

Combining these four integrals,

$$\langle E \rangle = \frac{\hbar^2}{m} \left(\frac{900}{a^5 b^5} \right) \left(\frac{a^3 b^5}{180} + \frac{a^5 b^3}{180} \right)$$

$$= \frac{5\hbar^2}{m} \left(\frac{1}{a^2} + \frac{1}{b^2} \right)$$

4–11. Evaluate the commutator $[\hat{A}, \hat{B}]$, where $\hat{A}$ and $\hat{B}$ are given below.

	$\hat{A}$	$\hat{B}$
(a)	$\dfrac{d^2}{dx^2}$	x
(b)	$\dfrac{d}{dx} - x$	$\dfrac{d}{dx} + x$
(c)	$\displaystyle\int_0^x dx$	$\dfrac{d}{dx}$
(d)	$\dfrac{d^2}{dx^2} - x$	$\dfrac{d}{dx} + x^2$

(a)

$$\hat{A}\hat{B}f = \frac{d^2}{dx^2}(xf)$$

$$= \frac{d}{dx}\left(f + x\frac{df}{dx} \right)$$

$$= \frac{df}{dx} + \frac{df}{dx} + x\frac{d^2 f}{dx^2}$$

$$= 2\frac{df}{dx} + x\frac{d^2 f}{dx^2}$$

$$\hat{B}\hat{A}f = x\frac{d^2 f}{dx^2}$$

Therefore,

$$\hat{A}\hat{B}f - \hat{B}\hat{A}f = 2\frac{df}{dx}$$

$$[\hat{A}, \hat{B}] = 2\frac{d}{dx}$$

(b)

$$\hat{A}\hat{B}f = \left(\frac{d}{dx} - x \right)\left(\frac{df}{dx} + xf \right)$$

$$= \frac{d^2 f}{dx^2} + f + x\frac{df}{dx} - x\frac{df}{dx} - x^2 f$$

$$= \frac{d^2 f}{dx^2} + f - x^2 f$$

$$\hat{B}\hat{A}f = \left(\frac{d}{dx} + x\right)\left(\frac{df}{dx} - xf\right)$$

$$= \frac{d^2 f}{dx^2} - f - x\frac{df}{dx} + x\frac{df}{dx} - x^2 f$$

$$= \frac{d^2 f}{dx^2} - f - x^2 f$$

Therefore,

$$\hat{A}\hat{B}f - \hat{B}\hat{A}f = 2f$$

$$[\hat{A}, \hat{B}] = 2$$

(c)

$$\hat{A}\hat{B}f = \int_0^x dx' \frac{df}{dx'}$$

$$= f(x) - f(0)$$

$$\hat{B}\hat{A}f = \frac{d}{dx}\int_0^x dx' f(x')$$

$$= f(x)$$

Therefore,

$$\hat{A}\hat{B}f - \hat{B}\hat{A}f = -f(0)$$

$$[\hat{A}, \hat{B}]f = -f(0)$$

(d)

$$\hat{A}\hat{B}f = \left(\frac{d^2}{dx^2} - x\right)\left(\frac{df}{dx} + x^2 f\right)$$

$$= \frac{d^2}{dx^2}\left(\frac{df}{dx} + x^2 f\right) - x\frac{df}{dx} - x^3 f$$

$$= \frac{d}{dx}\left(\frac{d^2 f}{dx^2} + 2xf + x^2\frac{df}{dx}\right) - x\frac{df}{dx} - x^3 f$$

$$= \frac{d^3 f}{dx^3} + 2f + 2x\frac{df}{dx} + 2x\frac{df}{dx} + x^2\frac{d^2 f}{dx^2} - x\frac{df}{dx} - x^3 f$$

$$= \frac{d^3 f}{dx^3} + x^2\frac{d^2 f}{dx^2} + 3x\frac{df}{dx} - x^3 f + 2f$$

$$\hat{B}\hat{A}f = \left(\frac{d}{dx} + x^2\right)\left(\frac{d^2 f}{dx^2} - xf\right)$$

$$= \frac{d^3 f}{dx^3} - f - x\frac{df}{dx} + x^2\frac{d^2 f}{dx^2} - x^3 f$$

Therefore,

$$\hat{A}\hat{B}f - \hat{B}\hat{A}f = 4x\frac{df}{dx} + 3f$$

$$[\hat{A}, \hat{B}] = 4x\frac{d}{dx} + 3$$

4–12. Referring to Table 4.1 for the operator expressions for angular momentum, show that

$$[\hat{L}_x, \hat{L}_y] = i\hbar\hat{L}_z$$

$$[\hat{L}_y, \hat{L}_z] = i\hbar\hat{L}_x$$

and

$$[\hat{L}_z, \hat{L}_x] = i\hbar\hat{L}_y$$

(Do you see a pattern here to help remember these commutation relations?) What do these expressions say about the ability to measure the components of angular momentum simultaneously?

$$\hat{L}_x\hat{L}_yf = \left[-i\hbar\left(y\frac{\partial}{\partial z} - z\frac{\partial}{\partial y}\right)\right]\left[-i\hbar\left(z\frac{\partial f}{\partial x} - x\frac{\partial f}{\partial z}\right)\right]$$

$$= -\hbar^2\left(y\frac{\partial f}{\partial x} + yz\frac{\partial^2 f}{\partial x\partial z} - xy\frac{\partial^2 f}{\partial z^2} - z^2\frac{\partial^2 f}{\partial x\partial y} + xz\frac{\partial^2 f}{\partial y\partial z}\right)$$

$$\hat{L}_y\hat{L}_xf = \left[-i\hbar\left(z\frac{\partial}{\partial x} - x\frac{\partial}{\partial z}\right)\right]\left[-i\hbar\left(y\frac{\partial f}{\partial z} - z\frac{\partial f}{\partial y}\right)\right]$$

$$= -\hbar^2\left(yz\frac{\partial^2 f}{\partial x\partial z} - z^2\frac{\partial^2 f}{\partial x\partial y} - xy\frac{\partial^2 f}{\partial z^2} + x\frac{\partial f}{\partial y} + xz\frac{\partial^2 f}{\partial y\partial z}\right)$$

Therefore,

$$\hat{L}_x\hat{L}_yf - \hat{L}_y\hat{L}_xf = -\hbar^2\left(y\frac{\partial f}{\partial x} - x\frac{\partial f}{\partial y}\right)$$

$$= -\hbar^2\left(\frac{\hat{L}_z}{i\hbar}\right)$$

$$= i\hbar\hat{L}_zf$$

$$\left[\hat{L}_x, \hat{L}_y\right] = i\hbar\hat{L}_z$$

The other two commutators are evaluated in a similar manner.

$$\hat{L}_y\hat{L}_z f = \left[-i\hbar\left(z\frac{\partial}{\partial x} - x\frac{\partial}{\partial z}\right)\right]\left[-i\hbar\left(x\frac{\partial f}{\partial y} - y\frac{\partial f}{\partial x}\right)\right]$$

$$= -\hbar^2\left(z\frac{\partial f}{\partial y} + xz\frac{\partial^2 f}{\partial x\partial y} - yz\frac{\partial^2 f}{\partial x^2} - x^2\frac{\partial^2 f}{\partial y\partial z} + xy\frac{\partial^2 f}{\partial x\partial z}\right)$$

$$\hat{L}_z\hat{L}_y f = \left[-i\hbar\left(x\frac{\partial}{\partial y} - y\frac{\partial}{\partial x}\right)\right]\left[-i\hbar\left(z\frac{\partial f}{\partial x} - x\frac{\partial f}{\partial z}\right)\right]$$

$$= -\hbar^2\left(xz\frac{\partial^2 f}{\partial x\partial y} - x^2\frac{\partial^2 f}{\partial y\partial z} - yz\frac{\partial^2 f}{\partial x^2} + y\frac{\partial f}{\partial z} + xy\frac{\partial^2 f}{\partial x\partial z}\right)$$

$$\left[\hat{L}_y, \hat{L}_z\right] f = -\hbar^2\left(z\frac{\partial f}{\partial y} - y\frac{\partial f}{\partial z}\right)$$

$$\left[\hat{L}_y, \hat{L}_z\right] = i\hbar\hat{L}_x$$

and

$$\hat{L}_z\hat{L}_x f = \left[-i\hbar\left(x\frac{\partial}{\partial y} - y\frac{\partial}{\partial x}\right)\right]\left[-i\hbar\left(y\frac{\partial f}{\partial z} - z\frac{\partial f}{\partial y}\right)\right]$$

$$= -\hbar^2\left(x\frac{\partial f}{\partial z} + xy\frac{\partial^2 f}{\partial y\partial z} - xz\frac{\partial^2 f}{\partial y^2} - y^2\frac{\partial^2 f}{\partial x\partial z} + yz\frac{\partial^2 f}{\partial x\partial y}\right)$$

$$\hat{L}_x\hat{L}_z f = \left[-i\hbar\left(y\frac{\partial}{\partial z} - z\frac{\partial}{\partial y}\right)\right]\left[-i\hbar\left(x\frac{\partial f}{\partial y} - y\frac{\partial f}{\partial x}\right)\right]$$

$$= -\hbar^2\left(xy\frac{\partial^2 f}{\partial y\partial z} - y^2\frac{\partial^2 f}{\partial x\partial z} - xz\frac{\partial^2 f}{\partial y^2} + z\frac{\partial f}{\partial x} + yz\frac{\partial^2 f}{\partial x\partial y}\right)$$

$$\left[\hat{L}_z, \hat{L}_x\right] f = -\hbar^2\left(x\frac{\partial f}{\partial z} - z\frac{\partial f}{\partial x}\right)$$

$$\left[\hat{L}_z, \hat{L}_x\right] = i\hbar\hat{L}_y$$

These commutation relations have a pattern that involves cyclic permutation of x, y, and z. Since no two operators for the components of angular momentum commute, their corresponding observable quantities do not have simultaneously well defined values. In other words, the components cannot be measured simultaneously to arbitrary precision.

4–13. Defining

$$\hat{L}^2 = \hat{L}_x^2 + \hat{L}_y^2 + \hat{L}_z^2$$

show that $\hat{L}^2$ commutes with each component separately. What does this result tell you about the ability to measure the square of the total angular momentum and its components simultaneously?

First consider the commutator $\left[\hat{L}^2, \hat{L}_x\right]$:

$$\left[\hat{L}^2, \hat{L}_x\right] = \left[\hat{L}_x^2 + \hat{L}_y^2 + \hat{L}_z^2, \hat{L}_x\right]$$

$$= \left[\hat{L}_x^2, \hat{L}_x\right] + \left[\hat{L}_y^2, \hat{L}_x\right] + \left[\hat{L}_z^2, \hat{L}_x\right]$$

Since $\hat{L}_x^2$ commutes with $\hat{L}_x$, the commutator $\left[\hat{L}_x^2, \hat{L}_x\right]$ is zero. Using, for example, $i\hbar\hat{L}_z = \hat{L}_x\hat{L}_y - \hat{L}_y\hat{L}_x = \left[\hat{L}_x, \hat{L}_y\right]$, the other two commutators in the last expression can be written as

$$\left[\hat{L}_y^2, \hat{L}_x\right] = \hat{L}_y^2\hat{L}_x - \hat{L}_x\hat{L}_y\hat{L}_y$$

$$= \hat{L}_y^2\hat{L}_x - \left(\hat{L}_y\hat{L}_x + i\hbar\hat{L}_z\right)\hat{L}_y$$

$$= \hat{L}_y\left(\hat{L}_y\hat{L}_x - \hat{L}_x\hat{L}_y\right) - i\hbar\hat{L}_z\hat{L}_y$$

$$= -i\hbar\hat{L}_y\hat{L}_z - i\hbar\hat{L}_z\hat{L}_y$$

$$\left[\hat{L}_z^2, \hat{L}_x\right] = \hat{L}_z^2\hat{L}_x - \hat{L}_x\hat{L}_z\hat{L}_z$$

$$= \hat{L}_z^2\hat{L}_x - \left(\hat{L}_z\hat{L}_x - i\hbar\hat{L}_y\right)\hat{L}_z$$

$$= \hat{L}_z\left(\hat{L}_z\hat{L}_x - \hat{L}_x\hat{L}_z\right) + i\hbar\hat{L}_y\hat{L}_z$$

$$= i\hbar\hat{L}_z\hat{L}_y + i\hbar\hat{L}_y\hat{L}_z$$

where the commutation relations in Problem 4–12 have been applied. Combining the three comutation relations between the square of the operator of each angular momentum component and $\hat{L}_x$,

$$\left[\hat{L}^2, \hat{L}_x\right] = 0$$

Similarly, $\left[\hat{L}^2, \hat{L}_y\right] = 0$ because

$$\left[\hat{L}_y^2, \hat{L}_y\right] = 0$$

$$\left[\hat{L}_x^2, \hat{L}_y\right] = -\left[\hat{L}_z^2, \hat{L}_y\right]$$

and $\left[\hat{L}^2, \hat{L}_z\right] = 0$ because

$$\left[\hat{L}_z^2, \hat{L}_z\right] = 0$$

$$\left[\hat{L}_x^2, \hat{L}_z\right] = -\left[\hat{L}_y^2, \hat{L}_z\right]$$

Since $\hat{L}^2$ commutes with the operator of each angular momentum component, the corresponding observable quantities for these operators have simultaneously well-defined values. It is therefore possible to measure the square of angular momentum and one of its components simultaneously to arbitrary precision.

4–14. The operators

$$\hat{L}_+ = \hat{L}_x + i\hat{L}_y \qquad \text{and} \qquad \hat{L}_- = \hat{L}_x - i\hat{L}_y$$

play a central role in the quantum-mechanical theory of angular momentum. (See the Appendix to Chapter 6.) Show that

$$\hat{L}_+\hat{L}_- = \hat{L}^2 - \hat{L}_z^2 + \hbar\hat{L}_z$$

$$[\hat{L}_z, \hat{L}_+] = \hbar\hat{L}_+$$

and that

$$[\hat{L}_z, \hat{L}_-] = -\hbar\hat{L}_-$$

$$\hat{L}_+\hat{L}_- = \left(\hat{L}_x + i\hat{L}_y\right)\left(\hat{L}_x - i\hat{L}_y\right)$$

$$= \hat{L}_x^2 - i\hat{L}_x\hat{L}_y + i\hat{L}_y\hat{L}_x + \hat{L}_y^2$$

$$= \hat{L}_x^2 + \hat{L}_y^2 - i\left(\hat{L}_x\hat{L}_y - \hat{L}_y\hat{L}_x\right)$$

$$= \hat{L}^2 - \hat{L}_z^2 - i\left(i\hbar\hat{L}_z\right)$$

$$= \hat{L}^2 - \hat{L}_z^2 + \hbar\hat{L}_z$$

where the relations $\hat{L}_x^2 + \hat{L}_y^2 = \hat{L}^2 - \hat{L}_z^2$ and $[\hat{L}_x, \hat{L}_y] = i\hbar\hat{L}_z$ have been applied.

The two commutation relations $\left[\hat{L}_z, \hat{L}_+\right]$ and $\left[\hat{L}_z, \hat{L}_-\right]$ can be obtained readily from the relations given in Problem 4–12:

$$\left[\hat{L}_z, \hat{L}_+\right] = \left[\hat{L}_z, \hat{L}_x + i\hat{L}_y\right]$$

$$= \hat{L}_z\hat{L}_x + i\hat{L}_z\hat{L}_y - \hat{L}_x\hat{L}_z - i\hat{L}_y\hat{L}_z$$

$$= \left[\hat{L}_z, \hat{L}_x\right] + i\left[\hat{L}_z, \hat{L}_y\right]$$

$$= i\hbar\hat{L}_y + i\left(-i\hbar\hat{L}_x\right)$$

$$= \hbar\hat{L}_+$$

$$\left[\hat{L}_z, \hat{L}_-\right] = \left[\hat{L}_z, \hat{L}_x - i\hat{L}_y\right]$$

$$= \hat{L}_z\hat{L}_x - i\hat{L}_z\hat{L}_y - \hat{L}_x\hat{L}_z + i\hat{L}_y\hat{L}_z$$

$$= \left[\hat{L}_z, \hat{L}_x\right] - i\left[\hat{L}_z, \hat{L}_y\right]$$

$$= i\hbar\hat{L}_y - i\left(-i\hbar\hat{L}_x\right)$$

$$= -\hbar\hat{L}_-$$

4–15. Consider a particle in a two-dimensional box. Determine $[\hat{X}, \hat{P}_y]$, $[\hat{X}, \hat{P}_x]$, $[\hat{Y}, \hat{P}_y]$, and $[\hat{Y}, \hat{P}_x]$.

As can be shown using any well behaved functions, the following relations apply to any system:

$$[\hat{X}, \hat{P}_y] = 0$$

$$[\hat{X}, \hat{P}_x] = i\hbar$$

$$[\hat{Y}, \hat{P}_y] = i\hbar$$

$$[\hat{Y}, \hat{P}_x] = 0$$

Here, the relations are shown explicitly using the wave functions for a particle in a two-dimensional box:

(a)

$$\hat{X}\hat{P}_y \psi_{n_x,n_y} = x\left(-i\hbar \frac{\partial}{\partial y}\right)\left[\left(\frac{4}{ab}\right)^{1/2} \sin\frac{n_x\pi x}{a} \sin\frac{n_y\pi y}{b}\right]$$

$$= -i\hbar x \left(\frac{4}{ab}\right)^{1/2} \frac{n_y\pi}{b} \sin\frac{n_x\pi x}{a} \cos\frac{n_y\pi y}{b}$$

$$\hat{P}_y\hat{X} \psi_{n_x,n_y} = \left(-i\hbar \frac{\partial}{\partial y}\right)\left[x\left(\frac{4}{ab}\right)^{1/2} \sin\frac{n_x\pi x}{a} \sin\frac{n_y\pi y}{b}\right]$$

$$= -i\hbar x \left(\frac{4}{ab}\right)^{1/2} \frac{n_y\pi}{b} \sin\frac{n_x\pi x}{a} \cos\frac{n_y\pi y}{b}$$

$$\hat{X}\hat{P}_y \psi_{n_x,n_y} - \hat{P}_y\hat{X} \psi_{n_x,n_y} = 0$$

$$[\hat{X}, \hat{P}_y] = 0$$

(b)

$$\hat{X}\hat{P}_x \psi_{n_x,n_y} = x\left(-i\hbar \frac{\partial}{\partial x}\right)\left[\left(\frac{4}{ab}\right)^{1/2} \sin\frac{n_x\pi x}{a} \sin\frac{n_y\pi y}{b}\right]$$

$$= -i\hbar x \left(\frac{4}{ab}\right)^{1/2} \frac{n_x\pi}{a} \cos\frac{n_x\pi x}{a} \sin\frac{n_y\pi y}{b}$$

$$\hat{P}_x\hat{X} \psi_{n_x,n_y} = \left(-i\hbar \frac{\partial}{\partial x}\right)\left[x\left(\frac{4}{ab}\right)^{1/2} \sin\frac{n_x\pi x}{a} \sin\frac{n_y\pi y}{b}\right]$$

$$= -i\hbar \left(\frac{4}{ab}\right)^{1/2}\left(\sin\frac{n_x\pi x}{a} \sin\frac{n_y\pi y}{b} + x\frac{n_x\pi}{a} \cos\frac{n_x\pi x}{a} \sin\frac{n_y\pi y}{b}\right)$$

$$\hat{X}\hat{P}_x \psi_{n_x,n_y} - \hat{P}_x\hat{X} \psi_{n_x,n_y} = i\hbar\psi_{n_x,n_y}$$

$$[\hat{X}, \hat{P}_x] = i\hbar$$

(c)

$$\hat{Y}\hat{P}_y \psi_{n_x,n_y} = y\left(-i\hbar\frac{\partial}{\partial y}\right)\left[\left(\frac{4}{ab}\right)^{1/2}\sin\frac{n_x\pi x}{a}\sin\frac{n_y\pi y}{b}\right]$$

$$= -i\hbar y\left(\frac{4}{ab}\right)^{1/2}\frac{n_y\pi}{b}\sin\frac{n_x\pi x}{a}\cos\frac{n_y\pi y}{b}$$

$$\hat{P}_y\hat{Y}\psi_{n_x,n_y} = \left(-i\hbar\frac{\partial}{\partial y}\right)\left[y\left(\frac{4}{ab}\right)^{1/2}\sin\frac{n_x\pi x}{a}\sin\frac{n_y\pi y}{b}\right]$$

$$= -i\hbar\left(\frac{4}{ab}\right)^{1/2}\left(\sin\frac{n_x\pi x}{a}\sin\frac{n_y\pi y}{b} + y\frac{n_y\pi}{b}\sin\frac{n_x\pi x}{a}\cos\frac{n_y\pi y}{b}\right)$$

$$\hat{Y}\hat{P}_y\psi_{n_x,n_y} - \hat{P}_y\hat{Y}\psi_{n_x,n_y} = i\hbar\psi_{n_x,n_y}$$

$$[\hat{Y},\hat{P}_y] = i\hbar$$

(d)

$$\hat{Y}\hat{P}_x\psi_{n_x,n_y} = y\left(-i\hbar\frac{\partial}{\partial x}\right)\left[\left(\frac{4}{ab}\right)^{1/2}\sin\frac{n_x\pi x}{a}\sin\frac{n_y\pi y}{b}\right]$$

$$= -i\hbar y\left(\frac{4}{ab}\right)^{1/2}\frac{n_x\pi}{a}\cos\frac{n_x\pi x}{a}\sin\frac{n_y\pi y}{b}$$

$$\hat{P}_x\hat{Y}\psi_{n_x,n_y} = \left(-i\hbar\frac{\partial}{\partial x}\right)\left[y\left(\frac{4}{ab}\right)^{1/2}\sin\frac{n_x\pi x}{a}\sin\frac{n_y\pi y}{b}\right]$$

$$= -i\hbar y\left(\frac{4}{ab}\right)^{1/2}\frac{n_x\pi}{a}\cos\frac{n_x\pi x}{a}\sin\frac{n_y\pi y}{b}$$

$$\hat{Y}\hat{P}_x\psi_{n_x,n_y} - \hat{P}_x\hat{Y}\psi_{n_x,n_y} = 0$$

$$[\hat{Y},\hat{P}_x] = 0$$

4–16. Can the position and kinetic energy of an electron be measured simultaneously to arbitrary precision? (See Problem 4–42.)

Consider first whether the kinetic energy and the x position of an electron can be measured simultaneously to arbitrary precision.

$$\left[\hat{T},x\right] = \left[\hat{T}_x,x\right] + \left[\hat{T}_y,x\right] + \left[\hat{T}_z,x\right]$$

The commutator $\left[\hat{T}_x,x\right]$ can be readily evaluated using the results from Problem 4–11a:

$$\left[\hat{T}_x, x\right] = -\frac{\hbar^2}{2m}\left[\frac{\partial^2}{\partial x^2}, x\right]$$

$$= -\frac{\hbar^2}{m}\frac{\partial}{\partial x} \neq 0$$

The commutators $\left[\hat{T}_y, x\right]$ and $\left[\hat{T}_z, x\right]$ are both zero since $\hat{T}_y$ and $\hat{T}_z$ contain partial derivatives that do not involve x. Combining the three commutators,

$$\left[\hat{T}, x\right] \neq 0$$

Similarly, $\left[\hat{T}, y\right] \neq 0$ and $\left[\hat{T}, z\right] \neq 0$. Therefore, it is not possible to measure the position and kinetic energy of any electron simultaneously to arbitrary precision.

4–17. Using the result of Problem 4–15, what are the "uncertainty relationships" $\Delta x \Delta p_y$ and $\Delta y \Delta p_x$ equal to?

Since $\hat{X}$ and $\hat{P}_y$ commute, $\Delta x \Delta p_y = 0$. Thus, the x coordinate and the y component of momentum can be measured simultaneously to arbitrary precision. Similarly, because $\hat{Y}$ and $\hat{P}_x$ commute, $\Delta y \Delta p_x = 0$ and the y coordinate and the x component of momentum can also be measured simultaneously to arbitrary precision.

4–18. Which of the following operators is Hermitian: d/dx, id/dx, d^2/dx^2, id^2/dx^2, xd/dx, and x? Assume that the functions on which these operators operate are appropriately well behaved at infinity.

An operator $\hat{A}$ is Hermitian if $\int_{-\infty}^{\infty} f^*\hat{A}g \, dx = \int_{-\infty}^{\infty} g\hat{A}^*f^* \, dx$. That the functions f and g are well-behaved means that their values at $x = \pm\infty$ are 0 and that their first derivatives at $x = \pm\infty$ are also 0.

(a) For $\hat{A} = d/dx$,

$$\int_{-\infty}^{\infty} f^*\hat{A}g \, dx = \int_{-\infty}^{\infty} f^*\frac{dg}{dx} \, dx$$

$$= f^*g\Big|_{-\infty}^{\infty} - \int_{-\infty}^{\infty} \frac{df^*}{dx}g \, dx$$

$$= -\int_{-\infty}^{\infty} g\hat{A}f^* \, dx$$

$$= -\int_{-\infty}^{\infty} g\hat{A}^*f^* \, dx$$

$$\neq \int_{-\infty}^{\infty} g\hat{A}^*f^* \, dx$$

The operator d/dx is not Hermitian.

(b) For $\hat{A} = i\,(d/dx)$,

$$\int_{-\infty}^{\infty} f^*\hat{A}g\,dx = i\int_{-\infty}^{\infty} f^*\frac{dg}{dx}\,dx$$

$$= i\,f^*g\Big|_{-\infty}^{\infty} - i\int_{-\infty}^{\infty}\frac{df^*}{dx}g\,dx$$

$$= \int_{-\infty}^{\infty} g\left(-i\frac{df^*}{dx}\right)dx$$

$$= \int_{-\infty}^{\infty} g\hat{A}^*f^*\,dx$$

The operator $i\,(d/dx)$ is Hermitian.

(c) For $\hat{A} = d^2/dx^2$,

$$\int_{-\infty}^{\infty} f^*\hat{A}g\,dx = \int_{-\infty}^{\infty} f^*\frac{d^2g}{dx^2}\,dx$$

$$= f^*\frac{dg}{dx}\Big|_{-\infty}^{\infty} - \int_{-\infty}^{\infty}\frac{df^*}{dx}\frac{dg}{dx}\,dx$$

$$= -\int_{-\infty}^{\infty}\frac{df^*}{dx}\frac{dg}{dx}\,dx$$

$$= -\left(\frac{df^*}{dx}g\Big|_{-\infty}^{\infty} - \int_{-\infty}^{\infty}\frac{d^2f^*}{dx}g\,dx\right)$$

$$= \int_{-\infty}^{\infty} g\hat{A}^*f^*\,dx$$

The operator d^2/dx^2 is Hermitian.

(d) For $\hat{A} = i\,(d^2/dx^2)$,

$$\int_{-\infty}^{\infty} f^*\hat{A}g\,dx = i\int_{-\infty}^{\infty} f^*\frac{d^2g}{dx^2}\,dx$$

$$= i\,f^*\frac{dg}{dx}\Big|_{-\infty}^{\infty} - i\int_{-\infty}^{\infty}\frac{df^*}{dx}\frac{dg}{dx}\,dx$$

$$= -i\int_{-\infty}^{\infty}\frac{df^*}{dx}\frac{dg}{dx}\,dx$$

$$= -i\left(\frac{df^*}{dx}g\Big|_{-\infty}^{\infty} - \int_{-\infty}^{\infty}\frac{d^2f^*}{dx}g\,dx\right)$$

$$= \int_{-\infty}^{\infty} g\left(i\frac{d^2f^*}{dx}\right)dx$$

$$\neq \int_{-\infty}^{\infty} g\hat{A}^*f^*\,dx$$

The operator $i\,(d^2/dx^2)$ is not Hermitian.

(e) For $\hat{A} = x \, (d/dx)$,

$$\int_{-\infty}^{\infty} f^* \hat{A} g \, dx = \int_{-\infty}^{\infty} f^* x \frac{dg}{dx} \, dx$$

$$= f^* x g \Big|_{-\infty}^{\infty} - \int_{-\infty}^{\infty} \left(x \frac{df^*}{dx} + f^* \right) g \, dx$$

$$= - \int_{-\infty}^{\infty} x \frac{df^*}{dx} g \, dx - \int_{-\infty}^{\infty} f^* g \, dx$$

$$= - \int_{-\infty}^{\infty} g \hat{A}^* f^* \, dx - \int_{-\infty}^{\infty} f^* g \, dx$$

$$\neq \int_{-\infty}^{\infty} g \hat{A}^* f^* \, dx$$

The operator $x \, (d/dx)$ is not Hermitian. Note that since f and g are well-behaved functions, as x approaches $\pm\infty$, they approach 0 faster than x diverges. Thus, $f^* x g \big|_{-\infty}^{\infty} = 0$.

(f) For $\hat{A} = x$,

$$\int_{-\infty}^{\infty} f^* \hat{A} g \, dx = \int_{-\infty}^{\infty} f^* x g \, dx$$

$$= \int_{-\infty}^{\infty} g x f^* \, dx$$

$$= \int_{-\infty}^{\infty} g \hat{A}^* f^* \, dx$$

The operator x is Hermitian.

4–19. Show that if $\hat{A}$ is Hermitian, then $\hat{A} - \langle a \rangle$ is Hermitian. Show that the sum of two Hermitian operators is Hermitian.

To show that $\hat{A} - \langle a \rangle$ is Hermitian, show that the two expressions $\int_{-\infty}^{\infty} f^* \left(\hat{A} - \langle a \rangle \right) g \, dx$ and $\int_{-\infty}^{\infty} g \left(\hat{A} - \langle a \rangle \right)^* f^* \, dx$ are equal.

$$\int_{-\infty}^{\infty} f^* \left(\hat{A} - \langle a \rangle \right) g \, dx = \int_{-\infty}^{\infty} f^* \hat{A} g \, dx - \int_{-\infty}^{\infty} f^* \langle a \rangle g \, dx$$

Since $\hat{A}$ is Hermitian, it follows that

$$\int_{-\infty}^{\infty} f^* \hat{A} g \, dx = \int_{-\infty}^{\infty} g \hat{A}^* f^* \, dx$$

Furthermore, $\langle a \rangle$ is real; thus, $\langle a \rangle^* = \langle a \rangle$. Finally, the order of multiplication among functions is irrelevant. Thus,

$$\int_{-\infty}^{\infty} f^* \left(\hat{A} - \langle a \rangle \right) g \, dx = \int_{-\infty}^{\infty} g \hat{A}^* f^* \, dx - \int_{-\infty}^{\infty} g \langle a \rangle^* f^* \, dx$$

$$= \int_{-\infty}^{\infty} g \left(\hat{A} - \langle a \rangle \right)^* f^* \, dx$$

Indeed, $\hat{A} - \langle a \rangle$ is Hermitian.

If $\hat{B}$ is also Hermitian, then

$$\int_{-\infty}^{\infty} f^* \hat{B} g \, dx = \int_{-\infty}^{\infty} g \hat{B}^* f^* \, dx$$

It follows that

$$\int_{-\infty}^{\infty} f^* \left(\hat{A} + \hat{B} \right) g \, dx = \int_{-\infty}^{\infty} f^* \hat{A}^* g \, dx + \int_{-\infty}^{\infty} f^* \hat{B}^* g^* \, dx$$

$$= \int_{-\infty}^{\infty} g \hat{A}^* f^* \, dx + \int_{-\infty}^{\infty} g \hat{B}^* f^* \, dx$$

$$= \int_{-\infty}^{\infty} g \left(\hat{A} + \hat{B} \right)^* f^* \, dx$$

Thus, the sum of two Hermitian operators is Hermitian.

4–20. To prove that Equation 4.30 follows from Equation 4.29, first write Equation 4.30 with f and with g:

$$\int f^* \hat{A} f \, dx = \int f \hat{A}^* f^* \, dx \qquad \text{and} \qquad \int g^* \hat{A} g \, dx = \int g \hat{A}^* g^* \, dx$$

Now let $\psi = c_1 f + c_2 g$, where c_1 and c_2 are arbitrary complex constants, to write

$$\int (c_1^* f^* + c_2^* g^*) \hat{A} \, (c_1 f + c_2 g) \, dx = \int (c_1 f + c_2 g) \hat{A}^* \, (c_1^* f^* + c_2^* g^*) \, dx$$

If we expand both sides and use the first two equations, we find that

$$c_1^* c_2 \int f^* \hat{A} g \, dx + c_2^* c_1 \int g^* \hat{A} f \, dx = c_1 c_2^* \int f \hat{A}^* g^* \, dx + c_1^* c_2 \int g \hat{A}^* f^* \, dx$$

Rearrange this into

$$c_1^* c_2 \int (f^* \hat{A} g - g \hat{A}^* f^*) \, dx = c_1 c_2^* \int (f \hat{A}^* g^* - g^* \hat{A} f) \, dx$$

Notice that the two sides of this equation are complex conjugates of each other. If $z = x + iy$ and $z = z^*$, then show that this implies that z is real. Thus, both sides of this equation are real. But because c_1 and c_2 are arbitrary complex constants, the only way for both sides to be real is for both integrals to equal zero. Show that this implies Equation 4.30.

Let $\psi = c_1 f + c_2 g$. Since $\hat{A}$ is Hermitian,

$$\int (c_1^* f^* + c_2^* g^*) \hat{A} \, (c_1 f + c_2 g) \, dx = \int (c_1 f + c_2 g) \hat{A}^* \, (c_1^* f^* + c_2^* g^*) \, dx$$

Expanding both sides of this last equation,

$$c_1^* c_1 \int f^* \hat{A} f \, dx + c_1^* c_2 \int f^* \hat{A} g \, dx + c_2^* c_1 \int g^* \hat{A} f \, dx + c_2^* c_2 \int g^* \hat{A} g \, dx$$

$$= c_1 c_1^* \int f \hat{A}^* f^* \, dx + c_1 c_2^* \int f \hat{A}^* g^* \, dx + c_2 c_1^* \int g \hat{A}^* f^* \, dx + c_2 c_2^* \int g \hat{A}^* g^* \, dx$$

Simplifying using $\int f^* \hat{A} f \, dx = \int f \hat{A}^* f^* \, dx$ and $\int g^* \hat{A} g \, dx = \int g \hat{A}^* g^* \, dx$, the above expression becomes

$$c_1^* c_2 \int f^* \hat{A} g \, dx + c_2^* c_1 \int g^* \hat{A} f \, dx = c_1 c_2^* \int f \hat{A}^* g^* \, dx + c_2 c_1^* \int g \hat{A}^* f^* \, dx$$

$$c_1^* c_2 \int (f^* \hat{A} g - g \hat{A}^* f^*) \, dx = c_1 c_2^* \int (f \hat{A}^* g^* - g^* \hat{A} f) \, dx$$

The two sides are complex conjugates of each other. That is, if one side is denoted by $z = x + iy$, the above expression can be represented by $z = z^*$, or $x + iy = x - iy$. Thus, y must be zero. Additionally, because c_1 and c_2 are arbitrary constants, both integrals in the last expression must equal 0.

$$\int (f^* \hat{A} g - g \hat{A}^* f^*) \, dx = \int (f \hat{A}^* g^* - g^* \hat{A} f) \, dx = 0$$

$$\int f^* \hat{A} g \, dx = \int g \hat{A}^* f^* \, dx \text{ or equivalently } \int f \hat{A}^* g^* \, dx = \int g^* \hat{A} f \, dx$$

which is equivalent to Equation 4.30.

4–21. Show that if $\hat{A}$ is Hermitian, then

$$\int \hat{A}^* \psi^* \hat{B} \psi \, dx = \int \psi^* \hat{A} \hat{B} \psi \, dx$$

Hint: Use Equation 4.30.

$\hat{A}^* \psi^*$ and $\hat{B} \psi$ are two functions, and their multiplication commutes:

$$\int \left(\hat{A}^* \psi^* \right) \left(\hat{B} \psi \right) \, dx = \int \left(\hat{B} \psi \right) \left(\hat{A}^* \psi^* \right) \, dx$$

Since $\hat{A}$ is Hermitian,

$$\int \left(\hat{B} \psi \right) \hat{A}^* \psi^* \, dx = \int \psi^* \hat{A} \left(\hat{B} \psi \right) \, dx$$

Thus, $\int \hat{A}^* \psi^* \hat{B} \psi \, dx = \int \psi^* \hat{A} \hat{B} \psi \, dx$.

4–22.   Show that

$$\psi_0(x) = \pi^{-1/4}e^{-x^2/2}$$

$$\psi_1(x) = (4/\pi)^{1/4}xe^{-x^2/2}$$

$$\psi_2(x) = (4\pi)^{-1/4}(2x^2-1)e^{-x^2/2}$$

are orthonormal over the interval $-\infty < x < \infty$.

First show that the functions are normalized, that is, $\int_{-\infty}^{\infty}\psi_n^*\psi_n\,dx = 1$:

$$\int_{-\infty}^{\infty}\psi_0^*\psi_0\,dx = \frac{1}{\pi^{1/2}}\int_{-\infty}^{\infty}e^{-x^2}\,dx$$

$$= \frac{2}{\pi^{1/2}}\int_{0}^{\infty}e^{-x^2}\,dx$$

$$= 1$$

$$\int_{-\infty}^{\infty}\psi_1^*\psi_1\,dx = \left(\frac{4}{\pi}\right)^{1/2}\int_{-\infty}^{\infty}x^2e^{-x^2}\,dx$$

$$= \frac{4}{\pi^{1/2}}\int_{0}^{\infty}x^2e^{-x^2}\,dx$$

$$= 1$$

$$\int_{-\infty}^{\infty}\psi_2^*\psi_2\,dx = \left(\frac{1}{4\pi}\right)^{1/2}\int_{-\infty}^{\infty}\left(2x^2-1\right)^2e^{-x^2}\,dx$$

$$= \left(\frac{1}{4\pi}\right)^{1/2}\int_{0}^{\infty}\left(8x^4e^{-x^2} - 8x^2e^{-x^2} + 2e^{-x^2}\right)\,dx$$

$$= \left(\frac{1}{4\pi}\right)^{1/2}\left(8\frac{3\pi^{1/2}}{8} - 8\frac{\pi^{1/2}}{4} + 2\frac{\pi^{1/2}}{2}\right)$$

$$= 1$$

Now show that the functions are orthogonal, that is, for $m \neq n$, $\int_{-\infty}^{\infty}\psi_n^*\psi_m\,dx = 0$:

$$\int_{-\infty}^{\infty}\psi_0^*\psi_1\,dx = \frac{1}{\pi^{1/4}}\left(\frac{4}{\pi}\right)^{1/4}\int_{-\infty}^{\infty}xe^{-x^2}\,dx$$

$$= 0$$

$$\int_{-\infty}^{\infty}\psi_0^*\psi_2\,dx = \frac{1}{\pi^{1/4}}\left(\frac{1}{4\pi}\right)^{1/4}\int_{-\infty}^{\infty}\left(2x^2-1\right)e^{-x^2}\,dx$$

$$= \frac{1}{\pi^{1/4}}\left(\frac{1}{4\pi}\right)^{1/4}\int_{0}^{\infty}\left(4x^2e^{-x^2} - 2e^{-x^2}\right)\,dx$$

$$= \frac{1}{\pi^{1/4}}\left(\frac{1}{4\pi}\right)^{1/4}\left(4\frac{\pi^{1/2}}{4} - 2\frac{\pi^{1/2}}{2}\right)$$

$$= 0$$

$$\int_{-\infty}^{\infty} \psi_1^* \psi_2 \, dx = \left(\frac{4}{\pi}\right)^{1/4} \left(\frac{1}{4\pi}\right)^{1/4} \int_{-\infty}^{\infty} x \left(2x^2 - 1\right) e^{-x^2} \, dx$$

$$= \left(\frac{4}{\pi}\right)^{1/4} \left(\frac{1}{4\pi}\right)^{1/4} \int_{-\infty}^{\infty} \left(2x^3 - x\right) e^{-x^2} \, dx$$

$$= 0$$

The first and third integrals are simply evaluated as 0 because they involve odd functions integrated over an interval symmetric about $x = 0$.

4–23. Show that the polynomials

$$P_0(x) = 1, \qquad P_1(x) = x, \qquad P_2(x) = \frac{1}{2}(3x^2 - 1) \qquad \text{and} \qquad P_3(x) = \frac{1}{2}(5x^3 - 3x)$$

satisfy the orthogonality relation

$$\int_{-1}^{1} P_l(x) P_n(x) dx = \frac{2\delta_{ln}}{2l + 1} \qquad l = 0, 1, 2, 3$$

If the polynomials satisfy the orthogonality relation, then for $l \neq n$, $\int_{-1}^{1} P_l(x) P_n(x) \, dx = 0$ and for

$$l = n = 0, \qquad \int_{-1}^{1} P_l(x) P_n(x) \, dx = 2$$

$$l = n = 1, \qquad \int_{-1}^{1} P_l(x) P_n(x) \, dx = 2/3$$

$$l = n = 2, \qquad \int_{-1}^{1} P_l(x) P_n(x) \, dx = 2/5$$

$$l = n = 3, \qquad \int_{-1}^{1} P_l(x) P_n(x) \, dx = 2/7$$

Explicit integration shows that the above results are indeed obtained, although some simplification is possible by noting that P_n is an odd function when n is odd, and is an even function when n is even, and recalling that the integral of an odd function over an interval symmetric about $x = 0$ is 0. Thus, $\int_{-1}^{1} P_l(x) P_n(x) \, dx = 0$ when $l + n$ is odd. For $l \neq n$,

$$\int_{-1}^{1} P_0(x) P_1(x)\, dx = 0$$

$$\int_{-1}^{1} P_0(x) P_2(x)\, dx = \frac{1}{2} \int_{-1}^{1} \left(3x^2 - 1\right)\, dx$$

$$= \frac{1}{2} \left(x^3 - x\right)\Big|_{-1}^{1}$$

$$= 0$$

$$\int_{-1}^{1} P_0(x) P_3(x)\, dx = 0$$

$$\int_{-1}^{1} P_1(x) P_2(x)\, dx = 0$$

$$\int_{-1}^{1} P_1(x) P_3(x)\, dx = \frac{1}{2} \int_{-1}^{1} \left(5x^4 - 3x^2\right)\, dx$$

$$= \frac{1}{2} \left(x^5 - x^3\right)\Big|_{-1}^{1}$$

$$= 0$$

$$\int_{-1}^{1} P_2(x) P_3(x)\, dx = 0$$

For $l = n$,

$$\int_{-1}^{1} P_0(x) P_0(x)\, dx = \int_{-1}^{1} dx$$

$$= 2$$

$$\int_{-1}^{1} P_1(x) P_1(x)\, dx = \int_{-1}^{1} x^2\, dx$$

$$= \frac{x^3}{3}\Big|_{-1}^{1}$$

$$= \frac{2}{3}$$

$$\int_{-1}^{1} P_2(x) P_2(x)\, dx = \frac{1}{4} \int_{-1}^{1} \left(9x^4 - 6x^2 + 1\right)\, dx$$

$$= \frac{1}{4} \left(\frac{9x^5}{5} - \frac{6x^3}{3} + x\right)\Big|_{-1}^{1}$$

$$= \frac{2}{5}$$

$$\int_{-1}^{1} P_3(x) P_3(x)\, dx = \frac{1}{4} \int_{-1}^{1} \left(25x^6 - 30x^4 + 9x^2\right)\, dx$$

$$= \frac{1}{4} \left(\frac{25x^7}{7} - \frac{30x^5}{5} + \frac{9x^3}{3}\right)\Big|_{-1}^{1}$$

$$= \frac{2}{7}$$

4–24. Show that the set of functions $\{(2/a)^{1/2} \cos(n\pi x/a)\}$, $n = 0, 1, 2, \ldots$, is orthonormal over the interval $0 \leq x \leq a$.

The orthonormality relation can be readily obtained using the results from Problem 3–23.

$$\int_0^a \psi_n^* \psi_m \, dx = \frac{2}{a} \int_0^a \cos\frac{n\pi x}{a} \cos\frac{m\pi x}{a} \, dx$$

$$= \frac{2}{a}\left(\frac{a}{2}\right)\delta_{nm}$$

$$= \delta_{nm}$$

4–25. Generate an orthogonal set of polynomials $\{\phi_j(x), \ j = 1, 2, 3\}$ over the interval $-1 \leq x \leq 1$ starting with $f_0(x) = 1$, $f_1(x) = x$, and $f_2(x) = x^2$. Instead of normalizing the final result, choose a multiplicative constant such that $\phi_j(1) = 1$. Compare these polynomials to those in Problem 4–23.

Choose the functions as linear combinations of $f_0(x)$, $f_1(x)$, and $f_2(x)$:

$$\phi_1(x) = f_0(x) = 1$$
$$\phi_2(x) = f_1(x) + a_0 f_0(x) = x + a_0$$
$$\phi_3(x) = f_2(x) + b_1 f_1(x) + b_0 f_0(x) = x^2 + b_1 x + b_0$$

To determine $\phi_2(x)$, use the fact that $\phi_2(x)$ and $\phi_1(x)$ are orthogonal:

$$\int_{-1}^1 \phi_2^*(x)\phi_1(x) \, dx = 0 = \int_{-1}^1 \left(x + a_0\right) \, dx$$

$$= \left(\frac{x^2}{2} + a_0 x\right)\Bigg|_{-1}^1$$

$$= 2a_0$$

$$a_0 = 0$$

Thus, $\phi_2(x) = x$, and it satisfies the condition $\phi_2(1) = 1$.

To determine $\phi_3(x)$, use the fact that $\phi_3(x)$ is orthogonal to $\phi_1(x)$:

$$\int_{-1}^{1} \phi_3^*(x)\phi_1(x)\,dx = 0 = \int_{-1}^{1} \left(x^2 + b_1 x + b_0\right)\,dx$$

$$= \left(\frac{x^3}{3} + \frac{b_1 x^2}{2} + b_0 x\right)\Bigg|_{-1}^{1}$$

$$= \frac{2}{3} + 2b_0$$

$$b_0 = -\frac{1}{3}$$

and also is orthogonal to $\phi_2(x)$:

$$\int_{-1}^{1} \phi_3^*(x)\phi_2(x)\,dx = 0 = \int_{-1}^{1} \left(x^3 + b_1 x^2 + b_0 x\right)\,dx$$

$$= \left(\frac{x^4}{4} + \frac{b_1 x^3}{3} + \frac{b_0 x^2}{2}\right)\Bigg|_{-1}^{1}$$

$$= \frac{2b_1}{3}$$

$$b_1 = 0$$

Thus, $\phi_3(x) = x^2 - 1/3$. This expression has to be multiplied by 3/2 to satisfy the condition $\phi_3(1) = 1$. That is,

$$\phi_3(x) = \frac{1}{2}\left(3x^2 - 1\right)$$

These orthogonal polynomials $\phi_j(x)$ are the same as those in Problem 4–23.

4–26. Prove that if δ_{nm} is the Kronecker delta

$$\delta_{nm} = \begin{cases} 1 & n = m \\ 0 & n \neq m \end{cases}$$

then

$$\sum_{n=1}^{\infty} c_n \delta_{nm} = c_m \qquad \text{and} \qquad \sum_{n}\sum_{m} a_n b_m \delta_{nm} = \sum_{n} a_n b_n$$

These results will be used often.

Separate the sum over $c_n \delta_{nm}$ into two terms:

$$\sum_{n=1}^{\infty} c_n \delta_{nm} = c_m \delta_{mm} + \sum_{\substack{n=1 \\ n \neq m}}^{\infty} c_n \delta_{nm}$$

$$= c_m + \sum_{\substack{n=1 \\ n \neq m}} c_n \, (0)$$

$$= c_m$$

Similarly, separate the sum $\sum_m a_n b_m \delta_{nm}$ into two terms:

$$\sum_n \sum_m a_n b_m \delta_{nm} = \sum_n \left(a_n b_n \delta_{nn} + \sum_{m \neq n} a_n b_m \delta_{nm} \right)$$

$$= \sum_n a_n b_n$$

4–27. Express the orthonormality of the set of functions $\{\psi_n(x)\}$ in Dirac notation. Express the eigenfunction expansion $\phi(x) = \sum_n c_n \psi_n(x)$ and the coefficients c_n in Dirac notation.

In Dirac notation, the orthonormality relation $\int_{-\infty}^{\infty} \psi_n^*(x) \psi_m(x) \, dx = \delta_{nm}$ is

$$\langle n \mid m \rangle = \delta_{nm}$$

The eigenfunction expansion $\phi(x) = \sum_n c_n \psi_n(x)$ is

$$\mid \phi \rangle = \sum_n c_n \mid n \rangle$$

and for the coefficient, $c_n = \int_{-\infty}^{\infty} \psi_n^*(x) \phi(x) \, dx$ is

$$c_n = \langle n \mid \phi \rangle$$

4–28. A general state function, expressed in the form of a ket vector $\mid \phi \rangle$, can be written as a superposition of the eigenstates $\mid 1 \rangle$, $\mid 2 \rangle$, ... of an operator $\hat{A}$ with eigenvalues $a_1, a_2, \ldots$ (in other words, $\hat{A} \mid n \rangle = a_n \mid n \rangle$):

$$\mid \phi \rangle = c_1 \mid 1 \rangle + c_2 \mid 2 \rangle + \cdots = \sum_n c_n \mid n \rangle$$

Show that $c_n = \langle n \mid \phi \rangle$. This quantity is called the *amplitude* of measuring a_n if a measurement of $\hat{A}$ is made in the state $\mid \phi \rangle$. The probability of obtaining a_n is $c_n^* c_n$. Show that $\mid \phi \rangle$ can be written as

$$\mid \phi \rangle = \sum_n \mid n \rangle \langle n \mid \phi \rangle$$

Similarly, the corresponding bra vector of $|\phi\rangle$ can be written in terms of the corresponding bra vectors of the $|n\rangle$ as

$$\langle\phi| = \sum_n c_n^* \langle n|$$

Show that $c_n^* = \langle\phi|n\rangle$. Show that $\langle\phi|$ can be written as

$$\langle\phi| = \sum_n \langle\phi|n\rangle\langle n|$$

Now show that if $\langle\phi|$ is normalized, then

$$\langle\phi|\phi\rangle = 1 = \sum_n \langle\phi|n\rangle\langle n|\phi\rangle$$

and use this result to argue that

$$\sum_n |n\rangle\langle n| = 1$$

is a unit operator.

_____

Since the summation variable is arbitrary,

$$|\phi\rangle = \sum_m c_m |m\rangle$$

then

$$\langle n|\phi\rangle = \sum_m c_m \langle n|m\rangle$$

$$= \sum_m c_m \delta_{nm}$$

$$= c_n$$

Using this relation,

$$|\phi\rangle = \sum_n c_n |n\rangle$$

$$= \sum_n \langle n|\phi\rangle |n\rangle$$

$$= \sum_n |n\rangle\langle n|\phi\rangle$$

Similarly,

$$\langle \phi | = \sum_m c_m^* \langle m |$$

$$\langle \phi | n \rangle = \sum_m c_m^* \langle m | n \rangle$$

$$= \sum_m c_m^* \delta_{mn}$$

$$= c_n^*$$

Furthermore,

$$\langle \phi | = \sum_n c_n^* \langle n |$$

$$= \sum_n \langle \phi | n \rangle \langle n |$$

Combining the expressions for $| \phi \rangle$ and $\langle \phi |$ and using the normalization relation,

$$\langle \phi | \phi \rangle = 1 = \sum_n \sum_m \langle \phi | n \rangle \langle n | m \rangle \langle m | \phi \rangle$$

$$= \sum_n \sum_m \langle \phi | n \rangle \delta_{nm} \langle m | \phi \rangle$$

$$= \sum_n \langle \phi | n \rangle \langle n | \phi \rangle$$

which can be written as

$$\langle \phi | \phi \rangle = \langle \phi \left(\sum_n | n \rangle \langle n | \right) \phi \rangle$$

For the equality to hold,

$$\sum_n | n \rangle \langle n | = 1$$

4–29. Given the three polynomials $f_0(x) = a_0$, $f_1(x) = a_1 + b_1 x$, and $f_2(x) = a_2 + b_2 x + c_2 x^2$, find the constants such that the f's form an orthonormal set over the interval $0 \le x \le 1$.

Take the constants a_n and b_n to be real.

Since $f_0(x)$ is normalized:

$$\int_0^1 f_0^*(x) f_0(x) \, dx = 1$$

$$\int_0^1 a_0^2 \, dx = 1$$

$$a_0^2 = 1$$

$$a_0 = 1$$

where a_0 is taken to be positive. Thus, $f_0(x) = 1$.

Because $f_1(x)$ is orthogonal to $f_0(x)$:

$$\int_0^1 f_1^*(x) f_0(x)\, dx = 0$$

$$\int_0^1 (a_1 + b_1 x)\, dx = 0$$

$$\left(a_1 x + \frac{b_1 x^2}{2}\right)\Big|_0^1 = 0$$

$$b_1 = -2a_1$$

Furthermore, $f_1(x)$ is normalized:

$$\int_0^1 f_1^*(x) f_1(x)\, dx = 1$$

$$\int_0^1 \left(a_1^2 + 2a_1 b_1 x + b_1^2 x^2\right)\, dx = 1$$

$$\left(a_1^2 x + a_1 b_1 x^2 + \frac{b_1^2 x^3}{3}\right)\Big|_0^1 = 1$$

$$a_1^2 + a_1 b_1 + \frac{b_1^2}{3} = 1$$

These expressions for the normality and orthogonality of $f_1(x)$ can be combined to solve for a_1 and b_1. Specifically, substitute the expression $b_1 = -2a_1$ into the normalization condition:

$$a_1^2 - 2a_1^2 + \frac{4a_1^2}{3} = 1$$

$$a_1 = \sqrt{3}$$

where a_1 is taken to be positive. Thus, $b_1 = -2a_1 = -2\sqrt{3}$ and

$$f_1(x) = \sqrt{3}\,(1 - 2x)$$

Since $f_2(x)$ is orthogonal to $f_0(x)$:

$$\int_0^1 f_2^*(x) f_0(x)\, dx = 0$$

$$\int_0^1 \left(a_2 + b_2 x + c_2 x^2\right)\, dx = 0$$

$$\left(a_2 x + \frac{b_2 x^2}{2} + \frac{c_2 x^3}{3}\right)\Big|_0^1 = 0$$

$$a_2 + \frac{b_2}{2} + \frac{c_2}{3} = 0$$

and because $f_2(x)$ is orthogonal to $f_1(x)$:

$$\int_0^1 f_2^*(x) f_1(x)\, dx = 0$$

$$\sqrt{3} \int_0^1 \left[a_2 + (b_2 - 2a_2)\, x + (c_2 - 2b_2)\, x^2 - 2c_2 x^3 \right] dx = 0$$

$$\left[a_2 x + \frac{(b_2 - 2a_2)\, x^2}{2} + \frac{(c_2 - 2b_2)\, x^3}{3} - \frac{c_2 x^4}{2} \right]\Bigg|_0^1 = 0$$

$$a_2 + \frac{(b_2 - 2a_2)}{2} + \frac{(c_2 - 2b_2)}{3} - \frac{c_2}{2} = 0$$

$$b_2 = -c_2$$

Thus, the orthogonality condition between $f_2(x)$ and $f_0(x)$ gives

$$a_2 - \frac{c_2}{2} + \frac{c_2}{3} = 0$$

$$a_2 = \frac{c_2}{6}$$

$f_2(x)$ can now be expressed in terms of c_2:

$$f_2(x) = a_2 + b_2 x + c_2 x^2$$

$$= \frac{c_2}{6} - c_2 x + c_2 x^2$$

To obtain c_2, use the fact that $f_2(x)$ is normalized:

$$\int_0^1 f_2^*(x) f_2(x)\, dx = 1$$

$$\int_0^1 \left(\frac{c_2^2}{36} - \frac{c_2^2 x}{3} + \frac{4c_2^2 x^2}{3} - 2c_2^2 x^3 + c_2^2 x^4 \right) dx = 1$$

$$c_2^2 \left(\frac{x}{36} - \frac{x^2}{6} + \frac{4x^3}{9} - \frac{x^4}{2} + \frac{x^5}{5} \right)\Bigg|_0^1 = 1$$

$$\frac{c_2^2}{180} = 1$$

$$c_2 = 6\sqrt{5}$$

where c_2 is taken to be positive. Thus,

$$a_2 = \frac{c_2}{6} = \sqrt{5}$$

$$b_2 = -c_2 = -6\sqrt{5}$$

and $f_2(x) = \sqrt{5}\left(1 - 6x + 6x^2\right)$.

4–30. Using the orthogonality of the set $\{\sin(n\pi x/a)\}$ over the interval $0 \le x \le a$, show that if

$$f(x) = \sum_{n=1}^{\infty} b_n \sin \frac{n\pi x}{a}$$

then

$$b_n = \frac{2}{a} \int_0^a f(x) \sin \frac{n\pi x}{a} \, dx \qquad n = 1, 2, \ldots$$

Use this to show that the Fourier expansion of $f(x) = x$, $0 \le x \le a$, is

$$x = \frac{2a}{\pi} \sum_{n=1}^{\infty} \frac{(-1)^{n+1}}{n} \sin \frac{n\pi x}{a}$$

Multiply $f(x)$ by $\sin(m\pi x/a)$ and integrate over the interval $0 \le x \le a$:

$$\int_0^a f(x) \sin \frac{m\pi x}{a} \, dx = \sum_{n=1}^{\infty} b_n \int_0^a \sin \frac{n\pi x}{a} \sin \frac{m\pi x}{a} \, dx$$

The integral on the right-hand side was evaluated in Problem 3–23. Thus,

$$\int_0^a f(x) \sin \frac{m\pi x}{a} \, dx = \sum_{n=1}^{\infty} b_n \frac{a}{2} \delta_{nm}$$

$$= \frac{a}{2} b_m$$

Rearranging this equation and changing the index variable from m to n,

$$b_n = \frac{2}{a} \int_0^a f(x) \sin \frac{n\pi x}{a} \, dx$$

For $f(x) = x$,

$$b_n = \frac{2}{a} \int_0^a x \sin \frac{n\pi x}{a} \, dx$$

$$= \frac{2}{a} \left(\frac{a^2}{n^2\pi^2} \sin \frac{n\pi x}{a} - \frac{ax}{n\pi} \cos \frac{n\pi x}{a} \right) \Big|_0^a$$

$$= -\frac{2a}{n\pi} \cos(n\pi)$$

$$= -\frac{2a}{n\pi} (-1)^n$$

$$= \frac{2a}{n\pi} (-1)^{n+1}$$

Thus,

$$f(x) = \sum_{n=1}^{\infty} b_n \sin \frac{n\pi x}{a}$$

$$= \frac{2a}{\pi} \sum_{n=1}^{\infty} \frac{(-1)^{n+1}}{n} \sin \frac{n\pi x}{a}$$

4–31. We can define functions of operators through their Maclaurin series (MathChapter D). For example, we define the operator $\exp(\hat{S})$ by

$$e^{\hat{S}} = \sum_{n=0}^{\infty} \frac{(\hat{S})^n}{n!}$$

Under what conditions does the equality $e^{\hat{A}+\hat{B}} \overset{?}{=} e^{\hat{A}} e^{\hat{B}}$ hold?

Write the Maclaurin series expansion of $e^{\hat{A}+\hat{B}}$ to second order:

$$e^{\hat{A}+\hat{B}} = \sum_{n=0}^{\infty} \frac{(\hat{A}+\hat{B})^n}{n!}$$

$$= \hat{I} + \hat{A} + \hat{B} + \frac{(\hat{A}+\hat{B})^2}{2} + O\left[(\hat{A}+\hat{B})^3\right]$$

$$= \hat{I} + \hat{A} + \hat{B} + \frac{\hat{A}^2}{2} + \frac{\hat{B}^2}{2} + \frac{\hat{A}\hat{B}}{2} + \frac{\hat{B}\hat{A}}{2} + O\left[(\hat{A}+\hat{B})^3\right]$$

Now consider $e^{\hat{A}} e^{\hat{B}}$:

$$e^{\hat{A}} e^{\hat{B}} = \left[\hat{I} + \hat{A} + \frac{\hat{A}^2}{2} + O\left(\hat{A}^3\right)\right]\left[\hat{I} + \hat{B} + \frac{\hat{B}^2}{2} + O\left(\hat{B}^3\right)\right]$$

$$= \hat{I} + \hat{A} + \hat{B} + \frac{\hat{A}^2}{2} + \frac{\hat{B}^2}{2} + \hat{A}\hat{B} + O\left[(\hat{A}+\hat{B})^3\right]$$

The equality $e^{\hat{A}+\hat{B}} = e^{\hat{A}} e^{\hat{B}}$ holds only if $\hat{A}\hat{B} = \hat{B}\hat{A}$, that is, if $\hat{A}$ and $\hat{B}$ commute.

4–32. In this chapter, we learned that if ψ_n is an eigenfunction of the time-independent Schrödinger equation, then

$$\Psi_n(x, t) = \psi_n(x) e^{-iE_n t/\hbar}$$

Show that if $\psi_m(x)$ and $\psi_n(x)$ are both stationary states of $\hat{H}$, then the state

$$\Psi(x, t) = c_m \psi_m(x) e^{-iE_m t/\hbar} + c_n \psi_n(x) e^{-iE_n t/\hbar}$$

satisfies the time-dependent Schrödinger equation.

If $\psi_m(x)$ and $\psi_n(x)$ are both stationary states of $\hat{H}$, then

$$\hat{H}\psi_m(x) = E_m\psi_m(x)$$

$$\hat{H}\psi_n(x) = E_n\psi_n(x)$$

To show that $\Psi(x, t)$ satisfies the time-dependent Schrödinger equation, show that $\hat{H}\Psi_n$ and $i\hbar\partial\Psi/\partial t$ are equal.

$$\hat{H}\Psi(x, t) = c_m\hat{H}\psi_m(x)e^{-iE_mt/\hbar} + c_n\hat{H}\psi_n(x)e^{-iE_nt/\hbar}$$

$$= c_m E_m\psi_m(x)e^{-iE_mt/\hbar} + c_n E_n\psi_n(x)e^{-iE_nt/\hbar}$$

$$i\hbar\frac{\partial\Psi}{\partial t} = i\hbar\left(c_m\psi_m(x)\frac{\partial}{\partial t}e^{-iE_mt/\hbar} + c_n\psi_n\frac{\partial}{\partial t}e^{-iE_nt/\hbar}\right)$$

$$= i\hbar\left[c_m\psi_m(x)\left(\frac{-iE_m}{\hbar}\right)e^{-iE_mt/\hbar} + c_n\psi_n\left(\frac{-iE_n}{\hbar}\right)e^{-iE_nt/\hbar}\right]$$

$$= c_m E_m\psi_m(x)e^{-iE_mt/\hbar} + c_n E_n\psi_n(x)e^{-iE_nt/\hbar}$$

$$= \hat{H}\Psi(x, t)$$

Indeed, $\Psi(x, t)$ satisfies the time-dependent Schrödinger equation.

4–33. Show that $\Psi(x, t)$ given by Equation 4.78 is normalized.

The wave function is given by Equation 4.78:

$$\Psi(x, t) = \left(\frac{1}{a}\right)^{1/2}e^{-iE_1t/\hbar}\sin\frac{\pi x}{a} + \left(\frac{1}{a}\right)^{1/2}e^{-iE_2t/\hbar}\sin\frac{2\pi x}{a}$$

Now show that it is normalized:

$$\int_0^a \Psi^*(x, t)\Psi(x, t)\, dx = \int_0^a\left[\left(\frac{1}{a}\right)^{1/2}e^{iE_1t/\hbar}\sin\frac{\pi x}{a} + \left(\frac{1}{a}\right)^{1/2}e^{iE_2t/\hbar}\sin\frac{2\pi x}{a}\right]$$

$$\times\left[\left(\frac{1}{a}\right)^{1/2}e^{-iE_1t/\hbar}\sin\frac{\pi x}{a} + \left(\frac{1}{a}\right)^{1/2}e^{-iE_2t/\hbar}\sin\frac{2\pi x}{a}\right]dx$$

$$= \frac{1}{a}\left[\int_0^a\sin^2\frac{\pi x}{a}\, dx + e^{i(E_1-E_2)t/\hbar}\int_0^a\sin\frac{\pi x}{a}\sin\frac{2\pi x}{a}\, dx\right.$$

$$\left. + e^{i(E_2-E_1)t/\hbar}\int_0^a\sin\frac{\pi x}{a}\sin\frac{2\pi x}{a}\, dx + \int_0^a\sin^2\frac{2\pi x}{a}\, dx\right]$$

As shown in Problem 3–23,

$$\int_0^a\sin\frac{n\pi x}{a}\sin\frac{m\pi x}{a}\, dx = \frac{a}{2}\delta_{nm}$$

Thus,

$$\int_0^a \Psi^*(x, t)\Psi(x, t)\, dx = \frac{1}{a}\left(\frac{a}{2} + \frac{a}{2}\right)$$

$$= 1$$

4–34. Verify Equation 4.79.

The wave function is given by Equation 4.78:

$$\Psi(x, t) = \left(\frac{1}{a}\right)^{1/2} e^{-iE_1 t/\hbar} \sin\frac{\pi x}{a} + \left(\frac{1}{a}\right)^{1/2} e^{-iE_2 t/\hbar} \sin\frac{2\pi x}{a}$$

The average position of x is

$$\langle x \rangle = \int_0^a \left[\left(\frac{1}{a}\right)^{1/2} e^{iE_1 t/\hbar} \sin\frac{\pi x}{a} + \left(\frac{1}{a}\right)^{1/2} e^{iE_2 t/\hbar} \sin\frac{2\pi x}{a}\right] x$$

$$\times \left[\left(\frac{1}{a}\right)^{1/2} e^{-iE_1 t/\hbar} \sin\frac{\pi x}{a} + \left(\frac{1}{a}\right)^{1/2} e^{-iE_2 t/\hbar} \sin\frac{2\pi x}{a}\right] dx$$

$$= \frac{1}{a}\left[\int_0^a x \sin^2\frac{\pi x}{a}\, dx + e^{i(E_1 - E_2)t/\hbar}\int_0^a x \sin\frac{\pi x}{a}\sin\frac{2\pi x}{a}\, dx\right.$$

$$\left. + e^{i(E_2 - E_1)t/\hbar}\int_0^a x \sin\frac{\pi x}{a}\sin\frac{2\pi x}{a}\, dx + \int_0^a x \sin^2\frac{2\pi x}{a}\, dx\right]$$

$$= \frac{1}{a}\int_0^a x \sin^2\frac{\pi x}{a}\, dx + \frac{1}{a}\int_0^a x \sin^2\frac{2\pi x}{a}\, dx$$

$$+ \frac{1}{a}\left[e^{-i(E_2 - E_1)t/\hbar} + e^{i(E_2 - E_1)t/\hbar}\right]\int_0^a x \sin\frac{\pi x}{a}\sin\frac{2\pi x}{a}\, dx$$

$$= \frac{1}{a}\int_0^a x \sin^2\frac{\pi x}{a}\, dx + \frac{1}{a}\int_0^a x \sin^2\frac{2\pi x}{a}\, dx$$

$$+ \frac{2\cos\omega_{12}t}{a}\int_0^a x \sin\frac{\pi x}{a}\sin\frac{2\pi x}{a}\, dx$$

where the relation $\cos\theta = (e^{i\theta} + e^{-i\theta})/2$ and the definition of $\omega_{12} = (E_2 - E_1)/\hbar$ have been applied. The first two integrals have been evaluated in Problem 3–14:

$$\int_0^a x \sin^2\frac{n\pi x}{a}\, dx = \frac{a^2}{4}$$

The third integral can be evaluated by using the trigonometric identity (Problem 3–21):

$$\sin\alpha \sin\beta = \frac{1}{2}\cos(\alpha - \beta) - \frac{1}{2}\cos(\alpha + \beta)$$

Thus,

$$
\int_0^a x \sin \frac{\pi x}{a} \sin \frac{2\pi x}{a} \, dx = \frac{1}{2} \int_0^a \left(x \cos \frac{\pi x}{a} - x \cos \frac{3\pi x}{a} \right) dx
$$

$$
= \frac{1}{2} \left(\frac{a^2}{\pi^2} \cos \frac{\pi x}{a} + \frac{ax}{\pi} \sin \frac{\pi x}{a} - \frac{a^2}{9\pi^2} \cos \frac{3\pi x}{a} - \frac{ax}{3\pi} \sin \frac{3\pi x}{a} \right) \Big|_0^a
$$

$$
= \frac{1}{2} \left[-\frac{a^2}{\pi^2} + \frac{a^2}{9\pi^2} - \left(\frac{a^2}{\pi^2} - \frac{a^2}{9\pi^2} \right) \right]
$$

$$
= -\frac{8a^2}{9\pi^2}
$$

Combining all the integrals,

$$
\langle x \rangle = \frac{1}{a} \frac{a^2}{4} + \frac{1}{a} \frac{a^2}{4} - \frac{2 \cos \omega_{12} t}{a} \frac{8a^2}{9\pi^2}
$$

$$
= \frac{a}{2} - \frac{16a}{9\pi^2} \cos \omega_{12} t
$$

4–35. What is the normalization constant for $\Psi(x, t) = \displaystyle\sum_{n=1}^{N} \psi_n(x) e^{-i E_n t/\hbar}$ if the $\psi_n(x)$ are normalized?

$$
\int_{-\infty}^{\infty} \Psi^*(x, t) \Psi(x, t) \, dx = \sum_{n=1}^{N} \sum_{m=1}^{N} e^{i E_n t/\hbar} e^{-i E_m t/\hbar} \int_{-\infty}^{\infty} \psi_n^*(x) \psi_m(x) \, dx
$$

$$
= \sum_{n=1}^{N} \sum_{m=1}^{N} e^{i E_n t/\hbar} e^{-i E_m t/\hbar} \delta_{nm}
$$

$$
= \sum_{n=1}^{N} e^{i E_n t/\hbar} e^{-i E_n t/\hbar}
$$

$$
= N
$$

The normalization constant is $(1/N)^{1/2}$.

4–36. Superimpose the behavior of $\langle x \rangle$ for a classical particle moving with the same period onto Figure 4.3.

A classical particle is equally likely to be found anywhere in the box. Thus, $\langle x \rangle = a/2$, and is represented by the dashed line in the figure below. The solid line represents the average position of the quantum mechanical particle whose state is described by Equation 4.78.

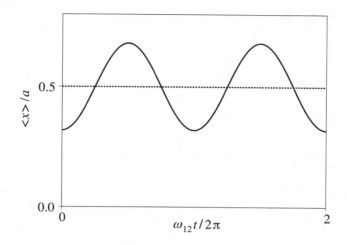

4–37. Show that the average energy of a particle described by Equation 4.78 is a constant.

The particle is described by

$$\Psi(x, t) = \left(\frac{1}{a}\right)^{1/2} e^{-iE_1 t/\hbar} \sin\frac{\pi x}{a} + \left(\frac{1}{a}\right)^{1/2} e^{-iE_2 t/\hbar} \sin\frac{2\pi x}{a}$$

$$= \frac{1}{\sqrt{2}}\left[\psi_1(x)e^{-iE_1 t/\hbar} + \psi_2(x)e^{-iE_2 t/\hbar}\right]$$

The average energy is

$$\langle E \rangle = \int_0^a \Psi^*(x, t)\hat{H}\Psi(x, t)\, dx$$

$$= \frac{1}{2}\int_0^a \left[\psi_1^*(x)e^{iE_1 t/\hbar} + \psi_2^*(x)e^{iE_2 t/\hbar}\right]\left[E_1\psi_1(x)e^{-iE_1 t/\hbar} + E_2\psi_2(x)e^{-iE_2 t/\hbar}\right] dx$$

$$= \frac{1}{2}\left[E_1\int_0^a \psi_1^*(x)\psi_1(x)\, dx + e^{i(E_1-E_2)t/\hbar}E_2\int_0^a \psi_1(x)^*\psi_2(x)\, dx\right.$$

$$\left. + E_1 e^{i(E_2-E_1)t/\hbar}\int_0^a \psi_2(x)^*\psi_1(x)\, dx + E_2\int_0^a \psi_2^*(x)\psi_2(x)\, dx\right]$$

$$= \frac{1}{2}\left(E_1 + E_2\right)$$

$$= \frac{5h^2}{16ma^2}$$

The average energy is a constant. The orthonormality of the wave functions is used in deriving this expression.

4–38. What would be the form of $\Psi(x, t)$ in Equation 4.78 if it were a superposition of the three lowest states instead of two?

As shown in Problem 4–35, the normalization constant for a wave function formed by a super-position of three states is $(1/3)^{1/2}$. Specifically,

$$\Psi(x, t) = \left(\frac{1}{3}\right)^{1/2} \left[\psi_1(x)e^{-iE_1t/\hbar} + \psi_2(x)e^{-iE_2t/\hbar} + \psi_3(x)e^{-iE_3t/\hbar}\right]$$

$$= \left(\frac{2}{3a}\right)^{1/2} e^{-iE_1t/\hbar} \sin\frac{\pi x}{a} + \left(\frac{2}{3a}\right)^{1/2} e^{-iE_2t/\hbar} \sin\frac{2\pi x}{a} + \left(\frac{2}{3a}\right)^{1/2} e^{-iE_3t/\hbar} \sin\frac{3\pi x}{a}$$

4–39. Derive an expression for the average position of a particle in a box in a state described by

$$\Psi(x, t) = \left(\frac{1}{a}\right)^{1/2} e^{-iE_2t/\hbar} \sin\frac{2\pi x}{a} + \left(\frac{1}{a}\right)^{1/2} e^{-iE_3t/\hbar} \sin\frac{3\pi x}{a}$$

With what frequency does the particle oscillate about the midpoint of the box?

This problem is similar to Problem 4–34. The average position of x is

$$\langle x \rangle = \int_0^a \left[\left(\frac{1}{a}\right)^{1/2} e^{iE_2t/\hbar} \sin\frac{2\pi x}{a} + \left(\frac{1}{a}\right)^{1/2} e^{iE_3t/\hbar} \sin\frac{3\pi x}{a}\right] x$$

$$\times \left[\left(\frac{1}{a}\right)^{1/2} e^{-iE_2t/\hbar} \sin\frac{2\pi x}{a} + \left(\frac{1}{a}\right)^{1/2} e^{-iE_3t/\hbar} \sin\frac{3\pi x}{a}\right] dx$$

$$= \frac{1}{a}\left[\int_0^a x \sin^2\frac{2\pi x}{a} dx + e^{i(E_2-E_3)t/\hbar} \int_0^a x \sin\frac{2\pi x}{a} \sin\frac{3\pi x}{a} dx\right.$$

$$\left. + e^{i(E_3-E_2)t/\hbar} \int_0^a x \sin\frac{2\pi x}{a} \sin\frac{3\pi x}{a} dx + \int_0^a x \sin^2\frac{3\pi x}{a} dx\right]$$

$$= \frac{1}{a}\int_0^a x \sin^2\frac{2\pi x}{a} dx + \frac{1}{a}\int_0^a x \sin^2\frac{3\pi x}{a} dx$$

$$+ \frac{1}{a}\left[e^{-i(E_3-E_2)t/\hbar} + e^{i(E_3-E_2)t/\hbar}\right]\int_0^a x \sin\frac{2\pi x}{a} \sin\frac{3\pi x}{a} dx$$

$$= \frac{1}{a}\int_0^a x \sin^2\frac{2\pi x}{a} dx + \frac{1}{a}\int_0^a x \sin^2\frac{3\pi x}{a} dx$$

$$+ \frac{2\cos\omega_{23}t}{a}\int_0^a x \sin\frac{2\pi x}{a} \sin\frac{3\pi x}{a} dx$$

where the relation $\cos\theta = (e^{i\theta} + e^{-i\theta})/2$ and the definition of $\omega_{23} = (E_3 - E_2)/\hbar$ have been applied. The first two integrals have been evaluated in Problem 3–14:

$$\int_0^a x \sin^2\frac{n\pi x}{a} dx = \frac{a^2}{4}$$

The third integral can be evaluated by using the trigonometric identity (Problem 3–21):

$$\sin\alpha \sin\beta = \frac{1}{2}\cos(\alpha - \beta) - \frac{1}{2}\cos(\alpha + \beta)$$

Thus,

$$
\int_0^a x \sin \frac{2\pi x}{a} \sin \frac{3\pi x}{a}\, dx = \frac{1}{2}\int_0^a \left(x \cos \frac{\pi x}{a} - x \cos \frac{5\pi x}{a} \right) dx
$$

$$
= \frac{1}{2}\left(\frac{a^2}{\pi^2} \cos \frac{\pi x}{a} + \frac{ax}{\pi}\sin\frac{\pi x}{a} - \frac{a^2}{25\pi^2}\cos\frac{5\pi x}{a} - \frac{ax}{5\pi}\sin\frac{5\pi x}{a}\right)\Bigg|_0^a
$$

$$
= \frac{1}{2}\left[-\frac{a^2}{\pi^2} + \frac{a^2}{25\pi^2} - \left(\frac{a^2}{\pi^2} - \frac{a^2}{25\pi^2} \right) \right]
$$

$$
= -\frac{24a^2}{25\pi^2}
$$

Combining all the integrals,

$$
\langle x \rangle = \frac{1}{a}\frac{a^2}{4} + \frac{1}{a}\frac{a^2}{4} - \frac{2\cos\omega_{23}t}{a}\frac{24a^2}{25\pi^2}
$$

$$
= \frac{a}{2} - \frac{48a}{25\pi^2}\cos\omega_{23}t
$$

The particle oscillates about the midpoint of the box with an angular frequency of ω_{23} (or a frequency of $\omega_{23}/2\pi$).

4–40. Calculate the amplitude associated with the oscillation of a particle in a box in a state described by

$$
\Psi(x, t) = \left(\frac{1}{a}\right)^{1/2} e^{-iE_1 t/\hbar}\sin\frac{\pi x}{a} + \left(\frac{1}{a}\right)^{1/2} e^{-iE_4 t/\hbar}\sin\frac{4\pi x}{a}
$$

What is the frequency? Compare the amplitude here with that in the previous problem.

By analogy to Problem 3–34 and 3–39, the average position of the particle in this case is

$$
\langle x \rangle = \frac{1}{a}\int_0^a x \sin^2\frac{\pi x}{a}\, dx + \frac{1}{a}\int_0^a x \sin^2\frac{4\pi x}{a}\, dx
$$

$$
+ \frac{2\cos\omega_{14}t}{a}\int_0^a x \sin\frac{\pi x}{a}\sin\frac{4\pi x}{a}\, dx
$$

$$
= \frac{1}{a}\frac{a^2}{4} + \frac{1}{a}\frac{a^2}{4} - \frac{2\cos\omega_{14}t}{a}\frac{16a^2}{225\pi^2}
$$

$$
= \frac{a}{2} - \frac{32a}{225\pi^2}\cos\omega_{14}t
$$

where $\omega_{14} = (E_4 - E_1)/\hbar$ is the angular frequency of the oscillation. The amplitude of oscillation is $32a/(225\pi^2)$, which is a small fraction $(2/27)$ of that in the last problem, $48a/(25\pi^2)$.

4–41. Use a program such as *MathCad* or *Mathematica* to plot the time evolution of the probability density for a particle in a box in a state described in the previous problem. Plot your result through one cycle.

The probability density is

$$\Psi^*(x, t)\Psi(x, t) = \frac{1}{a}\sin^2\frac{\pi x}{a} + \frac{1}{a}\sin^2\frac{4\pi x}{a} + \frac{2}{a}\sin\frac{\pi x}{a}\sin\frac{4\pi x}{a}\cos\omega_{14}t$$

The following plots show the probability density as a function of x for $t = 0$, $\pi/2\omega$, π/ω, $3\pi/2\omega$, and $2\pi/\omega$.

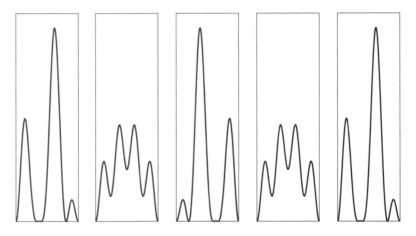

4–42. In this problem, we shall develop the consequence of measuring the position of a particle in a box. If we find that the particle is located between $a/2 - \epsilon/2$ and $a/2 + \epsilon/2$, then its wave function may be ideally represented by

$$\phi_\epsilon = \begin{cases} 0 & x < a/2 - \epsilon/2 \\ 1/\sqrt{\epsilon} & a/2 - \epsilon/2 < x < a/2 + \epsilon/2 \\ 0 & x > a/2 + \epsilon/2 \end{cases}$$

Plot $\phi_\epsilon(x)$ and show that it is normalized. The parameter ϵ is in a sense a gauge of the accuracy of the measurement; the smaller the value of ϵ, the more accurate the measurement. Now let's suppose we measure the energy of the particle. The probability that we observe the value E_n is given by the value of $|c_n|^2$ in the expansion

$$\phi_\epsilon(x) = \sum_{n=1}^{\infty} c_n\psi_n(x)e^{-iE_nt/\hbar}$$

where $\psi_n(x) = (2/a)^{1/2}\sin n\pi x/a$ and $E_n = n^2h^2/8ma^2$. Multiply both sides of this equation by $\psi_m(x)$ and integrate over x from 0 to a to get

$$c_m = e^{iE_mt/\hbar}\int_0^a \phi_\epsilon(x)\psi_m(x)\,dx = \frac{2^{3/2}a^{1/2}e^{iE_mt/\hbar}}{\epsilon^{1/2}m\pi}\sin\frac{m\pi}{2}\sin\frac{m\pi\epsilon}{2a}$$

Now show that the probability of observing E_n is given by

$$p(E_n) = \begin{cases} 0 & \text{if } n \text{ is even} \\ \dfrac{8a}{\epsilon}\left(\dfrac{1}{n\pi}\right)^2 \sin^2 \dfrac{n\pi\epsilon}{2a} & \text{if } n \text{ is odd} \end{cases}$$

Plot $p(E_n)$ against n for $\epsilon/a = 0.10$, 0.050, and 0.010. Interpret the result in terms of the uncertainty principle.

The wave function is plotted below:

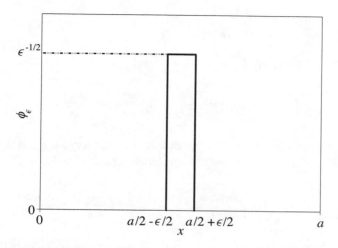

Now show that it is normalized:

$$\int_0^a \phi_\epsilon^* \phi_\epsilon \, dx = \int_{a/2-\epsilon/2}^{a/2+\epsilon/2} \frac{1}{\epsilon} \, dx$$

$$= \frac{1}{\epsilon}\left(\frac{a}{2} + \frac{\epsilon}{2} - \frac{a}{2} + \frac{\epsilon}{2}\right)$$

$$= 1$$

Multiplying $\phi_\epsilon(x)$ by $\psi_m(x)$ and integrating over x from 0 to a:

$$\int_0^a \phi_\epsilon(x)\psi_m(x) \, dx = \sum_{n=1}^\infty c_n e^{-iE_n t/\hbar} \int_0^a \psi_n(x)\psi_m(x) \, dx$$

$$= \sum_{n=1}^\infty c_n e^{-iE_n t/\hbar} \delta_{nm}$$

$$= c_m e^{-iE_m t/\hbar}$$

Thus,

$$c_m = e^{iE_m t/\hbar} \int_0^a \phi_\epsilon(x)\psi_m(x) \, dx$$

$$= e^{iE_m t/\hbar}\left(\frac{2}{a\epsilon}\right)^{1/2} \int_{a/2-\epsilon/2}^{a/2+\epsilon/2} \sin\frac{m\pi x}{a} \, dx$$

$$= e^{iE_m t/\hbar}\left(\frac{2}{a\epsilon}\right)^{1/2}\left(-\frac{a}{m\pi}\right)\left\{\cos\left[\frac{m\pi}{a}\left(\frac{a}{2}+\frac{\epsilon}{2}\right)\right] - \cos\left[\frac{m\pi}{a}\left(\frac{a}{2}-\frac{\epsilon}{2}\right)\right]\right\}$$

Using the trigonometric identity

$$\sin \alpha \sin \beta = \frac{1}{2} \cos (\alpha - \beta) - \frac{1}{2} \cos (\alpha + \beta)$$

the expression for c_m can be simplified:

$$c_m = e^{i E_m t/\hbar} \left(\frac{2}{a\epsilon} \right)^{1/2} \left(-\frac{a}{m\pi} \right) \left(-2 \sin \frac{m\pi}{2} \sin \frac{m\pi\epsilon}{2a} \right)$$

$$= \frac{2^{3/2} a^{1/2} e^{i E_m t/\hbar}}{\epsilon^{1/2} m\pi} \sin \frac{m\pi}{2} \sin \frac{m\pi\epsilon}{2a}$$

The probability of observing E_m is $|c_m|^2$:

$$|c_m|^2 = \frac{8a}{\epsilon m^2 \pi^2} \sin^2 \frac{m\pi}{2} \sin^2 \frac{m\pi\epsilon}{2a}$$

When m is even, $\sin^2(m\pi/2) = 0$ and the probability of observing E_m is 0. On the other hand, when m is odd, $\sin^2(m\pi/2) = 1$ and the probability of observing E_m is

$$|c_m|^2 = \frac{8a}{\epsilon m^2 \pi^2} \sin^2 \frac{m\pi\epsilon}{2a} = \frac{8}{\epsilon} a \left(\frac{1}{m\pi} \right)^2 \sin^2 \frac{m\pi\epsilon}{2a}$$

Replacing the index variable m with n gives the same expression as that in the question.

Plots of $p(E_n)$ against n are shown below for $\epsilon = 0.10$, 0.050, and 0.010 from left to right for $n = 1$ to 100.

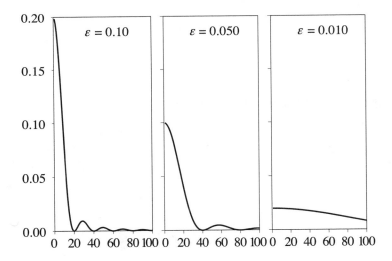

The figures show that the smaller the value of ϵ, the larger is the spread of energy. This is in accordance with the uncertainty principle in that the more accurate is the position measurement, the less accurate is the energy (or momentum) measurement of the particle.

4–43. Starting with

$$\langle x \rangle = \int \Psi^*(x, t) x \Psi(x, t) dx$$

and the time-dependent Schrödinger equation, show that

$$\frac{d\langle x \rangle}{dt} = \int \Psi^* \frac{i}{\hbar}[\hat{H}, x] \Psi \, dx = \frac{\langle p_x \rangle}{m}$$

Interpret this result.

Differentiate $\langle x \rangle$ with respect to t:

$$\frac{d\langle x \rangle}{dt} = \int \frac{\partial \Psi^*(x, t)}{\partial t} x \Psi(x, t) \, dx + \int \Psi^*(x, t) x \frac{\partial \Psi(x, t)}{\partial t} \, dx$$

$$= \int \left[\frac{1}{i\hbar} \hat{H} \Psi(x, t) \right]^* x \Psi(x, t) \, dx + \int \Psi^*(x, t) x \left[\frac{1}{i\hbar} \hat{H} \Psi(x, t) \right] \, dx$$

$$= -\frac{1}{i\hbar} \int x \Psi(x, t) \hat{H}^* \Psi^*(x, t) \, dx + \frac{1}{i\hbar} \int \Psi^*(x, t) x \hat{H} \Psi(x, t) \, dx$$

where the time dependent Schrödinger equation has been applied and the order of multiplication of $\hat{H}^* \Psi^*(x, t)$ and $x \Psi(x, t)$ has been reversed. Since $\hat{H}$ is Hermitian, the above expression becomes

$$\frac{d\langle x \rangle}{dt} = -\frac{1}{i\hbar} \int \Psi^*(x, t) \hat{H} x \Psi(x, t) \, dx + \frac{1}{i\hbar} \int \Psi^*(x, t) x \hat{H} \Psi(x, t) \, dx$$

$$= \frac{i}{\hbar} \int \Psi^*(x, t) \left(\hat{H} x - \hat{H} x \right) \Psi(x, t) \, dx$$

$$= \int \Psi^* \frac{i}{\hbar} [\hat{H}, x] \Psi \, dx$$

Now evaluate the commutator $[\hat{H}, x]$:

$$[\hat{H}, x] f(x) = \left[-\frac{\hbar^2}{2m} \frac{d^2}{dx^2} + V(x) \right] [x f(x)] - x \left[-\frac{\hbar^2}{2m} \frac{d^2 f}{dx^2} + V(x) f(x) \right]$$

$$= -\frac{\hbar^2}{2m} \frac{d}{dx} \left[f(x) + x \frac{df(x)}{dx} \right] + V(x) x f(x) + \frac{\hbar^2}{2m} x \frac{d^2 f}{dx^2} - x V(x) f(x)$$

$$= -\frac{\hbar^2}{2m} \left[2 \frac{df(x)}{dx} + x \frac{d^2 f(x)}{dx^2} \right] + \frac{\hbar^2}{2m} x \frac{d^2 f}{dx^2}$$

$$= -\frac{\hbar^2}{m} \frac{df(x)}{dx}$$

$$= -\frac{\hbar^2}{m} \frac{\hat{P}_x}{-i\hbar} f$$

$$= -\frac{i\hbar}{m} \hat{P}_x f$$

Using this commutation relationship, $d\langle x \rangle / dt$ becomes

$$\frac{d\langle x \rangle}{dt} = \int \Psi^* \frac{i}{\hbar} \left[-\frac{i\hbar}{m} \hat{P}_x \right] \Psi \, dx$$

$$= \frac{\langle p_x \rangle}{m}$$

The above equation shows that the average quantum mechanical values of position and momentum follow the same relation as their classical counterparts, $p_x = m \, (dx/dt)$.

4–44. Use Equation 4.84 to show that

$$\frac{d\langle \hat{P}_x \rangle}{dt} = \left\langle -\frac{d\hat{V}}{dx} \right\rangle$$

Interpret this result, which is known as *Ehrenfest's theorem*.

According to Equation 4.84,

$$\frac{d\langle \hat{P}_x \rangle}{dt} = \frac{i}{\hbar} \left\langle \Psi \left| \left[\hat{H}, \hat{P}_x \right] \right| \Psi \right\rangle + \left\langle \frac{\partial \hat{P}_x}{\partial t} \right\rangle$$

Since $\hat{P}_x$ is not time dependent, the second term on the right hand side of the equality sign in the above expression is 0. The commutator $\left[\hat{H}, \hat{P}_x \right]$ is

$$\left[\hat{H}, \hat{P}_x \right] f(x) = \left[-\frac{\hbar^2}{2m} \frac{d^2}{dx^2} + V(x) \right] (-i\hbar) \frac{df(x)}{dx} - (-i\hbar) \frac{d}{dx} \left[-\frac{\hbar^2}{2m} \frac{d^2 f(x)}{dx^2} + V(x) f(x) \right]$$

$$= \frac{i\hbar^3}{2m} \frac{d^3 f(x)}{dx^3} - i\hbar V(x) \frac{df(x)}{dx} - \frac{i\hbar^3}{2m} \frac{d^3 f(x)}{dx^3} + i\hbar V(x) \frac{df(x)}{dx} + i\hbar f(x) \frac{dV(x)}{dx}$$

$$= i\hbar f(x) \frac{dV(x)}{dx}$$

Thus,

$$\frac{d\langle \hat{P}_x \rangle}{dt} = \frac{i}{\hbar} \left\langle \Psi \left| i\hbar \frac{dV(x)}{dx} \right| \Psi \right\rangle$$

$$= \left\langle \Psi \left| -\frac{dV(x)}{dx} \right| \Psi \right\rangle$$

$$= \left\langle -\frac{dV}{dx} \right\rangle$$

This relation is equivalent to the classical relation between force and acceleration, $F = ma$, or equivalently, $-dV/dt = dp/dt$.

4–45. In this problem, we shall prove the Schwartz inequality, which says that if f and g are two suitably well-behaved functions, then

$$\left(\int |f|^2 \, dx \right) \left(\int |g|^2 \, dx \right) \geq \left| \int f^* g \, dx \right|^2$$

In order to prove the Schwartz inequality, we start with

$$\int (f + \lambda g)^* (f + \lambda g) \, dx \geq 0$$

where λ is an arbitrary complex number. Expand this to find

$$|\lambda|^2 \int g^* g \, dx + \lambda \int f^* g \, dx + \lambda^* \int g^* f \, dx + \int f^* f \, dx \geq 0$$

This inequality must be true for any complex λ and, in particular, choose

$$\lambda = -\frac{\int g^* f \, dx}{\int g^* g \, dx} = -\frac{\left(\int g f^* \, dx \right)^*}{\int g^* g \, dx}$$

Show that this choice of λ gives the Schwartz inequality:

$$\left(\int f^* f \, dx \right) \left(\int g^* g \, dx \right) \geq \left| \int f^* g \, dx \right|^2$$

When does the equality hold?

Expanding the inequality

$$\int (f + \lambda g)^* (f + \lambda g) \, dx \geq 0$$

results in four integrals:

$$|\lambda|^2 \int g^* g \, dx + \lambda^* \int f^* g \, dx + \lambda \int f g^* \, dx + \int f^* f \, dx \geq 0$$

Substituting the expression suggested for λ into the above inequality,

$$\frac{\left(\int f g^* \, dx \right) \left(\int f^* g \, dx \right)}{\left(\int g^* g \, dx \right)^2} \int g^* g \, dx - \frac{\int f g^* \, dx}{\int g^* g \, dx} \left(\int f^* g \, dx \right)$$

$$- \frac{\int f^* g \, dx}{\int g^* g \, dx} \left(\int f g^* \, dx \right) + \int f^* f \, dx \geq 0$$

$$\int f^* f \, dx - \frac{\left(\int f g^* \, dx \right) \left(\int f^* g \, dx \right)}{\int g^* g \, dx} \geq 0$$

Since $\int g^* g \, dx \geq 0$, multiplication of the above expression by this term does not change the inequality:

$$\left(\int f^*f \, dx\right)\left(\int g^*g \, dx\right) - \left|\int f^*g \, dx\right|^2 \geq 0$$

$$\left(\int f^*f \, dx\right)\left(\int g^*g \, dx\right) \geq \left|\int f^*g \, dx\right|^2$$

The equality holds when $f = g$.

4–46.   In this problem, we shall prove that if $f(x)$ and $g(x)$ are suitably behaved functions, then

$$\left[\int f^*(x)f(x) \, dx\right]\left[\int g^*(x)g(x) \, dx\right] \geq \frac{1}{4}\left[\int (f^*(x)g(x) + f(x)g^*(x)) \, dx\right]^2$$

We shall use this inequality to derive Equation 4.19 in the next problem. First, let

$$A = \int f^*(x)f(x) \, dx \qquad B = \int f^*(x)g(x) \, dx \qquad C = \int g^*(x)g(x) \, dx$$

Now argue that

$$\int [\lambda f^*(x) + g^*(x)][\lambda f(x) + g(x)] \, dx = A\lambda^2 + (B + B^*)\lambda + C$$

is equal to or greater than zero for real values of λ. Show that $A \geq 0$, $C \geq 0$, and then argue that the roots of the quadratic form $A\lambda^2 + (B + B^*)\lambda + C$ cannot be real. Show that this can be so only if

$$AC \geq \frac{1}{4}(B + B^*)^2$$

which is the same as the above inequality.

Define the function $h(x)$ as

$$h(x) = \lambda f(x) + g(x)$$

For real values of λ,

$$h^*(x) = \lambda f^*(x) + g^*(x)$$

Since $h^*(x)h(x)$ must be greater than or equal to zero at all x, the integral of this product over x must also be greater than or equal to zero. Thus,

$$\int [\lambda f^*(x) + g^*(x)][\lambda f(x) + g^*(x)] \, dx \geq 0$$

Expanding and then rewriting this expression in terms of A, B, and C,

$$\lambda^2 \int f^*(x)f(x) \, dx + \lambda \int f^*(x)g(x) \, dx + \lambda \int f(x)g^*(x) \, dx + \int g^*(x)g(x) \, dx \geq 0$$

$$A\lambda^2 + (B + B^*)\lambda + C \geq 0$$

If the function $A\lambda^2 + (B + B^*)\lambda + C$ had two real roots for λ, then a plot of the function versus λ would take the form of a parabola and intersect the λ axis in two places. Furthermore, for the values of λ between the two roots, the parabola would be negative. This is, however, not the case, as shown by the above inequality. Thus, the function can have at most one real root, which occurs at its tangent point to the λ axis. Using the quadratic formula, the root of the function is

$$\lambda = \frac{-(B + B^*) \pm \left[(B + B^*)^2 - 4AC\right]^{1/2}}{2A}$$

Since both roots cannot be real,

$$(B + B^*)^2 - 4AC \leq 0$$

As A and C each represents an integral of a product of a function and its complex conjugate, they must be greater or equal to zero. (The expression $B + B^*$ is real, but has no sign restriction.) It follows that

$$4AC \geq (B + B^*)^2$$

$$AC \geq \frac{1}{4}(B + B^*)^2$$

Writing A, B, and C in terms of their respective integrals,

$$\left[\int f^*(x)f(x)\,dx\right]\left[\int g^*(x)g(x)\,dx\right] \geq \frac{1}{4}\left\{\int \left[f^*(x)g(x) + f(x)g^*(x)\right]\,dx\right\}^2$$

4–47. We shall derive Equation 4.19 in this problem. You need the inequality that is derived in the previous problem to do this problem. Referring to the previous problem, let

$$f(x) = (\hat{A} - \langle a \rangle)\psi(x) \quad \text{and} \quad g(x) = i(\hat{B} - \langle b \rangle)\psi(x)$$

where $\psi(x)$ is any suitably behaved function. Substitute these into the left side of the inequality in the previous problem and use the fact that $\hat{A} - \langle a \rangle$ and $\hat{B} - \langle b \rangle$ are Hermitian (Problem 4–19) to write

$$\sigma_A^2 \sigma_B^2 \geq \frac{1}{4}\left[i\int dx\,(\hat{A} - \langle a \rangle)^*\psi^*(x)(\hat{B} - \langle b \rangle)\psi(x)\right.$$
$$\left. -i\int dx\,(\hat{B} - \langle b \rangle)^*\psi^*(x)(\hat{A} - \langle a \rangle)\psi(x)\right]^2$$

Use the Hermitian property of $\hat{A} - \langle a \rangle$ and $\hat{B} - \langle b \rangle$ again to write the right side as

$$-\frac{1}{4}\left\{\int dx\,\psi^*(x)\left[\hat{A}\hat{B} - \langle a \rangle\hat{B} - \hat{A}\langle b \rangle + \langle a \rangle\langle b \rangle\right.\right.$$
$$\left.\left. - \hat{B}\hat{A} + \langle b \rangle\hat{A} + \hat{B}\langle a \rangle - \langle a \rangle\langle b \rangle\right]\psi(x)\right\}^2$$

Now use the fact that $\langle a \rangle$ and $\langle b \rangle$ are just numbers to write

$$\sigma_A^2 \sigma_B^2 \geq -\frac{1}{4} \left\{ \int dx \ \psi^*(x) [\hat{A}, \hat{B}] \psi(x) \right\}^2$$

which is Equation 4.19.

$$\int f^*(x) f(x) \, dx = \int \left(\hat{A} - \langle a \rangle \right)^* \psi^*(x) \left(\hat{A} - \langle a \rangle \right) \psi(x) \, dx$$

$$= \int \left[\left(\hat{A} - \langle a \rangle \right) \psi(x) \right] \left(\hat{A} - \langle a \rangle \right)^* \psi^*(x) \, dx$$

$$\int g^*(x) g(x) \, dx = \int \left[i \left(\hat{B} - \langle b \rangle \right) \right]^* \psi^*(x) \left[i \left(\hat{B} - \langle b \rangle \right) \right] \psi(x) \, dx$$

$$= \int \left[\left(\hat{B} - \langle b \rangle \right) \psi(x) \right] \left(\hat{B} - \langle b \rangle \right)^* \psi^*(x) \, dx$$

Since $\hat{A}$ and $\hat{B}$ are Hermitian, $\hat{A} - \langle a \rangle$ and $\hat{B} - \langle b \rangle$ are also Hermitian (see Problem 4–19). Thus, the above expressions become

$$\int f^*(x) f(x) \, dx = \int \psi^*(x) \left(\hat{A} - \langle a \rangle \right)^2 \psi(x) \, dx = \sigma_A^2$$

$$\int g^*(x) g(x) \, dx = \int \psi^*(x) \left(\hat{B} - \langle b \rangle \right)^2 \psi(x) \, dx = \sigma_B^2$$

The inequality in Problem 4–46 becomes

$$\left[\int f^*(x) f(x) \, dx \right] \left[\int g^*(x) g(x) \, dx \right] \geq \frac{1}{4} \left\{ \int \left[f^*(x) g(x) + f(x) g^*(x) \right] \, dx \right\}^2$$

$$\sigma_A^2 \sigma_B^2 \geq \frac{1}{4} \left[i \int dx \, (\hat{A} - \langle a \rangle)^* \psi^*(x) (\hat{B} - \langle b \rangle) \psi(x) - i \int dx \, (\hat{B} - \langle b \rangle)^* \psi^*(x) (\hat{A} - \langle a \rangle) \psi(x) \right]^2$$

$$\sigma_A^2 \sigma_B^2 \geq \frac{1}{4} \left[i \int dx \, \left[\left(\hat{B} - \langle b \rangle \right) \psi(x) \right] (\hat{A} - \langle a \rangle)^* \psi^*(x) - i \int dx \, \left[\left(\hat{A} - \langle a \rangle \right) \psi(x) \right] (\hat{B} - \langle b \rangle)^* \psi^*(x) \right]^2$$

Using the Hermitian property of $\hat{A} - \langle a \rangle$ and $\hat{B} - \langle b \rangle$ again,

$$\sigma_A^2 \sigma_B^2 \geq \frac{1}{4} \left[i \int dx \, \psi^*(x) (\hat{A} - \langle a \rangle) \left(\hat{B} - \langle b \rangle \right) \psi(x) - i \int dx \, \psi^*(x) (\hat{B} - \langle b \rangle) \left(\hat{A} - \langle a \rangle \right) \psi(x) \right]^2$$

$$\sigma_A^2 \sigma_B^2 \geq -\frac{1}{4} \left\{ \int dx \, \psi^*(x) \left[\hat{A}\hat{B} - \langle a \rangle \hat{B} - \hat{A} \langle b \rangle + \langle a \rangle \langle b \rangle - \hat{B}\hat{A} + \langle b \rangle \hat{A} + \hat{B} \langle a \rangle - \langle a \rangle \langle b \rangle \right] \psi(x) \right\}^2$$

Since $\langle a \rangle$ and $\langle b \rangle$ are just numbers and $\hat{A}$ and $\hat{B}$ are linear operators, $\langle a \rangle \hat{B} = \hat{B} \langle a \rangle$ and $\langle b \rangle \hat{A} = \hat{A} \langle b \rangle$, and

$$\sigma_A^2 \sigma_B^2 \geq -\frac{1}{4} \left\{ \int dx \, \psi^*(x) \left[\hat{A}, \hat{B} \right] \psi(x) \right\}^2$$

4–48. Show that $\sin(n\pi x/a)$ is an even function of x about $a/2$ if n is odd and is an odd function about $a/2$ if n is even. Use this result to show that the c_n in Equation 4.60 are zero for even values of n.

Consider two points $x_1 = a/2 - k$ and $x_2 = a/2 + k$ that are symmetric about $x = a/2$,

$$\sin \frac{n\pi x_1}{a} = \sin \left(\frac{n\pi}{2} - \frac{n\pi k}{2} \right)$$

$$= \sin \frac{n\pi}{2} \cos \frac{n\pi k}{2} - \cos \frac{n\pi}{2} \sin \frac{n\pi k}{2}$$

$$\sin \frac{n\pi x_2}{a} = \sin \left(\frac{n\pi}{2} + \frac{n\pi k}{2} \right)$$

$$= \sin \frac{n\pi}{2} \cos \frac{n\pi k}{2} + \cos \frac{n\pi}{2} \sin \frac{n\pi k}{2}$$

If n is even,

$$\sin \frac{n\pi x_1}{a} = -\cos \frac{n\pi}{2} \sin \frac{n\pi k}{2}$$

$$\sin \frac{n\pi x_2}{a} = \cos \frac{n\pi}{2} \sin \frac{n\pi k}{2}$$

Since $\sin(n\pi x/a)$ is antisymmetric about $a/2$, it is odd. On the other hand, if n is odd,

$$\sin \frac{n\pi x_1}{a} = \sin \frac{n\pi}{2} \cos \frac{n\pi k}{2}$$

$$\sin \frac{n\pi x_2}{a} = \sin \frac{n\pi}{2} \cos \frac{n\pi k}{2}$$

Since $\sin(n\pi x/a)$ is symmetric about $a/2$, it is even.

The expression for c_n contains the function $x(a - x)$. This function also has a parity about $a/2$:

$$x_1(a - x_1) = \left(\frac{a}{2} - k \right) \left[a - \left(\frac{a}{2} - k \right) \right]$$

$$= \left(\frac{a}{2} - k \right) \left(\frac{a}{2} + k \right)$$

$$x_2(a - x_2) = \left(\frac{a}{2} + k \right) \left[a - \left(\frac{a}{2} + k \right) \right]$$

$$= \left(\frac{a}{2} + k \right) \left(\frac{a}{2} - k \right)$$

Since $x(a - x)$ is symmetric about $a/2$, it is even.

The expression for c_n is

$$c_n = \left(\frac{60}{a^6} \right)^{1/2} \int_0^a x(a - x) \sin \frac{n\pi x}{a} \, dx$$

When n is even, the integral involves an odd function [arising from a product of an even function, $x(a - x)$, and an odd function, $\sin(n\pi x/a)$] integrated over a range symmetric about $a/2$. Thus, $c_n = 0$.

Problems 4–49 through 4–54 deal with systems with piecewise constant potentials.

4–49. Consider a particle moving in the potential energy

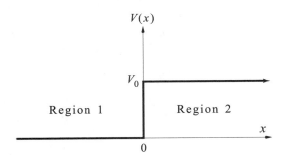

whose mathematical form is

$$V(x) = \begin{cases} 0 & x < 0 \\ V_0 & x > 0 \end{cases}$$

where V_0 is a constant. Show that if $E > V_0$, then the solutions to the Schrödinger equation in the two regions (1 and 2) are (see Problem 3–35)

$$\psi_1(x) = Ae^{ik_1x} + Be^{-ik_1x} \qquad x < 0 \tag{1}$$

and

$$\psi_2(x) = Ce^{ik_2x} + De^{-ik_2x} \qquad x > 0 \tag{2}$$

where

$$k_1 = \left(\frac{2mE}{\hbar^2}\right)^{1/2} \qquad \text{and} \qquad k_2 = \left[\frac{2m(E - V_0)}{\hbar^2}\right]^{1/2} \tag{3}$$

As we learned in Problem 3–35, e^{ikx} represents a particle traveling to the right and e^{-ikx} represents a particle traveling to the left. The physical problem we wish to set up is a particle of energy E traveling to the right and incident on a potential barrier of height V_0. If we wish to exclude the case of a particle traveling to the left in region 2, we set $D = 0$ in equation 2. The squares of the coefficients in equations 1 and 2 represent the probability that the particle is traveling in a certain direction in a given region. For example, $|A|^2$ is the probability that the particle is traveling with momentum $+\hbar k_1$ in the region $x < 0$. If we consider many particles, N_0, instead of just one, then we can interpret $|A|^2 N_0$ to be the number of particles with momentum $\hbar k_1$ in the region $x < 0$. The number of these particles that pass a given point per unit time is given by $v|A|^2 N_0$, where the velocity v is given by $\hbar k_1/m$.

Now apply the conditions that $\psi(x)$ and $d\psi/dx$ must be continuous at $x = 0$ (see Problem 4–4) to obtain

$$A + B = C$$

and

$$k_1(A - B) = k_2 C$$

Now define a quantity

$$r = \frac{\hbar k_1 |B|^2 N_0/m}{\hbar k_1 |A|^2 N_0/m} = \frac{|B|^2}{|A|^2}$$

and show that

$$r = \left(\frac{k_1 - k_2}{k_1 + k_2}\right)^2$$

Similarly, define

$$t = \frac{\hbar k_2 |C|^2 N_0/m}{\hbar k_1 |A|^2 N_0/m} = \frac{k_2 |C|^2}{k_1 |A|^2}$$

and show that

$$t = \frac{4k_1 k_2}{(k_1 + k_2)^2}$$

The symbols r and t stand for reflection coefficient and transmission coefficient, respectively. Give a physical interpretation of these designations. Show that $r + t = 1$. Would you have expected the particle to have been reflected even though its energy, E, is greater than the barrier height, V_0? Show that $r \to 0$ and $t \to 1$ as $V_0 \to 0$.

The Schrödinger equation in Region 1 is

$$-\frac{\hbar^2}{2m}\frac{d\psi_1}{dx} - E\psi_1 = 0$$

The general solution is a linear combination of two solutions (see Section 2.3):

$$\psi_1(x) = Ae^{ik_1 x} + Be^{-ik_1 x} \qquad \text{where } k_1 = \left(\frac{2mE}{\hbar^2}\right)^{1/2}$$

The Schrödinger equation in Region 2 is

$$-\frac{\hbar^2}{2m}\frac{d\psi_2}{dx} - (E - V_0)\psi_2 = 0$$

For $E > V_0$, the general solution is similar to that for Region 1 but with E replaced by $E - V_0$:

$$\psi_2(x) = Ce^{ik_2 x} + De^{-ik_2 x} \qquad \text{where } k_2 = \left[\frac{2m(E - V_0)}{\hbar^2}\right]^{1/2}$$

Excluding the possibility of a particle traveling to the left in Region 2, $D = 0$ and

$$\psi_2(x) = Ce^{ik_2 x}$$

The two wave functions must be continuous at $x = 0$,

$$\psi_1(0) = \psi_2(0)$$

$$A + B = C$$

and the continuity criterion must also be satisfied by the derivatives of the two wave functions:

$$\frac{d\psi_1}{dx}\bigg|_{x=0} = \frac{d\psi_2}{dx}\bigg|_{x=0}$$

$$Aik_1e^{ik_1x} - Bik_1e^{-ik_1x}\bigg|_{x=0} = Cik_2e^{ik_2x}\bigg|_{x=0}$$

$$k_1(A - B) = k_2C$$

These two boundary conditions can be combined to obtain the ratio B/A by multiplying the first condition by k_2 and subtracting from it the second condition:

$$A(k_2 - k_1) + B(k_2 + k_1) = 0$$

$$\frac{B}{A} = \frac{k_1 - k_2}{k_1 + k_2}$$

Alternatively, the conditions can be combined to obtain the ratio C/A by multiplying the first condition by k_1 and adding to it the second condition:

$$2k_1A = C(k_1 + k_2)$$

$$\frac{C}{A} = \frac{2k_1}{k_1 + k_2}$$

The quantity r becomes

$$r = \frac{\hbar k_1|B|^2N_0/m}{\hbar k_1|A|^2N_0/m} = \frac{|B|^2}{|A|^2}$$

$$= \left(\frac{k_1 - k_2}{k_1 + k_2}\right)^2$$

r represents the ratio of the number of particles with momentum $-\hbar k_1$ to the number of particles with momentum $\hbar k_1$ that pass a given point per unit time. In other words, r is the probability that particles traveling in region 1 towards region 2 are reflected at $x = 0$. Hence it is called the reflection coefficient.

The quantity t becomes

$$t = \frac{\hbar k_2|C|^2N_0/m}{\hbar k_1|A|^2N_0/m} = \frac{k_2|C|^2}{k_1|A|^2}$$

$$= \frac{4k_1k_2}{(k_1 + k_2)^2}$$

t represents the ratio of the number of particles with momentum $\hbar k_2$ to the number of particles with momentum $\hbar k_1$ that pass a given point per unit time. In other words, t is the probability that particles traveling in region 1 towards region 2 are transmitted to region 2 at $x = 0$. Hence it is called the transmission coefficient.

Since the particles are either reflected or transmitted, $r + t$ necessarily equals 1, which is verified below:

$$r + t = \left(\frac{k_1 - k_2}{k_1 + k_2}\right)^2 + \frac{4k_1 k_2}{(k_1 + k_2)^2}$$

$$= \frac{k_1^2 - 2k_1 k_2 + k_2^2 + 4k_1 k_2}{(k_1 + k_2)^2}$$

$$= 1$$

A classical particle would be transmitted and not be reflected if $E > V_0$. Thus, the results obtained here are due to quantum mechanical effects. As $V_0 \to 0$, $k_2 \to k_1$, giving $r \to 0$ and $t \to 1$.

4–50. Show that $r = 1$ for the system described in Problem 4–49 but with $E < V_0$. Discuss the physical interpretation of this result.

The Schrödinger equation in Region 1 is

$$-\frac{\hbar^2}{2m} \frac{d\psi_1}{dx} - E\psi_1 = 0$$

giving (see Section 2.3)

$$\psi_1(x) = Ae^{ik_1 x} + Be^{-ik_1 x} \qquad \text{where } k_1 = \left(\frac{2mE}{\hbar^2}\right)^{1/2}$$

The Schrödinger equation in Region 2 is

$$-\frac{\hbar^2}{2m} \frac{d\psi_2}{dx} - \left(E - V_0\right)\psi_2 = 0$$

For $E < V_0$, the general solution is (see Section 2.2):

$$\psi_2(x) = Ce^{\rho_2 x} + De^{-\rho_2 x} \qquad \text{where } \rho_2 = \left[\frac{2m\left(V_0 - E\right)}{\hbar^2}\right]^{1/2}$$

For the solution to remain finite when $x \to \infty$, C must be zero. Thus,

$$\psi_2(x) = De^{-\rho_2 x}$$

The two wave functions must be continuous at $x = 0$,

$$\psi_1(0) = \psi_2(0)$$

$$A + B = D$$

and the continuity criterion must also be satisfied by the derivatives of the two wave functions:

$$\left.\frac{d\psi_1}{dx}\right|_{x=0} = \left.\frac{d\psi_2}{dx}\right|_{x=0}$$

$$Aik_1e^{ik_1x} - Bik_1e^{-ik_1x}\Big|_{x=0} = -D\rho_2e^{-\rho_2x}\Big|_{x=0}$$

$$ik_1(A - B) = -\rho_2 D$$

These two boundary conditions can be combined to obtain the ratio B/A by multiplying the first condition by ρ_2 and adding to it the second condition:

$$A(\rho_2 + ik_1) + B(\rho_2 - ik_1) = 0$$

$$\frac{B}{A} = \frac{ik_1 + \rho_2}{ik_1 - \rho_2}$$

The reflection coefficient is

$$r = \frac{|B|^2}{|A|^2}$$

$$= \frac{(ik_1 + \rho_2)(-ik_1 + \rho_2)}{(ik_1 - \rho_2)(-ik_1 - \rho_2)}$$

$$= 1$$

The particle is always reflected at $x = 0$, as in the case for a classical particle.

4–51. In this problem, we introduce the idea of *quantum-mechanical tunneling*, which plays a central role in such diverse processes as the α decay of nuclei, electron-transfer reactions, and hydrogen bonding. Consider a particle in the potential energy regions as shown below.

Mathematically, we have

$$V(x) = \begin{cases} 0 & x < 0 \\ V_0 & 0 < x < a \\ 0 & x > a \end{cases}$$

Show that if $E < V_0$, the solution to the Schrödinger equation in each region is given by

$$\psi_1(x) = Ae^{ik_1x} + Be^{-ik_1x} \qquad x < 0 \tag{1}$$

$$\psi_2(x) = Ce^{k_2x} + De^{-k_2x} \qquad 0 < x < a \tag{2}$$

and

$$\psi_3(x) = Ee^{ik_1x} + Fe^{-ik_1x} \qquad x > a \tag{3}$$

where

$$k_1 = \left(\frac{2mE}{\hbar^2}\right)^{1/2} \quad \text{and} \quad k_2 = \left[\frac{2m(V_0 - E)}{\hbar^2}\right]^{1/2} \tag{4}$$

If we exclude the situation of the particle coming from large positive values of x, then $F = 0$ in equation 3. Following Problem 4–49, argue that the transmission coefficient, the probability the particle will get past the barrier, is given by

$$t = \frac{|E|^2}{|A|^2} \tag{5}$$

Now use the fact that $\psi(x)$ and $d\psi/dx$ must be continuous at $x = 0$ and $x = a$ to obtain

$$A + B = C + D \qquad ik_1(A - B) = k_2(C - D) \tag{6}$$

and

$$Ce^{k_2a} + De^{-k_2a} = Ee^{ik_1a} \qquad k_2Ce^{k_2a} - k_2De^{-k_2a} = ik_1Ee^{ik_1a} \tag{7}$$

Eliminate B from equations 6 to get A in terms of C and D. Then solve equations 7 for C and D in terms of E. Substitute these results into the equation for A in terms of C and D to get the intermediate result

$$2ik_1A = \left[(k_1^2 - k_2^2 + 2ik_1k_2)e^{k_2a} + (k_2^2 - k_1^2 + 2ik_1k_2)e^{-k_2a}\right]\frac{Ee^{ik_1a}}{2k_2}$$

Now use the relations $\sinh x = (e^x - e^{-x})/2$ and $\cosh x = (e^x + e^{-x})/2$ (Problem A–11) to get

$$\frac{E}{A} = \frac{4ik_1k_2e^{-ik_1a}}{2(k_1^2 - k_2^2)\sinh k_2a + 4ik_1k_2\cosh k_2a}$$

Now multiply the right side by its complex conjugate and use the relation $\cosh^2 x = 1 + \sinh^2 x$ to get

$$t = \left|\frac{E}{A}\right|^2 = \frac{4}{4 + \dfrac{(k_1^2 + k_2^2)^2}{k_1^2k_2^2}\sinh^2 k_2a}$$

Finally, use the definition of k_1 and k_2 to show that the probability the particle gets through the barrier (even though it does not have enough energy!) is

$$t = \frac{1}{1 + \dfrac{v_0^2}{4\varepsilon(v_0 - \varepsilon)}\sinh^2(v_0 - \varepsilon)^{1/2}} \tag{8}$$

or

$$t = \frac{1}{1 + \dfrac{\sinh^2[v_0^{1/2}(1 - \alpha)^{1/2}]}{4\alpha(1 - \alpha)}} \tag{9}$$

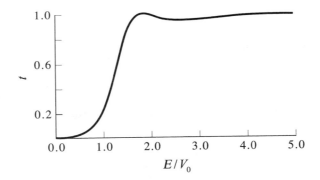

FIGURE 4.7
A plot of the probability that a particle of energy E will penetrate a barrier of height V_0 plotted against the ratio E/V_0 (equation 9 of Problem 4–51 with $v_0 = 10$).

where $v_0 = 2ma^2 V_0/\hbar^2$, $\varepsilon = 2ma^2 E/\hbar^2$, and $\alpha = E/V_0 = \varepsilon/v_0$. Figure 4.7 shows a plot of t versus α for $v_0 = 10$. To plot t versus α for values of $\alpha > 1$, you need to use the relation $\sinh ix = i \sin x$ (Problem A–12). What would the classical result look like?

The Schrödinger equation in Region 1 is

$$-\frac{\hbar^2}{2m}\frac{d\psi_1}{dx} - E\psi_1 = 0$$

giving (see Section 2.3)

$$\psi_1(x) = Ae^{ik_1x} + Be^{-ik_1x} \qquad x < 0 \tag{1}$$

The Schrödinger equation in Region 2 is

$$-\frac{\hbar^2}{2m}\frac{d\psi_2}{dx} - \left(E - V_0\right)\psi_2 = 0$$

For $E < V_0$, the general solution is (see Section 2.2):

$$\psi_2(x) = Ce^{k_2x} + De^{-k_2x} \qquad 0 < x < a \tag{2}$$

The Schrödinger equation in Region 3 is similar to that for Region 1:

$$-\frac{\hbar^2}{2m}\frac{d\psi_3}{dx} - E\psi_3 = 0$$

giving

$$\psi_3(x) = Ee^{ik_1x} + Fe^{-ik_1x} \qquad x > a \tag{3}$$

F is set to zero; thus,

$$\psi_3(x) = Ee^{ik_1x}$$

The expressions for k_1 and k_2 are

$$k_1 = \left(\frac{2mE}{\hbar^2}\right)^{1/2} \quad \text{and} \quad k_2 = \left[\frac{2m(V_0 - E)}{\hbar^2}\right]^{1/2} \tag{4}$$

The transmission coefficient represents the ratio of the number of particles with momentum hk_1 in Region 3 to the number of particles with momentum hk_1 in Region 1 that pass a given point per unit time. Thus,

$$t = \frac{\hbar k_1 |E|^2 N_0/m}{\hbar k_1 |A|^2 N_0/m} = \frac{|E|^2}{|A|^2}$$

Now relate the coefficients in the wave functions using boundary conditions. The wave functions ψ_1 and ψ_2 and their derivatives must be continuous at $x = 0$:

$$\psi_1(0) = \psi_2(0)$$

$$\frac{d\psi_1}{dx}\bigg|_{x=0} = \frac{d\psi_2}{dx}\bigg|_{x=0}$$

These conditions give

$$A + B = C + D \qquad ik_1(A - B) = k_2(C - D) \tag{6}$$

The wave functions ψ_2 and ψ_3 and their derivatives must be continuous at $x = a$:

$$\psi_2(a) = \psi_3(a)$$

$$\frac{d\psi_2}{dx}\bigg|_{x=a} = \frac{d\psi_3}{dx}\bigg|_{x=a}$$

to give

$$Ce^{k_2a} + De^{-k_2a} = Ee^{ik_1a} \qquad k_2Ce^{k_2a} - k_2De^{-k_2a} = ik_1Ee^{ik_1a} \tag{7}$$

Eliminate B from Equation 6 by adding together the equations

$$ik_1(A + B) = ik_1(C + D)$$

$$ik_1(A - B) = k_2(C - D)$$

to give

$$2ik_1A = C\left(k_2 + ik_1\right) - D\left(k_2 - ik_1\right) \tag{8}$$

Solve Equation 7 for C and D in terms of E by adding and subtracting the equations, respectively,

$$k_2\left(Ce^{k_2a} + De^{-k_2a}\right) = k_2Ee^{ik_1a}$$

$$k_2Ce^{k_2a} - k_2De^{-k_2a} = ik_1Ee^{ik_1a}$$

to give

$$C = \frac{(k_2 + ik_1)Ee^{ik_1a}}{2k_2e^{k_2a}} \tag{9}$$

$$D = \frac{(k_2 - ik_1)Ee^{ik_1a}}{2k_2e^{-k_2a}} \tag{10}$$

Substituting these two equations into Equation 8 gives

$$2ik_1A = \left[\frac{(k_2 + ik_1)Ee^{ik_1a}}{2k_2e^{k_2a}}(k_2 + ik_1) - \frac{(k_2 - ik_1)Ee^{ik_1a}}{2k_2e^{-k_2a}}(k_2 - ik_1)\right]$$

$$= \left[\frac{(k_2 + ik_1)^2}{e^{k_2a}} - \frac{(k_2 - ik_1)^2}{e^{-k_2a}}\right]\frac{Ee^{ik_1a}}{2k_2}$$

$$= \left[\frac{(k_2^2 - k_1^2 + 2ik_1k_2)e^{-k_2a} + (k_1^2 - k_2^2 + 2ik_1k_2)e^{k_2a}}{e^{-k_2a}e^{k_2a}}\right]\frac{Ee^{ik_1a}}{2k_2}$$

$$= [2k_1^2 \sinh k_2a - 2k_2^2 \sinh k_2a + 4ik_1k_2 \cosh k_2a]\frac{Ee^{ik_1a}}{2k_2}$$

$$\frac{E}{A} = \frac{4ik_1k_2e^{-ik_1a}}{2(k_1^2 - k_2^2) \sinh k_2a + 4ik_1k_2 \cosh k_2a}$$

The transmission coefficient is

$$t = \left|\frac{E}{A}\right|^2$$

$$= \frac{(4ik_1k_2e^{-ik_1a})(-4ik_1k_2e^{ik_1a})}{\left[2(k_1^2 - k_2^2) \sinh k_2a + 4ik_1k_2 \cosh k_2a\right]\left[2(k_1^2 - k_2^2) \sinh k_2a - 4ik_1k_2 \cosh k_2a\right]}$$

$$= \frac{16k_1^2k_2^2}{4(k_1^2 - k_2^2)^2 \sinh^2 k_2a + 16k_1^2k_2^2 \cosh^2 k_2a}$$

$$= \frac{16k_1^2k_2^2}{4(k_1^2 - k_2^2)^2 \sinh^2 k_2a + 16k_1^2k_2^2\left(1 + \sinh^2 k_2a\right)}$$

$$= \frac{16k_1^2k_2^2}{4(k_1^2 + k_2^2)^2 \sinh^2 k_2a + 16k_1^2k_2^2}$$

$$= \frac{4}{4 + \frac{(k_1^2 + k_2^2)^2}{k_1^2k_2^2} \sinh^2 k_2a}$$

Using the definitions of k_1 and k_2 in Equation 4 and letting $v_0 = 2ma^2V_0/\hbar^2$, $\varepsilon = 2ma^2E/\hbar$, and $\alpha = E/V_0 = \varepsilon/v_0$, the transmission coefficient becomes

$$t = \cfrac{4}{4 + \cfrac{\left[\frac{2mE}{\hbar^2} + \frac{2m(V_0-E)}{\hbar^2}\right]^2}{\left(\frac{2mE}{\hbar^2}\right)\left[\frac{2m(V_0-E)}{\hbar^2}\right]} \sinh^2\left[\left(\frac{2m(V_0-E)}{\hbar^2}\right)^{1/2} a\right]}$$

$$= \cfrac{4}{4 + \cfrac{v_0^2}{\varepsilon(v_0-\varepsilon)} \sinh^2(v_0-\varepsilon)^{1/2}}$$

$$= \cfrac{1}{1 + \cfrac{v_0^2}{4\varepsilon(v_0-\varepsilon)} \sinh^2(v_0-\varepsilon)^{1/2}}$$

$$= \cfrac{1}{1 + \cfrac{\sinh^2[v_0^{1/2}(1-\alpha)^{1/2}]}{4\alpha(1-\alpha)}}$$

A plot of t versus $\alpha = E/V_0$, as shown in Figure 4.7, indicates that the probability that a particle would penetrate a barrier increases as α increases. The probability is small, but finite even when $E < V_0$. This is not the case with a classical particle, which would not penetrate the barrier ($t = 0$) when $E \leq V_0$ but would transmit entirely ($t = 1$) when $E > V_0$. Thus, there is a discontinuity of the t versus α plot at $E = V_0$ for the classical particle.

4–52. Use the result of Problem 4–51 to determine the probability that an electron with a kinetic energy 8.0×10^{-21} J will tunnel through a 1.0 nm thick potential barrier with $V_0 = 12.0 \times 10^{-21}$ J.

$$\alpha = \frac{E}{V_0} = \frac{8.0 \times 10^{-21} \text{ J}}{12.0 \times 10^{-21} \text{ J}} = \frac{2}{3}$$

$$v_0 = \frac{2ma^2 V_0}{\hbar^2}$$

$$= \frac{2 \left(9.1094 \times 10^{-31} \text{ kg}\right) \left(1.0 \times 10^{-9} \text{ m}\right)^2 \left(12.0 \times 10^{-21} \text{ J}\right)}{\left(1.0546 \times 10^{-34} \text{ J·s}\right)^2}$$

$$= 1.9657$$

The probability that the electron will tunnel through the barrier is

$$t = \cfrac{1}{1 + \cfrac{\sinh^2[v_0^{1/2}(1-\alpha)^{1/2}]}{4\alpha(1-\alpha)}}$$

$$= \cfrac{1}{1 + \cfrac{\sinh^2[1.9657^{1/2}(1/3)^{1/2}]}{4(2/3)(1/3)}}$$

$$= 0.52$$

This is a significant probability.

4–53.   Problem 4–51 shows that the probability that a particle of relative energy E/V_0 will penetrate a rectangular potential barrier of height V_0 and thickness a is

$$t = \cfrac{1}{1 + \cfrac{\sinh^2[v_0^{1/2}(1-\alpha)^{1/2}]}{4\alpha(1-\alpha)}}$$

where $v_0 = 2mV_0a^2/\hbar^2$ and $\alpha = E/V_0$. What is the limit of t as $\alpha \to 1$? Plot t against α for $v_0 = 1/2$, 1, and 2. Interpret your results.

As $\alpha \to 1$, the argument of sinh approaches 0. Expand sinh x in a Maclaurin series:

$$\sinh x = x + \frac{x^3}{3!} + \frac{x^5}{5!} + \cdots$$

and keep only the first term, which is the most significant term, in the expansion and substitute into the transmission expression:

$$\lim_{\alpha \to 1} t = \cfrac{1}{1 + \cfrac{\left[v_0^{1/2}(1-\alpha)^{1/2}\right]^2}{4\alpha(1-\alpha)}}$$

$$= \frac{4\alpha(1-\alpha)}{4\alpha(1-\alpha) + v_0(1-\alpha)}$$

$$= \frac{4\alpha}{4\alpha + v_0}$$

$$= \frac{1}{1 + v_0/4}$$

A plot of t versus α below shows that the higher and thicker the potential barrier, the smaller the probability of transmission at a particular E.

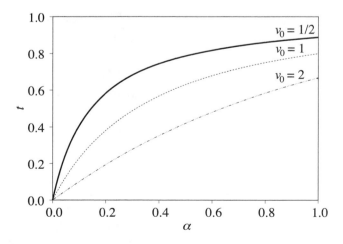

4–54.   In this problem, we will consider a particle in a *finite* potential well,

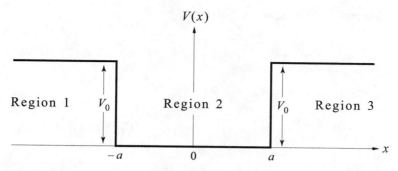

whose mathematical form is

$$V(x) = \begin{cases} V_0 & x < -a \\ 0 & -a < x < a \\ V_0 & x > a \end{cases} \tag{1}$$

Note that this potential describes what we have called a "particle in a box" if $V_0 \to \infty$. Show that if $0 < E < V_0$, the solution to the Schrödinger equation in each region is

$$\psi_1(x) = Ae^{k_1 x} \quad x < -a$$

$$\psi_2(x) = B \sin \alpha x + C \cos \alpha x \quad -a < x < a \tag{2}$$

$$\psi_3(x) = De^{-k_1 x} \quad x > a$$

where

$$k_1 = \left[\frac{2m(V_0 - E)}{\hbar^2} \right]^{1/2} \quad \text{and} \quad \alpha = \left(\frac{2mE}{\hbar^2} \right)^{1/2} \tag{3}$$

Now apply the conditions that $\psi(x)$ and $d\psi/dx$ must be continuous at $x = -a$ and $x = a$ to obtain

$$Ae^{-k_1 a} = -B \sin \alpha a + C \cos \alpha a \tag{4}$$

$$De^{-k_1 a} = B \sin \alpha a + C \cos \alpha a \tag{5}$$

$$k_1 Ae^{-k_1 a} = \alpha B \cos \alpha a + \alpha C \sin \alpha a \tag{6}$$

and

$$-k_1 De^{-k_1 a} = \alpha B \cos \alpha a - \alpha C \sin \alpha a \tag{7}$$

Add and subtract equations 4 and 5 and add and subtract equations 6 and 7 to obtain

$$2C \cos \alpha a = (A + D)e^{-k_1 a} \tag{8}$$

$$2B \sin \alpha a = (D - A)e^{-k_1 a} \tag{9}$$

$$2\alpha C \sin \alpha a = k_1(A + D)e^{-k_1 a} \tag{10}$$

and

$$2\alpha B \cos \alpha a = -k_1(D - A)e^{-k_1 a} \tag{11}$$

Now divide equation 10 by equation 8 to get

$$\frac{\alpha \sin \alpha a}{\cos \alpha a} = \alpha \tan \alpha a = k_1 \qquad (D \neq -A \text{ and } C \neq 0) \qquad (12)$$

and then divide equation 11 by equation 9 to get

$$\frac{\alpha \cos \alpha a}{\sin \alpha a} = \alpha \cot \alpha a = -k_1 \qquad (D \neq A \text{ and } B \neq 0) \qquad (13)$$

Referring back to equation 3, note that equations 12 and 13 give the allowed values of E in terms of V_0. It turns out that these two equations cannot be solved simultaneously, so we have two sets of equations:

$$\alpha \tan \alpha a = k_1 \qquad (14)$$

and

$$\alpha \cot \alpha a = -k_1 \qquad (15)$$

Let's consider equation 14 first. Multiply both sides by a and use the definitions of α and k_1 to get

$$\left(\frac{2ma^2 E}{\hbar^2}\right)^{1/2} \tan \left(\frac{2ma^2 E}{\hbar^2}\right)^{1/2} = \left[\frac{2ma^2}{\hbar^2}(V_0 - E)\right]^{1/2} \qquad (16)$$

Show that this equation simplifies to

$$\varepsilon^{1/2} \tan \varepsilon^{1/2} = (v_0 - \varepsilon)^{1/2} \qquad (17)$$

where $\varepsilon = 2ma^2 E/\hbar^2$ and $v_0 = 2ma^2 V_0/\hbar^2$. Thus, if we fix v_0 (actually $2ma^2 V_0/\hbar^2$), then we can use equation 17 to solve for the allowed values of ε (actually $2ma^2 E/\hbar^2$). Equation 17 cannot be solved analytically, but if we plot both $\varepsilon^{1/2} \tan \varepsilon^{1/2}$ and $(v_0 - \varepsilon)^{1/2}$ versus ε on the same graph, then the solutions are given by the intersections of the two curves. Figure 4.8a shows such a plot for $v_0 = 12$.

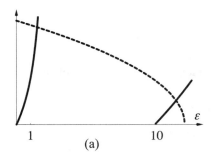
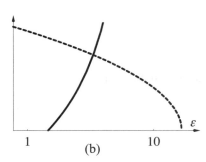

FIGURE 4.8
(a) Plots of both $\varepsilon^{1/2} \tan \varepsilon^{1/2}$ (solid curve) and $(12 - \varepsilon)^{1/2}$ (dotted curve) versus ε. The intersections of the curves give the allowed values of ε for a one-dimensional potential well of depth $V_0 = 12\hbar^2/2ma^2$. (b) Plots of both $-\varepsilon^{1/2} \cot \varepsilon^{1/2}$ (solid curve) and $(12 - \varepsilon)^{1/2}$ (dotted curve) plotted against ε. The intersection gives an allowed value of ε for a one-dimensional potential well of depth $V_0 = 12\hbar^2/2ma^2$.

The intersections occur at $\varepsilon = 2ma^2 E/\hbar^2 = 1.47$ and 11.37. The other value(s) of ε are given by the solutions to equation 15, which are obtained by finding the intersection of $-\varepsilon^{1/2} \cot \varepsilon^{1/2}$ and $(v_0 - \varepsilon)^{1/2}$ plotted against ε. Such a plot is shown in Figure 4.8b for $v_0 = 12$, giving $\varepsilon = 2ma^2 E/\hbar^2 = 5.68$. Thus, we see there are only three bound states for a well of depth $V_0 = 12\hbar^2/2ma^2$. The important point here is not the numerical values of E, but the fact that there is only a finite number of bound states. Show that there are only two bound states for $v_0 = 2ma^2 V_0/\hbar^2 = 4$.

The Schrödinger equation in Region 1 is

$$-\frac{\hbar^2}{2m} \frac{d\psi_1}{dx} - (E - V_0)\,\psi_1 = 0$$

giving

$$\psi_1(x) = Ae^{k_1 x}$$

The $e^{-k_1 x}$ term is not included because only the case of a particle traveling from Region 1 to 2 is considered. Furthermore, it is not finite as $x \to -a$. The Schrödinger equation in Region 2 is

$$-\frac{\hbar^2}{2m} \frac{d\psi_2}{dx} - E\psi_2 = 0$$

giving

$$\psi_2(x) = B \sin \alpha x + C \cos \alpha x$$

Here the solution is written in terms of trignometric functions instead of the equivalent forms for $e^{-i\alpha x}$ and $e^{i\alpha x}$. The Schrödinger equation in Region 3 is similar to that for Region 1:

$$-\frac{\hbar^2}{2m} \frac{d\psi_3}{dx} - (E - V_0)\,\psi_3 = 0$$

giving

$$\psi_3(x) = De^{-k_1 x}$$

The $e^{k_1 x}$ term is not included because it is not finite as $x \to \infty$ and furthermore, the case of a particle traveling to the left in Region 3 is not considered here. The expressions for k_1 and k_2 are

$$k_1 = \left[\frac{2m(V_0 - E)}{\hbar^2}\right]^{1/2} \quad \text{and} \quad \alpha = \left(\frac{2mE}{\hbar^2}\right)^{1/2} \tag{3}$$

Now relate the coefficients in the wave functions using boundary conditions. The wave functions ψ_1 and ψ_2 must be continuous at $x = -a$ and the wave functions ψ_2 and ψ_3 must be continuous at $x = a$, respectively:

$$Ae^{-k_1 a} = B \sin(-\alpha a) + C \cos(-\alpha a) = -B \sin \alpha a + C \cos \alpha a \tag{4}$$

$$De^{-k_1 a} = B \sin \alpha a + C \cos \alpha a \tag{5}$$

The derivatives of these wave functions must be continuous at $x = -a$ and $x = a$, respectively:

$$k_1 A e^{-k_1 a} = \alpha B \cos(-\alpha a) - \alpha C \sin(-\alpha a) = \alpha B \cos \alpha a + \alpha C \sin \alpha a \tag{6}$$

$$-k_1 D e^{-k_1 a} = \alpha B \cos \alpha a - \alpha C \sin \alpha a \tag{7}$$

Adding and subtracting Equations 4 and 5 gives

$$2C \cos \alpha a = (A + D)e^{-k_1 a} \tag{8}$$

$$2B \sin \alpha a = (D - A)e^{-k_1 a} \tag{9}$$

Subtracting and adding Equations 6 and 7 gives

$$2\alpha C \sin \alpha a = k_1 (A + D)e^{-k_1 a} \tag{10}$$

$$2\alpha B \cos \alpha a = -k_1 (D - A)e^{-k_1 a} \tag{11}$$

Dividing Equation 10 by Equation 8,

$$\frac{\alpha \sin \alpha a}{\cos \alpha a} = \alpha \tan \alpha a = k_1 \qquad (D \neq -A \text{ and } C \neq 0) \tag{12}$$

and dividing Equation 11 by Equation 9,

$$\frac{\alpha \cos \alpha a}{\sin \alpha a} = \alpha \cot \alpha a = -k_1 \qquad (D \neq A \text{ and } B \neq 0) \tag{13}$$

These two sets of equations:

$$\alpha \tan \alpha a = k_1 \tag{14}$$

$$\alpha \cot \alpha a = -k_1 \tag{15}$$

relate the allowed values of E to V_0. Multiplying Equation 14 by a and using the definitions of α and k_1,

$$\left(\frac{2ma^2 E}{\hbar^2}\right)^{1/2} \tan \left(\frac{2ma^2 E}{\hbar^2}\right)^{1/2} = \left[\frac{2ma^2}{\hbar^2}(V_0 - E)\right]^{1/2} \tag{16}$$

Letting $\varepsilon = 2ma^2 E/\hbar^2$ and $v_0 = 2ma^2 V_0/\hbar^2$, Equation 16 becomes

$$\varepsilon^{1/2} \tan \varepsilon^{1/2} = (v_0 - \varepsilon)^{1/2} \tag{17}$$

Multiplying Equation 15 by a followed by similar substitutions,

$$-\varepsilon^{1/2} \cot \varepsilon^{1/2} = (v_0 - \varepsilon)^{1/2} \tag{18}$$

For $v_0 = 4$, the curve $\varepsilon^{1/2} \tan \varepsilon^{1/2}$ versus ε intersects the curve $(v_0 - \varepsilon)^{1/2}$ versus ε once (see the left graph below) while the curve $-\varepsilon^{1/2} \cot \varepsilon^{1/2}$ versus ε also intersects the curve $(v_0 - \varepsilon)^{1/2}$ versus ε once (right graph). Therefore, there are only two bound states for $v_0 = 4$.

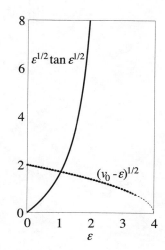
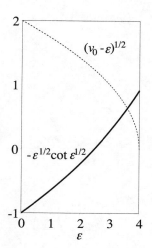

Series and Limits

PROBLEMS AND SOLUTIONS

D–1. Calculate the percentage difference between e^x and $1 + x$ for $x = 0.0050, 0.0100, 0.0150,$ ..., 0.1000.

With % difference defined as $\left| \dfrac{e^x - (1 + x)}{e^x} \right| \times 100$, the desired results are in the following table.

x	e^x	$1 + x$	% difference
0.0050	1.005 0125	1.0050	0.001 25%
0.0100	1.010 0502	1.0100	0.004 97%
0.0150	1.015 1131	1.0150	0.011 14%
0.0200	1.020 2013	1.0200	0.019 74%
0.0250	1.025 3151	1.0250	0.030 73%
0.0300	1.030 4545	1.0300	0.044 11%
0.0350	1.035 6197	1.0350	0.059 84%
0.0400	1.040 8108	1.0400	0.077 90%
0.0450	1.046 0279	1.0450	0.098 26%
0.0500	1.051 2711	1.0500	0.120 91%
0.0550	1.056 5406	1.0550	0.145 82%
0.0600	1.061 8365	1.0600	0.172 96%
0.0650	1.067 1590	1.0650	0.202 32%
0.0700	1.072 5082	1.0700	0.233 86%
0.0750	1.077 8842	1.0750	0.267 58%
0.0800	1.083 2871	1.0800	0.303 43%
0.0850	1.088 7171	1.0850	0.341 42%
0.0900	1.094 1743	1.0900	0.381 50%
0.0950	1.099 6589	1.0950	0.423 66%
0.1000	1.105 1709	1.1000	0.467 88%

D–2. Calculate the percentage difference between $\ln(1 + x)$ and x for $x = 0.0050, 0.0100, 0.0150,$ ..., 0.1000.

With % difference defined as $\left| \dfrac{\ln(1+x) - x}{\ln(1+x)} \right| \times 100$, the desired results are in the following table.

x	$\ln(1+x)$	x	% difference
0.0050	0.004 9875	0.0050	0.250%
0.0100	0.009 9503	0.0100	0.499%
0.0150	0.014 8886	0.0150	0.748%
0.0200	0.019 8026	0.0200	0.997%
0.0250	0.024 6926	0.0250	1.245%
0.0300	0.029 5588	0.0300	1.493%
0.0350	0.034 4014	0.0350	1.740%
0.0400	0.039 2207	0.0400	1.987%
0.0450	0.044 0169	0.0450	2.233%
0.0500	0.048 7902	0.0500	2.480%
0.0550	0.053 5408	0.0550	2.725%
0.0600	0.058 2689	0.0600	2.971%
0.0650	0.062 9748	0.0650	3.216%
0.0700	0.067 6586	0.0700	3.461%
0.0750	0.072 3207	0.0750	3.705%
0.0800	0.076 9610	0.0800	3.949%
0.0850	0.081 5800	0.0850	4.192%
0.0900	0.086 1777	0.0900	4.435%
0.0950	0.090 7544	0.0950	4.678%
0.1000	0.095 3102	0.1000	4.921%

D–3. Write out the expansion of $(1+x)^{1/2}$ through the quadratic term.

Use Equation D.13 with $n = 1/2$.

$$(1+x)^{1/2} = 1 + \frac{x}{2} + \frac{\frac{1}{2}\left(\frac{1}{2} - 1\right)}{2!}x^2 + O(x^3)$$

$$= 1 + \frac{x}{2} - \frac{x^2}{8} + O(x^3)$$

D–4. Write out the expansion of $(1+x)^{-1/2}$ through the quadratic term.

Use Equation D.13 with $n = -1/2$.

$$(1+x)^{-1/2} = 1 - \frac{x}{2} + \frac{-\frac{1}{2}\left(-\frac{1}{2} - 1\right)}{2!}x^2 + O(x^3)$$

$$= 1 - \frac{x}{2} + \frac{3x^2}{8} + O(x^3)$$

D–5. Show that

$$\frac{1}{(1-x)^2} = 1 + 2x + 3x^2 + 4x^3 + \cdots$$

Realize that

$$\frac{1}{(1-x)^2} = \frac{d}{dx}\left(\frac{1}{1-x}\right)$$

Then use Equation D.3

$$\frac{1}{(1-x)^2} = \frac{d}{dx}\left(\frac{1}{1-x}\right)$$

$$= \frac{d}{dx}\sum_{n=0}^{\infty} x^n$$

$$= \sum_{n=0}^{\infty} \frac{d}{dx}x^n$$

$$= \sum_{n=0}^{\infty} nx^{n-1}$$

$$= 1 + 2x + 3x^2 + 4x^3 + \cdots$$

Notice that the term with $n = 0$ does not contribute to the final summation.

D–6. Evaluate the series

$$S = \frac{1}{2} + \frac{1}{4} + \frac{1}{8} + \frac{1}{16} + \cdots$$

S is recognized as a geometric series with $x = 1/2$, but missing its first term. Equation D.3 is used to find the desired result.

$$S = \frac{1}{2} + \frac{1}{4} + \frac{1}{8} + \frac{1}{16} + \cdots$$

$$= \left[\sum_{n=0}^{\infty}\left(\frac{1}{2}\right)^n\right] - 1$$

$$= \frac{1}{1-\frac{1}{2}} - 1$$

$$= 1$$

D–7. Evaluate the series

$$S = \sum_{n=0}^{\infty} \frac{1}{3^n}$$

S is a geometric series with $x = 1/3$. Using Equation D.3,

$$S = \sum_{n=0}^{\infty} \frac{1}{3^n} = \frac{1}{1-\frac{1}{3}} = \frac{3}{2}$$

D–8. Evaluate the series

$$S = \sum_{n=1}^{\infty} \frac{(-1)^{n+1}}{2^n}$$

With a little manipulation, S can be put into the form of the geometric series in Equation D.3 with $x = -1/2$.

$$S = \sum_{n=1}^{\infty} \frac{(-1)^{n+1}}{2^n}$$

$$= -\sum_{n=1}^{\infty} \left(-\frac{1}{2}\right)^n$$

$$= 1 - \sum_{n=0}^{\infty} \left(-\frac{1}{2}\right)^n$$

$$= 1 - \frac{1}{1-\left(-\frac{1}{2}\right)}$$

$$= \frac{1}{3}$$

D–9. Numbers whose decimal formula are recurring decimals such as $0.272\,727\ldots$ are rational numbers, meaning that they can be expressed as the ratio of two numbers (in other words, as a fraction). Show that $0.272\,727\ldots = 27/99$.

Let $s = 0.272\,727\ldots$. Then,

$$99s = 100s - s = 27.272\,727\ldots - 0.272\,727\ldots = 27$$

or $s = 27/99$.

D–10. Show that $0.142\,857\,142\,857\,142\,857\ldots = 1/7$. (See the previous problem.)

Let $s = 0.142\,857\,142\,857\,142\,857\ldots$. Then,

$$999\,999s = 1\,000\,000s - s$$
$$= 142\,857.142\,857\,142\,857\,142\,857\ldots - 0.142\,857\,142\,857\,142\,857\ldots$$
$$= 142\,857$$

Thus, $s = 142\,857/999\,999 = 1/7$.

D–11. Series of the form

$$S(x) = \sum_{n=0}^{\infty} nx^n$$

occur frequently in physical problems. To find a closed expression for $S(x)$, we start with

$$\frac{1}{1-x} = \sum_{n=0}^{\infty} x^n$$

Notice now that $S(x)$ can be expressed as

$$x\frac{d}{dx}\sum_{n=0}^{\infty} x^n = \sum_{n=0}^{\infty} nx^n$$

and show that $S(x) = x/(1-x)^2$.

Since differentiation is a linear operation,

$$x\frac{d}{dx}\sum_{n=0}^{\infty} x^n = x\sum_{n=0}^{\infty}\frac{d}{dx}x^n = x\sum_{n=0}^{\infty} nx^{n-1} = \sum_{n=0}^{\infty} nx^n$$

The summation on the extreme left hand side may be evaluated using Equation D.3, which gives the desired result.

$$S(x) = \sum_{n=0}^{\infty} nx^n = x\frac{d}{dx}\left(\frac{1}{1-x}\right) = \frac{x}{(1-x)^2}$$

D–12. Using the method introduced in the previous problem, show that

$$S(x) = \sum_{n=0}^{\infty} n^2 x^n = \frac{x(1+x)}{(1-x)^3}$$

Start with the result from the previous problem,

$$\sum_{n=0}^{\infty} nx^n = \frac{x}{(1-x)^2}$$

differentiate both sides with respect to x,

$$\frac{d}{dx}\sum_{n=0}^{\infty} nx^n = \frac{d}{dx}\left[\frac{x}{(1-x)^2}\right]$$

$$\sum_{n=0}^{\infty} n^2 x^{n-1} = \frac{1}{(1-x)^2} + \frac{2x}{(1-x)^3} = \frac{1+x}{(1-x)^3}$$

and finally multiply this last result by x:

$$x\sum_{n=0}^{\infty} n^2 x^{n-1} = \sum_{n=0}^{\infty} n^2 x^n = \frac{x(1+x)}{(1-x)^3}$$

D–13. Use Equation D.9 to derive Equations D.10 and D.11.

Equation D.9 requires $f(0)$ and $d^n f/dx^n$ at $x = 0$, in this case for $f(x) = \sin x$ and $f(x) = \cos x$. For $f(x) = \sin x$, $f(0) = 0$ and

$$\frac{d^n f}{dx^n} = \begin{cases} (-1)^{(n-1)/2}\cos x & n \text{ odd} \\ (-1)^{n/2}\sin x & n \text{ even} \end{cases}$$

Evaluating at $x = 0$ gives,

$$\left(\frac{d^n f}{dx^n}\right)_{x=0} = \begin{cases} (-1)^{(n-1)/2} & n \text{ odd} \\ 0 & n \text{ even} \end{cases}$$

Thus, all the terms in Equation D.9 with even values of n vanish, leaving

$$\sin x = x + \frac{1}{3!}(-1)^1 x^3 + \frac{1}{5!}(-1)^2 x^5 + \frac{1}{7!}(-1)^3 x^7 + \cdots$$

$$= x - \frac{x^3}{3!} + \frac{x^5}{5!} - \frac{x^7}{7!} + \cdots$$

which is Equation D.10.

For $f(x) = \cos x$, $f(0) = 1$ and

$$\frac{d^n f}{dx^n} = \begin{cases} (-1)^{(n+1)/2}\sin x & n \text{ odd} \\ (-1)^{n/2}\cos x & n \text{ even} \end{cases}$$

Evaluating at $x = 0$ gives,

$$\left(\frac{d^n f}{dx^n}\right)_{x=0} = \begin{cases} 0 & n \text{ odd} \\ (-1)^{n/2} & n \text{ even} \end{cases}$$

This time, all the terms in Equation D.9 with odd values of n vanish, leaving

$$\cos x = 1 + \frac{1}{2!}(-1)^1 x^2 + \frac{1}{4!}(-1)^2 x^4 + \frac{1}{6!}(-1)^3 x^6 + \cdots$$

$$= 1 - \frac{x^2}{2!} + \frac{x^4}{4!} - \frac{x^6}{6!} + \cdots$$

which is Equation D.11.

D–14. Show that Equations D.2, D.10, and D.11 are consistent with the relation $e^{ix} = \cos x + i \sin x$.

Substitute ix for x in Equation D.2 and recall that

$$i^n = \begin{cases} i(-1)^{(n-1)/2} & n \text{ odd} \\ (-1)^{n/2} & n \text{ even} \end{cases}$$

to get

$$e^{ix} = \sum_{n=0}^{\infty} \frac{(ix)^n}{n!}$$

$$= \sum_{\substack{n=0 \\ n \text{ even}}}^{\infty} \frac{(-1)^{n/2} x^n}{n!} + \sum_{\substack{n=1 \\ n \text{ odd}}}^{\infty} \frac{i(-1)^{(n-1)/2} x^n}{n!}$$

$$= \sum_{\substack{n=0 \\ n \text{ even}}}^{\infty} (-1)^{n/2} \frac{x^n}{n!} + i \sum_{\substack{n=1 \\ n \text{ odd}}}^{\infty} (-1)^{(n-1)/2} \frac{x^n}{n!}$$

$$= \cos x + i \sin x$$

where Equations D.10 and D.11 are used in the last step.

D–15. Use Equation D.2 and the definitions

$$\sinh x = \frac{e^x - e^{-x}}{2} \quad \text{and} \quad \cosh x = \frac{e^x + e^{-x}}{2}$$

to show that

$$\sinh x = x + \frac{x^3}{3!} + \frac{x^5}{5!} + \cdots$$

$$\cosh x = 1 + \frac{x^2}{2!} + \frac{x^4}{4!} + \cdots$$

Use the series expansion of Equation D.2 for both e^x and e^{-x} in the defintion of sinh x.

$$\sinh x = \frac{e^x - e^{-x}}{2}$$

$$= \frac{1}{2} \sum_{n=0}^{\infty} \frac{x^n}{n!} - \frac{1}{2} \sum_{n=0}^{\infty} \frac{(-x)^n}{n!}$$

$$= \frac{1}{2} \sum_{n=0}^{\infty} \frac{x^n - (-x)^n}{n!}$$

$$= \frac{1}{2} \sum_{n=0}^{\infty} \frac{x^n}{n!} \left[1 - (-1)^n\right]$$

Since $\left[1 - (-1)^n\right] = 2$ for odd values of n, but is equal to 0 for even values of n, the last summation can be rewritten to give

$$\sinh x = \sum_{\substack{n=1 \\ n \text{ odd}}}^{\infty} \frac{x^n}{n!}$$

$$= x + \frac{x^3}{3!} + \frac{x^5}{5!} + \cdots$$

The argument is similar for $\cosh x$.

$$\cosh x = \frac{e^x + e^{-x}}{2}$$

$$= \frac{1}{2} \sum_{n=0}^{\infty} \frac{x^n}{n!} + \frac{1}{2} \sum_{n=0}^{\infty} \frac{(-x)^n}{n!}$$

$$= \frac{1}{2} \sum_{n=0}^{\infty} \frac{x^n + (-x)^n}{n!}$$

$$= \frac{1}{2} \sum_{n=0}^{\infty} \frac{x^n}{n!} \left[1 + (-1)^n\right]$$

This time the sum contains the term $\left[1 + (-1)^n\right]$, which is 0 for odd values of n and equal to 2 for even values of n, so that

$$\cosh x = \sum_{\substack{n=0 \\ n \text{ even}}}^{\infty} \frac{x^n}{n!}$$

$$= 1 + \frac{x^2}{2!} + \frac{x^4}{4!} + \cdots$$

D–16. Show that Equations D.10 and D.11 and the results of the previous problem are consistent with the relations

$$\sin ix = i \sinh x \qquad \cos ix = \cosh x$$

$$\sinh ix = i \sin x \qquad \cosh ix = \cos x$$

To show $\sin ix = i \sinh x$, start with Equation D.10 for $\sin ix$ and make appropriate substitutions for the powers of i.

$$\sin ix = ix - \frac{(ix)^3}{3!} + \frac{(ix)^5}{5!} - \frac{(ix)^7}{7!} + \cdots$$

$$= ix + i\frac{x^3}{3!} + i\frac{x^5}{5!} + i\frac{x^7}{7!} + \cdots$$

$$= i\left(x + \frac{x^3}{3!} + \frac{x^5}{5!} + \frac{x^7}{7!} + \cdots\right)$$

$$= i \sinh x$$

using the series expression for $\sinh x$ of the previous problem.

A similar method starting with Equation D.11 is used for $\cos ix$.

$$\cos ix = 1 - \frac{(ix)^2}{2!} + \frac{(ix)^4}{4!} - \frac{(ix)^6}{6!} + \cdots$$

$$= 1 + \frac{x^2}{2!} + \frac{x^4}{4!} + \frac{x^6}{6!} + \cdots$$

$$= \cosh x$$

from the series expression for $\cosh x$ found in the previous problem.

The results for hyperbolic sine and hyperbolic cosine of ix are handled in much the same manner, but starting with the results of the last problem and arriving at either Equation D.10 or D.11.

$$\sinh ix = ix + \frac{(ix)^3}{3!} + \frac{(ix)^5}{5!} + \frac{(ix)^7}{7!} + \cdots$$

$$= ix - i\frac{x^3}{3!} + i\frac{x^5}{5!} - i\frac{x^7}{7!} + \cdots$$

$$= i\left(x - \frac{x^3}{3!} + \frac{x^5}{5!} - \frac{x^7}{7!} + \cdots\right)$$

$$= i \sin x$$

$$\cosh ix = 1 + \frac{(ix)^2}{2!} + \frac{(ix)^4}{4!} + \frac{(ix)^6}{6!} + \cdots$$

$$= 1 - \frac{x^2}{2!} + \frac{x^4}{4!} - \frac{x^6}{6!} + \cdots$$

$$= \cos x$$

D–17. Evaluate the limit of

$$f(x) = \frac{e^{-x}\sin^2 x}{x^2}$$

as $x \to 0$.

Expand the two functions in the numerator of the fraction using Equations D.2 and D.10, respectively:

$$\lim_{x\to 0} f(x) = \lim_{x\to 0} \frac{e^{-x}\sin^2 x}{x^2}$$

$$= \lim_{x\to 0} \frac{\left[1-x+O(x^2)\right]\left[x+O(x^3)\right]^2}{x^2}$$

$$= \lim_{x\to 0} \frac{\left[1-x+O(x^2)\right]\left[x^2+O(x^4)\right]}{x^2}$$

$$= \lim_{x\to 0} \frac{x^2-x^3+O(x^4)}{x^2}$$

$$= \lim_{x\to 0} 1-x+O(x^2)$$

$$= 1$$

D–18. Evaluate the integral

$$I = \int_0^a x^2 e^{-x} \cos^2 x\, dx$$

for small values of a by expanding I in powers of a through quadratic terms.

Use Equations D.2 and D.11 to expand e^{-x} and $\cos x$, respectively.

$$I = \int_0^a x^2 e^{-x} \cos^2 x\, dx$$

$$= \int_0^a x^2 \left[1-x+\frac{x^2}{2}+O(x^3)\right]\left[1-\frac{x^2}{2}+O(x^4)\right]^2 dx$$

$$= \int_0^a x^2 \left[1-x+\frac{x^2}{2}+O(x^3)\right]\left[1-x^2+O(x^4)\right] dx$$

$$= \int_0^a \left[x^2-x^3+O(x^4)\right]\left[1-x^2+O(x^4)\right] dx$$

$$= \int_0^a \left[x^2-x^3+O(x^4)\right] dx$$

$$= \frac{x^3}{3}-\frac{x^4}{4}+O(x^5)\Big|_0^a$$

$$= \frac{a^3}{3}-\frac{a^4}{4}+O(a^5)$$

D–19. Prove that the series for $\sin x$ converges for all values of x.

Use the ratio test, Equation D.5, to test for convergence. Since,

$$\sin x = \sum_{n=0}^{\infty} (-1)^n \frac{x^{2n+1}}{(2n+1)!}$$

then

$$r = \lim_{n \to \infty} \left| \frac{u_{n+1}}{u_n} \right|$$

$$= \lim_{n \to \infty} \left| \frac{x^{2n+3}}{(2n+3)!} \cdot \frac{(2n+1)!}{x^{2n+1}} \right|$$

$$= \lim_{n \to \infty} \left| \frac{x^2}{(2n+3)(2n+2)} \right| = 0$$

Because $r < 1$, the series converges for all values of x.

D–20. A Maclaurin series is an expansion about the point $x = 0$. A series of the form

$$f(x) = c_0 + c_1(x - x_0) + c_2(x - x_0)^2 + \cdots$$

is an expansion about the point x_0 and is called a Taylor series. First show that $c_0 = f(x_0)$. Now differentiate both sides of the above expansion with respect to x and then let $x = x_0$ to show that $c_1 = (df/dx)_{x=x_0}$. Now show that

$$c_n = \frac{1}{n!} \left(\frac{d^n f}{dx^n} \right)_{x=x_0}$$

and so

$$f(x) = f(x_0) + \left(\frac{df}{dx} \right)_{x=x_0} (x - x_0) + \frac{1}{2} \left(\frac{d^2 f}{dx^2} \right)_{x=x_0} (x - x_0)^2 + \cdots$$

At $x = x_0$, the Taylor series becomes

$$f(x) = c_0 + c_1(x - x_0) + c_2(x - x_0)^2 + \cdots$$

$$f(x_0) = c_0$$

Differentiating the Taylor series with respect to x gives

$$f'(x) = c_1 + 2c_2(x - x_0) + \cdots$$

$$f'(x_0) = c_1$$

Likewise, the second derivative of $f(x)$ at $x = x_0$ is $2c_2$, the third derivative of $f(x)$ at $x = x_0$ is $3!c_3$, and generally,

$$\left(\frac{d^n f}{dx^n}\right)_{x=x_0} = n!c_n$$

$$c_n = \frac{1}{n!}\left(\frac{d^n f}{dx^n}\right)_{x=x_0}$$

Substitution into the Taylor series for c_n gives the desired result.

$$f(x) = f(x_0) + \left(\frac{df}{dx}\right)_{x=x_0}(x - x_0) + \frac{1}{2}\left(\frac{d^2 f}{dx^2}\right)_{x=x_0}(x - x_0)^2 + \cdots$$

D–21. Show that l'Hôpital's rule amounts to forming a Taylor expansion of both the numerator and the denominator. Evaluate the limit

$$\lim_{x \to 0} \frac{\ln(1+x) - x}{x^2}$$

both ways.

Let the numerator be $f(x)$ and the denominator $g(x)$, and expand both in a Taylor series about the point $x = x_0$ where the limit is sought.

$$f(x) = a_0 + a_1(x - x_0) + a_2(x - x_0)^2 + \cdots$$

$$g(x) = b_0 + b_1(x - x_0) + b_2(x - x_0)^2 + \cdots$$

So that

$$\frac{f(x)}{g(x)} = \frac{a_0 + a_1(x - x_0) + a_2(x - x_0)^2 + \cdots + a_n(x - x_0)^n + \cdots}{b_0 + b_1(x - x_0) + b_2(x - x_0)^2 + \cdots + b_n(x - x_0)^n + \cdots}$$

and

$$\frac{f'(x)}{g'(x)} = \frac{a_1 + 2a_2(x - x_0) + 3a_3(x - x_0)^2 + \cdots + na_n(x - x_0)^{n-1} + \cdots}{b_1 + 2b_2(x - x_0) + 3b_3(x - x_0)^2 + \cdots + nb_n(x - x_0)^{n-1} + \cdots}$$

L'Hôpital's rule states that if both $f(x)$ and $g(x)$ approach 0 as x approaches x_0, then

$$\lim_{x \to x_0} \frac{f(x)}{g(x)} = \lim_{x \to x_0} \frac{f'(x)}{g'(x)}$$

If both $f(x_0) = 0$ and $g(x_0) = 0$, then in the Taylor series expansions of $f(x)$ and $g(x)$, $a_0 = 0$ and $b_0 = 0$, so that

$$\frac{f(x)}{g(x)} = \frac{a_1(x - x_0) + a_2(x - x_0)^2 + \cdots + a_n(x - x_0)^n + \cdots}{b_1(x - x_0) + b_2(x - x_0)^2 + \cdots + b_n(x - x_0)^n + \cdots}$$

$$= \frac{a_1 + a_2(x - x_0) + \cdots + a_n(x - x_0)^{n-1} + \cdots}{b_1 + b_2(x - x_0) + \cdots + b_n(x - x_0)^{n-1} + \cdots}$$

and

$$\lim_{x \to x_0} \frac{f(x)}{g(x)} = \frac{a_1}{b_1}$$

From the Taylor series expressions for $f'(x)$ and $g'(x)$,

$$\lim_{x \to x_0} \frac{f'(x)}{g'(x)} = \lim_{x \to x_0} \frac{a_1 + 2a_2(x - x_0) + 3a_3(x - x_0)^2 + \cdots + na_n(x - x_0)^{n-1} + \cdots}{b_1 + 2b_2(x - x_0) + 3b_3(x - x_0)^2 + \cdots + nb_n(x - x_0)^{n-1} + \cdots}$$

$$= \frac{a_1}{b_1} = \lim_{x \to x_0} \frac{f(x)}{g(x)}$$

as l'Hôpital's rule states. If the first derivatives are both equal to zero at $x = x_0$, then both $a_1 = 0$ and $b_1 = 0$, and the process can be continued indefinitely until a nonzero numerator or denominator is found.

Applying l'Hôpital's rule,

$$\lim_{x \to 0} \frac{\ln(1 + x) - x}{x^2} = \lim_{x \to 0} \frac{(1 + x)^{-1} - 1}{2x}$$

Both the numerator and denominator still approach zero as $x \to 0$, so apply l'Hôpital's rule again.

$$\lim_{x \to 0} \frac{\ln(1 + x) - x}{x^2} = \lim_{x \to 0} \frac{-(1 + x)^{-2}}{2}$$

$$= -\frac{1}{2}$$

Using the expansion for $\ln(1 + x)$ in Equation D.12 (actually a Maclaurin expansion, since $x_0 = 0$),

$$f(x) = \ln(1 + x) - x = 0 + 0x - \frac{x^2}{2} + \cdots$$

$$g(x) = 0 + 0x + x^2$$

$$\lim_{x \to 0} \frac{f(x)}{g(x)} = \lim_{x \to 0} \frac{0 + 0x - x^2/2 + \cdots}{0 + 0x + x^2}$$

$$= -\frac{1}{2}$$

as above.

D–22. Start with

$$\frac{1}{1 - x} = 1 + x + x^2 + \cdots$$

Now let $x = 1/x$ to write

$$\frac{1}{1 - \dfrac{1}{x}} = \frac{x}{x - 1} = 1 + \frac{1}{x} + \frac{1}{x^2} + \cdots$$

Now add these two expressions to get

$$1 = \cdots + \frac{1}{x^2} + \frac{1}{x} + 2 + x + x^2 + \cdots$$

Does this make sense? What went wrong?

_____

Adding the left hand sides of the two expressions indeed gives 1, as seen below

$$\frac{1}{1-x} + \frac{x}{x-1} = \frac{1}{1-x} - \frac{x}{1-x} = \frac{1-x}{1-x} = 1$$

Yet it does not make sense when the right hand sides are added, since 1 is obviously not equal to

$$\cdots + \frac{1}{x^2} + \frac{1}{x} + 2 + x + x^2 + \cdots$$

The problem is that the expansion for $1/(1-x)$ is only valid for $|x| < 1$ and the expansion for $1/[1 - (1/x)]$ is only valid for $|1/x| < 1$, or $|x| > 1$. Consequently, the sum of the two expansions is valid for no value of x.

_____

D–23. The energy of a quantum-mechanical harmonic oscillator is given by $\varepsilon_n = (n + \frac{1}{2})hv$, $n = 0, 1, 2, \ldots$, where h is the Planck constant and v is the fundamental frequency of the oscillator. The average vibrational energy of a harmonic oscillator in an ideal gas is given by

$$\varepsilon_{\text{vib}} = (1 - e^{-hv/k_B T}) \sum_{n=0}^{\infty} \varepsilon_n e^{-nhv/k_B T}$$

where k_B is the Boltzmann constant and T is the kelvin temperature. Show that

$$\varepsilon_{\text{vib}} = \frac{hv}{2} + \frac{hve^{-hv/k_B T}}{1 - e^{-hv/k_B T}}$$

_____

Let $x = e^{-hv/k_B T}$, then $e^{-nhv/k_B T} = x^n$, and

$$\varepsilon_{\text{vib}} = \left(1 - e^{-hv/k_B T}\right) \sum_{n=0}^{\infty} \varepsilon_n e^{-nhv/k_B T}$$

$$= (1 - x) hv \sum_{n=0}^{\infty} \left(n + \frac{1}{2}\right) x^n$$

$$= (1 - x) hv \left(\sum_{n=0}^{\infty} nx^n + \frac{1}{2} \sum_{n=0}^{\infty} x^n\right)$$

One of these last two sums is a geometric series, the other was evaluated in Problem D–11.

$$\varepsilon_{\text{vib}} = (1 - x) hv \left[\frac{x}{(1-x)^2} + \frac{1}{2}\left(\frac{1}{1-x}\right)\right]$$

$$= hv \left(\frac{1}{2} + \frac{x}{1-x}\right)$$

Substitution for x completes the derivation.

$$\varepsilon_{\text{vib}} = \frac{hv}{2} + \frac{hve^{-hv/k_B T}}{1 - e^{-hv/k_B T}}$$

CHAPTER 5

The Harmonic Oscillator and Vibrational Spectroscopy

PROBLEMS AND SOLUTIONS

5–1. Show that the equation $Md^2X/dt^2 = 0$ (Equation 5.26) implies that the motion is uniform.

There is no force term in Equation 5.26, hence there is no acceleration of X ($F = ma = 0$ gives $a = 0$). Consequently, the center of mass moves with constant velocity, as can be seen by integrating once.

$$\int \frac{d^2X}{dt^2}\, dt = \frac{dX}{dt} = v = \text{constant} = v_0$$

A second integration gives $X(t)$,

$$X(t) = v_0 t + X_0$$

The center of mass is then seen to move with constant, uniform velocity.

5–2. Verify that $x(t) = A \sin \omega t + B \cos \omega t$, where $\omega = (k/m)^{1/2}$ is a solution to Newton's equation for a harmonic oscillator.

Newton's equation of motion for a harmonic oscillator is

$$\frac{d^2x}{dt^2} + \frac{k}{m}x = 0$$

Substitution of $x(t) = A \sin \omega t + B \cos \omega t$ into Newton's equation gives

$$-\omega^2 A \sin \omega t - \omega^2 B \cos \omega t + \frac{k}{m}\,(A \sin \omega t + B \cos \omega t)$$

$$= -\frac{k}{m}\,(A \sin \omega t + B \cos \omega t) + \frac{k}{m}\,(A \sin \omega t + B \cos \omega t) = 0$$

where $\omega = (k/m)^{1/2}$ has been used.

5–3. Verify that $x(t) = C \sin(\omega t + \phi)$ is a solution to Newton's equation for a harmonic oscillator.

Newton's equation of motion for a harmonic oscillator is

$$\frac{d^2x}{dt^2} + \frac{k}{m}x = 0$$

Substitution of $x(t) = C \sin(\omega t + \phi)$ into Newton's equation gives

$$-\omega^2 C \sin(\omega t + \phi) + \frac{k}{m}C \sin(\omega t + \phi) = -\frac{k}{m}C \sin(\omega t + \phi) + \frac{k}{m}C \sin(\omega t + \phi) = 0$$

where $\omega = (k/m)^{1/2}$ has been used.

5–4. The general solution for the classical harmonic oscillator is $x(t) = C \sin(\omega t + \phi)$. Show that the displacement oscillates between $+C$ and $-C$ with a frequency of ω radian·s^{-1} or $\nu = \omega/2\pi$ cycle·s^{-1}. What is the period of the oscillations; that is, how long does it take to undergo one cycle?

Examining the general solution $x(t) = C \sin(\omega t + \phi)$, the value of the sine function varies from $+1$ to -1, so the displacement oscillates between $+C$ and $-C$. The period of oscillation is the smallest time period τ that satisfies the condition

$$\begin{aligned} \sin(\omega t + \phi) &= \sin[\omega(t + \tau) + \phi] \\ &= \sin(\omega t + \phi + \omega\tau) \\ &= \sin(\omega t + \phi)\cos \omega\tau + \cos(\omega t + \phi)\sin \omega\tau \end{aligned}$$

This condition is met when τ satisfies the two conditions

$$\cos \omega\tau = 1 \qquad \sin \omega\tau = 0$$

or

$$\tau = \frac{2n\pi}{\omega} \qquad n = 1, 2, \ldots$$

The smallest value of τ that meets the condition is $\tau = 2\pi/\omega$, which is the time it takes for the oscillator to undergo one cycle. The frequency of the oscillator is the reciprocal of the period, or

$$\nu = \frac{1}{\tau} = \frac{\omega}{2\pi}$$

5–5. From Problem 5–4, we see that the period of a harmonic vibration is $\tau = 1/\nu$. The average of the kinetic energy over one cycle is given by

$$\langle T \rangle = \frac{1}{\tau}\int_0^\tau \frac{m\omega^2 C^2}{2}\cos^2(\omega t + \phi)dt$$

Show that $\langle T \rangle = E/2$, where E is the total energy. Show also that $\langle V \rangle = E/2$, where the instantaneous potential energy is given by

$$V = \frac{kC^2}{2} \sin^2(\omega t + \phi)$$

Interpret the result $\langle T \rangle = \langle V \rangle$.

The kinetic energy of a harmonic oscillator is (see Example 5–2)

$$T = \frac{m\omega^2 C^2}{2} \cos^2 (\omega t + \phi)$$

Averaging this quantity over one cycle gives

$$\langle T \rangle = \frac{1}{\tau} \int_0^\tau \frac{m\omega^2 C^2}{2} \cos^2 (\omega t + \phi) \; dt$$

$$= \frac{m\omega C^2}{2\tau} \int_{x=\phi}^{x=\omega\tau+\phi} \cos^2 x \; dx$$

$$= \frac{m\omega C^2}{2\tau} \left(\frac{x}{2} + \frac{\sin 2x}{4} \right) \Big|_{x=\phi}^{x=\omega\tau+\phi}$$

$$= \frac{m\omega C^2}{2\tau} \left[\frac{\omega\tau}{2} + \frac{\sin 2 (\omega\tau + \phi) - \sin 2\phi}{4} \right]$$

$$= \frac{m\omega C^2}{2\tau} \left(\frac{\omega\tau}{2} + \frac{\sin 2\omega\tau \cos 2\phi + \cos 2\omega\tau \sin 2\phi - \sin 2\phi}{4} \right)$$

$$= \frac{m\omega C^2}{2\tau} \left(\frac{\omega\tau}{2} + \frac{\sin 4\pi \cos 2\phi + \cos 4\pi \sin 2\phi - \sin 2\phi}{4} \right)$$

$$= \frac{m\omega^2 C^2}{4} = \frac{kC^2}{4} = \frac{E}{2}$$

where $\omega\tau = 2\pi$ (Problem 5–4), $k = m\omega^2$ (Equation 5.5), and $E = kC^2/2$ (Example 5–2) have been used.

Likewise,

$$\langle V \rangle = \frac{1}{\tau} \int_0^\tau \frac{kC^2}{2} \sin^2 (\omega t + \phi) \; dt$$

$$= \frac{kC^2}{2\tau} \int_0^\tau \left[1 - \cos^2 (\omega t + \phi) \right] dt$$

$$= \frac{kC^2}{2\tau} \left(\tau - \frac{\tau}{2} \right)$$

$$= \frac{kC^2}{4} = \frac{E}{2}$$

Thus, for a harmonic oscillator, on average there are equal amounts of kinetic and potential energy.

5–6. Consider two masses m_1 and m_2 in one dimension, interacting through a potential that depends only upon their relative separation $(x_1 - x_2)$, so that $V(x_1, x_2) = V(x_1 - x_2)$. Given that the force acting upon the jth particle is $f_j = -(\partial V/\partial x_j)$, show that $f_1 = -f_2$. What law is this?

Newton's equations for m_1 and m_2 are

$$m_1\frac{d^2x_1}{dt^2} = -\frac{\partial V}{\partial x_1} \quad \text{and} \quad m_2\frac{d^2x_2}{dt^2} = -\frac{\partial V}{\partial x_2}$$

Now introduce center-of-mass and relative coordinates by

$$X = \frac{m_1x_1 + m_2x_2}{M} \qquad x = x_1 - x_2$$

where $M = m_1 + m_2$, and solve for x_1 and x_2 to obtain

$$x_1 = X + \frac{m_2}{M}x \quad \text{and} \quad x_2 = X - \frac{m_1}{M}x$$

Show that Newton's equations in these coordinates are

$$m_1\frac{d^2X}{dt^2} + \frac{m_1m_2}{M}\frac{d^2x}{dt^2} = -\frac{\partial V}{\partial x}$$

and

$$m_2\frac{d^2X}{dt^2} - \frac{m_1m_2}{M}\frac{d^2x}{dt^2} = +\frac{\partial V}{\partial x}$$

Now add these two equations to find

$$M\frac{d^2X}{dt^2} = 0$$

Interpret this result. Now divide the first equation by m_1 and the second by m_2 and subtract to obtain

$$\frac{d^2x}{dt^2} = -\left(\frac{1}{m_1} + \frac{1}{m_2}\right)\frac{\partial V}{\partial x}$$

or

$$\mu\frac{d^2x}{dt^2} = -\frac{\partial V}{\partial x}$$

where $\mu = m_1m_2/(m_1 + m_2)$ is the reduced mass. Interpret this result, and discuss how the original two-body problem has been reduced to two one-body problems.

With the definition $x = x_1 - x_2$, $V(x_1 - x_2) = V(x)$, and the forces on the two particles are

$$f_1 = -\frac{\partial V}{\partial x_1} = -\frac{\partial V}{\partial x}\frac{\partial x}{\partial x_1} = -\frac{\partial V}{\partial x} \qquad f_2 = -\frac{\partial V}{\partial x_2} = -\frac{\partial V}{\partial x}\frac{\partial x}{\partial x_2} = \frac{\partial V}{\partial x}$$

This is Newton's third law: for every action there is an equal and opposite reaction.

Now use the definitions for the center-of-mass, X, and the relative coordinate, x,

$$X = \frac{m_1 x_1 + m_2 x_2}{M} \tag{1}$$

$$x = x_1 - x_2 \tag{2}$$

Multiply Equation 1 by M and Equation 2 by m_2 and add to find

$$x_1 = X + \frac{m_2}{M}x \tag{3}$$

Then multiply Equation 1 by M and Equation 2 by m_1 and subtract to get

$$x_2 = X - \frac{m_1}{M}x \tag{4}$$

Substitute Equations 3 and 4 into Newton's equations to obtain

$$m_1 \frac{d^2 x_1}{dt^2} = m_1 \frac{d^2 \left(X + \frac{m_2}{M}x \right)}{dt^2} = m_1 \frac{d^2 X}{dt^2} + \frac{m_1 m_2}{M} \frac{d^2 x}{dt^2} = -\frac{\partial V}{\partial x} \tag{5}$$

and

$$m_2 \frac{d^2 x_2}{dt^2} = m_2 \frac{d^2 \left(X - \frac{m_1}{M}x \right)}{dt^2} = m_2 \frac{d^2 X}{dt^2} - \frac{m_1 m_2}{M} \frac{d^2 x}{dt^2} = \frac{\partial V}{\partial x} \tag{6}$$

Adding Equations 5 and 6 and recalling $M = m_1 + m_2$ give,

$$M \frac{d^2 X}{dt^2} = 0$$

As discussed in Problem 5–1, the physical interpretation of this result is that the center of mass moves with constant, uniform velocity. Next, divide Equation 5 by m_1 and Equation 6 by m_2 and subtract to obtain

$$\frac{d^2 x}{dt^2} = -\left(\frac{1}{m_1} + \frac{1}{m_2} \right) \frac{\partial V}{\partial x} \tag{7}$$

Defining the reduced mass $\mu = m_1 m_2 / (m_1 + m_2)$ allows Equation 7 to be rewritten as

$$\mu \frac{d^2 x}{dt^2} = -\frac{\partial V}{\partial x}$$

This is the equation of motion for a body of mass μ moving under the influence of a force $-\partial V / \partial x$. Because the force acting on body 1 could be expressed in terms of the force acting on body 2, the original two-body problem is reduced to that of a single body.

5–7. Extend the results of Problem 5–6 to three dimensions. Realize that in three dimensions the relative separation is given by

$$r_{12} = [(x_1 - x_2)^2 + (y_1 - y_2)^2 + (z_1 - z_2)^2]^{1/2}$$

The x-, y-, and z-dimensions can be treated individually because these directions are orthogonal. The equations in the x-direction are the same as those in Problem 5–6. To find the equations in the y and z-directions, simply substitute y or z in place of x (and Y or Z in place of X) in the equations of Problem 5–6. The results are

$$\mu \frac{d^2 x}{dt^2} = -\frac{\partial V}{\partial x} \qquad \mu \frac{d^2 y}{dt^2} = -\frac{\partial V}{\partial y} \qquad \mu \frac{d^2 z}{dt^2} = -\frac{\partial V}{\partial z}$$

or

$$\mu \frac{d^2 \mathbf{r}}{dt^2} = -\nabla V$$

which is the three-dimensional extension of Problem 5–6.

5–8. Show that the reduced mass of two equal masses, m, is $m/2$.

If both m_1 and m_2 are equal to the same value m, the expression for the reduced mass becomes

$$\mu = \frac{m_1 m_2}{m_1 + m_2} = \frac{m^2}{2m} = \frac{m}{2}$$

5–9. Example 5–3 shows that a Maclaurin expansion of a Morse potential leads to

$$V(x) = D\beta^2 x^2 + \cdots$$

Given that $D = 7.31 \times 10^{-19}$ J·molecule^{-1} and $\beta = 1.81 \times 10^{10}$ m^{-1} for HCl, calculate the force constant of HCl. Plot the Morse potential for HCl, and plot the corresponding harmonic oscillator potential on the same graph (cf. Figure 5.5).

Since $V(x) = kx^2/2$ for a harmonic oscillator,

$$k = 2D\beta^2 = 2 \left(7.31 \times 10^{-19} \text{ J} \right) \left(1.81 \times 10^{10} \text{ m}^{-1} \right)^2 = 479 \text{ N·m}^{-1}$$

The graph below shows the Morse potential $V(x) = D \left(1 - e^{-\beta x} \right)^2$ (solid line) and the harmonic potential $V(x) = kx^2/2$ (dashed line) for HCl.

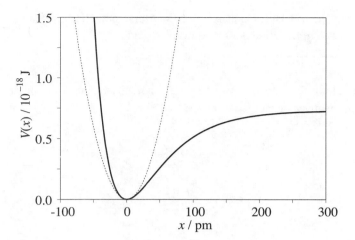

5–10. Use the result of Example 5–3 and Equation 5.39 to show that

$$\beta = 2\pi c \tilde{\omega}_{\text{obs}} \left(\frac{\mu}{2D}\right)^{1/2}$$

Given that $\tilde{\omega}_{\text{obs}} = 2886 \text{ cm}^{-1}$ and $D = 440.2 \text{ kJ·mol}^{-1}$ for $H^{35}Cl$, calculate β. Compare your result with that in Problem 5–9.

Substitution of the result of Example 5–3, $k = 2D\beta^2$, into Equation 5.39 gives

$$\tilde{\omega}_{\text{obs}} = \frac{1}{2\pi c} \left(\frac{k}{\mu}\right)^{1/2} = \frac{\beta}{2\pi c} \left(\frac{2D}{\mu}\right)^{1/2}$$

which can be rearranged to obtain

$$\beta = 2\pi c \tilde{\omega}_{\text{obs}} \left(\frac{\mu}{2D}\right)^{1/2}$$

For $H^{35}Cl$,

$$\mu = \frac{\left(34.97 \times 10^{-3} \text{ kg·mol}^{-1}\right)\left(1.008 \times 10^{-3} \text{ kg·mol}^{-1}\right)}{34.97 \times 10^{-3} \text{ kg·mol}^{-1} + 1.008 \times 10^{-3} \text{ kg·mol}^{-1}} = 0.9798 \times 10^{-3} \text{ kg·mol}^{-1}$$

and

$$\beta = 2\pi c \tilde{\omega}_{\text{obs}} \left(\frac{\mu}{2D}\right)^{1/2}$$

$$= 2\pi \left(2.997\,924\,58 \times 10^{10} \text{ cm·s}^{-1}\right)\left(2886 \text{ cm}^{-1}\right)\left[\frac{0.9798 \times 10^{-3} \text{ kg·mol}^{-1}}{2\left(440.2 \times 10^3 \text{ J·mol}^{-1}\right)}\right]^{1/2}$$

$$= 1.81 \times 10^{10} \text{ m}^{-1}$$

which is the same as the value given for β in Problem 5–9.

5–11. Carry out the Maclaurin expansion of the Morse potential in Example 5–3 through terms in x^4. Express γ_3 in Equation 5.30 in terms of D and β.

Continuing along the lines of Example 5–3, the subsequent terms in Equation 5.29 require additional derivatives of $V(x) = D\left(1 - e^{-\beta x}\right)^2$.

$$\left(\frac{d^3V}{dx^3}\right)_{x=0} = -2D\beta\left(-\beta^2 e^{-\beta x} + 4\beta^2 e^{-2\beta x}\right)_{x=0}$$

$$= -6D\beta^3$$

and

$$\left(\frac{d^4V}{dx^4}\right)_{x=0} = -2D\beta\left(\beta^3 e^{-\beta x} - 8\beta^3 e^{-2\beta x}\right)_{x=0}$$

$$= 14D\beta^4$$

Thus,

$$V(x) = D\beta^2 x^2 - D\beta^3 x^3 + \frac{7}{12}D\beta^4 x^4 + O(x^5)$$

Comparison with Equation 5.30 gives $\gamma_3 = -6D\beta^3$.

5–12. It turns out that the solution of the Schrödinger equation for the Morse potential can be expressed as

$$G(v) = \tilde{\omega}_e\left(v + \frac{1}{2}\right) - \tilde{\omega}_e\tilde{x}_e\left(v + \frac{1}{2}\right)^2$$

where

$$\tilde{x}_e = \frac{hc\tilde{\omega}_e}{4D}$$

Given that $\tilde{\omega}_e = 2886$ cm^{-1} and $D = 440.2$ kJ·mol^{-1} for H^{35}Cl, calculate $\tilde{x}_e$ and $\tilde{\omega}_e\tilde{x}_e$.

$$\tilde{x}_e = \frac{hc\tilde{\omega}_e}{4D}$$

$$= \frac{(6.626\,0755 \times 10^{-34}\text{ J·s})(2.997\,924\,58 \times 10^{10}\text{ cm·s}^{-1})(2886\text{ cm}^{-1})}{4\left(\dfrac{440.2 \times 10^3\text{ J·mol}^{-1}}{6.022\,1367 \times 10^{23}\text{ mol}^{-1}}\right)}$$

$$= 0.019\,61$$

$$\tilde{x}_e\tilde{\omega}_e = \left(2886\text{ cm}^{-1}\right)(0.01961) = 56.59\text{ cm}^{-1}$$

5–13. In the infrared spectrum of $H^{127}I$, there is an intense line at 2309 cm^{-1}. Calculate the force constant of $H^{127}I$ and the period of vibration of $H^{127}I$.

Equation 5.39 can be rearranged to give

$$k = \left(2\pi c\tilde{\omega}_{obs}\right)^2 \mu$$

The reduced mass for $H^{127}I$ is

$$\mu = \frac{(1.008 \text{ amu}) \, (126.9 \text{ amu})}{1.008 \text{ amu} + 126.9 \text{ amu}} = 1.000 \text{ amu}$$

so

$$k = \left[2\pi \left(2.997\,924\,58 \times 10^{10} \text{ cm·s}^{-1}\right) \left(2309 \text{ cm}^{-1}\right)\right]^2 (1.000 \text{ amu}) \left(1.660\,5402 \times 10^{-27} \text{ kg·amu}^{-1}\right)$$

$$= 314.1 \text{ N·m}^{-1}$$

The period of vibration is $\tau = 2\pi/\omega = 2\pi \, (\mu/k)^{1/2}$ (Problem 5–4).

$$\tau = 2\pi \left[(1.000 \text{ amu}) \left(1.660\,5402 \times 10^{-27} \text{ kg·amu}^{-1}\right) /314.1 \text{ N·m}^{-1}\right]^{1/2} = 1.445 \times 10^{-14} \text{ s}$$

5–14. The force constant of $^{35}Cl^{35}Cl$ is 319 $N·m^{-1}$. Calculate the fundamental vibrational frequency and the zero-point energy of $^{35}Cl^{35}Cl$.

From Problem 5–8, the reduced mass for $^{35}Cl^{35}Cl$ is $\mu = 34.97$ amu/2. This is used in Equation 5.39,

$$\tilde{\omega}_{obs} = \frac{1}{2\pi c} \left(\frac{k}{\mu}\right)^{1/2}$$

$$= \frac{1}{2\pi \left(2.997\,924\,58 \times 10^{10} \text{ cm·s}^{-1}\right)} \left[\frac{319 \text{ N·m}^{-1}}{\dfrac{34.97 \text{ amu}}{2} \left(1.660\,5402 \times 10^{-27} \text{ kg·amu}^{-1}\right)}\right]^{1/2}$$

$$= 556.5 \text{ cm}^{-1}$$

The zero-point energy is

$$E_0 = \frac{1}{2} hc\tilde{\omega}_{obs}$$

$$= \frac{1}{2} \left(6.626\,0755 \times 10^{-34} \text{ J·s}\right) \left(2.997\,924\,58 \times 10^{10} \text{ cm·s}^{-1}\right) \left(556.5 \text{ cm}^{-1}\right)$$

$$= 5.527 \times 10^{-21} \text{ J}$$

5–15. The fundamental line in the infrared spectrum of $^{12}C^{16}O$ occurs at 2143.0 cm^{-1}, and the first overtone occurs at 4260.0 cm^{-1}. Calculate the values of $\tilde{\omega}_e$ and $\tilde{x}_e\tilde{\omega}_e$ for $^{12}C^{16}O$.

Equation 5.43 gives $\tilde{\omega}_{obs} = \tilde{\omega}_e v - \tilde{x}_e\tilde{\omega}_e v\,(v+1)$.

Thus, for the two lines in this problem,

$$\text{Fundamental:} \quad \tilde{\omega}_e - 2\tilde{x}_e\tilde{\omega}_e = 2143.0 \text{ cm}^{-1}$$
$$\text{First overtone:} \quad 2\tilde{\omega}_e - 6\tilde{x}_e\tilde{\omega}_e = 4260.0 \text{ cm}^{-1}$$

Multiply the fundamental frequency by 3 and subtract the overtone to get

$$\tilde{\omega}_e = 3(2143.0 \text{ cm}^{-1}) - 4260.0 \text{ cm}^{-1} = 2169.0 \text{ cm}^{-1}$$

Multiply the fundamental frequency by 2 and subtract the overtone to get

$$2\tilde{x}_e\tilde{\omega}_e = 26.0 \text{ cm}^{-1}$$

or

$$\tilde{x}_e\tilde{\omega}_e = 13.0 \text{ cm}^{-1}$$

5–16. Using the parameters given in Table 5.1, calculate the fundamental and the first three overtones of H^{79}Br.

Use Equation 5.43

$$\tilde{\omega}_{obs} = \tilde{\omega}_e v - \tilde{x}_e\tilde{\omega}_e v\,(v+1)$$

with $\tilde{\omega}_e = 2648.97$ cm^{-1} and $\tilde{x}_e\tilde{\omega}_e = 45.218$ cm^{-1} to arrive at the results in the following table:

Line	v	Wavenumber / cm^{-1}
Fundamental	1	2558.53
1st overtone	2	5026.63
2nd overtone	3	7404.29
3rd overtone	4	9691.52

5–17. The frequencies of the vibrational transitions in the anharmonic-oscillator approximation are given by Equation 5.43. Show how the values of both $\tilde{\omega}_e$ and $\tilde{x}_e\tilde{\omega}_e$ may be obtained by plotting $\tilde{\omega}_{obs}/v$ versus $(v+1)$. Use this method and the data in Table 5.2 to determine the values of $\tilde{\omega}_e$ and $\tilde{x}_e\tilde{\omega}_e$ for H^{35}Cl.

Dividing both sides of Equation 5.43 by v gives

$$\frac{\tilde{\omega}_{obs}}{v} = \tilde{\omega}_e - \tilde{x}_e\tilde{\omega}_e\,(v+1)$$

Thus, a plot of $\tilde{\omega}_{obs}/v$ versus $v + 1$ will give a straight line graph that has a slope of $-\tilde{x}_e\tilde{\omega}_e$ and an intercept of $\tilde{\omega}_e$. Using the data from Table 5.2 results in the following plot:

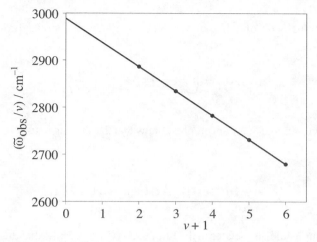

The best fit line to the graph has an intercept of $\tilde{\omega}_e = 2989.0 \text{ cm}^{-1}$ and a slope of $-\tilde{x}_e\tilde{\omega}_e = -51.64 \text{ cm}^{-1}$, so $\tilde{x}_e\tilde{\omega}_e = 51.64 \text{ cm}^{-1}$.

5–18. The following data are obtained from the infrared spectrum of $^{127}I^{35}Cl$. Using the method of Problem 5–17, determine the values of $\tilde{\omega}_e$ and $\tilde{x}_e\tilde{\omega}_e$ from these data.

Transition	Frequency/cm^{-1}
$0 \rightarrow 1$	381.20
$0 \rightarrow 2$	759.60
$0 \rightarrow 3$	1135.00
$0 \rightarrow 4$	1507.40
$0 \rightarrow 5$	1877.00

As in the previous problem, plot $\tilde{\omega}_{obs}/v$ versus $v + 1$.

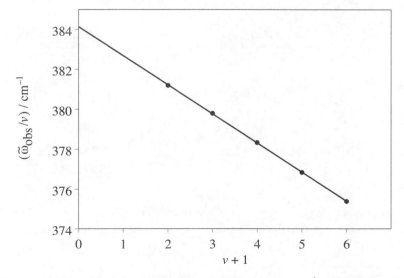

The best fit line to the graph has an intercept of $\tilde{\omega}_e = 384.14 \text{ cm}^{-1}$ and a slope of $-\tilde{x}_e\tilde{\omega}_e = -1.445 \text{ cm}^{-1}$, so $\tilde{x}_e\tilde{\omega}_e = 1.445 \text{ cm}^{-1}$

5–19. The vibrational term of a diatomic molecule is given to a good approximation by

$$G(v) = \left(v + \frac{1}{2}\right)\tilde{\omega}_e - \left(v + \frac{1}{2}\right)^2 \tilde{x}_e\tilde{\omega}_e$$

where v is the vibrational quantum number. Show that the spacing between the adjacent levels ΔG is given by

$$\Delta G = G(v+1) - G(v) = \tilde{\omega}_e\{1 - 2\tilde{x}_e(v+1)\} \tag{1}$$

The diatomic molecule dissociates in the limit that $\Delta G \to 0$. Show that the maximum vibrational quantum number, v_{max}, is given by

$$v_{max} = \frac{1}{2\tilde{x}_e} - 1$$

Use this result to show that the dissociation energy D_e of the diatomic molecule can be written as

$$\tilde{D}_e = \frac{\tilde{\omega}_e(1 - \tilde{x}_e^2)}{4\tilde{x}_e} \approx \frac{\tilde{\omega}_e}{4\tilde{x}_e} \tag{2}$$

Referring to equation 1, explain how the constants $\tilde{\omega}_e$ and $\tilde{x}_e$ can be evaluated from a plot of ΔG versus $v + 1$. This type of plot is called a *Birge–Sponer plot*. Once the values of $\tilde{\omega}_e$ and $\tilde{x}_e$ are known, equation 2 can be used to determine the dissociation energy of the molecule. Use the following experimental data for H_2 to calculate the dissociation energy, $\tilde{D}_e$.

v	$G(v)/\text{cm}^{-1}$	v	$G(v)/\text{cm}^{-1}$
0	4161.12	7	26 830.97
1	8087.11	8	29 123.93
2	11 782.35	9	31 150.19
3	15 250.36	10	32 886.85
4	18 497.92	11	34 301.83
5	21 505.65	12	35 351.01
6	24 287.83	13	35 972.97

Explain why your Birge–Sponer plot is not linear for high values of v. How does the value of $\tilde{D}_e$ obtained from the Birge–Sponer analysis compare with the experimental value of $38\,269.48\ \text{cm}^{-1}$?

Use the expression given for $G(v)$ in the problem to find ΔG,

$$\Delta G = G(v+1) - G(v)$$

$$= \left(v + \frac{3}{2}\right)\tilde{\omega}_e - \left(v + \frac{3}{2}\right)^2 \tilde{x}_e\tilde{\omega}_e - \left(v + \frac{1}{2}\right)\tilde{\omega}_e + \left(v + \frac{1}{2}\right)^2 \tilde{x}_e\tilde{\omega}_e$$

$$= \tilde{\omega}_e + \left(v^2 + v + \frac{1}{4} - v^2 - 3v - \frac{9}{4}\right)\tilde{x}_e\tilde{\omega}_e$$

$$= \tilde{\omega}_e - 2(v+1)\tilde{x}_e\tilde{\omega}_e$$

$$= \tilde{\omega}_e \left[1 - 2\tilde{x}_e(v+1)\right]$$

In the limit $\Delta G \to 0$, $v \to v_{max}$. Solving for v_{max},

$$0 = \tilde{\omega}_e \left[1 - 2\tilde{x}_e\left(v_{max} + 1\right)\right]$$

$$2\tilde{x}_e\left(v_{max} + 1\right) = 1$$

$$v_{max} = \frac{1}{2\tilde{x}_e} - 1$$

The molecule dissociates as $\Delta G \to 0$, so

$$\tilde{D}_e = G(v_{max}) = \left(\frac{1}{2\tilde{x}_e} - \frac{1}{2}\right)\tilde{\omega}_e - \left(\frac{1}{2\tilde{x}_e} - \frac{1}{2}\right)^2 \tilde{x}_e\tilde{\omega}_e$$

$$= \frac{\tilde{\omega}_e}{2\tilde{x}_e}\left(1 - \tilde{x}_e\right) - \frac{\tilde{\omega}_e}{4\tilde{x}_e}\left(1 - \tilde{x}_e\right)^2$$

$$= \frac{\tilde{\omega}_e}{4\tilde{x}_e}\left(2 - 2\tilde{x}_e - 1 + 2\tilde{x}_e - \tilde{x}_e^2\right)$$

$$= \frac{\tilde{\omega}_e}{4\tilde{x}_e}\left(1 - \tilde{x}_e^2\right) \approx \frac{\tilde{\omega}_e}{4\tilde{x}_e}$$

because $\tilde{x}_e$ is very small compared to one.

Writing Equation 1 as

$$\Delta G = \tilde{\omega}_e - 2\tilde{x}_e\tilde{\omega}_e(v+1)$$

shows that a plot of ΔG versus $v + 1$ should result in a straight line graph with a slope of $-2\tilde{x}_e\tilde{\omega}_e$ and an intercept of $\tilde{\omega}_e$. The experimental points for H_2 are plotted below:

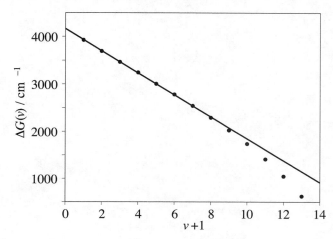

The best fit line to the first eight data points has an intercept of $\tilde{\omega}_e = 4164 \text{ cm}^{-1}$ and a slope of $-2\tilde{x}_e\tilde{\omega}_e = -232 \text{ cm}^{-1}$. Then $\tilde{x}_e = 0.0279$, and so

$$\tilde{D}_e = \frac{\tilde{\omega}_e}{4\tilde{x}_e} = \frac{4164 \text{ cm}^{-1}}{4(0.0279)} = 37\,300 \text{ cm}^{-1}$$

The Birge-Sponer plot is not linear for large values of v because the potential curve is not well described by the anharmonic potential energy function $G(v)$ given in the statement of the problem for vibrational levels close to the dissociation of the molecule. The result from the Birge-Sponer analysis underestimates the experimental value of $\tilde{D}_e$ by 2.5%.

5–20.   An analysis of the vibrational spectrum of the ground-electronic-state homonuclear diatomic molecule C_2 gives $\tilde{\omega}_e = 1854.71 \text{ cm}^{-1}$ and $\tilde{\omega}_e\tilde{x}_e = 13.34 \text{ cm}^{-1}$. Suggest an experimental method that can be used to determine these spectroscopic parameters. Use the expression derived in Problem 5–19 to determine the number of vibrational levels for the ground state of C_2.

The homonuclear diatomic molecule C_2 has no dipole moment regardless of its internuclear separation, and consequently its dipole moment does not change as the molecule vibrates. Thus, it does not have an infrared absorption (or emission) spectrum. It is possible to determine $\tilde{\omega}_e$ and $\tilde{\omega}_e\tilde{x}_e$ from an electronic emission spectrum, where the lines correspond to transitions between a specific vibrational state in the upper electronic state and the various vibrational states in the ground electronic state. The number of bound vibrational levels, v_{max}, is given by (see Problem 5–19)

$$v_{max} = \frac{1}{2\tilde{x}_e} - 1 = \frac{\tilde{\omega}_e}{2\tilde{\omega}_e\tilde{x}_e} - 1 = \frac{1854.71 \text{ cm}^{-1}}{2(13.34 \text{ cm}^{-1})} - 1 = 68.52$$

Thus, there are about 69 bound vibrational levels for the ground state of C_2, since the lowest level is $v = 0$.

5–21.   Verify that $\psi_1(x)$ and $\psi_2(x)$ given in Table 5.4 satisfy the Schrödinger equation for a harmonic oscillator.

The Schrödinger equation for a harmonic oscillator is given by Equation 5.31:

$$\frac{d^2\psi_v}{dx^2} + \frac{2\mu}{\hbar^2}\left(E_v - \frac{1}{2}kx^2\right)\psi_v = 0$$

where $E_v = \frac{h}{2\pi}\left(\frac{k}{\mu}\right)^{1/2}\left(v + \frac{1}{2}\right)$. From Table 5.4,

$$\psi_1(x) = \left(\frac{4\alpha^3}{\pi}\right)^{1/4} xe^{-\alpha x^2/2}$$

$$\psi_2(x) = \left(\frac{\alpha}{4\pi}\right)^{1/4}(2\alpha x^2 - 1)e^{-\alpha x^2/2}$$

where $\alpha = (k\mu)^{1/2}/\hbar$. Substituting ψ_1 into the Schrödinger equation with $v = 1$ gives

$$\frac{d^2\psi_1}{dx^2} + \frac{2\mu}{\hbar^2}\left(E_1 - \frac{1}{2}kx^2\right)\psi_1$$

$$= \left(\frac{4\alpha^3}{\pi}\right)^{1/4}\frac{d}{dx}\left(e^{-\alpha x^2/2} - \alpha x^2 e^{-\alpha x^2/2}\right) + \frac{2\mu}{\hbar^2}\left[\frac{3h}{4\pi}\left(\frac{k}{\mu}\right)^{1/2} - \frac{1}{2}kx^2\right]\left(\frac{4\alpha^3}{\pi}\right)^{1/4}xe^{-\alpha x^2/2}$$

$$= \left(\frac{4\alpha^3}{\pi}\right)^{1/4}\left(-\alpha x - 2\alpha x + \alpha^2 x^3\right)e^{-\alpha x^2/2} + \left(\frac{4\alpha^3}{\pi}\right)^{1/4}\left[\frac{3(\mu k)^{1/2}}{\hbar} - \frac{\mu k}{\hbar^2}x^2\right]xe^{-\alpha x^2/2}$$

$$= \left(\frac{4\alpha^3}{\pi}\right)^{1/4}\left(-3\alpha x + \alpha^2 x^3 + 3\alpha x - \alpha^2 x^3\right)$$

$$= 0$$

Substituting ψ_2 into the Schrödinger equation with $v = 2$ gives

$$\frac{d^2\psi_2}{dx^2} + \frac{2\mu}{\hbar^2}\left(E_2 - \frac{1}{2}kx^2\right)\psi_2$$

$$= \left(\frac{\alpha}{4\pi}\right)^{1/4}\frac{d}{dx}\left[4\alpha xe^{-\alpha x^2/2} - \alpha x\left(2\alpha x^2 - 1\right)e^{-\alpha x^2/2}\right]$$

$$\quad + \frac{2\mu}{\hbar^2}\left[\frac{5h}{4\pi}\left(\frac{k}{\mu}\right)^{1/2} - \frac{1}{2}kx^2\right]\left(\frac{\alpha}{4\pi}\right)^{1/4}e^{-\alpha x^2/2}$$

$$= \left(\frac{\alpha}{4\pi}\right)^{1/4}\left(5\alpha - 11\alpha^2 x^2 + 2\alpha^3 x^4\right)e^{-\alpha x^2/2}$$

$$\quad + \left[\frac{5(\mu k)^{1/2}}{\hbar} - \frac{\mu k}{\hbar^2}x^2\right]\left(\frac{\alpha}{4\pi}\right)^{1/4}\left(2\alpha x^2 - 1\right)e^{-\alpha x^2/2}$$

$$= \left(\frac{\alpha}{4\pi}\right)^{1/4}\left[5\alpha - 11\alpha^2 x^2 + 2\alpha^3 x^4 + \left(5\alpha - \alpha^2 x^2\right)\left(2\alpha x^2 - 1\right)\right]e^{-\alpha x^2/2}$$

$$= \left(\frac{\alpha}{4\pi}\right)^{1/4}\left(5\alpha - 11\alpha^2 x^2 + 2\alpha^3 x^4 - 2\alpha^3 x^4 + 11\alpha^2 x^2 - 5\alpha\right)e^{-\alpha x^2/2}$$

$$= 0$$

Both ψ_1 and ψ_2 are solutions to the Schrödinger equation.

5–22. Show explicitly for a harmonic oscillator that $\psi_0(\xi)$ is orthogonal to $\psi_1(\xi)$, $\psi_2(\xi)$, and $\psi_3(\xi)$ and that $\psi_1(\xi)$ is orthogonal to $\psi_2(\xi)$ and $\psi_3(\xi)$ (see Table 5.4).

From Table 5.4,

$$\psi_0(x) = \left(\frac{\alpha}{\pi}\right)^{1/4} e^{-\alpha x^2/2}$$

$$\psi_1(x) = \left(\frac{4\alpha^3}{\pi}\right)^{1/4} x e^{-\alpha x^2/2}$$

$$\psi_2(x) = \left(\frac{\alpha}{4\pi}\right)^{1/4} (2\alpha x^2 - 1) e^{-\alpha x^2/2}$$

$$\psi_3(x) = \left(\frac{\alpha^3}{9\pi}\right)^{1/4} (2\alpha x^3 - 3x) e^{-\alpha x^2/2}$$

Of the five integrals that must be evaluated to show orthogonality, three have integrands that are odd functions of x, and so are zero.

$$\int_{-\infty}^{\infty} \psi_0(x)\psi_1(x)\, dx = \int_{-\infty}^{\infty} \psi_0(x)\psi_3(x)\, dx = \int_{-\infty}^{\infty} \psi_1(x)\psi_2(x)\, dx = 0$$

This leaves the integrals with even integrands to be evaluated explicitly.

$$\int_{-\infty}^{\infty} \psi_0(x)\psi_2(x)\, dx = 2\left(\frac{\alpha}{\pi}\right)^{1/4}\left(\frac{\alpha}{4\pi}\right)^{1/4} \int_0^{\infty} (2\alpha x^2 - 1) e^{-\alpha x^2}\, dx$$

$$= 2\left(\frac{\alpha}{\pi}\right)^{1/4}\left(\frac{\alpha}{4\pi}\right)^{1/4} \left[2\alpha\left(\frac{1}{4\alpha}\right)\left(\frac{\pi}{\alpha}\right)^{1/2} - \left(\frac{\pi}{4\alpha}\right)^{1/2}\right]$$

$$= 0$$

and

$$\int_{-\infty}^{\infty} \psi_1(x)\psi_3(x)\, dx = 2\left(\frac{4\alpha^3}{\pi}\right)^{1/4}\left(\frac{\alpha^3}{9\pi}\right)^{1/4} \int_0^{\infty} (2\alpha x^4 - 3x^2) e^{-\alpha x^2}\, dx$$

$$= 2\left(\frac{4\alpha^3}{\pi}\right)^{1/4}\left(\frac{\alpha^3}{9\pi}\right)^{1/4} \left[2\alpha\left(\frac{3}{8\alpha^2}\right)\left(\frac{\pi}{\alpha}\right)^{1/2} - 3\left(\frac{1}{4\alpha}\right)\left(\frac{\pi}{\alpha}\right)^{1/2}\right]$$

$$= 0$$

5–23. To normalize the harmonic-oscillator wave functions and calculate various expectation values, we must be able to evaluate integrals of the form

$$I_v(a) = \int_{-\infty}^{\infty} x^{2v} e^{-ax^2}\, dx \qquad v = 0, 1, 2, \ldots$$

We can simply either look them up in a table of integrals or continue this problem. First, show that

$$I_v(a) = 2\int_0^{\infty} x^{2v} e^{-ax^2}\, dx$$

The case $v = 0$ can be handled by the following trick. Show that the square of $I_0(a)$ can be written in the form

$$I_0^2(a) = 4\int_0^{\infty}\int_0^{\infty} dx\, dy\, e^{-a(x^2+y^2)}$$

Now convert to plane polar coordinates, letting

$$r^2 = x^2 + y^2 \quad \text{and} \quad dx\,dy = r\,dr\,d\theta$$

Show that the appropriate limits of integration are $0 \leq r < \infty$ and $0 \leq \theta \leq \pi/2$ and that

$$I_0^2(a) = 4 \int_0^{\pi/2} d\theta \int_0^\infty dr\, r\, e^{-ar^2}$$

which is elementary and gives

$$I_0^2(a) = 4 \cdot \frac{\pi}{2} \cdot \frac{1}{2a} = \frac{\pi}{a}$$

or

$$I_0(a) = \left(\frac{\pi}{a}\right)^{1/2}$$

Now prove that the $I_v(a)$ may be obtained by repeated differentiation of $I_0(a)$ with respect to a and, in particular, that

$$\frac{d^v I_0(a)}{da^v} = (-1)^v I_v(a)$$

Use this result and the fact that $I_0(a) = (\pi/a)^{1/2}$ to generate $I_1(a)$, $I_2(a)$, and so forth.

The procedure of Problem B–4 can be used, since the integrand is an even function of x.

$$I_v(a) = \int_{-\infty}^{\infty} x^{2v} e^{-ax^2}\, dx$$

$$= \int_{-\infty}^{0} x^{2v} e^{-ax^2}\, dx + \int_0^\infty x^{2v} e^{-ax^2}\, dx$$

Let $z = -x$ in the first integral and $z = x$ in the second.

$$I_v(a) = -\int_\infty^0 z^{2v} e^{-az^2}\, dz + \int_0^\infty z^{2v} e^{-az^2}\, dz$$

$$= 2 \int_0^\infty z^{2v} e^{-az^2}\, dz$$

$$= 2 \int_0^\infty x^{2v} e^{-ax^2}\, dx$$

Since $I_v(a)$ depends only on a,

$$I_0^2(a) = 4 \left(\int_0^\infty e^{-ax^2}\, dx \right) \left(\int_0^\infty e^{-ay^2}\, dy \right) = 4 \int_0^\infty \int_0^\infty dx\, dy\, e^{-a(x^2+y^2)}$$

The integration is over the entire first quadrant, $0 \leq x \leq \infty$ and $0 \leq y \leq \infty$ so that in polar coordinates, the limits of integration are $0 \leq r < \infty$ and $0 \leq \theta \leq \pi/2$. Using the substitutions

$r^2 = x^2 + y^2$ and $dx\,dy = r\,dr\,d\theta$, the integral becomes

$$I_0^2(a) = 4 \int_0^{\pi/2} d\theta \int_0^\infty dr\, r e^{-ar^2}$$

$$= 4 \left(\frac{\pi}{2}\right) \frac{1}{2a} = \frac{\pi}{a}$$

Taking the square root of both sides of the equation leads directly to the desired result, $I_0(a) = (\pi/a)^{1/2}$.

Now, differentiate $I_0(a)$ with respect to a:

$$\frac{dI_0(a)}{da} = -\int_{-\infty}^\infty x^2 e^{-ax^2}\,dx = -I_1(a)$$

$$\frac{d^2 I_0(a)}{da^2} = \int_{-\infty}^\infty x^4 e^{-ax^2}\,dx = I_2(a)$$

This can be repeated, and the pattern is obvious, although the general result could be proved via induction if desired.

$$\frac{d^v I_0(a)}{da^v} = (-1)^v \int_{-\infty}^\infty x^{2v} e^{-ax^2}\,dx = (-1)^v I_v(a)$$

Starting with $I_0(a) = (\pi/a)^{1/2}$, application of the formula gives $I_1(a) = (\pi/a)^{1/2}/2a$, $I_2(a) = 3\,(\pi/a)^{1/2}/4a^2$, and so forth.

5–24. Prove that the product of two even functions is even, that the product of two odd functions is even, and that the product of an even and an odd function is odd.

Recall that an even function is one for which $f(x) = f(-x)$ and an odd function is one for which $f(x) = -f(-x)$. Let $P(x)$ be the product of two functions $f(x)$ and $g(x)$. For two even functions,

$$P(x) = f(x)g(x) = f(-x)g(-x) = P(-x)$$

so the product of two even functions is even. For two odd functions,

$$P(x) = f(x)g(x) = -f(-x)\,[-g(-x)] = f(-x)g(-x) = P(-x)$$

so the product of two odd functions is also even. For one odd and one even function,

$$P(x) = f(x)g(x) = [-f(-x)]g(-x) = -f(-x)g(-x) = -P(-x)$$

so the product of one odd and one even function is odd.

5–25. Prove that the derivative of an even (odd) function is odd (even).

If $f(x)$ is even, it can be represented by a power series of the form

$$f(x) = c_0 + c_2 x^2 + c_4 x^4 + O(x^6)$$

where the only allowed values of n in x^n are even. The derivative of this function is

$$f'(x) = 2c_2 x + 4c_4 x^3 + O(x^5)$$

which is an odd function expressed in a power series.

Similarly, if $g(x)$ is odd, it can be represented by

$$g(x) = c_1 x + c_3 x^3 + c_5 x^5 + O(x^7)$$

where the only allowed values of n in x^n are odd, and its derivative is

$$g'(x) = c_1 + 3c_3 x^2 + 5c_5 x^4 + O(x^6)$$

which is an even function.

5–26. Show that

$$\langle x^2 \rangle = \int_{-\infty}^{\infty} \psi_2(x) x^2 \psi_2(x) dx = \frac{5}{2} \frac{\hbar}{(\mu k)^{1/2}}$$

for a harmonic oscillator. Note that $\langle x^2 \rangle^{1/2}$ is the square root of the mean of the square of the displacement (the *root-mean-square displacement*) of the oscillator.

From Table 5.4, $\psi_2(x) = \left(\dfrac{\alpha}{4\pi}\right)^{1/4} \left(2\alpha x^2 - 1\right) e^{-\alpha x^2/2}$. So,

$$\langle x^2 \rangle = \int_{-\infty}^{\infty} \psi_2(x) x^2 \psi_2(x)\, dx$$

$$= 2 \left(\frac{\alpha}{4\pi}\right)^{1/2} \int_0^{\infty} x^2 \left(2\alpha x^2 - 1\right)^2 e^{-\alpha x^2}\, dx$$

$$= 2 \left(\frac{\alpha}{4\pi}\right)^{1/2} \int_0^{\infty} \left(4\alpha^2 x^6 - 4\alpha x^4 + x^2\right) e^{-\alpha x^2}\, dx$$

$$= 2 \left(\frac{\alpha}{4\pi}\right)^{1/2} \left[4\alpha^2 \left(\frac{15}{16\alpha^3}\right) - 4\alpha \left(\frac{3}{8\alpha^2}\right) + \frac{1}{4\alpha}\right] \left(\frac{\pi}{\alpha}\right)^{1/2}$$

$$= \frac{5}{2\alpha} = \frac{5}{2} \frac{\hbar}{(\mu k)^{1/2}}$$

5–27. Show that $\langle p \rangle = 0$ and that

$$\langle p^2 \rangle = \int_{-\infty}^{\infty} \psi_2(x) \hat{P}^2 \psi_2(x) dx = \frac{5}{2} \hbar (\mu k)^{1/2}$$

for a harmonic oscillator.

Recall that $\hat{P} = -i\hbar\dfrac{d}{dx}$ and $\hat{P}^2 = -\hbar^2\dfrac{d^2}{dx^2}$. Then from Table 5.4,

$$\psi_2(x) = \left(\frac{\alpha}{4\pi}\right)^{1/4}\left(2\alpha x^2 - 1\right)e^{-\alpha x^2/2}.$$

Note that $\psi_2(x)$ is an even function of x. Problem 5–25 showed that the derivative of an even function is odd, and Problem 5–24 showed that the product of an even function and an odd function is odd, so in evaluating

$$\langle p \rangle = \int_{-\infty}^{\infty} \psi_2(x)\hat{P}\psi_2(x)\,dx = -i\hbar \int_{-\infty}^{\infty} \psi_2(x)\frac{d}{dx}\psi_2(x)\,dx$$

the integrand is seen to be overall an odd function. Consequently, $\langle p \rangle = 0$.

For $\langle p^2 \rangle$,

$$\langle p^2 \rangle = \int_{-\infty}^{\infty} \psi_2(x)\hat{P}^2\psi_2(x)\,dx$$

$$= -2\hbar^2\left(\frac{\alpha}{4\pi}\right)^{1/2}\int_0^{\infty}\left(2\alpha x^2 - 1\right)e^{-\alpha x^2/2}\frac{d}{dx}\left[5\alpha x e^{-\alpha x^2/2} - 2\alpha^2 x^3 e^{-\alpha x^2/2}\right]dx$$

$$= -2\hbar^2\left(\frac{\alpha}{4\pi}\right)^{1/2}\int_0^{\infty}\left(2\alpha x^2 - 1\right)e^{-\alpha x^2}\left(5\alpha - 11\alpha^2 x^2 + 2\alpha^3 x^4\right)dx$$

$$= -2\hbar^2\left(\frac{\alpha}{4\pi}\right)^{1/2}\int_0^{\infty}\left(4\alpha^4 x^6 - 24\alpha^3 x^4 + 21\alpha^2 x^2 - 5\alpha\right)e^{-\alpha x^2}dx$$

$$= -2\hbar^2\left(\frac{\alpha}{4\pi}\right)^{1/2}\left[4\alpha^4\left(\frac{15}{16\alpha^3}\right) - 24\alpha^3\left(\frac{3}{8\alpha^2}\right) + 21\alpha^2\left(\frac{1}{4\alpha}\right) - \frac{5\alpha}{2}\right]\left(\frac{\pi}{\alpha}\right)^{1/2}$$

$$= \frac{5}{2}\hbar^2\alpha = \frac{5}{2}\hbar\left(\mu k\right)^{1/2}$$

5–28.   Use the result of the previous two problems to show that $E_2 = 5h\nu/2$.

The Schrödinger equation, $\hat{H}\psi_2 = E_2\psi_2$, and the normalization of the harmonic oscillator wave functions gives

$$E_2 = \int_{-\infty}^{\infty} \psi_2(x) \hat{H} \psi_2(x)\, dx$$

$$= \int_{-\infty}^{\infty} \psi_2(x) \left(-\frac{\hbar^2}{2\mu} \frac{d^2}{dx^2} + \frac{1}{2} k x^2 \right) \psi_2(x)\, dx$$

$$= \frac{\langle p^2 \rangle}{2\mu} + \frac{k \langle x^2 \rangle}{2}$$

Now use the results of the previous two problems.

$$E_2 = \frac{5}{2} \left[\frac{\hbar (\mu k)^{1/2}}{2\mu} \right] + \frac{5}{2} \left[\frac{k\hbar}{2 (\mu k)^{1/2}} \right]$$

$$= \frac{5}{2} \left(\frac{k}{\mu} \right)^{1/2} = \frac{5}{2} h\nu$$

5–29.   Prove that

$$\langle T \rangle = \langle V(x) \rangle = \frac{E_v}{2}$$

for a one-dimensional harmonic oscillator for $v = 0$ and $v = 1$.

Recall that $\hat{T} = -\dfrac{\hbar^2}{2\mu} \dfrac{d^2}{dx^2}$ and that $\hat{V}(x) = -\dfrac{1}{2} k x^2$ and find the expectation values of these

operators for the harmonic oscillator wave functions from Table 5.4: $\psi_0(x) = \left(\dfrac{\alpha}{\pi} \right)^{1/4} e^{-\alpha x^2/2}$

and $\psi_1(x) = \left(\dfrac{4\alpha^3}{\pi} \right)^{1/4} x e^{-\alpha x^2/2}$.

$$\langle T \rangle_0 = \int_{-\infty}^{\infty} \psi_0(x) \left(-\frac{\hbar^2}{2\mu} \right) \frac{d^2}{dx^2} \psi_0(x)\, dx$$

$$= -\frac{\hbar^2}{\mu} \left(\frac{\alpha}{\pi} \right)^{1/2} \int_0^{\infty} e^{-\alpha x^2/2} \frac{d}{dx} \left(-\alpha x e^{-\alpha x^2/2} \right)\, dx$$

$$= -\frac{\hbar^2}{\mu} \left(\frac{\alpha}{\pi} \right)^{1/2} \int_0^{\infty} \left(\alpha^2 x^2 - \alpha \right) e^{-\alpha x^2}\, dx$$

$$= -\frac{\hbar^2}{\mu} \left(\frac{\alpha}{\pi} \right)^{1/2} \left[\alpha^2 \left(\frac{1}{4\alpha} \right) - \frac{\alpha}{2} \right] \left(\frac{\pi}{\alpha} \right)^{1/2}$$

$$= \frac{\hbar^2 \alpha}{4\mu} = \frac{\hbar}{4} \left(\frac{k}{\mu} \right)^{1/2} = \frac{E_0}{2}$$

$$\langle V(x)\rangle_0 = \int_{-\infty}^{\infty} \psi_0(x) \left(\frac{1}{2}kx^2\right) \psi_0(x)\, dx$$

$$= k\left(\frac{\alpha}{\pi}\right)^{1/2} \int_0^{\infty} x^2 e^{-\alpha x^2}\, dx$$

$$= k\left(\frac{\alpha}{\pi}\right)^{1/2} \frac{1}{4\alpha} \left(\frac{\pi}{\alpha}\right)^{1/2}$$

$$= \frac{k}{4\alpha} = \frac{\hbar}{4}\left(\frac{k}{\mu}\right)^{1/2} = \frac{E_0}{2}$$

$$\langle T\rangle_1 = \int_{-\infty}^{\infty} \psi_1(x)\left(-\frac{\hbar^2}{2\mu}\right)\frac{d^2}{dx^2}\psi_1(x)\, dx$$

$$= -\frac{\hbar^2}{\mu}\left(\frac{4\alpha^3}{\pi}\right)^{1/2} \int_0^{\infty} xe^{-\alpha x^2/2}\frac{d}{dx}\left(e^{-\alpha x^2/2} - \alpha x^2 e^{-\alpha x^2/2}\right)\, dx$$

$$= -\frac{\hbar^2}{\mu}\left(\frac{4\alpha^3}{\pi}\right)^{1/2} \int_0^{\infty} \left(\alpha^2 x^4 - 3\alpha x^2\right) e^{-\alpha x^2}\, dx$$

$$= -\frac{\hbar^2}{\mu}\left(\frac{4\alpha^3}{\pi}\right)^{1/2} \left[\alpha^2\left(\frac{3}{8\alpha^2}\right) - 3\alpha\left(\frac{1}{4\alpha}\right)\right]\left(\frac{\pi}{\alpha}\right)^{1/2}$$

$$= \frac{3\hbar^2\alpha}{4\mu} = \frac{3\hbar}{4}\left(\frac{k}{\mu}\right)^{1/2} = \frac{E_1}{2}$$

$$\langle V(x)\rangle_1 = \int_{-\infty}^{\infty} \psi_1(x)\left(\frac{1}{2}kx^2\right)\psi_1(x)\, dx$$

$$= k\left(\frac{4\alpha^3}{\pi}\right)^{1/2} \int_0^{\infty} x^4 e^{-\alpha x^2}\, dx$$

$$= k\left(\frac{4\alpha^3}{\pi}\right)^{1/2} \frac{3}{8\alpha^2}\left(\frac{\pi}{\alpha}\right)^{1/2}$$

$$= \frac{3k}{4\alpha} = \frac{3\hbar}{4}\left(\frac{k}{\mu}\right)^{1/2} = \frac{E_1}{2}$$

5–30. Show that the eigenfunctions and eigenvalues of a three-dimensional harmonic oscillator whose potential energy is

$$V(x, y, z) = \frac{1}{2}k_x x^2 + \frac{1}{2}k_y y^2 + \frac{1}{2}k_z z^2$$

are

$$\psi_{v_x, v_y, v_z}(x, y, z) = \psi_{v_x}(x)\psi_{v_y}(y)\psi_{v_z}(z)$$

where

$$\psi_{v_u}(u) = \left[\frac{(\alpha_u/\pi)^{1/2}}{2^{v_u}v_u!}\right]^{1/2} H_{v_u}(\alpha_u^{1/2}u)e^{-\alpha_u u^2/2} \qquad (u = x, y, \text{ or } z)$$

$$\alpha_u^2 = \frac{\mu k_u}{\hbar^2}$$

and

$$E_{v_x v_y v_z} = h\nu_x(v_x + \tfrac{1}{2}) + h\nu_y(v_y + \tfrac{1}{2}) + h\nu_z(v_z + \tfrac{1}{2})$$

where

$$\nu_u = \frac{1}{2\pi}\left(\frac{k_u}{\mu}\right)^{1/2}$$

Discuss the degeneracy of this system when the oscillator is isotropic—that is, when $k_x = k_y = k_z$. Make a diagram like that shown in Figure 3.6.

The Schrödinger equation in three dimensions for the harmonic oscillator is

$$-\frac{\hbar^2}{2\mu}\left(\frac{\partial^2\psi}{\partial x^2} + \frac{\partial^2\psi}{\partial y^2} + \frac{\partial^2\psi}{\partial z^2}\right) + V(x, y, z)\psi(x, y, z) = E\psi(x, y, z)$$

Let $\psi(x, y, z) = X(x)Y(y)Z(z)$, substitute this into the Schrödinger equation, divide by XYZ and multiply by $2\mu/\hbar^2$ to obtain

$$-\frac{\hbar^2}{2\mu}\left(\frac{\partial^2\psi}{\partial x^2} + \frac{\partial^2\psi}{\partial y^2} + \frac{\partial^2\psi}{\partial z^2}\right) + V(x, y, z)\psi(x, y, z) = E\psi(x, y, z)$$

$$-\frac{\hbar^2}{2\mu}\left(YZ\frac{d^2X}{dx^2} + XZ\frac{d^2Y}{dy^2} + XY\frac{d^2Z}{dz^2}\right) + V(x, y, z)X(x)Y(y)Z(z) = EX(x)Y(y)Z(z)$$

$$\left[-\frac{1}{X}\frac{d^2X}{dx^2} + \frac{2\mu}{\hbar^2}\left(\frac{1}{2}k_x x^2\right)\right] + \left[-\frac{1}{Y}\frac{d^2Y}{dy^2} + \frac{2\mu}{\hbar^2}\left(\frac{1}{2}k_y y^2\right)\right] + \left[-\frac{1}{Z}\frac{d^2Z}{dz^2} + \frac{2\mu}{\hbar^2}\left(\frac{1}{2}k_z z^2\right)\right] = \frac{2\mu}{\hbar^2}E$$

As in Problem 2–15, the three-dimensional Schrödinger equation is reduced to three one-dimensional problems, for example

$$\frac{d^2X}{dx^2} + \frac{2\mu}{\hbar^2}\left(E_x - \frac{1}{2}k_x x^2\right)X(x) = 0$$

The solutions to this Schrödinger equation are discussed in Sections 5.5 and 5.8 of the text and are the one-dimensional harmonic oscillator wave functions,

$$\psi_{v_u}(u) = \left[\frac{(\alpha_u/\pi)^{1/2}}{2^{v_u} v_u!} \right]^{1/2} H_{v_u}(\alpha_u^{1/2} u) e^{-\alpha_u u^2/2} \qquad (u = x, y, \text{ or } z)$$

$$\alpha_u^2 = \frac{\mu k_u}{\hbar^2}$$

with

$$E_u = h\nu_u \left(v_u + \tfrac{1}{2} \right)$$

and

$$\nu_u = \frac{1}{2\pi} \left(\frac{k_u}{\mu} \right)^{1/2}$$

Thus,

$$\psi(x, y, z) = X(x)Y(y)Z(z) = \psi_{v_x}(x)\psi_{v_y}(y)\psi_{v_z}(z)$$

and

$$E_{v_x v_y v_z} = E_x + E_y + E_z = h\nu_x(v_x + \tfrac{1}{2}) + h\nu_y(v_y + \tfrac{1}{2}) + h\nu_z(v_z + \tfrac{1}{2})$$

If the oscillator is isotropic, $k_x = k_y = k_z$, and $\nu_x = \nu_y = \nu_z = \nu$, so that

$$E_{v_x v_y v_z} = h\nu \left(v_x + v_y + v_z + \tfrac{3}{2} \right) = h\nu \left(v + \tfrac{3}{2} \right)$$

with

$$v = v_x + v_y + v_z = 0, 1, 2, \ldots$$

There are many degeneracies for the three-dimensional isotropic oscillator. In fact the degree of degeneracy is given by the number of ways three integers (v_x, v_y, v_z) can sum to a given fourth integer (v). Namely, the energy level with quantum number v is $\frac{1}{2}(v+1)(v+2)$-fold degenerate, as seen in the figure below.

	$(v_x \, v_y \, v_z)$	Degeneracy
4	(400) (040) (004) (310) (301) (031) (130) (103) (013) (211) (121) (112) (220) (202) (022)	15
3	(300) (030) (003) (210) (120) (201) (102) (120) (210) (111)	10
2	(200) (020) (002) (110) (101) (011)	6
1	(100) (010) (001)	3
0	(000)	1

vertical axis: $v = v_x + v_y + v_z$

5–31. There are a number of general relations between the Hermite polynomials and their derivatives (which we will not derive). Some of these are

$$\frac{dH_v(\xi)}{d\xi} = 2\xi H_v(\xi) - H_{v+1}(\xi)$$

$$H_{v+1}(\xi) - 2\xi H_v(\xi) + 2v H_{v-1}(\xi) = 0$$

and

$$\frac{dH_v(\xi)}{d\xi} = 2v H_{v-1}(\xi)$$

Such connecting relations are called *recursion formulas*. Verify these formulas explicitly using the first few Hermite polynomials given in Table 5.3.

The first recursion relation, $\dfrac{dH_v(\xi)}{d\xi} = 2\xi H_v(\xi) - H_{v+1}(\xi)$, is verified in the table below through comparison of the final two columns.

v	$H_v(\xi)$	$2\xi H_v(\xi)$	$H_{v+1}(\xi)$	$2\xi H_v(\xi) - H_{v+1}(\xi)$	$\dfrac{dH_v(\xi)}{d\xi}$
0	1	2ξ	2ξ	0	0
1	2ξ	$4\xi^2$	$4\xi^2 - 2$	2	2
2	$4\xi^2 - 2$	$8\xi^3 - 4\xi$	$8\xi^3 - 12\xi$	8ξ	8ξ
3	$8\xi^3 - 12\xi$	$16\xi^4 - 24\xi^2$	$16\xi^4 - 48\xi^2 + 12$	$24\xi^2 - 12$	$24\xi^2 - 12$

The second recursion relation $H_{v+1}(\xi) - 2\xi H_v(\xi) + 2v H_{v-1}(\xi) = 0$ is verified by noting that the rows of the following table sum to zero.

v	$H_{v+1}(\xi)$	$-2\xi H_v(\xi)$	$2v H_{v-1}(\xi)$
1	$4\xi^2 - 2$	$-4\xi^2$	2
2	$8\xi^3 - 12\xi$	$-8\xi^3 + 4\xi$	8ξ
3	$16\xi^4 - 48\xi^2 + 12$	$-16\xi^4 + 24\xi^2$	$24\xi^2 - 12$
4	$32\xi^5 - 160\xi^3 + 120\xi$	$-32\xi^5 + 96\xi^3 - 24\xi$	$64\xi^3 - 96\xi$

Finally, $\dfrac{dH_v(\xi)}{d\xi} = 2v H_{v-1}(\xi)$ is verified by comparing the two rightmost columns in the following table.

v	$H_v(\xi)$	$2v H_{v-1}(\xi)$	$\dfrac{dH_v(\xi)}{d\xi}$
1	2ξ	2	2
2	$4\xi^2 - 2$	8ξ	8ξ
3	$8\xi^3 - 12\xi$	$24\xi^2 - 12$	$24\xi^2 - 12$
4	$16\xi^4 - 48\xi^2 + 12$	$64\xi^3 - 96\xi$	$64\xi^3 - 96\xi$
5	$32\xi^5 - 160\xi^3 + 120\xi$	$160\xi^4 - 480\xi^2 + 120$	$160\xi^4 - 480\xi^2 + 120$

5–32. Use the recursion formulas for the Hermite polynomials given in Problem 5–31 to show that $\langle p \rangle = 0$ and $\langle p^2 \rangle = \hbar(\mu k)^{1/2}(v + \frac{1}{2})$. Remember that the momentum operator involves a differentiation with respect to x, not ξ.

The easiest way to show that $\langle p \rangle = 0$ for the harmonic oscillator is to use the parity of the wave function and of its derivative as discussed in the text following Equation 5.52. Nevertheless, the result can also be obtained using the recursion properties of the Hermite polynomials. From Equation 5.52 and using Equations 5.44 through 5.46,

$$\langle p \rangle = \int_{-\infty}^{\infty} \psi_v(x) \left(-i\hbar \frac{d}{dx} \right) \psi_v(x)\, dx$$

$$= -i\hbar N_v^2 \int_{-\infty}^{\infty} H_v(\xi) e^{-\xi^2/2} \frac{d}{d\xi} \left[H_v(\xi) e^{-\xi^2/2} \right] d\xi$$

$$= -i\hbar N_v^2 \int_{-\infty}^{\infty} H_v(\xi) e^{-\xi^2/2} \left[\frac{dH_v(\xi)}{d\xi} - \xi H_v(\xi) \right] e^{-\xi^2/2}\, dx$$

where $\dfrac{d}{dx} = \dfrac{d\xi}{dx}\dfrac{d}{d\xi} = \alpha^{1/2}\dfrac{d}{d\xi}$, and $d\xi = \alpha^{1/2}\, dx$. Now use the recursion formulas, $\dfrac{dH_v(\xi)}{d\xi} = 2v H_{v-1}(\xi)$ and $\xi H_v(\xi) = v H_{v-1}(\xi) + \dfrac{1}{2} H_{v+1}(\xi)$.

$$\langle p \rangle = -i\hbar N_v^2 \int_{-\infty}^{\infty} H_v(\xi) \left[v H_{v-1}(\xi) - \frac{1}{2} H_{v+1}(\xi) \right] e^{-\xi^2}\, d\xi$$

$$= 0$$

since the two integrals vanish by the orthogonality properties of the Hermite polynomials with respect to the weighting function $e^{-\xi^2}$.

To evaluate $\langle p^2 \rangle$ using the recursion properties of the Hermite polynomials, consider the integral

$$\langle p^2 \rangle = \int_{-\infty}^{\infty} \psi_v(x) \left(-\hbar^2 \frac{d^2}{dx^2} \right) \psi_v(x)\, dx$$

$$= -\alpha^{1/2}\hbar^2 N_v^2 \int_{-\infty}^{\infty} H_v(\xi) e^{-\xi^2/2} \frac{d^2}{d\xi^2} \left[H_v(\xi) e^{-\xi^2/2} \right] d\xi \tag{1}$$

where $\dfrac{d^2}{dx^2} = \left(\dfrac{d\xi}{dx} \right)^2 \dfrac{d^2}{d\xi^2} = \alpha \dfrac{d^2}{d\xi^2}$ and again $d\xi = \alpha^{1/2}\, dx$. Continue by evaluating,

$$\frac{d^2}{d\xi^2} \left[H_v(\xi) e^{-\xi^2/2} \right] = \frac{d}{d\xi} \left[\frac{dH_v(\xi)}{d\xi} e^{-\xi^2/2} - \xi H_v(\xi) e^{-\xi^2/2} \right]$$

$$= \left[\frac{d^2 H_v(\xi)}{d\xi^2} - 2\xi \frac{dH_v(\xi)}{d\xi} + \left(\xi^2 - 1 \right) H_v(\xi) \right] e^{-\xi^2/2}$$

Now use $\dfrac{dH_v(\xi)}{d\xi} = 2v H_{v-1}(\xi)$, which can be repeated to give $\dfrac{d^2 H_v(\xi)}{d\xi^2} = 4v(v-1) H_{v-2}(\xi)$, so

$$\frac{d^2}{d\xi^2} \left[H_v(\xi) e^{-\xi^2/2} \right] = \left[4v(v-1) H_{v-2}(\xi) - 4v\xi H_{v-1}(\xi) + \left(\xi^2 - 1 \right) H_v(\xi) \right] e^{-\xi^2/2}$$

Then use the two recursion relations

$$\xi H_v(\xi) = v H_{v-1}(\xi) + \frac{1}{2} H_{v+1}(\xi)$$

$$\xi^2 H_v(\xi) = v(v-1) H_{v-2}(\xi) + \left(v + \frac{1}{2}\right) H_v(\xi) + \frac{1}{4} H_{v+2}(\xi)$$

(see the equation at the top of page 233 of the text for the second of these two), and arrive at

$$\frac{d^2}{d\xi^2}\left[H_v(\xi)e^{-\xi^2/2}\right] = \left[v(v-1) H_{v-2}(\xi) - \left(v + \frac{1}{2}\right) H_v(\xi) + \frac{1}{4} H_{v+2}(\xi)\right] e^{-\xi^2/2}$$

This result can be substituted into Equation 1 for $\langle p^2 \rangle$, to obtain

$$\langle p^2 \rangle = -\alpha^{1/2} \hbar^2 N_v^2 \int_{-\infty}^{\infty} H_v(\xi) \left[v(v-1) H_{v-2}(\xi) - \left(v + \frac{1}{2}\right) H_v(\xi) + \frac{1}{4} H_{v+2}(\xi)\right] e^{-\xi^2} \, d\xi$$

$$= -\alpha \hbar^2 N_v^2 \int_{-\infty}^{\infty} H_v(\alpha^{1/2}x) \left[v(v-1) H_{v-2}(\alpha^{1/2}x) - \left(v + \frac{1}{2}\right) H_v(\alpha^{1/2}x) + \frac{1}{4} H_{v+2}(\alpha^{1/2}x)\right] e^{-\alpha x^2} \, dx$$

$$= \alpha \hbar^2 \left(v + \frac{1}{2}\right) \int_{-\infty}^{\infty} \psi_v(x)\psi_v(x) \, dx$$

$$= \alpha \hbar^2 \left(v + \frac{1}{2}\right) = \hbar (\mu k)^{1/2} \left(v + \frac{1}{2}\right)$$

by the orthogonality of the Hermite polynomials and the normality of the harmonic oscillator wavefunctions.

5–33. It can be proved generally that

$$\langle x^2 \rangle = \frac{1}{\alpha} \left(v + \frac{1}{2}\right) = \frac{\hbar}{(\mu k)^{1/2}} \left(v + \frac{1}{2}\right)$$

and that

$$\langle x^4 \rangle = \frac{3}{4\alpha^2} (2v^2 + 2v + 1) = \frac{3\hbar^2}{4\mu k} (2v^2 + 2v + 1)$$

for a harmonic oscillator. Verify these formulas explicitly for the first two states of a harmonic oscillator.

Use the wave functions from Table 5.4 to evaluate the two quantities,

$$\psi_0(x) = \left(\frac{\alpha}{\pi}\right)^{1/4} e^{-\alpha x^2/2}$$

$$\psi_1(x) = \left(\frac{4\alpha^3}{\pi}\right)^{1/4} x e^{-\alpha x^2/2}$$

For $\langle x^2 \rangle$,

$$\langle x^2 \rangle_0 = \int_{-\infty}^{\infty} \psi_0(x) x^2 \psi_0(x)\, dx$$

$$= 2 \left(\frac{\alpha}{\pi} \right)^{1/2} \int_0^{\infty} x^2 e^{-\alpha x^2}\, dx$$

$$= 2 \left(\frac{\alpha}{\pi} \right)^{1/2} \frac{1}{4\alpha} \left(\frac{\pi}{\alpha} \right)^{1/2}$$

$$= \frac{1}{2\alpha} = \frac{\hbar}{2 \left(\mu k \right)^{1/2}}$$

$$\langle x^2 \rangle_1 = \int_{-\infty}^{\infty} \psi_1(x) x^2 \psi_1(x)\, dx$$

$$= 2 \left(\frac{4\alpha^3}{\pi} \right)^{1/2} \int_0^{\infty} x^4 e^{-\alpha x^2}\, dx$$

$$= 2 \left(\frac{4\alpha^3}{\pi} \right)^{1/2} \frac{3}{8\alpha^2} \left(\frac{\pi}{\alpha} \right)^{1/2}$$

$$= \frac{3}{2\alpha} = \frac{3\hbar}{2 \left(\mu k \right)^{1/2}}$$

Since $v + 1/2$ equals $1/2$ and $3/2$ for $v = 0$, 1, respectively, the formula for $\langle x^2 \rangle$ is verified for these two quantum numbers.

For $\langle x^4 \rangle$,

$$\langle x^4 \rangle_0 = \int_{-\infty}^{\infty} \psi_0(x) x^4 \psi_0(x)\, dx$$

$$= 2 \left(\frac{\alpha}{\pi} \right)^{1/2} \int_0^{\infty} x^4 e^{-\alpha x^2}\, dx$$

$$= 2 \left(\frac{\alpha}{\pi} \right)^{1/2} \frac{3}{8\alpha^2} \left(\frac{\pi}{\alpha} \right)^{1/2}$$

$$= \frac{3}{4\alpha^2} = \frac{3\hbar^2}{4\mu k}$$

$$\langle x^4 \rangle_1 = \int_{-\infty}^{\infty} \psi_1(x) x^4 \psi_1(x)\, dx$$

$$= 2 \left(\frac{4\alpha^3}{\pi} \right)^{1/2} \int_0^{\infty} x^6 e^{-\alpha x^2}\, dx$$

$$= 2 \left(\frac{4\alpha^3}{\pi} \right)^{1/2} \frac{15}{16\alpha^3} \left(\frac{\pi}{\alpha} \right)^{1/2}$$

$$= \frac{15}{4\alpha^2} = \frac{15\hbar^2}{4\mu k}$$

Since $2v^2 + 2v + 1$ equals 1 and 5 for $v = 0$, 1, respectively, the formula for $\langle x^4 \rangle$ is verified for these two quantum numbers.

5–34. This problem is similar to Problem 3–39. Show that the harmonic-oscillator wave functions are alternately even and odd functions of x because the Hamiltonian operator obeys $\hat{H}(x) = \hat{H}(-x)$. Define a reflection operator $\hat{R}$ by

$$\hat{R}u(x) = u(-x)$$

Show that $\hat{R}$ is linear and that it commutes with $\hat{H}$. Show also that the eigenvalues of $\hat{R}$ are ± 1. What are its eigenfunctions? Show that the harmonic-oscillator wave functions are eigenfunctions of $\hat{R}$. Note that they are eigenfunctions of both $\hat{H}$ and $\hat{R}$. What does this observation say about $\hat{H}$ and $\hat{R}$?

Consider the Schrödinger equation of a harmonic oscillator

$$\hat{H}(x)\psi_v(x) = E_v\psi_v(x)$$

Replace x by $-x$ and use the fact that $\hat{H}(x) = \hat{H}(-x)$ to obtain

$$\hat{H}\psi_v(-x) = E_v\psi_v(-x)$$

Both $\psi_v(-x)$ and $\psi_v(x)$ are eigenfunctions of $\hat{H}(x)$ corresponding to the eigenvalue E_v. Since the system is nondegenerate, these eigenfunctions can differ by only a multiplicative constant c, or $\psi_v(x) = c\psi_v(-x)$. But $\psi_v(-x) = c\psi_v(x)$, and so $c = \pm 1$ (as in Problem 3–39). Thus ψ_v is always either even or odd. Moreover, since $H_v(x)$ is even when v is even and odd when v is odd, and

$$\psi_v(x) = N_v H_v(\alpha^{1/2}x)e^{-\alpha x^2/2}$$

$\psi_v(x)$ is even when v is even and odd when v is odd.

Now define $\hat{R}$ as $\hat{R}u(x) = u(-x)$. $\hat{R}$ is linear because

$$\hat{R}\left[c_1u_1(x) + c_2u_2(x)\right] = c_1u_1(-x) + c_2u_2(-x)$$

$$= c_1\hat{R}u_1(x) + c_2\hat{R}u_2(x)$$

Because $\hat{R}\psi_v(x) = \psi_v(-x) = \pm\psi_v(x)$, we see that the eigenvalues of $\hat{R}$ are ± 1 and the eigenfunctions are $\psi_v(x)$. Because $\hat{H}$ and $\hat{R}$ have mutual eigenfunctions, they commute. Since the two operators commute, then it is possible to determine definite values simultaneously for both the energy and parity of a harmonic oscillator.

5–35. Use Ehrenfest's theorem (Problem 4–44) to show that $\langle p_x\rangle$ does not depend upon time for a one-dimensional harmonic oscillator.

From Ehrenfest's theorem

$$\frac{d\langle p_x\rangle}{dt} = \left\langle -\frac{dV}{dx}\right\rangle = \langle -kx\rangle = -k\int_{-\infty}^{\infty}\psi_v(x)x\psi_v(x)\,dx = 0$$

because x is an odd function. Since $d\langle p_x\rangle/dt = 0$, then $\langle p_x\rangle$ does not depend on time.

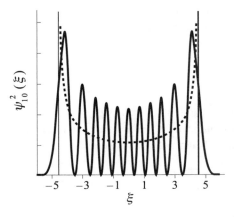

FIGURE 5.13
The probability distribution function of a harmonic oscillator in the $v = 10$ state. The dashed line is that for a classical harmonic oscillator with the same energy. The vertical lines at $\xi \approx \pm 4.6$ represent the extreme limits of the classical harmonic motion.

5–36. Figure 5.13 compares the probability distribution associated with $\psi_{10}(\xi)$ to the classical distribution. (See also Problem 3–18.) This problem illustrates what is meant by the classical distribution. Consider

$$x(t) = A \sin(\omega t + \phi)$$

which can be written as

$$\omega t = \sin^{-1}\left(\frac{x}{A}\right) - \phi$$

Now

$$dt = \frac{\omega^{-1} dx}{\sqrt{A^2 - x^2}} \tag{1}$$

This equation gives the time that the oscillator spends between x and $x + dx$. We can convert equation 1 to a probability distribution in x by dividing by the time that it takes for the oscillator to go from $-A$ to A. Show that this time is π/ω and that the probability distribution in x is

$$p(x)dx = \frac{dx}{\pi\sqrt{A^2 - x^2}} \tag{2}$$

Show that $p(x)$ is normalized. Why does $p(x)$ achieve its maximum value at $x = \pm A$? Now use the fact that $\xi = \alpha^{1/2}x$, where $\alpha = (k\mu/\hbar^2)^{1/2}$, to show that

$$p(\xi)d\xi = \frac{d\xi}{\pi\sqrt{\alpha A^2 - \xi^2}} \tag{3}$$

Show that the limits of ξ are $\pm(\alpha A^2)^{1/2} = \pm(21)^{1/2}$, and compare this result to the vertical lines shown in Figure 5.13. *Hint:* You need to use the fact that $kA^2/2 = E_{10}$ $(n = 10)$. Finally, plot equation 3 and compare your result with the curve in Figure 5.13.

─────────────

In going from $-A$ to A, the oscillator goes through 1/2 cycle, so the time, t, this takes is 1/2 the period of the oscillator, or from Problem 5–4, $t = \tau/2 = \pi/\omega$. Thus,

$$\frac{dt}{t} = p(x)\,dx = \frac{dx}{\pi\sqrt{A^2 - x^2}}$$

To show that this expression is normalized, integrate over the observed time period:

$$\int_{-A}^{A} \frac{dx}{\pi\sqrt{A^2 - x^2}} = \frac{1}{\pi}\sin^{-1}\frac{x}{A}\Big|_{-A}^{A} = 1$$

The maximum values of $p(x)$ are at $x = \pm A$ because these are the points at which the classical harmonic oscillator has zero velocity. Substituting $\xi = \alpha^{1/2}x$ and $d\xi = \alpha^{1/2}dx$,

$$p(\xi)d\xi = \frac{d\xi}{\pi\sqrt{\alpha A^2 - \xi^2}}$$

Since the limits of x are $\pm A$, the limits of ξ are $\pm\alpha^{1/2}A = \pm\sqrt{\alpha A^2}$. But $kA^2/2 = E_{10} = \frac{21}{2}\hbar\omega$, so $A^2 = 21\hbar/(\mu k)^{1/2}$. Also, $\alpha = (\mu k)^{1/2}/\hbar$, so $(\alpha A^2)^{1/2} = (21)^{1/2} = 4.58$. Using this value for A to plot Equation 3 gives the dashed curve in Figure 5.13.

5–37. We can use the harmonic-oscillator wave function to illustrate tunneling, a strictly quantum-mechancial property. The probability that the displacement of a harmonic oscillator in its ground state lies between x and $x + dx$ is given by

$$P(x)\,dx = \psi_0^2(x)\,dx = \left(\frac{\alpha}{\pi}\right)^{1/2}e^{-\alpha x^2}\,dx \qquad (1)$$

The energy of the oscillator in its ground state is $(\hbar/2)(k/\mu)^{1/2}$. Show that the greatest displacement that this oscillator can have classically is its amplitude

$$A = \left[\frac{\hbar}{(k\mu)^{1/2}}\right]^{1/2} = \frac{1}{\alpha^{1/2}} \qquad (2)$$

According to equation 1, however, there is a nonzero probability that the displacement of the oscillator will exceed this classical value and *tunnel* into the classically forbidden region. Show that this probability is given by

$$\int_{-\infty}^{-\alpha^{1/2}} P(x)\,dx + \int_{+\alpha^{1/2}}^{\infty} P(x)\,dx = \frac{2}{\pi^{1/2}}\int_{1}^{\infty}e^{-z^2}\,dz \qquad (3)$$

The integral here cannot be evaluated in closed form but occurs so frequently in a number of different fields (such as the kinetic theory of gases and statistics) that it is well tabulated under the name *complementary error function*, erfc(x), which is defined as

$$\text{erfc}(x) = \frac{2}{\pi^{1/2}}\int_{x}^{\infty}e^{-z^2}\,dz \qquad (4)$$

By referring to tables, it can be found that the probability that the displacement of the molecule will exceed its classical amplitude is 0.16.

Classically, the greatest displacement for an oscillator occurs when all the energy of the oscillator is potential energy and the kinetic energy equals zero, that is when $x = \pm A$, $kA^2/2 = E$. Taking $E = (\hbar/2)(k/\mu)^{1/2}$ gives

$$\frac{kA^2}{2} = \frac{\hbar}{2}\left(\frac{k}{\mu}\right)^{1/2}$$

$$A = \left[\frac{\hbar}{(k\mu)^{1/2}}\right]^{1/2} = \frac{1}{\alpha^{1/2}}$$

Classically, the oscillator is restricted to the interval $-\alpha^{-1/2} \le x \le \alpha^{-1/2}$, and is forbidden from the regions $|x| > \alpha^{-1/2}$ on either side. The probability that the quantum mechanical displacement will extend into this *classically forbidden region* is the sum of the probabilities for the two intervals $-\infty < x \le -\alpha^{-1/2}$ and $\alpha^{-1/2} \le x < \infty$, or since the probability is an even function of x,

$$\int_{-\infty}^{-\alpha^{1/2}} P(x)\,dx + \int_{\alpha^{1/2}}^{\infty} P(x)\,dx = 2\left(\frac{\alpha}{\pi}\right)^{1/2} \int_{\alpha^{1/2}}^{\infty} e^{-\alpha x^2}\,dx$$

$$= \frac{2}{\pi^{1/2}} \int_{1}^{\infty} e^{-z^2}\,dz$$

$$= \operatorname{erfc}(1) = 0.16$$

where $z = \alpha^{1/2}x$.

5–38. Problem 5–37 shows that although a classical harmonic oscillator has a fixed amplitude, a quantum-mechanical harmonic oscillator does not. We can, however, use $\langle x^2 \rangle$ as a measure of the square of the amplitude, and, in particular, we can use the square root of $\langle x^2 \rangle$, the root-mean-square value of x, $A_{\mathrm{rms}} = x_{\mathrm{rms}} = \langle x^2 \rangle^{1/2}$, as a measure of the amplitude. Show that

$$A_{\mathrm{rms}} = x_{\mathrm{rms}} = \left[\frac{\hbar}{2(\mu k)^{1/2}}\right]^{1/2} = \left(\frac{\hbar}{4\pi c\tilde{\omega}_{\mathrm{obs}}\mu}\right)^{1/2}$$

for the ground state of a harmonic oscillator. Thus, we can calculate $\langle x^2 \rangle^{1/2}$ in terms of $\tilde{\omega}_{\mathrm{obs}}$. Show that $A_{\mathrm{rms}} = 7.58$ pm for $H^{35}Cl$. The bond length of HCl is 127 pm, and so we can see that the root-mean-square amplitude is only about 5% of the bond length. This is typical of diatomic molecules.

From Problem 5–33, $\langle x^2 \rangle = \hbar/2\,(\mu k)^{1/2}$ for the ground state of a harmonic oscillator. Thus,

$$x_{\mathrm{rms}} = \langle x^2 \rangle^{1/2} = \left[\frac{\hbar}{2\,(\mu k)^{1/2}}\right]^{1/2} = \left(\frac{\hbar}{4\pi c\tilde{\omega}_{\mathrm{obs}}\mu}\right)^{1/2}$$

using Equation 5.39 for $\tilde{\omega}_{\mathrm{obs}}$.

For $H^{35}Cl$, $\tilde{\omega}_{\mathrm{obs}} = 2990.94$ cm^{-1} from Table 5.1, and the reduced mass was found in Problem 5–10, $\mu = 1.6270 \times 10^{-27}$ kg, so

$$x_{rms} = \left(\frac{\hbar}{4\pi c \tilde{\omega}_{obs} \mu} \right)^{1/2}$$

$$= \left[\frac{1.054\,572\,66 \times 10^{-34} \text{ J·s}}{4\pi \left(2.997\,924\,58 \times 10^{10} \text{ cm·s}^{-1} \right) \left(2990.94 \text{ cm}^{-1} \right) \left(1.6270 \times 10^{-27} \text{ kg} \right)} \right]^{1/2}$$

$$= 7.5845 \text{ pm}$$

5–39. The fundamental vibrational frequency of $H^{79}Br$ is 2.63×10^3 cm^{-1} and $\bar{l}$ is 141 pm. Calculate the root-mean-square displacement in the ground state and compare your result to $\bar{l}$.

For $H^{79}Br$, the reduced mass is calculated to be

$$\mu = \frac{(1.0078 \text{ amu}) \,(78.918 \text{ amu})}{1.0078 \text{ amu} + 78.918 \text{ amu}} = 0.99509 \text{ amu} = 1.6524 \times 10^{-27} \text{ kg}$$

Then using the value of $\tilde{\omega}_{obs}$ given in the problem,

$$x_{rms} = \left(\frac{\hbar}{4\pi c \tilde{\omega}_{obs} \mu} \right)^{1/2}$$

$$= \left[\frac{1.054\,572\,66 \times 10^{-34} \text{ J·s}}{4\pi \left(2.997\,924\,58 \times 10^{10} \text{ cm·s}^{-1} \right) \left(2.63 \times 10^3 \text{ cm}^{-1} \right) \left(1.6524 \times 10^{-27} \text{ kg} \right)} \right]^{1/2}$$

$$= 8.03 \text{ pm}$$

which is 5.69% of $\bar{l}$.

5–40. In this problem, we'll calculate the fraction of diatomic molecules in a particular vibrational state at a temperature T using the harmonic-oscillator approximation. A fundamental equation of physical chemistry is the *Boltzmann distribution*, which says that the number of molecules with an energy E_j is proportional to $e^{-E_j/k_B T}$, where k_B is the Boltzmann constant and T is the kelvin temperature. Thus, we write

$$N_j \propto e^{-E_j/k_B T}$$

Using the fact that $\sum_j N_j = N$, show that the fraction of molecules with an energy E_j is given by

$$f_j = \frac{e^{-E_j/k_B T}}{\sum_j e^{-E_j/k_B T}} = \frac{e^{-E_j/k_B T}}{Q(T)}$$

The denominator here, $Q(T)$, is called a partition function. Show that the partition function of a harmonic oscillator $[E_j = (j + \frac{1}{2})h\nu]$ is

$$Q(T) = e^{-h\nu/2k_B T} (1 - e^{-h\nu/k_B T})^{-1}$$

Hint: You need to use the geometric series $\sum_{j=0}^{\infty} x^j = (1-x)^{-1}$ for $|x| < 1$. Now show that

$$f_j = (1 - e^{-h\nu/k_\mathrm{B}T})e^{-jh\nu/k_\mathrm{B}T}$$

and, in particular, that

$$f_0 = 1 - e^{-h\nu/k_\mathrm{B}T}$$

Now show that $f_0 \approx 1$ at 300 K for a typical molecule, with $\tilde{\omega}_\mathrm{e} = 1000$ cm^{-1}. Plot the fraction of molecules in the jth vibrational state at $T = 300$ K.

─────────────────────────

The fraction of molecules with an energy E_j is the number with that energy divided by the total number of molecules, or

$$f_j = \frac{N_j}{N} = \frac{N_j}{\sum_j N_j} = \frac{e^{-E_j/k_\mathrm{B}T}}{\sum_j e^{-E_j/k_\mathrm{B}T}}$$

Using the expression for the energy of the jth level, the partition function becomes

$$Q(T) = \sum_{j=0}^{\infty} e^{-E_j/k_\mathrm{B}T}$$

$$= \sum_{j=0}^{\infty} e^{-(j+1/2)h\nu/k_\mathrm{B}T}$$

$$= e^{-h\nu/2k_\mathrm{B}T} \sum_{j=0}^{\infty} e^{-jh\nu/k_\mathrm{B}T}$$

The last sum has the form of a geometric series with $x = e^{-h\nu/k_\mathrm{B}T}$, so

$$Q(T) = e^{-h\nu/2k_\mathrm{B}T} \left(1 - e^{-h\nu/k_\mathrm{B}T}\right)^{-1}$$

This result for $Q(T)$ can be used in the expression for f_j,

$$f_j = \frac{e^{-E_j/k_\mathrm{B}T}}{Q(T)} = \frac{\left(1 - e^{-h\nu/k_\mathrm{B}T}\right) e^{-(j+1/2)h\nu/k_\mathrm{B}T}}{e^{-h\nu/2k_\mathrm{B}T}} = \left(1 - e^{-h\nu/k_\mathrm{B}T}\right) e^{-jh\nu/k_\mathrm{B}T}$$

which for $j = 0$ becomes

$$f_0 = 1 - e^{-h\nu/k_\mathrm{B}T}$$

For a typical molecule with $h\nu = \tilde{\omega}_\mathrm{e} = 1000$ cm^{-1}, the fraction of molecules in the ground vibrational state at $T = 300$ K is

$$f_0 = 1 - e^{-1000\ \mathrm{cm}^{-1}/\left(0.695\,038\ \mathrm{cm}^{-1}\cdot\mathrm{K}^{-1}\right)(300\ \mathrm{K})} = 1 - e^{-4.80} = 0.992 \approx 1$$

A plot of f_j versus j for $T = 300$ K and $\tilde{\omega}_e = 1000$ cm^{-1} is given below and shows quite dramatically that only very small fractions of the total number of molecules are in excited vibrational states at room temperature.

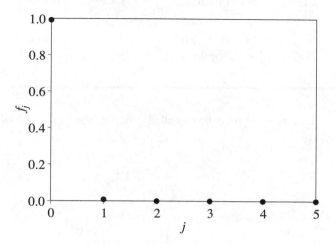

5–41. Calculate the fraction of HBr molecules in the ground vibrational state at 300 K and 2000 K. Take $\tilde{\omega}_e$ to be 2650 cm^{-1}.

From Problem 5–40, the fraction of molecules in the ground vibrational state is $f_0 = 1 - e^{-h\nu/k_B T}$. For HBr at 300 K,

$$f_0 = 1 - e^{-2650 \text{ cm}^{-1}/\left(0.695\,038 \text{ cm}^{-1}\cdot\text{K}^{-1}\right)(300 \text{ K})} = 1 - e^{-12.7} = 1.00$$

At 300 K, practically 100% of HBr molecules are in the ground vibrational state. At 2000 K,

$$f_0 = 1 - e^{-2650 \text{ cm}^{-1}/\left(0.695\,038 \text{ cm}^{-1}\cdot\text{K}^{-1}\right)(2000 \text{ K})} = 1 - e^{-1.91} = 0.851$$

Even at this higher temperature over 85% of the HBr molecules are in the ground vibrational state.

5–42. Determine the number of translational, rotational, and vibrational degrees of freedom in

(a) CH_3Cl (b) OCS

(c) C_6H_6 (d) H_2CO

All molecules have 3 translational degrees of freedom, linear molecules have 2 rotational degrees of freedom and non-linear molecules have 3 rotational degrees of freedom. If the molecule has N atoms, this leaves $3N - 5$ vibrational degrees of freedom if it is linear or $3N - 6$ vibrational degrees of freedom if it is non-linear.

	Molecule	N		Degrees of Freedom		
				Translational	Rotational	Vibrational
(a)	CH_3Cl	5	non-linear	3	3	9
(b)	OCS	3	linear	3	2	4
(c)	C_6H_6	12	non-linear	3	3	30
(d)	H_2CO	4	non-linear	3	3	6

5–43. In this problem, we will prove the so-called *quantum-mechanical virial theorem*. Start with $\hat{H}\psi = E\psi$, where

$$\hat{H} = -\frac{\hbar^2}{2m}\nabla^2 + V(x, y, z)$$

Using the fact that $\hat{H}$ is a Hermitian operator, show that

$$\int \psi^*[\hat{H}, \hat{A}]\psi \, d\tau = 0 \tag{1}$$

where $\hat{A}$ is any linear operator. Choose $\hat{A}$ to be

$$\hat{A} = -i\hbar \left(x\frac{\partial}{\partial x} + y\frac{\partial}{\partial y} + z\frac{\partial}{\partial z} \right) \tag{2}$$

and show that

$$[\hat{H}, \hat{A}] = i\hbar \left(x\frac{\partial V}{\partial x} + y\frac{\partial V}{\partial y} + z\frac{\partial V}{\partial z} \right) - \frac{i\hbar}{m}(\hat{P}_x^2 + \hat{P}_y^2 + \hat{P}_z^2)$$

$$= i\hbar \left(x\frac{\partial V}{\partial x} + y\frac{\partial V}{\partial y} + z\frac{\partial V}{\partial z} \right) - 2i\hbar\hat{T}$$

where $\hat{T}$ is the kinetic energy operator. Now use equation 1 and show that

$$\left\langle x\frac{\partial V}{\partial x} + y\frac{\partial V}{\partial y} + z\frac{\partial V}{\partial z} \right\rangle = 2\langle \hat{T} \rangle \tag{3}$$

Equation 3 is the quantum-mechanical *virial theorem*.

Start by writing out the commutator in Equation 1

$$\int \psi^*[\hat{H}, \hat{A}]\psi \, d\tau = \int \psi^*\hat{H}\hat{A}\psi \, d\tau - \int \psi^*\hat{A}\hat{H}\psi \, d\tau$$

Using the Hermitian property of $\hat{H}$ (see, for example, Problem 4–21) gives

$$\int \psi^*[\hat{H}, \hat{A}]\psi \, d\tau = \int \left(\hat{H}\psi\right)^* \hat{A}\psi \, d\tau - \int \psi^*\hat{A}\hat{H}\psi \, d\tau$$

then, since $\hat{H}\psi = E\psi$ and E is real

$$\int \psi^*[\hat{H}, \hat{A}]\psi \, d\tau = E\int \psi^*\hat{A}\psi \, d\tau - E\int \psi^*\hat{A}\psi \, d\tau = 0$$

Now let

$$\hat{A} = -i\hbar \left(x\frac{\partial}{\partial x} + y\frac{\partial}{\partial y} + z\frac{\partial}{\partial z} \right)$$

and use

$$\hat{H} = -\frac{\hbar^2}{2m}\nabla^2 + V(x, y, z)$$

Then

$$\hat{H}\hat{A}f = -i\hbar \left[-\frac{\hbar^2}{2m}\nabla^2 + V(x, y, z) \right] \left[x\frac{\partial f}{\partial x} + y\frac{\partial f}{\partial y} + z\frac{\partial f}{\partial z} \right]$$

$$= \frac{i\hbar^3}{2m} \left(2\frac{\partial^2 f}{\partial x^2} + x\frac{\partial^3 f}{\partial x^3} + y\frac{\partial^3 f}{\partial x^2 \partial y} + z\frac{\partial^3 f}{\partial x^2 \partial z} + x\frac{\partial^3 f}{\partial x \partial y^2} \right.$$

$$+ 2\frac{\partial^2 f}{\partial y^2} + y\frac{\partial^3 f}{\partial y^3} + z\frac{\partial^3 f}{\partial y^2 \partial z} + x\frac{\partial^3 f}{\partial x \partial z^2} + y\frac{\partial^3 f}{\partial y \partial z^2}$$

$$\left. + 2\frac{\partial^2 f}{\partial z^2} + z\frac{\partial^3 f}{\partial z^3} \right) - i\hbar \left[V \left(x\frac{\partial f}{\partial x} + y\frac{\partial f}{\partial y} + z\frac{\partial f}{\partial z} \right) \right]$$

and

$$\hat{A}\hat{H}f = \frac{i\hbar^3}{2m} \left(x\frac{\partial}{\partial x} + y\frac{\partial}{\partial y} + z\frac{\partial}{\partial z} \right) \left(\frac{\partial^2 f}{\partial x^2} + \frac{\partial^2 f}{\partial y^2} + \frac{\partial^2 f}{\partial z^2} \right)$$

$$- i\hbar \left(x\frac{\partial}{\partial x} + y\frac{\partial}{\partial y} + z\frac{\partial}{\partial z} \right) Vf$$

$$= \frac{i\hbar^3}{2m} \left(x\frac{\partial^3 f}{\partial x^3} + x\frac{\partial^3 f}{\partial x \partial y^2} + x\frac{\partial^3 f}{\partial x \partial z^2} + y\frac{\partial^3 f}{\partial x^2 \partial y} + y\frac{\partial^3 f}{\partial y^3} \right.$$

$$\left. + y\frac{\partial^3 f}{\partial y \partial z^2} + z\frac{\partial^3 f}{\partial x^2 \partial z} + z\frac{\partial^3 f}{\partial y^2 \partial z} + z\frac{\partial^3 f}{\partial z^3} \right) - i\hbar \left[\left(x\frac{\partial V}{\partial x} + y\frac{\partial V}{\partial y} + z\frac{\partial V}{\partial z} \right) f \right.$$

$$\left. + \left(x\frac{\partial f}{\partial x} + y\frac{\partial f}{\partial y} + z\frac{\partial f}{\partial z} \right) V \right]$$

Therefore,

$$\left[\hat{H}, \hat{A} \right] = \frac{i\hbar^3}{2m} \left(2\frac{\partial^2}{\partial x^2} + 2\frac{\partial^2}{\partial y^2} + 2\frac{\partial^2}{\partial z^2} \right) + ih \left(x\frac{\partial V}{\partial x} + y\frac{\partial V}{\partial y} + z\frac{\partial V}{\partial z} \right)$$

$$= i\hbar \left(x\frac{\partial V}{\partial x} + y\frac{\partial V}{\partial y} + z\frac{\partial V}{\partial z} \right) - \frac{i\hbar}{m} \left(\hat{P}_x^2 + \hat{P}_y^2 + \hat{P}_z^2 \right)$$

$$= i\hbar \left(x\frac{\partial V}{\partial x} + y\frac{\partial V}{\partial y} + z\frac{\partial V}{\partial z} \right) - 2i\hbar\hat{T}$$

where the operators $\hat{P}$ and $\hat{T}$ are recognized from Table 4.1 Now, substitute this result into Equation 1 to obtain

$$\int \psi^* \left[i\hbar \left(x\frac{\partial V}{\partial x} + y\frac{\partial V}{\partial y} + z\frac{\partial V}{\partial z} \right) - 2i\hbar \hat{T} \right] \psi d\tau = 0$$

$$i\hbar \int \psi^* \left(x\frac{\partial V}{\partial x} + y\frac{\partial V}{\partial y} + z\frac{\partial V}{\partial z} \right) \psi d\tau = \int \psi^* 2i\hbar \hat{T} \psi d\tau$$

$$\left\langle x\frac{\partial V}{\partial x} + y\frac{\partial V}{\partial y} + z\frac{\partial V}{\partial z} \right\rangle = 2\langle \hat{T} \rangle$$

5–44. Use the virial theorem (Problem 5–43) to prove that $\langle \hat{T} \rangle = \langle V \rangle = E/2$ for a harmonic oscillator.

For a three-dimensional harmonic oscillator,

$$V(x, y, z) = \frac{k_x x^2}{2} + \frac{k_y y^2}{2} + \frac{k_z z^2}{2}$$

Therefore,

$$x\frac{\partial V}{\partial x} + y\frac{\partial V}{\partial y} + z\frac{\partial V}{\partial z} = k_x x^2 + k_y y^2 + k_z z^2 = 2V$$

then from Problem 5–43, $2\langle V \rangle = 2\langle \hat{T} \rangle$. Because $\langle \hat{T} \rangle + \langle V \rangle = E$, this can be written as

$$\langle \hat{T} \rangle = \langle V \rangle = \tfrac{1}{2}E$$

5–45. Use the fact that

$$\hat{a}_+ \psi_v = \frac{1}{\sqrt{2}} \left[\left(\frac{\mu\omega}{\hbar} \right)^{1/2} x - \left(\frac{\hbar}{\mu\omega} \right)^{1/2} \frac{d}{dx} \right] \psi_v \propto \psi_{v+1}$$

and that $\psi_0(x) = (\alpha/\pi)^{1/4} e^{-\alpha x^2/2}$ to generate $\psi_1(x)$ and $\psi_2(x)$.

From Equation 5 of the Appendix,

$$\hat{a}_+ = \frac{1}{\sqrt{2}} \left(\hat{x} - i\hat{p} \right)$$

with $\hat{x} = (\mu\omega/\hbar)^{1/2} \hat{X}$ and $\hat{p} = (\mu\hbar\omega)^{-1/2} \hat{P}$. Thus, since $\hat{X} = x$ and $\hat{P} = -i\hbar (d/dx)$,

$$\hat{a}_+ = \frac{1}{\sqrt{2}}\left[\left(\frac{\mu\omega}{\hbar}\right)^{1/2}x - \left(\frac{\hbar}{\mu\omega}\right)^{1/2}\frac{d}{dx}\right]$$

$$= \frac{1}{\sqrt{2}}\left\{\left[\frac{(\mu k)^{1/2}}{\hbar}\right]^{1/2}x - \left[\frac{\hbar}{(\mu k)^{1/2}}\right]^{1/2}\frac{d}{dx}\right\}$$

$$= \frac{1}{\sqrt{2}}\left(\alpha^{1/2}x - \alpha^{-1/2}\frac{d}{dx}\right)$$

where $\omega = (k/\mu)^{1/2}$ and $\alpha = (\mu k)^{1/2}/\hbar$.

Now apply the raising operator to $\psi_0(x) = (\alpha/\pi)^{1/4}e^{-\alpha x^2/2}$,

$$\hat{a}_+\psi_0(x) = \frac{1}{\sqrt{2}}\left(\alpha^{1/2}x - \alpha^{-1/2}\frac{d}{dx}\right)\left(\frac{\alpha}{\pi}\right)^{1/4}e^{-\alpha x^2/2}$$

$$= \frac{1}{\sqrt{2}}\left(\frac{\alpha}{\pi}\right)^{1/4}\left(\alpha^{1/2}xe^{-\alpha x^2/2} + \alpha^{1/2}xe^{-\alpha x^2/2}\right)$$

$$= \left(\frac{4\alpha^3}{\pi}\right)^{1/4}xe^{-\alpha x^2/2}$$

$$= \psi_1(x)$$

and then to $\psi_1(x)$,

$$\hat{a}_+\psi_1(x) = \frac{1}{\sqrt{2}}\left(\alpha^{1/2}x - \alpha^{-1/2}\frac{d}{dx}\right)\left(\frac{4\alpha^3}{\pi}\right)^{1/4}xe^{-\alpha x^2/2}$$

$$= \frac{1}{\sqrt{2}}\left(\frac{4\alpha^3}{\pi}\right)^{1/4}\left[\alpha^{1/2}x^2e^{-\alpha x^2/2} - \alpha^{-1/2}\left(e^{-\alpha x^2/2} - \alpha x^2e^{-\alpha x^2/2}\right)\right]$$

$$= \left(\frac{\alpha}{\pi}\right)^{1/4}\left(2\alpha x^2 - 1\right)e^{-\alpha x^2/2}$$

$$= \psi_2(x)$$

Spherical Coordinates

PROBLEMS AND SOLUTIONS

E–1. Derive Equations E.2 from Equations E.1.

Equations E.1 are

$$x = r \sin \theta \cos \phi \qquad y = r \sin \theta \sin \phi \qquad z = r \cos \theta$$

Then, using $\sin^2 \alpha + \cos^2 \alpha = 1$ three times,

$$
\begin{aligned}
r^2 &= r^2 \left(\sin^2 \theta + \cos^2 \theta \right) \left(\sin^2 \phi + \cos^2 \phi \right) \\
&= r^2 \sin^2 \theta \sin^2 \phi + r^2 \sin^2 \theta \cos^2 \phi + r^2 \cos^2 \theta \sin^2 \phi + r^2 \cos^2 \theta \cos^2 \phi \\
&= (r \sin \theta \cos \phi)^2 + (r \sin \theta \sin \phi)^2 + (r \cos \theta)^2 \left(\sin^2 \phi + \cos^2 \phi \right) \\
&= x^2 + y^2 + z^2 \\
r &= \left(x^2 + y^2 + z^2 \right)^{1/2}
\end{aligned}
\tag{1}
$$

from which the equation for $\cos \theta$ follows,

$$\cos \theta = \frac{z}{r} = \frac{z}{\left(x^2 + y^2 + z^2 \right)^{1/2}} \tag{2}$$

and finally,

$$\tan \phi = \frac{\sin \phi}{\cos \phi} = \frac{r \sin \theta}{r \sin \theta} \left(\frac{\sin \phi}{\cos \phi} \right) = \frac{y}{x} \tag{3}$$

Equations 1, 2, and 3 are Equations E.2.

E–2. Express the following points given in cartesian coordinates in terms of spherical coordinates:
(x, y, z): $(1, 0, 0)$; $(0, 1, 0)$; $(0, 0, 1)$; $(0, 0, -1)$.

Use the equations derived in the previous problem (Equations E.2).

(a) (1, 0, 0)

$$r = \left(x^2 + y^2 + z^2 \right)^{1/2} = 1$$

$$\theta = \cos^{-1} \left(\frac{z}{r} \right) = \cos^{-1} 0 = \frac{\pi}{2}$$

$$\phi = \tan^{-1} \left(\frac{y}{x} \right) = \tan^{-1} 0 = 0$$

Spherical coordinates: $(1, \frac{\pi}{2}, 0)$

(b) (0, 1, 0)

$$r = \left(x^2 + y^2 + z^2 \right)^{1/2} = 1$$

$$\theta = \cos^{-1} \left(\frac{z}{r} \right) = \cos^{-1} 0 = \frac{\pi}{2}$$

$$\phi = \tan^{-1} \left(\frac{y}{x} \right) = \tan^{-1} \left(\frac{1}{0} \right) = \frac{\pi}{2}$$

Spherical coordinates: $(1, \frac{\pi}{2}, \frac{\pi}{2})$

(c) (0, 0, 1)

$$r = \left(x^2 + y^2 + z^2 \right)^{1/2} = 1$$

$$\theta = \cos^{-1} \left(\frac{z}{r} \right) = \cos^{-1}(1) = 0$$

$$\phi = \tan^{-1} \left(\frac{y}{x} \right) = \tan^{-1} \left(\frac{0}{0} \right)$$

(This means that the value of ϕ is undetermined, or that its value is irrelevant.)

Spherical coordinates: $(1, 0, \phi)$

(d)   (0, 0, −1)

$$r = \left(x^2 + y^2 + z^2 \right)^{1/2} = 1$$

$$\theta = \cos^{-1} \left(\frac{z}{r} \right) = \cos^{-1}(-1) = \pi$$

$$\phi = \tan^{-1} \left(\frac{y}{x} \right) = \tan^{-1} \left(\frac{0}{0} \right)$$

Spherical coordinates: $(1, \pi, \phi)$

E–3. Describe the graphs of the following equations:

(a) $r = 5$ (b) $\theta = \pi/4$ (c) $\phi = \pi/2$

(a) The surface of a sphere of radius 5 centered at the origin
(b) A cone about the z-axis with its point at the origin and an internal angle of $\frac{\pi}{4}$
(c) A plane containing the y and z axes.

E–4. Use Equation E.3 to determine the volume of a hemisphere of radius a.

A hemisphere of radius corresponds to $0 \le r \le a$, $0 \le \theta \le \frac{\pi}{2}$, and $0 \le \phi \le 2\pi$. (There are other choices; this is the upper hemisphere of a sphere cut by the xy plane.) Now integrate Equation E.3 over these limits.

$$dV = r^2 \sin\theta \, dr \, d\theta \, d\phi$$

$$V = \int_0^a dr \, r^2 \int_0^{\pi/2} d\theta \, \sin\theta \int_0^{2\pi} d\phi = \frac{2\pi a^3}{3}$$

E–5. Use Equation E.5 to determine the surface area of a hemisphere of radius a.

Integrate Equation E.5 over the angular coordinates with fixed $r = a$ using the same hemisphere of the previous problem.

$$dA = r^2 \sin\theta \, d\theta \, d\phi$$

$$V = a^2 \int_0^{\pi/2} d\theta \, \sin\theta \int_0^{2\pi} d\phi = 2\pi a^2$$

E–6. Evaluate the integral

$$I = \int_0^\pi \cos^2\theta \, \sin^3\theta \, d\theta$$

by letting $x = \cos\theta$.

If $x = \cos\theta$, then $dx = -\sin\theta \, d\theta$, and the limits of integration become $x = 1$ for $\theta = 0$ and $x = -1$ for $\theta = \pi$. Then,

$$I = \int_0^\pi \cos^2\theta \, \sin^3\theta \, d\theta = \int_0^\pi \cos^2\theta \left(1 - \cos^2\theta\right) \sin\theta \, d\theta$$

$$= -\int_1^{-1} x^2 \left(1 - x^2\right) dx = \int_{-1}^1 x^2 \, dx - \int_{-1}^1 x^4 \, dx$$

$$= \frac{1}{3} + \frac{1}{3} - \frac{1}{5} - \frac{1}{5} = \frac{4}{15}$$

E–7.   We will learn in Chapter 7 that a $2p_y$ hydrogen atom orbital is given by

$$\psi_{2p_y} = \frac{1}{4\sqrt{2\pi}}\, re^{-r/2} \sin\theta \sin\phi$$

Show that ψ_{2p_y} is normalized. (Don't forget to square ψ_{2p_y} first.)

Normalization requires that

$$I = \int_0^\infty \int_0^\pi \int_0^{2\pi} \psi_{2p_y}^* \psi_{2p_y} r^2 \sin\theta\, dr\, d\theta\, d\phi = 1$$

The square of ψ_{2p_y} is

$$\psi_{2p_y}^* \psi_{2p_y} = \frac{1}{32\pi} r^2 e^{-r} \sin^2\theta \sin^2\phi$$

Then

$$I = \frac{1}{32\pi}\int_0^\infty dr\, r^4 e^{-r} \int_0^\pi d\theta\, \sin^3\theta \int_0^{2\pi} d\phi\, \sin^2\phi$$

$$= \frac{1}{32\pi}\,(4!)\left(\frac{4}{3}\right)\left(\frac{2\pi}{2}\right) = 1$$

and consequently ψ_{2p_y} is normalized.

E–8.   We will learn in Chapter 7 that a $2s$ hydrogen atomic orbital is given by

$$\psi_{2s} = \frac{1}{4\sqrt{2\pi}}(2-r)e^{-r/2}$$

Show that ψ_{2s} is normalized.

As in the previous problem, evaluate

$$I = \int_0^\infty \int_0^\pi \int_0^{2\pi} \psi_{2s}^* \psi_{2s} r^2 \sin\theta\, dr\, d\theta\, d\phi$$

The square of ψ_{2s} is

$$\psi_{2s}^* \psi_{2s} = \frac{1}{32\pi}(2-r)^2\, e^{-r}$$

Then

$$I = \frac{1}{32\pi}\int_0^\infty dr\, r^2 (2-r)^2\, e^{-r} \int_0^\pi d\theta\, \sin\theta \int_0^{2\pi} d\phi$$

$$= \frac{1}{32\pi}\left(4\int_0^\infty dr\, r^2 e^{-r} - 4\int_0^\infty dr\, r^3 e^{-r} + \int_0^\infty dr\, r^4 e^{-r}\right)(2)\,(2\pi)$$

$$= \frac{1}{8}\,[4\,(2!) - 4\,(3!) + (4!)] = 1$$

and consequently ψ_{2s} is normalized.

E–9. Show that

$$Y_1^0(\theta, \phi) = \left(\frac{3}{4\pi}\right)^{1/2} \cos\theta$$

$$Y_1^1(\theta, \phi) = -\left(\frac{3}{8\pi}\right)^{1/2} e^{i\phi} \sin\theta$$

and

$$Y_1^{-1}(\theta, \phi) = \left(\frac{3}{8\pi}\right)^{1/2} e^{-i\phi} \sin\theta$$

are orthonormal over the surface of a sphere.

A set of functions defined for the angular coordinates (θ, ϕ) is orthonormal when members of the set, f_i and f_j, satisfy the condition

$$\int_0^\pi \int_0^{2\pi} f_i^* f_j \sin\theta \, d\theta \, d\phi = \delta_{ij}$$

For the functions above,

$$\int Y_1^{0*} Y_1^0 = \frac{3}{4\pi} \int_0^\pi d\theta \, \sin\theta \cos^2\theta \int_0^{2\pi} d\phi$$

$$= \frac{3}{4\pi} \left(\frac{-\cos^3\theta}{3} \bigg|_0^\pi\right)(2\pi) = \frac{3}{2}\left(\frac{2}{3}\right) = 1$$

$$\int Y_1^{1*} Y_1^1 = \frac{3}{8\pi} \int_0^\pi d\theta \, \sin^3\theta \int_0^{2\pi} d\phi$$

$$= \frac{3}{8\pi} \left(\frac{4}{3}\right)(2\pi) = 1$$

$$\int Y_1^{-1*} Y_1^{-1} = \frac{3}{8\pi} \int_0^\pi d\theta \, \sin^3\theta \int_0^{2\pi} d\phi$$

$$= \frac{3}{8\pi} \left(\frac{4}{3}\right)(2\pi) = 1$$

$$\int Y_1^{0*} Y_1^1 = -\frac{3}{4\sqrt{2}\pi} \int_0^\pi d\theta \, \cos\theta \sin^2\theta \int_0^{2\pi} e^{i\phi} d\phi = 0$$

$$\int Y_1^{0*} Y_1^{-1} = \frac{3}{4\sqrt{2}\pi} \int_0^\pi d\theta \, \cos\theta \sin^2\theta \int_0^{2\pi} e^{-i\phi} d\phi = 0$$

$$\int Y_1^{1*} Y_1^{-1} = -\frac{3}{8\pi} \int_0^\pi d\theta \, \sin^3\theta \int_0^{2\pi} e^{2i\phi} d\phi = 0$$

Recall from Problem A–9 that $\int_0^{2\pi} e^{-im\phi} d\phi = \int_0^{2\pi} e^{im\phi} d\phi = 0$.

E–10. Evaluate the average of $\cos\theta$ and $\cos^2\theta$ over the surface of a sphere.

The average of a function $f(\theta)$ [or more generally $f(\theta, \phi)$] over the surface of a sphere is

$$\langle f(\theta)\rangle = \frac{\int_0^{\pi}\int_0^{2\pi} f(\theta)\sin\theta\, d\theta\, d\phi}{\int_0^{\pi}\int_0^{2\pi}\sin\theta\, d\theta\, d\phi} = \frac{\int_0^{\pi}\int_0^{2\pi} f(\theta)\sin\theta\, d\theta\, d\phi}{4\pi}$$

Thus,

$$\langle\cos\theta\rangle = \frac{1}{4\pi}\int_0^{\pi} d\theta\,\sin\theta\cos\theta\int_0^{2\pi} d\phi = \frac{1}{4}\int_0^{\pi}\sin 2\theta\, d\theta = 0$$

and

$$\langle\cos^2\theta\rangle = \frac{1}{4\pi}\int_0^{\pi} d\theta\,\sin\theta\cos^2\theta\int_0^{2\pi} d\phi = \frac{1}{2}\int_{-1}^{1} x^2\, dx = \frac{1}{3}$$

where the substitution $x = \cos\theta$ (see Problem E–6) is used.

E–11. We shall frequently use the notation $d\mathbf{r}$ to represent the volume element in spherical coordinates. Evaluate the integral

$$I = \int d\mathbf{r} e^{-r}\cos^2\theta$$

where the integral is over all space (in other words, over all possible values of r, θ, and ϕ).

$$I = \int d\mathbf{r}\, e^{-r}\cos^2\theta$$

$$= \int_0^{\infty} dr\, r^2 e^{-r}\int_0^{\pi} d\theta\,\sin\theta\cos^2\theta\int_0^{2\pi} d\phi$$

$$= (2!)\left(\frac{2}{3}\right)(2\pi) = \frac{8\pi}{3}$$

E–12. Show that the two functions

$$f_1(r) = e^{-r}\cos\theta \quad \text{and} \quad f_2(r) = (2 - r)e^{-r/2}\cos\theta$$

are orthogonal over all space (in other words, over all possible values of r, θ, and ϕ).

If $f_1(\mathbf{r})$ and $f_2(\mathbf{r})$ are orthogonal, then

$$I = \int d\mathbf{r}\, f_1^*(\mathbf{r}) f_2(\mathbf{r}) = 0$$

Evaluating this integral gives

$$I = \int_0^\infty dr\, r^2 e^{-r}(2-r)e^{-r/2} \int_0^\pi d\theta\, \cos^2\theta \sin\theta \int_0^{2\pi} d\phi$$

$$= \int_0^\infty dr\, (2r^2 - r^3)e^{-3r/2}\left(\frac{4\pi}{3}\right)$$

$$= \frac{8\pi}{3}\int_0^\infty dr\, r^2 e^{-3r/2} - \frac{4\pi}{3}\int_0^\infty dr\, r^3 e^{-3r/2}$$

$$= \left(\frac{8\pi}{3}\right)\left(\frac{16}{27}\right) - \left(\frac{4\pi}{3}\right)\left(\frac{32}{27}\right) = 0$$

and the two functions are indeed orthogonal.

E–13. Consider the transformation from cartesian coordinates to plane polar coordinates,

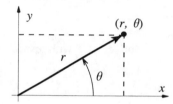

where

$$x = r\cos\theta \qquad r = (x^2 + y^2)^{1/2}$$

$$y = r\sin\theta \qquad \theta = \tan^{-1}\left(\frac{y}{x}\right) \tag{1}$$

If a function $f(r,\theta)$ depends upon the polar coordinates r and θ, then the chain rule of partial differentiation says that

$$\left(\frac{\partial f}{\partial x}\right)_y = \left(\frac{\partial f}{\partial r}\right)_\theta \left(\frac{\partial r}{\partial x}\right)_y + \left(\frac{\partial f}{\partial \theta}\right)_r \left(\frac{\partial \theta}{\partial x}\right)_y \tag{2}$$

and that

$$\left(\frac{\partial f}{\partial y}\right)_x = \left(\frac{\partial f}{\partial r}\right)_\theta \left(\frac{\partial r}{\partial y}\right)_x + \left(\frac{\partial f}{\partial \theta}\right)_r \left(\frac{\partial \theta}{\partial y}\right)_x \tag{3}$$

For simplicity, we will assume that r is equal to a constant, l, so that we can ignore terms involving derivatives with respect to r. In other words, we will consider a particle that is constrained to move on the circumference of a circle. This system is sometimes called a *particle on a ring*. Using equations 1 and 2, show that

$$\left(\frac{\partial f}{\partial x}\right)_y = -\frac{\sin\theta}{l}\left(\frac{\partial f}{\partial \theta}\right)_r \qquad \text{and} \qquad \left(\frac{\partial f}{\partial y}\right)_x = \frac{\cos\theta}{l}\left(\frac{\partial f}{\partial \theta}\right)_r \tag{4}$$

Now apply equation 2 again to show that

$$\left(\frac{\partial^2 f}{\partial x^2}\right)_y = \left[\frac{\partial}{\partial x}\left(\frac{\partial f}{\partial x}\right)_y\right] = \left[\frac{\partial}{\partial \theta}\left(\frac{\partial f}{\partial x}\right)_y\right]_l \left(\frac{\partial \theta}{\partial x}\right)_y$$

$$= \left\{\frac{\partial}{\partial \theta}\left[-\frac{\sin \theta}{l}\left(\frac{\partial f}{\partial \theta}\right)_r\right]\right\}_l \left(-\frac{\sin \theta}{l}\right)$$

$$= \frac{\sin \theta \cos \theta}{l^2}\left(\frac{\partial f}{\partial \theta}\right) + \frac{\sin^2 \theta}{l^2}\left(\frac{\partial^2 f}{\partial \theta^2}\right)$$

Similarly, show that

$$\left(\frac{\partial^2 f}{\partial y^2}\right)_x = -\frac{\sin \theta \cos \theta}{l^2}\left(\frac{\partial f}{\partial \theta}\right) + \frac{\cos^2 \theta}{l^2}\left(\frac{\partial^2 f}{\partial \theta^2}\right)$$

and that

$$\nabla^2 f = \frac{\partial^2 f}{\partial x^2} + \frac{\partial^2 f}{\partial y^2} \longrightarrow \frac{1}{l^2}\left(\frac{\partial^2 f}{\partial \theta^2}\right)$$

If r is a constant, then the first derivatives of f with respect to x and y are

$$\left(\frac{\partial f}{\partial x}\right)_y = \left(\frac{\partial f}{\partial r}\right)_\theta \left(\frac{\partial r}{\partial x}\right)_y + \left(\frac{\partial f}{\partial \theta}\right)_r \left(\frac{\partial \theta}{\partial x}\right)_y$$

$$= \left(\frac{\partial f}{\partial \theta}\right)_r \left(\frac{\partial \theta}{\partial x}\right)_y$$

$$\left(\frac{\partial f}{\partial y}\right)_x = \left(\frac{\partial f}{\partial r}\right)_\theta \left(\frac{\partial r}{\partial y}\right)_x + \left(\frac{\partial f}{\partial \theta}\right)_r \left(\frac{\partial \theta}{\partial y}\right)_x$$

$$= \left(\frac{\partial f}{\partial \theta}\right)_r \left(\frac{\partial \theta}{\partial y}\right)_x$$

To evaluate these derivatives, first determine the derivatives of θ with respect to x and y:

$$\left(\frac{\partial \theta}{\partial x}\right)_y = \left[\frac{\partial}{\partial x}\tan^{-1}\left(\frac{y}{x}\right)\right]_y$$

$$= \frac{1}{1 + (y/x)^2}\left(-\frac{y}{x^2}\right)$$

$$= -\frac{y}{x^2 + y^2}$$

$$= -\frac{l \sin \theta}{l^2}$$

$$= -\frac{\sin \theta}{l}$$

$$\left(\frac{\partial \theta}{\partial y}\right)_x = \left[\frac{\partial}{\partial y} \tan^{-1}\left(\frac{y}{x}\right)\right]_x$$

$$= \frac{1}{1 + (y/x)^2}\left(\frac{1}{x}\right)$$

$$= \frac{x}{x^2 + y^2}$$

$$= \frac{l\cos\theta}{l^2}$$

$$= \frac{\cos\theta}{l}$$

Thus,

$$\left(\frac{\partial f}{\partial x}\right)_y = -\frac{\sin\theta}{l}\left(\frac{\partial f}{\partial \theta}\right)_r$$

$$\left(\frac{\partial f}{\partial y}\right)_x = \frac{\cos\theta}{l}\left(\frac{\partial f}{\partial \theta}\right)_r$$

Applying the chain rule again, the second derivative of f with respect to x is

$$\left(\frac{\partial^2 f}{\partial x^2}\right)_y = \left[\frac{\partial}{\partial x}\left(\frac{\partial f}{\partial x}\right)_y\right]_y = \left[\frac{\partial}{\partial \theta}\left(\frac{\partial f}{\partial x}\right)_y\right]_r\left(\frac{\partial \theta}{\partial x}\right)_y$$

$$= \left\{\frac{\partial}{\partial \theta}\left[-\frac{\sin\theta}{l}\left(\frac{\partial f}{\partial \theta}\right)_r\right]\right\}_r\left(-\frac{\sin\theta}{l}\right)$$

$$= \left[-\frac{\cos\theta}{l}\left(\frac{\partial f}{\partial \theta}\right)_r - \frac{\sin\theta}{l}\left(\frac{\partial^2 f}{\partial \theta^2}\right)_r\right]\left(-\frac{\sin\theta}{l}\right)$$

$$= \frac{\sin\theta\cos\theta}{l^2}\left(\frac{\partial f}{\partial \theta}\right)_r + \frac{\sin^2\theta}{l^2}\left(\frac{\partial^2 f}{\partial \theta^2}\right)_r$$

Similarly, the second derivative of f with respect to y is

$$\left(\frac{\partial^2 f}{\partial y^2}\right)_x = \left[\frac{\partial}{\partial y}\left(\frac{\partial f}{\partial y}\right)_x\right]_x = \left[\frac{\partial}{\partial \theta}\left(\frac{\partial f}{\partial y}\right)_x\right]_r\left(\frac{\partial \theta}{\partial y}\right)_x$$

$$= \left\{\frac{\partial}{\partial \theta}\left[\frac{\cos\theta}{l}\left(\frac{\partial f}{\partial \theta}\right)_r\right]\right\}_r\left(\frac{\cos\theta}{l}\right)$$

$$= \left[-\frac{\sin\theta}{l}\left(\frac{\partial f}{\partial \theta}\right)_r + \frac{\cos\theta}{l}\left(\frac{\partial^2 f}{\partial \theta^2}\right)_r\right]\left(\frac{\cos\theta}{l}\right)$$

$$= -\frac{\sin\theta\cos\theta}{l^2}\left(\frac{\partial f}{\partial \theta}\right)_r + \frac{\cos^2\theta}{l^2}\left(\frac{\partial^2 f}{\partial \theta^2}\right)_r$$

Add these two second derivatives to give the Laplacian:

$$\nabla^2 f = \left(\frac{\partial^2 f}{\partial x^2}\right)_y + \left(\frac{\partial^2 f}{\partial y^2}\right)_x$$

$$= \frac{\sin^2\theta + \cos^2\theta}{l^2}\left(\frac{\partial^2 f}{\partial\theta^2}\right)_r$$

$$= \frac{1}{l^2}\left(\frac{\partial^2 f}{\partial\theta^2}\right)_r$$

E–14. Generalize Problem E–13 to the case of a particle moving in a plane under the influence of a central force; in other words, convert

$$\nabla^2 = \frac{\partial^2}{\partial x^2} + \frac{\partial^2}{\partial y^2}$$

to plane polar coordinates, this time without assuming that r is a constant. Use the method of separation of variables to separate the equation for this problem. Solve the angular equation.

If r is not constant, then the derivatives of r with respect to x and y need to be included.

$$\left(\frac{\partial r}{\partial x}\right)_y = \left[\frac{\partial}{\partial x}\left(x^2 + y^2\right)^{1/2}\right]_y$$

$$= \frac{x}{\left(x^2 + y^2\right)^{1/2}}$$

$$= \frac{r\cos\theta}{r}$$

$$= \cos\theta$$

$$\left(\frac{\partial r}{\partial y}\right)_x = \left[\frac{\partial}{\partial y}\left(x^2 + y^2\right)^{1/2}\right]_x$$

$$= \frac{y}{\left(x^2 + y^2\right)^{1/2}}$$

$$= \frac{r\sin\theta}{r}$$

$$= \sin\theta$$

The first derivatives of f with respect to x and y are then

$$\left(\frac{\partial f}{\partial x}\right)_y = \left(\frac{\partial f}{\partial r}\right)_\theta \left(\frac{\partial r}{\partial x}\right)_y + \left(\frac{\partial f}{\partial \theta}\right)_r \left(\frac{\partial \theta}{\partial x}\right)_y$$

$$= \cos\theta \left(\frac{\partial f}{\partial r}\right)_\theta - \frac{\sin\theta}{r}\left(\frac{\partial f}{\partial \theta}\right)_r$$

$$\left(\frac{\partial f}{\partial y}\right)_x = \left(\frac{\partial f}{\partial r}\right)_\theta \left(\frac{\partial r}{\partial y}\right)_x + \left(\frac{\partial f}{\partial \theta}\right)_r \left(\frac{\partial \theta}{\partial y}\right)_x$$

$$= \sin\theta \left(\frac{\partial f}{\partial r}\right)_\theta + \frac{\cos\theta}{r}\left(\frac{\partial f}{\partial \theta}\right)_r$$

Apply the chain rule to obtain the second derivatives of f with respect to x:

$$\left(\frac{\partial^2 f}{\partial x^2}\right)_y = \left[\frac{\partial}{\partial r}\left(\frac{\partial f}{\partial x}\right)_y\right]_\theta \left(\frac{\partial r}{\partial x}\right)_y + \left[\frac{\partial}{\partial \theta}\left(\frac{\partial f}{\partial x}\right)_y\right]_r \left(\frac{\partial \theta}{\partial x}\right)_y$$

$$= \left\{\frac{\partial}{\partial r}\left[\cos\theta \left(\frac{\partial f}{\partial r}\right)_\theta - \frac{\sin\theta}{r}\left(\frac{\partial f}{\partial \theta}\right)_r\right]\right\}_\theta (\cos\theta)$$

$$+ \left\{\frac{\partial}{\partial \theta}\left[\cos\theta \left(\frac{\partial f}{\partial r}\right)_\theta - \frac{\sin\theta}{r}\left(\frac{\partial f}{\partial \theta}\right)_r\right]\right\}_r \left(-\frac{\sin\theta}{r}\right)$$

$$= \left[\cos\theta \left(\frac{\partial^2 f}{\partial r^2}\right)_\theta + \frac{\sin\theta}{r^2}\left(\frac{\partial f}{\partial \theta}\right)_r - \frac{\sin\theta}{r}\left(\frac{\partial^2 f}{\partial r\partial\theta}\right)\right](\cos\theta)$$

$$+ \left[-\sin\theta \left(\frac{\partial f}{\partial r}\right)_\theta + \cos\theta \left(\frac{\partial^2 f}{\partial \theta\partial r}\right) - \frac{\cos\theta}{r}\left(\frac{\partial f}{\partial \theta}\right)_r - \frac{\sin\theta}{r}\left(\frac{\partial^2 f}{\partial \theta^2}\right)_r\right]\left(-\frac{\sin\theta}{r}\right)$$

$$= \cos^2\theta \left(\frac{\partial^2 f}{\partial r^2}\right)_\theta + \frac{\sin^2\theta}{r}\left(\frac{\partial f}{\partial r}\right)_\theta + \frac{2\sin\theta\cos\theta}{r^2}\left(\frac{\partial f}{\partial \theta}\right)_r$$

$$- \frac{2\sin\theta\cos\theta}{r}\left(\frac{\partial^2 f}{\partial r\partial\theta}\right) + \frac{\sin^2\theta}{r^2}\left(\frac{\partial^2 f}{\partial \theta^2}\right)_r$$

and with respect to y:

$$\left(\frac{\partial^2 f}{\partial y^2}\right)_x = \left[\frac{\partial}{\partial r}\left(\frac{\partial f}{\partial y}\right)_x\right]_\theta \left(\frac{\partial r}{\partial y}\right)_x + \left[\frac{\partial}{\partial \theta}\left(\frac{\partial f}{\partial y}\right)_x\right]_r \left(\frac{\partial \theta}{\partial y}\right)_x$$

$$= \left\{\frac{\partial}{\partial r}\left[\sin\theta \left(\frac{\partial f}{\partial r}\right)_\theta + \frac{\cos\theta}{r}\left(\frac{\partial f}{\partial \theta}\right)_r\right]\right\}_\theta (\sin\theta)$$

$$+ \left\{\frac{\partial}{\partial \theta}\left[\sin\theta \left(\frac{\partial f}{\partial r}\right)_\theta + \frac{\cos\theta}{r}\left(\frac{\partial f}{\partial \theta}\right)_r\right]\right\}_r \left(\frac{\cos\theta}{r}\right)$$

$$
= \left[\sin\theta \left(\frac{\partial^2 f}{\partial r^2} \right)_\theta - \frac{\cos\theta}{r^2} \left(\frac{\partial f}{\partial\theta} \right)_r + \frac{\cos\theta}{r} \left(\frac{\partial^2 f}{\partial r\partial\theta} \right) \right] (\sin\theta)
$$

$$
+ \left[\cos\theta \left(\frac{\partial f}{\partial r} \right)_\theta + \sin\theta \left(\frac{\partial^2 f}{\partial\theta\partial r} \right) - \frac{\sin\theta}{r} \left(\frac{\partial f}{\partial\theta} \right)_r + \frac{\cos\theta}{r} \left(\frac{\partial^2 f}{\partial\theta^2} \right)_r \right] \left(\frac{\cos\theta}{r} \right)
$$

$$
= \sin^2\theta \left(\frac{\partial^2 f}{\partial r^2} \right)_\theta + \frac{\cos^2\theta}{r} \left(\frac{\partial f}{\partial r} \right)_\theta - \frac{2\cos\theta\sin\theta}{r^2} \left(\frac{\partial f}{\partial\theta} \right)_r
$$

$$
+ \frac{2\cos\theta\sin\theta}{r} \left(\frac{\partial^2 f}{\partial r\partial\theta} \right) + \frac{\cos^2\theta}{r^2} \left(\frac{\partial^2 f}{\partial\theta^2} \right)_r
$$

Now add these two derivatives together to give

$$
\nabla^2 f = \left(\frac{\partial^2 f}{\partial x^2} \right)_y + \left(\frac{\partial^2 f}{\partial y^2} \right)_x
$$

$$
= \left(\frac{\partial^2 f}{\partial r^2} \right)_\theta + \frac{1}{r} \left(\frac{\partial f}{\partial r} \right)_\theta + \frac{1}{r^2} \left(\frac{\partial^2 f}{\partial\theta^2} \right)_r
$$

Multiplying the ∇^2 operator by $-\hbar^2/2m$ gives the kinetic energy operator in plane polar coordinates, which when combined with $V(r)$, the potential energy for a central force, will give the Schrödinger equation for a particle moving in a plane under the influence of a central force:

$$
-\frac{\hbar^2}{2m} \left[\frac{\partial^2\psi}{\partial r^2} + \frac{1}{r}\frac{\partial\psi}{\partial r} + \frac{1}{r^2}\frac{\partial^2\psi}{\partial\theta^2} \right] + V(r)\psi(r,\theta) = E\psi(r,\theta)
$$

Letting $\psi(r,\theta) = R(r)\Theta(\theta)$ and substituting it into the Schrödinger equation,

$$
-\frac{\hbar^2}{2m} \left[\Theta(\theta)\frac{d^2 R}{dr^2} + \frac{1}{r}\Theta(\theta)\frac{dR}{dr} + \frac{1}{r^2}R(r)\frac{d^2\Theta}{d\theta^2} \right] + V(r)R(r)\Theta(\theta) = ER(r)\Theta(\theta)
$$

Dividing the equation by $\psi(r,\theta) = R(r)\Theta(\theta)$,

$$
-\frac{\hbar^2}{2m} \left[\frac{1}{R(r)}\frac{d^2 R}{dr^2} + \frac{1}{rR(r)}\frac{dR}{dr} + \frac{1}{r^2\Theta(\theta)}\frac{d^2\Theta}{d\theta^2} \right] + V(r) = E
$$

Multiply by $-2mr^2/\hbar^2$ to give two sets of terms that depend on either r or θ:

$$
\frac{r^2}{R(r)}\frac{d^2 R}{dr^2} + \frac{r}{R(r)}\frac{dR}{dr} + \frac{1}{\Theta(\theta)}\frac{d^2\Theta}{d\theta^2} - \frac{2mr^2}{\hbar^2}V(r) = -\frac{2mr^2}{\hbar^2}E
$$

$$
\left\{ \frac{r^2}{R(r)}\frac{d^2 R}{dr^2} + \frac{r}{R(r)}\frac{dR}{dr} - \frac{2mr^2}{\hbar^2}[V(r) - E] \right\} + \left[\frac{1}{\Theta(\theta)}\frac{d^2\Theta}{d\theta^2} \right] = 0
$$

Since r and θ are independent, the two sets of terms in the above equation can be separated into two equations:

$$
\frac{r^2}{R(r)}\frac{d^2 R}{dr^2} + \frac{r}{R(r)}\frac{dR}{dr} - \frac{2mr^2}{\hbar^2}[V(r) - E] = -p^2
$$

$$
\frac{1}{\Theta(\theta)}\frac{d^2\Theta}{d\theta^2} = -q^2
$$

where $p^2 + q^2 = 0$. The angular equation is similar to that in Equation 2.15 in the text, with $\Theta(\theta)$ replacing $X(x)$ and q replacing β. The solution is

$$\Theta(\theta) = \frac{1}{(2\pi)^{1/2}} e^{iq\theta}$$

The boundary condition for this angular problem, $\Theta(\theta) = \Theta(\theta + 2\pi)$, was discussed in Problem 3–33 and gives $q = 0, \pm 1, \pm 2, \cdots$.

E–15. Show that $u(r, \theta, \phi) = r \sin \theta \cos \phi$ satisfies Laplace's equation, $\nabla^2 u = 0$.

Laplace's equation in spherical polar coordinates is (see Equation E.17)

$$\nabla^2 u = \left[\frac{1}{r^2} \frac{\partial}{\partial r} \left(r^2 \frac{\partial}{\partial r} \right) + \frac{1}{r^2 \sin \theta} \frac{\partial}{\partial \theta} \left(\sin \theta \frac{\partial}{\partial \theta} \right) + \frac{1}{r^2 \sin^2 \theta} \frac{\partial^2}{\partial^2 \phi} \right] u = 0$$

For the function $u(r, \theta, \phi) = r \sin \theta \cos \phi$,

$$
\begin{aligned}
\nabla^2 \left(r \sin \theta \cos \phi \right) &= \frac{\sin \theta \cos \phi}{r^2} \frac{\partial}{\partial r} \left(r^2 \right) + \frac{\cos \phi}{r \sin \theta} \frac{\partial}{\partial \theta} \left(\sin \theta \cos \theta \right) - \frac{1}{r \sin \theta} \cos \phi \\
&= \frac{2 \sin \theta \cos \phi}{r} + \frac{\cos \phi}{r \sin \theta} \left(\cos^2 \theta - \sin^2 \theta \right) - \frac{\cos \phi}{r \sin \theta} \\
&= \frac{2 \sin \theta \cos \phi}{r} + \frac{\left(1 - 2 \sin^2 \theta \right) \cos \phi}{r \sin \theta} - \frac{\cos \phi}{r \sin \theta} \\
&= \frac{2 \sin \theta \cos \phi}{r} + \frac{\cos \phi}{r \sin \theta} - \frac{2 \sin \theta \cos \phi}{r} - \frac{\cos \phi}{r \sin \theta} \\
&= 0
\end{aligned}
$$

Thus, the function satisfies Laplace's equation.

E–16. Show that $u(r, \theta, \phi) = r \sin^2 \theta \cos 2\phi$ satisfies Laplace's equation, $\nabla^2 u = 0$.

Apply the operator ∇^2 in spherical polar coordinates (see Equation E.17) to the function $u(r, \theta, \phi) = r^2 \sin^2 \theta \cos 2\phi$ to obtain

Apply the operator ∇^2 in spherical polar coordinates (see Equation E.17) to the function $u(r, \theta, \phi) = r^2 \sin^2 \theta \cos 2\phi$ to obtain

$$\nabla^2 \left(r^2 \sin^2 \theta \cos 2\phi \right) = \frac{\sin^2 \theta \cos 2\phi}{r^2} \frac{\partial}{\partial r} \left(2r^3 \right) + \frac{2\cos 2\phi}{\sin \theta} \frac{\partial}{\partial \theta} \left(\sin^2 \theta \cos \theta \right) - 4\cos 2\phi$$

$$= 6\sin^2 \theta \cos 2\phi + 2\cos 2\phi \left(2\cos^2 \theta - \sin^2 \theta \right) - 4\cos 2\phi$$

$$= 6\sin^2 \theta \cos 2\phi + 2 \left(2 - 3\sin^2 \theta \right) \cos 2\phi - 4\cos 2\phi$$

$$= 6\sin^2 \theta \cos 2\phi + 4\cos 2\phi - 6\sin^2 \theta \cos 2\phi - 4\cos 2\phi$$

$$= 0$$

Thus, the function satisfies Laplace's equation.

CHAPTER 6

The Rigid Rotator and Rotational Spectroscopy

PROBLEMS AND SOLUTIONS

6–1. Show that the moment of inertia for a rigid rotator can be written as $I = \mu l^2$, where $l = l_1 + l_2$ (the fixed separation of the two masses) and μ is the reduced mass.

The center of mass condition for a rigid rotator consisting of two masses is

$$m_1 l_1 = m_2 l_2$$

Substituting $l_2 = l - l_1$ into the above equation and solving for l_1 in terms of l,

$$m_1 l_1 = m_2 \left(l - l_1 \right)$$

$$l_1 = \frac{m_2}{m_1 + m_2} l$$

l_2 can also be written in terms of l,

$$l_2 = l - l_1$$

$$= l - \frac{m_2}{m_1 + m_2} l$$

$$= \frac{m_1}{m_1 + m_2} l$$

Substitute the expressions for l_1 and l_2 into the expression for moment of inertia to give

$$I = m_1 l_1^2 + m_2 l_2^2$$

$$= m_1 \left(\frac{m_2}{m_1 + m_2} l \right)^2 + m_2 \left(\frac{m_1}{m_1 + m_2} l \right)^2$$

$$= \frac{m_1 m_2 l^2 \left(m_2 + m_1 \right)}{\left(m_1 + m_2 \right)^2}$$

$$= \frac{m_1 m_2}{m_1 + m_2} l^2$$

$$= \mu l^2$$

279

6–2. Consider the transformation from cartesian coordinates to plane polar coordinates,

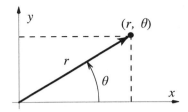

where

$$x = r \cos \theta \qquad r = (x^2 + y^2)^{1/2}$$

$$y = r \sin \theta \qquad \theta = \tan^{-1}\left(\frac{y}{x}\right) \tag{1}$$

If a function $f(r, \theta)$ depends upon the polar coordinates r and θ, then the chain rule of partial differentiation says that

$$\left(\frac{\partial f}{\partial x}\right)_y = \left(\frac{\partial f}{\partial r}\right)_\theta \left(\frac{\partial r}{\partial x}\right)_y + \left(\frac{\partial f}{\partial \theta}\right)_r \left(\frac{\partial \theta}{\partial x}\right)_y \tag{2}$$

and that

$$\left(\frac{\partial f}{\partial y}\right)_x = \left(\frac{\partial f}{\partial r}\right)_\theta \left(\frac{\partial r}{\partial y}\right)_x + \left(\frac{\partial f}{\partial \theta}\right)_r \left(\frac{\partial \theta}{\partial y}\right)_x \tag{3}$$

For simplicity, we will assume that r is equal to a constant, l, so that we can ignore terms involving derivatives with respect to r. In other words, we will consider a particle that is constrained to move on the circumference of a circle. This system is sometimes called a *particle on a ring*. Using equations 1 and 2, show that

$$\left(\frac{\partial f}{\partial x}\right)_y = -\frac{\sin \theta}{l} \left(\frac{\partial f}{\partial \theta}\right)_r \qquad \text{and} \qquad \left(\frac{\partial f}{\partial y}\right)_x = \frac{\cos \theta}{l} \left(\frac{\partial f}{\partial \theta}\right)_r \tag{4}$$

Now apply equation 2 again to show that

$$\left(\frac{\partial^2 f}{\partial x^2}\right)_y = \left[\frac{\partial}{\partial x}\left(\frac{\partial f}{\partial x}\right)_y\right] = \left[\frac{\partial}{\partial \theta}\left(\frac{\partial f}{\partial x}\right)_y\right]_r \left(\frac{\partial \theta}{\partial x}\right)_y$$

$$= \left\{\frac{\partial}{\partial \theta}\left[-\frac{\sin \theta}{l}\left(\frac{\partial f}{\partial \theta}\right)_r\right]\right\}_r \left(-\frac{\sin \theta}{l}\right)$$

$$= \frac{\sin \theta \cos \theta}{l^2}\left(\frac{\partial f}{\partial \theta}\right) + \frac{\sin^2 \theta}{l^2}\left(\frac{\partial^2 f}{\partial \theta^2}\right)$$

Similarly, show that

$$\left(\frac{\partial^2 f}{\partial y^2}\right)_x = -\frac{\sin\theta\cos\theta}{l^2}\left(\frac{\partial f}{\partial\theta}\right) + \frac{\cos^2\theta}{l^2}\left(\frac{\partial^2 f}{\partial\theta^2}\right)$$

and that

$$\nabla^2 f = \frac{\partial^2 f}{\partial x^2} + \frac{\partial^2 f}{\partial y^2} \longrightarrow \frac{1}{l^2}\left(\frac{\partial^2 f}{\partial\theta^2}\right)$$

Now show that the Schrödinger equation for a particle of mass m constrained to move on a circle of radius r is (see Problem 3–34)

$$-\frac{\hbar^2}{2I}\frac{\partial^2\psi(\theta)}{\partial\theta^2} = E\psi(\theta) \qquad 0 \le \theta \le 2\pi$$

where $I = ml^2$ is the moment of inertia. Solve this equation and determine the allowed values of E.

_____

If r is a constant, then the first derivatives of f with respect to x and y are

$$\left(\frac{\partial f}{\partial x}\right)_y = \left(\frac{\partial f}{\partial r}\right)_\theta\left(\frac{\partial r}{\partial x}\right)_y + \left(\frac{\partial f}{\partial\theta}\right)_r\left(\frac{\partial\theta}{\partial x}\right)_y$$

$$= \left(\frac{\partial f}{\partial\theta}\right)_r\left(\frac{\partial\theta}{\partial x}\right)_y$$

$$\left(\frac{\partial f}{\partial y}\right)_x = \left(\frac{\partial f}{\partial r}\right)_\theta\left(\frac{\partial r}{\partial y}\right)_x + \left(\frac{\partial f}{\partial\theta}\right)_r\left(\frac{\partial\theta}{\partial y}\right)_x$$

$$= \left(\frac{\partial f}{\partial\theta}\right)_r\left(\frac{\partial\theta}{\partial y}\right)_x$$

To evaluate these derivatives, first determine the derivatives of θ with respect to x and y:

$$\left(\frac{\partial \theta}{\partial x}\right)_y = \left[\frac{\partial}{\partial x} \tan^{-1}\left(\frac{y}{x}\right)\right]_y$$

$$= \frac{1}{1 + (y/x)^2}\left(-\frac{y}{x^2}\right)$$

$$= -\frac{y}{x^2 + y^2}$$

$$= -\frac{l \sin \theta}{l^2}$$

$$= -\frac{\sin \theta}{l}$$

$$\left(\frac{\partial \theta}{\partial y}\right)_x = \left[\frac{\partial}{\partial y} \tan^{-1}\left(\frac{y}{x}\right)\right]_x$$

$$= \frac{1}{1 + (y/x)^2}\left(\frac{1}{x}\right)$$

$$= \frac{x}{x^2 + y^2}$$

$$= \frac{l \cos \theta}{l^2}$$

$$= \frac{\cos \theta}{l}$$

Thus,

$$\left(\frac{\partial f}{\partial x}\right)_y = -\frac{\sin \theta}{l}\left(\frac{\partial f}{\partial \theta}\right)_r$$

$$\left(\frac{\partial f}{\partial y}\right)_x = \frac{\cos \theta}{l}\left(\frac{\partial f}{\partial \theta}\right)_r$$

Applying the chain rule again, the second derivative of f with respect to x is

$$\left(\frac{\partial^2 f}{\partial x^2}\right)_y = \left[\frac{\partial}{\partial x}\left(\frac{\partial f}{\partial x}\right)_y\right]_y = \left[\frac{\partial}{\partial \theta}\left(\frac{\partial f}{\partial x}\right)_y\right]_r\left(\frac{\partial \theta}{\partial x}\right)_y$$

$$= \left\{\frac{\partial}{\partial \theta}\left[-\frac{\sin \theta}{l}\left(\frac{\partial f}{\partial \theta}\right)_r\right]\right\}_r\left(-\frac{\sin \theta}{l}\right)$$

$$= \left[-\frac{\cos \theta}{l}\left(\frac{\partial f}{\partial \theta}\right)_r - \frac{\sin \theta}{l}\left(\frac{\partial^2 f}{\partial \theta^2}\right)_r\right]\left(-\frac{\sin \theta}{l}\right)$$

$$= \frac{\sin \theta \cos \theta}{l^2}\left(\frac{\partial f}{\partial \theta}\right)_r + \frac{\sin^2 \theta}{l^2}\left(\frac{\partial^2 f}{\partial \theta^2}\right)_r$$

Similarly, the second derivative of f with respect to y is

$$\left(\frac{\partial^2 f}{\partial y^2}\right)_x = \left[\frac{\partial}{\partial y}\left(\frac{\partial f}{\partial y}\right)_x\right]_x = \left[\frac{\partial}{\partial \theta}\left(\frac{\partial f}{\partial y}\right)_x\right]_r \left(\frac{\partial \theta}{\partial y}\right)_x$$

$$= \left\{\frac{\partial}{\partial \theta}\left[\frac{\cos \theta}{l}\left(\frac{\partial f}{\partial \theta}\right)_r\right]\right\}_r \left(\frac{\cos \theta}{l}\right)$$

$$= \left[-\frac{\sin \theta}{l}\left(\frac{\partial f}{\partial \theta}\right)_r + \frac{\cos \theta}{l}\left(\frac{\partial^2 f}{\partial \theta^2}\right)_r\right]\left(\frac{\cos \theta}{l}\right)$$

$$= -\frac{\sin \theta \cos \theta}{l^2}\left(\frac{\partial f}{\partial \theta}\right)_r + \frac{\cos^2 \theta}{l^2}\left(\frac{\partial^2 f}{\partial \theta^2}\right)_r$$

Add these two second derivatives to give the Laplacian:

$$\nabla^2 f = \left(\frac{\partial^2 f}{\partial x^2}\right)_y + \left(\frac{\partial^2 f}{\partial y^2}\right)_x$$

$$= \frac{\sin^2 \theta + \cos^2 \theta}{l^2}\left(\frac{\partial^2 f}{\partial \theta^2}\right)_r$$

$$= \frac{1}{l^2}\left(\frac{\partial^2 f}{\partial \theta^2}\right)_r$$

The Schrödinger equation for a particle of mass m constrained to move on a circle of radius r is

$$-\frac{\hbar^2}{2m}\nabla^2 \psi(\theta) = E\psi(\theta)$$

$$-\frac{\hbar^2}{2ml^2}\frac{\partial^2 \psi(\theta)}{\partial \theta^2} = E\psi(\theta)$$

$$-\frac{\hbar^2}{2I}\frac{\partial^2 \psi(\theta)}{\partial \theta^2} = E\psi(\theta)$$

Rearranging the equation as

$$\frac{\partial^2 \psi(\theta)}{\partial \theta^2} + \frac{2IE}{\hbar^2}\psi(\theta) = 0$$

the solution of this equation is readily seen to be

$$\psi(\theta) = Ae^{\pm in\theta}$$

where

$$n = \frac{(2IE)^{1/2}}{\hbar}$$

giving

$$E = \frac{n^2\hbar^2}{2I}$$

Application of the boundary condition $\psi(\theta) = \psi(\theta + 2\pi)$ gives $n = 0, \pm 1, \pm 2, \cdots$ and normalization of the wave function gives $A = 1/(2\pi)^{1/2}$. See Problem 3–33 for details.

6–3.   Generalize Problem 6–2 to the case of a particle moving in a plane under the influence of a central force; in other words, convert

$$\nabla^2 = \frac{\partial^2}{\partial x^2} + \frac{\partial^2}{\partial y^2}$$

to plane polar coordinates, this time without assuming that r is a constant. Use the method of separation of variables to separate the equation for this problem. Solve the angular equation.

If r is not constant, then the derivatives of r with respect to x and y need to be included.

$$\left(\frac{\partial r}{\partial x}\right)_y = \left[\frac{\partial}{\partial x}\left(x^2 + y^2\right)^{1/2}\right]_y$$

$$= \frac{x}{\left(x^2 + y^2\right)^{1/2}}$$

$$= \frac{r\cos\theta}{r}$$

$$= \cos\theta$$

$$\left(\frac{\partial r}{\partial y}\right)_x = \left[\frac{\partial}{\partial y}\left(x^2 + y^2\right)^{1/2}\right]_x$$

$$= \frac{y}{\left(x^2 + y^2\right)^{1/2}}$$

$$= \frac{r\sin\theta}{r}$$

$$= \sin\theta$$

The first derivatives of f with respect to x and y are then

$$\left(\frac{\partial f}{\partial x}\right)_y = \left(\frac{\partial f}{\partial r}\right)_\theta \left(\frac{\partial r}{\partial x}\right)_y + \left(\frac{\partial f}{\partial \theta}\right)_r \left(\frac{\partial \theta}{\partial x}\right)_y$$

$$= \cos\theta \left(\frac{\partial f}{\partial r}\right)_\theta - \frac{\sin\theta}{r}\left(\frac{\partial f}{\partial \theta}\right)_r$$

$$\left(\frac{\partial f}{\partial y}\right)_x = \left(\frac{\partial f}{\partial r}\right)_\theta \left(\frac{\partial r}{\partial y}\right)_x + \left(\frac{\partial f}{\partial \theta}\right)_r \left(\frac{\partial \theta}{\partial y}\right)_x$$

$$= \sin\theta \left(\frac{\partial f}{\partial r}\right)_\theta + \frac{\cos\theta}{r}\left(\frac{\partial f}{\partial \theta}\right)_r$$

Apply the chain rule to obtain the second derivatives of f with respect to x:

$$\left(\frac{\partial^2 f}{\partial x^2}\right)_y = \left[\frac{\partial}{\partial r}\left(\frac{\partial f}{\partial x}\right)_y\right]_\theta \left(\frac{\partial r}{\partial x}\right)_y + \left[\frac{\partial}{\partial \theta}\left(\frac{\partial f}{\partial x}\right)_y\right]_r \left(\frac{\partial \theta}{\partial x}\right)_y$$

$$= \left\{\frac{\partial}{\partial r}\left[\cos\theta\left(\frac{\partial f}{\partial r}\right)_\theta - \frac{\sin\theta}{r}\left(\frac{\partial f}{\partial \theta}\right)_r\right]\right\}_\theta (\cos\theta)$$

$$+ \left\{\frac{\partial}{\partial \theta}\left[\cos\theta\left(\frac{\partial f}{\partial r}\right)_\theta - \frac{\sin\theta}{r}\left(\frac{\partial f}{\partial \theta}\right)_r\right]\right\}_r \left(-\frac{\sin\theta}{r}\right)$$

$$= \left[\cos\theta\left(\frac{\partial^2 f}{\partial r^2}\right)_\theta + \frac{\sin\theta}{r^2}\left(\frac{\partial f}{\partial \theta}\right)_r - \frac{\sin\theta}{r}\left(\frac{\partial^2 f}{\partial r\partial\theta}\right)\right](\cos\theta)$$

$$+ \left[-\sin\theta\left(\frac{\partial f}{\partial r}\right)_\theta + \cos\theta\left(\frac{\partial^2 f}{\partial \theta\partial r}\right) - \frac{\cos\theta}{r}\left(\frac{\partial f}{\partial \theta}\right)_r - \frac{\sin\theta}{r}\left(\frac{\partial^2 f}{\partial \theta^2}\right)_r\right]\left(-\frac{\sin\theta}{r}\right)$$

$$= \cos^2\theta\left(\frac{\partial^2 f}{\partial r^2}\right)_\theta + \frac{\sin^2\theta}{r}\left(\frac{\partial f}{\partial r}\right)_\theta + \frac{2\sin\theta\cos\theta}{r^2}\left(\frac{\partial f}{\partial \theta}\right)_r$$

$$- \frac{2\sin\theta\cos\theta}{r}\left(\frac{\partial^2 f}{\partial r\partial\theta}\right) + \frac{\sin^2\theta}{r^2}\left(\frac{\partial^2 f}{\partial \theta^2}\right)_r$$

and with respect to y:

$$\left(\frac{\partial^2 f}{\partial y^2}\right)_x = \left[\frac{\partial}{\partial r}\left(\frac{\partial f}{\partial y}\right)_x\right]_\theta \left(\frac{\partial r}{\partial y}\right)_x + \left[\frac{\partial}{\partial \theta}\left(\frac{\partial f}{\partial y}\right)_x\right]_r \left(\frac{\partial \theta}{\partial y}\right)_x$$

$$= \left\{\frac{\partial}{\partial r}\left[\sin\theta\left(\frac{\partial f}{\partial r}\right)_\theta + \frac{\cos\theta}{r}\left(\frac{\partial f}{\partial \theta}\right)_r\right]\right\}_\theta (\sin\theta)$$

$$+ \left\{\frac{\partial}{\partial \theta}\left[\sin\theta\left(\frac{\partial f}{\partial r}\right)_\theta + \frac{\cos\theta}{r}\left(\frac{\partial f}{\partial \theta}\right)_r\right]\right\}_r \left(\frac{\cos\theta}{r}\right)$$

$$= \left[\sin\theta\left(\frac{\partial^2 f}{\partial r^2}\right)_\theta - \frac{\cos\theta}{r^2}\left(\frac{\partial f}{\partial \theta}\right)_r + \frac{\cos\theta}{r}\left(\frac{\partial^2 f}{\partial r\partial\theta}\right)\right](\sin\theta)$$

$$+ \left[\cos\theta\left(\frac{\partial f}{\partial r}\right)_\theta + \sin\theta\left(\frac{\partial^2 f}{\partial \theta\partial r}\right) - \frac{\sin\theta}{r}\left(\frac{\partial f}{\partial \theta}\right)_r + \frac{\cos\theta}{r}\left(\frac{\partial^2 f}{\partial \theta^2}\right)_r\right]\left(\frac{\cos\theta}{r}\right)$$

$$= \sin^2\theta\left(\frac{\partial^2 f}{\partial r^2}\right)_\theta + \frac{\cos^2\theta}{r}\left(\frac{\partial f}{\partial r}\right)_\theta - \frac{2\cos\theta\sin\theta}{r^2}\left(\frac{\partial f}{\partial \theta}\right)_r$$

$$+ \frac{2\cos\theta\sin\theta}{r}\left(\frac{\partial^2 f}{\partial r\partial\theta}\right) + \frac{\cos^2\theta}{r^2}\left(\frac{\partial^2 f}{\partial \theta^2}\right)_r$$

Now add these two derivatives together to give

$$\nabla^2 f = \left(\frac{\partial^2 f}{\partial x^2}\right)_y + \left(\frac{\partial^2 f}{\partial y^2}\right)_x$$

$$= \left(\frac{\partial^2 f}{\partial r^2}\right)_\theta + \frac{1}{r}\left(\frac{\partial f}{\partial r}\right)_\theta + \frac{1}{r^2}\left(\frac{\partial^2 f}{\partial \theta^2}\right)_r$$

The Schrödinger equation for a particle moving in a plane under the influence of a central force is

$$-\frac{\hbar^2}{2m}\left[\frac{\partial^2\psi}{\partial r^2}+\frac{1}{r}\frac{\partial\psi}{\partial r}+\frac{1}{r^2}\frac{\partial^2\psi}{\partial\theta^2}\right]+V(r)\psi(r,\theta)=E\psi(r,\theta)$$

Letting $\psi(r,\theta)=R(r)\Theta(\theta)$ and substituting it into the Schrödinger equation,

$$-\frac{\hbar^2}{2m}\left[\Theta(\theta)\frac{d^2R}{dr^2}+\frac{1}{r}\Theta(\theta)\frac{dR}{dr}+\frac{1}{r^2}R(r)\frac{d^2\Theta}{d\theta^2}\right]+V(r)R(r)\Theta(\theta)=ER(r)\Theta(\theta)$$

Dividing the equation by $\psi(r,\theta)=R(r)\Theta(\theta)$,

$$-\frac{\hbar^2}{2m}\left[\frac{1}{R(r)}\frac{d^2R}{dr^2}+\frac{1}{rR(r)}\frac{dR}{dr}+\frac{1}{r^2\Theta(\theta)}\frac{d^2\Theta}{d\theta^2}\right]+V(r)=E$$

Multiply by $-2mr^2/\hbar^2$ to give two sets of terms that depend on either r or θ:

$$\frac{r^2}{R(r)}\frac{d^2R}{dr^2}+\frac{r}{R(r)}\frac{dR}{dr}+\frac{1}{\Theta(\theta)}\frac{d^2\Theta}{d\theta^2}-\frac{2mr^2}{\hbar^2}V(r)=-\frac{2mr^2}{\hbar^2}E$$

$$\left\{\frac{r^2}{R(r)}\frac{d^2R}{dr^2}+\frac{r}{R(r)}\frac{dR}{dr}-\frac{2mr^2}{\hbar^2}[V(r)-E]\right\}+\left[\frac{1}{\Theta(\theta)}\frac{d^2\Theta}{d\theta^2}\right]=0$$

Since r and θ are independent, the two sets of terms in the above equation can be separated into two equations:

$$\frac{r^2}{R(r)}\frac{d^2R}{dr^2}+\frac{r}{R(r)}\frac{dR}{dr}-\frac{2mr^2}{\hbar^2}[V(r)-E]=-p^2$$

$$\frac{1}{\Theta(\theta)}\frac{d^2\Theta}{d\theta^2}=-q^2$$

where $p^2+q^2=0$. The angular equation is similar to that in Equation 6–17 in the text, with $\Theta(\theta)$ replacing $\Phi(\phi)$ and q replacing m. The solution is

$$\Theta(\theta)=\frac{1}{(2\pi)^{1/2}}e^{iq\theta}$$

Application of the boundary condition $\Theta(\theta)=\Theta(\theta+2\pi)$ gives $q=0,\pm1,\pm2,\dots$.

6–4. Show that rotational transitions of a diatomic molecule occur in the microwave region or the far infrared region of the spectrum.

As stated in Section 6–2 in the text, the typical reduced mass for a diatomic molecule of 10^{-25} to 10^{-26} kg and the typical bond distance of 10^{-10} m give a moment of inertia ranging from 10^{-45} to 10^{-46} kg·m^2. Taking I as 5×10^{-46} kg·m^2,

$$\nu_{\text{obs}}\approx 2B(J+1)$$

$$=\frac{h}{4\pi^2I}(J+1)$$

$$=\frac{6.626\,0755\times10^{-34}\text{ J·s}}{4\pi^2\left(5\times10^{-46}\text{ kg·m}^2\right)}(J+1)$$

$$=3.4\times10^{10}(J+1)\text{ Hz}$$

The rotational transitions occur at about 3.4×10^{10} Hz for $J = 0$ and at higher frequencies for higher J values. These frequencies correspond to the microwave or far infrared region of the electromagnetic radiation spectrum.

6–5. In the far infrared spectrum of $H^{79}Br$, there is a series of lines separated by 16.72 cm^{-1}. Calculate the values of the moment of inertia and the internuclear separation in $H^{79}Br$.

The lines are separated by $2\tilde{B}$, from which the moment of inertia can be determined:

$$2\tilde{B} = \frac{h}{4\pi^2 cI} = 16.72 \text{ cm}^{-1}$$

$$I = \frac{6.626\,0755 \times 10^{-34} \text{ J·s}}{4\pi^2 \left(2.997\,924\,58 \times 10^{10} \text{ cm·s}^{-1}\right) \left(16.72 \text{ cm}^{-1}\right)}$$

$$= 3.348 \times 10^{-47} \text{ kg·m}^2$$

The reduced mass of $H^{79}Br$ is

$$\mu = \frac{(1.008 \text{ amu})\,(78.918 \text{ amu})}{1.008 \text{ amu} + 78.918 \text{ amu}} \left(1.660\,5402 \times 10^{-27} \text{ kg·amu}^{-1}\right)$$

$$= 1.653 \times 10^{-27} \text{ kg}$$

Combining the moment of inertia and the reduced mass, the internuclear separation is

$$r = \left(\frac{I}{\mu}\right)^{1/2}$$

$$= \left(\frac{3.348 \times 10^{-47} \text{ kg·m}^2}{1.653 \times 10^{-27} \text{ kg}}\right)^{1/2}$$

$$= 1.42 \times 10^{-10} \text{ m} = 1.42 \text{ Å}$$

6–6. The $J = 0$ to $J = 1$ transition for carbon monoxide ($^{12}C^{16}O$) occurs at 1.153×10^5 MHz. Calculate the value of the bond length in carbon monoxide.

The $J = 0$ to $J = 1$ transition occurs at $\nu_{obs} = 2B(J + 1) = 2B$. From the observed frequency, the rotational constant, and therefore, the moment of inertia, of carbon monoxide can be determined:

$$2B = \frac{h}{4\pi^2 I} = 1.153 \times 10^{11} \text{ Hz}$$

$$I = \frac{6.626\,0755 \times 10^{-34} \text{ J·s}}{4\pi^2 \left(1.153 \times 10^{11} \text{ Hz}\right)}$$

$$= 1.4557 \times 10^{-46} \text{ kg·m}^2$$

The reduced mass of $^{12}C^{16}O$ is

$$\mu = \frac{(12.000 \text{ amu}) (15.995 \text{ amu})}{12.000 \text{ amu} + 15.995 \text{ amu}} \left(1.660\,5402 \times 10^{-27} \text{ kg·amu}^{-1}\right)$$

$$= 1.1385 \times 10^{-26} \text{ kg}$$

Combining the moment of inertia and the reduced mass, the bond length in carbon monoxide is

$$r = \left(\frac{I}{\mu}\right)^{1/2}$$

$$= \left(\frac{1.4557 \times 10^{-46} \text{ kg·m}^2}{1.1385 \times 10^{-26} \text{ kg}}\right)^{1/2}$$

$$= 1.131 \times 10^{-10} \text{ m} = 1.131 \text{ Å}$$

6–7. The spacing between the lines in the microwave spectrum of $H^{35}Cl$ is 6.350×10^{11} Hz. Calculate the bond length of $H^{35}Cl$.

The lines are separated by $2B$, from which the moment of inertia can be determined:

$$2B = \frac{h}{4\pi^2 I} = 6.350 \times 10^{11} \text{ Hz}$$

$$I = \frac{6.626\,0755 \times 10^{-34} \text{ J·s}}{4\pi^2 \left(6.350 \times 10^{11} \text{ Hz}\right)}$$

$$= 2.6432 \times 10^{-47} \text{ kg·m}^2$$

The reduced mass of $H^{35}Cl$ is

$$\mu = \frac{(1.008 \text{ amu}) (34.969 \text{ amu})}{1.008 \text{ amu} + 34.969 \text{ amu}} \left(1.660\,5402 \times 10^{-27} \text{ kg·amu}^{-1}\right)$$

$$= 1.6269 \times 10^{-27} \text{ kg}$$

Combining the moment of inertia and the reduced mass, the bond length of $H^{35}Cl$ is

$$r = \left(\frac{I}{\mu}\right)^{1/2}$$

$$= \left(\frac{2.6432 \times 10^{-47} \text{ kg·m}^2}{1.6269 \times 10^{-27} \text{ kg}}\right)^{1/2}$$

$$= 1.275 \times 10^{-10} \text{ m} = 1.275 \text{ Å}$$

6–8. The microwave spectrum of $^{39}K^{127}I$ consists of a series of lines whose spacing is almost constant at 3634 MHz. Calculate the bond length of $^{39}K^{127}I$.

The lines are separated by $2B$, from which the moment of inertia can be determined:

$$2B = \frac{h}{4\pi^2 I} = 3634 \times 10^6 \text{ Hz}$$

$$I = \frac{6.626\,0755 \times 10^{-34} \text{ J·s}}{4\pi^2 \left(3634 \times 10^6 \text{ Hz}\right)}$$

$$= 4.6186 \times 10^{-45} \text{ kg·m}^2$$

The reduced mass of $^{39}\text{K}^{127}\text{I}$ is

$$\mu = \frac{(38.964 \text{ amu})\,(126.904 \text{ amu})}{38.964 \text{ amu} + 126.904 \text{ amu}} \left(1.660\,5402 \times 10^{-27} \text{ kg·amu}^{-1}\right)$$

$$= 4.9502 \times 10^{-26} \text{ kg}$$

Combining the moment of inertia and the reduced mass, the bond length of $^{39}\text{K}^{127}\text{I}$ is

$$r = \left(\frac{I}{\mu}\right)^{1/2}$$

$$= \left(\frac{4.6186 \times 10^{-45} \text{ kg·m}^2}{4.9502 \times 10^{-26} \text{ kg}}\right)^{1/2}$$

$$= 3.055 \times 10^{-10} \text{ m} = 3.055 \text{ Å}$$

6–9. The equilibrium internuclear distance of H^{127}I is 160.4 pm. Calculate the value of B in wave numbers and megahertz.

The reduced mass of H^{127}I is

$$\mu = \frac{(1.008 \text{ amu})\,(126.904 \text{ amu})}{1.008 \text{ amu} + 126.904 \text{ amu}} \left(1.660\,5402 \times 10^{-27} \text{ kg·amu}^{-1}\right)$$

$$= 1.6606 \times 10^{-27} \text{ kg}$$

The moment of inertia of the molecule is

$$I = \mu r^2$$

$$= \left(1.6606 \times 10^{-27} \text{ kg}\right) \left(160.4 \times 10^{-12} \text{ m}\right)^2$$

$$= 4.2725 \times 10^{-47} \text{ kg·m}^2$$

from which the rotational constant can be determined:

$$B = \frac{h}{8\pi^2 I}$$

$$= \frac{6.626\,0755 \times 10^{-34}\ \text{J·s}}{8\pi^2\,(4.2725 \times 10^{-47}\ \text{kg·m}^2)}$$

$$= 1.964 \times 10^{11}\ \text{Hz}$$

$$= 1.964 \times 10^{5}\ \text{MHz}$$

$$\tilde{B} = B\left(\frac{1}{2.997\,924\,58 \times 10^{10}\ \text{cm·s}^{-1}}\right)$$

$$= 6.552\ \text{cm}^{-1}$$

6–10.   Assuming the rotation of a diatomic molecule in the $J = 10$ state may be approximated by classical mechanics, calculate how many revolutions per second $^{23}\text{Na}^{35}\text{Cl}$ makes in the $J = 10$ rotational state. The rotational constant of $^{23}\text{Na}^{35}\text{Cl}$ is 6500 MHz.

The rotational energy in the $J = 10$ state is

$$E = \frac{h^2}{8\pi^2 I} J\,(J+1)$$

Classically, this energy is

$$E = \frac{I\omega^2}{2}$$

where ω is the angular velocity. Its value can be determined by equating these two expressions:

$$\frac{I\omega^2}{2} = \frac{h^2}{8\pi^2 I} J\,(J+1)$$

$$\omega = \frac{h}{2\pi I}\,[J\,(J+1)]^{1/2}$$

$$= 4\pi B\,[J\,(J+1)]^{1/2}$$

$$= 4\pi\,\left(6500 \times 10^6\ \text{Hz}\right)(110)^{1/2}$$

$$= 8.567 \times 10^{11}\ \text{radians·s}^{-1}$$

It follows that

$$\nu = \omega\left(\frac{1\ \text{revolution}}{2\pi\ \text{radians}}\right)$$

$$= 1.36 \times 10^{11}\ \text{revolution·s}^{-1}$$

6–11. The results we derived for a rigid rotator apply to linear polyatomic molecules as well as to diatomic molecules. Given that the moment of inertia I for $H^{12}C^{14}N$ is 1.89×10^{-46} kg·m^2 (cf. Problem 6–12), predict the microwave spectrum of $H^{12}C^{14}N$.

The rotational constant of $H^{12}C^{14}N$ is

$$\tilde{B} = \frac{h}{8\pi^2 cI}$$

$$= \frac{6.626\,0755 \times 10^{-34}\,\text{J·s}}{8\pi^2 \left(2.997\,924\,58 \times 10^{10}\,\text{cm·s}^{-1}\right)\left(1.89 \times 10^{-46}\,\text{kg·m}^2\right)}$$

$$= 1.481\,\text{cm}^{-1}$$

The spectral lines prediced by the rigid rotator model are equally spaced by $2\tilde{B} = 2.96$ cm^{-1}.

6–12. This problem involves the calculation of the moment of inertia of a linear triatomic molecule such as $H^{12}C^{14}N$ (see Problem 6–11). The moment of inertia of any set of point masses is

$$I = \sum_j m_j l_j^2$$

where l_j is the distance of the jth mass from the center of mass. Thus, the moment of inertia of $H^{12}C^{14}N$ is

$$I = m_H l_H^2 + m_C l_C^2 + m_N l_N^2 \tag{1}$$

Show that equation 1 can be written as

$$I = \frac{m_H m_C r_{HC}^2 + m_H m_N r_{HN}^2 + m_C m_N r_{CN}^2}{m_H + m_C + m_N}$$

where the r's are the various internuclear distances. Given that $r_{HC} = 106.8$ pm and $r_{CN} = 115.6$ pm, calculate the value of I and compare the result with that given in Problem 6–11.

To introduce cross terms of masses into the moment of inertia expression, multiply and divide it by $M = m_H + m_C + m_N$:

$$I = m_H l_H^2 + m_C l_C^2 + m_N l_N^2$$

$$= \frac{m_H + m_C + m_N}{M}\left(m_H l_H^2 + m_C l_C^2 + m_N l_N^2\right)$$

$$= \frac{1}{M}\left(m_H^2 l_H^2 + m_H m_C l_C^2 + m_H m_N l_N^2 + m_C m_H l_H^2 + m_C^2 l_C^2 + m_C m_N l_N^2 \right.$$

$$\left. + m_N m_H l_H^2 + m_N m_C l_C^2 + m_N^2 l_N^2\right)$$

$$= \frac{1}{M}\left[m_H m_C \left(l_C^2 + l_H^2\right) + m_C m_N \left(l_C^2 + l_N^2\right) + m_H m_N \left(l_H^2 + l_N^2\right) + m_H^2 l_H^2 + m_C^2 l_C^2 + m_N^2 l_N^2\right]$$

The center of mass of HCN is located between the C and N atoms, giving the following relations between internuclear distances and the three l_j:

$$r_{HC} = l_H - l_C$$

$$r_{HN} = l_H + l_N$$

$$r_{CN} = l_C + l_N$$

Introducing these relations into the moment of inertia expression:

$$I = \frac{1}{M}\left[m_H m_C\left(l_H - l_C\right)^2 + 2m_H m_C l_H l_C + m_C m_N\left(l_C + l_N\right)^2 - 2m_C m_N l_C l_N + \right.$$

$$\left. m_H m_N\left(l_H + l_N\right)^2 - 2m_H m_N l_H l_N + m_H^2 l_H^2 + m_C^2 l_C^2 + m_N^2 l_N^2\right]$$

$$= \frac{1}{M}\left[m_H m_C r_{HC}^2 + m_C m_N r_{CN}^2 + m_H m_N r_{HN}^2 \right.$$

$$\left. + 2m_H m_C l_H l_C - 2m_C m_N l_C l_N - 2m_H m_N l_H l_N + m_H^2 l_H^2 + m_C^2 l_C^2 + m_N^2 l_N^2\right]$$

$$= \frac{1}{M}\left[m_H m_C r_{HC}^2 + m_C m_N r_{CN}^2 + m_H m_N r_{HN}^2 \right.$$

$$\left. + \left(m_H l_H + m_C l_C\right)^2 + m_N^2 l_N^2 - 2m_N l_N\left(m_C l_C + m_H l_H\right)\right]$$

Applying the center of mass condition, $m_N l_N = m_H l_H + m_C l_C$, the above expression becomes

$$I = \frac{1}{M}\left[m_H m_C r_{HC}^2 + m_C m_N r_{CN}^2 + m_H m_N r_{HN}^2 + m_N^2 l_N^2 + m_N^2 l_N^2 - 2m_N^2 l_N^2\right]$$

$$= \frac{m_H m_C r_{HC}^2 + m_H m_N r_{HN}^2 + m_C m_N r_{CN}^2}{m_H + m_C + m_N}$$

For $H^{12}C^{14}N$, $r_{HC} = 106.8$ pm, $r_{CN} = 115.6$ pm, and $r_{HN} = r_{HC} + r_{CN} = 222.4$ pm. Thus,

$$I = \frac{1}{1.008\text{ amu} + 12.000\text{ amu} + 14.003\text{ amu}}\left[(1.008\text{ amu})(12.000\text{ amu})(106.8\text{ pm})^2 \right.$$

$$\left. + (1.008\text{ amu})(14.003\text{ amu})(222.4\text{ pm})^2 + (12.000\text{ amu})(14.003\text{ amu})(115.6\text{ pm})^2\right]$$

$$= \left(1.1409 \times 10^5\text{ amu·pm}^2\right)\left(1.660\,5402 \times 10^{-27}\text{ kg·amu}^{-1}\right)\left(10^{-12}\text{ m·pm}^{-1}\right)^2$$

$$= 1.894 \times 10^{-46}\text{ kg·m}^2$$

which is the same as that given in Problem 6–12.

6–13. The following lines were observed in the microwave absorption spectrum of $H^{127}I$ and $D^{127}I$ between 60 cm^{-1} and 90 cm^{-1}.

	$\tilde{\nu}_{obs}/\text{cm}^{-1}$			
$H^{127}I$	64.275	77.130	89.985	
$D^{127}I$	65.070	71.577	78.094	84.591

Use the rigid-rotator approximation to determine the values of $\tilde{B}$, I, and $l(v = 0)$ for each molecule. Take the mass of ^{127}I to be 126.904 amu and the mass of D to be 2.013 amu.

Using the rigid rotor approximation,

$$\tilde{\nu}_{obs} = 2\tilde{B} \, (J + 1)$$

where J is the lower state quantum number. Since more than one transition is given for each molecule, it is best to determine the value of $\tilde{B}$ graphically. A plot of $\tilde{\nu}_{obs}$ versus $J + 1$ for each molecule gives a slope of $2\tilde{B}$.

To determine the lower state J value for the first line of each spectrum, rearrange the above equation to give

$$J = \frac{\tilde{\nu}_{obs}}{2\tilde{B}} - 1$$

where $2\tilde{B}$ can be approximated by the spacing between adjacent lines. For $H^{127}I$,

$$J = \frac{64.275 \text{ cm}^{-1}}{77.130 \text{ cm}^{-1} - 64.275 \text{ cm}^{-1}} - 1 = 4$$

Thus, the three transitions for $H^{127}I$ in the table arise from $J = 4$, 5, 6, respectively.

For $D^{127}I$,

$$J = \frac{65.070 \text{ cm}^{-1}}{71.577 \text{ cm}^{-1} - 65.070 \text{ cm}^{-1}} - 1 = 9$$

Thus, the four transitions for $H^{127}I$ in the table arise from $J = 9$, 10, 11, 12, respectively.

A plot of $\tilde{\nu}_{obs}$ versus $J + 1$ for each molecule is shown below.

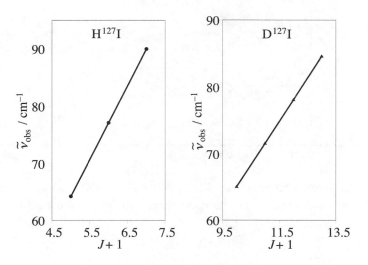

Linear least squares fit to give the best straight line for each plot gives a value of $2\tilde{B}$ of 12.8550 cm^{-1} for $H^{127}I$ and 6.5080 cm^{-1} for $D^{127}I$. Thus, for $H^{127}I$,

$$\tilde{B} = 6.4275 \text{ cm}^{-1}$$

$$I = \frac{h}{8\pi^2 c \tilde{B}}$$

$$= \frac{6.626\,0755 \times 10^{-34} \text{ J·s}}{8\pi^2 \left(2.997\,924\,58 \times 10^{10} \text{ cm·s}^{-1}\right) \left(6.4275 \text{ cm}^{-1}\right)}$$

$$= 4.3552 \times 10^{-47} \text{ kg·m}^2$$

$$\mu = \frac{(1.008 \text{ amu}) (126.904 \text{ amu})}{1.008 \text{ amu} + 126.904 \text{ amu}} \left(1.660\,5402 \times 10^{-27} \text{ kg·amu}^{-1}\right)$$

$$= 1.6606 \times 10^{-27} \text{ kg}$$

$$l = \left(\frac{I}{\mu}\right)^{1/2}$$

$$= \left(\frac{4.3552 \times 10^{-47} \text{ kg·m}^2}{1.6606 \times 10^{-27}}\right)^{1/2}$$

$$= 1.619 \times 10^{-10} \text{ m} = 1.619 \text{ Å}$$

For D^{127}I,

$$\tilde{B} = 3.2540 \text{ cm}^{-1} = 3.254 \text{ cm}^{-1}$$

$$I = \frac{h}{8\pi^2 c \tilde{B}}$$

$$= \frac{6.626\,0755 \times 10^{-34} \text{ J·s}}{8\pi^2 \left(2.997\,924\,58 \times 10^{10} \text{ cm·s}^{-1}\right) \left(3.2540 \text{ cm}^{-1}\right)}$$

$$= 8.6026 \times 10^{-47} \text{ kg·m}^2$$

$$\mu = \frac{(2.013 \text{ amu}) (126.904 \text{ amu})}{2.013 \text{ amu} + 126.904 \text{ amu}} \left(1.660\,5402 \times 10^{-27} \text{ kg·amu}^{-1}\right)$$

$$= 3.2905 \times 10^{-27} \text{ kg}$$

$$l = \left(\frac{I}{\mu}\right)^{1/2}$$

$$= \left(\frac{8.6026 \times 10^{-47} \text{ kg·m}^2}{3.2905 \times 10^{-27}}\right)^{1/2}$$

$$= 1.617 \times 10^{-10} \text{ m} = 1.617 \text{ Å}$$

6–14. Given that $B = 56\,000$ MHz and $\tilde{\omega} = 2143.0$ cm^{-1} for CO, calculate the frequencies of the first few lines of the R and P branches in the rotation-vibration spectrum of CO.

Converting B to cm^{-1},

$$\tilde{B} = \left(5.6 \times 10^{10}\,\text{Hz}\right)\left(\frac{1}{2.997\,924\,58 \times 10^{10}\,\text{cm·s}^{-1}}\right) = 1.87\,\text{cm}^{-1}$$

For the R branch,

$$\tilde{\nu}_{\text{obs}} = \tilde{\omega} + 2\tilde{B}\,(J+1)$$

$$= 2143.0\,\text{cm}^{-1} + 3.74\,\text{cm}^{-1}\,(J+1)$$

For the P branch,

$$\tilde{\nu}_{\text{obs}} = \tilde{\omega} - 2\tilde{B}\,J$$

$$= 2143.0\,\text{cm}^{-1} - 3.74\,\text{cm}^{-1}J$$

The first few lines in the rotation-vibration spectrum of CO are listed in the following table:

J	$R(J)/\text{cm}^{-1}$	$P(J)/\text{cm}^{-1}$
0	2146.74	–
1	2150.48	2139.26
2	2154.22	2135.52
3	2157.96	2131.78

6–15. Given that $l = 156\,\text{pm}$ and $k = 250\,\text{N·m}^{-1}$ for ^6LiF, use the rigid rotator–harmonic oscillator approximation to construct to scale an energy-level diagram for the first five rotational levels in the $v=0$ and $v=1$ vibrational states. Indicate the allowed transitions in an absorption experiment, and calculate the frequencies of the first few lines in the R and P branches of the rotation-vibration spectrum of ^6LiF.

Calculate the reduced mass and the moment of inertia for the molecule:

$$\mu = \frac{(6.015\,\text{amu})\,(18.998\,\text{amu})}{6.015\,\text{amu} + 18.998\,\text{amu}}\left(1.660\,5402 \times 10^{-27}\,\text{kg·amu}^{-1}\right)$$

$$= 7.5862 \times 10^{-27}\,\text{kg}$$

$$I = \mu l^2$$

$$= \left(7.5862 \times 10^{-27}\,\text{kg}\right)\left(156 \times 10^{-12}\,\text{m}\right)^2$$

$$= 1.8462 \times 10^{-46}\,\text{kg·m}^2$$

Now calculate $\tilde{B}$ and $\tilde{\omega}$:

$$\tilde{B} = \frac{h}{8\pi^2 cI}$$

$$= \frac{6.626\,0755 \times 10^{-34}\,\text{J}\cdot\text{s}}{8\pi^2\left(2.997\,924\,58 \times 10^{10}\,\text{cm}\cdot\text{s}^{-1}\right)\left(1.8462 \times 10^{-46}\,\text{kg}\cdot\text{m}^2\right)}$$

$$= 1.52\,\text{cm}^{-1}$$

$$\tilde{\omega} = \frac{1}{2\pi c}\left(\frac{k}{\mu}\right)^{1/2}$$

$$= \frac{1}{2\pi\left(2.997\,924\,58 \times 10^{10}\,\text{cm}\cdot\text{s}^{-1}\right)}\left(\frac{250\,\text{N}\cdot\text{m}^{-1}}{7.5862 \times 10^{-27}\,\text{kg}}\right)^{1/2}$$

$$= 963.7\,\text{cm}^{-1}$$

The energy for each rotational level in the v vibrational state is

$$E_{v,J} = \tilde{\omega}\left(v + \frac{1}{2}\right) + \tilde{B}J\,(J+1)$$

$$= \left(963.7\,\text{cm}^{-1}\right)\left(v + \frac{1}{2}\right) + \left(1.52\,\text{cm}^{-1}\right)J\,(J+1)$$

and those for the first five rotational levels in the $v = 0$ and $v = 1$ vibrational states are tabulated below:

J	$E(v = 0)/\text{cm}^{-1}$	$E(v = 1)/\text{cm}^{-1}$
0	481.85	1445.55
1	484.89	1448.59
2	490.97	1454.67
3	500.09	1463.79
4	512.25	1475.95

The energy-level diagram is shown below, with the allowed transitions marked as vertical lines. In an absorption experiment, the transitions originate from the $v = 0$ state. (See Problem 5–40.)

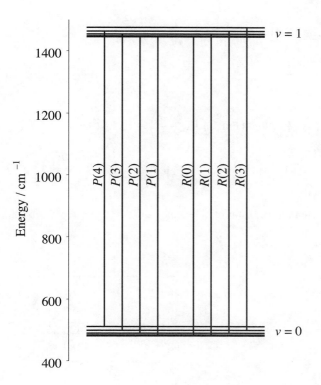

The frequencies for the transitions indicated can be calculated by subtracting the energy of the lower state from that of the upper state:

J	$R(J)/\text{cm}^{-1}$	$P(J)/\text{cm}^{-1}$
0	966.74	–
1	969.78	960.66
2	972.82	957.62
3	975.86	954.58
4	–	951.54

6–16. Using the values of $\tilde{\omega}_e$, $\tilde{x}_e\tilde{\omega}_e$, $\tilde{B}_e$, and $\tilde{\alpha}_e$ given in Tables 5.1 and 6.1, construct to scale an energy-level diagram for the first five rotational levels in the $v = 0$ and $v = 1$ vibrational states for $H^{35}Cl$. Indicate the allowed transitions in an absorption experiment, and calculate the frequencies of the first few lines in the R and P branches.

For $H^{35}Cl$, $\tilde{\omega}_e = 2990.94 \text{ cm}^{-1}$, $\tilde{x}_e\tilde{\omega}_e = 52.819 \text{ cm}^{-1}$, $\tilde{B}_e = 10.5934 \text{ cm}^{-1}$, and $\tilde{\alpha}_e = 0.3072 \text{ cm}^{-1}$. The energy for each rotational level in the vth vibrational state is

$$E_{v,J} = \tilde{\omega}_e \left(v + \frac{1}{2} \right) - \tilde{x}_e \tilde{\omega}_e \left(v + \frac{1}{2} \right)^2 + \tilde{B}_v J \left(J + 1 \right)$$

$$= \tilde{\omega}_e \left(v + \frac{1}{2} \right) - \tilde{x}_e \tilde{\omega}_e \left(v + \frac{1}{2} \right)^2 + \left[\tilde{B}_e - \tilde{\alpha}_e \left(v + \frac{1}{2} \right) \right] J \left(J + 1 \right)$$

$$= \left(2990.94 \ \text{cm}^{-1} \right) \left(v + \frac{1}{2} \right) - \left(52.819 \ \text{cm}^{-1} \right) \left(v + \frac{1}{2} \right)^2$$

$$+ \left[10.5934 \ \text{cm}^{-1} - \left(0.3072 \ \text{cm}^{-1} \right) \left(v + \frac{1}{2} \right) \right] J \left(J + 1 \right)$$

and those for the first five rotational levels in the $v = 0$ and $v = 1$ vibrational states are tabulated below:

J	$E(v = 0)/\text{cm}^{-1}$	$E(v = 1)/\text{cm}^{-1}$
0	1482.27	4367.57
1	1503.14	4387.83
2	1544.90	4428.36
3	1607.54	4489.16
4	1691.06	4570.22

The energy-level diagram is shown below, with the allowed transitions marked as vertical lines. In an absorption experiment, the transitions originate from the $v = 0$ state. (See Problem 5–40.)

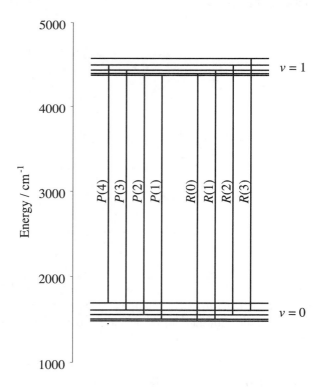

The transition frequencies for the transitions indicated can be calculated by subtracting the energy of the lower state from that of the upper state:

J	$R(J)/\text{cm}^{-1}$	$P(J)/\text{cm}^{-1}$
0	2905.56	–
1	2925.22	2864.43
2	2944.26	2842.93
3	2962.68	2820.82
4	–	2798.10

6–17. The following data are obtained for the rotation-vibration spectrum of $H^{127}I$. Determine $\tilde{B}_0$, $\tilde{B}_1$, $\tilde{B}_e$, and $\tilde{\alpha}_e$ from these data.

Line	Frequency/cm^{-1}
$R(0)$	2242.6
$R(1)$	2254.8
$P(1)$	2217.1
$P(2)$	2203.8

The transition frequencies for the R branch are given by Equation 6–41,

$$\tilde{\nu}_{\text{obs}} = \tilde{\omega}_0 + 2\tilde{B}_1 + \left(3\tilde{B}_1 - \tilde{B}_0\right) J + \left(\tilde{B}_1 - \tilde{B}_0\right) J^2$$

and those for the P branch are given by Equation 6–42,

$$\tilde{\nu}_{\text{obs}} = \tilde{\omega}_0 - \left(\tilde{B}_1 + \tilde{B}_0\right) J + \left(\tilde{B}_1 - \tilde{B}_0\right) J^2$$

For the four given transitions, these equations become

$$R(0): \quad 2242.6 \text{ cm}^{-1} = \tilde{\omega}_0 + 2\tilde{B}_1 \tag{1}$$

$$R(1): \quad 2254.8 \text{ cm}^{-1} = \tilde{\omega}_0 + 6\tilde{B}_1 - 2\tilde{B}_0 \tag{2}$$

$$P(1): \quad 2217.1 \text{ cm}^{-1} = \tilde{\omega}_0 - 2\tilde{B}_0 \tag{3}$$

$$P(2): \quad 2203.8 \text{ cm}^{-1} = \tilde{\omega}_0 + 2\tilde{B}_1 - 6\tilde{B}_0 \tag{4}$$

Subtracting Equation 3 from Equation 2:

$$6\tilde{B}_1 = 37.7 \text{ cm}^{-1}$$

$$\tilde{B}_1 = 6.28 \text{ cm}^{-1}$$

Subtracting Equation 4 from Equation 1:

$$6\tilde{B}_0 = 38.8 \text{ cm}^{-1}$$

$$\tilde{B}_0 = 6.47 \text{ cm}^{-1}$$

The values of $\tilde{B}_0$ and $\tilde{B}_1$ are related to the values of $\tilde{B}_e$ and α_e:

$$\tilde{B}_0 = \tilde{B}_e - \frac{\alpha_e}{2} \tag{5}$$

$$\tilde{B}_1 = \tilde{B}_e - \frac{3\alpha_e}{2} \tag{6}$$

Subtracting Equation 6 from Equation 5:

$$\alpha_e = \tilde{B}_0 - \tilde{B}_1$$

$$= 6.47 \text{ cm}^{-1} - 6.28 \text{ cm}^{-1}$$

$$= 0.19 \text{ cm}^{-1}$$

Multiplying Equation 5 by 3 and then subtracting Equation 6 from the result:

$$2\tilde{B}_e = 3\tilde{B}_0 - \tilde{B}_1$$

$$= 3 \left(6.47 \text{ cm}^{-1} \right) - 6.28 \text{ cm}^{-1}$$

$$B_e = 6.57 \text{ cm}^{-1}$$

6–18. The following spectroscopic constants were determined for pure samples of $^{74}\text{Ge}^{32}\text{S}$ and $^{72}\text{Ge}^{32}\text{S}$:

Molecule	B_e/MHz	α_e/MHz	D/kHz	$l(v=0)$/pm
$^{74}\text{Ge}^{32}\text{S}$	5593.08	22.44	2.349	0.20120
$^{72}\text{Ge}^{32}\text{S}$	5640.06	22.74	2.388	0.20120

Determine the frequency of the $J = 0$ to $J = 1$ transition for $^{74}\text{Ge}^{32}\text{S}$ and $^{72}\text{Ge}^{32}\text{S}$ in their ground vibrational state. The width of a microwave absorption line is on the order of 1 kHz. Could you distinguish a pure sample of $^{74}\text{Ge}^{32}\text{S}$ from a 50/50 mixture of $^{74}\text{Ge}^{32}\text{S}$ and $^{72}\text{Ge}^{32}\text{S}$ using microwave spectroscopy?

Since the rotational transition takes place in the ground vibrational state for each sample, the value of $B_0 = B_e - \alpha_e/2$ is necessary:

$$^{74}\text{Ge}^{32}\text{S}: \quad B_0 = 5593.08 \text{ MHz} - \frac{22.44 \text{ MHz}}{2} = 5581.86 \text{ MHz}$$

$$^{72}\text{Ge}^{32}\text{S}: \quad B_0 = 5640.06 \text{ MHz} - \frac{22.74 \text{ MHz}}{2} = 5628.69 \text{ MHz}$$

The frequency of the $J = 0$ to $J = 1$ transition is

$$\nu_{\text{obs}} = 2B(J+1) - 4D(J+1)^3$$

$$= 2B - 4D$$

For $^{74}\text{Ge}^{32}\text{S}$,

$$\nu_{\text{obs}} = 2(5581.86 \text{ MHz}) - 4\left(2.349 \times 10^{-3} \text{ MHz}\right) = 11163.71 \text{ MHz}$$

For $^{72}Ge^{32}S$,

$$\nu_{obs} = 2\,(5628.69 \text{ MHz}) - 4\left(2.388 \times 10^{-3} \text{ MHz}\right) = 11257.37 \text{ MHz}$$

These two frequencies differ by 94 MHz, and can easily be distinguished.

6–19. An analysis of the rotational spectrum of $^{12}C^{32}S$ gives the following results:

v	$\tilde{B}_v/cm^{-1}$
0	0.81708
1	0.81116
2	0.80524
3	0.79932

Determine the values of $\tilde{B}_e$ and $\tilde{\alpha}_e$ from these data.

Since

$$\tilde{B}_v = \tilde{B}_e - \alpha_e\left(v + \frac{1}{2}\right)$$

a plot of $\tilde{B}_v$ versus $v + 1/2$ gives a slope of $-\alpha_e$ and an intercept of $\tilde{B}_e$.

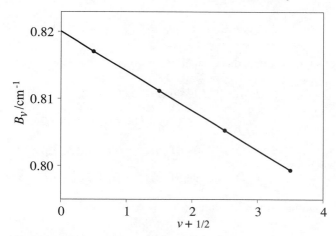

The best fit line of the above plot gives $y = -0.005\,92x + 0.820\,04$. Thus,

$$\tilde{B}_e = 0.820\,04 \text{ cm}^{-1}$$

$$\alpha_e = 0.005\,92 \text{ cm}^{-1}$$

6–20. How many degrees of vibrational freedom are there for NH_3, CO_2, CH_4, and C_2H_6?

According to Section 5–11, a linear molecule has $3N - 5$ vibrational degrees of freedom while a nonlinear molecule has $3N - 6$ vibrational degrees of freedom. For the molecules given, only CO_2 is linear. Thus,

Molecule	N	vibrational degrees of freedom
NH_3	4	6
CO_2	3	4
CH_4	5	9
C_2H_6	8	18

6–21. The following data are obtained for the rotation-vibration spectrum of $H^{79}Br$. Determine $\tilde{B}_0$, $\tilde{B}_1$, $\tilde{B}_e$, and $\tilde{\alpha}_e$ from these data.

Line	Frequency/cm^{-1}
$R(0)$	2642.60
$R(1)$	2658.36
$P(1)$	2609.67
$P(2)$	2592.51

The transition frequencies for the R branch are given by Equation 6–41,

$$\tilde{\nu}_{obs} = \tilde{\omega}_0 + 2\tilde{B}_1 + \left(3\tilde{B}_1 - \tilde{B}_0\right) J + \left(\tilde{B}_1 - \tilde{B}_0\right) J^2$$

and those for the P branch are given by Equation 6–42,

$$\tilde{\nu}_{obs} = \tilde{\omega}_0 - \left(\tilde{B}_1 + \tilde{B}_0\right) J + \left(\tilde{B}_1 - \tilde{B}_0\right) J^2$$

For the four given transitions, these equations become

$$R(0): \quad 2642.60 \text{ cm}^{-1} = \tilde{\omega}_0 + 2\tilde{B}_1 \tag{1}$$

$$R(1): \quad 2658.36 \text{ cm}^{-1} = \tilde{\omega}_0 + 6\tilde{B}_1 - 2\tilde{B}_0 \tag{2}$$

$$P(1): \quad 2609.67 \text{ cm}^{-1} = \tilde{\omega}_0 - 2\tilde{B}_0 \tag{3}$$

$$P(2): \quad 2592.51 \text{ cm}^{-1} = \tilde{\omega}_0 + 2\tilde{B}_1 - 6\tilde{B}_0 \tag{4}$$

Subtracting Equation 3 from Equation 2:

$$6\tilde{B}_1 = 48.69 \text{ cm}^{-1}$$

$$\tilde{B}_1 = 8.115 \text{ cm}^{-1}$$

Subtracting Equation 4 from Equation 1:

$$6\tilde{B}_0 = 50.09 \text{ cm}^{-1}$$

$$\tilde{B}_0 = 8.348 \text{ cm}^{-1}$$

The values of $\tilde{B}_0$ and $\tilde{B}_1$ are related to the values of $\tilde{B}_e$ and α_e:

$$\tilde{B}_0 = \tilde{B}_e - \frac{\alpha_e}{2} \tag{5}$$

$$\tilde{B}_1 = \tilde{B}_e - \frac{3\alpha_e}{2} \tag{6}$$

Subtracting Equation 6 from Equation 5:

$$\alpha_e = \tilde{B}_0 - \tilde{B}_1$$

$$= 8.348 \text{ cm}^{-1} - 8.115 \text{ cm}^{-1}$$

$$= 0.233 \text{ cm}^{-1}$$

Multiplying Equation 5 by 3 and then subtracting Equation 6 from the result:

$$2\tilde{B}_e = 3\tilde{B}_0 - \tilde{B}_1$$

$$= 3\left(8.348 \text{ cm}^{-1}\right) - 8.115 \text{ cm}^{-1}$$

$$B_e = 8.465 \text{ cm}^{-1}$$

6–22. The frequencies of the rotational transitions in the nonrigid-rotator approximation are given by Equation 6.45. Show how both $\tilde{B}$ and $\tilde{D}$ may be obtained by plotting $\tilde{\nu}/(J + 1)$ versus $(J + 1)^2$. Use this method and the data in Table 6.2 to determine both $\tilde{B}$ and $\tilde{D}$ for $H^{35}Cl$.

Dividing both sides of Equation 6.45, $\tilde{\nu} = 2\tilde{B}(J + 1) - 4\tilde{D}(J + 1)^3$, by $J + 1$,

$$\frac{\tilde{\nu}}{J + 1} = 2\tilde{B} - 4\tilde{D}(J + 1)^2$$

This equation describes a straight line: when $\tilde{\nu}/(J + 1)$ is plotted against $(J + 1)^2$, the slope is $-4\tilde{D}$ and the intercept is $2\tilde{B}$.

The data in Table 6.2 is plotted in the following graph.

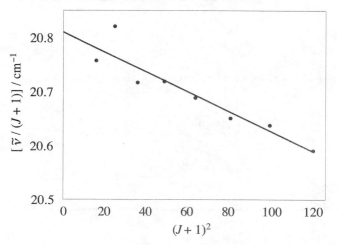

The straight line is described by $y = -1.82 \times 10^{-3}x + 20.81$. Thus,

$$\tilde{B} = \frac{20.81 \text{ cm}^{-1}}{2} = 10.40 \text{ cm}^{-1}$$

$$\tilde{D} = \frac{-1.82 \times 10^{-3} \text{ cm}^{-1}}{-4} = 4.6 \times 10^{-4} \text{ cm}^{-1}$$

6–23. The following data are obtained in the microwave spectrum of $^{12}C^{16}O$. Use the method of Problem 6–22 to determine the values of $\tilde{B}$ and $\tilde{D}$ from these data.

Transitions	Frequency/cm^{-1}
$0 \rightarrow 1$	3.84540
$1 \rightarrow 2$	7.69060
$2 \rightarrow 3$	11.53550
$3 \rightarrow 4$	15.37990
$4 \rightarrow 5$	19.22380
$5 \rightarrow 6$	23.06685

The plot of $\tilde{\nu}/(J+1)$ against $(J+1)^2$ is shown below.

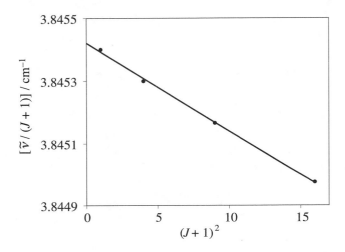

The straight line is described by $y = -2.61 \times 10^{-5}x + 3.845\,409$. Thus,

$$\tilde{B} = \frac{3.845\,409 \text{ cm}^{-1}}{2} = 1.922\,705 \text{ cm}^{-1}$$

$$\tilde{D} = \frac{-2.61 \times 10^{-5} \text{ cm}^{-1}}{-4} = 6.5 \times 10^{-6} \text{ cm}^{-1}$$

6–24. Using the parameters given in Table 6.1, calculate the frequencies (in cm^{-1}) of the $0 \rightarrow 1$, $1 \rightarrow 2$, $2 \rightarrow 3$, and $3 \rightarrow 4$ rotational transitions in the ground vibrational state of $H^{35}Cl$ in the nonrigid-rotator approximation.

For $H^{35}Cl$, $\tilde{B}_e = 10.5934$ cm^{-1}, $\tilde{\alpha}_e = 0.3072$ cm^{-1}, and $\tilde{D} = 5.319 \times 10^{-4}$ cm^{-1}. Transition frequencies can be determined using

$$\tilde{\nu} = 2\tilde{B}_0 (J + 1) - 4\tilde{D} (J + 1)^3$$

$$= 2 \left(\tilde{B}_e - \frac{\tilde{\alpha}_e}{2} \right) (J + 1) - 4\tilde{D} (J + 1)^3$$

$$= 2 \left(10.5934 \text{ cm}^{-1} - \frac{0.3072 \text{ cm}^{-1}}{2} \right) (J + 1) - 4 \left(5.319 \times 10^{-4} \text{ cm}^{-1} \right) (J + 1)^3$$

$$= 20.8796 (J + 1) - 2.1276 \times 10^{-3} (J + 1)^3$$

where J is the lower state quantum number. The frequencies for the first 4 rotational transitions in the ground vibrational state are tabulated below:

Transition	Frequency/cm^{-1}
$0 \rightarrow 1$	20.8775
$1 \rightarrow 2$	41.7422
$2 \rightarrow 3$	62.5814
$3 \rightarrow 4$	83.3822

6–25. Use the data in Table 6.1 to calculate the ratio of centrifugal distortion energy to the total rotational energy of $H^{35}Cl$ and $^{35}Cl^{35}Cl$ in the $J = 10$ state.

The energy of the $J = 10$ state is in the ground vibrational state is

$$E = \tilde{B}_0 J (J + 1) - D J^2 (J + 1)^2$$

$$= \tilde{B}_0 (10) (11) - D (10)^2 (11)^2$$

$$= 110\tilde{B}_0 - 12\,100\tilde{D}$$

The rotational constant in the ground vibrational state is $\tilde{B}_0 = \tilde{B}_e - \alpha_e/2$. For $H^{35}Cl$,

$$\tilde{B}_0 = 10.5934 \text{ cm}^{-1} - \frac{0.3072 \text{ cm}^{-1}}{2} = 10.4398 \text{ cm}^{-1}$$

and for $^{35}Cl^{35}Cl$,

$$\tilde{B}_0 = 0.2440 \text{ cm}^{-1} - \frac{0.001\,49 \text{ cm}^{-1}}{2} = 0.243\,26 \text{ cm}^{-1}$$

The ratio of centrifugal distortion energy to the total rotational energy is $12\,100\tilde{D}/\left(110\tilde{B}_0 - 12\,100\tilde{D}\right)$. For $H^{35}Cl$, the ratio is

$$\frac{12\,100\tilde{D}}{110\tilde{B}_0 - 12\,100\tilde{D}} = \frac{12\,100 \left(5.319 \times 10^{-4} \text{ cm}^{-1}\right)}{110 \left(10.4398 \text{ cm}^{-1}\right) - 12\,100 \left(5.319 \times 10^{-4} \text{ cm}^{-1}\right)} = 5.64 \times 10^{\pm 3}$$

For $^{35}Cl^{35}Cl$, the ratio is

$$\frac{12\,100\tilde{D}}{110\tilde{B}_0 - 12\,100\tilde{D}} = \frac{12\,100 \left(1.86 \times 10^{-7} \text{ cm}^{-1}\right)}{110 \left(0.243\,26 \text{ cm}^{-1}\right) - 12\,100 \left(1.86 \times 10^{-7} \text{ cm}^{-1}\right)} = 8.41 \times 10^{-5}$$

Centrifugal distortion plays a much greater role in H^{35}Cl than in ^{35}Cl^{35}Cl.

6–26. In this problem, we shall develop a semiclassical derivation of Equation 6.45. As usual, we'll consider one atom to be fixed at the origin with the other atom rotating about it with a reduced mass μ. Because the molecule is nonrigid, as the molecule rotates faster and faster (increasing J), the centrifugal force causes the bond to stretch. The extent of the stretching of the bond can be determined by balancing the Hooke's law force [$k(l - l_0)$] and the centrifugal force ($\mu v^2 / l = \mu l \omega^2$, Equation 1.20):

$$k(l - l_0) = \mu l \omega^2 \tag{1}$$

In equation 1, l_0 is the bond length when there is no rotation ($J = 0$). The total energy of the rotator is made up of a kinetic energy part and a potential energy part, which we write as

$$E = \frac{1}{2} I \omega^2 + \frac{1}{2} k (l - l_0)^2 \tag{2}$$

Substitute equation 1 into equation 2 and show that the result can be written as

$$E = \frac{L^2}{2\mu l^2} + \frac{L^4}{2\mu^2 k l^6} \tag{3}$$

where $L = I\omega$. Now use the quantum condition $L^2 = \hbar^2 J(J + 1)$ to obtain

$$E_J = \frac{\hbar^2}{2\mu l^2} J(J + 1) + \frac{\hbar^4 J^2 (J + 1)^2}{2\mu^2 k l^6} \tag{4}$$

Finally, solve equation 1 for $1/l^2$, and eliminate ω by using $\omega = L/I = L/\mu l^2$ to obtain

$$\frac{1}{l^2} = \frac{1}{l_0^2} \left(1 - \frac{2L^2}{\mu k l^4} + \frac{L^4}{\mu^2 k^2 l^8} \right)$$

Now substitute this result into equation 4 to obtain

$$E_J = \frac{\hbar^2}{2\mu l_0^2} J(J + 1) - \frac{\hbar^4}{2\mu^2 k l_0^6} J^2 (J + 1)^2$$

Rewrite Equation 1,

$$l - l_0 = \frac{\mu l \omega^2}{k}$$

and substitute it into Equation 2,

$$E = \frac{1}{2} I \omega^2 + \frac{1}{2} k \left(\frac{\mu l \omega^2}{k} \right)^2$$

$$= \frac{I^2 \omega^2}{2I} + \frac{\mu^2 l^2 \omega^4}{2k}$$

$$= \frac{I^2 \omega^2}{2I} + \frac{I^4 \omega^4}{2\mu^2 k l^6}$$

$$= \frac{L^2}{2\mu l^2} + \frac{L^4}{2\mu^2 k l^6}$$

where the relations $I = \mu l^2$ and $L = I\omega$ have been applied. Using the quantum condition $L^2 = \hbar^2 J(J+1)$, the above expression becomes

$$E_J = \frac{\hbar^2}{2\mu l^2} J(J+1) + \frac{\hbar^4 J^2 (J+1)^2}{2\mu^2 k l^6}$$

l can be expressed in terms of l_0 by rearranging Equation 1:

$$kl - kl_0 = \mu l \omega^2$$

$$\frac{1}{l} = \frac{k - \mu \omega^2}{k l_0}$$

$$\frac{1}{l^2} = \frac{k^2 - 2\mu \omega^2 k + \mu^2 \omega^4}{k^2 l_0^2}$$

Since $\omega = L/\mu l^2$ and $L^2 = \hbar^2 J(J+1)$,

$$\frac{1}{l^2} = \frac{1}{l_0^2} - \frac{2L^2}{\mu l^4 k l_0^2} + \frac{L^4}{\mu^2 l^8 k^2 l_0^2}$$

$$= \frac{1}{l_0^2} \left(1 - \frac{2L^2}{\mu k l^4} + \frac{L^4}{\mu^2 k^2 l^8} \right)$$

$$= \frac{1}{l_0^2} \left[1 - \frac{2\hbar^2}{\mu k l^4} J(J+1) + \frac{\hbar^4}{\mu^2 k^2 l^8} J^2 (J+1)^2 \right]$$

Substituting this expression into the E_J expression, and keeping only terms up to $J^2 (J+1)^2$,

$$E_J = \frac{\hbar^2}{2\mu} J(J+1) \left\{ \frac{1}{l_0^2} \left[1 - \frac{2\hbar^2}{\mu k l^4} J(J+1) + \frac{\hbar^4}{\mu^2 k^2 l^8} J^2 (J+1)^2 \right] \right\}$$

$$+ \frac{\hbar^4 J^2 (J+1)^2}{2\mu^2 k} \left\{ \frac{1}{l_0^2} \left[1 - \frac{2\hbar^2}{\mu k l^4} J(J+1) + \frac{\hbar^4}{\mu^2 k^2 l^8} J^2 (J+1)^2 \right] \right\}^3$$

$$= \frac{\hbar^2}{2\mu l_0^2} J(J+1) - \frac{\hbar^4}{\mu^2 k l_0^2 l^4} J^2 (J+1)^2 + \frac{\hbar^4}{2\mu^2 k l_0^6} J^2 (J+1)^2$$

$$= \frac{\hbar^2}{2\mu l_0^2} J(J+1) - \frac{\hbar^4}{2\mu^2 k l_0^6} J^2 (J+1)^2$$

The last step is obtained by approximating l as l_0. This is justified because if $l - l_0 = \delta$, then

$$\frac{1}{l^4} = \frac{1}{\left[l_0 \left(1 + \delta / l_0 \right) \right]^4}$$

$$= \frac{1}{l_0^4} \left(1 + \delta / l_0 \right)^{-4}$$

$$= \frac{1}{l_0^4} \left(1 - \frac{4\delta}{l_0} + \cdots \right)$$

$$= \frac{1}{l_0^4} - \frac{4\delta}{l_0^5} + \cdots$$

Since $\delta \ll l_0$, the terms after the first term in the expansion will be small (see Problems 5–38 and 5–39). In addition, they yield terms with orders higher than $J^2(J+1)^2$.

6–27. In terms of the variable θ, Legendre's equation is

$$\sin \theta \frac{d}{d\theta} \left[\sin \theta \frac{d\Theta(\theta)}{d\theta} \right] + (\beta \sin^2 \theta - m^2)\Theta(\theta) = 0$$

Let $x = \cos \theta$ and $P(x) = \Theta(\theta)$ and show that

$$(1 - x^2)\frac{d^2 P(x)}{dx^2} - 2x\frac{dP(x)}{dx} + \left(\beta - \frac{m^2}{1 - x^2} \right) P(x) = 0$$

Legendre's equation is

$$\sin \theta \frac{d}{d\theta} \left[\sin \theta \frac{d\Theta(\theta)}{d\theta} \right] + (\beta \sin^2 \theta - m^2)\Theta(\theta) = 0$$

$$\sin \theta \cos \theta \frac{d\Theta}{d\theta} + \sin^2 \theta \frac{d^2\Theta}{d\theta^2} + (\beta \sin^2 \theta - m^2)\Theta(\theta) = 0$$

Let $x = \cos \theta$ and $P(x) = \Theta(\theta)$. Then

$$\frac{d\Theta}{d\theta} = \frac{dP}{dx}\frac{dx}{d\theta}$$

$$= -\sin\theta\frac{dP}{dx}$$

$$= -(1-x^2)^{1/2}\frac{dP}{dx}$$

$$\frac{d^2\Theta}{d\theta^2} = \frac{d}{d\theta}\left(\frac{d\Theta}{d\theta}\right)$$

$$= \frac{dx}{d\theta}\frac{d}{dx}\left[-(1-x^2)^{1/2}\frac{dP}{dx}\right]$$

$$= -(1-x^2)^{1/2}\left[\frac{x}{(1-x^2)^{1/2}}\frac{dP}{dx} - (1-x^2)^{1/2}\frac{d^2P}{dx^2}\right]$$

$$= -x\frac{dP}{dx} + (1-x^2)\frac{d^2P}{dx^2}$$

Substituting these derivatives into Legendre's equation gives

$$\sin\theta\cos\theta\frac{d\Theta}{d\theta} + \sin^2\theta\frac{d^2\Theta}{d\theta^2} + (\beta\sin^2\theta - m^2)\Theta(\theta) = 0$$

$$(1-x^2)^{1/2}x\left[-(1-x^2)^{1/2}\frac{dP}{dx}\right] + (1-x^2)\left[-x\frac{dP}{dx} + (1-x^2)\frac{d^2P}{dx^2}\right] + \left[\beta\left(1-x^2\right) - m^2\right]P(x) = 0$$

$$(1-x^2)^2\frac{d^2P}{dx^2} - 2x(1-x^2)\frac{dP}{dx} + \left[\beta(1-x^2) - m^2\right]P(x) = 0$$

$$(1-x^2)\frac{d^2P}{dx^2} - 2x\frac{dP}{dx} + \left(\beta - \frac{m^2}{1-x^2}\right)P(x) = 0$$

6–28. Show that the Legendre polynomials given in Table 6.3 satisfy Equation 6.48 with $m = 0$.

Equation 6.48 is

$$\left(1-x^2\right)\frac{d^2P}{dx^2} - 2x\frac{dP}{dx} + \left[J(J+1) - \frac{m^2}{1-x^2}\right]P(x) = 0$$

For $m = 0$, the equation becomes

$$\left(1-x^2\right)\frac{d^2P}{dx^2} - 2x\frac{dP}{dx} + J(J+1)P(x) = 0$$

Showing that the Legendre polynomials satisfy this equation requires substituting a polynomial into the left hand side of the equation and showing the result is 0. The polynomials and their first two derivatives are shown in the table below:

J	$P_J(x)$	$\dfrac{dP_J}{dx}$	$\dfrac{d^2 P_J}{dx^2}$
0	1	0	0
1	x	1	0
2	$\dfrac{1}{2}\left(3x^2 - 1\right)$	$3x$	3
3	$\dfrac{1}{2}\left(5x^3 - 3x\right)$	$\dfrac{1}{2}\left(15x^2 - 3\right)$	$15x$
4	$\dfrac{1}{8}\left(35x^4 - 30x^2 + 3\right)$	$\dfrac{1}{8}\left(140x^3 - 60x\right)$	$\dfrac{1}{8}\left(420x^2 - 60\right)$

For $J = 0$,

$$\left(1 - x^2\right)\frac{d^2 P}{dx^2} - 2x\frac{dP}{dx} + J(J+1)P(x) = \left(1 - x^2\right)(0) - 2x(0) + 0(1)(1) = 0$$

For $J = 1$,

$$\left(1 - x^2\right)\frac{d^2 P}{dx^2} - 2x\frac{dP}{dx} + J(J+1)P(x) = \left(1 - x^2\right)(0) - 2x(1) + 1(2)(x) = 0$$

For $J = 2$,

$$\left(1 - x^2\right)\frac{d^2 P}{dx^2} - 2x\frac{dP}{dx} + J(J+1)P(x) = \left(1 - x^2\right)(3) - 2x(3x) + 2(3)\left(\frac{3x^2 - 1}{2}\right)$$

$$= 3 - 3x^2 - 6x^2 + 9x^2 - 3 = 0$$

For $J = 3$,

$$\left(1 - x^2\right)\frac{d^2 P}{dx^2} - 2x\frac{dP}{dx} + J(J+1)P(x)$$

$$= \left(1 - x^2\right)(15x) - 2x\left(\frac{15x^2 - 3}{2}\right) + 3(4)\left(\frac{5x^3 - 3x}{2}\right)$$

$$= 15x - 15x^3 - 15x^3 + 3x + 30x^3 - 18x = 0$$

For $J = 4$,

$$\left(1 - x^2\right)\frac{d^2 P}{dx^2} - 2x\frac{dP}{dx} + J(J+1)P(x)$$

$$= \left(1 - x^2\right)\left(\frac{420x^2 - 60}{8}\right) - 2x\left(\frac{140x^3 - 60x}{8}\right) + 4(5)\left(\frac{35x^4 - 30x^2 + 3}{8}\right)$$

$$= \frac{105}{2}x^2 - \frac{15}{2} - \frac{105}{2}x^4 + \frac{15}{2}x^2 - 35x^4 + 15x^2 + \frac{175}{2}x^4 - 75x^2 + \frac{15}{2} = 0$$

6–29. Show that the orthogonality integral for the Legendre polynomials, Equation 6.49, is equivalent to

$$\int_0^\pi P_l(\cos\theta)P_n(\cos\theta)\sin\theta\,d\theta = 0 \qquad l \neq n$$

Equation 6.49 is

$$\int_{-1}^{1} P_J(x) P_{J'}(x)\, dx = 0$$

for $J \neq J'$. Let $x = \cos\theta$. Then

$$dx = -\sin\theta\, d\theta$$

The limits in terms of θ are π when $x = -1$ and 0 when $x = 1$. Rewriting J and J' as l and n, respectively, the orthogonality integral becomes, for $l \neq n$,

$$\int_{\pi}^{0} P_l(\cos\theta) P_n(\cos\theta)\, (-\sin\theta)\, d\theta = 0$$

Switching the limits, the integral becomes

$$\int_{0}^{\pi} P_l(\cos\theta) P_n(\cos\theta)\, \sin\theta\, d\theta = 0$$

6–30. Show that the Legendre polynomials given in Table 6.3 satisfy the orthogonality and normalization conditions given by Equations 6.49 and 6.50.

The orthogonality relation is

$$\int_{-1}^{1} P_J(x) P_{J'}(x)\, dx = 0$$

for $J \neq J'$, whereas the normalization relation is

$$\int_{-1}^{1} \left[P_J(x) \right]^2\, dx = \frac{2}{2J+1}$$

Explicit integration shows that the above results are indeed obtained, although some simplification is possible by noting that P_J is an odd function when J is odd, and is an even function when J is even, and recalling that the integral of an odd function over an interval symmetric about $x = 0$ is 0. Thus, $\int_{-1}^{1} P_l(x) P_n(x)\, dx = 0$ when $J + J'$ is odd. For $J \neq J'$,

$$\int_{-1}^{1} P_0(x)P_1(x)\,dx = 0$$

$$\int_{-1}^{1} P_0(x)P_2(x)\,dx = \frac{1}{2}\int_{-1}^{1}\left(3x^2 - 1\right)\,dx = \frac{1}{2}\left(x^3 - x\right)\Big|_{-1}^{1} = 0$$

$$\int_{-1}^{1} P_0(x)P_3(x)\,dx = 0$$

$$\int_{-1}^{1} P_0(x)P_4(x)\,dx = 0 = \frac{1}{8}\int_{-1}^{1}\left(35x^4 - 30x^2 + 3\right)\,dx = \frac{1}{8}\left(7x^5 - 10x^3 + 3x\right)\Big|_{-1}^{1} = 0$$

$$\int_{-1}^{1} P_1(x)P_2(x)\,dx = 0$$

$$\int_{-1}^{1} P_1(x)P_3(x)\,dx = \frac{1}{2}\int_{-1}^{1}\left(5x^4 - 3x^2\right)\,dx = \frac{1}{2}\left(x^5 - x^3\right)\Big|_{-1}^{1} = 0$$

$$\int_{-1}^{1} P_1(x)P_4(x)\,dx = 0$$

$$\int_{-1}^{1} P_2(x)P_3(x)\,dx = 0$$

$$\int_{-1}^{1} P_2(x)P_4(x)\,dx = \frac{1}{16}\int_{-1}^{1}\left(105x^6 - 125x^4 + 39x^2 - 3\right)\,dx$$

$$= \frac{1}{16}\left(15x^7 - 25x^5 + 13x^3 - 3x\right)\Big|_{-1}^{1} = 0$$

$$\int_{-1}^{1} P_3(x)P_4(x)\,dx = 0$$

For $l = n$,

$$\int_{-1}^{1} \left[P_0(x)\right]^2\,dx = \int_{-1}^{1} dx = 2$$

$$\int_{-1}^{1} \left[P_1(x)\right]^2\,dx = \int_{-1}^{1} x^2\,dx = \frac{x^3}{3}\Big|_{-1}^{1} = \frac{2}{3}$$

$$\int_{-1}^{1} \left[P_2(x)\right]^2\,dx = \frac{1}{4}\int_{-1}^{1}\left(9x^4 - 6x^2 + 1\right)\,dx = \frac{1}{4}\left(\frac{9x^5}{5} - 2x^3 + x\right)\Big|_{-1}^{1} = \frac{2}{5}$$

$$\int_{-1}^{1} \left[P_3(x)\right]^2\,dx = \frac{1}{4}\int_{-1}^{1}\left(25x^6 - 30x^4 + 9x^2\right)\,dx = \frac{1}{4}\left(\frac{25x^7}{7} - 6x^5 + 3x^3\right)\Big|_{-1}^{1} = \frac{2}{7}$$

$$\int_{-1}^{1} \left[P_4(x)\right]^2\,dx = \frac{1}{64}\int_{-1}^{1}\left(1225x^8 - 2100x^6 + 1110x^4 - 180x^2 + 9\right)\,dx$$

$$= \frac{1}{64}\left(\frac{1225x^9}{9} - 300x^7 + 222x^5 - 60x^3 + 9x\right)\Big|_{-1}^{1} = \frac{2}{9}$$

6–31. Use Equation 6.51 to generate the associated Legendre functions in Table 6.4.

According to Equation 6.51,

$$P_J^{|m|}(x) = (1 - x^2)^{|m|/2} \frac{d^{|m|}}{dx^{|m|}} P_J(x)$$

For $m = 0$,

$$\frac{d^0}{dx^0} P_J(x) = P_J(x)$$

Thus,

$$P_J^0(x) = P_J(x)$$

which can be verified by comparing the associated Legendre functions, $P_J^0(x)$, in Table 6.4, with the Legendre polynomials, $P_J(x)$, in Table 6.3. For $m = 1$, $P_J^1(x) = (1 - x^2)^{1/2} d/dx \left[P_J(x) \right]$. Using the first derivatives of the Legendre polynomials calculated in Problem 6–28,

$$P_1^1(x) = (1 - x^2)^{1/2}$$

$$P_2^1(x) = 3x(1 - x^2)^{1/2}$$

$$P_3^1(x) = (1 - x^2)^{1/2} \left[\frac{1}{2} \left(15x^2 - 3 \right) \right] = \frac{3}{2} \left(5x^2 - 1 \right) (1 - x^2)^{1/2}$$

For $m = 2$, $P_J^2(x) = (1 - x^2) d^2/dx^2 \left[P_J(x) \right]$. Using the second derivatives of the Legendre polynomials calculated in Problem 6–28,

$$P_2^2(x) = 3(1 - x^2)$$

$$P_3^2(x) = 15x(1 - x^2)$$

Lastly,

$$P_3^3(x) = (1 - x^2)^{3/2} \frac{d^3}{dx^3} P_3(x)$$

$$= 15(1 - x^2)^{3/2}$$

6–32. Show that the first few associated Legendre functions given in Table 6.4 are solutions to Equation 6.48 and that they satisfy the orthonormality condition (Equation 6.53).

To show that the functions are solutions to Equation 6.48, show that the expression

$$\left(1 - x^2 \right) \frac{d^2 P}{dx^2} - 2x \frac{dP}{dx} + \left[J \left(J + 1 \right) - \frac{m^2}{1 - x^2} \right] P(x)$$

equals 0. For the function $P_0^0(x) = 1$,

$$\left(1 - x^2\right) \frac{d^2 P}{dx^2} - 2x \frac{dP}{dx} + \left[J\,(J+1) - \frac{m^2}{1-x^2}\right] P(x)$$

$$= \left(1 - x^2\right)(0) - 2x\,(0) + \left[0\,(1) - \frac{0}{1-x^2}\right](1) = 0$$

For the function $P_1^0(x) = x$,

$$\left(1 - x^2\right) \frac{d^2 P}{dx^2} - 2x \frac{dP}{dx} + \left[J\,(J+1) - \frac{m^2}{1-x^2}\right] P(x)$$

$$= \left(1 - x^2\right)(0) - 2x\,(1) + \left[1\,(2) - \frac{0}{1-x^2}\right](x) = 0$$

For the function $P_1^1(x) = (1 - x^2)^{1/2}$,

$$\left(1 - x^2\right) \frac{d^2 P}{dx^2} - 2x \frac{dP}{dx} + \left[J\,(J+1) - \frac{m^2}{1-x^2}\right] P(x)$$

$$= \left(1 - x^2\right)\left[-\frac{1}{(1-x^2)^{1/2}} - \frac{x^2}{(1-x^2)^{3/2}}\right] - 2x\left[-\frac{x}{(1-x^2)^{1/2}}\right] + \left[1\,(2) - \frac{1}{1-x^2}\right](1-x^2)^{1/2}$$

$$= -\left(1 - x^2\right)^{1/2} - \frac{x^2}{(1-x^2)^{1/2}} + \frac{2x^2}{(1-x^2)^{1/2}} + 2\left(1 - x^2\right)^{1/2} - \frac{1}{(1-x^2)^{1/2}}$$

$$= \frac{x^2 - 1}{(1-x^2)^{1/2}} + (1-x^2)^{1/2} = 0$$

The orthonormality condition for the associated Legendre functions is

$$\int_{-1}^{1} P_J^{|m|}(x) P_{J'}^{|m|}(x)\, dx = \frac{2}{2J+1} \frac{(J+|m|)!}{(J-|m|)!} \delta_{JJ'}$$

Use P_1^1 and P_2^1 as examples to show that this condition is satisfied.

$$\int_{-1}^{1} \left[P_1^1(x)\right]^2 dx = \int_{-1}^{1} \left(1 - x^2\right) dx = x - \frac{x^3}{3}\bigg|_{-1}^{1} = \frac{4}{3} = \frac{2}{3}\left(\frac{2!}{0!}\right)$$

$$\int_{-1}^{1} \left[P_2^1(x)\right]^2 dx = \int_{-1}^{1} \left(9x^2 - 9x^4\right) dx = 3x^3 - \frac{9x^5}{5}\bigg|_{-1}^{1} = \frac{12}{5} = \frac{2}{5}\left(\frac{3!}{1!}\right)$$

$$\int_{-1}^{1} P_1^1(x) P_2^1(x)\, dx = \int_{-1}^{1} \left(3x - 3x^3\right) dx = 0$$

The last integral vanishes because the integrand is an odd function, and the integration limits are symmetric about 0.

6–33. There are a number of recursion formulas for the associated Legendre functions. Show that the first few associated Legendre functions in Table 6.4 satisfy the recursion formula (Equation 6.66)

$$(2l + 1)x P_l^{|m|}(x) = (l - |m| + 1) P_{l+1}^{|m|}(x) + (l + |m|) P_{l-1}^{|m|}(x)$$

For $l = 1$ and $m = 0$,

$$(2l + 1)x P_l^{|m|}(x) = 3x P_1^0(x) = 3x^2$$

$$(l - |m| + 1) P_{l+1}^{|m|}(x) + (l + |m|) P_{l-1}^{|m|}(x) = 2P_2^0(x) + P_0^0 = 3x^2 - 1 + 1 = 3x^2$$

For $l = 1$ and $m = 1$,

$$(2l + 1)x P_l^{|m|}(x) = 3x P_1^1(x) = 3x(1 - x^2)^{1/2}$$

$$(l - |m| + 1) P_{l+1}^{|m|}(x) + (l + |m|) P_{l-1}^{|m|}(x) = P_2^1(x) + 2P_0^1 = 3x(1 - x^2)^{1/2}$$

Note that $P_0^1 = 0$, since m cannot be greater than l.

For $l = 2$ and $m = 0$,

$$(2l + 1)x P_l^{|m|}(x) = 5x P_2^0(x) = \frac{5}{2} \left(3x^3 - x\right)$$

$$(l - |m| + 1) P_{l+1}^{|m|}(x) + (l + |m|) P_{l-1}^{|m|}(x) = 3P_3^0(x) + 2P_1^0 = \frac{3}{2} \left(5x^3 - 3x\right) + 2x = \frac{5}{2} \left(3x^3 - x\right)$$

For $l = 2$ and $m = 1$,

$$(2l + 1)x P_l^{|m|}(x) = 5x P_2^1(x) = 15x^2 \left(1 - x^2\right)^{1/2}$$

$$(l - |m| + 1) P_{l+1}^{|m|}(x) + (l + |m|) P_{l-1}^{|m|}(x) = 2P_3^1(x) + 3P_1^1$$

$$= 3 \left(5x^2 - 1\right) \left(1 - x^2\right)^{1/2} + 3 \left(1 - x^2\right)^{1/2} = 15x^2 \left(1 - x^2\right)^{1/2}$$

For $l = 2$ and $m = 2$,

$$(2l + 1)x P_l^{|m|}(x) = 5x P_2^2(x) = 15x \left(1 - x^2\right)$$

$$(l - |m| + 1) P_{l+1}^{|m|}(x) + (l + |m|) P_{l-1}^{|m|}(x) = P_3^2(x) + 4P_1^2 = 15x \left(1 - x^2\right)$$

Note that $P_1^2 = 0$, since m cannot be greater than l.

6–34. Show that the factor $i^{m+|m|} = 1$ when m is odd and negative and that it equals -1 when m is odd and positive.

When m is odd and negative, $|m| = -m$. Therefore,

$$i^{m+|m|} = i^0 = 1$$

On the other hand, when m is odd and positive, $|m| = m$, giving

$$i^{m+|m|} = i^{2m}$$

Substitute $m = 2n + 1$ into the above equation, where $n = 0, 1, 2, \cdots$,

$$i^{m+|m|} = i^{4n+2} = i^{4n}i^2 = (1)(-1) = -1$$

6–35. Show that the integral of $\sin\theta \, d\theta d\phi$ over the surface of a sphere is equal to 4π.

The surface of a sphere is described by $0 \le \theta \le \pi$ and $0 \le \phi \le 2\pi$.

$$\int_0^{2\pi} d\phi \int_0^\pi d\theta \, \sin\theta = (\phi)|_0^{2\pi} \, (-\cos\theta)|_0^\pi$$

$$= (2\pi)(1+1)$$

$$= 4\pi$$

6–36. Show that the first few spherical harmonics in Table 6.5 satisfy the orthonormality condition (Equation 6.57).

The orthonormality relation is

$$\int_0^\pi d\theta \, \sin\theta \int_0^{2\pi} d\phi \, Y_J^m(\theta, \phi)^* Y_{J'}^k(\theta, \phi) = \delta_{JJ'}\delta_{mk}$$

The normalization condition is

$$\int_0^{2\pi} d\phi \int_0^\pi d\theta \, \sin\theta \left| Y_0^0(\theta, \phi) \right|^2 = \frac{1}{4\pi} \int_0^{2\pi} d\phi \int_0^\pi d\theta \, \sin\theta$$

$$= \frac{1}{4\pi} (2\pi)(2) = 1$$

$$\int_0^{2\pi} d\phi \int_0^\pi d\theta \, \sin\theta \left| Y_1^0(\theta, \phi) \right|^2 = \frac{3}{4\pi} \int_0^{2\pi} d\phi \int_0^\pi d\theta \, \sin\theta \cos^2\theta$$

$$= \left(\frac{3}{4\pi}\right)(2\pi)\left(-\frac{\cos^3\theta}{3}\right)\Bigg|_0^\pi = 1$$

$$\int_0^{2\pi} d\phi \int_0^\pi d\theta \, \sin\theta \left| Y_1^1(\theta, \phi) \right|^2 = \int_0^{2\pi} d\phi \int_0^\pi d\theta \, \sin\theta \left| Y_1^{-1}(\theta, \phi) \right|^2$$

$$= \frac{3}{8\pi} \int_0^{2\pi} d\phi \int_0^\pi d\theta \, \sin\theta \sin^2\theta$$

$$= \left(\frac{3}{8\pi}\right)(2\pi)\left[-\frac{1}{3}\cos\theta\left(\sin^2\theta + 2\right)\right]\Bigg|_0^\pi = 1$$

The orthogonality condition is

$$\int_0^{2\pi} d\phi \int_0^{\pi} d\theta \sin\theta\, Y_0^0(\theta, \phi)^* Y_1^0(\theta, \phi) = \frac{3^{1/2}}{4\pi} \int_0^{2\pi} d\phi \int_0^{\pi} d\theta \sin\theta \cos\theta$$

$$= \left(\frac{3^{1/2}}{4\pi}\right)(2\pi)\left(\frac{\sin^2\theta}{2}\right)\Bigg|_0^{\pi} = 0$$

$$\int_0^{2\pi} d\phi \int_0^{\pi} d\theta \sin\theta\, Y_0^0(\theta, \phi)^* Y_1^{\pm 1}(\theta, \phi) = \mp\left(\frac{3}{32\pi^2}\right)^{1/2}\int_0^{2\pi} e^{\pm i\phi}d\phi \int_0^{\pi} d\theta \sin^2\theta = 0$$

$$\int_0^{2\pi} d\phi \int_0^{\pi} d\theta \sin\theta\, Y_1^0(\theta, \phi)^* Y_1^{\pm 1}(\theta, \phi) = \mp\left(\frac{9}{32\pi^2}\right)^{1/2}\int_0^{2\pi} e^{\pm i\phi}d\phi \int_0^{\pi} d\theta \sin^2\theta \cos\theta$$

$$= 0$$

The last two integrals vanish, since $\displaystyle\int_0^{2\pi} d\phi\, e^{\pm i\phi} = 0$.

6–37. Using explicit expressions for $Y_l^m(\theta, \phi)$, show that

$$|Y_1^1(\theta, \phi)|^2 + |Y_1^0(\theta, \phi)|^2 + |Y_{-1}^1(\theta, \phi)|^2 = \text{constant}$$

This is a special case of the general theorem

$$\sum_{m=-l}^{+l} |Y_l^m(\theta, \phi)|^2 = \text{constant}$$

known as Unsöld's theorem. What is the physical significance of this result?

$$|Y_1^1(\theta, \phi)|^2 + |Y_1^0(\theta, \phi)|^2 + |Y_{-1}^1(\theta, \phi)|^2 = \frac{3}{8\pi}\sin^2\theta + \frac{3}{4\pi}\cos^2\theta + \frac{3}{8\pi}\sin^2\theta$$

$$= \frac{3}{4\pi}\left(\sin^2\theta + \cos^2\theta\right)$$

$$= \frac{3}{4\pi}$$

The magnitudes of spherical harmonics $Y_l^m(\theta, \phi)^* Y_l^m(\theta, \phi)$ describe the probability densities of locating the system on a spherical surface. Thus, the sum of these probability densities over the sphere at a fixed l must be a constant if the potential energy is isotropic.

6–38. In cartesian coordinates,

$$\hat{L}_z = -i\hbar\left(x\frac{\partial}{\partial y} - y\frac{\partial}{\partial x}\right)$$

Convert this equation to spherical coordinates, showing that

$$\hat{L}_z = -i\hbar\frac{\partial}{\partial\phi}$$

Write the spherical coordinates in terms of cartesian coordinates,

$$r = \left(x^2 + y^2 + z^2\right)^{1/2}$$

$$\theta = \cos^{-1}\left[\frac{z}{\left(x^2 + y^2 + z^2\right)^{1/2}}\right]$$

$$\phi = \tan^{-1}\frac{y}{x}$$

Then carry out the conversion using the chain rule of partial differentiation:

$$\frac{\partial f}{\partial x} = \left(\frac{\partial f}{\partial r}\right)\left(\frac{\partial r}{\partial x}\right) + \left(\frac{\partial f}{\partial \theta}\right)\left(\frac{\partial \theta}{\partial x}\right) + \left(\frac{\partial f}{\partial \phi}\right)\left(\frac{\partial \phi}{\partial x}\right)$$

$$\frac{\partial f}{\partial y} = \left(\frac{\partial f}{\partial r}\right)\left(\frac{\partial r}{\partial y}\right) + \left(\frac{\partial f}{\partial \theta}\right)\left(\frac{\partial \theta}{\partial y}\right) + \left(\frac{\partial f}{\partial \phi}\right)\left(\frac{\partial \phi}{\partial y}\right)$$

Applying the relations between cartesian and spherical coordinates, the above expressions become

$$\frac{\partial f}{\partial x} = \sin\theta\cos\phi\left(\frac{\partial f}{\partial r}\right) + \frac{\cos\theta\cos\phi}{r}\left(\frac{\partial f}{\partial \theta}\right) - \frac{\sin\phi}{r\sin\theta}\left(\frac{\partial f}{\partial \phi}\right)$$

$$\frac{\partial f}{\partial y} = \sin\theta\sin\phi\left(\frac{\partial f}{\partial r}\right) + \frac{\cos\theta\sin\phi}{r}\left(\frac{\partial f}{\partial \theta}\right) + \frac{\cos\phi}{r\sin\theta}\left(\frac{\partial f}{\partial \phi}\right)$$

Substitution into the $\hat{L}_z$ expression gives

$$\hat{L}_z f = -i\hbar\left(x\frac{\partial f}{\partial y} - y\frac{\partial f}{\partial x}\right)$$

$$= -i\hbar\left[r\sin\theta\cos\phi\left(\frac{\partial f}{\partial y}\right) - r\sin\theta\sin\phi\left(\frac{\partial f}{\partial x}\right)\right]$$

$$= -i\hbar\left[\left(r\sin^2\theta\cos\phi\sin\phi - r\sin^2\theta\cos\phi\sin\phi\right)\left(\frac{\partial f}{\partial r}\right)\right.$$

$$+ \left(\sin\theta\cos\theta\cos\phi\sin\phi - \sin\theta\cos\theta\sin\phi\cos\phi\right)\left(\frac{\partial f}{\partial \theta}\right)$$

$$+ \left.\left(\cos^2\phi + \sin^2\phi\right)\left(\frac{\partial f}{\partial \phi}\right)\right]$$

$$= -i\hbar\frac{\partial f}{\partial \phi}$$

Thus, $\hat{L}_z = -i\hbar\,\partial/\partial\phi$.

6–39. Convert $\hat{L}_x$ and $\hat{L}_y$ from cartesian coordinates to spherical coordinates.

In addition to the expressions for $\partial f/\partial x$ and $\partial f/\partial y$ derived in the last problem, the following expression is also required:

$$\frac{\partial f}{\partial z} = \left(\frac{\partial f}{\partial r}\right)\left(\frac{\partial r}{\partial z}\right) + \left(\frac{\partial f}{\partial \theta}\right)\left(\frac{\partial \theta}{\partial z}\right) + \left(\frac{\partial f}{\partial \phi}\right)\left(\frac{\partial \phi}{\partial z}\right)$$

$$= \cos\theta\left(\frac{\partial f}{\partial r}\right) - \frac{\sin\theta}{r}\left(\frac{\partial f}{\partial \theta}\right)$$

Therefore

$$\hat{L}_x f = -i\hbar\left[y\frac{\partial f}{\partial z} - z\frac{\partial f}{\partial y}\right]$$

$$= -i\hbar\left[(r\sin\theta\cos\theta\sin\phi - r\sin\theta\cos\theta\sin\phi)\frac{\partial f}{\partial r} + (-\sin^2\theta\sin\phi - \cos^2\theta\sin\phi)\frac{\partial f}{\partial \theta}\right.$$

$$\left. - \cot\theta\cos\phi\frac{\partial f}{\partial \phi}\right]$$

$$= -i\hbar\left(-\sin\phi\frac{\partial f}{\partial \theta} - \cot\theta\cos\phi\frac{\partial f}{\partial \phi}\right)$$

$$\hat{L}_x = -i\hbar\left(-\sin\phi\frac{\partial}{\partial \theta} - \cot\theta\cos\phi\frac{\partial}{\partial \phi}\right)$$

and

$$\hat{L}_y f = -i\hbar\left[z\frac{\partial f}{\partial x} - x\frac{\partial f}{\partial z}\right]$$

$$= -i\hbar\left[(r\cos\theta\sin\theta\cos\phi - r\cos\theta\sin\theta\cos\phi)\frac{\partial f}{\partial r} + (\cos^2\theta\cos\phi - \sin^2\theta\cos\phi)\frac{\partial f}{\partial \theta}\right.$$

$$\left. - \cot\theta\sin\phi\frac{\partial f}{\partial \phi}\right]$$

$$= -i\hbar\left(\cos\phi\frac{\partial f}{\partial \theta} - \cot\theta\sin\phi\frac{\partial f}{\partial \phi}\right)$$

$$\hat{L}_y = -i\hbar\left(\cos\phi\frac{\partial}{\partial \theta} - \cot\theta\sin\phi\frac{\partial}{\partial \phi}\right)$$

6–40. Compute the value of $\hat{L}^2 Y(\theta, \phi)$ for the following functions:

(a) $1/(4\pi)^{1/2}$ (b) $(3/4\pi)^{1/2}\cos\theta$

(c) $(3/8\pi)^{1/2}\sin\theta e^{i\phi}$ (d) $(3/8\pi)^{1/2}\sin\theta e^{-i\phi}$

Do you find anything interesting about the results?

The $\hat{L}^2$ operator is

$$\hat{L}^2 = -\hbar^2 \left[\frac{1}{\sin\theta} \frac{\partial}{\partial\theta} \left(\sin\theta \frac{\partial}{\partial\theta} \right) + \frac{1}{\sin^2\theta} \frac{\partial^2}{\partial\phi^2} \right]$$

(a)

$$\hat{L}^2 \left[\frac{1}{(4\pi)^{1/2}} \right] = -\hbar^2 \left[\frac{1}{\sin\theta} \frac{\partial}{\partial\theta} \left(\sin\theta \frac{\partial}{\partial\theta} \right) + \frac{1}{\sin^2\theta} \frac{\partial^2}{\partial\phi^2} \right] \left[\frac{1}{(4\pi)^{1/2}} \right]$$

$$= 0 = 0 \cdot \left[\frac{1}{(4\pi)^{1/2}} \right]$$

(b)

$$\hat{L}^2 \left[\left(\frac{3}{4\pi} \right)^{1/2} \cos\theta \right] = -\hbar^2 \left[\frac{1}{\sin\theta} \frac{\partial}{\partial\theta} \left(\sin\theta \frac{\partial}{\partial\theta} \right) + \frac{1}{\sin^2\theta} \frac{\partial^2}{\partial\phi^2} \right] \left[\left(\frac{3}{4\pi} \right)^{1/2} \cos\theta \right]$$

$$= -\hbar^2 \left(\frac{3}{4\pi} \right)^{1/2} \left[\frac{1}{\sin\theta} \frac{\partial}{\partial\theta} \left(-\sin^2\theta \right) \right]$$

$$= 2\hbar^2 \left[\left(\frac{3}{4\pi} \right)^{1/2} \cos\theta \right]$$

(c)

$$\hat{L}^2 \left[\left(\frac{3}{8\pi} \right)^{1/2} \sin\theta e^{i\phi} \right] = -\hbar^2 \left[\frac{1}{\sin\theta} \frac{\partial}{\partial\theta} \left(\sin\theta \frac{\partial}{\partial\theta} \right) + \frac{1}{\sin^2\theta} \frac{\partial^2}{\partial\phi^2} \right] \left[\left(\frac{3}{8\pi} \right)^{1/2} \sin\theta e^{i\phi} \right]$$

$$= -\hbar^2 \left(\frac{3}{8\pi} \right)^{1/2} \left[e^{i\phi} \frac{1}{\sin\theta} \frac{\partial}{\partial\theta} (\sin\theta \cos\theta) + \frac{1}{\sin\theta} \left(-e^{i\phi} \right) \right]$$

$$= -\hbar^2 \left(\frac{3}{8\pi} \right)^{1/2} \left[\frac{1}{\sin\theta} \left(\cos^2\theta - \sin^2\theta \right) - \frac{1}{\sin\theta} \right] e^{i\phi}$$

$$= -\hbar^2 \left(\frac{3}{8\pi} \right)^{1/2} \left[\frac{1}{\sin\theta} \left(1 - 2\sin^2\theta \right) - \frac{1}{\sin\theta} \right] e^{i\phi}$$

$$= 2\hbar^2 \left[\left(\frac{3}{8\pi} \right)^{1/2} \sin\theta e^{i\phi} \right]$$

(d)

$$\hat{L}^2 \left[\left(\frac{3}{8\pi} \right)^{1/2} \sin\theta e^{-i\phi} \right] = -\hbar^2 \left[\frac{1}{\sin\theta} \frac{\partial}{\partial\theta} \left(\sin\theta \frac{\partial}{\partial\theta} \right) + \frac{1}{\sin^2\theta} \frac{\partial^2}{\partial\phi^2} \right] \left[\left(\frac{3}{8\pi} \right)^{1/2} \sin\theta e^{-i\phi} \right]$$

$$= -\hbar^2 \left(\frac{3}{8\pi} \right)^{1/2} \left[e^{-i\phi} \frac{1}{\sin\theta} \frac{\partial}{\partial\theta} (\sin\theta \cos\theta) + \frac{1}{\sin\theta} \left(-e^{-i\phi} \right) \right]$$

$$= -\hbar^2 \left(\frac{3}{8\pi} \right)^{1/2} \left[\frac{1}{\sin\theta} \left(\cos^2\theta - \sin^2\theta \right) - \frac{1}{\sin\theta} \right] e^{-i\phi}$$

$$= -\hbar^2 \left(\frac{3}{8\pi} \right)^{1/2} \left[\frac{1}{\sin\theta} \left(1 - 2\sin^2\theta \right) - \frac{1}{\sin\theta} \right] e^{-i\phi}$$

$$= 2\hbar^2 \left[\left(\frac{3}{8\pi} \right)^{1/2} \sin\theta e^{-i\phi} \right]$$

These functions are spherical harmonics. Specifically, (a) is Y_0^0, (b) is Y_1^0, (c) is $-Y_1^1$ (the multiplicative constant -1 changes only the phase of the function), and (d) is Y_1^{-1}. As expected, they are eigenfunctions of $\hat{L}^2$, with eigenvalues $J(J+1)\hbar^2$.

6–41. Prove that $\hat{L}^2$ commutes with $\hat{L}_x$, $\hat{L}_y$, and $\hat{L}_z$ but that

$$[\hat{L}_x, \hat{L}_y] = i\hbar\hat{L}_z \qquad [\hat{L}_y, \hat{L}_z] = i\hbar\hat{L}_x \qquad [\hat{L}_z, \hat{L}_x] = i\hbar\hat{L}_y$$

(*Hint:* Use cartesian coordinates.) Do you see a pattern in these formulas?

First consider the commutator $\left[\hat{L}^2, \hat{L}_x\right]$:

$$\left[\hat{L}^2, \hat{L}_x\right] = \left[\hat{L}_x^2 + \hat{L}_y^2 + \hat{L}_z^2, \hat{L}_x\right]$$

$$= \left[\hat{L}_x^2, \hat{L}_x\right] + \left[\hat{L}_y^2, \hat{L}_x\right] + \left[\hat{L}_z^2, \hat{L}_x\right]$$

Since $\hat{L}_x^2$ commutes with $\hat{L}_x$, the commutator $\left[\hat{L}_x^2, \hat{L}_x\right]$ is zero. Using, for example, $i\hbar\hat{L}_z = \hat{L}_x\hat{L}_y - \hat{L}_y\hat{L}_x = \left[\hat{L}_x, \hat{L}_y\right]$, the other two commutators in the last expression can be written as

$$\left[\hat{L}_y^2, \hat{L}_x\right] = \hat{L}_y^2\hat{L}_x - \hat{L}_x\hat{L}_y\hat{L}_y$$

$$= \hat{L}_y^2\hat{L}_x - \left(\hat{L}_y\hat{L}_x + i\hbar\hat{L}_z\right)\hat{L}_y$$

$$= \hat{L}_y\left(\hat{L}_y\hat{L}_x - \hat{L}_x\hat{L}_y\right) - i\hbar\hat{L}_z\hat{L}_y$$

$$= -i\hbar\hat{L}_y\hat{L}_z - i\hbar\hat{L}_z\hat{L}_y$$

$$\left[\hat{L}_z^2, \hat{L}_x\right] = \hat{L}_z^2\hat{L}_x - \hat{L}_x\hat{L}_z\hat{L}_z$$

$$= \hat{L}_z^2\hat{L}_x - \left(\hat{L}_z\hat{L}_x - i\hbar\hat{L}_y\right)\hat{L}_z$$

$$= \hat{L}_z\left(\hat{L}_z\hat{L}_x - \hat{L}_x\hat{L}_z\right) + i\hbar\hat{L}_y\hat{L}_z$$

$$= i\hbar\hat{L}_z\hat{L}_y + i\hbar\hat{L}_y\hat{L}_z$$

where the commutation relations in Problem 4–12 have been applied. Combining the three comutation relations between the square of the operator of each angular momentum component and $\hat{L}_x$,

$$\left[\hat{L}^2, \hat{L}_x\right] = 0$$

Similarly, $\left[\hat{L}^2, \hat{L}_y\right] = 0$ because

$$\left[\hat{L}_y^2, \hat{L}_y\right] = 0$$

$$\left[\hat{L}_x^2, \hat{L}_y\right] = -\left[\hat{L}_z^2, \hat{L}_y\right]$$

and $\left[\hat{L}^2, \hat{L}_z\right] = 0$ because

$$\left[\hat{L}_z^2, \hat{L}_z\right] = 0$$

$$\left[\hat{L}_x^2, \hat{L}_z\right] = -\left[\hat{L}_y^2, \hat{L}_z\right]$$

Thus, $\hat{L}^2$ commutes with $\hat{L}_x$, $\hat{L}_y$, and $\hat{L}_z$.

Now consider the commutation relation beween two components of the angular momentum operator.

$$\hat{L}_x\hat{L}_y f = \left[-i\hbar\left(y\frac{\partial}{\partial z} - z\frac{\partial}{\partial y}\right)\right]\left[-i\hbar\left(z\frac{\partial f}{\partial x} - x\frac{\partial f}{\partial z}\right)\right]$$

$$= -\hbar^2\left(y\frac{\partial f}{\partial x} + yz\frac{\partial^2 f}{\partial x \partial z} - xy\frac{\partial^2 f}{\partial z^2} - z^2\frac{\partial^2 f}{\partial x \partial y} + xz\frac{\partial^2 f}{\partial y \partial z}\right)$$

$$\hat{L}_y\hat{L}_x f = \left[-i\hbar\left(z\frac{\partial}{\partial x} - x\frac{\partial}{\partial z}\right)\right]\left[-i\hbar\left(y\frac{\partial f}{\partial z} - z\frac{\partial f}{\partial y}\right)\right]$$

$$= -\hbar^2\left(yz\frac{\partial^2 f}{\partial x \partial z} - z^2\frac{\partial^2 f}{\partial x \partial y} - xy\frac{\partial^2 f}{\partial z^2} + x\frac{\partial f}{\partial y} + xz\frac{\partial^2 f}{\partial y \partial z}\right)$$

Therefore,

$$\hat{L}_x\hat{L}_y f - \hat{L}_y\hat{L}_x f = -\hbar^2\left(y\frac{\partial f}{\partial x} - x\frac{\partial f}{\partial y}\right)$$

$$= -\hbar^2\left(\frac{\hat{L}_z}{i\hbar}\right)$$

$$= i\hbar\hat{L}_z f$$

$$\left[\hat{L}_x, \hat{L}_y\right] = i\hbar\hat{L}_z$$

The other two commutators are evaluated in a similar manner.

$$\hat{L}_y \hat{L}_z f = \left[-i\hbar \left(z\frac{\partial}{\partial x} - x\frac{\partial}{\partial z} \right) \right] \left[-i\hbar \left(x\frac{\partial f}{\partial y} - y\frac{\partial f}{\partial x} \right) \right]$$

$$= -\hbar^2 \left(z\frac{\partial f}{\partial y} + xz\frac{\partial^2 f}{\partial x \partial y} - yz\frac{\partial^2 f}{\partial x^2} - x^2\frac{\partial^2 f}{\partial y \partial z} + xy\frac{\partial^2 f}{\partial x \partial z} \right)$$

$$\hat{L}_z \hat{L}_y f = \left[-i\hbar \left(x\frac{\partial}{\partial y} - y\frac{\partial}{\partial x} \right) \right] \left[-i\hbar \left(z\frac{\partial f}{\partial x} - x\frac{\partial f}{\partial z} \right) \right]$$

$$= -\hbar^2 \left(xz\frac{\partial^2 f}{\partial x \partial y} - x^2\frac{\partial^2 f}{\partial y \partial z} - yz\frac{\partial^2 f}{\partial x^2} + y\frac{\partial f}{\partial z} + xy\frac{\partial^2 f}{\partial x \partial z} \right)$$

$$\left[\hat{L}_y, \hat{L}_z \right] f = -\hbar^2 \left(z\frac{\partial f}{\partial y} - y\frac{\partial f}{\partial z} \right)$$

$$\left[\hat{L}_y, \hat{L}_z \right] = i\hbar \hat{L}_x$$

and

$$\hat{L}_z \hat{L}_x f = \left[-i\hbar \left(x\frac{\partial}{\partial y} - y\frac{\partial}{\partial x} \right) \right] \left[-i\hbar \left(y\frac{\partial f}{\partial z} - z\frac{\partial f}{\partial y} \right) \right]$$

$$= -\hbar^2 \left(x\frac{\partial f}{\partial z} + xy\frac{\partial^2 f}{\partial y \partial z} - xz\frac{\partial^2 f}{\partial y^2} - y^2\frac{\partial^2 f}{\partial x \partial z} + yz\frac{\partial^2 f}{\partial x \partial y} \right)$$

$$\hat{L}_x \hat{L}_z f = \left[-i\hbar \left(y\frac{\partial}{\partial z} - z\frac{\partial}{\partial y} \right) \right] \left[-i\hbar \left(x\frac{\partial f}{\partial y} - y\frac{\partial f}{\partial x} \right) \right]$$

$$= -\hbar^2 \left(xy\frac{\partial^2 f}{\partial y \partial z} - y^2\frac{\partial^2 f}{\partial x \partial z} - xz\frac{\partial^2 f}{\partial y^2} + z\frac{\partial f}{\partial x} + yz\frac{\partial^2 f}{\partial x \partial y} \right)$$

$$\left[\hat{L}_z, \hat{L}_x \right] f = -\hbar^2 \left(x\frac{\partial f}{\partial z} - z\frac{\partial f}{\partial x} \right)$$

$$\left[\hat{L}_z, \hat{L}_x \right] = i\hbar \hat{L}_y$$

These commutation relations have a pattern that involves cyclic permutation of x, y, and z.

6–42. It is a somewhat advanced exercise to prove generally that $\langle L_x \rangle = \langle L_y \rangle = 0$, but prove that they are zero at least for the first few l, m states by using the spherical harmonics given in Table 6.5.

The $\hat{L}_x$ and $\hat{L}_y$ operators are

$$\hat{L}_x = -i\hbar \left(-\sin\phi\frac{\partial}{\partial \theta} - \cot\theta\cos\phi\frac{\partial}{\partial \phi} \right)$$

$$\hat{L}_y = -i\hbar \left(\cos\phi\frac{\partial}{\partial \theta} - \cot\theta\sin\phi\frac{\partial}{\partial \phi} \right)$$

For the spherical harmonic $Y_0^0(\theta, \phi) = 1/(4\pi)^{1/2}$,

$$\hat{L}_x \left[\frac{1}{(4\pi)^{1/2}} \right] = -i\hbar \left(-\sin\phi \frac{\partial}{\partial\theta} - \cot\theta \cos\phi \frac{\partial}{\partial\phi} \right) \left[\frac{1}{(4\pi)^{1/2}} \right] = 0$$

$$\hat{L}_y \left[\frac{1}{(4\pi)^{1/2}} \right] = -i\hbar \left(\cos\phi \frac{\partial}{\partial\theta} - \cot\theta \sin\phi \frac{\partial}{\partial\phi} \right) \left[\frac{1}{(4\pi)^{1/2}} \right] = 0$$

Thus,

$$\langle L_x \rangle = \int_0^\pi d\theta \sin\theta \int_0^{2\pi} d\phi \, Y_0^0(\theta, \phi)^* \hat{L}_x Y_0^0(\theta, \phi) = 0$$

$$\langle L_y \rangle = \int_0^\pi d\theta \sin\theta \int_0^{2\pi} d\phi \, Y_0^0(\theta, \phi)^* \hat{L}_y Y_0^0(\theta, \phi) = 0$$

For the spherical harmonic $Y_1^0(\theta, \phi) = [3/(4\pi)]^{1/2} \cos\theta$,

$$\hat{L}_x \left[\left(\frac{3}{4\pi} \right)^{1/2} \cos\theta \right] = -i\hbar \left(-\sin\phi \frac{\partial}{\partial\theta} - \cot\theta \cos\phi \frac{\partial}{\partial\phi} \right) \left[\left(\frac{3}{4\pi} \right)^{1/2} \cos\theta \right]$$

$$= -i\hbar \left(\frac{3}{4\pi} \right)^{1/2} (\sin\phi \sin\theta)$$

$$\hat{L}_y \left[\left(\frac{3}{4\pi} \right)^{1/2} \cos\theta \right] = -i\hbar \left(\cos\phi \frac{\partial}{\partial\theta} - \cot\theta \sin\phi \frac{\partial}{\partial\phi} \right) \left[\left(\frac{3}{4\pi} \right)^{1/2} \cos\theta \right]$$

$$= -i\hbar \left(\frac{3}{4\pi} \right)^{1/2} (-\cos\phi \sin\theta)$$

Thus,

$$\langle L_x \rangle = \int_0^\pi d\theta \sin\theta \int_0^{2\pi} d\phi \, Y_1^0(\theta, \phi)^* \hat{L}_x Y_1^0(\theta, \phi)$$

$$= -i\hbar \left(\frac{3}{4\pi} \right) \int_0^\pi d\theta \sin^2\theta \cos\theta \int_0^{2\pi} d\phi \, \sin\phi = 0$$

$$\langle L_y \rangle = \int_0^\pi d\theta \sin\theta \int_0^{2\pi} d\phi \, Y_1^0(\theta, \phi)^* \hat{L}_y Y_1^0(\theta, \phi)$$

$$= i\hbar \left(\frac{3}{4\pi} \right) \int_0^\pi d\theta \sin^2\theta \cos\theta \int_0^{2\pi} d\phi \, \cos\phi = 0$$

Note that $\int_0^{2\pi} d\phi \, \sin\phi = \int_0^{2\pi} d\phi \, \cos\phi = 0$.

For the spherical harmonic $Y_1^{\pm 1}(\theta, \phi) = [\mp 3/(8\pi)]^{1/2} \sin\theta e^{\pm i\phi}$,

$$\hat{L}_x\left[\mp\left(\frac{3}{8\pi}\right)^{1/2}\sin\theta e^{\pm i\phi}\right]=-i\hbar\left(-\sin\phi\frac{\partial}{\partial\theta}-\cot\theta\cos\phi\frac{\partial}{\partial\phi}\right)\left[\mp\left(\frac{3}{8\pi}\right)^{1/2}\sin\theta e^{\pm i\phi}\right]$$

$$=\pm i\hbar\left(\frac{3}{8\pi}\right)^{1/2}\left(-\sin\phi\cos\theta e^{\pm i\phi}\mp i\cot\theta\cos\phi\sin\theta e^{\pm i\phi}\right)$$

$$=\pm i\hbar\left(\frac{3}{8\pi}\right)^{1/2}e^{\pm i\phi}\left(-\sin\phi\cos\theta\mp i\cos\theta\cos\phi\right)$$

$$\hat{L}_y\left[\mp\left(\frac{3}{8\pi}\right)^{1/2}\sin\theta e^{\pm i\phi}\right]=-i\hbar\left(\cos\phi\frac{\partial}{\partial\theta}-\cot\theta\sin\phi\frac{\partial}{\partial\phi}\right)\left[\mp\left(\frac{3}{8\pi}\right)^{1/2}\sin\theta e^{\pm i\phi}\right]$$

$$=\pm i\hbar\left(\frac{3}{8\pi}\right)^{1/2}\left(-\cos\phi\cos\theta e^{\pm i\phi}\mp i\cot\theta\sin\phi\sin\theta e^{\pm i\phi}\right)$$

$$=\pm i\hbar\left(\frac{3}{8\pi}\right)^{1/2}e^{\pm i\phi}\left(-\cos\phi\cos\theta\mp i\cos\theta\sin\phi\right)$$

Thus,

$$\langle L_x\rangle=\int_0^\pi d\theta\sin\theta\int_0^{2\pi}d\phi\,Y_1^\pm(\theta,\phi)^*\hat{L}_x Y_1^\pm(\theta,\phi)$$

$$=\mp i\hbar\left(\frac{3}{8\pi}\right)\int_0^\pi d\theta\sin\theta\int_0^{2\pi}d\phi\,(-\sin\theta\sin\phi\cos\theta\mp i\sin\theta\cos\theta\cos\phi)$$

$$=\mp i\hbar\left(\frac{3}{8\pi}\right)\left(-\int_0^\pi d\theta\sin^2\theta\cos\theta\int_0^{2\pi}d\phi\,\sin\phi\mp i\int_0^\pi d\theta\sin^2\theta\cos\theta\int_0^{2\pi}d\phi\,\cos\phi\right)$$

$$=0$$

$$\langle L_y\rangle=\int_0^\pi d\theta\sin\theta\int_0^{2\pi}d\phi\,Y_1^\pm(\theta,\phi)^*\hat{L}_y Y_1^\pm(\theta,\phi)$$

$$=\mp i\hbar\left(\frac{3}{8\pi}\right)\int_0^\pi d\theta\sin\theta\int_0^{2\pi}d\phi\,(-\sin\theta\cos\phi\cos\theta\mp i\sin\theta\cos\theta\sin\phi)$$

$$=\mp i\hbar\left(\frac{3}{8\pi}\right)\left(-\int_0^\pi d\theta\sin^2\theta\cos\theta\int_0^{2\pi}d\phi\,\cos\phi\mp i\int_0^\pi d\theta\sin^2\theta\cos\theta\int_0^{2\pi}d\phi\,\sin\phi\right)$$

$$=0$$

These integrals vanish because $\displaystyle\int_0^{2\pi}d\phi\,\sin\phi=\int_0^{2\pi}d\phi\,\cos\phi=0$.

6–43. Calculate the ratio of the dipole transition moments for the $0\to1$ and $1\to2$ rotational transitions in the rigid-rotator approximation.

The dipole transition moment for the $0\to1$ transition is

$$I_{0\to1}=\mu_0\int_0^{2\pi}\int_0^\pi Y_1^m(\theta,\phi)^*Y_0^0(\theta,\phi)\cos\theta\sin\theta\,d\theta\,d\phi$$

For the integration over ϕ to be nonzero, $m = 0$. Therefore,

$$
\begin{aligned}
I_{0\to1} &= \mu_0 \frac{(3)^{1/2}}{4\pi} \int_0^{2\pi} d\phi \int_0^\pi \cos^2\theta \sin\theta \, d\theta \\
&= \mu_0 \frac{(3)^{1/2}}{4\pi} (2\pi) \left(-\frac{\cos^3\theta}{3} \right)\Big|_0^\pi \\
&= \mu_0 \frac{(3)^{1/2}}{2} \left(\frac{2}{3} \right) \\
&= \frac{\mu_0}{3^{1/2}}
\end{aligned}
$$

The dipole transition moment for the $1 \to 2$ transition is

$$
I_{1\to2} = \mu_0 \int_0^{2\pi} \int_0^\pi Y_2^{m'}(\theta, \phi)^* Y_1^m(\theta, \phi) \cos\theta \sin\theta \, d\theta \, d\phi
$$

For the integration over ϕ to be nonzero, $m = m'$. Choose $m = 0$ in the following calculation.

$$
\begin{aligned}
I_{1\to2} &= \mu_0 \frac{(15)^{1/2}}{8\pi} \int_0^{2\pi} d\phi \int_0^\pi \left(3\cos^4\theta \sin\theta - \cos^2\theta \sin\theta \right) d\theta \\
&= \mu_0 \frac{(15)^{1/2}}{8\pi} (2\pi) \left(-\frac{3\cos^5\theta}{5} + \frac{\cos^3\theta}{3} \right)\Big|_0^\pi \\
&= \mu_0 \frac{(15)^{1/2}}{4} \left(\frac{8}{15} \right) \\
&= \frac{2\mu_0}{15^{1/2}}
\end{aligned}
$$

Therefore, the ratio of the dipole transition moments for the $0 \to 1$ and $1 \to 2$ rotational transitions is

$$
\frac{I_{0\to1}}{I_{1\to2}} = \frac{\mu_0/3^{1/2}}{2\mu_0/15^{1/2}} = \frac{5^{1/2}}{2}
$$

If m and m' are chosen to be 1, then $I_{1\to2} = \mu_0/5^{1/2}$ (the same result is obtained for $m = m' = -1$), giving $I_{0\to1}/I_{1\to2} = (5/3)^{1/2}$.

6–44. In this problem, we'll calculate the fraction of diatomic molecules in a particular rotational level at a temperature T using the rigid-rotator approximation. A fundamental equation of physical chemistry is the *Boltzmann distribution*, which says that the number of molecules with an energy E_J is proportional to $e^{-E_J/k_B T}$, where k_B is the Boltzmann constant and T is the kelvin temperature. Furthermore, because the degeneracy of the Jth rotational level is $2J + 1$, we write

$$
N_J \propto (2J + 1)e^{-E_J/k_B T} = (2J + 1)e^{-BJ(J+1)/k_B T}
$$

or

$$
N_J = c(2J + 1)e^{-BJ(J+1)/k_B T}
$$

where c is a proportionality constant. Plot N_J/N_0 versus J for $H^{35}Cl$ ($\tilde{B} = 10.60 \text{ cm}^{-1}$) and $^{127}I^{35}Cl$ ($\tilde{B} = 0.114 \text{ cm}^{-1}$) at 300 K. Treating J as a continuous parameter, show that the value of J in the most populated rotational state is the nearest integer to

$$J_{max} = \frac{1}{2}\left[\left(\frac{2k_BT}{\tilde{B}}\right)^{1/2} - 1\right]$$

Calculate J_{max} for $H^{35}Cl$ ($\tilde{B} = 10.60 \text{ cm}^{-1}$) and $^{127}I^{35}Cl$ ($\tilde{B} = 0.114 \text{ cm}^{-1}$) at 300 K.

From $N_J = c(2J+1)e^{-BJ(J+1)/k_BT}$, $N_0 = c$ for $J = 0$. Thus, $N_J/N_0 = (2J+1)e^{-BJ(J+1)/k_BT}$.
For $H^{35}Cl$,

$$\frac{N_J}{N_0} = (2J+1)\, e^{-J(J+1)\left(10.60 \text{ cm}^{-1}\right)/\left[\left(0.695\,038 \text{ cm}^{-1}\cdot K^{-1}\right)(300 \text{ K})\right]}$$

$$= (2J+1)\, e^{-5.0837\times10^{-2}J(J+1)}$$

For $^{127}I^{35}Cl$,

$$\frac{N_J}{N_0} = (2J+1)\, e^{-J(J+1)\left(0.114 \text{ cm}^{-1}\right)/\left[\left(0.695\,038 \text{ cm}^{-1}\cdot K^{-1}\right)(300 \text{ K})\right]}$$

$$= (2J+1)\, e^{-5.467\times10^{-4}J(J+1)}$$

The graphs below show N_J/N_0 versus J for $H^{35}Cl$ (left) and $^{127}I^{35}Cl$ (right). Note that the scales are not the same for the plots. $H^{35}Cl$ has a smaller moment of inertia, resulting in a larger rotational constant. The lower J states have the majority of the population. On the other hand, $^{127}I^{35}Cl$ has a larger moment of inertia and therefore a smaller rotational constant. The population spreads over many J states.

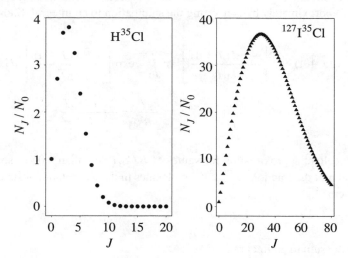

The value of J for the most populated rotational state can be obtained by differentiating N_J/N_0 with respect to J

$$\frac{dN_J/N_0}{dJ} = 2e^{-BJ(J+1)/k_BT} + (2J+1)\left(-\frac{2BJ}{k_BT} - \frac{B}{k_BT}\right)e^{-BJ(J+1)/k_BT}$$

and setting the expression to 0, followed by dividing the entire equation by $e^{-BJ(J+1)/k_BT}$:

$$2 - (2J_{max} + 1)\left(\frac{2BJ_{max}}{k_BT} + \frac{B}{k_BT}\right) = 0$$

$$(2J_{max} + 1)^2 = \frac{2k_BT}{B}$$

$$J_{max} = \frac{1}{2}\left[\left(\frac{2k_BT}{B}\right)^{1/2} - 1\right]$$

For $H^{35}Cl$,

$$J_{max} = \frac{1}{2}\left\{\left[\frac{2\,(0.695\,038\ \text{cm}^{-1}\cdot\text{K}^{-1})\,(300\ \text{K})}{10.60\ \text{cm}^{-1}}\right]^{1/2} - 1\right\} = 2.64$$

Rounding to the nearest integer, $J_{max} = 3$, which agrees with the figure for $H^{35}Cl$ above. For $^{127}I^{35}Cl$,

$$J_{max} = \frac{1}{2}\left\{\left[\frac{2\,(0.695\,038\ \text{cm}^{-1}\cdot\text{K}^{-1})\,(300\ \text{K})}{0.114\ \text{cm}^{-1}}\right]^{1/2} - 1\right\} = 29.74$$

Rounding to the nearest integer, $J_{max} = 30$, which agrees with the figure for $^{127}I^{35}Cl$ above.

6–45. The summation that occurs in the rotational Boltzmann distribution (previous problem) can be evaluated approximately by converting the summation to an integral. Show that

$$\sum_{J=0}^{\infty}(2J + 1)\exp\left[\frac{-\tilde{B}J(J+1)}{k_BT}\right] \approx \int_0^{\infty}\exp\left[\frac{-\tilde{B}J(J+1)}{k_BT}\right]d[J(J+1)]$$

$$= \frac{k_BT}{\tilde{B}} = \frac{8\pi^2cIk_BT}{h}$$

This is an excellent approximation for values of $\tilde{B}/k_BT$ less than 0.05 or so. Using this result, calculate and plot the fraction of $^{127}I^{35}Cl$ molecules in the Jth rotational state versus J at 25°C. ($\tilde{B} = 0.114\ \text{cm}^{-1}$.)

Converting the sum to an integral,

$$\sum_{J=0}^{\infty}(2J + 1)\exp\left[\frac{-\tilde{B}J(J+1)}{k_BT}\right] \approx \int_0^{\infty}(2J + 1)\exp\left[\frac{-\tilde{B}J(J+1)}{k_BT}\right]dJ$$

Since $d[J(J + 1)] = (2J + 1)dJ$, the above integral becomes

$$\sum_{J=0}^{\infty}(2J+1)\exp\left[\frac{-\tilde{B}J(J+1)}{k_BT}\right] \approx \int_0^\infty \exp\left[\frac{-\tilde{B}J(J+1)}{k_BT}\right]d[J(J+1)]$$

$$= -\frac{k_BT}{\tilde{B}}\exp\left[\frac{-\tilde{B}J(J+1)}{k_BT}\right]\Bigg|_{J(J+1)=0}^{\infty}$$

$$= \frac{k_BT}{\tilde{B}}$$

$$= \frac{8\pi^2 c I k_B T}{h}$$

For $^{127}I^{35}Cl$ at 25°C,

$$\sum_{J=0}^{\infty}(2J+1)\exp\left[\frac{-\tilde{B}J(J+1)}{k_BT}\right] \approx \frac{k_BT}{\tilde{B}}$$

$$= \frac{(0.695\,038\ \text{cm}^{-1}\cdot\text{K}^{-1})\,(298\ \text{K})}{0.114\ \text{cm}^{-1}}$$

$$= 1.817 \times 10^3$$

The fraction of $^{127}I^{35}Cl$ in the Jth rotational state is

$$\frac{(2J+1)\exp\left[-BJ(J+1)/k_BT\right]}{\displaystyle\sum_{J=0}^{\infty}(2J+1)\exp\left[-\tilde{B}J(J+1)/k_BT\right]}$$

$$= \frac{(2J+1)\exp\left\{\left[-J(J+1)\left(0.114\ \text{cm}^{-1}\right)\right]/\left[\left(0.695\,038\ \text{cm}^{-1}\cdot\text{K}^{-1}\right)(298\ \text{K})\right]\right\}}{1.817 \times 10^3}$$

$$= \frac{(2J+1)\,e^{-5.504\times10^{-4}J(J+1)}}{1.817 \times 10^3}$$

and a plot of this fraction versus J is shown below.

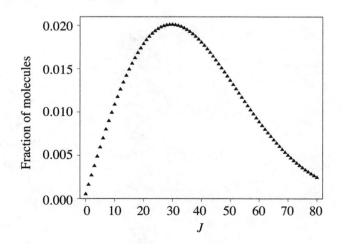

6–46. Can you use the result of the previous problem to rationalize the envelope of the lines in the P and R branches in the rotation-vibration spectrum in Figure 6.4?

The intensity in the P and R branches in the vibration-rotation spectrum is proportional to the lower state population and thus have similar envelope structure as the population of molecules as a function of J.

6–47. Many students are mystified when they see that $0! = 1$. The standard formula $n! = n(n-1)\cdots 1$ is applicable *only* for $n = 1, 2, 3, \ldots$. Euler showed that we could extend the idea of a factorial to other numbers through the definite integral

$$I_n = \int_0^\infty x^n e^{-x}\, dx$$

Integration repeatedly by parts shows that $I_n = n(n-1)\cdots 1$ when $n = 1, 2, 3, \ldots$, so that I_n gives our "standard" result in that case. However, there is no reason for n to be restricted in the above integral, and we can use it to *define* $n!$ for other values of n. Use this extended definition to show that $0! = 1$. What about $(1/2)!$? Even $(-1/2)!$? For a more complete discussion of $n!$ for a general value of n, read about the gamma function in any applied mathematics book.

Since $I_n = n!$,

$$0! = I_0 = \int_0^\infty x^0 e^{-x}\, dx$$

$$= \int_0^\infty e^{-x}\, dx$$

$$= -e^{-x}\Big|_0^\infty$$

$$= 1$$

Similarly, $\frac{1}{2}! = I_{1/2} = \int_0^\infty x^{1/2} e^{-x}\, dx$. Let $u = x^{1/2}$, then

$$x = u^2$$
$$dx = 2u\, du$$

Therefore,

$$\frac{1}{2}! = 2 \int_0^\infty u^2 e^{-u^2}\, du$$

$$= \frac{\pi^{1/2}}{2}$$

Finally, $-\frac{1}{2}! = I_{-1/2} = \int_0^\infty x^{-1/2} e^{-x}\, dx$. Once again let $u = x^{1/2}$, then

$$-\frac{1}{2}! = 2 \int_0^\infty e^{-u^2}\, du$$

$$= \pi^{1/2}$$

Determinants

PROBLEMS AND SOLUTIONS

F–1. Evaluate the determinant

$$D = \begin{vmatrix} 2 & 1 & 1 \\ -1 & 3 & 2 \\ 2 & 0 & 1 \end{vmatrix}$$

Add column 2 to column 1 to get

$$\begin{vmatrix} 3 & 1 & 1 \\ 2 & 3 & 2 \\ 2 & 0 & 1 \end{vmatrix}$$

and evaluate it. Compare your result with the value of D. Now add row 2 to row 1 of D to get

$$\begin{vmatrix} 1 & 4 & 3 \\ -1 & 3 & 2 \\ 2 & 0 & 1 \end{vmatrix}$$

and evaluate it. Compare your result with the value of D above.

To evaluate D, expand about the second column of elements:

$$D = -\begin{vmatrix} -1 & 2 \\ 2 & 1 \end{vmatrix} + 3 \begin{vmatrix} 2 & 1 \\ 2 & 1 \end{vmatrix}$$

$$= -(-1 - 4) + 3(2 - 2) = 5$$

Adding column 2 to column 1 of D gives the second determinant. To evaluate it, expand about the second column of elements:

$$\begin{vmatrix} 3 & 1 & 1 \\ 2 & 3 & 2 \\ 2 & 0 & 1 \end{vmatrix} = -\begin{vmatrix} 2 & 2 \\ 2 & 1 \end{vmatrix} + 3 \begin{vmatrix} 3 & 1 \\ 2 & 1 \end{vmatrix}$$

$$= -(2 - 4) + 3(3 - 2) = 5$$

331

The result is the same as D. This is expected as stated by rule 6: the value of a determinant is unchanged if one row or column is added to or subtracted from another.

Adding row 2 to row 1 of D gives the third determinant. Expand this determinant about the second column of elements:

$$\begin{vmatrix} 1 & 4 & 3 \\ -1 & 3 & 2 \\ 2 & 0 & 1 \end{vmatrix} = -4 \begin{vmatrix} -1 & 2 \\ 2 & 1 \end{vmatrix} + 3 \begin{vmatrix} 1 & 3 \\ 2 & 1 \end{vmatrix}$$

$$= -4(-1-4) + 3(1-6) = 5$$

As expected, this determinant is the same as D according to rule 6.

F–2. Interchange columns 1 and 3 in D in Problem F–1 and evaluate the resulting determinant. Compare your result with the value of D. Interchange rows 1 and 2 and do the same.

Interchanging columns 1 and 3 in D and expanding about the second column of elements:

$$\begin{vmatrix} 1 & 1 & 2 \\ 2 & 3 & -1 \\ 1 & 0 & 2 \end{vmatrix} = - \begin{vmatrix} 2 & -1 \\ 1 & 2 \end{vmatrix} + 3 \begin{vmatrix} 1 & 2 \\ 1 & 2 \end{vmatrix}$$

$$= -(4+1) + 3(2-2) = -5$$

Interchanging rows 1 and 2 in D and expanding about the second column of elements:

$$\begin{vmatrix} -1 & 3 & 2 \\ 2 & 1 & 1 \\ 2 & 0 & 1 \end{vmatrix} = -3 \begin{vmatrix} 2 & 1 \\ 2 & 1 \end{vmatrix} + \begin{vmatrix} -1 & 2 \\ 2 & 1 \end{vmatrix}$$

$$= -3(2-2) + (-1-4) = -5$$

These results are consistent with rule 3: if any two rows or columns are interchanged, the sign of the determinant is changed.

F–3. Evaluate the determinant

$$D = \begin{vmatrix} 1 & 6 & 1 \\ -2 & 4 & -2 \\ 1 & -3 & 1 \end{vmatrix}$$

Can you determine its value by inspection? What about

$$D = \begin{vmatrix} 2 & 6 & 1 \\ -4 & 4 & -2 \\ 2 & -3 & 1 \end{vmatrix}$$

Expanding the first determinant about the first column of elements:

$$\begin{vmatrix} 1 & 6 & 1 \\ -2 & 4 & -2 \\ 1 & -3 & 1 \end{vmatrix} = \begin{vmatrix} 4 & -2 \\ -3 & 1 \end{vmatrix} + 2\begin{vmatrix} 6 & 1 \\ -3 & 1 \end{vmatrix} + \begin{vmatrix} 6 & 1 \\ 4 & -2 \end{vmatrix}$$

$$= (4 - 6) + 2(6 + 3) + (-12 - 4) = 0$$

By inspection, this determinant is zero according to rule 2: if two rows or columns are the same, the value of the determinant is zero.

According to rule 4, the two determinants given are related:

$$\begin{vmatrix} 2 & 6 & 1 \\ -4 & 4 & -2 \\ 2 & -3 & 1 \end{vmatrix} = 2\begin{vmatrix} 1 & 6 & 1 \\ -2 & 4 & -2 \\ 1 & -3 & 1 \end{vmatrix} = 2(0) = 0$$

F–4. Find the values of x that satisfy the following determinantal equation:

$$\begin{vmatrix} x & 1 & 1 & 1 \\ 1 & x & 0 & 0 \\ 1 & 0 & x & 0 \\ 1 & 0 & 0 & x \end{vmatrix} = 0$$

Expand the determinant about the third column of elements:

$$\begin{vmatrix} 1 & x & 0 \\ 1 & 0 & 0 \\ 1 & 0 & x \end{vmatrix} + x\begin{vmatrix} x & 1 & 1 \\ 1 & x & 0 \\ 1 & 0 & x \end{vmatrix} = 0$$

Now expand the first determinant about its second column and the second determinant about its third column:

$$-x\begin{vmatrix} 1 & 0 \\ 1 & x \end{vmatrix} + x\left(\begin{vmatrix} 1 & x \\ 1 & 0 \end{vmatrix} + x\begin{vmatrix} x & 1 \\ 1 & x \end{vmatrix}\right) = 0$$

$$-x^2 + x\left(-x + x^3 - x\right) = 0$$

$$x^4 - 3x^2 = 0$$

$$x^2(x^2 - 3) = 0$$

$$x = 0, 0, \pm\sqrt{3}$$

F–5. Find the values of x that satisfy the following determinantal equation:

$$\begin{vmatrix} x & 1 & 0 & 1 \\ 1 & x & 1 & 0 \\ 0 & 1 & x & 1 \\ 1 & 0 & 1 & x \end{vmatrix} = 0$$

Expand the determinant about the first column of elements:

$$x \begin{vmatrix} x & 1 & 0 \\ 1 & x & 1 \\ 0 & 1 & x \end{vmatrix} - \begin{vmatrix} 1 & 0 & 1 \\ 1 & x & 1 \\ 0 & 1 & x \end{vmatrix} - \begin{vmatrix} 1 & 0 & 1 \\ x & 1 & 0 \\ 1 & x & 1 \end{vmatrix} = 0$$

Now expand each of the first two determinants about its first column and the third about its third column.

$$x \left(x \begin{vmatrix} x & 1 \\ 1 & x \end{vmatrix} - \begin{vmatrix} 1 & 0 \\ 1 & x \end{vmatrix} \right) - \left(\begin{vmatrix} x & 1 \\ 1 & x \end{vmatrix} - \begin{vmatrix} 0 & 1 \\ 1 & x \end{vmatrix} \right) - \left(\begin{vmatrix} x & 1 \\ 1 & x \end{vmatrix} + \begin{vmatrix} 1 & 0 \\ x & 1 \end{vmatrix} \right) = 0$$

$$x(x^3 - x - x) - (x^2 - 1 + 1) - (x^2 - 1 + 1) = 0$$

$$x^4 - 4x^2 = 0$$

$$x^2(x^2 - 4) = 0$$

$$x = 0,\ 0,\ \pm 2$$

F–6.   Show that

$$\begin{vmatrix} \cos\theta & -\sin\theta & 0 \\ \sin\theta & \cos\theta & 0 \\ 0 & 0 & 1 \end{vmatrix} = 1$$

Expand the determinant about its first column:

$$\begin{vmatrix} \cos\theta & -\sin\theta & 0 \\ \sin\theta & \cos\theta & 0 \\ 0 & 0 & 1 \end{vmatrix} = \cos\theta \begin{vmatrix} \cos\theta & 0 \\ 0 & 1 \end{vmatrix} - \sin\theta \begin{vmatrix} -\sin\theta & 0 \\ 0 & 1 \end{vmatrix}$$

$$= \cos^2\theta + \sin^2\theta = 1$$

F–7.   Find the three roots of the determinantal equation

$$\begin{vmatrix} 1-\lambda & 1 & 0 \\ 1 & 1-\lambda & 1 \\ 0 & 1 & 1-\lambda \end{vmatrix} = 0$$

Expand the determinant about its first column:

$$(1-\lambda)\begin{vmatrix} 1-\lambda & 1 \\ 1 & 1-\lambda \end{vmatrix} - \begin{vmatrix} 1 & 0 \\ 1 & 1-\lambda \end{vmatrix} = 0$$

$$(1-\lambda)^3 - (1-\lambda) - (1-\lambda) = 0$$

$$(1-\lambda)\left[(1-\lambda)^2 - 2\right] = 0$$

$$\lambda = 1, 1 \pm \sqrt{2}$$

F–8. Solve the following set of equations using Cramer's rule:

$$x + y = 2$$

$$3x - 2y = 5$$

$$x = \frac{\begin{vmatrix} 2 & 1 \\ 5 & -2 \end{vmatrix}}{\begin{vmatrix} 1 & 1 \\ 3 & -2 \end{vmatrix}} = \frac{-9}{-5} = \frac{9}{5}$$

$$y = \frac{\begin{vmatrix} 1 & 2 \\ 3 & 5 \end{vmatrix}}{\begin{vmatrix} 1 & 1 \\ 3 & -2 \end{vmatrix}} = \frac{-1}{-5} = \frac{1}{5}$$

F–9. Solve the following set of equations using Cramer's rule:

$$x + 2y + 3z = -5$$

$$-x - 3y + z = -14$$

$$2x + y + z = 1$$

$$x = \frac{\begin{vmatrix} -5 & 2 & 3 \\ -14 & -3 & 1 \\ 1 & 1 & 1 \end{vmatrix}}{\begin{vmatrix} 1 & 2 & 3 \\ -1 & -3 & 1 \\ 2 & 1 & 1 \end{vmatrix}} = \frac{17}{17} = 1$$

$$y = \frac{\begin{vmatrix} 1 & -5 & 3 \\ -1 & -14 & 1 \\ 2 & 1 & 1 \end{vmatrix}}{\begin{vmatrix} 1 & 2 & 3 \\ -1 & -3 & 1 \\ 2 & 1 & 1 \end{vmatrix}} = \frac{51}{17} = 3$$

$$z = \frac{\begin{vmatrix} 1 & 2 & -5 \\ -1 & -3 & -14 \\ 2 & 1 & 1 \end{vmatrix}}{\begin{vmatrix} 1 & 2 & 3 \\ -1 & -3 & 1 \\ 2 & 1 & 1 \end{vmatrix}} = \frac{-68}{17} = -4$$

F–10. Determine the values of x for which the following equations will have a nontrivial solution.

$$xc_1 + c_2 + c_4 = 0$$

$$c_1 + xc_2 + c_3 = 0$$

$$c_2 + xc_3 + c_4 = 0$$

$$c_1 + c_3 + xc_4 = 0$$

A nontrivial solution is obtained when

$$\begin{vmatrix} x & 1 & 0 & 1 \\ 1 & x & 1 & 0 \\ 0 & 1 & x & 1 \\ 1 & 0 & 1 & x \end{vmatrix} = 0$$

This equation is the same as that in Problem F–5. Thus,

$$x^2(x^2 - 4) = 0$$

$$x = 0, 0, \pm 2$$

CHAPTER 7

The Hydrogen Atom

PROBLEMS AND SOLUTIONS

7–1. Show that both $\hbar^2 \nabla^2 / 2m_e$ and $e^2 / 4\pi \epsilon_0 r$ have the units of energy (joules).

Because ∇^2 has units of m^{-2},

$$\frac{\hbar^2 \nabla^2}{2m_e} \sim J^2 \cdot s^2 \cdot m^{-2} \cdot kg^{-1} = J$$

and

$$\frac{e^2}{4\pi \epsilon_0 r} \sim \frac{C^2}{C^2 \cdot J^{-1} \cdot m^{-1} \cdot m} = J$$

7–2. Show that the substitution $R(r)Y_l^{m_l}(\theta, \phi)$ into Equation 7.7 yields Equation 7.10.

Start with Equation 7.7 and make the suggested substitution, recalling that $\hat{L}^2$ operates on only the angular coordinates.

$$-\hbar^2 \frac{\partial}{\partial r} \left(r^2 \frac{\partial \psi}{\partial r} \right) + \hat{L}^2 \psi + 2m_e r^2 \left[V(r) - E \right] \psi(r, \theta, \phi) = 0$$

$$-\hbar^2 \frac{\partial}{\partial r} \left[r^2 \frac{\partial}{\partial r} R(r) Y_l^{m_l}(\theta, \phi) \right] + \hat{L}^2 R(r) Y_l^{m_l}(\theta, \phi) + 2m_e r^2 \left[V(r) - E \right] R(r) Y_l^{m_l}(\theta, \phi) = 0$$

$$-\hbar^2 Y_l^{m_l}(\theta, \phi) \frac{d}{dr} \left(r^2 \frac{dR}{dr} \right) + R(r) \hat{L}^2 Y_l^{m_l}(\theta, \phi) + 2m_e r^2 \left[V(r) - E \right] R(r) Y_l^{m_l}(\theta, \phi) = 0$$

$$-\hbar^2 Y_l^{m_l}(\theta, \phi) \frac{d}{dr} \left(r^2 \frac{dR}{dr} \right) + R(r) \hbar^2 l (l + 1) Y_l^{m_l}(\theta, \phi) + 2m_e r^2 \left[V(r) - E \right] R(r) Y_l^{m_l}(\theta, \phi) = 0$$

Now divide by $2m_e r^2 Y_l^{m_l}(\theta, \phi)$ to arrive at Equation 7.10.

$$-\frac{\hbar^2}{2m_e r^2} \frac{d}{dr} \left(r^2 \frac{dR}{dr} \right) + \left[\frac{\hbar^2 l (l + 1)}{2m_e r^2} + V(r) - E \right] R(r) = 0$$

7–3. Show that the radial functions given in Table 7.2 are normalized.

Normalization requires that

$$I = \int_0^\infty r^2 R_{nl}^2(r)\, dr = \left(\frac{a_0}{Z}\right)^3 \int_0^\infty \rho^2 R_{nl}^2(\rho)\, d\rho = 1$$

The integral

$$\int_0^\infty x^n e^{-ax}\, dx = \frac{n!}{a^{n+1}}$$

is used many times in this problem.

For $R_{10} = 2\left(\dfrac{Z}{a_0}\right)^{3/2} e^{-\rho}$:

$$I = 4\int_0^\infty \rho^2 e^{-2\rho}\, d\rho = 4\left(\frac{2!}{2^3}\right) = 1$$

For $R_{20} = \left(\dfrac{Z}{2a_0}\right)^{3/2}(2 - \rho)\, e^{-\rho/2}$:

$$I = \frac{1}{8}\int_0^\infty \rho^2 (2 - \rho)^2 e^{-\rho}\, d\rho$$

$$= \frac{1}{8}\int_0^\infty \left(4\rho^2 - 4\rho^3 + \rho^4\right) e^{-\rho}\, d\rho$$

$$= \frac{1}{8}\left[4\cdot 2! - 4\cdot 3! + 4!\right] = 1$$

For $R_{21} = \dfrac{1}{\sqrt{3}}\left(\dfrac{Z}{2a_0}\right)^{3/2} \rho e^{-\rho/2}$:

$$I = \frac{1}{24}\int_0^\infty \rho^4 e^{-\rho}\, d\rho = \frac{1}{24}(4!) = 1$$

For $R_{30} = \dfrac{2}{27}\left(\dfrac{Z}{3a_0}\right)^{3/2}\left(27 - 18\rho + 2\rho^2\right) e^{-\rho/3}$:

$$I = \frac{4}{3^9}\int_0^\infty \rho^2 \left(27 - 18\rho + 2\rho^2\right)^2 e^{-2\rho/3}\, d\rho$$

$$= \frac{4}{3^9}\int_0^\infty \left[3^6 \rho^2 - 4\left(3^5\right)\rho^3 + 16\left(3^3\right)\rho^4 - 8\left(3^2\right)\rho^5 + 4\rho^6\right] e^{-2\rho/3}\, d\rho$$

$$= \frac{4}{3^9}\left[3^6\cdot 2!\left(\frac{3}{2}\right)^3 - 4\cdot 3^5\cdot 3!\left(\frac{3}{2}\right)^4 + 16\cdot 3^3\cdot 4!\left(\frac{3}{2}\right)^5 - 8\cdot 3^2\cdot 5!\left(\frac{3}{2}\right)^6 + 4\cdot 6!\left(\frac{3}{2}\right)^7\right]$$

$$= 1$$

For $R_{31} = \dfrac{1}{27}\left(\dfrac{2Z}{3a_0}\right)^{3/2} \rho\,(6 - \rho)\, e^{-\rho/3}$:

$$I = \frac{8}{3^9} \int_0^\infty \rho^4 \, (6 - \rho)^2 \, e^{-2\rho/3} \, d\rho$$

$$= \frac{8}{3^9} \int_0^\infty \left[4 \left(3^2 \right) \rho^4 - 4 \, (3) \, \rho^5 + \rho^6 \right] e^{-2\rho/3} \, d\rho$$

$$= \frac{8}{3^9} \left[4 \cdot 3^2 \cdot 4! \left(\frac{3}{2} \right)^5 - 4 \cdot 3 \cdot 5! \left(\frac{3}{2} \right)^6 + 6! \left(\frac{3}{2} \right)^7 \right]$$

$$= 1$$

For $R_{32} = \dfrac{4}{27\sqrt{10}} \left(\dfrac{Z}{3a_0} \right)^{3/2} \rho^2 e^{-\rho/3}$:

$$I = \frac{16}{10 \cdot 3^9} \int_0^\infty \rho^6 e^{-2\rho/3} \, d\rho$$

$$= \frac{16}{10 \cdot 3^9} \cdot 6! \left(\frac{3}{2} \right)^7$$

$$= 1$$

7–4. Referring to Table 7.2, show that $R_{10}(r)$ and $R_{20}(r)$ are orthonormal.

To show orthonormality, evaluate the integral,

$$\int_0^\infty r^2 R_{10}(r) R_{20}(r) \, dr = \left(\frac{a_0}{Z} \right)^3 \int_0^\infty \rho^2 R_{10}(\rho) R_{20}(\rho) \, d\rho$$

$$= \frac{1}{\sqrt{2}} \int_0^\infty \rho^2 \, (2 - \rho) \, e^{-3\rho/2} \, d\rho$$

$$= \frac{1}{\sqrt{2}} \int_0^\infty \left(2\rho^2 - \rho^3 \right) e^{-3\rho/2} \, d\rho$$

$$= \frac{1}{\sqrt{2}} \left[2 \cdot 2! \left(\frac{2}{3} \right)^3 - 3! \left(\frac{2}{3} \right)^4 \right]$$

$$= 0$$

Since the integral is equal to zero, the functions are orthogonal.

7–5. Referring to Table 7.3, show that the first few hydrogen atomic wave functions are orthonormal.

The orthonormality condition for hydrogen atomic wave functions is

$$\int_0^\infty dr \, r^2 \int_0^\pi d\theta \, \sin \theta \int_0^{2\pi} d\phi \, \psi_{n'l'm'}^*(r, \theta, \phi) \psi_{nlm}(r, \theta, \phi) = \delta_{nn'} \delta_{ll'} \delta_{mm'}$$

Begin by showing that the first few hydrogen atomic wave functions are normalized. For ψ_{100},

$$\int d\tau\, \psi_{100}^* \psi_{100} = \int_0^{2\pi} d\phi \int_0^\pi d\theta\, \sin\theta \int_0^\infty dr\, r^2 \left[\frac{1}{\pi} \left(\frac{Z}{a_0}\right)^3 e^{-2\rho} \right]$$

$$= 4\left(\frac{Z}{a_0}\right)^3 \int_0^\infty dr\, r^2 e^{-2\rho}$$

$$= 4\int_0^\infty d\rho\, \rho^2 e^{-2\rho} = 4\left(\frac{2!}{2^3}\right) = 1$$

For ψ_{200},

$$\int d\tau\, \psi_{200}^* \psi_{200} = \int_0^{2\pi} d\phi \int_0^\pi d\theta\, \sin\theta \int_0^\infty dr\, r^2 \left[\frac{1}{32\pi} \left(\frac{Z}{a_0}\right)^3 \left(4 - 4\rho + \rho^2\right) e^{-\rho} \right]$$

$$= \frac{1}{8}\left(\frac{Z}{a_0}\right)^3 \int_0^\infty dr\, r^2 \left(4 - 4\rho + \rho^2\right) e^{-\rho}$$

$$= \frac{1}{8}\int_0^\infty d\rho \left(4\rho^2 - 4\rho^3 + \rho^4\right) e^{-\rho}$$

$$= \frac{1}{8}\left(4\cdot 2! - 4\cdot 3! + 4!\right) = 1$$

For ψ_{210},

$$\int d\tau\, \psi_{210}^* \psi_{210} = \int_0^{2\pi} d\phi \int_0^\pi d\theta\, \sin\theta \int_0^\infty dr\, r^2 \left[\frac{1}{32\pi} \left(\frac{Z}{a_0}\right)^3 \rho^2 \cos^2\theta\, e^{-\rho} \right]$$

$$= \frac{1}{16}\left(\frac{Z}{a_0}\right)^3 \int_0^\pi d\theta\, \sin\theta \cos^2\theta \int_0^\infty dr\, r^2 \rho^2 e^{-\rho}$$

$$= \frac{1}{16}\left(\frac{2}{3}\right) \int_0^\infty d\rho\, \rho^4 e^{-\rho}$$

$$= \frac{1}{24}(4!) = 1$$

For ψ_{211} or ψ_{21-1},

$$\int d\tau\, \psi_{21\pm1}^* \psi_{21\pm1} = \int_0^{2\pi} d\phi \int_0^\pi d\theta\, \sin\theta \int_0^\infty dr\, r^2 \left[\frac{1}{64\pi} \left(\frac{Z}{a_0}\right)^3 \rho^2 \sin^2\theta\, e^{-\rho} \right]$$

$$= \frac{1}{32}\left(\frac{Z}{a_0}\right)^3 \int_0^\pi d\theta\, \sin^3\theta \int_0^\infty dr\, r^2 \rho^2 e^{-\rho}$$

$$= \frac{1}{32}\left(\frac{4}{3}\right) \int_0^\infty d\rho\, \rho^4 e^{-\rho} = \frac{1}{24}(4!) = 1$$

Then show that the first few hydrogen atomic wave functions are orthogonal:

$$\int d\tau\,\psi^*_{100}\psi_{200} = \int_0^{2\pi} d\phi \int_0^\pi d\theta\,\sin\theta \int_0^\infty dr\,r^2 \left[\frac{1}{\pi^{1/2}}\left(\frac{Z}{a_0}\right)^{3/2} e^{-\rho}\right]\left[\frac{1}{(32\pi)^{1/2}}\left(\frac{Z}{a_0}\right)^{3/2}(2-\rho)e^{-\rho/2}\right]$$

$$= \frac{1}{\sqrt{2}}\left(\frac{Z}{a_0}\right)^3 \int_0^\infty dr\,r^2(2-\rho)e^{-3\rho/2}$$

$$= \frac{1}{\sqrt{2}} \int_0^\infty d\rho\,(2\rho^2 - \rho^3)e^{-3\rho/2}$$

$$= \frac{1}{\sqrt{2}}\left[2\cdot 2!\left(\frac{2}{3}\right)^3 - 3!\left(\frac{2}{3}\right)^4\right] = 0$$

In showing that either ψ_{100} or ψ_{200} is orthogonal to ψ_{210}, the integral over θ is

$$\int_0^\pi d\theta\,\sin\theta\cos\theta = 0$$

and so these orbitals are orthogonal. In fact, this result can be used to conclude that any $l = 0$ orbital is orthogonal to any $l = 1$, $m_l = 0$ orbital.

Likewise, in showing that either ψ_{100}, ψ_{200} or ψ_{210} is orthogonal to ψ_{211} or ψ_{21-1}, the integral over ϕ is

$$\int_0^{2\pi} d\phi\,e^{\pm i\phi} = 0$$

and so these orbitals are orthogonal, as would any two orbitals with different values of m_l.

7–6. Show explicitly that

$$\hat{H}\psi = -\frac{m_e e^4}{8\epsilon_0^2 h^2}\psi$$

for the ground state of a hydrogen atom.

The Hamiltonian operator for a hydrogen atom is (Equations 7.2 and 7.4)

$$\hat{H} = -\frac{\hbar^2}{2m_e}\left[\frac{1}{r^2}\frac{\partial}{\partial r}\left(r^2\frac{\partial}{\partial r}\right) + \frac{1}{r^2\sin\theta}\frac{\partial}{\partial\theta}\left(\sin\theta\frac{\partial}{\partial\theta}\right) + \frac{1}{r^2\sin^2\theta}\frac{\partial^2}{\partial\phi^2}\right] - \frac{e^2}{4\pi\epsilon_0 r}$$

and from Table 7.3 the ground state wave function of a hydrogen atom is

$$\psi_0 = \frac{1}{\pi^{1/2}a_0^{3/2}}e^{-r/a_0} \qquad a_0 = \frac{\epsilon_0 h^2}{\pi m_e e^2}$$

Since the derivatives with respect to the angular coordinates vanish for ψ_0,

$$\hat{H}\psi_0 = -\frac{\hbar^2}{2m_er^2}\frac{\partial}{\partial r}\left(r^2\frac{\partial\psi_0}{\partial r}\right) - \frac{e^2\psi_0}{4\pi\epsilon_0 r}$$

$$= -\frac{h^2}{(4\pi^2)2m_er^2}\frac{1}{\pi^{1/2}a_0^{3/2}}\left(-\frac{2r}{a_0}e^{-r/a_0} + \frac{r^2}{a_0^2}e^{-r/a_0}\right) - \frac{e^2}{4\pi\epsilon_0 r}\frac{1}{\pi^{1/2}a_0^{3/2}}e^{-r/a_0}$$

$$= \left(\frac{e^2}{4\pi\epsilon_0 r} - \frac{m_ee^4}{8\epsilon_0^2 h^2} - \frac{e^2}{4\pi\epsilon_0 r}\right)\psi_0$$

$$= -\frac{m_ee^4}{8\epsilon_0^2 h^2}\psi_0$$

7–7. Show explicitly that

$$\hat{H}\psi = -\frac{m_ee^4}{32\epsilon_0^2 h^2}\psi$$

for a $2p_0$ state of a hydrogen atom.

The Hamiltonian operator of a hydrogen atom is the same as in Problem 7–6, and the wave function of the $2p_0$ state is

$$\psi_{2p_0} = \frac{1}{(32\pi)^{1/2}a_0^{5/2}}re^{-r/2a_0}\cos\theta \qquad a_0 = \frac{\epsilon_0 h^2}{\pi m_ee^2}$$

Thus, realizing that the derivative with respect to ϕ vanishes,

$$\hat{H}\psi_{2p_0} = -\frac{\hbar^2}{2m_e}\left[\frac{1}{r^2}\frac{\partial}{\partial r}\left(r^2\frac{\partial\psi_{2p_0}}{\partial r}\right) + \frac{1}{r^2\sin\theta}\frac{\partial}{\partial\theta}\left(\sin\theta\frac{\partial\psi_{2p_0}}{\partial\theta}\right)\right] - \frac{e^2}{4\pi\epsilon_0 r}\psi_{2p_0}$$

$$= -\frac{\hbar^2}{2m_er^2}\frac{\cos\theta}{(32\pi)^{1/2}a_0^{5/2}}\frac{\partial}{\partial r}\left[\left(r^2 - \frac{r^3}{2a_0}\right)e^{-r/2a_0}\right] + \frac{\hbar^2}{2m_er^2}\frac{re^{-r/2a_0}}{(32\pi)^{1/2}a_0^{5/2}}\frac{1}{\sin\theta}\frac{\partial}{\partial\theta}\sin^2\theta$$

$$\quad - \frac{e^2}{4\pi\epsilon_0 r}\frac{re^{-r/2a_0}}{(32\pi)^{1/2}a_0^{5/2}}\cos\theta$$

$$= -\frac{\hbar^2}{2m_er^2}\frac{1}{(32\pi)^{1/2}a_0^{5/2}}\left[\left(2 - \frac{r}{2a_0} - \frac{3r}{2a_0} + \frac{r^2}{4a_0^2}\right)re^{-r/2a_0}\cos\theta\right]$$

$$\quad + \frac{\hbar^2}{2m_er^2}\frac{2re^{-r/2a_0}\cos\theta}{(32\pi)^{1/2}a_0^{5/2}} - \frac{e^2}{4\pi\epsilon_0 r}\frac{re^{-r/2a_0}\cos\theta}{(32\pi)^{1/2}a_0^{5/2}}$$

$$= -\frac{\hbar^2}{2m_e} \frac{1}{(32\pi)^{1/2}a_0^{5/2}} \left(\frac{r}{4a_0^2} e^{-r/2a_0} \cos\theta \right)$$

$$+ \frac{1}{(32\pi)^{1/2}a_0^{5/2}} \left(\frac{\hbar^2}{m_e r a_0} - \frac{e^2}{4\pi\epsilon_0 r} \right) r e^{-r/2a_0} \cos\theta$$

$$= -\frac{\hbar^2}{8m_e a_0^2} \psi_{2p_0} = -\frac{m_e e^4}{32\epsilon_0^2 h^2} \psi_{2p_0}$$

7–8. Show that all of the following expressions for the ground state of a hydrogen atom are equivalent:

$$E_0 = -\frac{\hbar^2}{2m_e a_0^2} = -\frac{e^2}{8\pi\epsilon_0 a_0} = -\frac{m_e e^4}{32\pi^2\epsilon_0^2 \hbar^2} = -\frac{m_e e^4}{8\epsilon_0^2 h^2}$$

Since $\hat{H}\psi_0 = E_0\psi_0$ for the hydrogen atom, Problem 7–6 shows that

$$E_0 = -\frac{m_e e^4}{8\epsilon_0^2 h^2}$$

Then, using $a_0 = \dfrac{\epsilon_0 h^2}{\pi m_e e^2} = \dfrac{4\pi\epsilon_0 \hbar^2}{m_e e^2}$ and $\hbar = \dfrac{h}{2\pi}$,

$$E_0 = -\frac{m_e e^4}{8\epsilon_0^2 h^2} = -\frac{m_e e^4}{(2\pi)^2\, 8\epsilon_0^2 \hbar^2} = -\frac{m_e e^4}{32\pi^2\epsilon_0^2 \hbar^2} = -\frac{e^2}{8\pi\epsilon_0}\frac{m_e e^2}{4\pi\epsilon_0 \hbar^2}$$

$$= -\frac{e^2}{8\pi\epsilon_0 a_0} = -\frac{\hbar^2}{2m_e a_0}\frac{m_e e^2}{4\pi\epsilon_0 \hbar^2} = -\frac{\hbar^2}{2m_e a_0^2}$$

7–9. Calculate the probability that a hydrogen $1s$ electron will be found within a distance $2a_0$ from the nucleus.

This problem is similar to Example 7–2. The wave function for the $1s$ orbital of hydrogen is

$$\psi_{100} = \frac{1}{\sqrt{\pi}} \left(\frac{1}{a_0} \right)^{3/2} e^{-\rho}$$

where $\rho = r/a_0$, and the probability that the electron will be found within a distance $2a_0$ from the nucleus is

$$\text{prob} = \int_0^{2\pi} d\phi \int_0^{\pi} d\theta \, \sin\theta \int_0^{2a_0} dr \, r^2 \frac{1}{\pi}\left(\frac{1}{a_0}\right)^3 e^{-2\rho}$$

$$= 4\left(\frac{1}{a_0}\right)^3 \int_0^{2a_0} dr \, r^2 e^{-2\rho} = 4\int_0^{2} d\rho \, \rho^2 e^{-2\rho}$$

$$= 4\left(-\frac{13}{4}e^{-4} + \frac{1}{4}\right) = 1 - 13e^{-4} = 0.762$$

7–10. Calculate the radius of the sphere that encloses a 50% probability of finding a hydrogen $1s$ electron. Repeat the calculation for a 90% probability.

The probability that a $1s$ electron will be found within a distance Da_0 of the nucleus is given by

$$\text{prob}(D) = \int_0^{2\pi} d\phi \int_0^{\pi} d\theta \, \sin\theta \int_0^{Da_0} dr \, r^2 \frac{1}{\pi}\left(\frac{1}{a_0}\right)^3 e^{-2\rho}$$

$$= 4\int_0^{D} d\rho \, \rho^2 e^{-2\rho} = 1 - e^{-2D}(2D^2 + 2D + 1)$$

This equation may be numerically solved for D to find that $D = 1.34$ for $\text{prob}(D) = 0.50$ and $D = 2.66$ for $\text{prob}(D) = 0.90$, so the 50% and 90% probability spheres have radii of $1.34a_0$ and $2.66a_0$, respectively.

7–11. Many problems involving the calculation of average values for the hydrogen atom require doing integrals of the form

$$I_n = \int_0^{\infty} r^n e^{-\beta r} dr$$

This integral can be evaluated readily by starting with the elementary integral

$$I_0(\beta) = \int_0^{\infty} e^{-\beta r} dr = \frac{1}{\beta}$$

Show that the derivatives of $I(\beta)$ are

$$\frac{dI_0}{d\beta} = -\int_0^{\infty} r e^{-\beta r} dr$$

$$\frac{d^2 I_0}{d\beta^2} = \int_0^{\infty} r^2 e^{-\beta r} dr$$

and so on. Using the fact that $I_0(\beta) = 1/\beta$, show that the values of these two integrals are $-1/\beta^2$ and $2/\beta^3$, respectively. Show that, in general,

$$\frac{d^n I_0}{d\beta^n} = (-1)^n \int_0^{\infty} r^n e^{-\beta r} dr = (-1)^n \frac{n!}{\beta^{n+1}}$$

and that

$$I_n = \frac{n!}{\beta^{n+1}}$$

Differentiating $I_0(\beta)$ with respect to β gives

$$I_0(\beta) = \int_0^\infty e^{-\beta r}\, dr$$

$$\frac{dI_0}{d\beta} = \int_0^\infty \frac{d}{d\beta}\left(e^{-\beta r}\right) dr = -\int_0^\infty r e^{-\beta r}\, dr = -I_1$$

$$\frac{d^2 I_0}{d\beta^2} = \int_0^\infty \frac{d}{d\beta}\left(-r e^{-\beta r}\right) dr = \int_0^\infty r^2 e^{-\beta r}\, dr = I_2$$

Alternatively, since $I_0(\beta) = 1/\beta$,

$$\frac{dI_0}{d\beta} = \frac{d}{d\beta}\left(\frac{1}{\beta}\right) = -\frac{1}{\beta^2}$$

$$\frac{d^2 I_0}{d\beta^2} = \frac{d}{d\beta}\left(-\frac{1}{\beta^2}\right) = \frac{2}{\beta^3}$$

Generally,

$$\frac{d^n I_0}{d\beta^n} = (-1)^n \int_0^\infty r^n e^{-\beta r}\, dr = (-1)^n I_n$$

$$= (-1)^n \frac{n!}{\beta^{n+1}}$$

and so

$$I_n = \frac{n!}{\beta^{n+1}}$$

7–12. Prove that the average value of r in the $1s$ and $2s$ states for a hydrogen-like atom is $3a_0/2Z$ and $6a_0/Z$, respectively. (Compare your result to Table 7.4.)

The average value of r, $\langle r \rangle$, is given by

$$\langle r \rangle = \int_0^{2\pi} d\phi \int_0^\pi d\theta\, \sin\theta \int_0^\infty dr\, r^3 \psi_{nlm}(r, \theta, \phi)^* \psi_{nlm}(r, \theta, \phi)$$

The wave functions for the $1s$ and $2s$ states are (Table 7.3)

$$\psi_{100} = \frac{1}{\sqrt{\pi}} \left(\frac{Z}{a_0}\right)^{3/2} e^{-\rho}$$

$$\psi_{200} = \frac{1}{\sqrt{32\pi}} \left(\frac{Z}{a_0}\right)^{3/2} (2 - \rho)e^{-\rho/2}$$

Because these are s orbitals, the integrals over the two angular coordinates give a simple factor of 4π, and

$$\langle r \rangle_{1s} = 4\left(\frac{Z}{a_0}\right)^3 \int_0^\infty dr\, r^3 e^{-2\rho}$$

$$= \frac{4a_0}{Z}\int_0^\infty d\rho\, \rho^3 e^{-2\rho} = \frac{4a_0}{Z}\left(\frac{3!}{16}\right) = \frac{3}{2}\left(\frac{a_0}{Z}\right)$$

$$\langle r \rangle_{2s} = \frac{1}{8}\left(\frac{Z}{a_0}\right)^3 \int_0^\infty dr\, r^3(4 - 4\rho + \rho^2)e^{-\rho}$$

$$= \frac{a_0}{8Z}\int_0^\infty d\rho\,(4\rho^3 - 4\rho^4 + \rho^5)e^{-\rho}$$

$$= \frac{a_0}{8Z}(4\cdot3! - 4\cdot4! + 5!) = \frac{6a_0}{Z}$$

For s orbitals, which have $l = 0$ and $m_l = 0$, Table 7.4 gives $\langle r \rangle_{n00} = 3a_0 n^2/2Z$, which agrees with the results found here for $n = 1, 2$.

7–13. Prove that $\langle V \rangle = 2\langle E \rangle$ and, consequently, that $\langle V \rangle/\langle \hat{T} \rangle = -2$, for a $2p_0$ electron.

The average potential energy of a hydrogen-like $2p_0$ electron is (Equation 7.1)

$$\langle V \rangle = \left\langle -Z\frac{e^2}{4\pi\epsilon_0 r}\right\rangle$$

$$= \int_0^{2\pi} d\phi \int_0^\pi d\theta\, \sin\theta \int_0^\infty dr\, r^2 \psi_{2p_0}^* \left(-\frac{Ze^2}{4\pi\epsilon_0 r}\right)\psi_{2p_0}$$

$$= -\frac{Ze^2}{64\pi\epsilon_0}\left(\frac{Z}{a_0}\right)^3 \int_0^\pi d\theta\, \sin\theta \cos^2\theta \int_0^\infty dr\, r\rho^2 e^{-\rho}$$

$$= -\frac{Ze^2}{96\pi\epsilon_0}\left(\frac{Z}{a_0}\right) \int_0^\infty d\rho\, \rho^3 e^{-\rho}$$

$$= -\frac{Z^2 e^2}{16\pi\epsilon_0 a_0}$$

The total energies of a hydrogen-like atom are (Problem 7–29)

$$E_n = -\frac{Z^2 e^2}{8\pi\epsilon_0 a_0 n^2} = -\frac{Z^2 e^2}{32\pi\epsilon_0 a_0}$$

for $n = 2$, so $\langle V \rangle = 2\langle E \rangle$.

Because $\langle \hat{T} \rangle + \langle V \rangle = \langle E \rangle$, $\langle \hat{T} \rangle = -\langle E \rangle$, or $\langle V \rangle/\langle \hat{T} \rangle = -2$.

7–14. By evaluating the appropriate integrals, compute $\langle r \rangle$ in the $2s$, $2p$, and $3s$ states of the hydrogen atom; compare your results with Table 7.4 and the general formula

$$\langle r \rangle_{nl} = \frac{a_0}{2Z}[3n^2 - l(l+1)]$$

In Problem 7–12 the average value of r in a hydrogen $2s$ orbital was found to be $\langle r \rangle = 6a_0$. The wave functions for the $2p$ and $3s$ states are (Table 7.3)

$$\psi_{210} = \frac{1}{\sqrt{32\pi}}\left(\frac{Z}{a_0}\right)^{3/2} \rho\, e^{-\rho/2}\cos\theta$$

$$\psi_{300} = \frac{1}{81\sqrt{3\pi}}\left(\frac{Z}{a_0}\right)^{3/2}(27 - 18\rho + 2\rho^2)e^{-\rho/3}$$

Since the average value of r does not depend on m_l, the choice of $2p$ orbital is arbitrary. For hydrogen, $Z = 1$, and

$$\langle r \rangle_{2p} = \frac{1}{32\pi a_0^3}\int_0^{2\pi} d\phi \int_0^\pi d\theta\, \sin\theta \cos^2\theta \int_0^\infty dr\, r^3\rho^2 e^{-\rho}$$

$$= \frac{a_0}{16}\left(\frac{2}{3}\right)\int_0^\infty d\rho\, \rho^5 e^{-\rho} = \frac{a_0}{24}(5!) = 5a_0$$

$$\langle r \rangle_{3s} = \frac{1}{3^9\pi a_0^3}\int_0^{2\pi} d\phi \int_0^\pi d\theta\, \sin\theta \int_0^\infty dr\, r^3(27 - 18\rho + 2\rho^2)^2 e^{-2\rho/3}$$

$$= \frac{4a_0}{3^9}\int_0^\infty d\rho\, \rho^3\left[3^6 + 4(3^4\rho^2) + 4\rho^4 - 4(3^5\rho) + 4(3^3\rho^2) - 8(3^2\rho^3)\right]e^{-2\rho/3}$$

$$= \frac{4a_0}{3^9}\left[3^6\cdot 3!\left(\frac{3}{2}\right)^4 + 4\cdot 3^4\cdot 5!\left(\frac{3}{2}\right)^6 + 4\cdot 7!\left(\frac{3}{2}\right)^8 - 4\cdot 3^5\cdot 4!\left(\frac{3}{2}\right)^5 + 4\cdot 3^3\cdot 5!\left(\frac{3}{2}\right)^6 - 8\cdot 3^2\cdot 6!\left(\frac{3}{2}\right)^7\right]$$

$$= \frac{27a_0}{2}$$

The general formula

$$\langle r \rangle_{nl} = \frac{a_0}{2}[3n^2 - l(l+1)]$$

gives $6a_0$, $5a_0$ and $27a_0/2$ for the values of $\langle r \rangle_{nl}$ for the $2s$, $2p$ and $3s$ orbitals, in agreement with the above calculations.

7–15. By evaluating the appropriate integrals, compute $\langle 1/r \rangle$ in the $2s$, $2p$, and $3s$ states of the hydrogen atom; compare your results with Table 7.4 and the general formula

$$\left\langle \frac{1}{r} \right\rangle_{n,l} = \frac{Z}{a_0 n^2}$$

The average value of $1/r$, $\langle 1/r \rangle_{nl}$, is given by

$$\left\langle \frac{1}{r} \right\rangle_{nl} = \int_0^{2\pi} d\phi \int_0^{\pi} d\theta \ \sin\theta \int_0^{\infty} dr \ r\psi_{nlm}(r,\theta,\phi)^* \psi_{nlm}(r,\theta,\phi)$$

The wave functions for the $2s$, $2p$, and $3s$ states are (Table 7.3)

$$\psi_{200} = \frac{1}{\sqrt{32\pi}} \left(\frac{Z}{a_0}\right)^{3/2} (2-\rho)e^{-\rho/2}$$

$$\psi_{210} = \frac{1}{\sqrt{32\pi}} \left(\frac{Z}{a_0}\right)^{3/2} \rho e^{-\rho/2} \cos\theta$$

$$\psi_{300} = \frac{1}{81\sqrt{3\pi}} \left(\frac{Z}{a_0}\right)^{3/2} (27 - 18\rho + 2\rho^2)e^{-\rho/3}$$

The choice of one of the three p orbitals can be made arbitrarily. For hydrogen, $Z = 1$, and

$$\left\langle \frac{1}{r} \right\rangle_{2s} = \frac{1}{8a_0^3} \int_0^{\infty} dr \ r(4 - 4\rho + \rho^2)e^{-\rho}$$

$$= \frac{1}{8a_0} \int_0^{\infty} d\rho \ (4\rho - 4\rho^2 + \rho^3)e^{-\rho}$$

$$= \frac{1}{8a_0}(4 - 4\cdot2! + 3!) = \frac{1}{4a_0}$$

$$\left\langle \frac{1}{r} \right\rangle_{2p} = \frac{1}{32\pi a_0^3} \int_0^{2\pi} d\phi \int_0^{\pi} d\theta \ \sin\theta \cos^2\theta \int_0^{\infty} dr \ r\rho^2 e^{-\rho}$$

$$= \frac{a_0}{16}\left(\frac{2}{3}\right) \int_0^{\infty} d\rho \ \rho^3 e^{-\rho} = \frac{a_0}{24}(3!) = \frac{1}{4a_0}$$

$$\left\langle \frac{1}{r} \right\rangle_{3s} = \frac{1}{3^9\pi a_0^3} \int_0^{2\pi} d\phi \int_0^{\pi} d\theta \ \sin\theta \int_0^{\infty} dr \ r(27 - 18\rho + 2\rho^2)^2 e^{-2\rho/3}$$

$$= \frac{4}{3^9 a_0} \int_0^{\infty} d\rho \ \rho \left[3^6 + 4(3^4\rho^2) + 4\rho^4 - 4(3^5\rho) + 4(3^3\rho^2) - 8(3^2\rho^3)\right]e^{-2\rho/3}$$

$$= \frac{4}{3^9 a_0}\left[3^6\cdot1!\left(\frac{3}{2}\right)^2 + 4\cdot3^4\cdot3!\left(\frac{3}{2}\right)^4 + 4\cdot5!\left(\frac{3}{2}\right)^6 - 4\cdot3^5\cdot2!\left(\frac{3}{2}\right)^3 \right.$$

$$\left. + 4\cdot3^3\cdot3!\left(\frac{3}{2}\right)^4 - 8\cdot3^2\cdot4!\left(\frac{3}{2}\right)^5\right]$$

$$= \frac{1}{9a_0}$$

The general formula

$$\left\langle \frac{1}{r} \right\rangle_{nl} = \frac{Z}{a_0 n^2}$$

gives $1/4a_0$, $1/4a_0$ and $1/9a_0$ for the values of $\langle 1/r \rangle_{nl}$ for the $2s$, $2p$ and $3s$ orbitals, in agreement with the above calculations.

7–16. Use the results in Table 7.4 to show that $1/\langle r \rangle \neq \langle 1/r \rangle$.

From Table 7.4,

$$\langle r \rangle_{nl} = \frac{a_0 n^2}{Z} \left\{ 1 + \frac{1}{2} \left[1 - \frac{l(l+1)}{n^2} \right] \right\} = \frac{a_0}{2Z} \left[3n^2 - l(l+1) \right]$$

$$\left\langle \frac{1}{r} \right\rangle_{nl} = \frac{Z}{a_0 n^2}$$

Thus,

$$\frac{1}{\langle r \rangle_{nl}} = \frac{2Z}{a_0 \left[3n^2 - l(l+1) \right]} \neq \frac{Z}{a_0 n^2} = \left\langle \frac{1}{r} \right\rangle_{nl}$$

A comparison of the results from Problems 7–14 and 7–15 shows this indeed to be the case for the $2s$, $2p$, and $3s$ orbitals of the hydrogen atom.

7–17. Show that the two maxima in the plot of $r^2 R_{20}^2(r)$ against r occur at $(3 \pm \sqrt{5})a_0$. (See Figure 7.2.)

Figure 7.2 is for the hydrogen atom, so in the radial function from Table 7.2, set $Z = 1$ and make the substitution, $\rho = r/a_0$,

$$R_{20}(r) = \left(\frac{1}{2a_0} \right)^{3/2} \left(2 - \frac{r}{a_0} \right) e^{-r/2a_0}$$

$$r^2 R_{20}^2(r) = \left(\frac{1}{2a_0} \right)^3 r^2 \left(2 - \frac{r}{a_0} \right)^2 e^{-r/a_0}$$

To find the extrema of this function, take the derivative with respect to r and set equal to zero:

$$\frac{d}{dr} \left[r^2 R_{20}^2(r) \right] = \frac{1}{8a_0^3} \left[2r \left(2 - \frac{r}{a_0} \right)^2 e^{-r/a_0} - \frac{2r^2}{a_0} \left(2 - \frac{r}{a_0} \right) e^{-r/a_0} - \frac{r^2}{a_0} \left(2 - \frac{r}{a_0} \right)^2 e^{-r/a_0} \right]$$

$$0 = 2 \left(2 - \frac{r}{a_0} \right) - \frac{2r}{a_0} - \frac{r}{a_0} \left(2 - \frac{r}{a_0} \right)$$

$$= 4a_0^2 - 2a_0 r - 2a_0 r - 2a_0 r + r^2$$

$$= r^2 - 6a_0 r + 4a_0^2$$

$$r = (3 \pm \sqrt{5})a_0$$

These two roots are in addition to the roots at $r = 0$ and $r = 2a_0$. Evaluation of the second derivative would determine which extrema are maxima and which are minima, but it is easily seen that $r^2 R_{20}^2(r) = 0$ at $r = 0$ and $r = 2a_0$. Since $r^2 R_{20}^2(r) \geq 0$ for all values of r, $r = 0$ and $r = 2a_0$ are the minima of the function and $r = (3 \pm \sqrt{5})a_0$ correspond to the maxima.

7–18. Calculate the value of $\langle r \rangle$ for the $n = 2$, $l = 1$ state and the $n = 2$, $l = 0$ state of the hydrogen atom. Are you surprised by the answers? Explain.

The average value of r, $\langle r \rangle$, is given by

$$\langle r \rangle = \int d\tau \, \psi_{nl}(r, \theta, \phi)^* r \psi_{nl}(r, \theta, \phi)$$

Use the wave functions in Table 7.3, making the choice of $2p_z$ for the $n = 2$, $l = 1$ state, and with $Z = 1$ to find

$$\langle r \rangle_{20} = \frac{1}{32\pi a_0^3} \int_0^{2\pi} d\phi \int_0^{\pi} d\theta \, \sin\theta \int_0^{\infty} dr \, r^3 \, (2 - \rho)^2 \, e^{-\rho}$$

$$= \frac{a_0}{8} \int_0^{\infty} d\rho \, \rho^3 (2 - \rho)^2 e^{-\rho}$$

$$= \frac{a_0}{8} (4 \cdot 3! - 4 \cdot 4! + 5!)$$

$$= 6a_0$$

$$\langle r \rangle_{21} = \frac{1}{32\pi a_0^3} \int_0^{2\pi} d\phi \int_0^{\infty} d\theta \, \sin\theta \cos^2\theta \int_0^{\infty} dr \, r^3 \rho^2 e^{-\rho}$$

$$= \frac{a_0}{16} \left(\frac{2}{3}\right) \int_0^{\infty} d\rho \, \rho^5 e^{-\rho}$$

$$= \frac{a_0}{16} \left(\frac{2}{3}\right) (5!)$$

$$= 5a_0$$

Alternatively, the general formula in Table 7.4 could have been used. Either way, these results show that for the hydrogen atom, an electron in the $2s$ orbital is farther from the nucleus (on average) than an electron in the $2p$ orbital. This is surprising, the reverse is expected to be true from studying multi-electron systems in general chemistry; note, however, that a one-electron hydrogen-like wave function differs from multi-electron wave functions (Chapter 9). Thus, for a one-electron system, the greater the value of l, the closer, on average, is the electron to the nucleus, but for multi-electron atoms, the effects of electron-electron interactions lead to the reverse being true.

7–19. The average value of r for a hydrogen-like atom can be evaluated in general and is given by (see Table 7.4)

$$\langle r \rangle_{nl} = \frac{n^2 a_0}{Z} \left\{ 1 + \frac{1}{2} \left[1 - \frac{l(l + 1)}{n^2} \right] \right\}$$

Verify this formula explicitly for the ψ_{211} orbital.

First determine $\langle r \rangle_{21}$ directly:

$$\langle r \rangle_{21} = \int_0^{2\pi} d\phi \int_0^\pi d\theta \, \sin\theta \int_0^\infty dr \, r^2 \psi_{211}^* r \psi_{211}$$

$$= \frac{Z^3}{64\pi a_0^3} \int_0^{2\pi} d\phi \int_0^\pi d\theta \, \sin^3\theta \int_0^\infty dr \, r^3 \rho^2 e^{-\rho}$$

$$= \frac{a_0}{32Z} \left(\frac{4}{3}\right) \int_0^\infty d\rho \, \rho^5 e^{-\rho}$$

$$= \frac{a_0}{24Z}(5!) = \frac{5a_0}{Z}$$

Using the equation given in the problem,

$$\langle r \rangle_{21} = \frac{2^2 a_0}{Z} \left\{ 1 + \frac{1}{2}\left[1 - \frac{1(1+1)}{2^2}\right]\right\}$$

$$= \frac{4a_0}{Z}\frac{5}{4} = \frac{5a_0}{Z}$$

7–20. The average value of r^2 for a hydrogen-like atom is given in Table 7.4. Verify the entry explicitly for the ψ_{210} orbital.

First determine $\langle r^2 \rangle_{21}$ directly:

$$\langle r^2 \rangle_{21} = \int_0^{2\pi} d\phi \int_0^\pi \sin\theta \, d\theta \int_0^\infty dr \, r^2 \psi_{210}^* r^2 \psi_{210}$$

$$= \frac{Z^3}{32\pi a_0^3} \int_0^{2\pi} d\phi \int_0^\pi d\theta \, \sin\theta \cos^2\theta \int_0^\infty dr \, r^4 \rho^2 e^{-\rho}$$

$$= \frac{a_0^2}{16Z^2} \left(\frac{2}{3}\right) \int_0^\infty d\rho \, \rho^6 e^{-\rho}$$

$$= \frac{a_0^2}{24Z^2}(6!) = \frac{30a_0^2}{Z^2}$$

Using the equation given in Table 7.4,

$$\langle r^2 \rangle_{21} = \frac{2^4 a_0^2}{Z^2} \left\{ 1 + \frac{3}{2}\left[1 - \frac{1(1+1) - \frac{1}{3}}{2^2}\right]\right\}$$

$$= \frac{16a_0^2}{Z^2}\left[1 + \frac{3}{2}\left(\frac{7}{12}\right)\right]$$

$$= \frac{16a_0^2}{Z^2}\left(\frac{15}{8}\right) = \frac{30a_0^2}{Z^2}$$

7–21. The average values of $1/r$, $1/r^2$, and $1/r^3$ are given in Table 7.4. Verify these entries explicitly for the ψ_{210} orbital.

First determine these values directly:

$$\left\langle \frac{1}{r} \right\rangle_{21} = \int_0^{2\pi} d\phi \int_0^{\pi} \sin\theta d\theta \int_0^{\infty} dr\, r^2 \psi_{210}^* \left(\frac{1}{r} \right) \psi_{210}$$

$$= \frac{Z^3}{32\pi a_0^3} \int_0^{2\pi} d\phi \int_0^{\pi} d\theta\, \sin\theta \cos^2\theta \int_0^{\infty} dr\, r\rho^2 e^{-\rho}$$

$$= \frac{Z}{16a_0} \left(\frac{2}{3} \right) \int_0^{\infty} d\rho\, \rho^3 e^{-\rho}$$

$$= \frac{Z}{24a_0}(3!) = \frac{Z}{4a_0}$$

$$\left\langle \frac{1}{r^2} \right\rangle_{21} = \int_0^{2\pi} d\phi \int_0^{\pi} \sin\theta d\theta \int_0^{\infty} dr\, r^2 \psi_{210}^* \left(\frac{1}{r^2} \right) \psi_{210}$$

$$= \frac{Z^3}{32\pi a_0^3} \int_0^{2\pi} d\phi \int_0^{\pi} d\theta\, \sin\theta \cos^2\theta \int_0^{\infty} dr\, \rho^2 e^{-\rho}$$

$$= \frac{Z^2}{16a_0^2} \left(\frac{2}{3} \right) \int_0^{\infty} d\rho\, \rho^2 e^{-\rho}$$

$$= \frac{Z^2}{24a_0^2}(2!) = \frac{Z^2}{12a_0^2}$$

$$\left\langle \frac{1}{r^3} \right\rangle_{21} = \int_0^{2\pi} d\phi \int_0^{\pi} \sin\theta d\theta \int_0^{\infty} dr\, r^2 \psi_{210}^* \left(\frac{1}{r^3} \right) \psi_{210}$$

$$= \frac{Z^3}{32\pi a_0^3} \int_0^{2\pi} d\phi \int_0^{\pi} d\theta\, \sin\theta \cos^2\theta \int_0^{\infty} dr\, \frac{1}{r}\rho^2 e^{-\rho}$$

$$= \frac{Z^3}{16a_0^3} \left(\frac{2}{3} \right) \int_0^{\infty} d\rho\, \rho e^{-\rho}$$

$$= \frac{Z^3}{24a_0^3}$$

Using the equations given in Table 7.4,

$$\left\langle \frac{1}{r} \right\rangle_{21} = \frac{Z}{a_0(2)^2} = \frac{Z}{4a_0}$$

$$\left\langle \frac{1}{r^2} \right\rangle_{21} = \frac{Z^2}{a_0^2(2)^3(1+\frac{1}{2})} = \frac{Z^2}{12a_0^2}$$

$$\left\langle \frac{1}{r^3} \right\rangle_{21} = \frac{Z^3}{a_0^3(2)^3(1+\frac{1}{2})(1+1)} = \frac{Z^3}{24a_0^3}$$

7–22. In Chapter 4, we learned that if ψ_1 and ψ_2 are solutions of the Schrödinger equation that have the same energy E_n, then $c_1\psi_1 + c_2\psi_2$ is also a solution with that energy. Let $\psi_1 = \psi_{210}$ and $\psi_2 = \psi_{211}$ (see Table 7.3). What is the energy corresponding to $\psi = c_1\psi_1 + c_2\psi_2$, where

$c_1^2 + c_2^2 = 1$? What does this result tell you about the uniqueness of the three p orbitals, p_x, p_y, and p_z?

Recall that the energy of the hydrogen atom depends only on the value of n. Therefore, ψ_{211} and ψ_{210} have the same energy, E_2, and so (see Chapter 4) the energy corresponding to $\psi = c_1\psi_1 + c_2\psi_2$ where $c_1^2 + c_2^2 = 1$ is also E_2. The three p orbitals (p_x, p_y, and p_z), therefore, are not a unique representation of the three degenerate orbitals for $n = 2$ and $l = 1$.

7–23. Verify the linear combinations given in Equations 7.32.

There are four linear combinations to verify using the spherical harmonics from Table 7.1 and Euler's relations (Problem A–6).

$$\frac{1}{\sqrt{2}}\left(Y_2^{-1} - Y_2^{+1}\right) = \left(\frac{15}{16\pi}\right)^{1/2}\sin\theta\cos\theta e^{-i\phi} + \left(\frac{15}{16\pi}\right)^{1/2}\sin\theta\cos\theta e^{+i\phi}$$

$$= \left(\frac{15}{16\pi}\right)^{1/2}\sin\theta\cos\theta\left(e^{+i\phi} + e^{-i\phi}\right)$$

$$= \left(\frac{15}{4\pi}\right)^{1/2}\sin\theta\cos\theta\cos\phi$$

$$\frac{i}{\sqrt{2}}\left(Y_2^{-1} + Y_2^{+1}\right) = i\left(\frac{15}{16\pi}\right)^{1/2}\sin\theta\cos\theta e^{-i\phi} - i\left(\frac{15}{16\pi}\right)^{1/2}\sin\theta\cos\theta e^{+i\phi}$$

$$= -i\left(\frac{15}{16\pi}\right)^{1/2}\sin\theta\cos\theta\left(e^{+i\phi} - e^{-i\phi}\right)$$

$$= \left(\frac{15}{4\pi}\right)^{1/2}\sin\theta\cos\theta\sin\phi$$

$$\frac{1}{\sqrt{2}}\left(Y_2^{+2} + Y_2^{-2}\right) = \left(\frac{15}{64\pi}\right)^{1/2}\sin^2\theta e^{+2i\phi} + \left(\frac{15}{64\pi}\right)^{1/2}\sin^2\theta e^{-2i\phi}$$

$$= \left(\frac{15}{64\pi}\right)^{1/2}\sin^2\theta\left(e^{+2i\phi} + e^{-2i\phi}\right)$$

$$= \left(\frac{15}{16\pi}\right)^{1/2}\sin^2\theta\cos 2\phi$$

$$\frac{1}{\sqrt{2i}}\left(Y_2^{+2} - Y_2^{-2}\right) = \frac{1}{i}\left(\frac{15}{64\pi}\right)^{1/2}\sin^2\theta e^{+2i\phi} - \frac{1}{i}\left(\frac{15}{64\pi}\right)^{1/2}\sin^2\theta e^{-2i\phi}$$

$$= \frac{1}{i}\left(\frac{15}{64\pi}\right)^{1/2}\sin^2\theta\left(e^{+2i\phi} - e^{-2i\phi}\right)$$

$$= \left(\frac{15}{16\pi}\right)^{1/2}\sin^2\theta\sin 2\phi$$

7–24. Show that the total probability density of the $2p$ orbitals is spherically symmetric by evaluating $\sum\limits_{m_l=-1}^{1} \left|\psi_{21m_l}\right|^2$. (Use the wave functions in Table 7.3.)

Since the wave functions in Table 7.3 are complex valued, the probability density is $\psi_{21m_l}^* \psi_{21m_l} = \left|\psi_{21m_l}\right|^2$.

$$\sum_{m_l=-1}^{1} \left|\psi_{21m_l}\right|^2 = \frac{1}{32\pi}\left(\frac{Z}{a_0}\right)^3 \rho^2 e^{-\rho}\left(\cos^2\theta + \frac{1}{2}\sin^2\theta + \frac{1}{2}\sin^2\theta\right)$$

$$= \frac{Z^3\rho^2 e^{-\rho}}{32\pi a_0^3}\left(\cos^2\theta + \sin^2\theta\right)$$

$$= \frac{Z^3\rho^2 e^{-\rho}}{32\pi a_0^3}$$

The sum depends only on the variable r (through ρ), so the total probability density of the $2p$ orbitals is spherically symmetric.

7–25. Show that the total probability density of the $3d$ orbitals is spherically symmetric by evaluating $\sum\limits_{m_l=-2}^{2} \left|\psi_{32m_l}\right|^2$. (Use the wave functions in Table 7.3.)

Since the wave functions in Table 7.3 are complex valued, the probability density is $\psi_{32m_l}^* \psi_{32m_l} = \left|\psi_{32m_l}\right|^2$.

$$\sum_{m_l=-2}^{2} \left|\psi_{32m_l}\right|^2 = \frac{1}{81^2\pi}\left(\frac{Z}{a_0}\right)^3 \rho^4 e^{-2\rho/3}$$

$$\times \left[\frac{(3\cos^2\theta - 1)^2}{6} + \sin^2\theta\cos^2\theta + \sin^2\theta\cos^2\theta + \frac{\sin^4\theta}{4} + \frac{\sin^4\theta}{4}\right]$$

$$= \frac{Z^3\rho^4 e^{-2\rho/3}}{81^2\pi a_0^3}\left[\frac{(3\cos^2\theta - 1)^2}{6} + 2\sin^2\theta\cos^2\theta + \frac{\sin^4\theta}{2}\right]$$

$$= \frac{Z^3\rho^4 e^{-2\rho/3}}{(81)^2 6\pi a_0^3}\left[(3\cos^2\theta - 1)^2 + 12\sin^2\theta\cos^2\theta + 3\sin^4\theta\right]$$

Now substitute $\sin^2\theta = 1 - \cos^2\theta$ into the above expression to get

$$\sum_{m_l=-2}^{2} \left|\psi_{32m_l}\right|^2 = \frac{Z^3 \rho^4 e^{-2\rho/3}}{(81)^2 6\pi a_0^3} \left[9\cos^4\theta - 6\cos^2\theta + 1 + 12(1-\cos^2\theta)\cos^2\theta + 3(1-\cos^2\theta)^2\right]$$

$$= \frac{Z^3 \rho^4 e^{-2\rho/3}}{(81)^2 6\pi a_0^3} \left[9\cos^4\theta - 6\cos^2\theta + 1 + 12\cos^2\theta - 12\cos^4\theta\right.$$

$$\left. +3 - 6\cos^2\theta + 3\cos^4\theta\right]$$

$$= \frac{4Z^3 \rho^4 e^{-2\rho/3}}{(81)^2 6\pi a_0^3} = \frac{2Z^3 \rho^4 e^{-2\rho/3}}{(81)^2 3\pi a_0^3}$$

The sum depends only on the variable r (through ρ), so the total probability density of the $3d$ orbitals is spherically symmetric.

7–26. Show that the sum of the probability densities for the $n = 3$ states of a hydrogen atom is spherically symmetric. Do you expect this to be true for all values of n? Explain.

Problem 7–25 showed that the sum of the probability densities of the $3d$ orbitals is spherically symmetric. The probability density of the $3s$ orbital is spherically symmetric, and so only the sum of the probability densities of the $3p$ orbitals remains. The angular dependence, however, of the $3p$ orbitals is the same as that of the $2p$ orbitals, and in Problem 7–24 the sum of the squares of the $2p$ orbitals was shown to be spherically symmetric. Therefore, the same must be true for the $3p$ orbitals. Thus, the sum of the probability densities for the $n = 3$ states of the hydrogen atom is spherically symmetric. This will be the case for all values of n. Recall from Problem 6–37 that

$$\sum_{m=-l}^{+l} |Y_l^{m_l}(\theta, \phi)|^2 = \text{constant} \tag{1}$$

Summing all of the probability densities corresponding to any given n requires evaluating a sum similar to that in Equation 1. Because the sum is equal to a constant, it cannot have any angular dependence and will depend only on r. Such a sum is spherically symmetric.

7–27. The designations of the d orbitals can be rationalized in the following way. Equation 7.32 shows that d_{xz} goes as $\sin\theta\cos\theta\cos\phi$. Using the relation between cartesian and spherical coordinates, show that $\sin\theta\cos\theta\cos\phi$ is proportional to xz. Similarly, show that $\sin\theta\cos\theta\sin\phi$ (d_{yz}) is proportional to yz, that $\sin^2\theta\cos 2\phi$ ($d_{x^2-y^2}$) is proportional to $x^2 - y^2$, and that $\sin^2\theta\sin 2\phi$ (d_{xy}) is proportional to xy.

The relations between Cartesian and spherical coordinates are

$$x = r\sin\theta\cos\phi \qquad y = r\sin\theta\sin\phi \qquad z = r\cos\theta$$

Thus, for the d_{xz} and d_{yz} orbitals,

$$r^2 \sin \theta \cos \theta \cos \phi = xz$$

$$\sin \theta \cos \theta \cos \phi \propto xz$$

$$r^2 \sin \theta \cos \theta \sin \phi = yz$$

$$\sin \theta \cos \theta \sin \phi \propto yz$$

Likewise, for the $d_{x^2-y^2}$ and d_{xy} orbitals,

$$r^2 \sin^2 \theta \cos^2 \phi - r^2 \sin^2 \theta \sin^2 \phi = x^2 - y^2$$

$$r^2 \sin^2 \theta \left(\cos^2 \phi - \sin^2 \phi \right) = x^2 - y^2$$

$$r^2 \sin^2 \theta \cos 2\phi = x^2 - y^2$$

$$\sin^2 \theta \cos 2\phi \propto x^2 - y^2$$

$$r^2 \sin^2 \theta \cos \phi \sin \phi = xy$$

$$\tfrac{1}{2} r^2 \sin^2 \theta \sin 2\phi = xy$$

$$\sin^2 \theta \sin 2\phi \propto xy$$

7–28. What is the degeneracy of each of the hydrogen atomic energy levels (neglecting spin-orbit interactions)?

The energy depends only on the quantum number n:

$$E_n = -\frac{e^2}{8\pi \varepsilon_0 a_0 n^2} \qquad n = 1, 2, \ldots$$

For a given value of n, there are $n - 1$ allowed values of l. For each l there are $2l + 1$ allowed values of m_l. All of these combinations of l and m_l are degenerate, and so the total number of energy sublevels for one value of n is

$$\sum_{l=0}^{n-1}(2l + 1) = 2\sum_{l=0}^{n-1} l + \sum_{l=0}^{n-1} 1 = 2\frac{(n - 1)n}{2} + n = n^2$$

7–29. Set up the Hamiltonian operator for the system of an electron interacting with a fixed nucleus of atomic number Z. The simplest such system is singly ionized helium, where $Z = 2$. We will call this a hydrogen-like system. Observe that the only difference between this Hamiltonian operator and the hydrogen Hamiltonian operator is the correspondence that e^2 for the hydrogen atom becomes Ze^2 for the hydrogen-like ion. Consequently, show that the energy becomes (cf. Equation 7.11)

$$E_n = -\frac{m_e Z^2 e^4}{8\epsilon_0^2 h^2 n^2} \qquad n = 1, 2, \ldots$$

Furthermore, now show that the solutions to the radial equation, Equation 7.10, are

$$R_{nl}(r) = -\left\{\frac{(n-l-1)!}{2n[(n+l)!]^3}\right\}^{1/2}\left(\frac{2Z}{na_0}\right)^{l+3/2} r^l e^{-Zr/na_0} L_{n+l}^{2l+1}\left(\frac{2Zr}{na_0}\right)$$

Show that the $1s$ orbital for this system is

$$\psi_{1s} = \frac{1}{\sqrt{\pi}}\left(\frac{Z}{a_0}\right)^{3/2} e^{-Zr/a_0}$$

and show that it is normalized. Show that

$$\langle r \rangle = \frac{3\,a_0}{2Z} \qquad \text{and} \qquad r_{\text{mp}} = \frac{a_0}{Z}$$

Last, compare the ionization energy of a hydrogen atom and that of a singly ionized helium atom. Express your answer in kilojoules per mole.

The Hamiltonian operator for a hydrogen-like system is

$$\hat{H} = -\frac{\hbar^2}{2m_{\text{e}}}\nabla^2 - \frac{Ze^2}{4\pi\epsilon_0 r}$$

The only difference between this Hamiltonian operator and that of a hydrogen atom is that e^2 is replaced by Ze^2. Because e appears nowhere else, the results of this problem can be obtained by replacing e^2 wherever it appears in the hydrogen atom results by Ze^2. For example, the expression for the energy of a hydrogen atom is (Equation 7.11)

$$E_n = -\frac{m_{\text{e}}e^4}{8\epsilon_0^2 h^2 n^2} \qquad n = 1, 2, \ldots$$

and so the energy of a hydrogen-like atom is given by

$$E_n = -\frac{m_{\text{e}}(Ze^2)^2}{8\epsilon_0^2 h^2 n^2} = -\frac{m_{\text{e}}Z^2 e^4}{8\epsilon_0^2 h^2 n^2} \qquad n = 1, 2, \ldots \tag{1}$$

The expression for the Bohr radius, $\epsilon_0 h^2/\pi m_{\text{e}}e^2$, contains e^2, so likewise, wherever a_0 appears in an expression for a hydrogen-like atom it needs to be replaced by a_0/Z.

Consequently, Equation 7.14 becomes

$$R_{nl}(r) = -\left\{\frac{(n-l-1)!}{2n[(n+l)!]^3}\right\}^{1/2}\left(\frac{2Z}{na_0}\right)^{l+3/2} r^l e^{-Zr/na_0} L_{n+l}^{2l+1}\left(\frac{2Zr}{na_0}\right)$$

The hydrogen atomic $1s$ orbital for a hydrogen atom is

$$\psi_{1s} = \frac{1}{\sqrt{\pi}a_0^{3/2}} e^{-r/a_0}$$

and becomes, for a hydrogen-like atom of atomic number Z,

$$\psi_{1s} = \frac{1}{\sqrt{\pi}}\left(\frac{Z}{a_0}\right)^{3/2} e^{-Zr/a_0} \tag{2}$$

To see whether this function is normalized, evaluate

$$\int d\tau \, \psi_{1s}^* \psi_{1s} = \int_0^{2\pi} d\phi \int_0^{\pi} d\theta \, \sin\theta \int_0^{\infty} dr \, r^2 \frac{Z^3}{\pi a_0^3} e^{-2Zr/a_0}$$

$$= \frac{4Z^3}{a_0^3} \int_0^{\infty} dr \, r^2 e^{-2Zr/a_0}$$

$$= \frac{4Z^3}{a_0^3} \frac{2! a_0^3}{8Z^3} = 1$$

Now evaluate $\langle r \rangle_{1s}$ using Equation 2.

$$\langle r \rangle_{1s} = \int_0^{2\pi} d\phi \int_0^{\pi} d\theta \, \sin\theta \int_0^{\infty} dr \, r^3 \left(\frac{Z^3}{\pi a_0^3} \right) e^{-2Zr/a_0}$$

$$= \frac{4Z^3}{a_0^3} \int_0^{\infty} dr \, r^3 e^{-2Zr/a_0}$$

$$= \frac{4Z^3}{a_0^3} \frac{3! a_0^4}{16Z^4} = \frac{3a_0}{2Z}$$

The value of r_{mp} is found by determining the value of r where the derivative of the radial probability function is equal to zero:

$$\frac{d}{dr} \left(r^2 \psi_{1s}^* \psi_{1s} \right) = 0$$

$$\frac{d}{dr} \left(r^2 \frac{Z^3}{\pi a_0^3} e^{-2Zr/a_0} \right) = 0$$

$$2r e^{-2Zr/a_0} - \frac{2Zr^2}{a_0} e^{-2Zr/a_0} = 0$$

$$a_0 - Zr = 0$$

$$r = \frac{a_0}{Z}$$

The ionization energy is given by $-E_1$, and so from Equation 1,

$$IE = \frac{m_e Z^2 e^4}{8\epsilon_0^2 h^2 (1)^2} = \left(1313 \, \text{kJ} \cdot \text{mol}^{-1} \right) Z^2 = (13.61 \, \text{eV}) \, Z^2$$

Therefore,

$$IE_H = 1313 \, \text{kJ} \cdot \text{mol}^{-1} = 13.61 \, \text{eV} \quad \text{and} \quad IE_{He^+} = 5252 \, \text{kJ} \cdot \text{mol}^{-1} = 54.44 \, \text{eV}$$

7–30. What is the ratio of the ground-state energy of atomic hydrogen to that of atomic deuterium?

In the approximation of the fixed (or infinitely heavy) nucleus used in Chapter 7, the ratio is one, since the ground state energy (Equation 7.11, with $n = 1$) depends only on the mass of the electron:

$$E_1 = -\frac{m_e e^4}{8\epsilon_0^2 h^2}$$

Consequently, to this level of approximation, the two energies are equal.

The hydrogen atom, however, is a two-body system, as is the deuterium atom, and Problem 5–6 shows that the equations of motion for such a system can be reduced to a one-body problem by using the reduced mass,

$$\mu = \frac{m_1 m_2}{m_1 + m_2}$$

Thus, relaxing the fixed nucleus approximation for the hydrogen atom requires only replacing m_e by μ wherever it appears, and the ground state energy becomes

$$E_1 = -\frac{\mu e^4}{8\epsilon_0^2 h^2}$$

which will be slightly different for these two isotopes of hydrogen. In fact,

$$\frac{E_{\text{H}}}{E_{\text{D}}} = \frac{\mu_{\text{H}}}{\mu_{\text{D}}}$$

For hydrogen,

$$\mu_{\text{H}} = \frac{m_1 m_2}{m_1 + m_2} = \frac{\left(1.672\,6231 \times 10^{-27}\,\text{kg}\right)\left(9.109\,3897 \times 10^{-31}\,\text{kg}\right)}{\left(1.672\,6231 \times 10^{-27}\,\text{kg}\right) + \left(9.109\,3897 \times 10^{-31}\,\text{kg}\right)}$$

$$= 9.104\,431 \times 10^{-31}\,\text{kg}$$

For deuterium, the mass of the nucleus is the sum of the proton mass and the neutron mass, or $3.343\,586 \times 10^{-27}$ kg:

$$\mu_{\text{H}} = \frac{m_1 m_2}{m_1 + m_2} = \frac{\left(3.343\,586 \times 10^{-27}\,\text{kg}\right)\left(9.109\,3897 \times 10^{-31}\,\text{kg}\right)}{\left(3.343\,586 \times 10^{-27}\,\text{kg}\right) + \left(9.109\,3897 \times 10^{-31}\,\text{kg}\right)}$$

$$= 9.106\,908 \times 10^{-31}\,\text{kg}$$

Thus,

$$\frac{E_{\text{H}}}{E_{\text{D}}} = \frac{9.104\,431 \times 10^{-31}\,\text{kg}}{9.106\,908 \times 10^{-31}\,\text{kg}} = 0.999\,7279$$

7–31. The virial theorem is proved in general in Problem 5–43. Show that if $V(x, y, z)$ is a coulombic potential

$$V(x, y, z) = \frac{Ze^2}{4\pi\epsilon_0(x^2 + y^2 + z^2)^{1/2}}$$

then

$$\langle V \rangle = -2\langle \hat{T} \rangle = 2\langle E \rangle \tag{1}$$

where

$$\langle E \rangle = \langle \hat{T} \rangle + \langle V \rangle$$

We proved that this result is valid for a $1s$ electron in Section 7.2 and for a $2s$ electron in Example 7–4. Although we proved equation 1 only for the case of one electron in the field of one nucleus, equation 1 is valid for many-electron atoms and molecules. The proof is a straightforward extension of the proof developed in this problem.

The virial theorem is

$$\left\langle x\frac{\partial V}{\partial x} + y\frac{\partial V}{\partial y} + z\frac{\partial V}{\partial z} \right\rangle = 2\langle \hat{T} \rangle$$

and for a coulombic potential

$$x\frac{\partial V}{\partial x} + y\frac{\partial V}{\partial y} + z\frac{\partial V}{\partial z} = -\frac{Ze^2}{4\pi\epsilon_0}\left(\frac{x^2}{r^3} + \frac{y^2}{r^3} + \frac{z^2}{r^3}\right)$$

$$= \frac{Ze^2}{4\pi\epsilon_0 r} = -V$$

and so by the virial theorem, $\langle V \rangle = -2\langle \hat{T} \rangle$. Because $\langle \hat{T} \rangle + \langle V \rangle = \langle E \rangle$, then

$$\langle E \rangle = \langle \hat{T} \rangle - 2\langle \hat{T} \rangle = -\langle \hat{T} \rangle$$

and $\langle V \rangle = 2\langle E \rangle$ for a coulombic potential.

7–32. Show that Equation 7.36 reduces to Equation 7.35 in the case of a circular orbit.

In a circular orbit the two vectors $\mathbf{r}$ and $\mathbf{v}$ are orthogonal to each other. Therefore, $|\mathbf{r} \times \mathbf{v}| = rv\sin\theta = rv\sin 90° = rv$, and

$$m = |\mathbf{m}| = \frac{q\,|\mathbf{r} \times \mathbf{v}|}{2} = \frac{qrv}{2}$$

7–33. Show that V in Equation 7.42 has units of energy.

From Equation 7.40,

$$\mathbf{F} = q\,(\mathbf{v} \times \mathbf{B})$$

$$\text{kg·m·s}^{-2} = \text{C·m·s}^{-1}\text{·T}$$

$$\text{T} = \text{kg·C}^{-1}\text{·s}^{-1}$$

since $1\,\text{N} = 1\,\text{kg·m·s}^{-2}$. Therefore, in Equation 7.42,

$$\frac{|e|B_z}{2m_e}L_z \sim \frac{\text{C·T}}{\text{kg}}\,\text{kg·m}^2\text{·s}^{-1} = \frac{\text{C·kg·C}^{-1}\text{·s}^{-1}}{\text{kg}}\,\text{kg·m}^2\text{·s}^{-1} = \text{kg·m}^2\text{·s}^{-2} = \text{J}$$

7–34. Verify the value of a Bohr magneton, given in Equation 7.47.

$$\frac{|e|\hbar}{2m_e} = \frac{\left(1.602\,177\,33 \times 10^{-19}\,\text{C}\right)\left(1.054\,572\,66 \times 10^{-34}\,\text{J·s}\right)}{2\left(9.109\,3897 \times 10^{-31}\,\text{kg}\right)}$$

$$= 9.274\,015\,41 \times 10^{-24}\,\text{J·T}^{-1}$$

7–35. Superconducting magnets have magnetic field strengths of the order of 15 T. Calculate the magnitude of the splitting shown in Figure 7.9 for a magnetic field of 15 T. Compare your result with the energy difference between the unperturbed $1s$ and $2p$ levels. Show that the three distinct transitions shown in Figure 7.9 lie very close together.

This problem is very similar to Example 7–5, and as noted there, Equation 7.45 shows that the splitting of the 3 m_l levels in the $2p$ state, and consequently the splitting in the spectrum, is given by

$$\Delta E = \beta_B m_l B_z$$

$$= \left(9.274\,015\,41 \times 10^{-24}\,\text{J·T}^{-1}\right)(15\,\text{T})\,m_l$$

$$= \left(1.4 \times 10^{-22}\,\text{J}\right)m_l \qquad m_l = 0, \pm 1$$

The energy difference between the unperturbed $1s$ and $2p$ levels is

$$E_{2p} - E_{1s} = -\frac{m_e e^4}{8\epsilon_0^2 \hbar^2}\left(\frac{1}{4} - 1\right) = 1.635 \times 10^{-18}\,\text{J}$$

The magnitude of the splitting caused by the magnetic field is less than 0.01% of the energy difference between the two unperturbed levels, and the three lines in the spectrum lie very close togther.

7–36. Show that the force acting upon a magnetic dipole moment in a magnetic field B_z that varies in the z direction is given by $F_z = m_z \partial B_z / \partial z$.

In general, force is given as the negative gradient of the potential energy, and the potential energy of a magnetic dipole in a magnetic field is given by Equation 7.39,

$$\mathbf{F} = -\nabla V$$

$$= -\nabla \left(-\mathbf{m} \cdot \mathbf{B} \right)$$

$$= \left(\mathbf{i} \frac{\partial}{\partial x} m_x B_x + \mathbf{j} \frac{\partial}{\partial y} m_y B_y + \mathbf{k} \frac{\partial}{\partial z} m_z B_z \right)$$

The magnetic dipole is a constant, and the information given in the problem says that the magnetic field only varies in the z direction. Hence, only the partial derivative with respect to z is non-zero, and

$$F_z = m_z \frac{\partial B_z}{\partial z}$$

7–37. Explain why the spin functions α and β can be represented by $\mid \frac{1}{2} \frac{1}{2} \rangle$ and $\mid \frac{1}{2} \ -\frac{1}{2} \rangle$ in Dirac notation.

In the Dirac notation for angular momentum wave functions, two quantum numbers are specified. The first refers to the magnitude of the angular momentum and the second to its projection on the z-axis. Thus, for orbital angular momentum, $\mid l m_l \rangle$ specifies a wave function for which

$$\hat{L}^2 \mid l m_l \rangle = \hbar^2 l(l + 1) \mid l m_l \rangle$$

and

$$\hat{L}_z \mid l m_l \rangle = \hbar m_l \mid l m_l \rangle$$

For the spin functions α and β, $l \to s = 1/2$ and $m_l \to m_s = \pm 1/2$, so in Dirac notation the functions are $\mid s m_s \rangle$, or $\mid \frac{1}{2} \frac{1}{2} \rangle$ and $\mid \frac{1}{2} -\frac{1}{2} \rangle$, respectively, where

$$\hat{S}^2 \left| \frac{1}{2} \pm \frac{1}{2} \right\rangle = \hbar^2 \frac{3}{4} \left| \frac{1}{2} \pm \frac{1}{2} \right\rangle$$

and

$$\hat{S}_z \left| \frac{1}{2} \pm \frac{1}{2} \right\rangle = \pm \frac{\hbar}{2} \left| \frac{1}{2} \pm \frac{1}{2} \right\rangle$$

7–38. Townsend (see end-of-chapter references or Problem 7–43) gives the following formula for the shift of a hydrogen atom energy level due to spin-orbit coupling:

$$\Delta E_{so} = \frac{m_e c^2 Z^4 \alpha^4}{4 n^3 (l + \frac{1}{2}) l(l + 1)} \times \begin{cases} l & j = l + \frac{1}{2} \\ -(l + 1) & j = l - \frac{1}{2} \end{cases}$$

The only new quantity in this expression is α, which is called the *fine structure constant*, and is given by $e^2 / 4\pi \epsilon_0 \hbar c$. Use this formula for ΔE_{so} to show that the difference in energy between two states with the same value of n and l is

$$\text{diff} = \frac{m_e c^2 Z^4 \alpha^4}{2n^3 l(l+1)} = \frac{5.8437 Z^4 \text{ cm}^{-1}}{n^3 l(l+1)}$$

Calculate the difference in energies between the $2p\ ^2P_{1/2}$ and $2p\ ^2P_{3/2}$ states of a hydrogen atom and compare your results to what you obtain from Table 7.7. Do the same for the $3p\ ^2P_{1/2}$ and $3p\ ^2P_{3/2}$ states and the $4p\ ^2P_{1/2}$ and $4p\ ^2P_{3/2}$ states.

Two hydrogen atom states with the same values of n and l will have $j = l + 1/2$ and $j = l - 1/2$, respectively. Consequently, the difference in their spin-orbit energies will be

$$\text{diff} = \Delta E_{so}^{j=l+1/2} - \Delta E_{so}^{j=l-1/2}$$

$$= \frac{m_e c^2 Z^4 \alpha^4}{4n^3(l+\frac{1}{2})l(l+1)}\{l - [-(l+1)]\}$$

$$= \frac{m_e c^2 Z^4 \alpha^4}{4n^3(l+\frac{1}{2})l(l+1)}(2l+1)$$

$$= \frac{m_e c^2 Z^4 \alpha^4}{2n^3 l(l+1)}$$

$$= \frac{(9.109\,3897 \times 10^{-31}\text{ kg})(2.997\,924\,58 \times 10^8 \text{ m·s}^{-1})^2 (7.297\,353\,08 \times 10^{-3})^4\ Z^4}{2n^3 l(l+1)}$$

$$= \frac{(1.160\,812 \times 10^{-22}\text{ J})\ Z^4}{n^3 l(l+1)}$$

For the hydrogen atom, $Z = 1$ and converting to cm^{-1} ($1\text{ J} = 5.034\,11 \times 10^{22}$ cm^{-1}) gives

$$\text{diff} = \frac{5.8437 \text{ cm}^{-1}}{n^3 l(l+1)}$$

This formula is used to calculate the difference in energies for the requested states and compared with those determined from the experimental values of Table 7.7 in the following table:

States		n	l	Calculated / cm^{-1}	Experimental / cm^{-1}
$2p\ ^2P_{3/2}$	$2p\ ^2P_{1/2}$	2	1	0.3652	0.3659
$3p\ ^2P_{3/2}$	$3p\ ^2P_{1/2}$	3	1	0.1082	0.1084
$4p\ ^2P_{3/2}$	$4p\ ^2P_{1/2}$	4	1	0.0457	0.0457

7–39. Use the equation of the previous problem to calculate the difference in energy between the $3d\ ^2D_{3/2}$ and $3d\ ^2D_{5/2}$ states and the $4d\ ^2D_{3/2}$ and $4d\ ^2D_{5/2}$ states of a hydrogen atom. How about for the $4f\ ^2F_{5/2}$ and $4f\ ^2F_{7/2}$ states?

The equation of the previous problem is used along with values from Table 7.7 to give the following table:

States		n	l	Calculated / cm^{-1}	Experimental / cm^{-1}
$3d\,^2D_{5/2}$	$3d\,^2D_{3/2}$	3	2	0.0361	0.0362
$4d\,^2D_{5/2}$	$4d\,^2D_{3/2}$	4	2	0.0152	0.0153
$4f\,^2F_{7/2}$	$4f\,^2F_{5/2}$	4	3	0.0076	0.0077

7–40. Use the formula in Problem 7–38 to determine the value of the fine structure constant. Take its reciprocal.

From Problem 7–38,

$$\alpha = \frac{e^2}{4\pi\epsilon_0\hbar c}$$

$$= \frac{\left(1.602\,177\,33 \times 10^{-19}\,\text{C}\right)^2}{\left(1.112\,650\,056 \times 10^{-10}\,\text{C}^2\cdot\text{J}^{-1}\cdot\text{m}^{-1}\right)\left(1.054\,572\,66 \times 10^{-34}\,\text{J}\cdot\text{s}\right)\left(2.997\,924\,58 \times 10^8\,\text{m}\cdot\text{s}^{-1}\right)}$$

$$= 7.297\,353\,08 \times 10^{-3}$$

$$\frac{1}{\alpha} = 137.035\,89$$

7–41. Repeat Problem 7–38 for a singly ionized helium atom. Go to *http://physics.nist.gov/PhysRefData.ASD/levels_form.html* to see the energy levels of He$^+$, and compare your results to the experimental values.

Adapting the equation from Problem 7–38 to a one-electron ion of atomic number Z is simply a matter of keeping the factor of Z^4 in the final step. That is,

$$\text{diff} = \frac{\left(5.8437\,\text{cm}^{-1}\right)Z^4}{n^3 l(l+1)}$$

For the singly ionized helium atom, which has just one electron, $Z = 2$, and the following results are calculated and compared with the experimental values of the website.

States		n	l	Calculated / cm^{-1}	Experimental / cm^{-1}
$2p\,^2P_{3/2}$	$2p\,^2P_{1/2}$	2	1	5.8437	5.8571
$3p\,^2P_{3/2}$	$3p\,^2P_{1/2}$	3	1	1.7315	1.7355
$4p\,^2P_{3/2}$	$4p\,^2P_{1/2}$	4	1	0.7305	0.7321

7–42. Repeat Problem 7–39 for a singly ionized helium atom. Go to *http://physics.nist.gov/PhysRefData.ASD/levels_form.html* to see the energy levels of He$^+$, and compare your results to the experimental values.

Using the equation for the difference in spin-orbit energies as modified in the previous problem and the experimental values from the website gives the following table:

States		n	l	Calculated / cm^{-1}	Experimental / cm^{-1}
$3d\,^2\mathrm{D}_{5/2}$	$3d\,^2\mathrm{D}_{3/2}$	3	2	0.5772	0.5784
$4d\,^2\mathrm{D}_{5/2}$	$4d\,^2\mathrm{D}_{3/2}$	4	2	0.2435	0.2440
$4f\,^2\mathrm{F}_{7/2}$	$4f\,^2\mathrm{F}_{5/2}$	4	3	0.1217	0.1220

7–43. This problem develops a heuristic derivation of the formula for ΔE_{so}, the shift in the hydrogen atom energy level due to spin-orbit interaction (the first formula in Problem 7–38). We start with an explicit formula for the Hamiltonian operator for spin-orbit interaction (see Townsend, end-of–chapter references):

$$\hat{H}_{\mathrm{so}} = \frac{Ze^2}{2m_e^2 c^2 \kappa_0 r^3} \hat{\mathbf{L}} \cdot \hat{\mathbf{S}}$$

We want to take the average of this quantity. Show that $\hat{H}_{\mathrm{so}}$ can be written as

$$\hat{H}_{\mathrm{so}} = \frac{Ze^2}{4m_e^2 c^2 \kappa_0 r^3} (\hat{J}^2 - \hat{L}^2 - \hat{S}^2)$$

(Remember that $\hat{\mathbf{J}} = \hat{\mathbf{L}} + \hat{\mathbf{S}}$.) Although the inclusion of $\hat{H}_{\mathrm{so}}$ into the hydrogen atom Hamiltonian operator alters the wave functions, if the effect of $\hat{H}_{\mathrm{so}}$ is small (as we are assuming), then we can use the hydrogen atomic wave functions as a good approximation. The hydrogen atomic wave functions are eigenfunctions of $\hat{L}^2$, $\hat{L}_z$, $\hat{S}^2$, and $\hat{S}_z$, but it's possible to build linear combinations of them that are eigenfunctions of $\hat{J}^2$, $\hat{J}_z$, $\hat{L}^2$, and $\hat{S}^2$ (denote them by $| n, j, m_j, l, s \rangle$), in which case j, m_j, l, and s are good quantum numbers. Use these wave functions to argue that

$$\langle n, j, m_j, l, s \,|\, \hat{H}_{\mathrm{so}} |\, n, j, m_j, l, s \rangle$$

$$= \frac{Ze^2}{4m_e^2 c^2 \kappa_0} \left\langle \frac{1}{r^3} \right\rangle_{n,l} \langle j, m_j, l, s \,|\, \hat{J}^2 - \hat{L}^2 - \hat{S}^2 \,|\, j, m_j, l, s \rangle$$

Now use the fact that (see Table 7.4)

$$\left\langle \frac{1}{r^3} \right\rangle_{n,l} = \frac{Z^3}{a_0^3 n^3 l(l+1)(l+\frac{1}{2})}$$

and that

$$(\hat{J}^2 - \hat{L}^2 - \hat{S}^2) \,|\, j, m_j, l, s \rangle = \hbar^2 \left[j(j+1) - l(l+1) - s(s+1) \right] |\, j, m_j, l, s \rangle$$

to obtain

$$\Delta E_{\mathrm{so}} = \frac{Z^4 e^2 \hbar^2 [J(J+1) - L(L+1) - S(S+1)]}{4m_e^2 c^2 \kappa_0 a_0^3 n^3 l(l+1)(l+\frac{1}{2})}$$

Finally, show that this result becomes

$$\Delta E_{\mathrm{so}} = \frac{m_e c^2 Z^4 \alpha^4}{4n^3 l(l+1)(l+\frac{1}{2})} \times \begin{cases} l & j = l + \frac{1}{2} \\ -(l+1) & j = l - \frac{1}{2} \end{cases}$$

This is not a rigorous derivation, but it does give an idea of the "flavor" of the derivation.

Since $\hat{\mathbf{J}} = \hat{\mathbf{L}} + \hat{\mathbf{S}}$, then

$$\hat{J}^2 = \hat{\mathbf{J}} \cdot \hat{\mathbf{J}}$$

$$= \left(\hat{\mathbf{L}} + \hat{\mathbf{S}} \right) \cdot \left(\hat{\mathbf{L}} + \hat{\mathbf{S}} \right)$$

$$= \hat{L}^2 + \hat{S}^2 + \hat{\mathbf{L}} \cdot \hat{\mathbf{S}} + \hat{\mathbf{S}} \cdot \hat{\mathbf{L}}$$

The two operators $\hat{\mathbf{L}}$ and $\hat{\mathbf{S}}$ commute because they operate on different coordinates, so $\hat{\mathbf{L}} \cdot \hat{\mathbf{S}} = \hat{\mathbf{S}} \cdot \hat{\mathbf{L}}$, and therefore

$$\hat{\mathbf{L}} \cdot \hat{\mathbf{S}} = \frac{1}{2} \left(\hat{J}^2 - \hat{L}^2 - \hat{S}^2 \right)$$

Making this substitution in the Hamiltonian operator for the spin-obit interaction gives

$$\hat{H}_{so} = \frac{Ze^2}{2m_e^2 c^2 \kappa_0 r^3} \hat{\mathbf{L}} \cdot \hat{\mathbf{S}}$$

$$= \frac{Ze^2}{2m_e^2 c^2 \kappa_0 r^3} \frac{\left(\hat{J}^2 - \hat{L}^2 - \hat{S}^2 \right)}{2}$$

$$= \frac{Ze^2}{4m_e^2 c^2 \kappa_0 r^3} \left(\hat{J}^2 - \hat{L}^2 - \hat{S}^2 \right)$$

Under the approximation that the effect of $\hat{H}_{so}$ is small, ΔE_{so} is given by

$$\Delta E_{so} = \left\langle n, j, m_j, l, s \left| \hat{H}_{so} \right| n, j, m_j, l, s \right\rangle$$

The hydrogen atomic wave functions can be written as a product of a radial part and an angular part. That is,

$$\left| n, j, m_j, l, s \right\rangle = |n, l\rangle \left| j, m_j, l, s \right\rangle$$

The angular momentum operators in $\hat{H}_{so}$ affect only the angular part of the wave function, and the factor of r^{-3} affects only the radial portion. Of course, all the constants factor out of the integrals, and consequently,

$$\Delta E_{so} = \left\langle n, j, m_j, l, s \left| \hat{H}_{so} \right| n, j, m_j, l, s \right\rangle$$

$$= \left\langle n, j, m_j, l, s \left| \frac{Ze^2}{4m_e^2 c^2 \kappa_0 r^3} \left(\hat{J}^2 - \hat{L}^2 - \hat{S}^2 \right) \right| n, j, m_j, l, s \right\rangle$$

$$= \frac{Ze^2}{4m_e^2 c^2 \kappa_0} \left\langle n, l \left| \frac{1}{r^3} \right| n, l \right\rangle \left\langle j, m_j, l, s \left| \hat{J}^2 - \hat{L}^2 - \hat{S}^2 \right| j, m_j, l, s \right\rangle$$

$$= \frac{Ze^2}{4m_e^2 c^2 \kappa_0} \left\langle \frac{1}{r^3} \right\rangle_{nl} \left\langle j, m_j, l, s \left| \hat{J}^2 - \hat{L}^2 - \hat{S}^2 \right| j, m_j, l, s \right\rangle$$

Then, by the orthonormality of the angular portion of the wave function,

$$\left\langle j, m_j, l, s \left| \hat{J}^2 - \hat{L}^2 - \hat{S}^2 \right| j, m_j, l, s \right\rangle$$

$$= \hbar^2 \left[j(j+1) - l(l+1) - s(s+1) \right] \left\langle j, m_j, l, s \mid j, m_j, l, s \right\rangle$$

$$= \hbar^2 \left[j(j+1) - l(l+1) - s(s+1) \right]$$

Now make this substitution along with the one suggested for $\langle 1/r^3 \rangle_{nl}$ to obtain

$$\Delta E_{so} = \frac{Ze^2}{4m_e^2 c^2 \kappa_0} \left[\frac{Z^3}{a_0^3 n^3 l(l+1)\left(l+\frac{1}{2}\right)} \right] \hbar^2 \left[j(j+1) - l(l+1) - s(s+1) \right]$$

$$= \frac{Z^4 e^2 \hbar^2 \left[j(j+1) - l(l+1) - s(s+1) \right]}{4m_e^2 c^2 \kappa_0 a_0^3 n^3 l(l+1)\left(l+\frac{1}{2}\right)}$$

Using $a_0 = \dfrac{4\pi\epsilon_0 \hbar^2}{m_e e^2}$ and $\kappa_0 = 4\pi\epsilon_0$ (Table 9.1),

$$\Delta E_{so} = \frac{Z^4 e^2 \hbar^2 \left[j(j+1) - l(l+1) - s(s+1) \right]}{4m_e^2 c^2 \kappa_0 a_0^3 n^3 l(l+1)\left(l+\frac{1}{2}\right)}$$

$$= \frac{Z^4 e^2 \hbar^2 \left[j(j+1) - l(l+1) - s(s+1) \right]}{4m_e^2 c^2 n^3 l(l+1)\left(l+\frac{1}{2}\right)} \left[\frac{m_e^3 e^6}{(4\pi\epsilon_0)^4 \hbar^6} \right]$$

$$= \frac{Z^4 m_e c^2 \left[j(j+1) - l(l+1) - s(s+1) \right]}{4n^3 l(l+1)\left(l+\frac{1}{2}\right)} \left(\frac{e^2}{4\pi\epsilon_0 \hbar c} \right)^4$$

$$= \frac{Z^4 m_e c^2 \alpha^4 \left[j(j+1) - l(l+1) - s(s+1) \right]}{4n^3 l(l+1)\left(l+\frac{1}{2}\right)}$$

where $\alpha = e^2/4\pi\epsilon_0 \hbar c$ is the fine structure constant (Problem 7–38).

For hydrogen-like atoms, $s = 1/2$ and $j = l + 1/2$ or $j = l - 1/2$. For $j = l + 1/2$,

$$j(j+1) - l(l+1) - s(s+1) = \left(l+\frac{1}{2}\right)\left(l+\frac{3}{2}\right) - l(l+1) - \frac{3}{4}$$

$$= l^2 + 2l + \frac{3}{4} - l^2 - l - \frac{3}{4}$$

$$= l$$

while for $j = l - 1/2$,

$$j(j+1) - l(l+1) - s(s+1) = \left(l-\frac{1}{2}\right)\left(l+\frac{1}{2}\right) - l(l+1) - \frac{3}{4}$$

$$= l^2 - \frac{1}{4} - l^2 - l - \frac{3}{4}$$

$$= -(l+1)$$

This gives the final result,

$$\Delta E_{so} = \frac{Z^4 m_e c^2 \alpha^4}{4n^3 l \, (l+1) \left(l+\frac{1}{2}\right)} \times \begin{cases} l & j = l + \frac{1}{2} \\ -(l+1) & j = l - \frac{1}{2} \end{cases}$$

7-44.   Extend Table 7.6 to the case $l = 4$. How many states are there?

For $l = 4$, the possible values of j are $9/2$ and $7/2$, and for each value of j, m_j can take on $2j + 1$ possible values between $+j$ and $-j$.

$$l = 4 \quad \text{18 states}$$

$$m_j = \frac{9}{2}, \frac{7}{2}, \frac{5}{2}, \frac{3}{2}, \frac{1}{2}, -\frac{1}{2}, -\frac{3}{2}, -\frac{5}{2}, -\frac{7}{2}, -\frac{9}{2} \qquad j = \frac{9}{2}$$

$$m_j = \frac{7}{2}, \frac{5}{2}, \frac{3}{2}, \frac{1}{2}, -\frac{1}{2}, -\frac{3}{2}, -\frac{5}{2}, -\frac{7}{2}, \qquad j = \frac{7}{2}$$

7-45.   Can you deduce a general formula for the total number of states for each value of l in Table 7.6?

There are $2l + 1$ possible orbital angular-momentum states for a fixed value of l and $2s + 1 = 2$ possible spin angular-momentum states. Consequently, there are $2\,(2l + 1)$ total states possible.

7-46.   We'll see in Section 8.6 that the selection rule for electronic transitions in a hydrogen atom depends upon whether the integral

$$I = \langle \, n, l, m_l | \, z \, | \, n', l', m_l' \, \rangle$$

is zero or nonzero. If $I = 0$, then the transition is forbidden; otherwise, it may occur. Show that $I = 0$ unless $\Delta l = \pm 1$ and $\Delta m_l = 0$. Do you find any restriction on Δn?

The selection rules may be derived through consideration of the integral,

$$I = \langle \, n, l, m_l | z | n', l', m_l' \, \rangle$$

$$= \langle \, n, l, m_l | r \cos \theta | n', l', m_l' \, \rangle$$

$$= \int_0^{2\pi} d\phi \int_0^{\pi} d\theta \, \sin \theta \int_0^{\infty} dr \, r^2 \psi_{nlm_l}^* \, r \cos \theta \, \psi_{n'l'm_l'}$$

$$= \int_0^{2\pi} d\phi \int_0^{\pi} d\theta \ \sin\theta Y_l^{m_l}(\theta,\phi)^* \cos\theta Y_{l'}^{m_l'}(\theta,\phi) \int_0^{\infty} dr \ r^2 R_{nl}(r)^* r R_{n'l'}(r)$$

$$= i^{m'+|m'|}(-i)^{m+|m|} N_{lm_l} N_{l'm_l'} \int_0^{2\pi} d\phi \ e^{i(m'-m)\phi} \int_0^{\pi} d\theta \ \sin\theta P_l^{|m|}(\cos\theta) \cos\theta P_{l'}^{|m'|}(\cos\theta)$$

$$\times \int_0^{\infty} dr \ r^2 R_{nl}(r)^* r R_{n'l'}(r)$$

$$= i^{m'+|m'|}(-i)^{m+|m|} N_{lm_l} N_{l'm_l'} \int_0^{2\pi} d\phi \ e^{i(m'-m)\phi}$$

$$\times \int_0^{\pi} d\theta \ \sin\theta P_l^{|m|}(\cos\theta) \left[\frac{(l'-|m'|+1)}{2l'+1} P_{l'+1}^{|m'|}(\cos\theta) + \frac{(l'+|m'|)}{2l'+1} P_{l'-1}^{|m'|}(\cos\theta) \right]$$

$$\times \int_0^{\infty} dr \ r^3 R_{nl}(r)^* R_{n'l'}(r)$$

where Equation 6.63 and the recursion formula from Problem 6–33 for the associated Legendre functions have been used. It is seen that I_z factors into three integrals. The integral over ϕ is zero unless $m = m'$, or $\Delta m = 0$. The orthogonality of the associated Legendre functions (Equation 6.53) requires that the integral over θ vanishes unless $l = l' + 1$ or $l = l' - 1$, or $\Delta l = 0$. The integral over r governs any selection rules for n, and in fact gives no restrictions on the allowed values for Δn, as is seen experimentally in the hydrogen atom emission spectrum where all values of $n > n_f$ are seen in the various series of lines (see Chapter 1).

7–47. Using the data in Table 7.7, calculate the frequency (in cm^{-1}) of the extreme left-hand transition in Figure 7.13 in an external magnetic field of 1.00 tesla. Compare your result to the frequency that you would obtain neglecting spin-orbit interaction.

The values of l, j, and m_j in the upper state ($2p\ ^2P_{1/2}$) are 1, 1/2, and $-1/2$, respectively. This state has $j = l - 1/2$, so $g(j, l) = 2/3$ in Equation 7.75. Therefore,

$$\Delta E_{\text{upper}} = g(j, l)\beta_B m_j B_z$$

$$= \frac{2}{3}\left(9.274\,0154 \times 10^{-24}\ \text{J}\cdot\text{T}^{-1}\right)\left(-\frac{1}{2}\right)(1.00\ \text{T})$$

$$= -3.09 \times 10^{-24}\ \text{J} = -0.156\ \text{cm}^{-1}$$

The energy of the upper state is, using the data from Table 7.7

$$E_{\text{upper}} = 82\,258.9206\ \text{cm}^{-1} - 0.156\ \text{cm}^{-1} = 82\,258.765\ \text{cm}^{-1}$$

For the lower state ($1s\ ^2S_{1/2}$), the values of l, j, and m_j are 0, 1/2, and 1/2, respectively. This state has $j = l + 1/2$, therefore $g(j, l) = 2$ in Equation 7.75, which gives

$$\Delta E_{\text{lower}} = g(j, l)\beta_B m_j B_z$$

$$= 2\left(9.274\,0154 \times 10^{-24}\,\text{J}\cdot\text{T}^{-1}\right)\left(\frac{1}{2}\right)(1.00\,\text{T})$$

$$= 9.27 \times 10^{-24}\,\text{J} = 0.467\,\text{cm}^{-1}$$

and leads to

$$E_{\text{lower}} = 0\,\text{cm}^{-1} + 0.467\,\text{cm}^{-1} = 0.467\,\text{cm}^{-1}$$

The frequency of the transition is then

$$\Delta E = E_{\text{upper}} - E_{\text{lower}} = 82\,258.298\,\text{cm}^{-1}$$

Because the spin-orbit wave functions are linear combinations of the $|lm_l m_s\rangle$ wave functions (Problem 7–43), it is a subtle matter to decide to which of the three frequencies obtained by neglecting spin-orbit interaction this transition would correspond. Nevertheless, since this is the lowest frequency transition in Figure 7.13, it can be compared with the lowest frequency transition obtained in the absence of spin-orbit effects. From the results of Example 7–5, this transition would be $9.274 \times 10^{-24}\,\text{J} = 0.467\,\text{cm}^{-1}$ lower in energy than the unsplit $2p$ to $1s$ transition, or

$$\Delta E_{\text{no spin-orbit}} = \left(109\,677.58\,\text{cm}^{-1}\right)\left(1 - \frac{1}{4}\right) - 0.467\,\text{cm}^{-1} = 82\,257.718\,\text{cm}^{-1}$$

7–48. Using the data in Table 7.7, calculate the frequency (in cm^{-1}) of the second-from-right transition in Figure 7.13 in an external magnetic field of 1.00 tesla. Compare your result to the frequency that you would obtain neglecting spin-orbit interaction.

The values of l, j, and m_j in the upper state ($2p\,^2P_{3/2}$) are 1, 3/2, and $-1/2$, respectively. This state has $j = l + 1/2$, so $g(j, l) = 4/3$ in Equation 7.75. Therefore,

$$\Delta E_{\text{upper}} = g(j, l)\beta_B m_j B_z$$

$$= \frac{4}{3}\left(9.274\,0154 \times 10^{-24}\,\text{J}\cdot\text{T}^{-1}\right)\left(-\frac{1}{2}\right)(1.00\,\text{T})$$

$$= -6.18 \times 10^{-24}\,\text{J} = -0.311\,\text{cm}^{-1}$$

The energy of the upper state is, using the data from Table 7.7

$$E_{\text{upper}} = 82\,259.2865\,\text{cm}^{-1} - 0.311\,\text{cm}^{-1} = 82\,258.975\,\text{cm}^{-1}$$

For the lower state ($1s\,^2S_{1/2}$), the values of l, j, and m_j are 0, 1/2, and $-1/2$, respectively. This state has $j = l + 1/2$, therefore $g(j, l) = 2$ in Equation 7.75, which gives

$$\Delta E_{\text{lower}} = g(j, l)\beta_{\text{B}} m_j B_z$$

$$= 2 \left(9.274\,0154 \times 10^{-24} \text{ J} \cdot \text{T}^{-1}\right) \left(-\frac{1}{2}\right) (1.00 \text{ T})$$

$$= -9.27 \times 10^{-24} \text{ J} = -0.467 \text{ cm}^{-1}$$

and leads to

$$E_{\text{lower}} = 0 \text{ cm}^{-1} - 0.467 \text{ cm}^{-1} = -0.467 \text{ cm}^{-1}$$

The frequency of the transition is then

$$\Delta E = E_{\text{upper}} - E_{\text{lower}} = 82\,259.442 \text{ cm}^{-1}$$

As discussed in the previous problem, the correct choice of frequency for comparison if the spin-orbit interaction were neglected is not obvious. There are good reasons to choose the unsplit $2p$ to $1s$ transition, or

$$\Delta E_{\text{no spin}-\text{orbit}} = \left(109\,677.58 \text{ cm}^{-1}\right) \left(1 - \frac{1}{4}\right) = 82\,258.185 \text{ cm}^{-1}$$

Matrices

PROBLEMS AND SOLUTIONS

G–1. Given the two matrices

$$A = \begin{pmatrix} 1 & 0 & -1 \\ -1 & 2 & 0 \\ 0 & 1 & 1 \end{pmatrix} \quad \text{and} \quad B = \begin{pmatrix} -1 & 1 & 0 \\ 3 & 0 & 2 \\ 1 & 1 & 1 \end{pmatrix}$$

form the matrices $C = 2\,A - 3\,B$ and $D = 6\,B - A$.

(a) $C = 2\,A - 3\,B$

$$C = \begin{pmatrix} 2 & 0 & -2 \\ -2 & 4 & 0 \\ 0 & 2 & 2 \end{pmatrix} - \begin{pmatrix} -3 & 3 & 0 \\ 9 & 0 & 6 \\ 3 & 3 & 3 \end{pmatrix} = \begin{pmatrix} 5 & -3 & -2 \\ -11 & 4 & -6 \\ -3 & -1 & -1 \end{pmatrix}$$

(b) $D = 6\,B - A$

$$D = \begin{pmatrix} -6 & 6 & 0 \\ 18 & 0 & 12 \\ 6 & 6 & 6 \end{pmatrix} - \begin{pmatrix} 1 & 0 & -1 \\ -1 & 2 & 0 \\ 0 & 1 & 1 \end{pmatrix} = \begin{pmatrix} -7 & 6 & 1 \\ 19 & -2 & 12 \\ 6 & 5 & 5 \end{pmatrix}$$

G–2. Given the three matrices

$$A = \frac{1}{2}\begin{pmatrix} 0 & 1 \\ 1 & 0 \end{pmatrix} \qquad B = \frac{1}{2}\begin{pmatrix} 0 & -i \\ i & 0 \end{pmatrix} \qquad C = \frac{1}{2}\begin{pmatrix} 1 & 0 \\ 0 & -1 \end{pmatrix}$$

show that $A^2 + B^2 + C^2 = \frac{3}{4}I$, where I is a unit matrix. Also show that

$$AB - BA = iC$$

$$BC - CB = iA$$

$$CA - AC = iB$$

(a)

$$A^2 + B^2 + C^2 = \frac{1}{4}\begin{pmatrix} 0 & 1 \\ 1 & 0 \end{pmatrix}\begin{pmatrix} 0 & 1 \\ 1 & 0 \end{pmatrix} + \frac{1}{4}\begin{pmatrix} 0 & -i \\ i & 0 \end{pmatrix}\begin{pmatrix} 0 & -i \\ i & 0 \end{pmatrix}$$

$$+ \frac{1}{4}\begin{pmatrix} 1 & 0 \\ 0 & -1 \end{pmatrix}\begin{pmatrix} 1 & 0 \\ 0 & -1 \end{pmatrix}$$

$$= \frac{1}{4}\begin{pmatrix} 1 & 0 \\ 0 & 1 \end{pmatrix} + \frac{1}{4}\begin{pmatrix} 1 & 0 \\ 0 & 1 \end{pmatrix} + \frac{1}{4}\begin{pmatrix} 1 & 0 \\ 0 & 1 \end{pmatrix} = \frac{3}{4}I$$

(b)

$$AB - BA = \frac{1}{4}\begin{pmatrix} 0 & 1 \\ 1 & 0 \end{pmatrix}\begin{pmatrix} 0 & -i \\ i & 0 \end{pmatrix} - \frac{1}{4}\begin{pmatrix} 0 & -i \\ i & 0 \end{pmatrix}\begin{pmatrix} 0 & 1 \\ 1 & 0 \end{pmatrix}$$

$$= \frac{1}{4}\begin{pmatrix} i & 0 \\ 0 & -i \end{pmatrix} - \frac{1}{4}\begin{pmatrix} -i & 0 \\ 0 & i \end{pmatrix} = \frac{i}{2}\begin{pmatrix} 1 & 0 \\ 0 & -1 \end{pmatrix} = iC$$

(c)

$$BC - CB = \frac{1}{4}\begin{pmatrix} 0 & -i \\ i & 0 \end{pmatrix}\begin{pmatrix} 1 & 0 \\ 0 & -1 \end{pmatrix} - \frac{1}{4}\begin{pmatrix} 1 & 0 \\ 0 & -1 \end{pmatrix}\begin{pmatrix} 0 & -i \\ i & 0 \end{pmatrix}$$

$$= \frac{1}{4}\begin{pmatrix} 0 & i \\ i & 0 \end{pmatrix} - \frac{1}{4}\begin{pmatrix} 0 & -i \\ -i & 0 \end{pmatrix} = \frac{i}{2}\begin{pmatrix} 0 & 1 \\ 1 & 0 \end{pmatrix} = iA$$

(d)

$$CA - AC = \frac{1}{4}\begin{pmatrix} 1 & 0 \\ 0 & -1 \end{pmatrix}\begin{pmatrix} 0 & 1 \\ 1 & 0 \end{pmatrix} - \frac{1}{4}\begin{pmatrix} 0 & 1 \\ 1 & 0 \end{pmatrix}\begin{pmatrix} 1 & 0 \\ 0 & -1 \end{pmatrix}$$

$$= \frac{1}{4}\begin{pmatrix} 0 & 1 \\ -1 & 0 \end{pmatrix} - \frac{1}{4}\begin{pmatrix} 0 & -1 \\ 1 & 0 \end{pmatrix} = \frac{i}{2}\begin{pmatrix} 0 & -i \\ i & 0 \end{pmatrix} = iB$$

G–3.   Given the matrices

$$A = \frac{1}{\sqrt{2}}\begin{pmatrix} 0 & 1 & 0 \\ 1 & 0 & 1 \\ 0 & 1 & 0 \end{pmatrix} \qquad B = \frac{1}{\sqrt{2}}\begin{pmatrix} 0 & -i & 0 \\ i & 0 & -i \\ 0 & i & 0 \end{pmatrix} \qquad C = \begin{pmatrix} 1 & 0 & 0 \\ 0 & 0 & 0 \\ 0 & 0 & -1 \end{pmatrix}$$

show that

$$AB - BA = iC$$

$$BC - CB = iA$$

$$CA - AC = iB$$

and

$$A^2 + B^2 + C^2 = 2I$$

where I is a unit matrix.

———————————————

(a)

$$AB - BA = \frac{1}{2}\begin{pmatrix} 0 & 1 & 0 \\ 1 & 0 & 1 \\ 0 & 1 & 0 \end{pmatrix}\begin{pmatrix} 0 & -i & 0 \\ i & 0 & -i \\ 0 & i & 0 \end{pmatrix} - \frac{1}{2}\begin{pmatrix} 0 & -i & 0 \\ i & 0 & -i \\ 0 & i & 0 \end{pmatrix}\begin{pmatrix} 0 & 1 & 0 \\ 1 & 0 & 1 \\ 0 & 1 & 0 \end{pmatrix}$$

$$= \frac{1}{2}\begin{pmatrix} i & 0 & -i \\ 0 & 0 & 0 \\ i & 0 & -i \end{pmatrix} - \frac{1}{2}\begin{pmatrix} -i & 0 & -i \\ 0 & 0 & 0 \\ i & 0 & i \end{pmatrix} = \frac{1}{2}\begin{pmatrix} 2i & 0 & 0 \\ 0 & 0 & 0 \\ 0 & 0 & -2i \end{pmatrix}$$

$$= i\begin{pmatrix} 1 & 0 & 0 \\ 0 & 0 & 0 \\ 0 & 0 & -1 \end{pmatrix} = i\mathsf{C}$$

(b)

$$BC - CB = \frac{1}{\sqrt{2}}\begin{pmatrix} 0 & -i & 0 \\ i & 0 & -i \\ 0 & i & 0 \end{pmatrix}\begin{pmatrix} 1 & 0 & 0 \\ 0 & 0 & 0 \\ 0 & 0 & -1 \end{pmatrix} - \frac{1}{\sqrt{2}}\begin{pmatrix} 1 & 0 & 0 \\ 0 & 0 & 0 \\ 0 & 0 & -1 \end{pmatrix}\begin{pmatrix} 0 & -i & 0 \\ i & 0 & -i \\ 0 & i & 0 \end{pmatrix}$$

$$= \frac{1}{\sqrt{2}}\begin{pmatrix} 0 & 0 & 0 \\ i & 0 & i \\ 0 & 0 & 0 \end{pmatrix} - \frac{1}{\sqrt{2}}\begin{pmatrix} 0 & -i & 0 \\ 0 & 0 & 0 \\ 0 & -i & 0 \end{pmatrix} = \frac{1}{\sqrt{2}}\begin{pmatrix} 0 & i & 0 \\ i & 0 & i \\ 0 & i & 0 \end{pmatrix}$$

$$= \frac{i}{\sqrt{2}}\begin{pmatrix} 0 & 1 & 0 \\ 1 & 0 & 1 \\ 0 & 1 & 0 \end{pmatrix} = i\mathsf{A}$$

(c)

$$CA - AC = \frac{1}{\sqrt{2}}\begin{pmatrix} 1 & 0 & 0 \\ 0 & 0 & 0 \\ 0 & 0 & -1 \end{pmatrix}\begin{pmatrix} 0 & 1 & 0 \\ 1 & 0 & 1 \\ 0 & 1 & 0 \end{pmatrix} - \frac{1}{\sqrt{2}}\begin{pmatrix} 0 & 1 & 0 \\ 1 & 0 & 1 \\ 0 & 1 & 0 \end{pmatrix}\begin{pmatrix} 1 & 0 & 0 \\ 0 & 0 & 0 \\ 0 & 0 & -1 \end{pmatrix}$$

$$= \frac{1}{\sqrt{2}}\begin{pmatrix} 0 & 1 & 0 \\ 0 & 0 & 0 \\ 0 & -1 & 0 \end{pmatrix} - \frac{1}{\sqrt{2}}\begin{pmatrix} 0 & 0 & 0 \\ 1 & 0 & -1 \\ 0 & 0 & 0 \end{pmatrix} = \frac{1}{\sqrt{2}}\begin{pmatrix} 0 & 1 & 0 \\ -1 & 0 & 1 \\ 0 & -1 & 0 \end{pmatrix}$$

$$= \frac{i}{\sqrt{2}}\begin{pmatrix} 0 & -i & 0 \\ i & 0 & -i \\ 0 & i & 0 \end{pmatrix} = i\mathsf{B}$$

(d)

$$
A^2 + B^2 + C^2 = \frac{1}{2} \begin{pmatrix} 0 & 1 & 0 \\ 1 & 0 & 1 \\ 0 & 1 & 0 \end{pmatrix} \begin{pmatrix} 0 & 1 & 0 \\ 1 & 0 & 1 \\ 0 & 1 & 0 \end{pmatrix} + \frac{1}{2} \begin{pmatrix} 0 & -i & 0 \\ i & 0 & -i \\ 0 & i & 0 \end{pmatrix} \begin{pmatrix} 0 & -i & 0 \\ i & 0 & -i \\ 0 & i & 0 \end{pmatrix}
$$

$$
+ \begin{pmatrix} 1 & 0 & 0 \\ 0 & 0 & 0 \\ 0 & 0 & -1 \end{pmatrix} \begin{pmatrix} 1 & 0 & 0 \\ 0 & 0 & 0 \\ 0 & 0 & -1 \end{pmatrix}
$$

$$
= \frac{1}{2} \begin{pmatrix} 1 & 0 & 1 \\ 0 & 2 & 0 \\ 1 & 0 & 1 \end{pmatrix} + \frac{1}{2} \begin{pmatrix} 1 & 0 & -1 \\ 0 & 2 & 0 \\ -1 & 0 & 1 \end{pmatrix} + \begin{pmatrix} 1 & 0 & 0 \\ 0 & 0 & 0 \\ 0 & 0 & 1 \end{pmatrix}
$$

$$
= \frac{1}{2} \begin{pmatrix} 4 & 0 & 0 \\ 0 & 4 & 0 \\ 0 & 0 & 4 \end{pmatrix} = 2I
$$

G–4. Do you see any similarity between the results of Problems G–2 and G–3 and the commutation relations involving the components of angular momentum?

Yes, the commutation relations for these sets of matrices behave like those for the components of angular momentum. That is, if A, B, and C correspond to $\hat{J}_x$, $\hat{J}_y$, and $\hat{J}_z$, respectively, then the results are similar to the commutation relations for $\hat{J}_x$, $\hat{J}_y$, and $\hat{J}_z$. Additionally, the sum of the squares of the matrices behaves like the sum of the squares of the angular momentum components for $j = 1/2$ and $j = 1$, respectively, in Problems G–2 and G–3 giving $3/4 = (1/2)(1/2 + 1)$ and $2 = 1(1 + 1)$ times the identity.

G–5. A three-dimensional rotation about the z axis can be represented by the matrix

$$
R = \begin{pmatrix} \cos\theta & -\sin\theta & 0 \\ \sin\theta & \cos\theta & 0 \\ 0 & 0 & 1 \end{pmatrix}
$$

Show that

$$
\det R = |R| = 1
$$

Also show that

$$
R^{-1} = R(-\theta) = \begin{pmatrix} \cos\theta & \sin\theta & 0 \\ -\sin\theta & \cos\theta & 0 \\ 0 & 0 & 1 \end{pmatrix}
$$

Evaluate $|R|$ by expanding along the third column to obtain

$$|R| = \begin{vmatrix} \cos\theta & -\sin\theta & 0 \\ \sin\theta & \cos\theta & 0 \\ 0 & 0 & 1 \end{vmatrix} = 1 \begin{vmatrix} \cos\theta & -\sin\theta \\ \sin\theta & \cos\theta \end{vmatrix}$$

$$= \cos^2\theta + \sin^2\theta = 1$$

Since $\cos(-\theta) = \cos\theta$ and $\sin(-\theta) = -\sin\theta$, then

$$R(-\theta) = \begin{pmatrix} \cos\theta & \sin\theta & 0 \\ -\sin\theta & \cos\theta & 0 \\ 0 & 0 & 1 \end{pmatrix}$$

Now consider the product,

$$R(-\theta)R(\theta) = \begin{pmatrix} \cos\theta & \sin\theta & 0 \\ -\sin\theta & \cos\theta & 0 \\ 0 & 0 & 1 \end{pmatrix} \begin{pmatrix} \cos\theta & -\sin\theta & 0 \\ \sin\theta & \cos\theta & 0 \\ 0 & 0 & 1 \end{pmatrix} = \begin{pmatrix} 1 & 0 & 0 \\ 0 & 1 & 0 \\ 0 & 0 & 1 \end{pmatrix}$$

and also,

$$R(\theta)R(-\theta) = \begin{pmatrix} \cos\theta & -\sin\theta & 0 \\ \sin\theta & \cos\theta & 0 \\ 0 & 0 & 1 \end{pmatrix} \begin{pmatrix} \cos\theta & \sin\theta & 0 \\ -\sin\theta & \cos\theta & 0 \\ 0 & 0 & 1 \end{pmatrix} = \begin{pmatrix} 1 & 0 & 0 \\ 0 & 1 & 0 \\ 0 & 0 & 1 \end{pmatrix}$$

Therefore, $R(-\theta) = R^{-1}$.

G–6. Show that the matrix R in Problem G–5 is orthogonal.

Form the transpose of the matrix, R in Problem G–5, to find

$$R^T = \begin{pmatrix} \cos\theta & \sin\theta & 0 \\ -\sin\theta & \cos\theta & 0 \\ 0 & 0 & 1 \end{pmatrix} = R^{-1}$$

as shown in the previous problem. Since the inverse of R is equal to its transpose, R is an orthogonal matrix.

G–7. Given the matrices

$$C_3 = \begin{pmatrix} -\frac{1}{2} & -\frac{\sqrt{3}}{2} \\ \frac{\sqrt{3}}{2} & -\frac{1}{2} \end{pmatrix} \qquad \sigma_v = \begin{pmatrix} 1 & 0 \\ 0 & -1 \end{pmatrix}$$

$$\sigma_v' = \begin{pmatrix} -\frac{1}{2} & \frac{\sqrt{3}}{2} \\ \frac{\sqrt{3}}{2} & \frac{1}{2} \end{pmatrix} \qquad \sigma_v'' = \begin{pmatrix} -\frac{1}{2} & -\frac{\sqrt{3}}{2} \\ -\frac{\sqrt{3}}{2} & \frac{1}{2} \end{pmatrix}$$

show that

$$\sigma_v C_3 = \sigma_v'' \qquad C_3 \sigma_v = \sigma_v'$$

$$\sigma_v'' \sigma_v' = C_3 \qquad C_3 \sigma_v'' = \sigma_v$$

Calculate the determinant associated with each matrix. Calculate the trace of each matrix.

___

(a)

$$\sigma_v C_3 = \begin{pmatrix} 1 & 0 \\ 0 & -1 \end{pmatrix} \begin{pmatrix} -\frac{1}{2} & -\frac{\sqrt{3}}{2} \\ \frac{\sqrt{3}}{2} & -\frac{1}{2} \end{pmatrix} = \begin{pmatrix} -\frac{1}{2} & -\frac{\sqrt{3}}{2} \\ -\frac{\sqrt{3}}{2} & \frac{1}{2} \end{pmatrix} = \sigma_v''$$

(b)

$$C_3 \sigma_v = \begin{pmatrix} -\frac{1}{2} & -\frac{\sqrt{3}}{2} \\ \frac{\sqrt{3}}{2} & -\frac{1}{2} \end{pmatrix} \begin{pmatrix} 1 & 0 \\ 0 & -1 \end{pmatrix} = \begin{pmatrix} -\frac{1}{2} & \frac{\sqrt{3}}{2} \\ \frac{\sqrt{3}}{2} & \frac{1}{2} \end{pmatrix} = \sigma_v'$$

(c)

$$\sigma_v'' \sigma_v' = \begin{pmatrix} -\frac{1}{2} & -\frac{\sqrt{3}}{2} \\ -\frac{\sqrt{3}}{2} & \frac{1}{2} \end{pmatrix} \begin{pmatrix} -\frac{1}{2} & \frac{\sqrt{3}}{2} \\ \frac{\sqrt{3}}{2} & \frac{1}{2} \end{pmatrix} = \begin{pmatrix} -\frac{1}{2} & -\frac{\sqrt{3}}{2} \\ \frac{\sqrt{3}}{2} & -\frac{1}{2} \end{pmatrix} = C_3$$

(d)

$$C_3 \sigma_v'' = \begin{pmatrix} -\frac{1}{2} & -\frac{\sqrt{3}}{2} \\ \frac{\sqrt{3}}{2} & -\frac{1}{2} \end{pmatrix} \begin{pmatrix} -\frac{1}{2} & -\frac{\sqrt{3}}{2} \\ -\frac{\sqrt{3}}{2} & \frac{1}{2} \end{pmatrix} = \begin{pmatrix} 1 & 0 \\ 0 & -1 \end{pmatrix} = \sigma_v$$

The determinant and trace of each matrix are found below.

$$\begin{aligned}
|C_3| &= \frac{1}{4} + \frac{3}{4} = 1 & \text{Tr } C_3 &= -1 \\
|\sigma_v| &= -1 & \text{Tr } \sigma_v &= 0 \\
|\sigma_v'| &= -\frac{1}{4} - \frac{3}{4} = -1 & \text{Tr } \sigma_v' &= 0 \\
|\sigma_v''| &= -\frac{1}{4} - \frac{3}{4} = -1 & \text{Tr } \sigma_v'' &= 0
\end{aligned}$$

___

G–8. Which of the matrices in Problem G–7 are orthogonal?

___

A matrix R is orthogonal if $R^T = R^{-1}$. In other words, R is orthogonal if $R^{-1}R = R^T R = I$ where I is the identity matrix.

(a)

$$C_3 = \begin{pmatrix} -\frac{1}{2} & -\frac{\sqrt{3}}{2} \\ \frac{\sqrt{3}}{2} & -\frac{1}{2} \end{pmatrix} \qquad C_3^T = \begin{pmatrix} -\frac{1}{2} & \frac{\sqrt{3}}{2} \\ -\frac{\sqrt{3}}{2} & -\frac{1}{2} \end{pmatrix}$$

$$C_3^T C_3 = \begin{pmatrix} -\frac{1}{2} & \frac{\sqrt{3}}{2} \\ -\frac{\sqrt{3}}{2} & -\frac{1}{2} \end{pmatrix} \begin{pmatrix} -\frac{1}{2} & -\frac{\sqrt{3}}{2} \\ \frac{\sqrt{3}}{2} & -\frac{1}{2} \end{pmatrix} = \begin{pmatrix} 1 & 0 \\ 0 & 1 \end{pmatrix}$$

C_3 is orthogonal.

(b)

$$\sigma_v = \begin{pmatrix} 1 & 0 \\ 0 & -1 \end{pmatrix} \qquad \sigma_v^T = \begin{pmatrix} 1 & 0 \\ 0 & -1 \end{pmatrix}$$

$$\sigma_v^T \sigma_v = \begin{pmatrix} 1 & 0 \\ 0 & -1 \end{pmatrix} \begin{pmatrix} 1 & 0 \\ 0 & -1 \end{pmatrix} = \begin{pmatrix} 1 & 0 \\ 0 & 1 \end{pmatrix}$$

σ_v is orthogonal.

(c)

$$\sigma_v' = \begin{pmatrix} -\frac{1}{2} & \frac{\sqrt{3}}{2} \\ \frac{\sqrt{3}}{2} & \frac{1}{2} \end{pmatrix} \qquad \sigma_v'^T = \begin{pmatrix} -\frac{1}{2} & \frac{\sqrt{3}}{2} \\ \frac{\sqrt{3}}{2} & \frac{1}{2} \end{pmatrix}$$

$$\sigma_v'^T \sigma_v' = \begin{pmatrix} -\frac{1}{2} & \frac{\sqrt{3}}{2} \\ \frac{\sqrt{3}}{2} & \frac{1}{2} \end{pmatrix} \begin{pmatrix} -\frac{1}{2} & \frac{\sqrt{3}}{2} \\ \frac{\sqrt{3}}{2} & \frac{1}{2} \end{pmatrix} = \begin{pmatrix} 1 & 0 \\ 0 & 1 \end{pmatrix}$$

σ_v' is orthogonal.

(d)

$$\sigma_v'' = \begin{pmatrix} -\frac{1}{2} & -\frac{\sqrt{3}}{2} \\ -\frac{\sqrt{3}}{2} & \frac{1}{2} \end{pmatrix} \qquad \sigma_v''^T = \begin{pmatrix} -\frac{1}{2} & -\frac{\sqrt{3}}{2} \\ -\frac{\sqrt{3}}{2} & \frac{1}{2} \end{pmatrix}$$

$$\sigma_v''^T \sigma_v'' = \begin{pmatrix} -\frac{1}{2} & -\frac{\sqrt{3}}{2} \\ -\frac{\sqrt{3}}{2} & \frac{1}{2} \end{pmatrix} \begin{pmatrix} -\frac{1}{2} & -\frac{\sqrt{3}}{2} \\ -\frac{\sqrt{3}}{2} & \frac{1}{2} \end{pmatrix} = \begin{pmatrix} 1 & 0 \\ 0 & 1 \end{pmatrix}$$

σ_v'' is orthogonal.

All four matrices are orthogonal.

G–9. The inverse of a matrix A can be found by using the following procedure:

1. Replace each element of A by its cofactor in the corresponding determinant (see MathChapter F for a definition of a cofactor).
2. Take the transpose of the matrix obtained in step 1.
3. Divide each element of the matrix obtained in step 2 by the determinant of A.

For example, if

$$A = \begin{pmatrix} 1 & 2 \\ 3 & 4 \end{pmatrix}$$

then det A = −2 and

$$A^{-1} = -\frac{1}{2} \begin{pmatrix} 4 & -2 \\ -3 & 1 \end{pmatrix}$$

Show that $AA^{-1} = A^{-1}A = I$. Use the above procedure to find the inverse of

$$A = \begin{pmatrix} \frac{1}{2} & \frac{1}{\sqrt{2}} \\ \frac{1}{\sqrt{2}} & 0 \end{pmatrix} \qquad \text{and} \qquad B = \begin{pmatrix} 0 & 2 & 3 \\ 1 & 1 & 1 \\ 2 & 0 & 1 \end{pmatrix}$$

For the matrices

$$A = \begin{pmatrix} 1 & 2 \\ 3 & 4 \end{pmatrix} \qquad A^{-1} = -\frac{1}{2}\begin{pmatrix} 4 & -2 \\ -3 & 1 \end{pmatrix}$$

Form the two products,

$$AA^{-1} = -\frac{1}{2}\begin{pmatrix} 1 & 2 \\ 3 & 4 \end{pmatrix}\begin{pmatrix} 4 & -2 \\ -3 & 1 \end{pmatrix} = -\frac{1}{2}\begin{pmatrix} -2 & 0 \\ 0 & -2 \end{pmatrix} = \begin{pmatrix} 1 & 0 \\ 0 & 1 \end{pmatrix}$$

and

$$A^{-1}A = -\frac{1}{2}\begin{pmatrix} 4 & -2 \\ -3 & 1 \end{pmatrix}\begin{pmatrix} 1 & 2 \\ 3 & 4 \end{pmatrix} = -\frac{1}{2}\begin{pmatrix} -2 & 0 \\ 0 & -2 \end{pmatrix} = \begin{pmatrix} 1 & 0 \\ 0 & 1 \end{pmatrix}$$

Thus, $AA^{-1} = A^{-1}A = I$.

To find the inverse of

$$A = \begin{pmatrix} \frac{1}{2} & \frac{1}{\sqrt{2}} \\ \frac{1}{\sqrt{2}} & 0 \end{pmatrix}$$

Start by finding

$$|A| = -\frac{1}{2}$$

Recall from Mathchapter F that the cofactor, A_{ij}, of an element a_{ij} is a $(n-1) \times (n-1)$ determinant obtained by deleting the ith row and the jth column, multiplied by $(-1)^{i+j}$. The cofactors are then

$$\begin{aligned}
A_{11} &= 0 & A_{12} &= -\frac{1}{\sqrt{2}} \\
A_{21} &= -\frac{1}{\sqrt{2}} & A_{22} &= \frac{1}{2}
\end{aligned}$$

Form the transpose of these elements and divide by the determinant to give the matrix A^{-1}:

$$A^{-1} = -2\begin{pmatrix} 0 & -\frac{1}{\sqrt{2}} \\ -\frac{1}{\sqrt{2}} & \frac{1}{2} \end{pmatrix} = \begin{pmatrix} 0 & \sqrt{2} \\ \sqrt{2} & -1 \end{pmatrix}$$

For

$$B = \begin{pmatrix} 0 & 2 & 3 \\ 1 & 1 & 1 \\ 2 & 0 & 1 \end{pmatrix}$$

The determinant is

$$|B| = -2\begin{vmatrix} 1 & 1 \\ 2 & 1 \end{vmatrix} + 3\begin{vmatrix} 1 & 1 \\ 2 & 0 \end{vmatrix} = -2(-1) + 3(-2) = -4$$

and the cofactors are

$$\begin{aligned}
B_{11} &= 1 & B_{12} &= 1 & B_{13} &= -2 \\
B_{21} &= -2 & B_{22} &= -6 & B_{23} &= 4 \\
B_{31} &= -1 & B_{32} &= 3 & B_{33} &= -2
\end{aligned}$$

So that the inverse is

$$B^{-1} = -\frac{1}{4}\begin{pmatrix} 1 & -2 & 1 \\ 1 & -6 & 3 \\ -2 & 4 & -2 \end{pmatrix}$$

G–10. Recall that a singular matrix is one whose determinant is equal to zero. Referring to the procedure in Problem G–9, do you see why a singular matrix has no inverse?

To find the inverse of a matrix, the elements of the transpose of the matrix of cofactors is divided by the determinant of the original matrix. If the determinant of a matrix is equal to zero (that is, the matrix is singular), then this division process is not defined and the inverse cannot be obtained.

G–11. Consider the matrices A and S,

$$A = \begin{pmatrix} 1 & 0 & 1 \\ 0 & 1 & 0 \\ 1 & 0 & 1 \end{pmatrix} \qquad S = \begin{pmatrix} \frac{1}{\sqrt{2}} & 0 & \frac{1}{\sqrt{2}} \\ 0 & 1 & 0 \\ \frac{1}{\sqrt{2}} & 0 & -\frac{1}{\sqrt{2}} \end{pmatrix}$$

First, show that S is orthogonal. Then evaluate the matrix $D = S^{-1}AS = S^TAS$. What form does D have?

To show that S is orthogonal, show that $SS^T = S^TS = I$. That is, that $S^T = S^{-1}$.

$$S^T = \begin{pmatrix} \frac{1}{\sqrt{2}} & 0 & \frac{1}{\sqrt{2}} \\ 0 & 1 & 0 \\ \frac{1}{\sqrt{2}} & 0 & -\frac{1}{\sqrt{2}} \end{pmatrix}$$

$$SS^T = S^TS = \begin{pmatrix} \frac{1}{\sqrt{2}} & 0 & \frac{1}{\sqrt{2}} \\ 0 & 1 & 0 \\ \frac{1}{\sqrt{2}} & 0 & -\frac{1}{\sqrt{2}} \end{pmatrix}\begin{pmatrix} \frac{1}{\sqrt{2}} & 0 & \frac{1}{\sqrt{2}} \\ 0 & 1 & 0 \\ \frac{1}{\sqrt{2}} & 0 & -\frac{1}{\sqrt{2}} \end{pmatrix} = \begin{pmatrix} 1 & 0 & 0 \\ 0 & 1 & 0 \\ 0 & 0 & 1 \end{pmatrix}$$

Note that S is symmetric so that in addition, $S = S^T$. Finally, evaluate D:

$$D = S^T A S = \begin{pmatrix} \frac{1}{\sqrt{2}} & 0 & \frac{1}{\sqrt{2}} \\ 0 & 1 & 0 \\ \frac{1}{\sqrt{2}} & 0 & -\frac{1}{\sqrt{2}} \end{pmatrix} \begin{pmatrix} 1 & 0 & 1 \\ 0 & 1 & 0 \\ 1 & 0 & 1 \end{pmatrix} \begin{pmatrix} \frac{1}{\sqrt{2}} & 0 & \frac{1}{\sqrt{2}} \\ 0 & 1 & 0 \\ \frac{1}{\sqrt{2}} & 0 & -\frac{1}{\sqrt{2}} \end{pmatrix}$$

$$= \begin{pmatrix} \frac{1}{\sqrt{2}} & 0 & \frac{1}{\sqrt{2}} \\ 0 & 1 & 0 \\ \frac{1}{\sqrt{2}} & 0 & -\frac{1}{\sqrt{2}} \end{pmatrix} \begin{pmatrix} \sqrt{2} & 0 & 0 \\ 0 & 1 & 0 \\ \sqrt{2} & 0 & 0 \end{pmatrix}$$

$$= \begin{pmatrix} 2 & 0 & 0 \\ 0 & 1 & 0 \\ 0 & 0 & 0 \end{pmatrix}$$

D is a diagonal matrix, and its elements are the eigenvalues of A.

G–12. A matrix whose elements satisfy the relation $a_{ij} = a_{ji}^*$ is called *Hermitian*. You can think of a Hermitian matrix as a symmetric matrix in a complex space. Show that the eigenvalues of a Hermitian matrix are real. (Note the similarity between a Hermitian operator and a Hermitian matrix.) *Hint*: Start with $H\mathbf{x}_i = \lambda_i \mathbf{x}_i$ and $H^*\mathbf{x}_j^* = \lambda_j^* \mathbf{x}_j^*$ and multiply the first equation from the left by $\mathbf{x}_j^*$ and the second from the left by $\mathbf{x}_i$ and then use the Hermitian property of H.

Designating the nth component of the vector $\mathbf{x}_i$ by x_n^i, the matrix equation, $H\mathbf{x}_i = \lambda_i \mathbf{x}_i$, can be written as

$$\sum_n h_{mn} x_n^i = \lambda_i x_m^i$$

Left multiply by $\mathbf{x}_j^*$ (expressed as a row vector) to give

$$\mathbf{x}_j^* H \mathbf{x}_i = \lambda_i \mathbf{x}_j^* \cdot \mathbf{x}_i$$

$$\sum_m \sum_n x_m^{j*} h_{mn} x_n^i = \lambda_i \sum_m x_m^{j*} x_m^i$$

Similarly, multiply the second equation by $\mathbf{x}_i$ to give $\mathbf{x}_i H^* \mathbf{x}_j^* = \lambda_j^* \mathbf{x}_i \cdot \mathbf{x}_j^*$ which can be expressed as

$$\sum_m \sum_n x_m^i h_{mn}^* x_n^{j*} = \lambda_j^* \sum_m x_m^{j*} x_m^i$$

Since H is Hermitian,

$$\sum_m \sum_n x_m^i h_{mn}^* x_n^{j*} = \sum_m \sum_n x_m^i h_{nm} x_n^{j*}$$

then interchange the summation variables

$$\sum_m \sum_n x_m^i h_{mn}^* x_n^{j*} = \sum_n \sum_m x_n^i h_{mn} x_m^{j*}$$

$$= \lambda_i \sum_m x_m^{j*} x_m^i$$

using the result for $\mathbf{x}_j^* H \mathbf{x}_i$ from earlier. Thus,

$$\lambda_j^* \sum_m x_m^{j*} x_m^i = \lambda_i \sum_m x_m^{j*} x_m^i$$

$$\left(\lambda_j^* - \lambda_i\right) \mathbf{x}_j^* \cdot \mathbf{x}_i = 0$$

When $j = i$, this last equation becomes

$$\left(\lambda_i^* - \lambda_i\right) \mathbf{x}_i^* \cdot \mathbf{x}_i = 0$$

which, aside from the trivial case of $\mathbf{x}_i$ equal to the null vector, requires that $\lambda_i^* = \lambda_i$, since $\mathbf{x}_i^* \cdot \mathbf{x}_i \neq 0$. Thus, λ_i is real. Note that the case $i \neq j$ shows that if $\lambda_i \neq \lambda_j$, then the two vectors are orthogonal.

G–13. Show that $(AB)^T = B^T A^T$.

Let $C = AB$, so that

$$c_{ij} = \sum_k a_{ik} b_{kj}$$

Then, $(AB)^T = C^T$, and

$$c_{ij}^T = c_{ji} = \sum_k a_{jk} b_{ki} = \sum_k a_{kj}^T b_{ik}^T = \sum_k b_{ik}^T a_{kj}^T$$

Thus, using the definition of matrix multiplication, $C^T = (AB)^T = B^T A^T$.

G–14. Show that $(AB)^{-1} = B^{-1} A^{-1}$.

Consider the matrix product $B^{-1} A^{-1} AB$. Since matrix multiplication is associative,

$$B^{-1} \left(A^{-1} A\right) B = B^{-1} IB = B^{-1} B = I$$

Similarly,

$$ABB^{-1} A^{-1} = A \left(BB^{-1}\right) A^{-1} = AIA^{-1} = AA^{-1} = I$$

Thus, writing $C = AB$,

$$B^{-1} A^{-1} C = I \qquad \text{and} \qquad CB^{-1} A^{-1} = I$$

or

$$B^{-1} A^{-1} = C^{-1} = (AB)^{-1}$$

G–15.   Show that Tr $AB = $ Tr BA.

The trace of AB is the sum of the diagonal elements of the product of the two matrices, or

$$\text{Tr } AB = \sum_i \sum_k a_{ik}b_{ki} = \sum_k \sum_i b_{ik}a_{ki} = \text{Tr } BA$$

where the summation variables have been switched and because the order of summation is unimportant.

G–16.   Consider the simultaneous algebraic equations

$$x + y = 3$$
$$4x - 3y = 5$$

Show that this pair of equations can be written in the matrix form

$$Ax = c \tag{1}$$

where

$$x = \begin{pmatrix} x \\ y \end{pmatrix} \qquad c = \begin{pmatrix} 3 \\ 5 \end{pmatrix} \qquad \text{and} \qquad A = \begin{pmatrix} 1 & 1 \\ 4 & -3 \end{pmatrix}$$

Now multiply equation 1 from the left by A^{-1} to obtain

$$x = A^{-1}c \tag{2}$$

Now show that

$$A^{-1} = -\frac{1}{7}\begin{pmatrix} -3 & -1 \\ -4 & 1 \end{pmatrix}$$

and that

$$x = -\frac{1}{7}\begin{pmatrix} -3 & -1 \\ -4 & 1 \end{pmatrix}\begin{pmatrix} 3 \\ 5 \end{pmatrix} = \begin{pmatrix} 2 \\ 1 \end{pmatrix}$$

or that $x = 2$ and $y = 1$. Do you see how this procedure generalizes to any number of simultaneous equations?

Writing out the equation $Ax = c$ gives

$$\begin{pmatrix} 1 & 1 \\ 4 & -3 \end{pmatrix}\begin{pmatrix} x \\ y \end{pmatrix} = \begin{pmatrix} 3 \\ 5 \end{pmatrix}$$

After performing the matrix multiplication this becomes

$$x + y = 3$$
$$4x - 3y = 5$$

which are the original two equations. Multiplying $\mathbf{Ax} = \mathbf{c}$ from the left by $\mathbf{A}^{-1}$ gives

$$\mathbf{A}^{-1}\mathbf{Ax} = \mathbf{A}^{-1}\mathbf{c}$$

$$\mathbf{Ix} = \mathbf{A}^{-1}\mathbf{c}$$

$$\mathbf{x} = \mathbf{A}^{-1}\mathbf{c}$$

The procedure of Problem G–9 can be used to find $\mathbf{A}^{-1}$. First, $|\mathbf{A}| = -7$. Then the cofactors are found:

$$A_{11} = -3 \quad A_{12} = -4$$
$$A_{21} = -1 \quad A_{22} = 1$$

and

$$\mathbf{A}^{-1} = -\frac{1}{7}\begin{pmatrix} -3 & -1 \\ -4 & 1 \end{pmatrix}$$

Therefore,

$$\mathbf{x} = -\frac{1}{7}\begin{pmatrix} -3 & -1 \\ -4 & 1 \end{pmatrix}\begin{pmatrix} 3 \\ 5 \end{pmatrix} = \begin{pmatrix} 2 \\ 1 \end{pmatrix}$$

The procedure is readily generalized to a system of N equations in N unknowns, where $\mathbf{A}$ is the $N \times N$ matrix of coefficients from the left hand side of the system, and $\mathbf{c}$ is the $N \times 1$ column vector of constants from right side of the system.

G–17. Solve the following simultaneous algebraic equations by the matrix inverse method developed in Problem G–16:

$$x + y - z = 1$$
$$2x - 2y + z = 6$$
$$x + 3z = 0$$

First, show that

$$\mathbf{A}^{-1} = \frac{1}{13}\begin{pmatrix} 6 & 3 & 1 \\ 5 & -4 & 3 \\ -2 & -1 & 4 \end{pmatrix}$$

and evaluate $\mathbf{x} = \mathbf{A}^{-1}\mathbf{c}$.

The 3×3 matrix $\mathbf{A}$ is

$$\mathbf{A} = \begin{pmatrix} 1 & 1 & -1 \\ 2 & -2 & 1 \\ 1 & 0 & 3 \end{pmatrix}$$

with

$$|\mathbf{A}| = (1)\begin{vmatrix} 1 & -1 \\ -2 & 1 \end{vmatrix} + (3)\begin{vmatrix} 1 & 1 \\ 2 & -2 \end{vmatrix} = 1(-1) + 3(-4) = -13$$

and cofactors

$$
\begin{array}{lll}
A_{11} = -6 & A_{12} = -5 & A_{13} = 2 \\
A_{21} = -3 & A_{22} = 4 & A_{23} = 1 \\
A_{31} = -1 & A_{32} = -3 & A_{33} = -4
\end{array}
$$

so that

$$
\mathsf{A}^{-1} = -\frac{1}{13}
\begin{pmatrix}
-6 & -3 & -1 \\
-5 & 4 & -3 \\
2 & 1 & -4
\end{pmatrix}
= \frac{1}{13}
\begin{pmatrix}
6 & 3 & 1 \\
5 & -4 & 3 \\
-2 & -1 & 4
\end{pmatrix}
$$

Then,

$$
\mathbf{x} =
\begin{pmatrix}
x \\
y \\
z
\end{pmatrix}
= \mathsf{A}^{-1}
\begin{pmatrix}
1 \\
6 \\
0
\end{pmatrix}
= \frac{1}{13}
\begin{pmatrix}
6 & 3 & 1 \\
5 & -4 & 3 \\
-2 & -1 & 4
\end{pmatrix}
\begin{pmatrix}
1 \\
6 \\
0
\end{pmatrix}
= \frac{1}{13}
\begin{pmatrix}
24 \\
-19 \\
-8
\end{pmatrix}
$$

Approximation Methods

PROBLEMS AND SOLUTIONS

8–1. This problem involves the proof of the variational principle (Equation 8.4). Let $\hat{H}\psi_n = E_n\psi_n$ be the problem of interest, and let ϕ be our approximation to ψ_0. Even though we do not know the ψ_n, we can express ϕ formally as

$$\phi = \sum_n c_n \psi_n$$

where the c_n are constants. Using the fact that the ψ_n are orthonormal, show that

$$c_n = \int \psi_n^* \phi \, d\tau$$

Substitute ϕ as given above into

$$E_\phi = \frac{\int \phi^* \hat{H} \phi \, d\tau}{\int \phi^* \phi \, d\tau}$$

to obtain

$$E_\phi = \frac{\sum_n c_n^* c_n E_n}{\sum_n c_n^* c_n}$$

Subtract E_0 from the left side of the above equation and $E_0 \sum_n c_n^* c_n / \sum_n c_n^* c_n$ from the right side to obtain

$$E_\phi - E_0 = \frac{\sum_n c_n^* c_n (E_n - E_0)}{\sum_n c_n^* c_n}$$

Now explain why every term on the right side is either zero or positive, proving that $E_\phi \geq E_0$.

Writing ϕ as a linear combination of the eigenfunctions of $\hat{H}$ gives

$$\phi = \sum_m c_m \psi_m$$

Left multiply both sides of this equation by ψ_n^* and integrate over all space:

$$\int \psi_n^* \phi \, d\tau = \sum_m c_m \int \psi_n^* \psi_m \, d\tau$$

$$= \sum_m c_m \delta_{nm}$$

$$= c_n$$

Now substitute ϕ into the E_ϕ expression and use the facts that the functions ψ_n are eigenfunctions of $\hat{H}$ and are orthonormal.

$$E_\phi = \frac{\int \phi^* \hat{H} \phi \, d\tau}{\int \phi^* \phi \, d\tau}$$

$$= \frac{\sum_m \sum_n c_m^* c_n \int \psi_m^* \hat{H} \psi_n \, d\tau}{\sum_m \sum_n c_m^* c_n \int \psi_m^* \psi_n \, d\tau}$$

$$= \frac{\sum_m \sum_n c_m^* c_n E_n \int \psi_m^* \psi_n \, d\tau}{\sum_m \sum_n c_m^* c_n \delta_{mn}}$$

$$= \frac{\sum_m \sum_n c_m^* c_n E_n \delta_{mn}}{\sum_n c_n^* c_n}$$

$$= \frac{\sum_n c_n^* c_n E_n}{\sum_n c_n^* c_n}$$

Subtract E_0 from both sides of the above equation:

$$E_\phi - E_0 = \frac{\sum_n c_n^* c_n E_n}{\sum_n c_n^* c_n} - E_0 \frac{\sum_n c_n^* c_n}{\sum_n c_n^* c_n}$$

$$= \frac{\sum_n c_n^* c_n (E_n - E_0)}{\sum_n c_n^* c_n}$$

Since E_0 is the ground-state energy, $E_n \geq E_0$. In addition, $c_n^* c_n \geq 0$. Thus, $E_\phi - E_0 \geq 0$ or $E_\phi \geq E_0$.

8–2. Show that the expression for α in a trial function $e^{-\alpha r^2}$ for a hydrogen atom is given by $\alpha = 8/9\pi a_0^2$, where a_0 is the Bohr radius.

Since $E_\phi = \int \phi^* \hat{H} \phi \, d\tau / \int \phi^* \phi \, d\tau$, the integrals in the numerator and denominator must be evaluated. First evaluate $\hat{H}\phi$:

$$\hat{H}\phi = \left[-\frac{\hbar^2}{2m_e r^2} \frac{d}{dr}\left(r^2 \frac{d}{dr}\right) - \frac{e^2}{4\pi\epsilon_0 r} \right] e^{-\alpha r^2}$$

$$= -\frac{\hbar^2}{2m_e r^2} \frac{d}{dr}\left(-2\alpha r^3 e^{-\alpha r^2}\right) - \frac{e^2}{4\pi\epsilon_0 r} e^{-\alpha r^2}$$

$$= -\frac{\hbar^2}{2m_e r^2}\left(-6\alpha r^2 e^{-\alpha r^2} + 4\alpha^2 r^4 e^{-\alpha r^2}\right) - \frac{e^2}{4\pi\epsilon_0 r} e^{-\alpha r^2}$$

It follows that

$$\int \phi^* \hat{H}\phi \, d\tau = \int_0^{2\pi} d\phi \int_0^{\pi} d\theta \, \sin\theta \int_0^{\infty} dr \, r^2 e^{-\alpha r^2}$$

$$\times \left[-\frac{\hbar^2}{2m_e r^2}\left(-6\alpha r^2 e^{-\alpha r^2} + 4\alpha^2 r^4 e^{-\alpha r^2}\right) - \frac{e^2}{4\pi\epsilon_0 r} e^{-\alpha r^2} \right]$$

$$= (4\pi)\left[-\frac{\hbar^2}{2m_e} \int_0^{\infty} dr \left(-6\alpha r^2 e^{-2\alpha r^2} + 4\alpha^2 r^4 e^{-2\alpha r^2}\right) - \frac{e^2}{4\pi\epsilon_0} \int_0^{\infty} dr \, r e^{-2\alpha r^2} \right]$$

$$= (4\pi)\left\{ -\frac{\hbar^2}{2m_e}\left[-6\alpha\left(\frac{1}{8\alpha}\sqrt{\frac{\pi}{2\alpha}}\right) + 4\alpha^2\left(\frac{3}{32\alpha^2}\sqrt{\frac{\pi}{2\alpha}}\right) \right] - \frac{e^2}{4\pi\epsilon_0}\left(\frac{1}{4\alpha}\right) \right\}$$

$$= \frac{3\hbar^2 \pi^{3/2}}{32^{1/2} m_e \alpha^{1/2}} - \frac{e^2}{4\epsilon_0 \alpha}$$

Now evaluate $\int \phi^* \phi \, d\tau$:

$$\int \phi^* \phi \, d\tau = \int_0^{2\pi} d\phi \int_0^{\pi} d\theta \, \sin\theta \int_0^{\infty} dr \, r^2 e^{-2\alpha r^2}$$

$$= (4\pi)\left(\frac{1}{8\alpha}\sqrt{\frac{\pi}{2\alpha}}\right)$$

$$= \left(\frac{\pi}{2\alpha}\right)^{3/2}$$

Thus,

$$E_\phi = \frac{\int \phi^* \hat{H}\phi \, d\tau}{\int \phi^* \phi \, d\tau}$$

$$= \left(\frac{2\alpha}{\pi}\right)^{3/2}\left(\frac{3\hbar^2 \pi^{3/2}}{32^{1/2} m_e \alpha^{1/2}} - \frac{e^2}{4\epsilon_0 \alpha}\right)$$

$$= \frac{3\hbar^2 \alpha}{2m_e} - \frac{e^2 \alpha^{1/2}}{2^{1/2}\epsilon_0 \pi^{3/2}}$$

To minimize E_ϕ, differentiate it with respect to α and set the result equal to zero:

$$\frac{dE_\phi}{d\alpha} = \frac{3\hbar^2}{2m_e} - \frac{e^2}{(2\pi)^{3/2}\epsilon_0 \alpha^{1/2}} = 0$$

Solve for α to give

$$\frac{e^2}{(2\pi)^{3/2}\,\epsilon_0\alpha^{1/2}} = \frac{3\hbar^2}{2m_e}$$

$$\alpha = \frac{m_e^2 e^4}{18\pi^3\epsilon_0^2\hbar^4} = \frac{8}{9\pi a_0^2}$$

where $a_0 = 4\pi\epsilon_0\hbar^2/(m_e e^2)$.

8–3. Calculate the ground state of a hydrogen atom using a trial function of the form $e^{-\alpha r}$. Why does the result turn out to be so good?

Since $E_\phi = \int \phi^*\hat{H}\phi\,d\tau / \int \phi^*\phi\,d\tau$, the integrals in the numerator and denominator must be evaluated. First evaluate $\hat{H}\phi$:

$$\hat{H}\phi = \left[-\frac{\hbar^2}{2m_e r^2}\frac{d}{dr}\left(r^2\frac{d}{dr}\right) - \frac{e^2}{4\pi\epsilon_0 r}\right]e^{-\alpha r}$$

$$= -\frac{\hbar^2}{2m_e r^2}\frac{d}{dr}\left(-\alpha r^2 e^{-\alpha r}\right) - \frac{e^2}{4\pi\epsilon_0 r}e^{-\alpha r}$$

$$= -\frac{\hbar^2}{2m_e r^2}\left(-2\alpha r e^{-\alpha r^2} + \alpha^2 r^2 e^{-\alpha r}\right) - \frac{e^2}{4\pi\epsilon_0 r}e^{-\alpha r}$$

It follows that

$$\int \phi^*\hat{H}\phi\,d\tau = \int_0^{2\pi} d\phi \int_0^\pi d\theta\,\sin\theta \int_0^\infty dr\,r^2 e^{-\alpha r}$$

$$\times \left[-\frac{\hbar^2}{2m_e r^2}\left(-2\alpha r e^{-\alpha r^2} + \alpha^2 r^2 e^{-\alpha r}\right) - \frac{e^2}{4\pi\epsilon_0 r}e^{-\alpha r}\right]$$

$$= (4\pi)\left[-\frac{\hbar^2}{2m_e}\int_0^\infty dr\,\left(-2\alpha r e^{-2\alpha r} + \alpha^2 r^2 e^{-2\alpha r}\right) - \frac{e^2}{4\pi\epsilon_0}\int_0^\infty dr\,r e^{-2\alpha r}\right]$$

$$= (4\pi)\left\{-\frac{\hbar^2}{2m_e}\left[-2\alpha\left(\frac{1}{4\alpha^2}\right) + \alpha^2\left(\frac{1}{4\alpha^3}\right)\right] - \frac{e^2}{4\pi\epsilon_0}\left(\frac{1}{4\alpha^2}\right)\right\}$$

$$= \frac{\hbar^2\pi}{2m_e\alpha} - \frac{e^2}{4\epsilon_0\alpha^2}$$

Now evaluate $\int \phi^*\phi\,d\tau$:

$$\int \phi^*\phi\,d\tau = \int_0^{2\pi} d\phi \int_0^\pi d\theta\,\sin\theta \int_0^\infty dr\,r^2 e^{-2\alpha r}$$

$$= (4\pi)\left(\frac{1}{4\alpha^3}\right)$$

$$= \frac{\pi}{\alpha^3}$$

Thus,

$$E_\phi = \frac{\int \phi^* \hat{H} \phi \, d\tau}{\int \phi^* \phi \, d\tau}$$

$$= \left(\frac{\alpha^3}{\pi}\right) \left(\frac{\hbar^2 \pi}{2m_e \alpha} - \frac{e^2}{4\epsilon_0 \alpha^2}\right)$$

$$= \frac{\hbar^2 \alpha^2}{2m_e} - \frac{e^2 \alpha}{4\epsilon_0 \pi}$$

To minimize E_ϕ, differentiate it with respect to α and set the result equal to zero:

$$\frac{dE_\phi}{d\alpha} = \frac{\hbar^2 \alpha}{m_e} - \frac{e^2}{4\epsilon_0 \pi} = 0$$

Solve for α to give

$$\alpha = \frac{m_e e^2}{4\pi \epsilon_0 \hbar^2}$$

Substitute this expression into the E_ϕ expression to give the estimate of the ground-state energy from this trial function:

$$E_{\min} = \frac{\hbar^2}{2m_e} \left(\frac{m_e e^2}{4\pi \epsilon_0 \hbar^2}\right)^2 - \frac{e^2}{4\epsilon_0 \pi} \left(\frac{m_e e^2}{4\pi \epsilon_0 \hbar^2}\right)$$

$$= -\frac{m_e e^4}{32 \hbar^2 \pi^2 \epsilon_0^2} = -\frac{m_e e^4}{8h^2 \epsilon_0^2}$$

The value of $E_{\min}$ agrees with the exact ground-state energy. This is because the trial function has the same form as the exact ground-state wave function.

8–4. Use a trial function of the form $\phi(x) = 1/(1 + \beta x^2)$ to calculate the ground-state energy of a harmonic oscillator. The necessary integrals are

$$\int_{-\infty}^{\infty} \frac{dx}{(1 + \beta x^2)^2} = \frac{\pi}{2\beta^{1/2}}$$

$$\int_{-\infty}^{\infty} \frac{dx}{(1 + \beta x^2)^n} = \frac{(2n - 3)(2n - 5)(2n - 7) \cdots (1)}{(2n - 2)(2n - 4)(2n - 6) \cdots (2)} \frac{\pi}{\beta^{1/2}} \qquad n \geq 2$$

and

$$\int_{-\infty}^{\infty} \frac{x^2 dx}{(1 + \beta x^2)^n} = \frac{(2n - 5)(2n - 7) \cdots (1)}{(2n - 2)(2n - 4) \cdots (2)} \frac{\pi}{\beta^{3/2}} \qquad n \geq 3$$

Since $E_\phi = \int \phi^* \hat{H} \phi \, d\tau / \int \phi^* \phi \, d\tau$, the integrals in the numerator and denominator must be evaluated. First evaluate $\hat{H}\phi$:

$$\hat{H}\phi = \left(-\frac{\hbar^2}{2\mu}\frac{d^2}{dx^2} + \frac{1}{2}kx^2\right)\frac{1}{1+\beta x^2}$$

$$= -\frac{\hbar^2}{2\mu}\left[\frac{8\beta^2 x^2}{(1+\beta x^2)^3} - \frac{2\beta}{(1+\beta x^2)^2}\right] + \frac{kx^2}{2}\frac{1}{1+\beta x^2}$$

It follows that

$$\int \phi^* \hat{H}\phi \, d\tau = \int_{-\infty}^{\infty}\frac{1}{1+\beta x^2}\left\{-\frac{\hbar^2}{2\mu}\left[\frac{8\beta^2 x^2}{(1+\beta x^2)^3} - \frac{2\beta}{(1+\beta x^2)^2}\right] + \frac{kx^2}{2}\frac{1}{1+\beta x^2}\right\} dx$$

$$= -\frac{\hbar^2}{2\mu}\left[8\beta^2\left(\frac{3\pi}{48\beta^{3/2}}\right) - 2\beta\left(\frac{3\pi}{8\beta^{1/2}}\right)\right] + \frac{k}{2}\frac{\pi}{2\beta^{3/2}}$$

$$= \frac{\pi\hbar^2\beta^{1/2}}{8\mu} + \frac{\pi k}{4\beta^{3/2}}$$

Now evaluate $\int \phi^*\phi \, d\tau$:

$$\int \phi^*\phi \, d\tau = \int_{-\infty}^{\infty}\left(\frac{1}{1+\beta x^2}\right)^2 dx$$

$$= \frac{\pi}{2\beta^{1/2}}$$

Thus,

$$E_\phi = \frac{\int \phi^* \hat{H}\phi \, d\tau}{\int \phi^*\phi \, d\tau}$$

$$= \left(\frac{2\beta^{1/2}}{\pi}\right)\left(\frac{\pi\hbar^2\beta^{1/2}}{8\mu} + \frac{\pi k}{4\beta^{3/2}}\right)$$

$$= \frac{\hbar^2\beta}{4\mu} + \frac{k}{2\beta}$$

To minimize E_ϕ, differentiate it with respect to β and set the result equal to zero:

$$\frac{dE_\phi}{d\beta} = \frac{\hbar^2}{4\mu} - \frac{k}{2\beta^2} = 0$$

Solve for β to give

$$\beta = \frac{(2k\mu)^{1/2}}{\hbar}$$

The second root for β gives a negative value for E_ϕ, which is nonphysical for the harmonic oscillator. Substitute this expression into the E_ϕ expression to give the estimate of the ground-state energy from this trial function:

$$E_{min} = \frac{\hbar^2}{4\mu}\left[\frac{(2k\mu)^{1/2}}{\hbar}\right] + \frac{k}{2}\frac{\hbar}{(2k\mu)^{1/2}}$$

$$= \frac{\hbar}{2^{1/2}}\left(\frac{k}{\mu}\right)^{1/2} = 0.7071\hbar\left(\frac{k}{\mu}\right)^{1/2}$$

which is greater than the exact ground-state energy, $(\hbar/2)(k/\mu)^{1/2}$, by 41%.

8–5. Use a trial function $\phi(x) = 1/(1 + \beta x^2)^2$ to calculate the ground-state energy of a harmonic oscillator variationally. The necessary integrals are given in the previous problem.

Since $E_\phi = \int \phi^* \hat{H} \phi \, d\tau / \int \phi^* \phi \, d\tau$, the integrals in the numerator and denominator must be evaluated. First evaluate $\hat{H}\phi$:

$$\hat{H}\phi = \left(-\frac{\hbar^2}{2\mu}\frac{d^2}{dx^2} + \frac{1}{2}kx^2 \right) \frac{1}{(1 + \beta x^2)^2}$$

$$= -\frac{\hbar^2}{2\mu}\left[-\frac{4\beta}{(1 + \beta x^2)^3} + \frac{24\beta^2 x^2}{(1 + \beta x^2)^4} \right] + \frac{kx^2}{2(1 + \beta x^2)^2}$$

It follows that

$$\int \phi^* \hat{H} \phi \, d\tau = \int_{-\infty}^{\infty} \frac{1}{(1 + \beta x^2)^2} \left\{ -\frac{\hbar^2}{2\mu}\left[-\frac{4\beta}{(1 + \beta x^2)^3} + \frac{24\beta^2 x^2}{(1 + \beta x^2)^4} \right] + \frac{kx^2}{2(1 + \beta x^2)^2} \right\} dx$$

$$= \frac{2\beta\hbar^2}{\mu}\left(\frac{35\pi}{128\beta^{1/2}} \right) - \frac{12\beta^2\hbar^2}{\mu}\left(\frac{7\pi}{256\beta^{3/2}} \right) + \frac{k}{2}\left(\frac{\pi}{16\beta^{3/2}} \right)$$

$$= \frac{7\pi\beta^{1/2}\hbar^2}{32\mu} + \frac{k\pi}{32\beta^{3/2}}$$

Now evaluate $\int \phi^* \phi \, d\tau$:

$$\int \phi^* \phi \, d\tau = \int_{-\infty}^{\infty} \frac{1}{(1 + \beta x^2)^4} dx$$

$$= \frac{5\pi}{16\beta^{1/2}}$$

Thus,

$$E_\phi = \frac{\int \phi^* \hat{H} \phi \, d\tau}{\int \phi^* \phi \, d\tau}$$

$$= \left(\frac{16\beta^{1/2}}{5\pi} \right)\left(\frac{7\pi\beta^{1/2}\hbar^2}{32\mu} + \frac{k\pi}{32\beta^{3/2}} \right)$$

$$= \frac{7\beta\hbar^2}{10\mu} + \frac{k}{10\beta}$$

To minimize E_ϕ, differentiate it with respect to β and set the result equal to zero:

$$\frac{dE_\phi}{d\beta} = \frac{7\hbar^2}{10\mu} - \frac{k}{10\beta^2} = 0$$

Solve for β to give

$$\beta = \left(\frac{\mu k}{7\hbar^2} \right)^{1/2}$$

The second root for β gives a negative value for E_ϕ, which is nonphysical for the harmonic oscillator. Substitute this expression into the E_ϕ expression to give the estimate of the ground-state energy from this trial function:

$$E_{\min} = \frac{7\hbar^2}{10\mu}\left(\frac{\mu k}{7\hbar^2}\right)^{1/2} + \frac{k}{10}\left(\frac{7\hbar^2}{\mu k}\right)^{1/2}$$

$$= \frac{7^{1/2}\hbar}{5}\left(\frac{k}{\mu}\right)^{1/2} = 0.5292\,\hbar\left(\frac{k}{\mu}\right)^{1/2}$$

which is greater than the exact ground-state energy, $(\hbar/2)\,(k/\mu)^{1/2}$, by 6%.

8–6. If you were to use a trial function of the form $\phi(x) = (1 + c\alpha x^2)e^{-\alpha x^2/2}$, where $\alpha = (k\mu/\hbar^2)^{1/2}$ and c is a variational parameter, to calculate the ground-state energy of a harmonic oscillator, what do you think the value of c will come out to be? Why?

The exact ground-state wave function of a harmonic oscillator is of the form $e^{-\alpha x^2/2}$. Thus, the value of c will be zero.

8–7. Use a trial function of the form $\phi(r) = re^{-\alpha r}$ with α as a variational parameter to calculate the ground-state energy of a hydrogen atom.

Since $E_\phi = \int \phi^* \hat{H}\phi \, d\tau / \int \phi^*\phi \, d\tau$, the integrals in the numerator and denominator must be evaluated. First evaluate $\hat{H}\phi$:

$$\hat{H}\phi = \left[-\frac{\hbar^2}{2m_e r^2}\frac{d}{dr}\left(r^2\frac{d}{dr}\right) - \frac{e^2}{4\pi\epsilon_0 r}\right]re^{-\alpha r}$$

$$= -\frac{\hbar^2}{2m_e r^2}\frac{d}{dr}\left(r^2 e^{-\alpha r} - \alpha r^3 e^{-\alpha r}\right) - \frac{e^2}{4\pi\epsilon_0}e^{-\alpha r}$$

$$= -\frac{\hbar^2}{2m_e r^2}\left(2r - 4\alpha r^2 + \alpha^2 r^3\right)e^{-\alpha r} - \frac{e^2}{4\pi\epsilon_0}e^{-\alpha r}$$

It follows that

$$\int \phi^* \hat{H} \phi \, d\tau = \int_0^{2\pi} d\phi \int_0^{\pi} d\theta \, \sin\theta \int_0^{\infty} dr \, r^2 r e^{-\alpha r}$$

$$\times \left[-\frac{\hbar^2}{2m_e r^2} \left(2r - 4\alpha r^2 + \alpha^2 r^3 \right) e^{-\alpha r} - \frac{e^2}{4\pi\epsilon_0} e^{-\alpha r} \right]$$

$$= (4\pi) \left[-\frac{\hbar^2}{2m_e} \int_0^{\infty} dr \left(2r^2 - 4\alpha r^3 + \alpha^2 r^4 \right) e^{-2\alpha r} - \frac{e^2}{4\pi\epsilon_0} \int_0^{\infty} dr \, r^3 e^{-2\alpha r} \right]$$

$$= (4\pi) \left\{ -\frac{\hbar^2}{2m_e} \left[2\left(\frac{1}{4\alpha^3} \right) - 4\alpha \left(\frac{3}{8\alpha^4} \right) + \alpha^2 \left(\frac{3}{4\alpha^5} \right) \right] - \frac{e^2}{4\pi\epsilon_0} \left(\frac{3}{8\alpha^4} \right) \right\}$$

$$= \frac{\pi\hbar^2}{2m_e\alpha^3} - \frac{3e^2}{8\epsilon_0\alpha^4}$$

Now evaluate $\int \phi^* \phi \, d\tau$:

$$\int \phi^* \phi \, d\tau = \int_0^{2\pi} d\phi \int_0^{\pi} d\theta \, \sin\theta \int_0^{\infty} dr \, r^4 e^{-2\alpha r}$$

$$= (4\pi) \left(\frac{3}{4\alpha^5} \right)$$

$$= \frac{3\pi}{\alpha^5}$$

Thus,

$$E_\phi = \frac{\int \phi^* \hat{H} \phi \, d\tau}{\int \phi^* \phi \, d\tau}$$

$$= \left(\frac{\alpha^5}{3\pi} \right) \left(\frac{\pi\hbar^2}{2m_e\alpha^3} - \frac{3e^2}{8\epsilon_0\alpha^4} \right)$$

$$= \frac{\alpha^2 \hbar^2}{6m_e} - \frac{\alpha e^2}{8\epsilon_0\pi}$$

To minimize E_ϕ, differentiate it with respect to α and set the result equal to zero:

$$\frac{dE_\phi}{d\alpha} = \frac{\alpha\hbar^2}{3m_e} - \frac{e^2}{8\epsilon_0\pi} = 0$$

Solve for α to give

$$\alpha = \frac{3m_e e^2}{8\epsilon_0\pi\hbar^2}$$

Substitute this expression into the E_ϕ expression to give the estimate of the ground-state energy from this trial function:

$$E_{\min} = \frac{\hbar^2}{6m_e} \left(\frac{3m_e e^2}{8\epsilon_0\pi\hbar^2} \right)^2 - \frac{e^2}{8\epsilon_0\pi} \left(\frac{3m_e e^2}{8\epsilon_0\pi\hbar^2} \right)$$

$$= -\frac{3m_e e^4}{128\epsilon_0^2\pi^2\hbar^2}$$

The exact ground-state energy is $-m_e e^4/(8h^2\epsilon_0^2) = -m_e e^4/(32\pi^2\hbar^2\epsilon_0^2)$. Thus, the minimum energy is greater than the exact ground-state energy by 25%.

8–8. Suppose we were to use a trial function of the form $\phi = c_1 e^{-\alpha r} + c_2 e^{-\beta r^2}$ to carry out a variational calculation for the ground-state energy of a hydrogen atom. Can you guess without doing any calculations what c_1, c_2, α, and E_{min} will be? What about a trial function of the form $\phi = \sum_{k=1}^{5} c_k e^{-\alpha_k r - \beta_k r^2}$?

For each of the trial functions, the energy E_{min} will turn out to be the exact ground-state energy of the hydrogen atom, $-m_e e^4/(8h^2\epsilon_0^2)$ because the form of the exact ground-state wave function is contained in the trial function. Thus, for the function $\phi = c_1 e^{-\alpha r} + c_2 e^{-\beta r^2}$, $c_2 = 0$ (making β irrelevant), $c_1 = 1/(\pi^{1/2} a_0^{3/2})$ (normalization constant), and $\alpha = 1/a_0$.

For the function $\phi = \sum_{k=1}^{5} c_k e^{-\alpha_k r - \beta_k r^2}$, only one of the c_k coefficients is nonzero. Calling that coefficient c_i, it follows that $c_i = 1/(\pi^{1/2} a_0^{3/2})$, $\alpha_i = 1/a_0$, and $\beta_i = 0$. For $k \neq i$, $c_k = 0$ (making α_k and β_k irrelevant).

8–9. Use a trial function of the form $e^{-\beta x^2}$ with β as a variational parameter to calculate the ground-state energy of a harmonic oscillator. Compare your result with the exact energy $h\nu/2$. Why is the agreement so good?

Since $E_\phi = \int \phi^*\hat{H}\phi \, d\tau / \int \phi^*\phi \, d\tau$, the integrals in the numerator and denominator must be evaluated. First evaluate $\hat{H}\phi$:

$$\hat{H}\phi = \left(-\frac{\hbar^2}{2\mu}\frac{d^2}{dx^2} + \frac{1}{2}kx^2\right)e^{-\beta x^2}$$

$$= -\frac{\hbar^2}{2\mu}\frac{d}{dx}\left(-2\beta x e^{-\beta x^2}\right) + \frac{kx^2}{2}e^{-\beta x^2}$$

$$= -\frac{\hbar^2}{2\mu}\left(-2\beta e^{-\beta x^2} + 4\beta^2 x^2 e^{-\beta x^2}\right) + \frac{kx^2}{2}e^{-\beta x^2}$$

It follows that

$$\int \phi^*\hat{H}\phi \, d\tau = \int_{-\infty}^{\infty} e^{-\beta x^2}\left[-\frac{\hbar^2}{2\mu}\left(-2\beta e^{-\beta x^2} + 4\beta^2 x^2 e^{-\beta x^2}\right) + \frac{kx^2}{2}e^{-\beta x^2}\right] dx$$

$$= \frac{\hbar^2}{2\mu}\left[2\beta\left(\frac{\pi}{2\beta}\right)^{1/2} - 4\beta^2\left(\frac{1}{4\beta}\right)\left(\frac{\pi}{2\beta}\right)^{1/2}\right] + \frac{k}{2}\left(\frac{1}{4\beta}\right)\left(\frac{\pi}{2\beta}\right)^{1/2}$$

$$= \frac{\hbar^2\beta}{2\mu}\left(\frac{\pi}{2\beta}\right)^{1/2} + \frac{k}{8\beta}\left(\frac{\pi}{2\beta}\right)^{1/2}$$

Now evaluate $\int \phi^* \phi \, d\tau$:

$$\int \phi^* \phi \, d\tau = \int_{-\infty}^{\infty} e^{-2\beta x^2} \, dx$$

$$= \left(\frac{\pi}{2\beta} \right)^{1/2}$$

Thus,

$$E_\phi = \frac{\int \phi^* \hat{H} \phi \, d\tau}{\int \phi^* \phi \, d\tau}$$

$$= \left(\frac{2\beta}{\pi} \right)^{1/2} \left[\frac{\hbar^2 \beta}{2\mu} \left(\frac{\pi}{2\beta} \right)^{1/2} + \frac{k}{8\beta} \left(\frac{\pi}{2\beta} \right)^{1/2} \right]$$

$$= \frac{\hbar^2 \beta}{2\mu} + \frac{k}{8\beta}$$

To minimize E_ϕ, differentiate it with respect to β and set the result equal to zero:

$$\frac{dE_\phi}{d\beta} = \frac{\hbar^2}{2\mu} - \frac{k}{8\beta^2} = 0$$

Solve for β to give

$$\beta = \frac{(k\mu)^{1/2}}{2\hbar}$$

The second root for β gives a negative value for E_ϕ, which is nonphysical for the harmonic oscillator. Substitute this expression into the E_ϕ expression to give the estimate of the ground-state energy from this trial function:

$$E_{\min} = \frac{\hbar^2}{2\mu} \left[\frac{(k\mu)^{1/2}}{2\hbar} \right] + \frac{k}{8} \left[\frac{2\hbar}{(k\mu)^{1/2}} \right]$$

$$= \frac{\hbar}{2} \left(\frac{k}{\mu} \right)^{1/2}$$

which is the exact ground-state energy. The agreement is so good because the trial function has the same form as the exact ground-state wave function.

8–10. Consider a three-dimensional, spherically symmetric, isotropic harmonic oscillator with $V(r) = kr^2/2$. Using a trial function $e^{-\alpha r^2}$ with α as a variational parameter, calculate the ground-state energy. Do the same using $e^{-\alpha r}$. The Hamiltonian operator is

$$\hat{H} = -\frac{\hbar^2}{2\mu r^2} \frac{d}{dr} \left(r^2 \frac{d}{dr} \right) + \frac{k}{2} r^2$$

Compare these results with the exact ground-state energy, $E = \frac{3}{2} h\nu$. Why is one of these so much better than the other?

First use $\phi = e^{-\alpha r^2}$ as the trial function. Since $E_\phi = \int \phi^* \hat{H} \phi \, d\tau / \int \phi^* \phi \, d\tau$, the integrals in the numerator and denominator must be evaluated. First evaluate $\hat{H}\phi$:

$$\hat{H}\phi = \left[-\frac{\hbar^2}{2\mu r^2} \frac{d}{dr}\left(r^2 \frac{d}{dr} \right) + \frac{kr^2}{2} \right] e^{-\alpha r^2}$$

$$= -\frac{\hbar^2}{2\mu r^2} \frac{d}{dr} \left(-2\alpha r^3 e^{-\alpha r^2} \right) + \frac{kr^2}{2} e^{-\alpha r^2}$$

$$= -\frac{\hbar^2}{2\mu r^2} \left(-6\alpha r^2 e^{-\alpha r^2} + 4\alpha^2 r^4 e^{-\alpha r^2} \right) + \frac{kr^2}{2} e^{-\alpha r^2}$$

It follows that

$$\int \phi^* \hat{H} \phi \, d\tau = \int_0^{2\pi} d\phi \int_0^\pi d\theta \, \sin\theta \int_0^\infty dr \, r^2 e^{-\alpha r^2}$$

$$\times \left[-\frac{\hbar^2}{2\mu r^2} \left(-6\alpha r^2 e^{-\alpha r^2} + 4\alpha^2 r^4 e^{-\alpha r^2} \right) + \frac{kr^2}{2} e^{-\alpha r^2} \right]$$

$$= (4\pi) \left[-\frac{\hbar^2}{2\mu} \int_0^\infty dr \, \left(-6\alpha r^2 e^{-2\alpha r^2} + 4\alpha^2 r^4 e^{-2\alpha r^2} \right) + \frac{k}{2} \int_0^\infty dr \, r^4 e^{-2\alpha r^2} \right]$$

$$= (4\pi) \left\{ -\frac{\hbar^2}{2\mu} \left[-6\alpha \left(\frac{1}{8\alpha} \right) \left(\frac{\pi}{2\alpha} \right)^{1/2} + 4\alpha^2 \left(\frac{3}{32\alpha^2} \right) \left(\frac{\pi}{2\alpha} \right)^{1/2} \right] \right.$$

$$\left. + \frac{k}{2} \left(\frac{3}{32\alpha^2} \right) \left(\frac{\pi}{2\alpha} \right)^{1/2} \right\}$$

$$= \frac{3\pi \hbar^2}{4\mu} \left(\frac{\pi}{2\alpha} \right)^{1/2} + \left(\frac{3\pi k}{16\alpha^2} \right) \left(\frac{\pi}{2\alpha} \right)^{1/2}$$

Now evaluate $\int \phi^* \phi \, d\tau$:

$$\int \phi^* \phi \, d\tau = \int_0^{2\pi} d\phi \int_0^\pi d\theta \, \sin\theta \int_0^\infty dr \, r^2 e^{-2\alpha r^2}$$

$$= (4\pi) \left(\frac{1}{8\alpha} \right) \left(\frac{\pi}{2\alpha} \right)^{1/2}$$

$$= \left(\frac{\pi}{2\alpha} \right)^{3/2}$$

Thus,

$$E_\phi = \frac{\int \phi^* \hat{H} \phi \, d\tau}{\int \phi^* \phi \, d\tau}$$

$$= \left(\frac{2\alpha}{\pi} \right)^{3/2} \left[\frac{3\pi \hbar^2}{4\mu} \left(\frac{\pi}{2\alpha} \right)^{1/2} + \left(\frac{3\pi k}{16\alpha^2} \right) \left(\frac{\pi}{2\alpha} \right)^{1/2} \right]$$

$$= \frac{3\hbar^2 \alpha}{2\mu} + \frac{3k}{8\alpha}$$

To minimize E_ϕ, differentiate it with respect to α and set the result equal to zero:

$$\frac{dE_\phi}{d\alpha} = \frac{3\hbar^2}{2\mu} - \frac{3k}{8\alpha^2} = 0$$

Solve for α to give

$$\alpha = \frac{(k\mu)^{1/2}}{2\hbar}$$

The second root for α gives a negative E_ϕ, which is nonphysical for the three-dimensional spherically symmetric, isotropic harmonic oscillator. Substitute this expression into the E_ϕ expression to give the estimate of the ground-state energy from this trial function:

$$E_{\min} = \frac{3\hbar^2}{2\mu} \left[\frac{(k\mu)^{1/2}}{2\hbar} \right] + \frac{3k}{8} \left[\frac{2\hbar}{(k\mu)^{1/2}} \right]$$

$$= \frac{3\hbar}{2} \left(\frac{k}{\mu} \right)^{1/2}$$

which is the exact ground-state energy. This arises because the trial function has the same form as the exact ground-state wave function.

Following the same procedure for $\phi = e^{-\alpha r}$ as the trial function,

$$\int \phi^* \hat{H} \phi \, d\tau = \frac{\pi \hbar^2}{2\alpha\mu} + \frac{3\pi k}{2\alpha^5}$$

$$\int \phi^* \phi \, d\tau = \frac{\pi}{\alpha^3}$$

Therefore,

$$E_\phi = \frac{\int \phi^* \hat{H} \phi \, d\tau}{\int \phi^* \phi \, d\tau}$$

$$= \frac{h^2 \alpha^2}{2\mu} + \frac{3k}{2\alpha^2}$$

To minimize E_ϕ, differentiate it with respect to α and set the result equal to zero:

$$\frac{dE_\phi}{d\alpha} = \frac{h^2 \alpha}{\mu} - \frac{3k}{\alpha^3} = 0$$

Solve for α to give

$$\alpha = \pm \left(\frac{3k\mu}{\hbar^2} \right)^{1/4}$$

Only the positive value for α is physical. The negative value gives a wave function that diverges. The other two roots are imaginary and they give nonphysical, negative values for E_ϕ. Substitute α into the E_ϕ expression to give the estimate of the ground-state energy from this trial function:

$$E_{min} = \frac{h^2}{2\mu}\left(\frac{3k\mu}{\hbar^2}\right)^{1/2} + \frac{3k}{2}\left(\frac{\hbar^2}{3k\mu}\right)^{1/2}$$

$$= 3^{1/2}\hbar\left(\frac{k}{\mu}\right)^{1/2}$$

which is greater than the exact ground-state energy by 15%.

8–11. Use a trial function of the form $e^{-\alpha x^2/2}$ to calculate the ground-state energy of a quartic oscillator, whose potential is $V(x) = cx^4$.

Since $E_\phi = \int \phi^* \hat{H}\phi \, d\tau / \int \phi^*\phi \, d\tau$, the integrals in the numerator and denominator must be evaluated. First evaluate $\hat{H}\phi$:

$$\hat{H}\phi = \left(-\frac{\hbar^2}{2\mu}\frac{d^2}{dx^2} + cx^4\right)e^{-\alpha x^2/2}$$

$$= -\frac{\hbar^2}{2\mu}\frac{d}{dx}\left(-\alpha x e^{-\alpha x^2/2}\right) + cx^4 e^{-\alpha x^2/2}$$

$$= -\frac{\hbar^2}{2\mu}\left(-\alpha e^{-\alpha x^2/2} + \alpha^2 x^2 e^{-\alpha x^2/2}\right) + cx^4 e^{-\alpha x^2/2}$$

It follows that

$$\int \phi^* \hat{H}\phi \, d\tau = \int_{-\infty}^{\infty} e^{-\alpha x^2/2}\left[-\frac{\hbar^2}{2\mu}\left(-\alpha e^{-\alpha x^2} + \alpha^2 x^2 e^{-\alpha x^2/2}\right) + cx^4 e^{-\alpha x^2/2}\right] dx$$

$$= \frac{\hbar^2}{2\mu}\left[\alpha\left(\frac{\pi}{2}\right)^{1/2} - \alpha^2\left(\frac{1}{2\alpha}\right)\left(\frac{\pi}{\alpha}\right)^{1/2}\right] + c\left(\frac{3}{4\alpha^2}\right)\left(\frac{\pi}{\alpha}\right)^{1/2}$$

$$= \frac{\hbar^2\alpha}{4\mu}\left(\frac{\pi}{\alpha}\right)^{1/2} + \frac{3c}{4\alpha^2}\left(\frac{\pi}{\alpha}\right)^{1/2}$$

Now evaluate $\int \phi^*\phi \, d\tau$:

$$\int \phi^*\phi \, d\tau = \int_{-\infty}^{\infty} e^{-\alpha x^2} dx$$

$$= \left(\frac{\pi}{\alpha}\right)^{1/2}$$

Thus,

$$E_\phi = \frac{\int \phi^* \hat{H}\phi \, d\tau}{\int \phi^*\phi \, d\tau}$$

$$= \left(\frac{\alpha}{\pi}\right)^{1/2}\left[\frac{\hbar^2\alpha}{4\mu}\left(\frac{\pi}{\alpha}\right)^{1/2} + \frac{3c}{4\alpha^2}\left(\frac{\pi}{\alpha}\right)^{1/2}\right]$$

$$= \frac{\hbar^2\alpha}{4\mu} + \frac{3c}{4\alpha^2}$$

To minimize E_ϕ, differentiate it with respect to α and set the result equal to zero:

$$\frac{dE_\phi}{d\alpha} = \frac{\hbar^2}{4\mu} - \frac{3c}{2\alpha^3} = 0$$

Solve for α to give

$$\alpha = \left(\frac{6\mu c}{\hbar^2}\right)^{1/3}$$

The other roots for α give nonphysical, imaginary values for E_ϕ for a quartic oscillator. Substitute the above expression for α into the E_ϕ expression to give the estimate of the ground-state energy from this trial function:

$$E_{min} = \frac{\hbar^2}{4\mu}\left(\frac{6\mu c}{\hbar^2}\right)^{1/3} + \frac{3c}{4}\left(\frac{\hbar^2}{6\mu c}\right)^{2/3}$$

$$= \frac{3\hbar}{4}\left(\frac{3c\hbar}{4\mu^2}\right)^{1/3}$$

8–12. Use the variational method to calculate the ground-state energy of a particle constrained to move within the region $0 \le x \le a$ in a potential given by

$$V(x) = \begin{cases} V_0 x & 0 \le x \le \dfrac{a}{2} \\[2mm] V_0(a-x) & \dfrac{a}{2} \le x \le a \end{cases}$$

As a trial function, use a linear combination of the first two particle-in-a-box wave functions:

$$\phi(x) = c_1\left(\frac{2}{a}\right)^{1/2}\sin\frac{\pi x}{a} + c_2\left(\frac{2}{a}\right)^{1/2}\sin\frac{2\pi x}{a}$$

Determine the secular determinant, set it equal to zero to solve for the ground-state energy.

The trial function is a linear combination of two eigenfunctions, $\psi_1 = (2/a)^{1/2}\sin(\pi x/a)$ and $\psi_2 = (2/a)^{1/2}\sin(2\pi x/a)$, for the particle-in-a-box system. The calculations of the matrix elements H_{ij} are simplified when the following relations are used:

(1) If the particle-in-a-box Hamiltonian operator is written as $\hat{H}_0$, then $\hat{H}_0\psi_n = E_n\psi_n = [n^2 h^2/(8ma^2)]\psi_n$;

(2) the functions ψ_n are orthonormal;

(3) the potential energy for the present system is an even function about $x = a/2$;

(4) the functions ψ_n are even about $x = a/2$ when n is odd and odd about $x = a/2$ when n is even.

The H_{ij} matrix elements are

$$H_{11} = \int_0^{a/2} \psi_1 \left(-\frac{\hbar^2}{2m}\frac{d^2}{dx^2} + V_0 x \right) \psi_1 \, dx + \int_{a/2}^a \psi_1 \left[-\frac{\hbar^2}{2m}\frac{d^2}{dx^2} + V_0(a-x) \right] \psi_1 \, dx$$

$$= \int_0^a \psi_1 \hat{H}_0 \psi_1 \, dx + 2V_0 \int_0^{a/2} \left[\left(\frac{2}{a}\right)^{1/2} \sin\frac{\pi x}{a} \right] x \left[\left(\frac{2}{a}\right)^{1/2} \sin\frac{\pi x}{a} \right] dx$$

$$= \frac{h^2}{8ma^2} + aV_0 \left(\frac{1}{\pi^2} + \frac{1}{4} \right)$$

$$H_{22} = \int_0^{a/2} \psi_2 \left(-\frac{\hbar^2}{2m}\frac{d^2}{dx^2} + V_0 x \right) \psi_2 \, dx + \int_{a/2}^a \psi_2 \left[-\frac{\hbar^2}{2m}\frac{d^2}{dx^2} + V_0(a-x) \right] \psi_2 \, dx$$

$$= \int_0^a \psi_2 \hat{H}_0 \psi_2 \, dx + 2V_0 \int_0^{a/2} \left[\left(\frac{2}{a}\right)^{1/2} \sin\frac{2\pi x}{a} \right] x \left[\left(\frac{2}{a}\right)^{1/2} \sin\frac{2\pi x}{a} \right] dx$$

$$= \frac{h^2}{2ma^2} + \frac{aV_0}{4}$$

$$H_{12} = H_{21} = \int_0^{a/2} \psi_1 \left(-\frac{\hbar^2}{2m}\frac{d^2}{dx^2} + V_0 x \right) \psi_2 \, dx + \int_{a/2}^a \psi_1 \left[-\frac{\hbar^2}{2m}\frac{d^2}{dx^2} + V_0(a-x) \right] \psi_2 \, dx$$

$$= \int_0^a \psi_1 \hat{H}_0 \psi_2 \, dx + V_0 \int_0^{a/2} \left[\left(\frac{2}{a}\right)^{1/2} \sin\frac{\pi x}{a} \right] x \left[\left(\frac{2}{a}\right)^{1/2} \sin\frac{2\pi x}{a} \right] dx$$

$$+ V_0 \int_{a/2}^a \left[\left(\frac{2}{a}\right)^{1/2} \sin\frac{\pi x}{a} \right] (a-x) \left[\left(\frac{2}{a}\right)^{1/2} \sin\frac{2\pi x}{a} \right] dx$$

$$= 0$$

Note that the last two integrals in H_{12} sum to 0: the integrands have opposite signs but the same magnitude.

Since the functions ψ_n are orthonormal, the matrix elements S_{ij} are

$$S_{11} = \int_0^a \psi_1 \psi_1 \, dx = 1$$

$$S_{22} = \int_0^a \psi_2 \psi_2 \, dx = 1$$

$$S_{12} = S_{21} = \int_0^a \psi_1 \psi_2 \, dx = 0$$

The ijth element of the secular determinant is $H_{ij} - E S_{ij}$. Setting it equal to zero,

$$\begin{vmatrix} \frac{h^2}{8ma^2} + aV_0 \left(\frac{1}{\pi^2} + \frac{1}{4} \right) - E & 0 \\ 0 & \frac{h^2}{2ma^2} + \frac{aV_0}{4} - E \end{vmatrix} = 0$$

gives the secular equation

$$\left[\frac{h^2}{8ma^2} + aV_0 \left(\frac{1}{\pi^2} + \frac{1}{4} \right) - E \right] \left[\frac{h^2}{2ma^2} + \frac{aV_0}{4} - E \right] = 0$$

Solving for E results in two roots,

$$E = \frac{h^2}{8ma^2} + aV_0\left(\frac{1}{\pi^2} + \frac{1}{4}\right) \quad \text{and} \quad \frac{h^2}{2ma^2} + \frac{aV_0}{4}$$

The lower of the two is the estimate of the ground-state energy from the given trial function.

8–13. Consider a particle of mass m in the potential energy field described by

$$V(x) = \begin{cases} V_0 & x < -a \\ 0 & -a < x < a \\ V_0 & x > a \end{cases}$$

(See also the figure in Problem 4–55.) This problem describes a particle in a finite well. If $V_0 \to \infty$, then we have a particle in a box. Using $\phi(x) = l^2 - x^2$ for $-l < x < l$ and $\phi(x) = 0$ otherwise as a trial function with l as a variational parameter, calculate the ground-state energy of this system for $\alpha = 2mV_0a^2/\hbar^2 = 4$ and 12. The exact ground-state energies are $0.530\hbar^2/ma^2$ and $0.736\hbar^2/ma^2$, respectively (see Problem 4–55).

Since $E_\phi = \int \phi^*\hat{H}\phi \, d\tau / \int \phi^*\phi \, d\tau$, the integrals in the numerator and denominator must be evaluated. First evaluate $\hat{H}\phi$. For $x < -a$ and $x > a$,

$$\hat{H}\phi = \left(-\frac{\hbar^2}{2m}\frac{d^2}{dx^2} + V_0\right)\left(l^2 - x^2\right)$$

$$= \frac{\hbar^2}{m} + V_0\left(l^2 - x^2\right)$$

For $-a < x < a$,

$$\hat{H}\phi = -\frac{\hbar^2}{2m}\frac{d^2}{dx^2}\left(l^2 - x^2\right)$$

$$= \frac{\hbar^2}{m}$$

It follows that

$$\int \phi^*\hat{H}\phi \, d\tau = \int_{-l}^{-a}\left(l^2 - x^2\right)\left[\frac{\hbar^2}{m} + V_0\left(l^2 - x^2\right)\right] dx + \int_{-a}^{a}\left(l^2 - x^2\right)\left(\frac{\hbar^2}{m}\right) dx$$

$$+ \int_{a}^{l}\left(l^2 - x^2\right)\left[\frac{\hbar^2}{m} + V_0\left(l^2 - x^2\right)\right] dx$$

$$= \frac{\hbar^2}{m}\int_{-l}^{l}(l^2 - x^2)\, dx + 2V_0\int_{a}^{l}(l^2 - x^2)^2\, dx$$

Since $(l^2 - x^2)^2$ is even about $x = 0$, the integrals involving V_0 could be combined. Evaluating the integrals gives

$$\int \phi^*\hat{H}\phi \, d\tau = \frac{4\hbar^2 l^3}{3m} + 2V_0\left(\frac{8l^5}{15} - al^4 + \frac{2}{3}a^3l^2 - \frac{a^5}{5}\right)$$

Now evaluate $\int \phi^* \phi \, d\tau$:

$$\int \phi^* \phi \, d\tau = \int_{-l}^{l} \left(l^2 - x^2 \right)^2 dx$$

$$= \frac{16l^5}{15}$$

Thus,

$$E_\phi = \frac{\int \phi^* \hat{H} \phi \, d\tau}{\int \phi^* \phi \, d\tau}$$

$$= \left(\frac{15}{16l^5} \right) \left[\frac{4\hbar^2 l^3}{3m} + 2V_0 \left(\frac{8l^5}{15} - al^4 + \frac{2}{3}a^3 l^2 - \frac{a^5}{5} \right) \right]$$

$$= \frac{5\hbar^2}{4ml^2} + V_0 - \frac{15aV_0}{8l} + \frac{5a^3 V_0}{4l^3} - \frac{3a^5 V_0}{8l^5}$$

Introducing $V_0 = \hbar^2 \alpha / (2ma^2)$ into the above equation,

$$E_\phi = \frac{\hbar^2}{m} \left[\frac{5}{4l^2} + \frac{\alpha}{2a^2} - \frac{15\alpha}{16al} + \frac{5a\alpha}{8l^3} - \frac{3a^3\alpha}{16l^5} \right]$$

To write the solutions for l in terms of a, introduce a unitless quantity $s = l/a$ to give

$$E_\phi = \frac{5\hbar^2}{4ma^2} \left[\frac{1}{s^2} + \frac{2\alpha}{5} - \frac{3\alpha}{4s} + \frac{\alpha}{2s^3} - \frac{3\alpha}{20s^5} \right]$$

To minimize E_ϕ, differentiate it with respect to l, or equivalently, with respect to s and set the result equal to zero:

$$\frac{dE_\phi}{ds} = \frac{5\hbar^2}{4ma^2} \left(-\frac{2}{s^3} + \frac{3\alpha}{4s^2} - \frac{3\alpha}{2s^4} + \frac{3\alpha}{4s^6} \right) = 0$$

$$3\alpha s^4 - 8s^3 - 6\alpha s^2 + 3\alpha = 0$$

For $\alpha = 4$,

$$3s^4 - 2s^3 - 6s^2 + 3 = 0$$

There are four roots for s, two are imaginary and two are real and positive. Since E and s have an inverse relationship, the larger, real, positive root, $s = 1.6546$, gives the estimate of the ground-state energy from this trial function:

$$E_{\min} = E_\phi(\text{at } s = 1.6546) = 0.6816 \frac{\hbar^2}{ma^2}$$

This is greater than the exact ground-state energy by 29%.

For $\alpha = 12$, $dE_\phi/ds = 0$ becomes

$$36s^4 - 8s^3 - 72s^2 + 36 = 0$$

$$9s^4 - 2s^3 - 18s^2 + 9 = 0$$

Once again, there are four roots for s, two are imaginary and two are real and positive. Take the larger, real, positive root, $s = 1.3049$ to give the estimate of the ground-state energy from this trial function:

$$E_{\min} = E_\phi(\text{at } s = 1.3049) = 0.8935\frac{\hbar^2}{ma^2}$$

This is greater than the exact ground-state energy by 21%.

8–14. Repeat the calculation in the previous problem for a trial function $\phi(x) = \cos \lambda x$ for $-\pi/2\lambda < x < \pi/2\lambda$ and $\phi(x) = 0$ otherwise. Use λ as a variational parameter.

Since $E_\phi = \int \phi^* \hat{H} \phi \, d\tau / \int \phi^* \phi \, d\tau$, the integrals in the numerator and denominator must be evaluated. First evaluate $\hat{H}\phi$. For $x < -a$ and $x > a$,

$$\hat{H}\phi = \left(-\frac{\hbar^2}{2m}\frac{d^2}{dx^2} + V_0\right)\cos(\lambda x)$$

$$= \frac{\hbar^2\lambda^2}{2m}\cos(\lambda x) + V_0\cos(\lambda x)$$

For $-a < x < a$,

$$\hat{H}\phi = -\frac{\hbar^2}{2m}\frac{d^2}{dx^2}\cos(\lambda x)$$

$$= \frac{\hbar^2\lambda^2}{2m}\cos(\lambda x)$$

It follows that

$$\int \phi^* \hat{H} \phi \, d\tau = \int_{-\pi/(2\lambda)}^{-a} \cos(\lambda x)\left[\frac{\hbar^2\lambda^2}{2m}\cos(\lambda x) + V_0\cos(\lambda x)\right] dx + \int_{-a}^{a}\cos(\lambda x)\left[\frac{\hbar^2\lambda^2}{2m}\cos(\lambda x)\right] dx$$

$$+ \int_{a}^{\pi/(2\lambda)}\cos(\lambda x)\left[\frac{\hbar^2\lambda^2}{2m}\cos(\lambda x) + V_0\cos(\lambda x)\right] dx$$

$$= \frac{\hbar^2\lambda^2}{2m}\int_{-\pi/(2\lambda)}^{\pi/(2\lambda)}\cos^2(\lambda x)\,dx + 2V_0\int_{a}^{\pi/(2\lambda)}\cos^2(\lambda x)\,dx$$

Since $\cos^2(\lambda x)$ is even about $x = 0$, the integrals involving V_0 can be combined. Evaluation of the integrals gives

$$\int \phi^* \hat{H} \phi \, d\tau = \frac{\hbar^2\pi\lambda}{4m} + 2V_0\left[\frac{\pi}{4\lambda} - \frac{a}{2} - \frac{\sin(2\lambda a)}{4\lambda}\right]$$

Now evaluate $\int \phi^* \phi \, d\tau$:

$$\int \phi^* \phi \, d\tau = \int_{-\pi/(2\lambda)}^{\pi/(2\lambda)}\cos^2(\lambda x)\,dx$$

$$= \frac{\pi}{2\lambda}$$

Thus,

$$E_\phi = \frac{\int \phi^* \hat{H} \phi \, d\tau}{\int \phi^* \phi \, d\tau}$$

$$= \left(\frac{2\lambda}{\pi}\right)\left\{\frac{\hbar^2 \pi \lambda}{4m} + 2V_0\left[\frac{\pi}{4\lambda} - \frac{a}{2} - \frac{\sin(2\lambda a)}{4\lambda}\right]\right\}$$

$$= \frac{\hbar^2 \lambda^2}{2m} + V_0\left[1 - \frac{2\lambda a}{\pi} - \frac{\sin(2\lambda a)}{\pi}\right]$$

Introducing $V_0 = \hbar^2 \alpha/(2ma^2)$ into the above equation,

$$E_\phi = \frac{\hbar^2}{m}\left[\frac{\lambda^2}{2} + \frac{\alpha}{2a^2} - \frac{\lambda\alpha}{\pi a} - \frac{\alpha \sin(2\lambda a)}{2\pi a^2}\right]$$

To write the solutions for λ (which has a unit of length^{-1}) in terms of a, introduce a unitless quantity $s = \lambda a$ to give

$$E_\phi = \frac{\hbar^2}{ma^2}\left[\frac{s^2}{2} + \frac{\alpha}{2} - \frac{s\alpha}{\pi} - \frac{\alpha \sin(2s)}{2\pi}\right]$$

To minimize E_ϕ, differentiate it with respect to λ, or equivalently, with respect to s and set the result equal to zero:

$$\frac{dE_\phi}{ds} = \frac{\hbar^2}{ma^2}\left[s - \frac{\alpha}{\pi} - \frac{\alpha \cos(2s)}{\pi}\right] = 0$$

$$\pi s - \alpha - \alpha \cos(2s) = 0$$

For $\alpha = 4$,

$$\pi s - 4 - 4\cos(2s) = 0$$

Solving for s numerically yields only one real root, $s = 0.9242$. Thus, the estimate of the ground-state energy from this trial function is

$$E_{\min} = E_\phi(\text{at } s = 0.9242) = 0.6381\frac{\hbar^2}{ma^2}$$

This is greater than the exact ground-state energy by 20%.

For $\alpha = 12$, $dE_\phi/ds = 0$ becomes

$$\pi s - 12 - 12\cos(2s) = 0$$

Numerical solutions for s show that there are five real roots for s. The smallest one, $s = 1.1689$, gives the estimate of the ground-state energy from this trial function:

$$E_{\min} = E_\phi(\text{at } s = 1.1689) = 0.8432\frac{\hbar^2}{ma^2}$$

This is greater than the exact ground-state energy by 15%.

8–15. Consider a particle that is confined to a sphere of radius a. The Hamiltonian operator for this system is (see Equation 7.10)

$$\hat{H} = -\frac{\hbar^2}{2mr^2}\frac{d}{dr}\left(r^2\frac{d}{dr}\right) + \frac{\hbar^2 l(l+1)}{2mr^2} \qquad 0 < r \le a$$

In the ground state, $l = 0$ and so

$$\hat{H} = -\frac{\hbar^2}{2mr^2}\frac{d}{dr}\left(r^2\frac{d}{dr}\right) \qquad 0 < r \le a$$

As in the case of a particle in a rectangular box, $\phi(a) = 0$. Use $\phi(r) = a - r$ to calculate an upper bound to the ground-state energy of this system. There is no variational parameter in this case, but the calculated energy is still an upper bound to the ground-state energy. The exact ground-state energy is $\pi^2\hbar^2/2ma^2$ (see Problem 8–17).

Since $E_\phi = \int \phi^*\hat{H}\phi \, d\tau / \int \phi^*\phi \, d\tau$, the integrals in the numerator and denominator must be evaluated. The numerator is

$$\int \phi^*\hat{H}\phi \, d\tau = \int_0^{2\pi} d\phi \int_0^\pi d\theta \, \sin\theta \int_0^a dr \, r^2 (a-r)\left[-\frac{\hbar^2}{2mr^2}\frac{d}{dr}\left(r^2\frac{d}{dr}\right)\right](a-r)$$

$$= (4\pi)\left(\frac{\hbar^2}{m}\right)\int_0^a dr \left(ar - r^2\right)$$

$$= \frac{2\pi\hbar^2 a^3}{3m}$$

Now evaluate the denominator:

$$\int \phi^*\phi \, d\tau = \int_0^{2\pi} d\phi \int_0^\pi d\theta \, \sin\theta \int_0^a dr \, r^2(a-r)^2$$

$$= (4\pi)\left(\frac{a^5}{30}\right)$$

$$= \frac{2\pi a^5}{15}$$

The estimate of the ground-state energy from this trial function is

$$E_\phi = \frac{\int \phi^*\hat{H}\phi \, d\tau}{\int \phi^*\phi \, d\tau}$$

$$= \left(\frac{15}{2\pi a^5}\right)\left(\frac{2\pi\hbar^2 a^3}{3m}\right)$$

$$= \frac{5\hbar^2}{ma^2}$$

which is 1% greater than the exact ground-state energy.

8–16. Repeat the calculation in Problem 8–15 using $\phi(r) = (a - r)^2$ as a trial function. Compare your result to the one obtained in the previous problem. The exact (normalized) wave function is given in the next problem. Compare plots of $(1 - r)$ and $(1 - r)^2$ (after normalizing them) and the exact wave functions.

Since $E_\phi = \int \phi^* \hat{H} \phi \, d\tau / \int \phi^* \phi \, d\tau$, the integrals in the numerator and denominator must be evaluated. The numerator is

$$\int \phi^* \hat{H} \phi \, d\tau = \int_0^{2\pi} d\phi \int_0^{\pi} d\theta \, \sin\theta \int_0^a dr \, r^2 \, (a - r)^2 \left[-\frac{\hbar^2}{2mr^2} \frac{d}{dr} \left(r^2 \frac{d}{dr} \right) \right] (a - r)^2$$

$$= (4\pi) \left(\frac{\hbar^2}{m} \right) \int_0^a dr \, (a - r)^2 \left(2ar - 3r^2 \right)$$

$$= \frac{4\pi \hbar^2 a^5}{15m}$$

Now evaluate the denominator:

$$\int \phi^* \phi \, d\tau = \int_0^{2\pi} d\phi \int_0^{\pi} d\theta \, \sin\theta \int_0^a dr \, r^2 (a - r)^4$$

$$= (4\pi) \left(\frac{a^7}{105} \right)$$

$$= \frac{4\pi a^7}{105}$$

The estimate of the ground-state energy from this trial function is

$$E_\phi = \frac{\int \phi^* \hat{H} \phi \, d\tau}{\int \phi^* \phi \, d\tau}$$

$$= \left(\frac{105}{4\pi a^7} \right) \frac{4\pi \hbar^2 a^5}{15m}$$

$$= \frac{7\hbar^2}{ma^2}$$

which is 42% greater than the exact ground-state energy.

Taking $a = 1$, the normalized function $(1 - r)$ in the last problem is $[15/(2\pi)]^{1/2}(1 - r)$ and the normalized function $(1 - r)^2$ in this problem is $[105/(4\pi)]^{1/2}(1 - r)^2$. The exact ground-state wave function is given in Problem 8–17, and for $a = 1$, it is $(2\pi)^{-1/2}[\sin(\pi r)]/r$. Plots for these functions are shown below:

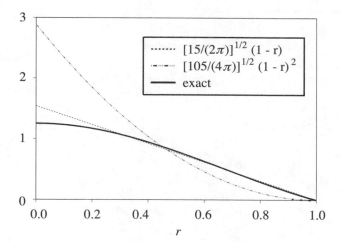

Note that the trial function $[15/(2\pi)]^{1/2}(1-r)$ is closer to the exact wave function. Thus, the ground-state energy predicted by this trial function is closer to the exact energy than that predicted by the function $[105/(4\pi)]^{1/2}(1-r)^2$, as shown in these two problems.

8–17. In this problem, we will solve the Schrödinger equation for the ground-state wave function and energy of a particle confined to a sphere of radius a. The Schrödinger equation is given by Equation 7.10 with $l = 0$ (ground state) and without the $e^2/4\pi\epsilon_0 r$ term:

$$-\frac{\hbar^2}{2mr^2}\frac{d}{dr}\left(r^2\frac{d\psi}{dr}\right) = E\psi$$

Substitute $u = r\psi$ into this equation to get

$$\frac{d^2u}{dr^2} + \frac{2mE}{\hbar^2}u = 0$$

The general solution to this equation is

$$u(r) = A\cos\alpha r + B\sin\alpha r$$

or

$$\psi(r) = \frac{A\cos\alpha r}{r} + \frac{B\sin\alpha r}{r}$$

where $\alpha = (2mE/\hbar^2)^{1/2}$. Which of these terms is finite at $r = 0$? Now use the fact that $\psi(a) = 0$ to prove that

$$\alpha a = \pi$$

for the ground state, or that the ground-state energy is

$$E = \frac{\pi^2\hbar^2}{2ma^2}$$

Show that the normalized ground-state wave function is

$$\psi(r) = (2\pi a)^{-1/2}\frac{\sin\pi r/a}{r}$$

Substitute $\psi = u/r$ into the Schrödinger equation:

$$-\frac{\hbar^2}{2mr^2}\frac{d}{dr}\left[r^2\frac{d}{dr}\left(\frac{u}{r}\right)\right] = E\left(\frac{u}{r}\right)$$

$$-\frac{\hbar^2}{2mr^2}\frac{d}{dr}\left[r^2\left(\frac{1}{r}\frac{du}{dr} - \frac{u}{r^2}\right)\right] = E\left(\frac{u}{r}\right)$$

$$-\frac{\hbar^2}{2mr^2}\frac{d}{dr}\left(r\frac{du}{dr} - u\right) = E\left(\frac{u}{r}\right)$$

$$-\frac{\hbar^2}{2mr^2}\left(r\frac{d^2u}{dr^2} + \frac{du}{dr} - \frac{du}{dr}\right) = E\left(\frac{u}{r}\right)$$

$$\frac{d^2u}{dr^2} + \frac{2mE}{\hbar^2}u = 0$$

In the general solution,

$$\psi(r) = \frac{A\cos\alpha r}{r} + \frac{B\sin\alpha r}{r}$$

the cosine term diverges as r goes to zero. Thus, $A = 0$. On the other hand, the sine term is finite as r goes to zero; specifically, $\lim_{r\to 0}[\sin(\alpha r)]/r = \alpha$. Since $\psi(a) = 0$,

$$\psi(a) = \frac{B\sin(\alpha a)}{a} = 0$$

Nontrivial solutions result when $\alpha a = n\pi$, with $n = 1, 2, 3, \cdots$. The ground-state energy is given by $n = 1$, that is, $\alpha a = \pi$:

$$\alpha = \frac{\pi}{a} = \left(\frac{2mE}{\hbar^2}\right)^{1/2}$$

$$E = \frac{\pi^2\hbar^2}{2ma^2}$$

The ground-state wave function is normalized:

$$\int_0^{2\pi} d\phi \int_0^{\pi} d\theta \, \sin\theta \int_0^a dr \, r^2 B^2\frac{\sin^2(\alpha r)}{r^2} = 1$$

$$(4\pi)\int_0^a dr \, B^2 \sin^2(\alpha r) = 1$$

$$2\pi a B^2 = 1$$

For a positive normalization constant, $B = (2\pi a)^{-1/2}$ and the wave function is

$$\psi(r) = (2\pi a)^{-1/2}\frac{\sin\pi r/a}{r}$$

8–18. This problem fills in the steps of the variational treatment of a helium atom. We use a trial function of the form

$$\phi(\mathbf{r}_1, \mathbf{r}_2) = \frac{Z^3}{a_0^3 \pi} e^{-Z(r_1+r_2)/a_0}$$

with Z as an adjustable parameter. The Hamiltonian operator of a helium atom is

$$\hat{H} = -\frac{\hbar^2}{2m_e}\nabla_1^2 - \frac{\hbar^2}{2m_e}\nabla_2^2 - \frac{2e^2}{4\pi\epsilon_0 r_1} - \frac{2e^2}{4\pi\epsilon_0 r_2} + \frac{e^2}{4\pi\epsilon_0 r_{12}}$$

We now evaluate

$$E(Z) = \int d\mathbf{r}_1 d\mathbf{r}_2 \, \phi^* \hat{H} \phi$$

The evaluation of this integral is greatly simplified if you recall that

$$\psi(r_j) = (Z^3/a_0^3\pi)^{1/2} e^{-Zr_j/a_0}$$

is an eigenfunction of a hydrogen-like Hamiltonian operator, one for which the nucleus has a charge Z. Show that the helium atom Hamiltonian operator can be written as

$$\hat{H} = -\frac{\hbar^2}{2m_e}\nabla_1^2 - \frac{Ze^2}{4\pi\epsilon_0 r_1} - \frac{\hbar^2}{2m_e}\nabla_2^2 - \frac{Ze^2}{4\pi\epsilon_0 r_2} + \frac{(Z-2)e^2}{4\pi\epsilon_0 r_1} + \frac{(Z-2)e^2}{4\pi\epsilon_0 r_2} + \frac{e^2}{4\pi\epsilon_0 r_{12}}$$

where

$$\left(-\frac{\hbar^2}{2m_e}\nabla^2 - \frac{Ze^2}{4\pi\epsilon_0 r}\right)\left(\frac{Z^3}{a_0^3\pi}\right)^{1/2} e^{-Zr/a_0} = -\frac{Z^2 e^2}{8\pi\epsilon_0 a_0}\left(\frac{Z^3}{a_0^3\pi}\right)^{1/2} e^{-Zr/a_0}$$

Show that

$$E(Z) = \frac{Z^6}{a_0^6 \pi^2} \iint d\mathbf{r}_1 d\mathbf{r}_2 e^{-Z(r_1+r_2)/a_0}\left[-\frac{Z^2 e^2}{8\pi\epsilon_0 a_0} - \frac{Z^2 e^2}{8\pi\epsilon_0 a_0} + \frac{(Z-2)e^2}{4\pi\epsilon_0 r_1}\right.$$

$$\left. +\frac{(Z-2)e^2}{4\pi\epsilon_0 r_2} + \frac{e^2}{4\pi\epsilon_0 r_{12}}\right] e^{-Z(r_1+r_2)/a_0}$$

The last integral is evaluated in Problem 8–39 or 8–40 and the others are elementary. Show that $E(Z)$, in units of $(m_e e^4/16\pi^2\epsilon_0^2\hbar^2)$, is given by

$$E(Z) = -Z^2 + 2(Z-2)\frac{Z^3}{\pi}\int d\mathbf{r}\frac{e^{-2Zr}}{r} + \frac{5}{8}Z$$

$$= -Z^2 + 2(Z-2)Z + \frac{5}{8}Z$$

$$= Z^2 - \frac{27}{8}Z$$

Now minimize E with respect to Z and show that

$$E = -\left(\frac{27}{16}\right)^2 = -2.8477$$

in units of $m_e e^4/16\pi^2\epsilon_0^2\hbar^2$. Interpret the value of Z that minimizes E.

Add to and subtract from the Hamiltonian operator,

$$\hat{H} = -\frac{\hbar^2}{2m_e}\nabla_1^2 - \frac{\hbar^2}{2m_e}\nabla_2^2 - \frac{2e^2}{4\pi\epsilon_0 r_1} - \frac{2e^2}{4\pi\epsilon_0 r_2} + \frac{e^2}{4\pi\epsilon_0 r_{12}}$$

the expression $\dfrac{Ze^2}{4\pi\epsilon_0 r_1} + \dfrac{Ze^2}{4\pi\epsilon_0 r_2}$ to give

$$\hat{H} = -\frac{\hbar^2}{2m_e}\nabla_1^2 - \frac{Ze^2}{4\pi\epsilon_0 r_1} - \frac{\hbar^2}{2m_e}\nabla_2^2 - \frac{Ze^2}{4\pi\epsilon_0 r_2} + \frac{(Z-2)e^2}{4\pi\epsilon_0 r_1} + \frac{(Z-2)e^2}{4\pi\epsilon_0 r_2} + \frac{e^2}{4\pi\epsilon_0 r_{12}}$$

For a hydrogenlike atom, the Hamiltonian operator is

$$\hat{H}_H = -\frac{\hbar^2}{2m_e}\nabla^2 - \frac{Ze^2}{4\pi\epsilon_0 r}$$

and the ground-state wave function and energy (see Problem 7–29) are

$$\psi_H(r) = \left(\frac{Z^3}{a_0^3\pi}\right)^{1/2} e^{-Zr/a_0}$$

$$E_H = -\frac{m_e Z^2 e^4}{8\epsilon_0^2 h^2} = -\frac{Z^2 e^2}{8\pi\epsilon_0 a_0}$$

(The equivalence of the two expressions for E_H is shown in Problem 7–8.) The eigenvalue equation for the ground-state of a hydrogenlike atom is then

$$\left(-\frac{\hbar^2}{2m_e}\nabla^2 - \frac{Ze^2}{4\pi\epsilon_0 r}\right)\left(\frac{Z^3}{a_0^3\pi}\right)^{1/2} e^{-Zr/a_0} = -\frac{Z^2 e^2}{8\pi\epsilon_0 a_0}\left(\frac{Z^3}{a_0^3\pi}\right)^{1/2} e^{-Zr/a_0}$$

The Hamiltonian operator for a helium atom is related to those for two hydrogenlike atoms (with the subscripts 1 and 2 labeling the two electrons):

$$\hat{H} = \hat{H}_{H_1} + \hat{H}_{H_2} + \frac{(Z-2)e^2}{4\pi\epsilon_0 r_1} + \frac{(Z-2)e^2}{4\pi\epsilon_0 r_2} + \frac{e^2}{4\pi\epsilon_0 r_{12}}$$

To obtain the ground-state energy of a helium atom variationally, determine both $\int \phi^*\hat{H}\phi\, d\tau$ and $\int \phi^*\phi\, d\tau$. The trial function is

$$\phi(\mathbf{r}_1, \mathbf{r}_2) = \psi_{H_1}(\mathbf{r}_1)\psi_{H_2}(\mathbf{r}_2) = \frac{Z^3}{a_0^3\pi} e^{-Z(r_1+r_2)/a_0}$$

Since ψ_{H_1} and ψ_{H_2} are normalized,

$$\int \phi^*\phi\, d\tau = \int\int d\mathbf{r}_1\, d\mathbf{r}_2\, \psi_{H_1}^*(\mathbf{r}_1)\psi_{H_2}^*(\mathbf{r}_2)\psi_{H_1}(\mathbf{r}_1)\psi_{H_2}(\mathbf{r}_2)$$

$$= \int d\mathbf{r}_2\, \psi_{H_2}^*(\mathbf{r}_2)\psi_{H_2}(\mathbf{r}_2) \int d\mathbf{r}_1\, \psi_{H_1}^*(\mathbf{r}_1)\psi_{H_1}(\mathbf{r}_1)$$

$$= 1$$

Operate on the trial function with the Hamiltonian operator to give the following expression:

$$\hat{H}\phi(\mathbf{r}_1, \mathbf{r}_2) = \left[\hat{H}_{H_1} + \hat{H}_{H_2} + \frac{(Z-2)e^2}{4\pi\epsilon_0 r_1} + \frac{(Z-2)e^2}{4\pi\epsilon_0 r_2} + \frac{e^2}{4\pi\epsilon_0 r_{12}}\right]\psi_{H_1}(\mathbf{r}_1)\psi_{H_2}(\mathbf{r}_2)$$

$$= \left[-\frac{Z^2 e^2}{8\pi\epsilon_0 a_0} - \frac{Z^2 e^2}{8\pi\epsilon_0 a_0} + \frac{(Z-2)e^2}{4\pi\epsilon_0 r_1} + \frac{(Z-2)e^2}{4\pi\epsilon_0 r_2} + \frac{e^2}{4\pi\epsilon_0 r_{12}}\right]\psi_{H_1}(\mathbf{r}_1)\psi_{H_2}(\mathbf{r}_2)$$

$$= \left[-\frac{Z^2 e^2}{8\pi\epsilon_0 a_0} - \frac{Z^2 e^2}{8\pi\epsilon_0 a_0} + \frac{(Z-2)e^2}{4\pi\epsilon_0 r_1} + \frac{(Z-2)e^2}{4\pi\epsilon_0 r_2} + \frac{e^2}{4\pi\epsilon_0 r_{12}}\right]\phi(\mathbf{r}_1, \mathbf{r}_2)$$

It follows that

$$\int \phi^*\hat{H}\phi \, d\tau = \int\int d\mathbf{r}_1 \, d\mathbf{r}_2 \, \phi(\mathbf{r}_1, \mathbf{r}_2)\left[-\frac{Z^2 e^2}{8\pi\epsilon_0 a_0} - \frac{Z^2 e^2}{8\pi\epsilon_0 a_0} + \frac{(Z-2)e^2}{4\pi\epsilon_0 r_1} + \frac{(Z-2)e^2}{4\pi\epsilon_0 r_2} + \frac{e^2}{4\pi\epsilon_0 r_{12}}\right]\phi(\mathbf{r}_1, \mathbf{r}_2)$$

Thus, the ground-state energy for a helium atom is

$$E(Z) = \frac{\int \phi^*\hat{H}\phi \, d\tau}{\int \phi^*\phi \, d\tau}$$

$$= \int\int d\mathbf{r}_1 \, d\mathbf{r}_2 \, \phi(\mathbf{r}_1, \mathbf{r}_2)\left[-\frac{Z^2 e^2}{8\pi\epsilon_0 a_0} - \frac{Z^2 e^2}{8\pi\epsilon_0 a_0} + \frac{(Z-2)e^2}{4\pi\epsilon_0 r_1} + \frac{(Z-2)e^2}{4\pi\epsilon_0 r_2} + \frac{e^2}{4\pi\epsilon_0 r_{12}}\right]\phi(\mathbf{r}_1, \mathbf{r}_2)$$

This last expression is the same as that given in the problem when each $\phi(\mathbf{r}_1, \mathbf{r}_2)$ is written explicitly in terms of r_1 and r_2 and its normalization factor is pulled out of the integrals. Proceeding with the evaluation of $E(Z)$ and using the results of Problem 8–39 or 8–40 to evaluate the last integral:

$$E(Z) = -\frac{Z^2 e^2}{4\pi\epsilon_0 a_0}\int\int d\mathbf{r}_1 \, d\mathbf{r}_2 \, \phi(\mathbf{r}_1, \mathbf{r}_2)\phi(\mathbf{r}_1, \mathbf{r}_2)$$

$$+ \frac{(Z-2)e^2}{4\pi\epsilon_0}\int d\mathbf{r}_2 \, \psi_{H_2}(\mathbf{r}_2)\psi_{H_2}(\mathbf{r}_2)\int d\mathbf{r}_1 \, \psi_{H_1}(\mathbf{r}_1)\frac{1}{r_1}\psi_{H_1}(\mathbf{r}_1)$$

$$+ \frac{(Z-2)e^2}{4\pi\epsilon_0}\int d\mathbf{r}_2 \, \psi_{H_2}(\mathbf{r}_2)\frac{1}{r_2}\psi_{H_2}(\mathbf{r}_2)\int d\mathbf{r}_1 \, \psi_{H_1}(\mathbf{r}_1)\psi_{H_1}(\mathbf{r}_1)$$

$$+ \frac{e^2}{4\pi\epsilon_0}\int\int d\mathbf{r}_1 \, d\mathbf{r}_2 \, \phi(\mathbf{r}_1, \mathbf{r}_2)\frac{1}{r_{12}}\phi(\mathbf{r}_1, \mathbf{r}_2)$$

$$= -\frac{Z^2 e^2}{4\pi\epsilon_0 a_0} + \frac{(Z-2)e^2}{4\pi\epsilon_0}\left(\frac{Z^3}{a_0^3\pi}\right)(4\pi)\int_0^\infty dr_1 \, r_1^2 \, e^{-2Zr_1/a_0}\frac{1}{r_1}$$

$$+ \frac{(Z-2)e^2}{4\pi\epsilon_0}\left(\frac{Z^3}{a_0^3\pi}\right)(4\pi)\int_0^\infty dr_2 \, r_2^2 \, e^{-2Zr_2/a_0}\frac{1}{r_2} + \frac{5}{8}Z\left(\frac{e^2}{4\pi\epsilon_0 a_0}\right)$$

$$= -\frac{Z^2 e^2}{4\pi\epsilon_0 a_0} + \frac{(Z-2)e^2}{4\pi\epsilon_0}\left(\frac{Z^3}{a_0^3\pi}\right)(4\pi)\left(\frac{a_0^2}{4Z^2}\right) + \frac{(Z-2)e^2}{4\pi\epsilon_0}\left(\frac{Z^3}{a_0^3\pi}\right)(4\pi)\left(\frac{a_0^2}{4Z^2}\right)$$

$$+ \frac{5}{8}Z\left(\frac{e^2}{4\pi\epsilon_0 a_0}\right)$$

$$= \left(\frac{e^2}{4\pi\epsilon_0 a_0}\right)\left[-Z^2 + 2Z(Z-2) + \frac{5}{8}Z\right]$$

Writing $E(Z)$ in units of $m_e e^4 / 16\pi^2 \epsilon_0^2 \hbar^2$, which is equivalent to $e^2 / 4\pi \epsilon_0 a_0$, gives

$$E(Z) = -Z^2 + 2Z(Z-2) + \frac{5}{8}Z$$

$$= Z^2 - \frac{27}{8}Z$$

To minimize Z, differentiate $E(Z)$ with respect to Z and set the expression equal to zero:

$$\frac{dE}{dZ} = 2Z - \frac{27}{8}$$

$$Z = \frac{27}{16}$$

where the positive root for Z is used. The estimate of the ground-state energy from the given trial function is then

$$E(Z) = \left(\frac{27}{16}\right)^2 - \frac{27}{8}\left(\frac{27}{16}\right)$$

$$= -\left(\frac{27}{16}\right)^2 = -2.8477$$

in units of $m_e e^4 / 16\pi^2 \epsilon_0^2 \hbar^2$. The value of Z that minimizes E is less than 2 (the nuclear charge of helium) because each electron partially screens the nucleus from the other.

8–19. Use the spectral data for He and He$^+$ from the website *http://physics.nist.gov/PhysRefData/ASD/levels_form.html* to determine the experimental ground-state energy of a helium atom.

The ground-state energy of a helium atom is the amount of energy released when He is formed from He^{2+}, or the negative of the sum of the first ionization and the second ionization energies of He. From the website, the first ionization energy of He is $198\,310.6672$ cm^{-1} and the second ionization energy of He (which is the same as the ionization energy of He$^+$) is $438\,908.8863$ cm^{-1}. Thus, the ground-state energy of He is $-637\,219.5535$ cm^{-1}.

8–20. Verify all the matrix elements in Equation 8.37.

The Hamiltonian operator is

$$\hat{H} = -\frac{\hbar^2}{2m}\frac{d^2}{dx^2}$$

and the trial function, for $0 < x < 1$, is the linear combinations of two functions:

$$f_1 = x\,(1-x)$$

$$f_2 = x^2\,(1-x)^2$$

The H_{ij} matrix elements are

$$H_{11} = \int_0^1 x(1-x)\left[-\frac{\hbar^2}{2m}\frac{d^2}{dx^2}\right]x(1-x)\ dx$$

$$= \frac{\hbar^2}{m}\int_0^1 x(1-x)\ dx$$

$$= \frac{\hbar^2}{6m}$$

$$H_{22} = \int_0^1 x^2(1-x)^2\left[-\frac{\hbar^2}{2m}\frac{d^2}{dx^2}\right]x^2(1-x)^2\ dx$$

$$= -\frac{\hbar^2}{2m}\int_0^1 x^2(1-x)^2\left(2-12x+12x^2\right)\ dx$$

$$= \frac{\hbar^2}{105m}$$

$$H_{12} = \int_0^1 x(1-x)\left[-\frac{\hbar^2}{2m}\frac{d^2}{dx^2}\right]x^2(1-x)^2\ dx$$

$$= -\frac{\hbar^2}{2m}\int_0^1 x(1-x)\left(2-12x+12x^2\right)\ dx$$

$$= \frac{\hbar^2}{30m}$$

$$H_{21} = \int_0^1 x^2(1-x)^2\left[-\frac{\hbar^2}{2m}\frac{d^2}{dx^2}\right]x(1-x)\ dx$$

$$= \frac{\hbar^2}{m}\int_0^1 x^2(1-x)^2\ dx$$

$$= \frac{\hbar^2}{30m}$$

Note that because $\hat{H}$ is Hermitian and real and $f_1(x)$ and $f_2(x)$ are real, $H_{12} = H_{21}$. The S_{ij} matrix elements are

$$S_{11} = \int_0^1 x^2(1-x)^2\ dx$$

$$= \frac{1}{30}$$

$$S_{22} = \int_0^1 x^4(1-x)^4\ dx$$

$$= \frac{1}{630}$$

$$S_{12} = S_{21} = \int_0^1 x^3(1-x)^3\ dx$$

$$= \frac{1}{140}$$

8–21.   Consider a system subject to the potential

$$V(x) = \frac{k}{2}x^2 + \frac{\gamma_3}{6}x^3 + \frac{\gamma_4}{24}x^4$$

Calculate the ground-state energy of this system using a trial function of the form

$$\phi = c_1\psi_0(x) + c_2\psi_2(x)$$

where $\psi_0(x)$ and $\psi_2(x)$ are the harmonic-oscillator wave functions.

Determine the secular determinant, and set it equal to zero to solve for the ground-state energy.

The Hamiltonian operator is

$$\hat{H} = -\frac{\hbar^2}{2\mu}\frac{d^2}{dx^2} + \frac{k}{2}x^2 + \frac{\gamma_3}{6}x^3 + \frac{\gamma_4}{24}x^4$$

$$= \hat{H}_{\mathrm{HO}} + \frac{\gamma_3}{6}x^3 + \frac{\gamma_4}{24}x^4$$

where $\hat{H}_{\mathrm{HO}}$ is the harmonic oscillator Hamiltonian operator. The trial function is a linear combination of two eigenfunctions of the harmonic oscillator:

$$\psi_0 = \left(\frac{\alpha}{\pi}\right)^{1/4} e^{-\alpha x^2/2}$$

$$\psi_2 = \left(\frac{\alpha}{4\pi}\right)^{1/4} (2\alpha x^2 - 1)e^{-\alpha x^2/2}$$

The calculation of the matrix elements H_{ij} is simplified when the following relations are used:

(1) $\hat{H}_{\mathrm{HO}}\psi_v = \left(v + \frac{1}{2}\right)\hbar\omega\psi_v$;

(2) the functions ψ_v are orthonormal;

(3) $\int_{-\infty}^{\infty} \psi_0 x^3 \psi_0\, dx = \int_{-\infty}^{\infty} \psi_2 x^3 \psi_2\, dx = \int_{-\infty}^{\infty} \psi_0 x^3 \psi_2\, dx = 0$ since each integrand is odd and the integration occurs over a range symmetric about $x = 0$.

The H_{ij} matrix elements are

$$H_{11} = \int_{-\infty}^{\infty} \psi_0 \left(\hat{H}_{\text{HO}} + \frac{\gamma_3}{6} x^3 + \frac{\gamma_4}{24} x^4 \right) \psi_0 \, dx$$

$$= \int_{-\infty}^{\infty} \psi_0 \hat{H}_{\text{HO}} \psi_0 \, dx + \frac{\gamma_3}{6} \int_{-\infty}^{\infty} \psi_0 x^3 \psi_0 \, dx + \frac{\gamma_4}{24} \int_{-\infty}^{\infty} \psi_0 x^4 \psi_0 \, dx$$

$$= \frac{\hbar\omega}{2} + \frac{\gamma_4}{24} \left(\frac{\alpha}{\pi} \right)^{1/2} \int_{-\infty}^{\infty} x^4 e^{-\alpha x^2} \, dx$$

$$= \frac{\hbar\omega}{2} + \frac{\gamma_4}{32\alpha^2}$$

$$H_{22} = \int_{-\infty}^{\infty} \psi_2 \left(\hat{H}_{\text{HO}} + \frac{\gamma_3}{6} x^3 + \frac{\gamma_4}{24} x^4 \right) \psi_2 \, dx$$

$$= \int_{-\infty}^{\infty} \psi_2 \hat{H}_{\text{HO}} \psi_2 \, dx + \frac{\gamma_3}{6} \int_{-\infty}^{\infty} \psi_2 x^3 \psi_2 \, dx + \frac{\gamma_4}{24} \int_{-\infty}^{\infty} \psi_2 x^4 \psi_2 \, dx$$

$$= \frac{5\hbar\omega}{2} + \frac{\gamma_4}{24} \left(\frac{\alpha}{4\pi} \right)^{1/2} \int_{-\infty}^{\infty} x^4 \left(2\alpha x^2 - 1 \right)^2 e^{-\alpha x^2} \, dx$$

$$= \frac{5\hbar\omega}{2} + \frac{13\gamma_4}{32\alpha^2}$$

$$H_{12} = \int_{-\infty}^{\infty} \psi_0 \left(\hat{H}_{\text{HO}} + \frac{\gamma_3}{6} x^3 + \frac{\gamma_4}{24} x^4 \right) \psi_2 \, dx$$

$$= \int_{-\infty}^{\infty} \psi_0 \hat{H}_{\text{HO}} \psi_2 \, dx + \frac{\gamma_3}{6} \int_{-\infty}^{\infty} \psi_0 x^3 \psi_2 \, dx + \frac{\gamma_4}{24} \int_{-\infty}^{\infty} \psi_0 x^4 \psi_2 \, dx$$

$$= \frac{\gamma_4}{24(2^{1/2})} \left(\frac{\alpha}{\pi} \right)^{1/2} \int_{-\infty}^{\infty} x^4 \left(2\alpha x^2 - 1 \right) e^{-\alpha x^2} \, dx$$

$$= \frac{\gamma_4}{2^{1/2} \left(8\alpha^2 \right)}$$

$$H_{21} = H_{12}^* = \frac{\gamma_4}{2^{1/2} \left(8\alpha^2 \right)}$$

Since the functions ψ_n are orthonormal, the matrix elements S_{ij} are

$$S_{11} = \int_{-\infty}^{\infty} \psi_0^2 \, dx = 1$$

$$S_{22} = \int_{-\infty}^{\infty} \psi_2^2 \, dx = 1$$

$$S_{12} = S_{21} = \int_{-\infty}^{\infty} \psi_0 \psi_2 \, dx = 0$$

Forming the secular determinant and setting it equal to zero,

$$\begin{vmatrix} \dfrac{\hbar\omega}{2} + \dfrac{\gamma_4}{32\alpha^2} - E & \dfrac{\gamma_4}{2^{1/2} \left(8\alpha^2 \right)} \\[2em] \dfrac{\gamma_4}{2^{1/2} \left(8\alpha^2 \right)} & \dfrac{5\hbar\omega}{2} + \dfrac{13\gamma_4}{32\alpha^2} - E \end{vmatrix} = 0$$

gives the secular equation

$$\left(\frac{\hbar\omega}{2} + \frac{\gamma_4}{32\alpha^2} - E\right)\left(\frac{5\hbar\omega}{2} + \frac{13\gamma_4}{32\alpha^2} - E\right) - \left[\frac{\gamma_4}{2^{1/2}\left(8\alpha^2\right)}\right]^2 = 0$$

which can be rearranged as

$$E^2 - \left(3\hbar\omega + \frac{7\gamma_4}{16\alpha^2}\right)E + \left(\frac{5\hbar^2\omega^2}{4} + \frac{9\gamma_4\hbar\omega}{32\alpha^2} + \frac{5\gamma_4^2}{1024\alpha^4}\right) = 0$$

Using the quadratic formula to solve for E and keeping only the smaller root, the estimate of the ground-state energy from the given trial function is

$$E = \frac{1}{2}\left(3\hbar\omega + \frac{7\gamma_4}{16\alpha^2}\right) - \frac{1}{2}\left[\left(3\hbar\omega + \frac{7\gamma_4}{16\alpha^2}\right)^2 - 4\left(\frac{5\hbar^2\omega^2}{4} + \frac{9\gamma_4\hbar\omega}{32\alpha^2} + \frac{5\gamma_4^2}{1024\alpha^4}\right)\right]^{1/2}$$

$$= \frac{3\hbar\omega}{2} + \frac{7\gamma_4}{32\alpha^2} - \frac{1}{2}\left(4\hbar^2\omega^2 + \frac{3\gamma_4\hbar\omega}{2\alpha^2} + \frac{11\gamma_4^2}{64\alpha^4}\right)^{1/2}$$

8–22. It is quite common to assume a trial function of the form

$$\phi = c_1\phi_1 + c_2\phi_2 + \cdots + c_n\phi_n$$

where the variational parameters and the ϕ_n may be complex. Using the simple, special case

$$\phi = c_1\phi_1 + c_2\phi_2$$

show that the variational method leads to

$$E_\phi = \frac{c_1^* c_1 H_{11} + c_1^* c_2 H_{12} + c_1 c_2^* H_{21} + c_2^* c_2 H_{22}}{c_1^* c_1 S_{11} + c_1^* c_2 S_{12} + c_1 c_2^* S_{21} + c_2^* c_2 S_{22}}$$

where

$$H_{ij} = \int \phi_i^* \hat{H} \phi_j \, d\tau = H_{ji}^*$$

and

$$S_{ij} = \int \phi_i^* \phi_j \, d\tau = S_{ji}^*$$

because $\hat{H}$ is a Hermitian operator. Now write the above equation for E_ϕ as

$$c_1^* c_1 H_{11} + c_1^* c_2 H_{12} + c_1 c_2^* H_{21} + c_2^* c_2 H_{22}$$
$$= E_\phi(c_1^* c_1 S_{11} + c_1^* c_2 S_{12} + c_1 c_2^* S_{21} + c_2^* c_2 S_{22})$$

and show that if we set

$$\frac{\partial E_\phi}{\partial c_1^*} = 0 \quad \text{and} \quad \frac{\partial E_\phi}{\partial c_2^*} = 0$$

we obtain

$$(H_{11} - E_\phi S_{11})c_1 + (H_{12} - E_\phi S_{12})c_2 = 0$$

and

$$(H_{21} - E_\phi S_{21})c_1 + (H_{22} - E_\phi S_{22})c_2 = 0$$

There is a nontrivial solution to this pair of equations if and only if the determinant

$$\begin{vmatrix} (H_{11} - E_\phi S_{11}) & (H_{21} - E_\phi S_{21}) \\ (H_{12} - E_\phi S_{12}) & (H_{22} - E_\phi S_{22}) \end{vmatrix} = 0$$

which gives a quadratic equation for E_ϕ. We choose the smaller solution as an approximation to the ground-state energy.

Since $E_\phi = \int \phi^* \hat{H} \phi \, d\tau / \int \phi^* \phi \, d\tau$, the integrals in the numerator and denominator must be evaluated. Using

$$\phi = c_1 \phi_1 + c_2 \phi_2$$

the numerator is

$$\int \phi^* \hat{H} \phi \, d\tau = \int (c_1^* \phi_1^* + c_2^* \phi_2^*) \hat{H}(c_1 \phi_1 + c_2 \phi_2) \, d\tau$$

$$= c_1^* c_1 \int \phi_1^* \hat{H} \phi_1 \, d\tau + c_1^* c_2 \int \phi_1^* \hat{H} \phi_2 \, d\tau + c_2^* c_1 \int \phi_2^* \hat{H} \phi_1 \, d\tau + c_2^* c_2 \int \phi_2^* \hat{H} \phi_2 \, d\tau$$

$$= c_1^* c_1 H_{11} + c_1^* c_2 H_{12} + c_2^* c_1 H_{21} + c_2^* c_2 H_{22}$$

and the denominator is

$$\int \phi^* \phi \, d\tau = \int (c_1^* \phi_1^* + c_2^* \phi_2^*) \hat{H}(c_1 \phi_1 + c_2 \phi_2) \, d\tau$$

$$= c_1^* c_1 \int \phi_1^* \phi_1 \, d\tau + c_1^* c_2 \int \phi_1^* \phi_2 \, d\tau + c_2^* c_1 \int \phi_2^* \phi_1 \, d\tau + c_2^* c_2 \int \phi_2^* \phi_2 \, d\tau$$

$$= c_1^* c_1 S_{11} + c_1^* c_2 S_{12} + c_2^* c_1 S_{21} + c_2^* c_2 S_{22}$$

Thus,

$$E_\phi = \frac{\int \phi^* \hat{H} \phi \, d\tau}{\int \phi^* \phi \, d\tau}$$

$$= \frac{c_1^* c_1 H_{11} + c_1^* c_2 H_{12} + c_2^* c_1 H_{21} + c_2^* c_2 H_{22}}{c_1^* c_1 S_{11} + c_1^* c_2 S_{12} + c_2^* c_1 S_{21} + c_2^* c_2 S_{22}}$$

Since $\hat{H}$ is a Hermitian operator,

$$H_{ij} = \int \phi_i^* \hat{H} \phi_j \, d\tau = \int \phi_j \hat{H}^* \phi_i^* \, d\tau = \left(\int \phi_j^* \hat{H} \phi_i \, d\tau \right)^* = H_{ji}^*$$

The S_{ij} and S_{ji} elements are also related:

$$S_{ij} = \int \phi_i^* \phi_j \, d\tau = \left(\int \phi_j^* \phi_i d\tau \right)^* = S_{ji}^*$$

Now multiply both sides of the E_ϕ expression by $\int d\tau \phi^* \phi$ to obtain

$$c_1^* c_1 H_{11} + c_1^* c_2 H_{12} + c_2^* c_1 H_{21} + c_2^* c_2 H_{22} = E_\phi (c_1^* c_1 S_{11} + c_1^* c_2 S_{12} + c_2^* c_1 S_{21} + c_2^* c_2 S_{22})$$

Differentiate with respect to c_1^* and set the derivative of E_ϕ equal to zero to find the minimum value of E_ϕ with respect to c_1^*:

$$c_1 H_{11} + c_2 H_{12} = \frac{\partial E_\phi}{\partial c_1^*} (c_1^* c_1 S_{11} + c_1^* c_2 S_{12} + c_2^* c_1 S_{21} + c_2^* c_2 S_{22}) + E_\phi \left(c_1 S_{11} + c_2 S_{12} \right)$$

$$c_1 H_{11} + c_2 H_{12} = E_\phi \left(c_1 S_{11} + c_2 S_{12} \right)$$

$$\left(H_{11} - E_\phi S_{11} \right) c_1 + \left(H_{12} - E_\phi S_{12} \right) c_2 = 0$$

Repeat the differentiation with respect to c_2^* to find the minimum value of E_ϕ with respect to c_2^*:

$$c_1 H_{21} + c_2 H_{22} = \frac{\partial E_\phi}{\partial c_2^*} (c_1^* c_1 S_{11} + c_1^* c_2 S_{12} + c_2^* c_1 S_{21} + c_2^* c_2 S_{22}) + E_\phi \left(c_1 S_{21} + c_2 S_{22} \right)$$

$$c_1 H_{21} + c_2 H_{22} = E_\phi \left(c_1 S_{21} + c_2 S_{22} \right)$$

$$\left(H_{21} - E_\phi S_{21} \right) c_1 + \left(H_{22} - E_\phi S_{22} \right) c_2 = 0$$

These two equations give nontrivial solutions only if the secular determinant shown in the question equals zero.

8–23. You may have noticed in Example 8–4 that only the diagonal elements of the secular determinant contain E. Do you see why? If not, see the next problem.

If the trial function chosen is a linear combination of orthogonal functions, then for the off-diagonal elements, $S_{ij} = 0$ and therefore $E S_{ij} = 0$. Thus, only the diagonal elements contain E. (In fact, if the trial function is a linear combination of orthonormal functions, then for the diagonal elements, $S_{ii} = 1$ and $E S_{ii} = E$.)

8–24. What does the secular determinant in Equation 8.39 look like if we expand ϕ in Equation 8.24 in an orthonormal set of functions?

If ϕ is expanded in an orthonormal set of functions, then the off-diagonal elements of the secular determinant do not contain E but the diagonal elements all contain E (see Problem 8–23). Specifically, the determinant is

$$\begin{vmatrix} H_{11} - E & H_{12} & \cdots & H_{1N} \\ H_{21} & H_{22} - E & \cdots & H_{2N} \\ \vdots & \vdots & \ddots & \vdots \\ H_{N1} & H_{N2} & \cdots & H_{NN} - E \end{vmatrix}$$

Because $\hat{H}$ is a Hermitian operator, $H_{ij} = H_{ji}^*$ (see Problem 8–22), or if H_{ij} is real, then $H_{ij} = H_{ji}$.

8–25. Determine the wave function corresponding to the energy in Example 8–4 for $v_0 = 10$ (see Figure 8.5).

For $v_0 = 10$, the estimate of the ground-state energy from the given trial function is

$$\epsilon = \frac{5 + v_0}{2} - \frac{1}{2}\left[9 + \left(\frac{32v_0}{9\pi^2}\right)^2\right]^{1/2}$$

$$= 5.155\,953$$

Use this and matrix elements in Example 8–4 to calculate the ratio c_2/c_1:

$$\frac{c_2}{c_1} = \frac{H_{11} - \epsilon S_{11}}{H_{12} - \epsilon S_{12}}$$

$$= \frac{1 + (v_0/2) - \epsilon}{-16v_0/(9\pi^2)}$$

$$= -0.468\,586$$

or $c_2 = -0.468\,586c_1$. The wave function can now be rewritten in terms of c_1:

$$\phi(x) = c_1\left(\frac{2}{a}\right)^{1/2}\sin\frac{\pi x}{a} + c_2\left(\frac{2}{a}\right)^{1/2}\sin\frac{2\pi x}{a}$$

$$= c_1\left[\left(\frac{2}{a}\right)^{1/2}\sin\frac{\pi x}{a} - 0.468\,586\left(\frac{2}{a}\right)^{1/2}\sin\frac{2\pi x}{a}\right]$$

Determine c_1 by requiring $\phi(x)$ to be normalized:

$$\int_0^a \phi^2(x)\,dx = 1$$

$$c_1^2\left(\frac{2}{a}\right)\left[\int_0^a \sin^2\frac{\pi x}{a}\,dx - 2\,(0.468\,586)\int_0^a \sin\frac{\pi x}{a}\sin\frac{2\pi x}{a}\,dx + (0.468\,586)^2\int_0^a \sin^2\frac{2\pi x}{a}\,dx\right] = 1$$

$$c_1^2\left(\frac{2}{a}\right)\left[\frac{a}{2} - 0 + (0.468\,586)^2\left(\frac{a}{2}\right)\right] = 1$$

$$c_1 = 0.905\,516$$

where c_1 is taken to be positive. Therefore, c_2 is

$$c_2 = -0.468\,586c_1 = -0.424\,312$$

which gives the wave function

$$\phi(x) = c_1 \left(\frac{2}{a}\right)^{1/2} \sin\frac{\pi x}{a} + c_2 \left(\frac{2}{a}\right)^{1/2} \sin\frac{2\pi x}{a}$$

$$= \frac{1}{a^{1/2}} \left(1.280\ 59 \sin\frac{\pi x}{a} - 0.600\ 07 \sin\frac{2\pi x}{a}\right)$$

8–26.   Repeat the calculation in Example 8–4 using a trial function of the form

$$\phi(x) = c_1 \left(\frac{2}{a}\right)^{1/2} \sin\frac{\pi x}{a} + c_2 \left(\frac{2}{a}\right)^{1/2} \sin\frac{3\pi x}{a}$$

What happened? What if we add $\sin 5\pi x/a$ to $\sin \pi x/a$?

The Hamiltonian operator is

$$\hat{H} = \hat{H}_0 + \frac{V_0 x}{a}$$

where $\hat{H}_0$ is the particle-in-a-box Hamiltonian operator. The trial function

$$\phi(x) = c_1 \left(\frac{2}{a}\right) \sin\frac{\pi x}{a} + c_2 \left(\frac{2}{a}\right) \sin\frac{3\pi x}{a}$$

is a linear combination of two eigenfunctions, ψ_1 and ψ_3, of $\hat{H}_0$.
The H_{ij} matrix elements are

$$H_{11} = \int_0^a \psi_1^* \left(\hat{H}_0 + \frac{V_0 x}{a}\right) \psi_1\, dx$$

$$= \frac{h^2}{8ma^2} + \frac{2V_0}{a^2}\int_0^a x \sin^2\frac{\pi x}{a}\, dx$$

$$= \frac{h^2}{8ma^2} + \frac{V_0}{2}$$

$$H_{22} = \int_0^a \psi_3^* \left(\hat{H}_0 + \frac{V_0 x}{a}\right) \psi_3\, dx$$

$$= \frac{9h^2}{8ma^2} + \frac{2V_0}{a^2}\int_0^a x \sin^2\frac{3\pi x}{a}\, dx$$

$$= \frac{9h^2}{8ma^2} + \frac{V_0}{2}$$

$$H_{12} = H_{21} = \int_0^a \psi_1^* \left(\hat{H}_0 + \frac{V_0 x}{a}\right) \psi_3\, dx$$

$$= \frac{2V_0}{a^2}\int_0^a x \sin\frac{\pi x}{a} \sin\frac{3\pi x}{a}\, dx$$

$$= 0$$

Since the ψ_n functions are orthonormal, the S_{ij} elements are

$$S_{11} = \int_0^a \psi_1^* \psi_1 \, dx = 1$$

$$S_{22} = \int_0^a \psi_3^* \psi_3 \, dx = 1$$

$$S_{12} = S_{21} = \int_0^a \psi_1^* \psi_3 \, dx = 0$$

Forming the secular determinant and setting it equal to zero,

$$\begin{vmatrix} \dfrac{h^2}{8ma^2} + \dfrac{V_0}{2} - E & 0 \\ 0 & \dfrac{9h^2}{8ma^2} + \dfrac{V_0}{2} - E \end{vmatrix} = 0$$

Let $\epsilon = 8ma^2 E / h^2$ and $v_0 = 8ma^2 V_0 / h^2$ to rewrite this determinant,

$$\begin{vmatrix} 1 + \dfrac{v_0}{2} - \epsilon & 0 \\ 0 & 9 + \dfrac{v_0}{2} - \epsilon \end{vmatrix} = 0$$

The secular equation is

$$\left(1 + \frac{v_0}{2} - \epsilon\right)\left(9 + \frac{v_0}{2} - \epsilon\right) = 0$$

Solving for ϵ results in two roots,

$$\epsilon = 1 + \frac{v_0}{2} \quad \text{and} \quad 9 + \frac{v_0}{2}$$

These energies are simply the H_{11} and H_{22} elements. Note that the off-diagonal elements $H_{ij} - E S_{ij}$ are equal to zero because H_{ij} consists of an odd (about $x = a/2$) integrand integrated over an interval symmetric about $x = a/2$. The functions in the linear combination, ψ_1 and ψ_3 do not mix with each other, and thus are not effective in the determination of ground-state energy variationally. The same is true if $\psi_5 = (2/a)^{1/2} \sin(5\pi x/a)$ or the sine of any odd multiple of $\pi x/a$ is added to ψ_1.

8–27. Repeat the calculation in Example 8–4 using a trial function of the form

$$\phi(x) = c_1 \left(\frac{2}{a}\right)^{1/2} \sin \frac{\pi x}{a} + c_2 \left(\frac{2}{a}\right)^{1/2} \sin \frac{4\pi x}{a}$$

Compare your result for the energy to that obtained in Example 8–4 for $v_0 = 10$. What do you think would happen if you used $\sin 6\pi x/a$ instead of $\sin 4\pi x/a$? See the next problem for an explanation of what's going on.

The Hamiltonian operator is

$$\hat{H} = \hat{H}_0 + \frac{V_0 x}{a}$$

where $\hat{H}_0$ is the particle-in-a-box Hamiltonian operator. The trial function

$$\phi(x) = c_1 \left(\frac{2}{a}\right)^{1/2} \sin \frac{\pi x}{a} + c_2 \left(\frac{2}{a}\right)^{1/2} \sin \frac{4\pi x}{a}$$

is a linear combination of two eigenfunctions, ψ_1 and ψ_4, of $\hat{H}_0$.

The H_{ij} matrix elements are

$$H_{11} = \int_0^a \psi_1^* \left(\hat{H}_0 + \frac{V_0 x}{a}\right) \psi_1 \, dx$$

$$= \frac{h^2}{8ma^2} + \frac{2V_0}{a^2} \int_0^a x \sin^2 \frac{\pi x}{a} \, dx$$

$$= \frac{h^2}{8ma^2} + \frac{V_0}{2}$$

$$H_{22} = \int_0^a \psi_4^* \left(\hat{H}_0 + \frac{V_0 x}{a}\right) \psi_4 \, dx$$

$$= \frac{16h^2}{8ma^2} + \frac{2V_0}{a^2} \int_0^a x \sin^2 \frac{4\pi x}{a} \, dx$$

$$= \frac{16h^2}{8ma^2} + \frac{V_0}{2}$$

$$H_{12} = H_{21} = \int_0^a \psi_1^* \left(\hat{H}_0 + \frac{V_0 x}{a}\right) \psi_4 \, dx$$

$$= \frac{2V_0}{a^2} \int_0^a x \sin \frac{\pi x}{a} \sin \frac{4\pi x}{a} \, dx$$

$$= -\frac{32V_0}{225\pi^2}$$

Since the ψ_n functions are orthonormal, the S_{ij} elements are

$$S_{11} = \int_0^a \psi_1^* \psi_1 \, dx = 1$$

$$S_{22} = \int_0^a \psi_4^* \psi_4 \, dx = 1$$

$$S_{12} = S_{21} = \int_0^a \psi_1^* \psi_4 \, dx = 0$$

Forming the secular determinant and setting it equal to zero,

$$\begin{vmatrix} \dfrac{h^2}{8ma^2} + \dfrac{V_0}{2} - E & -\dfrac{32V_0}{225\pi^2} \\ -\dfrac{32V_0}{225\pi^2} & \dfrac{16h^2}{8ma^2} + \dfrac{V_0}{2} - E \end{vmatrix} = 0$$

Let $\epsilon = 8ma^2 E / h^2$ and $v_0 = 8ma^2 V_0 / h^2$ to rewrite this determinant,

$$\begin{vmatrix} 1 + \dfrac{v_0}{2} - \epsilon & -\dfrac{32v_0}{225\pi^2} \\ -\dfrac{32v_0}{225\pi^2} & 16 + \dfrac{v_0}{2} - \epsilon \end{vmatrix} = 0$$

The secular equation is

$$\left(1 + \frac{v_0}{2} - \epsilon \right) \left(16 + \frac{v_0}{2} - \epsilon \right) - \left(\frac{32v_0}{225\pi^2} \right)^2 = 0$$

$$\epsilon^2 - (17 + v_0)\,\epsilon + \left(1 + \frac{v_0}{2} \right) \left(16 + \frac{v_0}{2} \right) - \left(\frac{32v_0}{225\pi^2} \right)^2 = 0$$

Solving for ϵ when $v_0 = 10$ results in two roots, and the smallest root corresponds to the calculated ground-state energy:

$$\epsilon = 5.998\,62$$

This is greater than the value of $\epsilon = 5.155\,95$ determined in Example 8–4 (see Problem 8–25).

If $\psi_6 = (2/a)^{1/2} \sin(6\pi x/a)$ is used instead of ψ_4, H_{11} remains the same, but H_{22}, H_{12} and H_{21} change.

$$H_{22} = \int_0^a \psi_6^* \left(\hat{H}_0 + \frac{V_0 x}{a} \right) \psi_6 \, dx$$

$$= \frac{36h^2}{8ma^2} + \frac{2V_0}{a^2} \int_0^a x \sin^2 \frac{3\pi x}{a} \, dx$$

$$= \frac{36h^2}{8ma^2} + \frac{V_0}{2}$$

$$H_{12} = H_{21} = \int_0^a \psi_1^* \left(\hat{H}_0 + \frac{V_0 x}{a} \right) \psi_6 \, dx$$

$$= \frac{2V_0}{a^2} \int_0^a x \sin \frac{\pi x}{a} \sin \frac{6\pi x}{a} \, dx$$

$$= -\frac{48V_0}{1225\pi^2}$$

Forming the secular determinant and setting it equal to zero,

$$\begin{vmatrix} \dfrac{h^2}{8ma^2} + \dfrac{V_0}{2} - E & -\dfrac{48V_0}{1225\pi^2} \\ -\dfrac{48V_0}{1225\pi^2} & \dfrac{36h^2}{8ma^2} + \dfrac{V_0}{2} - E \end{vmatrix} = 0$$

Once again, let $\epsilon = 8ma^2 E/h^2$ and $v_0 = 8ma^2 V_0/h^2$ to rewrite this determinant,

$$\begin{vmatrix} 1 + \dfrac{v_0}{2} - \epsilon & -\dfrac{48v_0}{1225\pi^2} \\ -\dfrac{48v_0}{1225\pi^2} & 36 + \dfrac{v_0}{2} - \epsilon \end{vmatrix} = 0$$

The secular equation is

$$\left(1 + \frac{v_0}{2} - \epsilon\right)\left(36 + \frac{v_0}{2} - \epsilon\right) - \left(\frac{48v_0}{1225\pi^2}\right)^2 = 0$$

$$\epsilon^2 - (37 + v_0)\,\epsilon + \left(1 + \frac{v_0}{2}\right)\left(36 + \frac{v_0}{2}\right) - \left(\frac{48v_0}{1225\pi^2}\right)^2 = 0$$

Solving for ϵ when $v_0 = 10$ results in two roots, and the smallest root corresponds to the calculated ground-state energy:

$$\epsilon = 5.999\,95$$

This is even greater than the value of ϵ calculated using ψ_4. This problem suggests that higher energy functions contribute less to the calculated ground-state energy.

8–28. This problem shows that terms in a trial function that correspond to progressively higher energies contribute progressively less to the ground-state energy. For algebraic simplicity, assume that the Hamiltonian operator can be written in the form

$$\hat{H} = \hat{H}^{(0)} + \hat{H}^{(1)}$$

and choose a trial function

$$\phi = c_1\psi_1 + c_2\psi_2$$

where

$$\hat{H}^{(0)}\psi_j = E_j^{(0)}\psi_j \quad j = 1, 2$$

Show that the secular equation associated with the trial function is

$$\begin{vmatrix} H_{11} - E & H_{12} \\ H_{12} & H_{22} - E \end{vmatrix} = \begin{vmatrix} E_1^{(0)} + E_1^{(1)} - E & H_{12} \\ H_{12} & E_2^{(0)} + E_2^{(1)} - E \end{vmatrix} = 0 \qquad (1)$$

where

$$E_j^{(1)} = \int \psi_j^* \hat{H}^{(1)}\psi_j \, d\tau \quad \text{and} \quad H_{12} = \int \psi_1^* \hat{H}^{(1)}\psi_2 \, d\tau$$

Solve equation 1 for E to obtain

$$E = \frac{E_1^{(0)} + E_1^{(1)} + E_2^{(0)} + E_2^{(1)}}{2} \pm \frac{1}{2}\left\{[E_1^{(0)} + E_1^{(1)} - E_2^{(0)} - E_2^{(1)}]^2 + 4H_{12}^2\right\}^{1/2} \qquad (2)$$

If we arbitrarily assume that $E_1^{(0)} + E_1^{(1)} < E_2^{(0)} + E_2^{(1)}$, then we take the positive sign in equation 2 and write

$$E = \frac{E_1^{(0)} + E_1^{(1)} + E_2^{(0)} + E_2^{(1)}}{2} + \frac{E_1^{(0)} + E_1^{(1)} - E_2^{(0)} - E_2^{(1)}}{2}$$

$$\times \left\{1 + \frac{4H_{12}^2}{[E_1^{(0)} + E_1^{(1)} - E_2^{(0)} - E_2^{(1)}]^2}\right\}^{1/2}$$

Use the expansion $(1 + x)^{1/2} = 1 + x/2 + \cdots$ to get

$$E = E_1^{(0)} + E_1^{(1)} + \frac{H_{12}^2}{E_1^{(0)} + E_1^{(1)} - E_2^{(0)} - E_2^{(1)}} + \cdots \tag{3}$$

Note that if $E_1^{(0)} + E_1^{(1)}$ and $E_2^{(0)} + E_2^{(1)}$ are widely separated, the term involving H_{12}^2 in equation 3 is small. Therefore, the energy is simply that calculated using ψ_1 alone; the ψ_2 part of the trial function contributes little to the overall energy. The general result is that terms in a trial function that correspond to higher and higher energies contribute less and less to the total ground-state energy.

Assume that the functions ψ_j are orthonormal. The H_{ij} matrix elements are

$$H_{11} = \int \psi_1^* \left[\hat{H}^{(0)} + \hat{H}^{(1)} \right] \psi_1 \, d\tau$$

$$= E_1^{(0)} + \int \psi_1^* \hat{H}^{(1)} \psi_1 \, d\tau$$

$$= E_1^{(0)} + E_1^{(1)}$$

$$H_{22} = \int \psi_2^* \left[\hat{H}^{(0)} + \hat{H}^{(1)} \right] \psi_2 \, d\tau$$

$$= E_2^{(0)} + \int \psi_2^* \hat{H}^{(1)} \psi_2 \, d\tau$$

$$= E_2^{(0)} + E_2^{(1)}$$

$$H_{12} = H_{21}^* = \int \psi_1^* \left[\hat{H}^{(0)} + \hat{H}^{(1)} \right] \psi_2 \, d\tau$$

$$= \int \psi_1^* \hat{H}^{(1)} \psi_2 \, d\tau$$

The relationship between H_{12} and H_{21} is shown in Problem 8–22. If H_{12} is real, then $H_{21} = H_{12}$. The S_{ij} elements are

$$S_{11} = S_{22} = 1$$
$$S_{12} = S_{21} = 0$$

The secular equation is

$$\begin{vmatrix} H_{11} - ES_{11} & H_{12} - ES_{12} \\ H_{21} - ES_{21} & H_{22} - ES_{22} \end{vmatrix} = \begin{vmatrix} E_1^{(0)} + E_1^{(1)} - E & H_{12} \\ H_{12} & E_2^{(0)} + E_2^{(1)} - E \end{vmatrix} = 0$$

Expanding the determinant gives

$$\left[E_1^{(0)} + E_1^{(1)} - E \right] \left[E_2^{(0)} + E_2^{(1)} - E \right] - H_{12}^2 = 0$$

$$E^2 - \left[E_1^{(0)} + E_1^{(1)} + E_2^{(0)} + E_2^{(1)} \right] E + \left[E_1^{(0)} + E_1^{(1)} \right] \left[E_2^{(0)} + E_2^{(1)} \right] - H_{12}^2 = 0$$

This is a quadratic equation in E. Solving for E gives

$$E = \frac{E_1^{(0)} + E_1^{(1)} + E_2^{(0)} + E_2^{(1)}}{2}$$

$$\pm \frac{1}{2} \left\{ \left[E_1^{(0)} + E_1^{(1)} + E_2^{(0)} + E_2^{(1)} \right]^2 - 4 \left[E_1^{(0)} + E_1^{(1)} \right] \left[E_2^{(0)} + E_2^{(1)} \right] + 4H_{12}^2 \right\}^{1/2}$$

$$= \frac{E_1^{(0)} + E_1^{(1)} + E_2^{(0)} + E_2^{(1)}}{2} \pm \frac{1}{2} \left\{ \left[E_1^{(0)} + E_1^{(1)} \right]^2 + \left[E_2^{(0)} + E_2^{(1)} \right]^2 \right.$$

$$\left. +2 \left[E_1^{(0)} + E_1^{(1)} \right] \left[E_2^{(0)} + E_2^{(1)} \right] - 4 \left[E_1^{(0)} + E_1^{(1)} \right] \left[E_2^{(0)} + E_2^{(1)} \right] + 4H_{12}^2 \right\}^{1/2}$$

$$= \frac{E_1^{(0)} + E_1^{(1)} + E_2^{(0)} + E_2^{(1)}}{2} \pm \frac{1}{2} \left\{ \left[E_1^{(0)} + E_1^{(1)} - E_2^{(0)} - E_2^{(1)} \right]^2 + 4H_{12}^2 \right\}^{1/2}$$

Assuming $E_1^{(0)} + E_1^{(1)} < E_2^{(0)} + E_2^{(1)}$ and taking the positive sign in the above equation,

$$E = \frac{E_1^{(0)} + E_1^{(1)} + E_2^{(0)} + E_2^{(1)}}{2} + \frac{E_1^{(0)} + E_1^{(1)} - E_2^{(0)} - E_2^{(1)}}{2}$$

$$\times \left\{ 1 + \frac{4H_{12}^2}{\left[E_1^{(0)} + E_1^{(1)} - E_2^{(0)} - E_2^{(1)} \right]^2} \right\}^{1/2}$$

$$= \frac{E_1^{(0)} + E_1^{(1)} + E_2^{(0)} + E_2^{(1)}}{2} + \frac{E_1^{(0)} + E_1^{(1)} - E_2^{(0)} - E_2^{(1)}}{2}$$

$$\times \left\{ 1 + \frac{2H_{12}^2}{\left[E_1^{(0)} + E_1^{(1)} - E_2^{(0)} - E_2^{(1)} \right]^2} + \cdots \right\}$$

$$= E_1^{(0)} + E_1^{(1)} + \frac{H_{12}^2}{E_1^{(0)} + E_1^{(1)} - E_2^{(0)} - E_2^{(1)}} + \cdots$$

where the series expansion $(1 + x)^{1/2} = 1 + x/2 + \cdots$ has been applied.

8–29. Use a program such as *MathCad* or *Mathematica* to repeat the calculation in Example 8–4 using a trial function of the form

$$\phi(x) = c_1 \sin \frac{\pi x}{a} + c_2 \sin \frac{2\pi x}{a} + c_3 \sin \frac{4\pi x}{a}$$

for $v_0 = 10$. Compare your answer to $\varepsilon = 5.155\,95$ from Example 8–4 and to $\varepsilon_{\text{exact}} = 5.0660$.

Since it is convenient to use orthonormal wave functions, write the trial function as

$$\phi(x) = N_1 \left(\frac{2}{a} \right)^{1/2} \sin \frac{\pi x}{a} + N_2 \left(\frac{2}{a} \right)^{1/2} \sin \frac{2\pi x}{a} + N_3 \left(\frac{2}{a} \right)^{1/2} \sin \frac{4\pi x}{a}$$

where $c_i = N_i \, (2/a)^{1/2}$. The Hamiltonian operator is

$$\hat{H} = -\frac{h^2}{8\pi^2 m}\frac{d^2}{dx^2} + \frac{V_0 x}{a}$$

Using a computation program, the H_{ij} matrix elements are

$$H_{11} = \frac{h^2}{8ma^2} + \frac{V_0}{2}$$

$$H_{22} = \frac{h^2}{2ma^2} + \frac{V_0}{2}$$

$$H_{33} = \frac{2h^2}{ma^2} + \frac{V_0}{2}$$

$$H_{12} = H_{21} = -\frac{16V_0}{9\pi^2}$$

$$H_{13} = H_{31} = -\frac{32V_0}{225\pi^2}$$

$$H_{23} = H_{32} = 0$$

Since the functions in the trial function are orthonormal,

$$S_{ij} = \delta_{ij}$$

Forming the secular determinant and setting it equal to zero results in the following expression:

$$\begin{vmatrix} \dfrac{h^2}{8ma^2} + \dfrac{V_0}{2} - E & -\dfrac{16V_0}{9\pi^2} & -\dfrac{32V_0}{225\pi^2} \\[2ex] -\dfrac{16V_0}{9\pi^2} & \dfrac{h^2}{2ma^2} + \dfrac{V_0}{2} - E & 0 \\[2ex] -\dfrac{32V_0}{225\pi^2} & 0 & \dfrac{2h^2}{ma^2} + \dfrac{V_0}{2} - E \end{vmatrix} = 0$$

Let $\epsilon = 8ma^2 E/h^2$ and $v_0 = 8ma^2 V_0/h^2$ to rewrite this determinant,

$$\begin{vmatrix} 1 + \dfrac{v_0}{2} - \epsilon & -\dfrac{16v_0}{9\pi^2} & -\dfrac{32v_0}{225\pi^2} \\[2ex] -\dfrac{16v_0}{9\pi^2} & 4 + \dfrac{v_0}{2} - \epsilon & 0 \\[2ex] -\dfrac{32v_0}{225\pi^2} & 0 & 16 + \dfrac{v_0}{2} - \epsilon \end{vmatrix} = 0$$

Setting $v_0 = 10$, the above equation gives $\epsilon_{min} = 5.154\,88$. This is lower than that obtained from Example 8–4 and thus more closely approximate the exact ground-state energy. In general, the use of more functions in the trial function gives an energy closer to the exact ground-state energy.

8–30. Use a program such as *MathCad* or *Mathematica* to verify the first few entries in Table 8.1.

As an illustration, use the trial function

$$\phi = c_1 e^{-\alpha_1 r^2} + c_2 e^{-\alpha_2 r^2}$$

The Hamiltonian operator is

$$\hat{H} = -\frac{\hbar^2}{2m_e r^2}\frac{d}{dr}\left(r^2\frac{d}{dr}\right) - \frac{e^2}{4\pi\epsilon_0 r}$$

The two necessary integrals and the integration results are shown below:

$$\int \phi^* \hat{H}\phi \, d\tau = 4\pi \int_0^\infty r^2 \phi \hat{H}\phi \, dr$$

$$= \frac{c_1^2\left(18^{1/2}\hbar^2\pi^{3/2}\epsilon_0\alpha_1^{1/2} - 2m_e e^2\right)}{8m_e\epsilon_0\alpha_1} + \frac{c_2^2\left(18^{1/2}\hbar^2\pi^{3/2}\epsilon_0\alpha_2^{1/2} - 2m_e e^2\right)}{8m_e\epsilon_0\alpha_2}$$

$$+ \frac{8c_1 c_2\left[6\hbar^2\pi^{3/2}\epsilon_0\alpha_1\alpha_2 - m_e e^2\left(\alpha_1 + \alpha_2\right)^{3/2}\right]}{8m_e\epsilon_0\left(\alpha_1 + \alpha_2\right)^{5/2}}$$

$$\int \phi^*\phi \, d\tau = 4\pi \int_0^\infty r^2\phi^2 \, dr$$

$$= \frac{\pi^{3/2}}{4}\left[\frac{2^{1/2}c_1^2}{\alpha_1^{3/2}} + \frac{2^{1/2}c_2^2}{\alpha_2^{3/2}} + \frac{8c_1 c_2}{\left(\alpha_1 + \alpha_2\right)^{3/2}}\right]$$

The energy expression is

$$E_\phi = \frac{\int \phi^* \hat{H}\phi \, d\tau}{\int \phi^*\phi \, d\tau}$$

Divide E_ϕ by $m_e e^4/(16\pi^2\epsilon_0^2\hbar^2)$ as done in Table 8.1 and substitute the numerical value of the constants m_e, e, $\hbar$, ϵ_0 into the resulting expression. Minimizing E_ϕ numerically gives $E_{min} = -0.485\,813$.

8–31. In this problem, we shall calculate the polarizability of a hydrogen atom using the variational method. The polarizability of an atom is a measure of the distortion of the electronic distribution of the atom when it is placed in an external electric field. When an atom is placed in an external electric field, the field induces a dipole moment in the atom. It is a good approximation to say that the magnitude of the induced dipole moment is proportional to the strength of the electric field. In an equation, we have

$$\mu = \alpha\mathcal{E} \tag{1}$$

where μ is the magnitude of the induced dipole moment, $\mathcal{E}$ is the strength of the electric field, and α is a proportionality constant called the *polarizability*. The value of α depends upon the particular atom.

The energy required to induce a dipole moment is given by

$$E = -\int_0^{\mathcal{E}} \mu \, d\mathcal{E}' = -\int_0^{\mathcal{E}} \alpha\mathcal{E}' \, d\mathcal{E}' = -\frac{\alpha\mathcal{E}^2}{2} \tag{2}$$

Equation 2 is the energy associated with a polarizable atom in an electric field.

Consider now a hydrogen atom for simplicity. In a hydrogen atom, there is an instantaneous dipole moment pointing from the electron to the nucleus. This instantaneous dipole moment is given by $-e\mathbf{r}$ and interacts with an external electric field according to

$$E = -\boldsymbol{\mu} \cdot \mathcal{E} = e\mathbf{r} \cdot \mathcal{E} \tag{3}$$

If $\mathcal{E}$ is taken to be in the z direction, then equation 3 introduces a perturbation term to the Hamiltonian operator of the hydrogen atom that is of the form

$$\hat{H}^{(1)} = e\mathcal{E}_z r \cos\theta$$

and so the complete Hamiltonian operator is

$$\hat{H} = \hat{H}^{(0)} + e\mathcal{E}_z r \cos\theta \tag{4}$$

We can solve this problem using perturbation theory, but we shall use the variational method here to calculate the ground-state energy of a hydrogen atom in an external electric field.

Problem 8–24 shows that it is convenient to write a trial function as a linear combination of orthonormal functions. In particular, in this case it is convenient to choose the orthonormal functions to be the eigenfunctions of the unperturbed system. Because the field induces a dipole in the z direction, let's take

$$\phi = c_1 \psi_{1s} + c_2 \psi_{2p_z} \tag{5}$$

as our trial function. Using the hydrogen atomic wave functions given in Table 7.5, show that

$$H_{11} = -\frac{e^2}{2\,\kappa_0\,a_0}$$

$$H_{22} = -\frac{e^2}{8\,\kappa_0\,a_0} \tag{6}$$

$$H_{12} = \frac{8}{\sqrt{2}} \left(\frac{2}{3}\right)^5 e\mathcal{E}_z a_0$$

where $\kappa_0 = 4\pi\epsilon_0$. Show that the two roots of the corresponding secular equation are

$$E = -\frac{5\,e^2}{16\,\kappa_0\,a_0} \pm \frac{3\,e^2}{16\,\kappa_0\,a_0}\left(1 + \frac{2^{23}}{3^{12}}\frac{\mathcal{E}_z^2\kappa_0^2 a_0^4}{e^2}\right)^{1/2} \tag{7}$$

Now use the expansion of $(1+x)^{1/2}$ given in Equation D.14 to obtain

$$E = -\frac{e^2}{2\,\kappa_0\,a_0} - 2.96\,\kappa_0\,a_0^3\frac{\mathcal{E}_z^2}{2} + \cdots \tag{8}$$

Compare this result to the macroscopic equation (equation 2) to show that the polarizability of a hydrogen atom is

$$\alpha = 2.96\,\kappa_0\,a_0^3 \tag{9}$$

The exact value for the hydrogen atom is $9\,\kappa_0\,a_0^3/2$. Although the numerical value is in error by 35%, we do see that the polarizability is proportional to a_0^3 (i.e., to a measure of the volume of the atom). This is a general result and can be used to estimate polarizabilities. Why is there no linear term in $\mathcal{E}_z$ in the above equation for E? What do you think a first-order perturbation calculation of the $1s$ state would give?

The ψ_{1s} and ψ_{2p_z} functions are

$$\psi_{1s} = \frac{1}{\sqrt{\pi}}\left(\frac{1}{a_0}\right)^{3/2} e^{-r/a_0}$$

$$\psi_{2p_z} = \frac{1}{4\sqrt{2\pi}}\left(\frac{1}{a_0}\right)^{3/2}\left(\frac{r}{a_0}\right) e^{-r/2a_0}\cos\theta$$

These are orthonormal eigenfunctions of $\hat{H}^{(0)}$ with eigenvalues of $-e^2/(8\pi\epsilon_0 a_0)$ and $-e^2/(32\pi\epsilon_0 a_0)$, respectively. Using $\kappa_0 = 4\pi\epsilon_0$, they become $-e^2/(2\kappa_0 a_0)$ and $-e^2/(8\kappa_0 a_0)$, respectively.

The H_{ij} elements are

$$H_{11} = \int \psi_{1s}\left[\hat{H}^{(0)} + e\mathcal{E}_z r \cos\theta\right]\psi_{1s}\,d\tau$$

$$= -\frac{e^2}{2\kappa_0 a_0} + \frac{e\mathcal{E}_z}{\pi a_0^3}\int_0^{2\pi} d\phi \int_0^{\pi} d\theta \,\sin\theta\cos\theta \int_0^{\infty} dr\, r^3 e^{-2r/a_0}$$

$$= -\frac{e^2}{2\kappa_0 a_0}$$

$$H_{22} = \int \psi_{2p_z}\left[\hat{H}^{(0)} + e\mathcal{E}_z r \cos\theta\right]\psi_{2p_z}\,d\tau$$

$$= -\frac{e^2}{8\kappa_0 a_0} + \frac{e\mathcal{E}_z}{32\pi a_0^5}\int_0^{2\pi} d\phi \int_0^{\pi} d\theta \,\sin\theta\cos^3\theta \int_0^{\infty} dr\, r^5 e^{-r/a_0}$$

$$= -\frac{e^2}{8\kappa_0 a_0}$$

$$H_{12} = H_{21} = \int \psi_{1s}\left[\hat{H}^{(0)} + e\mathcal{E}_z r \cos\theta\right]\psi_{2p_z}\,d\tau$$

$$= \frac{e\mathcal{E}_z}{4\sqrt{2\pi}a_0^4}\int_0^{2\pi} d\phi \int_0^{\pi} d\theta \,\sin\theta\cos^2\theta \int_0^{\infty} dr\, r^4 e^{-3r/2a_0}$$

$$= \frac{e\mathcal{E}_z}{4\sqrt{2\pi}a_0^4}\,(2\pi)\left(\frac{2}{3}\right)\left(\frac{256a_0^5}{81}\right)$$

$$= \frac{8}{\sqrt{2}}\left(\frac{2}{3}\right)^5 e\mathcal{E}_z a_0$$

The θ integrals in H_{11} and H_{22} vanish; thus, these matrix elements are simply the energies of the $1s$ and $2p_z$ states, respectively.

Since ψ_{1s} and ψ_{2p_z} are orthonormal,

$$S_{11} = S_{22} = 1$$

$$S_{12} = S_{21} = 0$$

Forming the secular determinant and setting it equal to zero results in the following expression:

$$\begin{vmatrix} -\dfrac{e^2}{2\kappa_0 a_0} - E & \dfrac{8}{\sqrt{2}} \left(\dfrac{2}{3}\right)^5 e\mathcal{E}_z a_0 \\ \dfrac{8}{\sqrt{2}} \left(\dfrac{2}{3}\right)^5 e\mathcal{E}_z a_0 & -\dfrac{e^2}{8\kappa_0 a_0} - E \end{vmatrix} = 0$$

which gives

$$\left(-\dfrac{e^2}{2\kappa_0 a_0} - E\right) \left(-\dfrac{e^2}{8\kappa_0 a_0} - E\right) - \left[\dfrac{8}{\sqrt{2}} \left(\dfrac{2}{3}\right)^5 e\mathcal{E}_z a_0\right]^2 = 0$$

$$E^2 + \left(\dfrac{5e^2}{8\kappa_0 a_0}\right) E + \dfrac{e^4}{16\kappa_0^2 a_0^2} - 32 \left(\dfrac{2}{3}\right)^{10} e^2 \mathcal{E}_z^2 a_0^2 = 0$$

Solving this quadratic equation for E,

$$E = -\dfrac{1}{2} \left(\dfrac{5e^2}{8\kappa_0 a_0}\right) \pm \dfrac{1}{2} \left\{\left(\dfrac{5e^2}{8\kappa_0 a_0}\right)^2 - 4\left[\dfrac{e^4}{16\kappa_0^2 a_0^2} - 32 \left(\dfrac{2}{3}\right)^{10} e^2 \mathcal{E}_z^2 a_0^2\right]\right\}^{1/2}$$

$$= -\dfrac{5e^2}{16\kappa_0 a_0} \pm \dfrac{1}{2} \left[\dfrac{9e^4}{64\kappa_0^2 a_0^2} + \dfrac{2^{17}}{3^{10}} e^2 \mathcal{E}_z^2 a_0^2\right]^{1/2}$$

$$= -\dfrac{5e^2}{16\kappa_0 a_0} \pm \dfrac{3e^2}{16\kappa_0 a_0} \left(1 + \dfrac{2^{23}}{3^{12}} \dfrac{\mathcal{E}_z^2 \kappa_0^2 a_0^4}{e^2}\right)^{1/2}$$

Expanding the last expression using $(1 + x)^{1/2} = 1 + x/2 + \cdots$ and taking the negative sign to give the minimum energy,

$$E = -\dfrac{5e^2}{16\kappa_0 a_0} - \dfrac{3e^2}{16\kappa_0 a_0} \left(1 + \dfrac{2^{22}}{3^{12}} \dfrac{\mathcal{E}_z^2 \kappa_0^2 a_0^4}{e^2} + \cdots\right)$$

$$= -\dfrac{e^2}{2\kappa_0 a_0} - 2.96\kappa_0 a_0^3 \dfrac{\mathcal{E}_z^2}{2}$$

The first term $-e^2/(2\kappa_0 a_0)$ in the E expression is the energy of the hydrogen atom in the $1s$ orbital. A comparison between the second term in this equation and $E = -\alpha \mathcal{E}^2/2$ gives

$$\alpha = 2.96\kappa_0 a_0^3$$

A linear term in $\mathcal{E}_z$ could only be present if the integrals involving this term in the diagonal H_{ii} elements are nonzero, which is not the case here. A first-order perturbation calculation for the energy of the $1s$ state therefore would give a value of 0 for the correction term.

8–32. It is instructive to redo the calculation of the polarizability of a hydrogen atom in the previous problem using a trial function of the form

$$\phi = c_1 \psi_{1s} + c_2 \psi_{3p_z}$$

This trial function has the same symmetry as equation 5 in the previous problem, but it involves the ψ_{3p_z} orbital instead of the ψ_{2p_z}. Show that in this case

$$E = -\frac{5\,e^2}{18\,\kappa_0\,a_0} \pm \frac{1}{2}\left(\frac{16}{81}\frac{e^4}{\kappa_0^2\,a_0^2} + \frac{3^6}{2^{11}}e^2\mathcal{E}_z^2 a_0^2\right)^{1/2}$$

or

$$E = -\frac{e^2}{2\,\kappa_0\,a_0} - 0.400\,\kappa_0\,a_0^3\frac{\mathcal{E}_z^2}{2}$$

for a polarizability, $\alpha = 0.400\,\kappa_0\,a_0^3$. Note that, in this case, the energy is quite a bit higher than that in equation 8 of the previous problem, and in fact it is not very far from the $1s$ energy, $-e^2/2\,\kappa_0\,a_0$. This result suggests that the $3p_z$ orbital somehow does not play much of a role in the trial function, particularly compared to the trial function involving the $1s$ and $2p_z$ orbitals. These two calculations of the polarizability of a hydrogen atom illustrate a general principle that we discuss in Problem 8–28.

_____

The ψ_{1s} and ψ_{3p_z} functions are

$$\psi_{1s} = \frac{1}{\sqrt{\pi}}\left(\frac{1}{a_0}\right)^{3/2} e^{-r/a_0}$$

$$\psi_{3p_z} = \frac{\sqrt{2}}{81\sqrt{\pi}}\left(\frac{1}{a_0}\right)^{3/2}\left(\frac{r}{a_0}\right)\left(6 - \frac{r}{a_0}\right)e^{-r/3a_0}\cos\theta$$

These are orthonormal eigenfunctions of $\hat{H}^{(0)}$ with eigenvalues of $-e^2/(8\pi\epsilon_0 a_0)$ and $-e^2/(72\pi\epsilon_0 a_0)$, respectively. Using $\kappa_0 = 4\pi\epsilon_0$, they become $-e^2/(2\kappa_0 a_0)$ and $-e^2/(18\kappa_0 a_0)$, respectively.

As in the last problem, each diagonal H_{ii} element is simply the energy of the hydrogen atom in the state described by the corresponding hydrogen-atom wave function:

$$H_{11} = -\frac{e^2}{2\kappa_0 a_0}$$

$$H_{22} = -\frac{e^2}{18\kappa_0 a_0}$$

The off-diagonal elements are

$$H_{12} = H_{21} = \int \psi_{1s}\left[\hat{H}^{(0)} + e\mathcal{E}_z r\cos\theta\right]\psi_{3p_z}\,d\tau$$

$$= \frac{e\mathcal{E}_z\sqrt{2}}{81\pi a_0^4}\int_0^{2\pi}d\phi\int_0^{\pi}d\theta\,\sin\theta\cos^2\theta\int_0^{\infty}dr\,r^4\left(6 - \frac{r}{a_0}\right)e^{-4r/3a_0}$$

$$= \frac{e\mathcal{E}_z\sqrt{2}}{81\pi a_0^4}(2\pi)\left(\frac{2}{3}\right)\left(\frac{6561 a_0^5}{512}\right)$$

$$= \frac{3^3}{2^{13/2}}e\mathcal{E}_z a_0$$

Since ψ_{1s} and ψ_{3p_z} are orthonormal,

$$S_{11} = S_{22} = 1$$

$$S_{12} = S_{21} = 0$$

Forming the secular determinant and setting it equal to zero results in the following expression:

$$\begin{vmatrix} -\dfrac{e^2}{2\kappa_0 a_0} - E & \dfrac{3^3}{2^{13/2}} e\mathcal{E}_z a_0 \\ \dfrac{3^3}{2^{13/2}} e\mathcal{E}_z a_0 & -\dfrac{e^2}{18\kappa_0 a_0} - E \end{vmatrix} = 0$$

which gives

$$\left(-\frac{e^2}{2\kappa_0 a_0} - E \right) \left(-\frac{e^2}{18\kappa_0 a_0} - E \right) - \left(\frac{3^3}{2^{13/2}} e\mathcal{E}_z a_0 \right)^2 = 0$$

$$E^2 + \left(\frac{10e^2}{18\kappa_0 a_0} \right) E + \frac{e^4}{36\kappa_0^2 a_0^2} - \frac{3^6}{2^{13}} e^2 \mathcal{E}_z^2 a_0^2 = 0$$

Solving this quadratic equation for E,

$$E = -\frac{1}{2} \left(\frac{10e^2}{18\kappa_0 a_0} \right) \pm \frac{1}{2} \left[\left(\frac{10e^2}{18\kappa_0 a_0} \right)^2 - 4 \left(\frac{e^4}{36\kappa_0^2 a_0^2} - \frac{3^6}{2^{13}} e^2 \mathcal{E}_z^2 a_0^2 \right) \right]^{1/2}$$

$$= -\frac{5e^2}{18\kappa_0 a_0} \pm \frac{1}{2} \left[\frac{16e^4}{81\kappa_0^2 a_0^2} + \frac{3^6}{2^{11}} e^2 \mathcal{E}_z^2 a_0^2 \right]^{1/2}$$

$$= -\frac{5e^2}{18\kappa_0 a_0} \pm \frac{2e^2}{9\kappa_0 a_0} \left(1 + \frac{3^{10}}{2^{15}} \frac{\mathcal{E}_z^2 \kappa_0^2 a_0^4}{e^2} \right)^{1/2}$$

Expanding the last expression using $(1+x)^{1/2} = 1 + x/2 + \cdots$ and taking the negative sign to give the minimum energy,

$$E = -\frac{5e^2}{18\kappa_0 a_0} - \frac{2e^2}{9\kappa_0 a_0} \left(1 + \frac{3^{10}}{2^{16}} \frac{\mathcal{E}_z^2 \kappa_0^2 a_0^4}{e^2} + \cdots \right)$$

$$= -\frac{e^2}{2\kappa_0 a_0} - 0.400\kappa_0 a_0^3 \frac{\mathcal{E}_z^2}{2}$$

A comparison between the $\mathcal{E}_z^2/2$ term in this equation and $E = -\alpha \mathcal{E}^2/2$ gives

$$\alpha = 0.400\kappa_0 a_0^3$$

8–33. Verify the expansion in Equation 8.57.

Start with

$$\left[\hat{H}^{(0)} + \lambda\hat{H}^{(1)}\right]\left[\psi_n^{(0)} + \lambda\psi_n^{(1)} + \lambda^2\psi_n^{(2)} + \cdots\right] = \left[E_n^{(0)} + \lambda E_n^{(1)} + \lambda^2 E_n^{(2)} + \cdots\right]$$

$$\times \left[\psi_n^{(0)} + \lambda\psi_n^{(1)} + \lambda^2\psi_n^{(2)} + \cdots\right]$$

Keeping terms up to second order in λ,

$$\hat{H}^{(0)}\psi_n^{(0)} + \lambda\hat{H}^{(0)}\psi_n^{(1)} + \lambda^2\hat{H}^{(0)}\psi_n^{(2)} + \lambda\hat{H}^{(1)}\psi_n^{(0)} + \lambda^2\hat{H}^{(1)}\psi_n^{(1)}$$

$$= E_n^{(0)}\psi_n^{(0)} + \lambda E_n^{(0)}\psi_n^{(1)} + \lambda^2 E_n^{(0)}\psi_n^{(2)} + \lambda E_n^{(1)}\psi_n^{(0)} + \lambda^2 E_n^{(1)}\psi_n^{(1)} + \lambda^2 E_n^{(2)}\psi_n^{(0)} + O(\lambda^3)$$

and grouping terms according to powers of λ:

$$\left[\hat{H}^{(0)}\psi_n^{(0)} - E_n^{(0)}\psi_n^{(0)}\right] + \left[\hat{H}^{(0)}\psi_n^{(1)} + \hat{H}^{(1)}\psi_n^{(0)} - E_n^{(0)}\psi_n^{(1)} - E_n^{(1)}\psi_n^{(0)}\right]\lambda$$

$$+ \left[\hat{H}^{(0)}\psi_n^{(2)} + \hat{H}^{(1)}\psi_n^{(1)} - E_n^{(0)}\psi_n^{(2)} - E_n^{(1)}\psi_n^{(1)} - E_n^{(2)}\psi_n^{(0)}\right]\lambda^2 + O(\lambda^3) = 0$$

This last expression is the same as Equation 8.57.

8–34. Use the series expansion of $(1 + x)^{1/2}$ (Equation D.14) to show that the variational result of Example 8–4 can be written as

$$\varepsilon = 1 + \frac{v_0}{2} - \frac{256}{243\pi^4}v_0^2 + O(v_0^3)$$

Rewriting the variational result,

$$\epsilon = \frac{5 + v_0}{2} \pm \frac{1}{2}\left[9 + \left(\frac{32v_0}{9\pi^2}\right)^2\right]^{1/2}$$

$$= \frac{5 + v_0}{2} \pm \frac{3}{2}\left[1 + \left(\frac{32v_0}{27\pi^2}\right)^2\right]^{1/2}$$

and using $(1 + x)^{1/2} = 1 + x/2 + \cdots$, the above equation becomes

$$\epsilon = \frac{5 + v_0}{2} \pm \frac{3}{2}\left[1 + \frac{1}{2}\left(\frac{32v_0}{27\pi^2}\right)^2 + O(v_0^3)\right]$$

Taking the negative sign,

$$\epsilon = 1 + \frac{v_0}{2} - \frac{256}{243\pi^4}v_0^2 + O(v_0^3)$$

8–35. Calculate the first-order correction to the first excited state of an anharmonic oscillator whose potential is given in Example 8–5.

The anharmonic oscillator potential is

$$V(x) = \frac{1}{2}kx^2 + \frac{1}{6}\gamma_3 x^3 + \frac{1}{24}\gamma_4 x^4$$

Take the unperturbed Hamiltonian operator as that of a harmonic oscillator,

$$\hat{H}^{(0)} = -\frac{\hbar^2}{2\mu}\frac{d^2}{dx^2} + \frac{1}{2}kx^2$$

then the perturbation is

$$\hat{H}^{(1)} = \frac{1}{6}\gamma_3 x^3 + \frac{1}{24}\gamma_4 x^4$$

The unperturbed first excited state is,

$$\psi_1^{(0)} = \left(\frac{4\alpha^3}{\pi}\right)^{1/4} x e^{-\alpha x^2/2}$$

and the first order correction to its energy is

$$E_1^{(1)} = \int \psi_1^{(0)} \hat{H}^{(1)} \psi_1^{(0)} \, d\tau$$

$$= \frac{\gamma_3}{6} \int_{-\infty}^{\infty} \psi_1^{(0)} x^3 \psi_1^{(0)} \, dx + \frac{\gamma_4}{24} \int_{-\infty}^{\infty} \psi_1^{(0)} x^4 \psi_1^{(0)} \, dx$$

The first integral is 0 since the integrand is odd and the limits of integration are symmetric about $x = 0$. It follows that

$$E_1^{(1)} = \frac{\gamma_4}{24}\left(\frac{4\alpha^3}{\pi}\right)^{1/2} \int_{-\infty}^{\infty} x^6 e^{-\alpha x^2} \, dx$$

$$= \frac{\gamma_4}{24}\left(\frac{4\alpha^3}{\pi}\right)^{1/2}\left(\frac{15\pi^{1/2}}{8\alpha^{7/2}}\right)$$

$$= \frac{5\gamma_4}{32\alpha^2}$$

8–36. Calculate the first-order correction to the energy of a particle constrained to move within the region $0 \le x \le a$ in the potential

$$V(x) = \begin{cases} V_0 x & 0 \le x \le \frac{a}{2} \\ V_0(a-x) & \frac{a}{2} \le x \le a \end{cases}$$

where V_0 is a constant.

Take the unperturbed system to be a particle in a box. Then the perturbed Hamiltonian operator, $\hat{H}^{(1)}$, is the potential given. Thus, the first order correction to the energy of a constrained particle with the potential described is

$$E_n^{(1)} = \int \psi_n^{(0)} \hat{H}^{(1)} \psi_n^{(0)} \, d\tau$$

$$= \int_0^{a/2} \psi_n^{(0)} V_0 x \psi_n^{(0)} \, dx + \int_{a/2}^a \psi_n^{(0)} V_0(a-x) \psi_n^{(0)} \, dx$$

$$= 2 \int_0^{a/2} \psi_n^{(0)} V_0 x \psi_n^{(0)} \, dx$$

$$= \frac{4V_0}{a} \int_0^{a/2} x \sin^2 \frac{n\pi x}{a} \, dx$$

$$= \frac{4V_0}{a} \left\{ \frac{a^2 \left[n^2\pi^2 + 2 - 2\cos(n\pi) \right]}{16n^2\pi^2} \right\}$$

$$= \frac{V_0 a}{2} \left[\frac{1}{2} + \frac{1 - \cos(n\pi)}{n^2\pi^2} \right]$$

8–37. Use first-order perturbation theory to calculate the first-order correction to the ground-state energy of a quartic oscillator whose potential energy is

$$V(x) = cx^4$$

In this case, use a harmonic oscillator as the unperturbed system. What is the perturbing potential?

If the unperturbed system is a harmonic oscillator, then

$$\hat{H}^{(0)} = -\frac{\hbar^2}{2\mu} \frac{d^2}{dx^2} + \frac{k}{2} x^2$$

Since the perturbed system has a potential energy of $U(x) = cx^4$ and $\hat{H} = \hat{H}^{(0)} + \hat{H}^{(1)}$, the perturbation, or the perturbing potential, is

$$\hat{H}^{(1)} = cx^4 - \frac{k}{2} x^2$$

The unperturbed ground-state wave function is

$$\psi_0 = \left(\frac{\alpha}{\pi} \right)^{1/4} e^{-\alpha x^2/2}$$

and the first order correction to the ground-state energy of a quartic oscillator is

$$E_0^{(1)} = \int \psi_0^{(0)} \hat{H}^{(1)} \psi_0^{(0)} \, d\tau$$

$$= \left(\frac{\alpha}{\pi}\right)^{1/2} \left(c \int_{-\infty}^{\infty} x^4 e^{-\alpha x^2} \, dx - \frac{k}{2} \int_{-\infty}^{\infty} x^2 e^{-\alpha x^2} \, dx \right)$$

$$= \left(\frac{\alpha}{\pi}\right)^{1/2} \left\{ c \left[\frac{3}{4\alpha^2} \left(\frac{\pi}{\alpha}\right)^{1/2} \right] + \frac{k}{2} \left[\frac{1}{2\alpha} \left(\frac{\pi}{\alpha}\right)^{1/2} \right] \right\}$$

$$= \frac{3c}{4\alpha^2} - \frac{k}{4\alpha}$$

8–38. In Example 5–2, we introduced the Morse potential

$$V(x) = D(1 - e^{-\beta x})^2$$

as a description of the internuclear potential energy of a diatomic molecule. First expand the Morse potential in a power series about x. (*Hint:* Use the expansion $e^x = 1 + x + \frac{x^2}{2} + \frac{x^3}{6} + \cdots$.) What is the Hamiltonian operator for the Morse potential? Show that the Hamiltonian operator can be written in the form

$$\hat{H} = -\frac{\hbar^2}{2\mu} \frac{d^2}{dx^2} + ax^2 + bx^3 + cx^4 + \cdots \tag{1}$$

How are the constants a, b, and c related to the constants D and β? What part of the Hamiltonian operator would you associate with $\hat{H}^{(0)}$, and what are the functions $\psi_n^{(0)}$ and energies $E_n^{(0)}$? Use perturbation theory to evaluate the first-order corrections to the energy of the first three states that arise from the cubic and quartic terms.

Using the power series expansion for e^x to rewrite the Morse potential:

$$V(x) = D(1 - e^{-\beta x})^2$$

$$= D \left\{ 1 - \left[1 - \beta x + \frac{\beta^2 x^2}{2} - \frac{\beta^3 x^3}{6} + O(x^4) \right] \right\}^2$$

$$= D \left[\beta x - \frac{\beta^2 x^2}{2} + \frac{\beta^3 x^3}{6} + O(x^4) \right]^2$$

$$= D\beta^2 x^2 \left[1 - \frac{\beta x}{2} + \frac{\beta^2 x^2}{6} + O(x^3) \right]^2$$

$$= D\beta^2 x^2 \left[1 - \beta x + \frac{7\beta^2 x^2}{12} + O(x^3) \right]$$

$$= D\beta^2 x^2 - D\beta^3 x^3 + \frac{7}{12} D\beta^4 x^4 + O(x^5)$$

The Hamiltonian operator for the system described by the Morse potential is

$$\hat{H} = -\frac{\hbar^2}{2\mu}\frac{d^2}{dx^2} + V(x)$$

$$= -\frac{\hbar^2}{2\mu}\frac{d^2}{dx^2} + ax^2 + bx^3 + cx^4 + \dots$$

where $a = D\beta^2$, $b = -D\beta^3$, and $c = 7D\beta^4/12$. If the unperturbed system is a harmonic oscillator, then

$$\hat{H}^{(0)} = -\frac{\hbar^2}{2\mu}\frac{d^2}{dx^2} + ax^2$$

$$\hat{H}^{(1)} = bx^3 + cx^4$$

$$\psi_v^{(0)} = N_v H_v\left(\alpha^{1/2}x\right)e^{-\alpha x^2/2}$$

$$E_v^{(0)} = \hbar\omega\left(v + \frac{1}{2}\right) \qquad v = 0, 1, 2, \dots$$

The first three unperturbed states are

$$\psi_0^{(0)}(x) = \left(\frac{\alpha}{\pi}\right)^{1/4}e^{-\alpha x^2/2}$$

$$\psi_1^{(0)}(x) = \left(\frac{4\alpha^3}{\pi}\right)^{1/4}xe^{-\alpha x^2/2}$$

$$\psi_2^{(0)}(x) = \left(\frac{\alpha}{4\pi}\right)^{1/4}(2\alpha x^2 - 1)e^{-\alpha x^2/2}$$

and the first-order corrections to the energy of the these states due to the cubic and quartic terms in the potential are

$$E_v^{(1)} = \int \psi_v^{(0)}\hat{H}^{(1)}\psi_v^{(0)}\,d\tau$$

$$= b\int_{-\infty}^{\infty}x^3\left[\psi_v^{(0)}\right]^2\,dx + c\int_{-\infty}^{\infty}x^4\left[\psi_v^{(0)}\right]^2\,dx$$

The first integral is 0 since the integrand is odd and the integration limits are symmetric about $x = 0$. Thus,

$$E_0^{(1)} = c\left(\frac{\alpha}{\pi}\right)^{1/2}\int_{-\infty}^{\infty}x^4 e^{-\alpha x^2}\,dx$$

$$= c\left(\frac{\alpha}{\pi}\right)^{1/2}\left[\frac{3}{4\alpha^2}\left(\frac{\pi}{\alpha}\right)^{1/2}\right]$$

$$= \frac{3c}{4\alpha^2}$$

$$E_1^{(1)} = c \left(\frac{4\alpha^3}{\pi} \right)^{1/2} \int_{-\infty}^{\infty} x^6 e^{-\alpha x^2} \, dx$$

$$= c \left(\frac{4\alpha^3}{\pi} \right)^{1/2} \left[\frac{15}{8\alpha^3} \left(\frac{\pi}{\alpha} \right)^{1/2} \right]$$

$$= \frac{15c}{4\alpha^2}$$

$$E_2^{(1)} = c \left(\frac{\alpha}{4\pi} \right)^{1/2} \int_{-\infty}^{\infty} x^4 \left(2\alpha x^2 - 1 \right)^2 e^{-\alpha x^2} \, dx$$

$$= c \left(\frac{\alpha}{4\pi} \right)^{1/2} \left[\frac{39}{2\alpha^2} \left(\frac{\pi}{\alpha} \right)^{1/2} \right]$$

$$= \frac{39c}{4\alpha^2}$$

Note that the perturbation becomes more significant with higher energy states. This is expected, since these states deviate more from a harmonic oscillator description.

8–39. In applying first-order perturbation theory to a helium atom, we must evaluate the integral (Equation 8.70)

$$E^{(1)} = \frac{e^2}{4\pi\epsilon_0} \iint d\mathbf{r}_1 d\mathbf{r}_2 \psi_{1s}^*(\mathbf{r}_1) \psi_{1s}^*(\mathbf{r}_2) \frac{1}{r_{12}} \psi_{1s}(\mathbf{r}_1) \psi_{1s}(\mathbf{r}_2)$$

where

$$\psi_{1s}(\mathbf{r}_j) = \left(\frac{Z^3}{a_0^3 \pi} \right)^{1/2} e^{-Z\mathbf{r}_j/a_0}$$

and $Z = 2$ for a helium atom. This same integral occurs in a variational treatment of a helium atom, where in that case the value of Z is left arbitrary. This problem proves that

$$E^{(1)} = \frac{5Z}{8} \left(\frac{m_e e^4}{16\pi^2 \epsilon_0^2 \hbar^2} \right)$$

Let $\mathbf{r}_1$ and $\mathbf{r}_2$ be the radius vectors of electrons 1 and 2, respectively, and let θ be the angle between these two vectors. Now this is generally *not* the θ of spherical coordinates, but if we choose one of the radius vectors, say $\mathbf{r}_1$, to be the z axis, then the two θ's are the same. Using the law of cosines,

$$r_{12} = (r_1^2 + r_2^2 - 2r_1 r_2 \cos\theta)^{1/2}$$

show that $E^{(1)}$ becomes

$$E^{(1)} = \frac{e^2}{4\pi\epsilon_0} \frac{Z^6}{a_0^6 \pi^2} \int_0^\infty dr_1 e^{-Zr_1/a_0} 4\pi r_1^2 \int_0^\infty dr_2 e^{-Zr_2/a_0} r_2^2$$

$$\times \int_0^{2\pi} d\phi \int_0^\pi \frac{d\theta \sin\theta}{(r_1^2 + r_2^2 - 2r_1 r_2 \cos\theta)^{1/2}}$$

Letting $x = \cos\theta$, show that the integrand over θ is

$$\int_0^\pi \frac{d\theta\,\sin\theta}{(r_1^2 + r_2^2 - 2r_1r_2\cos\theta)^{1/2}} = \int_{-1}^{1} \frac{dx}{(r_1^2 + r_2^2 - 2r_1r_2x)^{1/2}}$$

$$= \frac{2}{r_1} \quad r_1 > r_2$$

$$= \frac{2}{r_2} \quad r_1 < r_2$$

Substituting this result into $E^{(1)}$, show that

$$E^{(1)} = \frac{e^2}{4\pi\epsilon_0} \frac{16Z^6}{a_0^6} \int_0^\infty dr_1 e^{-2Zr_1/a_0} r_1^2 \left(\frac{1}{r_1} \int_0^{r_1} dr_2 e^{-2Zr_2/a_0} r_2^2 \right.$$

$$\left. + \int_{r_1}^\infty dr_2 e^{-2Zr_2/a_0} r_2 \right)$$

$$= \frac{e^2}{4\pi\epsilon_0} \frac{4Z^3}{a_0^3} \int_0^\infty dr_1 e^{-2Zr_1/a_0} r_1^2 \left[\frac{1}{r_1} - e^{-2Zr_1/a_0} \left(\frac{Z}{a_0} + \frac{1}{r_1} \right) \right]$$

$$= \frac{5}{8} Z \left(\frac{e^2}{4\pi\epsilon_0 a_0} \right) = \frac{5}{8} Z \left(\frac{m_e e^4}{16\pi^2\epsilon_0^2\hbar^2} \right)$$

Show that the energy through first order is

$$E^{(0)} + E^{(1)} = \left(-Z^2 + \frac{5}{8}Z \right) \left(\frac{m_e e^4}{16\pi^2\epsilon_0^2\hbar^2} \right) = -\frac{11}{4} \left(\frac{m_e e^4}{16\pi^2\epsilon_0^2\hbar^2} \right)$$

$$= -2.75 \left(\frac{m_e e^4}{16\pi^2\epsilon_0^2\hbar^2} \right)$$

compared with the exact result, $E_{\text{exact}} = -2.9037(m_e e^4/16\pi^2\epsilon_0^2\hbar^2)$.

Choose $\mathbf{r}_1$ to be the z-axis and use

$$r_{12} = (r_1^2 + r_2^2 - 2r_1r_2\cos\theta)^{1/2}$$

where θ is the angle between the two radius vectors. Substitute this result into the $E^{(1)}$ expression to give

$$E^{(1)} = \frac{e^2}{4\pi\epsilon_0} \int_0^{2\pi} d\phi_2 \int_0^{2\pi} d\phi_1 \int_0^{\pi} d\theta_1 \sin\theta_1 \int_0^{\infty} dr_2\, r_2^2 \psi_2^2 \int_0^{\infty} dr_1\, r_1^2 \psi_1^2$$

$$\times \int_0^{\pi} \frac{d\theta\, \sin\theta}{(r_1^2 + r_2^2 - 2r_1r_2\cos\theta)^{1/2}}$$

$$= \frac{e^2}{4\pi\epsilon_0} 8\pi^2 \left(\frac{Z^3}{a_0^3\pi}\right)^2 \int_0^{\infty} dr_2\, r_2^2 e^{-2Zr_2/a_0} \int_0^{\infty} dr_1\, r_1^2 e^{-2Zr_1/a_0}$$

$$\times \int_0^{\pi} \frac{d\theta\, \sin\theta}{(r_1^2 + r_2^2 - 2r_1r_2\cos\theta)^{1/2}}$$

where θ_2 becomes θ and θ_1 is integrated over all possible values.

Let $x = \cos\theta$. Then $dx = -\sin\theta\, d\theta$, and the limits of $\theta = 0$ and $\theta = \pi$ become $x = 1$ and $x = -1$, respectively. The last integral in the $E^{(1)}$ expression then becomes

$$\int_0^{\pi} \frac{d\theta\, \sin\theta}{(r_1^2 + r_2^2 - 2r_1r_2\cos\theta)^{1/2}} = \int_{-1}^{1} \frac{dx}{(r_1^2 + r_2^2 - 2r_1r_2x)^{1/2}}$$

$$= -\frac{(r_1^2 + r_2^2 - 2r_1r_2x)^{1/2}}{r_1r_2}\Bigg|_{-1}^{1}$$

$$= -\frac{(r_1^2 + r_2^2 - 2r_1r_2)^{1/2}}{r_1r_2} + \frac{(r_1^2 + r_2^2 + 2r_1r_2)^{1/2}}{r_1r_2}$$

The first term in this expression is $\pm(r_1 - r_2)/(r_1r_2)$ and the second term is equal to $\pm(r_1 + r_2)/(r_1r_2)$. Since $E^{(1)}$ is an intrinsically positive quantity, not all combinations of signs are physical. In particular, only the positive sign is permissible for the second term. To pick the sign for the first term when $r_1 > r_2$, consider the limit when r_1 is very large, then $(r_1^2 + r_2^2 - 2r_1r_2x)^{1/2} \to r_1$ and therefore is consistent with the result

$$\int_{-1}^{1} \frac{dx}{(r_1^2 + r_2^2 - 2r_1r_2x)^{1/2}} = \frac{r_2 - r_1 + r_1 + r_2}{r_1r_2} = \frac{2r_2}{r_1r_2} = \frac{2}{r_1}$$

On the other hand, for $r_2 > r_1$, when r_2 is very large, $(r_1^2 + r_2^2 - 2r_1r_2x)^{1/2} \to r_2$, which is consistent with the result

$$\int_{-1}^{1} \frac{dx}{(r_1^2 + r_2^2 - 2r_1r_2x)^{1/2}} = \frac{r_1 - r_2 + r_1 + r_2}{r_1r_2} = \frac{2r_1}{r_1r_2} = \frac{2}{r_2}$$

Substituting these results into the $E^{(1)}$ expression gives

$$E^{(1)} = \frac{e^2}{4\pi\epsilon_0} \frac{8Z^6}{a_0^6} \int_0^{\infty} dr_1\, r_1^2 e^{-2Zr_1/a_0} \left(\frac{2}{r_1} \int_0^{r_1} dr_2\, r_2^2 e^{-2Zr_2/a_0} + 2\int_{r_1}^{\infty} dr_2\, r_2 e^{-2Zr_2/a_0}\right)$$

$$= \frac{e^2}{4\pi\epsilon_0} \frac{16Z^6}{a_0^6} \int_0^{\infty} dr_1\, r_1^2 e^{-2Zr_1/a_0} \left(\frac{1}{r_1} \int_0^{r_1} dr_2\, r_2^2 e^{-2Zr_2/a_0} + \int_{r_1}^{\infty} dr_2\, r_2 e^{-2Zr_2/a_0}\right)$$

Evaluate the two integrals involving r_2

$$\int_0^{r_1} dr_2\, r_2^2 e^{-2Zr_2/a_0} = \left.\frac{r_2^2 a_0 e^{-2Zr_2/a_0}}{-2Z}\right|_0^{r_1} + \frac{2a_0}{2Z}\int_0^{r_1} dr_2\, r_2 e^{-2Zr_2/a_0}$$

$$= -\frac{r_1^2 a_0 e^{-2Zr_1/a_0}}{2Z} + \left.\frac{a_0^3 e^{-2Zr_2/a_0}}{4Z^3}\left(\frac{-2Zr_2}{a_0} - 1\right)\right|_0^{r_1}$$

$$= -\frac{r_1^2 a_0 e^{-2Zr_1/a_0}}{2Z} - \frac{a_0^3 e^{-2Zr_1/a_0}}{4Z^3}\left(\frac{2Zr_1}{a_0} + 1\right) + \frac{a_0^3}{4Z^3}$$

$$\int_{r_1}^{\infty} dr_2\, r_2 e^{-2Zr_2/a_0} = \left.\frac{a_0^2 e^{-2Zr_2/a_0}}{4Z^2}\left(\frac{-2Zr_2}{a_0} - 1\right)\right|_{r_1}^{\infty}$$

$$= \frac{a_0^2 e^{-2Zr_1/a_0}}{4Z^2}\left(\frac{2Zr_1}{a_0} + 1\right)$$

to give

$$\frac{1}{r_1}\int_0^{r_1} dr_2\, r_2^2 e^{-2Zr_2/a_0} + \int_{r_1}^{\infty} dr_2 r_2 e^{-2Zr_2/a_0} = \frac{a_0^2 e^{-2Zr_1/a_0}}{4Z^2} - \frac{a_0^2 e^{-2Zr_1/a_0}}{2Z^2}$$

$$- \frac{a_0^3 e^{-2Zr_1/a_0}}{4Z^3 r_1} + \frac{a_0^3}{4Z^3 r_1}$$

$$= \frac{a_0^3}{4Z^3}\left[\frac{1}{r_1} - e^{-2Zr_1/a_0}\left(\frac{Z}{a_0} + \frac{1}{r_1}\right)\right]$$

and the $E^{(1)}$ expression becomes

$$E^{(1)} = \frac{e^2}{4\pi\epsilon_0}\frac{4Z^3}{a_0^3}\int_0^{\infty} dr_1\, r_1^2 e^{-2Zr_1/a_0}\left[\frac{1}{r_1} - e^{-2Zr_1/a_0}\left(\frac{Z}{a_0} + \frac{1}{r_1}\right)\right]$$

$$= \frac{e^2}{4\pi\epsilon_0}\frac{4Z^3}{a_0^3}\left(\frac{a_0^2}{4Z^2} - \frac{Z}{a_0}\frac{a_0^3}{32Z^3} - \frac{a_0^2}{16Z^2}\right)$$

$$= \frac{e^2}{4\pi\epsilon_0}\frac{4Z^3}{a_0^3}\left(\frac{5a_0^2}{32Z^2}\right)$$

$$= \frac{5Z}{8a_0}\frac{e^2}{4\pi\epsilon_0} = \frac{5Z}{8}\left(\frac{m_e e^4}{16\pi^2\epsilon_0^2\hbar^2}\right)$$

The zero-order energy $E^{(0)}$ for this system is

$$E^{(0)} = -\frac{Z^2 m_e e^4}{16\pi^2\epsilon_0^2\hbar^2}$$

Thus, the energy through first order is

$$E^{(0)} + E^{(1)} = \left(-Z^2 + \frac{5}{8}Z\right)\left(\frac{m_e e^4}{16\pi^2\epsilon_0^2\hbar^2}\right) = -\frac{11}{4}\left(\frac{m_e e^4}{16\pi^2\epsilon_0^2\hbar^2}\right)$$

$$= -2.75\left(\frac{m_e e^4}{16\pi^2\epsilon_0^2\hbar^2}\right)$$

where $Z = 2$ has been used. The calculated energy is 5% greater than the exact value.

8–40. In the previous problem we evaluated the integral that occurs in the first-order perturbation theory treatment of a helium atom (see Equation 8.70). In this problem we will evaluate the integral by another method, one that uses an expansion for $1/r_{12}$ that is useful in many applications. We can write $1/r_{12}$ as an expansion in terms of spherical harmonics,

$$\frac{1}{r_{12}} = \frac{1}{|\mathbf{r}_1 - \mathbf{r}_2|} = \sum_{l=0}^{\infty}\sum_{m=-l}^{+l}\frac{4\pi}{2l+1}\frac{r_<^l}{r_>^{l+1}}Y_l^m(\theta_1, \phi_1)Y_l^{m*}(\theta_2, \phi_2)$$

where θ_i and ϕ_i are the angles that describe $\mathbf{r}_i$ in a spherical coordinate system and $r_<$ and $r_>$ are, respectively, the smaller and larger values of r_1 and r_2. In other words, if $r_1 < r_2$, then $r_< = r_1$ and $r_> = r_2$. Substitute $\psi_{1s}(r_i) = (Z^3/a_0^3\pi)^{1/2}e^{-Zr_i/a_0}$, and the above expansion for $1/r_{12}$ into Equation 8.70, integrate over the angles, and show that all the terms except for the $l = 0$, $m = 0$ term vanish. Show that

$$E^{(1)} = \frac{e^2}{4\pi\epsilon_0}\frac{16Z^6}{a_0^6}\int_0^{\infty}dr_1 r_1^2 e^{-2Zr_1/a_0}\int_0^{\infty}dr_2 r_2^2\frac{e^{-2Zr_2/a_0}}{r_>}$$

Now show that

$$E^{(1)} = \frac{e^2}{4\pi\epsilon_0}\frac{16Z^6}{a_0^6}\int_0^{\infty}dr_1 r_1 e^{-2Zr_1/a_0}\int_0^{r_1}dr_2 r_2^2 e^{-2Zr_2/a_0}$$

$$+ \frac{e^2}{4\pi\epsilon_0}\frac{16Z^6}{a_0^6}\int_0^{\infty}dr_1 r_1^2 e^{-2Zr_1/a_0}\int_{r_1}^{\infty}dr_2 r_2 e^{-2Zr_2/a_0}$$

$$= -\frac{e^2}{4\pi\epsilon_0}\frac{4Z^6}{a_0^6}\int_0^{\infty}dr_1 r_1 e^{-2Zr_1/a_0}\left[e^{-2Zr_1/a_0}\left(\frac{2Z^2 r_1^2}{a_0^2} + \frac{2Zr_1}{a_0} + 1\right) - 1\right]$$

$$+ \frac{e^2}{4\pi\epsilon_0}\frac{4Z^6}{a_0^6}\int_0^{\infty}dr_1 r_1^2 e^{-2Zr_1/a_0}\left[e^{-2Zr_1/a_0}\left(\frac{2Zr_1}{a_0} + 1\right)\right]$$

$$= -\frac{e^2}{4\pi\epsilon_0}\frac{4Z^6}{a_0^6}\int_0^{\infty}dr_1 e^{-4Zr_1/a_0}\left[\frac{r_1^2 a_0^2}{Z^2} + \frac{r_1 a_0^3}{Z^3}\right]$$

$$+ \frac{e^2}{4\pi\epsilon_0}\frac{4Z^3}{a_0^3}\int_0^{\infty}dr_1 r_1 e^{-2Zr_1/a_0}$$

$$= \frac{5}{8}Z\left(\frac{e^2}{4\pi\epsilon_0 a_0}\right)$$

as in Problem 8–39.

Substituting the expression for $\psi_{1s}(r_i)$ and the expansion of $1/r_{12}$ into Equation 8.70,

$$
\begin{aligned}
E^{(1)} &= \frac{e^2}{4\pi\epsilon_0} \int \int d\mathbf{r}_1 d\mathbf{r}_2 \, \psi_{1s}(\mathbf{r}_1)\psi_{1s}(\mathbf{r}_2)\frac{1}{r_{12}}\psi_{1s}(\mathbf{r}_1)\psi_{1s}(\mathbf{r}_2) \\
&= \frac{e^2}{4\pi\epsilon_0}\left(\frac{Z^6}{a_0^6\pi^2}\right)\int_0^\infty dr_1 \, r_1^2 e^{-2Zr_1/a_0} \int_0^\infty dr_2 \, r_2^2 e^{-2Zr_2/a_0} \int_0^{2\pi} d\phi_1 \int_0^\pi d\theta_1 \, \sin\theta_1 \\
&\quad \times \int_0^{2\pi} d\phi_2 \int_0^\pi d\theta_2 \, \sin\theta_2 \left[\sum_{l=0}^\infty \sum_{m=-l}^{+l} \frac{4\pi}{2l+1}\frac{r_<^l}{r_>^{l+1}} Y_l^m(\theta_1, \phi_1)Y_l^{m*}(\theta_2, \phi_2)\right]
\end{aligned}
$$

Consider the integrals over the angles and introduce $Y_0^0 = Y_0^{0*} = (4\pi)^{-1/2}$ into the expression:

$$
\begin{aligned}
&\int_0^{2\pi} d\phi_1 \int_0^\pi d\theta_1 \, \sin\theta_1 Y_l^m(\theta_1, \phi_1) \int_0^{2\pi} d\phi_2 \int_0^\pi d\theta_2 \, \sin\theta_2 Y_l^{m*}(\theta_2, \phi_2) \\
&= 4\pi \int_0^{2\pi} d\phi_1 \int_0^\pi d\theta_1 \, \sin\theta_1 Y_0^{0*}(\theta_1, \phi_1)Y_l^m(\theta_1, \phi_1) \int_0^{2\pi} d\phi_2 \int_0^\pi d\theta_2 \, \sin\theta_2 Y_l^{m*}(\theta_2, \phi_2)Y_0^0(\theta_2, \phi_2) \\
&= 4\pi \, \delta_{l0}\,\delta_{m0}
\end{aligned}
$$

The last equality arises from the orthonormality of spherical harmonics. Thus, all terms containing spherical harmonics vanish unless $l = m = 0$. The $E^{(1)}$ expression becomes

$$
\begin{aligned}
E^{(1)} &= \frac{e^2}{4\pi\epsilon_0}\left(\frac{Z^6}{a_0^6\pi^2}\right)\int_0^\infty dr_1 \, r_1^2 e^{-2Zr_1/a_0}\int_0^\infty dr_2 \, r_2^2 e^{-2Zr_2/a_0}\left(\frac{4\pi}{2\cdot 0+1}\right)\frac{r_<^0}{r_>^1}(4\pi) \\
&= \frac{e^2}{4\pi\epsilon_0}\left(\frac{16Z^6}{a_0^6}\right)\int_0^\infty dr_1 \, r_1^2 e^{-2Zr_1/a_0}\int_0^\infty dr_2 \, r_2^2 e^{-2Zr_2/a_0}\frac{1}{r_>}
\end{aligned}
$$

When $r_2 < r_1$, $r_> = r_1$, and when $r_2 > r_1$, $r_> = r_2$. Separating the integral over r_2 into two parts,

$$
\begin{aligned}
E^{(1)} &= \frac{e^2}{4\pi\epsilon_0}\left(\frac{16Z^6}{a_0^6}\right)\int_0^\infty dr_1 \, r_1 e^{-2Zr_1/a_0}\int_0^{r_1} dr_2 \, r_2^2 e^{-2Zr_2/a_0} \\
&\quad + \frac{e^2}{4\pi\epsilon_0}\left(\frac{16Z^6}{a_0^6}\right)\int_0^\infty dr_1 \, r_1^2 e^{-2Zr_1/a_0}\int_{r_1}^\infty dr_2 \, r_2 e^{-2Zr_2/a_0} \\
&= \frac{e^2}{4\pi\epsilon_0}\left(\frac{4Z^3}{a_0^3}\right)\int_0^\infty dr_1 \, r_1 e^{-2Zr_1/a_0}\left[1 - e^{-2Zr_1/a_0}\left(\frac{2Z^2r_1^2}{a_0^2} + \frac{2Zr_1}{a_0} + 1\right)\right] \\
&\quad + \frac{e^2}{4\pi\epsilon_0}\left(\frac{4Z^4}{a_0^4}\right)\int_0^\infty dr_1 \, r_1^2 e^{-2Zr_1/a_0}\left[e^{-2Zr_1/a_0}\left(\frac{2Zr_1}{a_0} + 1\right)\right]
\end{aligned}
$$

(These integrals have been evaluated in detail in Problem 8–39.) Finally, to evaluate the integrals over r_1, set $x = 2Zr_1/a_0$:

$$E^{(1)} = \frac{e^2}{4\pi\epsilon_0}\left(\frac{Z}{a_0}\right)\int_0^\infty dx\, xe^{-x}\left[1 - e^{-x}\left(\frac{x^2}{2} + x + 1\right)\right] + \frac{e^2}{4\pi\epsilon_0}\left(\frac{Z}{2a_0}\right)\int_0^\infty dx\, x^2 e^{-x}\left[e^{-x}(x+1)\right]$$

$$= \frac{e^2}{4\pi\epsilon_0}\left(\frac{Z}{a_0}\right)\int_0^\infty dx\left(xe^{-x} - xe^{-2x} - \frac{x^2}{2}e^{-2x}\right)$$

$$= \frac{e^2}{4\pi\epsilon_0}\left(\frac{Z}{a_0}\right)\left(1 - \frac{1}{4} - \frac{1}{8}\right) = \frac{5Z}{8}\left(\frac{e^2}{4\pi\epsilon_0 a_0}\right)$$

8–41. Consider a molecule with a dipole moment $\boldsymbol{\mu}$ in an electric field $\mathcal{E}$. We picture the dipole moment as a positive charge and a negative charge of magnitude q separated by a vector $\mathbf{l}$.

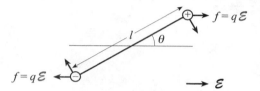

The field $\mathcal{E}$ causes the dipole to rotate into a direction parallel to $\mathcal{E}$. Therefore, work is required to rotate the dipole to an angle θ to $\mathcal{E}$. The force causing the molecule to rotate is actually a torque (torque is the angular analog of force) and is given by $l/2$ times the force perpendicular to $\mathbf{l}$ at each end of the vector $\mathbf{l}$. Show that this torque is equal to $\mu\mathcal{E}\sin\theta$ and that the energy required to rotate the dipole from some initial angle θ_0 to some arbitrary angle θ is

$$V = \int_{\theta_0}^\theta \mu\mathcal{E}\sin\theta'\, d\theta'$$

Given that θ_0 is customarily taken to be $\pi/2$, show that

$$V = -\mu\mathcal{E}\cos\theta = -\boldsymbol{\mu}\cdot\mathcal{E}$$

The force perpendicular to $\mathbf{l}$ is

$$f_\perp = f\sin\theta = q\mathcal{E}\sin\theta$$

giving a torque of

$$\tau = 2\left(\frac{l}{2}\right)q\mathcal{E}\sin\theta$$

$$= \mu\mathcal{E}\sin\theta$$

where the dipole moment relation $\mu = ql$ is used. The energy required to rotate the dipole from θ_0 to θ is

$$V = \int_{\theta_0}^\theta \tau\, d\theta$$

$$= \int_{\theta_0}^\theta \mu\mathcal{E}\sin\theta'\, d\theta'$$

For $\theta_0 = \pi/2$,

$$V = \mu\mathcal{E} \int_{\pi/2}^{\theta} \sin\theta' \, d\theta'$$

$$= -\mu\mathcal{E} \, \cos\theta' \Big|_{\pi/2}^{\theta}$$

$$= -\mu\mathcal{E} \, \cos\theta = -\boldsymbol{\mu} \cdot \boldsymbol{\mathcal{E}}$$

8–42. Derive Equation 8.93 using the second term in Equation 8.91 (cf. Problem A–6).

Equation 8.91 gives

$$a_2(t) = \frac{(\mu_z)_{12} E_{0z}}{2} \left\{ \frac{1 - \exp\left[i\left(E_2 - E_1 + \hbar\omega\right) t/\hbar\right]}{E_2 - E_1 + \hbar\omega} + \frac{1 - \exp\left[i\left(E_2 - E_1 - \hbar\omega\right) t/\hbar\right]}{E_2 - E_1 - \hbar\omega} \right\}$$

Taking $E_2 > E_1$, the first term in the braces becomes negligible compared to the second term; thus,

$$a_2(t) = \frac{(\mu_z)_{12} E_{0z}}{2} \left\{ \frac{1 - \exp\left[i\left(E_2 - E_1 - \hbar\omega\right) t/\hbar\right]}{E_2 - E_1 - \hbar\omega} \right\}$$

The probability of absorption is

$$P_{1\rightarrow 2}(\omega, t) = a_2^*(t) a_2(t)$$

$$= \frac{(\mu_z)_{12}^2 E_{0z}^2}{4} \left\{ \frac{1 - \exp\left[-i\left(E_2 - E_1 - \hbar\omega\right) t/\hbar\right]}{E_2 - E_1 - \hbar\omega} \right\} \left\{ \frac{1 - \exp\left[i\left(E_2 - E_1 - \hbar\omega\right) t/\hbar\right]}{E_2 - E_1 - \hbar\omega} \right\}$$

$$= \frac{(\mu_z)_{12}^2 E_{0z}^2}{4\left(E_2 - E_1 - \hbar\omega\right)^2} \left\{ 1 - \exp\left[-i\left(E_2 - E_1 - \hbar\omega\right) t/\hbar\right] \right\} \left\{ 1 - \exp\left[i\left(E_2 - E_1 - \hbar\omega\right) t/\hbar\right] \right\}$$

Writing $(E_2 - E_1 - \hbar\omega)t/\hbar$ as x,

$$\left\{ 1 - \exp\left[-i\left(E_2 - E_1 - \hbar\omega\right) t/\hbar\right] \right\} \left\{ 1 - \exp\left[i\left(E_2 - E_1 - \hbar\omega\right) t/\hbar\right] \right\}$$

$$= \left(1 - e^{-ix}\right)\left(1 - e^{ix}\right)$$

$$= 2 - e^{ix} - e^{-ix}$$

$$= 2 - 2\cos x$$

$$= 4\sin^2 \frac{x}{2}$$

where the relations $\cos x = (e^{ix} + e^{-ix})/2$ (see Problem A–6) and $\sin^2(x/2) = (1 - \cos x)/2$ have been applied. Thus,

$$P_{1 \to 2}(\omega, t) = \frac{(\mu_z)_{12}^2 E_{0z}^2}{4 \left(E_2 - E_1 - \hbar\omega\right)^2} \left[4 \sin^2 \frac{(E_2 - E_1 - \hbar\omega)t/\hbar}{2} \right]$$

$$= \frac{(\mu_z)_{12}^2 E_{0z}^2 \sin^2 \left[(E_2 - E_1 - \hbar\omega)t/2\hbar\right]}{\left(E_2 - E_1 - \hbar\omega\right)^2}$$

8–43. In this problem we shall derive the formulas of perturbation theory to second order in the perturbation. First, substitute Equation 8.61 into Equation 8.58 to obtain an equation in which the only unknown quantity is $\psi_n^{(1)}$. A standard way to solve the equation for $\psi_n^{(1)}$ is to expand the unknown $\psi_n^{(1)}$ in terms of the eigenfunctions of the unperturbed problem. Show that if we substitute

$$\psi_n^{(1)} = \sum_j a_{nj} \psi_j^{(0)} \tag{1}$$

into Equation 8.58, multiply by $\psi_k^{(0)*}$, and integrate, then we obtain

$$\sum_j a_{nj} \langle \psi_k^{(0)} \mid \hat{H}^{(0)} - E_n^{(0)} \mid \psi_j^{(0)} \rangle = E_n^{(1)} \langle \psi_k^{(0)} \mid \psi_n^{(0)} \rangle - \langle \psi_k^{(0)} \mid \hat{H}^{(1)} \mid \psi_n^{(0)} \rangle \tag{2}$$

There are two cases to consider here, $k = n$ and $k \neq n$. Show that when $k = n$, we obtain Equation 8.61 again and that when $k \neq n$, we obtain

$$a_{nk} = \frac{\langle \psi_k^{(0)} \mid \hat{H}^{(1)} \mid \psi_n^{(0)} \rangle}{E_n^{(0)} - E_k^{(0)}} = \frac{H_{kn}^{(1)}}{E_n^{(0)} - E_k^{(0)}} \tag{3}$$

where

$$H_{kn}^{(1)} = \langle \psi_k^{(0)} \mid \hat{H}^{(1)} \mid \psi_n^{(0)} \rangle = \int \psi_k^{(0)*} \hat{H}^{(1)} \psi_n^{(0)} \, d\tau \tag{4}$$

Thus, we have determined all the a's in equation 1 except for a_{nn}. We can determine a_{nn} by requiring that ψ_n in Equation 8.55 be normalized through first order, or through terms linear in λ. Show that this requirement is equivalent to requiring that $\psi_n^{(0)}$ be orthogonal to $\psi_n^{(1)}$ and that it gives $a_{nn} = 0$. The complete wave function to first order, then, is

$$\psi_n = \psi_n^{(0)} + \lambda \sum_{j \neq n} \frac{H_{jn}^{(1)} \psi_j^{(0)}}{E_n^{(0)} - E_j^{(0)}} = \psi_n^{(0)} + \lambda \psi_n^{(1)} \tag{5}$$

which defines $\psi_n^{(1)}$. Now that we have $\psi_n^{(1)}$, we can determine the second-order energy by setting the λ^2 term in Equation 8.57 equal to zero:

$$H^{(0)} \psi_n^{(2)} + \hat{H}^{(1)} \psi_n^{(1)} - E_n^{(0)} \psi_n^{(2)} - E_n^{(1)} \psi_n^{(1)} - E_n^{(2)} \psi_n^{(0)} = 0 \tag{6}$$

As with $\psi_n^{(1)}$, we write $\psi_n^{(2)}$ as

$$\psi_n^{(2)} = \sum_s b_{ns} \psi_s^{(0)} \tag{7}$$

Substitute this expression into equation 6, multiply from the left by $\psi_m^{(0)*}$, and integrate to obtain

$$b_{nm}E_m^{(0)} + \sum_{j \neq n} a_{nj}H_{mj}^{(1)} = b_{nm}E_n^{(0)} + a_{nm}E_n^{(1)} + \delta_{nm}E_n^{(2)} \tag{8}$$

Let $n = m$ to get (remember that $a_{nn} = 0$)

$$E_n^{(2)} = \sum_{j \neq n} \frac{H_{nj}^{(1)} H_{jn}^{(1)}}{E_n^{(0)} - E_j^{(0)}} = \sum_{j \neq n} \frac{|H_{nj}^{(1)}|^2}{E_n^{(0)} - E_j^{(0)}} \tag{9}$$

Thus, the energy through second order is

$$E_n = E_n^{(0)} + E_n^{(1)} + E_n^{(2)}$$

where $E_n^{(1)}$ is given by Equation 8.61 and $E_n^{(2)}$ is given by equation 9.

Substitute Equation 8.61,

$$E_n^{(1)} = \langle \psi_n^{(0)} | \hat{H}^{(1)} | \psi_n^{(0)} \rangle = \int d\tau \, \psi_n^{(0)*} \hat{H}^{(1)} \psi_n^{(0)}$$

into Equation 8.58,

$$\hat{H}^{(0)} \psi_n^{(1)} + \hat{H}^{(1)} \psi_n^{(0)} = E_n^{(0)} \psi_n^{(1)} + E_n^{(1)} \psi_n^{(0)}$$

to give

$$\hat{H}^{(0)} \psi_n^{(1)} + \hat{H}^{(1)} \psi_n^{(0)} = E_n^{(0)} \psi_n^{(1)} + \langle \psi_n^{(0)} | \hat{H}^{(1)} | \psi_n^{(0)} \rangle \psi_n^{(0)}$$

The only unknown in this equation is $\psi_n^{(1)}$, which can be written as a linear combination of the eigenfunctions of the unperturbed system:

$$\psi_n^{(1)} = \sum_j a_{nj} \psi_j^{(0)}$$

Substitute this expression into Equation 8.58, multiply by $\psi_k^{(0)*}$, and integrate:

$$\psi_k^{(0)*} \hat{H}^{(0)} \sum_j a_{nj} \psi_j^{(0)} + \psi_k^{(0)*} \hat{H}^{(1)} \psi_n^{(0)} = E_n^{(0)} \psi_k^{(0)*} \sum_j a_{nj} \psi_j^{(0)} + E_n^{(1)} \psi_k^{(0)*} \psi_n^{(0)}$$

$$\sum_j a_{nj} \langle \psi_k^{(0)} | \hat{H}^{(0)} | \psi_j^{(0)} \rangle + \langle \psi_k^{(0)} | \hat{H}^{(1)} | \psi_n^{(0)} \rangle = \sum_j a_{nj} E_n^{(0)} \langle \psi_k^{(0)} | \psi_j^{(0)} \rangle + E_n^{(1)} \langle \psi_k^{(0)} | \psi_n^{(0)} \rangle$$

$$\sum_j a_{nj} \langle \psi_k^{(0)} | \hat{H}^{(0)} - E_n^{(0)} | \psi_j^{(0)} \rangle = E_n^{(1)} \langle \psi_k^{(0)} | \psi_n^{(0)} \rangle - \langle \psi_k^{(0)} | \hat{H}^{(1)} | \psi_n^{(0)} \rangle$$

which can be simplified to give

$$\sum_j a_{nj} \left[E_j^{(0)} - E_n^{(0)} \right] \langle \psi_k^{(0)} | \psi_j^{(0)} \rangle = \sum_j a_{nj} \left[E_j^{(0)} - E_n^{(0)} \right] \delta_{kj} = E_n^{(1)} \delta_{kn} - \langle \psi_k^{(0)} | \hat{H}^{(1)} | \psi_n^{(0)} \rangle$$

$$a_{nk} \left[E_k^{(0)} - E_n^{(0)} \right] = E_n^{(1)} \delta_{kn} - \langle \psi_k^{(0)} | \hat{H}^{(1)} | \psi_n^{(0)} \rangle$$

When $k = n$, $E_k^{(0)} = E_n^{(0)}$ and $\delta_{kn} = 1$, giving

$$a_{nk}\left[E_k^{(0)} - E_n^{(0)}\right] = 0 = E_n^{(1)} - \langle\,\psi_n^{(0)}\mid\hat{H}^{(1)}\mid\psi_n^{(0)}\,\rangle$$

$$E_n^{(1)} = \langle\,\psi_n^{(0)}\mid\hat{H}^{(1)}\mid\psi_n^{(0)}\,\rangle$$

which is the same as Equation 8.61. On the other hand, when $k \neq n$, $E_k^{(0)} \neq E_n^{(0)}$ and $\delta_{nk} = 0$, and

$$a_{nk}\left[E_k^{(0)} - E_n^{(0)}\right] = 0 - \langle\,\psi_k^{(0)}\mid\hat{H}^{(1)}\mid\psi_n^{(0)}\,\rangle$$

$$a_{nk} = \frac{\langle\,\psi_k^{(0)}\mid\hat{H}^{(1)}\mid\psi_n^{(0)}\,\rangle}{E_n^{(0)} - E_k^{(0)}} = \frac{H_{kn}^{(1)}}{E_n^{(0)} - E_k^{(0)}}$$

To determine a_{nn}, require the wave function to be normalized through first order:

$$\langle\,\psi\mid\psi\,\rangle = 1 = \langle\,\psi_n^{(0)} + \lambda\psi_n^{(1)}\mid\psi_n^{(0)} + \lambda\psi_n^{(1)}\,\rangle$$

$$= \langle\,\psi_n^{(0)}\mid\psi_n^{(0)}\,\rangle + \lambda\langle\,\psi_n^{(0)}\mid\psi_n^{(1)}\,\rangle + \lambda\langle\,\psi_n^{(1)}\mid\psi_n^{(0)}\,\rangle + O(\lambda^2)$$

$$= 1 + \lambda\langle\,\psi_n^{(0)}\mid\psi_n^{(1)}\,\rangle + \lambda\langle\,\psi_n^{(1)}\mid\psi_n^{(0)}\,\rangle + O(\lambda^2)$$

To satisfy this relation, it must be true that

$$\langle\,\psi_n^{(0)}\mid\psi_n^{(1)}\,\rangle = \langle\,\psi_n^{(1)}\mid\psi_n^{(0)}\,\rangle = 0$$

In other words, that $\psi_n^{(1)}$ is orthogonal to $\psi_n^{(0)}$. Since

$$\psi_n^{(1)} = \sum_j a_{nj}\psi_j^{(0)}$$

it follows that

$$\langle\,\psi_n^{(0)}\mid\psi_n^{(1)}\,\rangle = \sum_j a_{nj}\langle\,\psi_n^{(0)}\mid\psi_j^{(0)}\,\rangle$$

$$0 = \sum_j a_{nj}\delta_{nj} = a_{nn}$$

The expressions for a define the wave function to first order and allow the second order energy to be determined. Setting the λ^2 term in Equation 8.57 equal to zero,

$$H^{(0)}\psi_n^{(2)} + \hat{H}^{(1)}\psi_n^{(1)} - E_n^{(0)}\psi_n^{(2)} - E_n^{(1)}\psi_n^{(1)} - E_n^{(2)}\psi_n^{(0)} = 0$$

and substituting into this expression

$$\psi_n^{(1)} = \sum_j a_{nj}\psi_j^{(0)}$$

$$\psi_n^{(2)} = \sum_s b_{ns}\psi_s^{(0)}$$

gives

$$H^{(0)} \sum_s b_{ns} \psi_s^{(0)} + \hat{H}^{(1)} \sum_{j \neq n} a_{nj} \psi_j^{(0)} = E_n^{(0)} \sum_s b_{ns} \psi_s^{(0)} + E_n^{(1)} \sum_{j \neq n} a_{nj} \psi_j^{(0)} + E_n^{(2)} \psi_n^{(0)}$$

Multiplying this expression by $\psi_k^{(0)*}$, and integrating,

$$\psi_m^{(0)*} H^{(0)} \sum_s b_{ns} \psi_s^{(0)} + \psi_m^{(0)*} \hat{H}^{(1)} \sum_{j \neq n} a_{nj} \psi_j^{(0)}$$

$$= E_n^{(0)} \psi_m^{(0)*} \sum_s b_{ns} \psi_s^{(0)} + E_n^{(1)} \psi_m^{(0)*} \sum_{j \neq n} a_{nj} \psi_j^{(0)} + E_n^{(2)} \psi_m^{(0)*} \psi_n^{(0)}$$

$$\sum_s b_{ns} \langle \psi_m^{(0)} | H^{(0)} | \psi_s^{(0)} \rangle + \sum_{j \neq n} a_{nj} \langle \psi_m^{(0)} | H^{(1)} | \psi_j^{(0)} \rangle$$

$$= E_n^{(0)} \sum_s b_{ns} \langle \psi_m^{(0)} | \psi_s^{(0)} \rangle + E_n^{(1)} \sum_{j \neq n} a_{nj} \langle \psi_m^{(0)} | \psi_j^{(0)} \rangle + E_n^{(2)} \langle \psi_m^{(0)} | \psi_n^{(0)} \rangle$$

$$\sum_s b_{ns} E_s^{(0)} \delta_{ms} + \sum_{j \neq n} a_{nj} H_{mj}^{(1)} = E_n^{(0)} \sum_s b_{ns} \delta_{ms} + E_n^{(1)} \sum_{j \neq n} a_{nj} \delta_{mj} + E_n^{(2)} \delta_{mn}$$

$$b_{nm} E_m^{(0)} + \sum_{j \neq n} a_{nj} H_{mj}^{(1)} = b_{nm} E_n^{(0)} + a_{nm} E_n^{(1)} + \delta_{mn} E_n^{(2)}$$

Letting $n = m$ and noting that $a_{nn} = 0$, the above equation becomes

$$b_{nn} E_n^{(0)} + \sum_{j \neq n} a_{nj} H_{nj}^{(1)} = b_{nn} E_n^{(0)} + E_n^{(2)}$$

Writing explicitly the expression for a_{nj},

$$E_n^{(2)} = \sum_{j \neq n} a_{nj} H_{nj}^{(1)}$$

$$= \sum_{j \neq n} \frac{H_{jn}^{(1)} H_{nj}^{(1)}}{E_n^{(0)} - E_j^{(0)}}$$

8–44. Derive the equation for E_n through second order by starting with

$$E = \frac{\langle \psi_n | \hat{H} | \psi_n \rangle}{\langle \psi_n | \psi_n \rangle}$$

Starting with a variational approach, an expression for E through second order can be derived by setting up a secular determinant. Write the Hamiltonian operator as

$$\hat{H} = \hat{H}^{(0)} + \hat{H}^{(1)}$$

and the wave function as a linear combination of zero-order wave functions,

$$\psi = \sum_j c_j \psi_j^{(0)}$$

where $\hat{H}^{(0)} \psi_j^{(0)} = E_j^{(0)} \psi_j^{(0)}$. The Schrödinger equation becomes

$$\hat{H}\psi = E\psi$$

$$\left[\hat{H}^{(0)} + \hat{H}^{(1)}\right] \sum_j c_j \psi_j^{(0)} = E \sum_j c_j \psi_j^{(0)}$$

$$\sum_j c_j \left[E_j^{(0)} \psi_j^{(0)} + \hat{H}^{(1)} \psi_j^{(0)}\right] = E \sum_j c_j \psi_j^{(0)}$$

Left multiply the above equation by $\psi_k^{(0)*}$ and integrate over all space:

$$\sum_j c_j \left[E_j^{(0)} \langle \psi_k^{(0)} | \psi_j^{(0)} \rangle + \langle \psi_k^{(0)} | \hat{H}^{(1)} | \psi_j^{(0)} \rangle\right] = E \sum_j c_j \langle \psi_k^{(0)} | \psi_j^{(0)} \rangle$$

$$\sum_j c_j \left[E_j^{(0)} \delta_{kj} + H_{kj}^{(1)} - E\delta_{kj}\right] = 0$$

where $H_{kj}^{(1)} = \langle \psi_k^{(0)} | \hat{H}^{(1)} | \psi_j^{(0)} \rangle$ and the zero-order wave functions are taken to be orthonormal. This set of equations gives a secular determinant, which has a solution if the determinant is 0:

$$\begin{vmatrix} E_1^{(0)} + H_{11}^{(1)} - E & H_{12}^{(1)} & H_{13}^{(1)} & \cdots \\ H_{21}^{(1)} & E_2^{(0)} + H_{22}^{(1)} - E & H_{23}^{(1)} & \cdots \\ H_{31}^{(1)} & H_{32}^{(1)} & E_3^{(0)} + H_{33}^{(1)} - E & \cdots \\ \vdots & \vdots & \vdots & \vdots \end{vmatrix} = 0$$

Note that in the second order approximation, perturbation theory gives an energy expression connecting only the desired state with other states. The mixing between states that do not include the desired state is ignored. Applying this approximation to the ground state means that all off-diagonal H_{kj} terms in the secular determinant are set equal to zero if $k \neq 1$ or $j \neq 1$:

$$\begin{vmatrix} E_1^{(0)} + H_{11}^{(1)} - E & H_{12}^{(1)} & H_{13}^{(1)} & \cdots \\ H_{21}^{(1)} & E_2^{(0)} + H_{22}^{(1)} - E & 0 & \cdots \\ H_{31}^{(1)} & 0 & E_3^{(0)} + H_{33}^{(1)} - E & \cdots \\ \vdots & \vdots & \vdots & \vdots \end{vmatrix} = 0$$

Now transform the determinant into a diagonal form. First, multiply the second row by $H_{12}^{(1)}/[E_2^{(0)} + H_{22}^{(1)} - E]$ and subtract from the first row to obtain a new determinant with the first element of the first row equal to

$$E_1^{(0)} + H_{11}^{(1)} - E - \frac{H_{21}^{(1)} H_{12}^{(1)}}{E_2^{(0)} + H_{22}^{(1)} - E}$$

and the second element of the first row becomes zero. Continue this process of multiplying the jth row by $H_{1j}^{(1)}/[E_j^{(0)} + H_{jj}^{(1)} - E]$ and subtracting from the first row to give a determinant where all but the first element is zero. The nonzero first element is then

$$E_1^{(0)} + H_{11}^{(1)} - E - \sum_{j=2} \frac{|H_{1j}|^2}{E_j^{(0)} + H_{jj}^{(1)} - E}$$

Expanding the determinant about the first column gives

$$\left[E_1^{(0)} + H_{11}^{(1)} - E - \sum_{j=2} \frac{\left|H_{1j}\right|^2}{E_j^{(0)} + H_{jj}^{(1)} - E} \right] \left[E_2^{(0)} + H_{22}^{(1)} - E \right] \left[E_3^{(0)} + H_{33}^{(1)} - E \right] + \cdots = 0$$

The smallest root for E is obtained by setting the first term equal to zero, which then gives

$$E = E_1^{(0)} + H_{11}^{(1)} + \sum_{j=2} \frac{\left|H_{1j}\right|^2}{E_j^{(0)} + H_{jj}^{(1)} - E}$$

A solution for E can be found using successive approximation. As a first approximation, set E in the denominator of the last equation equal to the first two terms of that expression, $E = E_1^{(0)} + H_{11}^{(1)}$, then

$$E = E_1^{(0)} + H_{11}^{(1)} + \sum_{j=2} \frac{\left|H_{1j}\right|^2}{E_j^{(0)} + H_{jj}^{(1)} - E_1^{(0)} - H_{11}^{(1)}}$$

Note that the first-order energy terms, $H_{jj}^{(1)}$ and $H_{11}^{(1)}$, are small compared to the zero order energies. If they are ignored in the denominator of the third term, then the expression is the same as that obtained by perturbation theory.

8–45. Problem 8–43 shows that $\psi_n^{(1)}$ has the form $\psi_n^{(1)} = \sum_{j \neq n} a_{nj} \psi_j^{(0)}$. Given that $\psi_n^{(2)}$ has a similar form, $\psi_n^{(2)} = \sum_{j \neq n} b_{nj} \psi_j^{(0)}$, show that a knowledge of the wave function through first order determines the energy through third order.

By expressing the Hamiltonian operator as a sum of unperturbed and perturbed operators, and the wave function ψ_n and the energy E_n as power series in λ, the Schrödinger equation can be written as

$$\left[\hat{H}^{(0)} + \lambda \hat{H}^{(1)} \right] \left[\psi_n^{(0)} + \lambda \psi_n^{(1)} + \lambda^2 \psi_n^{(2)} + \lambda^3 \psi_n^{(3)} + \cdots \right]$$

$$= \left[E_n^{(0)} + \lambda E_n^{(1)} + \lambda^2 E_n^{(2)} + \lambda^3 E_n^{(3)} + \cdots \right] \left[\psi_n^{(0)} + \lambda \psi_n^{(1)} + \lambda^2 \psi_n^{(2)} + \lambda^3 \psi_n^{(3)} + \cdots \right] \quad (1)$$

To obtain the third order energy correction, consider the λ^3 terms in the above equation:

$$\hat{H}^{(0)} \psi_n^{(3)} + \hat{H}^{(1)} \psi_n^{(2)} = E_n^{(0)} \psi_n^{(3)} + E_n^{(1)} \psi_n^{(2)} + E_n^{(2)} \psi_n^{(1)} + E_n^{(3)} \psi_n^{(0)}$$

Multiply this equation by $\psi_k^{(0)*}$ and integrate:

$$\langle \psi_k^{(0)} | \hat{H}^{(0)} | \psi_n^{(3)} \rangle + \langle \psi_k^{(0)} | \hat{H}^{(1)} | \psi_n^{(2)} \rangle$$

$$= E_n^{(0)} \langle \psi_k^{(0)} | \psi_n^{(3)} \rangle + E_n^{(1)} \langle \psi_k^{(0)} | \psi_n^{(2)} \rangle + E_n^{(2)} \langle \psi_k^{(0)} | \psi_n^{(1)} \rangle + E_n^{(3)} \langle \psi_k^{(0)} | \psi_n^{(0)} \rangle$$

Note that since $\hat{H}^{(0)}$ is Hermitian, $\langle \psi_k^{(0)} | \hat{H}^{(0)} | \psi_n^{(3)} \rangle = \langle \hat{H}^{(0)} \psi_k^{(0)} | \psi_n^{(3)} \rangle = E_n^{(0)} \langle \psi_k^{(0)} | \psi_n^{(3)} \rangle$. Thus, the first term on the left hand side of the last equation cancels the first one on the right hand side. When $k = n$,

$$\langle \psi_n^{(0)} \mid \hat{H}^{(1)} \mid \psi_n^{(2)} \rangle = E_n^{(1)} \langle \psi_n^{(0)} \mid \psi_n^{(2)} \rangle + E_n^{(2)} \langle \psi_n^{(0)} \mid \psi_n^{(1)} \rangle + E_n^{(3)}$$

Furthermore, it was shown in Problem 8.43 that $\langle \psi_n^{(0)} \mid \psi_n^{(1)} \rangle = \langle \psi_n^{(1)} \mid \psi_n^{(0)} \rangle = 0$. Therefore, the third-order energy correction is

$$E_n^{(3)} = \left\langle \psi_n^{(0)} \left| \left[\hat{H}^{(1)} - E_n^{(1)} \right] \right| \psi_n^{(2)} \right\rangle = \left\langle \left[\hat{H}^{(1)} - E_n^{(1)} \right] \psi_n^{(0)} \mid \psi_n^{(2)} \right\rangle \tag{2}$$

An expression for $\left[\hat{H}^{(1)} - E_n^{(1)} \right] \psi_n^{(0)}$ can be obtained by considering the λ terms in Equation 1:

$$\hat{H}^{(0)} \psi_n^{(1)} + \hat{H}^{(1)} \psi_n^{(0)} = E_n^{(0)} \psi_n^{(1)} + E_n^{(1)} \psi_n^{(0)}$$

$$\left[\hat{H}^{(1)} - E_n^{(1)} \right] \psi_n^{(0)} = - \left[\hat{H}^{(0)} - E_n^{(0)} \right] \psi_n^{(1)} \tag{3}$$

Substitute Equation 3 into Equation 2 to give

$$E_n^{(3)} = - \left\langle \left[\hat{H}^{(0)} - E_n^{(0)} \right] \psi_n^{(1)} \mid \psi_n^{(2)} \right\rangle$$

$$= - \left\langle \psi_n^{(1)} \left| \left[\hat{H}^{(0)} - E_n^{(0)} \right] \right| \psi_n^{(2)} \right\rangle \tag{4}$$

Now use the λ^2 terms in Equation 1 to rewrite $\left[\hat{H}^{(0)} - E_n^{(0)} \right] \psi_n^{(2)}$:

$$\hat{H}^{(0)} \psi_n^{(2)} + \hat{H}^{(1)} \psi_n^{(1)} = E_n^{(0)} \psi_n^{(2)} + E_n^{(1)} \psi_n^{(1)} + E_n^{(2)} \psi_n^{(0)}$$

$$\left[\hat{H}^{(0)} - E_n^{(0)} \right] \psi_n^{(2)} = - \left[\hat{H}^{(1)} - E_n^{(1)} \right] \psi_n^{(1)} + E_n^{(2)} \psi_n^{(0)} \tag{5}$$

Substitute Equation 5 into Equation 4 to give

$$E_n^{(3)} = \left\langle \psi_n^{(1)} \left| \left[\hat{H}^{(1)} - E_n^{(1)} \right] \right| \psi_n^{(1)} \right\rangle + E_n^{(2)} \langle \psi_n^{(1)} \mid \psi_n^{(0)} \rangle$$

$$= \left\langle \psi_n^{(1)} \left| \left[\hat{H}^{(1)} - E_n^{(1)} \right] \right| \psi_n^{(1)} \right\rangle$$

where the relation $\langle \psi_n^{(1)} \mid \psi_n^{(0)} \rangle = 0$ has been applied. Note that the third-order energy depends only on the first-order wave function.

8–46. In this problem we'll calculate the ground-state energy of a particle in a gravitational well (Example 8–4 and Equation 8.68) through second order in perturbation theory using the results of Problem 8–43. From equation 9 of Problem 8–43, we see that the second-order correction to the ground-state ($n = 1$) energy is given by

$$E_1^{(2)} = \sum_{j \neq 1} \frac{[H_{1j}^{(1)}]^2}{E_1^{(0)} - E_j^{(0)}}$$

where

$$H_{1j}^{(1)} = \left\langle \psi_1^{(0)} \left| \frac{V_0 x}{a} \right| \psi_j^{(0)} \right\rangle$$

and $\psi_k^{(0)} = (2/a)^{1/2} \sin k\pi x/a$. Show that

$$H_{1j}^{(1)} = \frac{2V_0}{a^2} \int_0^a dx \, x \sin \frac{\pi x}{a} \sin \frac{j\pi x}{a} = \begin{cases} -\dfrac{8jV_0}{\pi^2(1-j^2)^2} & j \text{ even} \\ 0 & j \text{ odd} \end{cases}$$

for $j \geq 2$. Now show that

$$E_1^{(2)} = \frac{64V_0^2}{\pi^4} \frac{8ma^2}{h^2} \sum_{\substack{j \geq 2 \\ (j \text{ even})}}^{\infty} \frac{j^2}{(1-j^2)^5}$$

$$= \frac{64v_0^2}{\pi^4} \frac{h^2}{8ma^2} \sum_{\substack{j \geq 2 \\ (j \text{ even})}}^{\infty} \frac{j^2}{(1-j^2)^5}$$

where $v_0 = 8mV_0a^2/h^2$.

Show that the energy through second order is

$$\varepsilon = \varepsilon_1^{(0)} + \varepsilon_1^{(1)} + \varepsilon_1^{(2)} = 1 + \frac{v_0}{2} + v_0^2 \left[\frac{64}{\pi^4} \sum_{\substack{j \geq 2 \\ (j \text{ even})}}^{\infty} \frac{j^2}{(1-j^2)^5} \right]$$

where $\varepsilon = 8mEa^2/h^2$.

The summation here converges very rapidly; two terms give 0.01648, which is accurate to four significant figures. Therefore,

$$\varepsilon = 1 + \frac{v_0}{2} + 0.01083 \, v_0^2 + O(v_0^3)$$

Compare this result to that obtained in Problem 8–34. Comment on the comparison. The following figure compares this result to the exact energy as a function of v_0. Note that the two sets of values agree for small values of v_0, but diverge as v_0 increases.

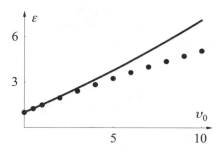

To determine the second-order correction to the ground-state energy, first evaluate $H_{1j}^{(1)}$:

$$H_{1j}^{(1)} = \int_0^a \psi_1^{(0)} \left(V_0 \frac{x}{a} \right) \psi_j^{(0)} \, dx$$

$$= \frac{2V_0}{a^2} \int_0^a x \sin \frac{\pi x}{a} \sin \frac{j\pi x}{a} \, dx$$

$$= \frac{V_0}{a^2} \int_0^a x \left[\cos \frac{(j-1)\pi x}{a} - \cos \frac{(j+1)\pi x}{a} \right] dx$$

$$= \frac{V_0}{a^2} \left\{ \frac{a^2}{(j-1)^2\pi^2} \cos \frac{(j-1)\pi x}{a} + \frac{ax}{(j-1)\pi} \sin \frac{(j-1)\pi x}{a} \right.$$

$$\left. - \left[\frac{a^2}{(j+1)^2\pi^2} \cos \frac{(j+1)\pi x}{a} + \frac{ax}{(j+1)\pi} \sin \frac{(j+1)\pi x}{a} \right] \right\} \Bigg|_0^a$$

$$= \frac{V_0}{a^2} \left\{ \frac{a^2}{(j-1)^2\pi^2} \cos(j-1)\pi + \frac{ax}{(j-1)\pi} \sin(j-1)\pi - \frac{a^2}{(j+1)^2\pi^2} \cos(j+1)\pi \right.$$

$$\left. - \frac{ax}{(j+1)\pi} \sin(j+1)\pi - \left[\frac{a^2}{(j-1)^2\pi^2} - \frac{a^2}{(j+1)^2\pi^2} \right] \right\}$$

If $j \geq 2$ and is even,

$$\cos(j-1)\pi = \cos(j+1)\pi = -1$$

$$\sin(j-1)\pi = \sin(j+1)\pi = 0$$

Therefore,

$$H_{1j}^{(1)} = \frac{V_0}{a^2} \left\{ -\frac{a^2}{(j-1)^2\pi^2} + \frac{a^2}{(j+1)^2\pi^2} - \left[\frac{a^2}{(j-1)^2\pi^2} - \frac{a^2}{(j+1)^2\pi^2} \right] \right\}$$

$$= \frac{2V_0}{\pi^2} \left[\frac{1}{(j+1)^2} - \frac{1}{(j-1)^2} \right]$$

$$= -\frac{8jV_0}{\pi^2(1-j^2)^2}$$

If $j \geq 2$ and is odd,

$$\cos(j-1)\pi = \cos(j+1)\pi = 1$$

$$\sin(j-1)\pi = \sin(j+1)\pi = 0$$

and

$$H_{1j}^{(1)} = \frac{V_0}{a^2} \left\{ \frac{a^2}{(j-1)^2\pi^2} - \frac{a^2}{(j+1)^2\pi^2} - \left[\frac{a^2}{(j-1)^2\pi^2} - \frac{a^2}{(j+1)^2\pi^2} \right] \right\}$$

$$= 0$$

The difference between $E_1^{(0)}$ and $E_j^{(0)}$ is

$$E_1^{(0)} - E_j^{(0)} = \frac{h^2}{8ma^2} - \frac{j^2 h^2}{8ma^2} = \frac{(1-j^2)h^2}{8ma^2}$$

Combining the above expressions, the second-order correction to the ground-state energy is

$$E_1^{(2)} = \sum_{j \neq 1} \frac{[H_{1j}^{(1)}]^2}{E_1^{(0)} - E_j^{(0)}}$$

$$= \sum_{\substack{j \geq 2 \\ (j \text{ even})}}^{\infty} \frac{\{-8jV_0/[\pi^2(1-j^2)^2]\}^2}{(1-j^2)h^2/(8ma^2)}$$

$$= \frac{64V_0^2}{\pi^4} \frac{8ma^2}{h^2} \sum_{\substack{j \geq 2 \\ (j \text{ even})}}^{\infty} \frac{j^2}{(1-j^2)^5}$$

$$= \frac{64v_0^2}{\pi^4} \frac{h^2}{8ma^2} \sum_{\substack{j \geq 2 \\ (j \text{ even})}}^{\infty} \frac{j^2}{(1-j^2)^5}$$

$$\epsilon_1^{(2)} = \frac{64v_0^2}{\pi^4} \sum_{\substack{j \geq 2 \\ (j \text{ even})}}^{\infty} \frac{j^2}{(1-j^2)^5}$$

where $v_0 = 8mV_0a^2/h^2$ and $\epsilon = 8mEa^2/h^2$.

Combining the present result with Equation 8.68 which describes the ground-state energy through first order, the expression for the ground-state energy through second order is

$$\epsilon = 1 + \frac{v_0}{2} + v_0^2 \left[\frac{64}{\pi^4} \sum_{\substack{j \geq 2 \\ (j \text{ even})}}^{\infty} \frac{j^2}{(1-j^2)^5} \right]$$

Taking just the first two terms of the sum gives

$$\epsilon = 1 + \frac{v_0}{2} - 0.010\,83 v_0^2 + O(v_0^3)$$

The variational result (see Problem 8–34) is

$$\epsilon = 1 + \frac{v_0}{2} - \frac{256}{243\pi^4} v_0^2 + O(v_0^3)$$

$$= 1 + \frac{v_0}{2} - 0.010\,82 v_0^2 + O(v_0^3)$$

The two approximation methods are comparable.

Matrix Eigenvalue Problems

PROBLEMS AND SOLUTIONS

H–1. Determine the eigenvalues and eigenvectors of $A = \begin{pmatrix} 1 & 1 \\ 1 & 1 \end{pmatrix}$.

The eigenvalues are found by setting the determinant of $A - \lambda I$ equal to zero,

$$\begin{vmatrix} 1 - \lambda & 1 \\ 1 & 1 - \lambda \end{vmatrix} = (1 - \lambda)^2 - 1 = \lambda^2 - 2\lambda = 0$$

which gives $\lambda = 2$ and 0.

The equations for the eigenvectors are

$$(1 - \lambda)\, c_1 + c_2 = 0 \tag{1}$$
$$c_1 + (1 - \lambda)\, c_2 = 0 \tag{2}$$

Substituting $\lambda = 2$ into these equations gives

$$-c_1 + c_2 = 0$$
$$c_1 - c_2 = 0$$

Thus, $c_2 = c_1$. The eigenvector associated with $\lambda = 2$ can be taken as

$$\begin{pmatrix} 1 \\ 1 \end{pmatrix}$$

or, if the eigenvector is required to be normalized, that is, $c_1^2 + c_2^2 = 1$, it becomes

$$\begin{pmatrix} 1/\sqrt{2} \\ 1/\sqrt{2} \end{pmatrix}$$

The other eigenvector is obtained by substituting $\lambda = 0$ into Equations 1 and 2, both give

$$c_1 + c_2 = 0$$

Thus, $c_2 = -c_1$. The eigenvector associated with $\lambda = 0$ can be taken as

$$\begin{pmatrix} 1 \\ -1 \end{pmatrix}$$

which upon normalization becomes

$$\begin{pmatrix} 1/\sqrt{2} \\ -1/\sqrt{2} \end{pmatrix}$$

H–2.   Determine the eigenvalues and eigenvectors of $A = \begin{pmatrix} 1 & -2 \\ -2 & 1 \end{pmatrix}$.

The eigenvalues are found by setting the determinant of $A - \lambda I$ equal to zero,

$$\begin{vmatrix} 1 - \lambda & -2 \\ -2 & 1 - \lambda \end{vmatrix} = (1 - \lambda)^2 - 4 = \lambda^2 - 2\lambda - 3 = (\lambda - 3)(\lambda + 1) = 0$$

which gives $\lambda = 3$ and -1.

The equations for the eigenvectors are

$$(1 - \lambda) c_1 - 2c_2 = 0 \tag{1}$$
$$-2c_1 + (1 - \lambda) c_2 = 0 \tag{2}$$

Substituting $\lambda = 3$ into these equations gives

$$-2c_1 - 2c_2 = 0$$

Thus, $c_2 = -c_1$. The eigenvector associated with $\lambda = 3$ can be taken as

$$\begin{pmatrix} 1 \\ -1 \end{pmatrix}$$

or, if the eigenvector is required to be normalized, that is, $c_1^2 + c_2^2 = 1$, it becomes

$$\begin{pmatrix} 1/\sqrt{2} \\ -1/\sqrt{2} \end{pmatrix}$$

The other eigenvector is obtained by substituting $\lambda = -1$ into Equations 1 and 2 to give

$$2c_1 - 2c_2 = 0$$
$$-2c_1 + 2c_2 = 0$$

Thus, $c_2 = c_1$. The eigenvector associated with $\lambda = -1$ can be taken as

$$\begin{pmatrix} 1 \\ 1 \end{pmatrix}$$

which upon normalization becomes

$$\begin{pmatrix} 1/\sqrt{2} \\ 1/\sqrt{2} \end{pmatrix}$$

H–3. Determine the eigenvalues and eigenvectors of $A = \begin{pmatrix} 1 & 0 & 1 \\ 0 & 1 & 0 \\ 1 & 0 & 0 \end{pmatrix}$.

The eigenvalues are found by setting the determinant of $A - \lambda I$ equal to 0,

$$\begin{vmatrix} 1-\lambda & 0 & 1 \\ 0 & 1-\lambda & 0 \\ 1 & 0 & -\lambda \end{vmatrix} = 0$$

and expanding the determinant about the second column:

$$(1-\lambda) \begin{vmatrix} 1-\lambda & 1 \\ 1 & -\lambda \end{vmatrix} = (1-\lambda)[-\lambda(1-\lambda)-1] = \lambda^3 - 2\lambda^2 + 1 = 0$$

which gives $\lambda = (1+\sqrt{5})/2$, 1, and $(1-\sqrt{5})/2$.

The equations for the eigenvectors are

$$(1-\lambda)c_1 + c_3 = 0 \tag{1}$$

$$(1-\lambda)c_2 = 0 \tag{2}$$

$$c_1 - \lambda c_3 = 0 \tag{3}$$

Substituting $\lambda = (1+\sqrt{5})/2$ into these equations gives

$$\frac{1-\sqrt{5}}{2}c_1 + c_3 = 0$$

$$\frac{1-\sqrt{5}}{2}c_2 = 0$$

$$c_1 - \frac{1+\sqrt{5}}{2}c_3 = 0$$

which in turn give

$$c_2 = 0$$

$$c_1 = \frac{1+\sqrt{5}}{2}c_3$$

The eigenvector can be taken as

$$\begin{pmatrix} (1+\sqrt{5})/2 \\ 0 \\ 1 \end{pmatrix}$$

or, if the eigenvector is required to be normalized, that is, $c_1^2 + c_2^2 + c_3^2 = 1$, it becomes

$$\left(\frac{2}{5+\sqrt{5}}\right)^{1/2} \begin{pmatrix} (1+\sqrt{5})/2 \\ 0 \\ 1 \end{pmatrix}$$

The second eigenvector is obtained by substituting $\lambda = 1$ into Equations 1, 2, and 3 to give

$$c_3 = 0$$

$$0 \cdot c_2 = 0$$

$$c_1 - c_3 = 0$$

Thus, $c_1 = c_3 = 0$ and c_2 can have any value. Taking $c_2 = 1$, the eigenvector is

$$\begin{pmatrix} 0 \\ 1 \\ 0 \end{pmatrix}$$

and is normalized.

The last eigenvector is obtained by substituting $\lambda = (1 - \sqrt{5})/2$ into Equations 1, 2, and 3 to give

$$\frac{1+\sqrt{5}}{2}c_1 + c_3 = 0$$

$$\frac{1+\sqrt{5}}{2}c_2 = 0$$

$$c_1 - \frac{1-\sqrt{5}}{2}c_3 = 0$$

which in turn give

$$c_2 = 0$$

$$c_1 = \frac{1-\sqrt{5}}{2}c_3$$

The eigenvector can be taken as

$$\begin{pmatrix} (1-\sqrt{5})/2 \\ 0 \\ 1 \end{pmatrix}$$

which upon nomalization becomes

$$\left(\frac{2}{5-\sqrt{5}}\right)^{1/2} \begin{pmatrix} (1-\sqrt{5})/2 \\ 0 \\ 1 \end{pmatrix}$$

H–4. Determine the eigenvalues and eigenvectors of $A = \begin{pmatrix} 1 & 0 & -1 \\ 0 & 1 & 0 \\ -1 & 0 & 1 \end{pmatrix}$.

The eigenvalues are found by setting the determinant of $A - \lambda I$ equal to 0,

$$\begin{vmatrix} 1-\lambda & 0 & -1 \\ 0 & 1-\lambda & 0 \\ -1 & 0 & 1-\lambda \end{vmatrix} = 0$$

and expanding the determinant about the second column:

$$(1-\lambda)\begin{vmatrix} 1-\lambda & -1 \\ -1 & 1-\lambda \end{vmatrix} = (1-\lambda)\left[(1-\lambda)^2 - 1\right] = 0$$

which gives $\lambda = 2$, 1, and 0.

The equations for the eigenvectors are

$$(1-\lambda)\,c_1 - c_3 = 0 \tag{1}$$

$$(1-\lambda)\,c_2 = 0 \tag{2}$$

$$-c_1 + (1-\lambda)\,c_3 = 0 \tag{3}$$

Substituting $\lambda = 2$ into these equations gives

$$-c_1 - c_3 = 0$$

$$-c_2 = 0$$

$$-c_1 - c_3 = 0$$

The eigenvector can be taken as

$$\begin{pmatrix} 1 \\ 0 \\ -1 \end{pmatrix}$$

or, if the eigenvector is required to be normalized, that is, $c_1^2 + c_2^2 + c_3^2 = 1$, it becomes

$$\frac{1}{\sqrt{2}}\begin{pmatrix} 1 \\ 0 \\ -1 \end{pmatrix}$$

The second eigenvector is obtained by substituting $\lambda = 1$ into Equations 1, 2, and 3 to give

$$c_3 = 0$$

$$0 \cdot c_2 = 0$$

$$c_1 = 0$$

Thus, c_2 can have any value. Taking $c_2 = 1$, the eigenvector is

$$\begin{pmatrix} 0 \\ 1 \\ 0 \end{pmatrix}$$

and is normalized.

The last eigenvector is obtained by substituting $\lambda = 0$ into Equations 1, 2, and 3 to give

$$c_1 - c_3 = 0$$

$$c_2 = 0$$

$$-c_1 + c_3 = 0$$

The eigenvector can be taken as

$$\begin{pmatrix} 1 \\ 0 \\ 1 \end{pmatrix}$$

which upon normalization becomes

$$\frac{1}{\sqrt{2}} \begin{pmatrix} 1 \\ 0 \\ 1 \end{pmatrix}$$

H–5. Show that the matrix $\mathsf{A} = \begin{pmatrix} 1 & i & 1-i \\ -i & 0 & -1+i \\ 1+i & -1-i & 3 \end{pmatrix}$ is Hermitian.

The matrix A is Hermitian if $A_{ij} = A_{ji}^*$. For the diagonal elements, this requirement becomes $A_{ii} = A_{ii}^*$, which can be satisfied only if A_{ii} is real, and is indeed the case here. For the off-diagonal elements,

$$A_{21}^* = (-i)^* = i = A_{12}$$

$$A_{31}^* = (1+i)^* = 1-i = A_{13}$$

$$A_{32}^* = (-1-i)^* = -1+i = A_{23}$$

Thus, A is Hermitian.

H–6. Verify that $(\lambda_1 c_1, \lambda_2 c_2, \ldots, \lambda_N c_N) = \mathsf{SD}$ in Equation H.16.

$$\mathsf{SD} = (c_1 \quad c_2 \quad \cdots \quad c_N) \begin{pmatrix} \lambda_1 & 0 & 0 & \cdots & 0 \\ 0 & \lambda_2 & 0 & \cdots & 0 \\ \vdots & \vdots & \vdots & \ddots & \vdots \\ 0 & 0 & 0 & \cdots & \lambda_N \end{pmatrix} = (\lambda_1 c_1 \quad \lambda_2 c_2 \quad \cdots \quad \lambda_N c_N)$$

H–7. The three eigenvectors of A in Problem H–4 are $c_1(-1, 0, 1)$, $c_2(0, 1, 0)$, and $c_3(1, 0, 1)$, where c_1, c_2, and c_3 are arbitrary. Choose them so that the three eigenvectors are normalized. Now form the matrix S whose columns consist of the three normalized eigenvectors. Find the inverse of S and then show explicitly that $S^{-1} = S^T$, or that S is indeed orthogonal.

An eigenvector is normalized when the sum of the squares of each component in the eigenvector equals 1. Thus, choosing positive values for the normalization constants, the three normalized vectors are

$$\frac{1}{\sqrt{2}} \begin{pmatrix} -1 \\ 0 \\ 1 \end{pmatrix}, \quad \begin{pmatrix} 0 \\ 1 \\ 0 \end{pmatrix}, \quad \frac{1}{\sqrt{2}} \begin{pmatrix} 1 \\ 0 \\ 1 \end{pmatrix}$$

The matrix S is then

$$S = \begin{pmatrix} -1/\sqrt{2} & 0 & 1/\sqrt{2} \\ 0 & 1 & 0 \\ 1/\sqrt{2} & 0 & 1/\sqrt{2} \end{pmatrix}$$

The matrix S^{-1} is determined by forming the transpose of the matrix of cofactors of S and dividing by the determinant of S. The cofactors of S are

$$\begin{array}{lll} S_{11} = 1/\sqrt{2} & S_{12} = 0 & S_{13} = -1/\sqrt{2} \\ S_{21} = 0 & S_{22} = -1 & S_{23} = 0 \\ S_{31} = -1/\sqrt{2} & S_{32} = 0 & S_{33} = -1/\sqrt{2} \end{array}$$

The determinant of S is

$$|S| = 1 \begin{vmatrix} -1/\sqrt{2} & 1/\sqrt{2} \\ 1/\sqrt{2} & 1/\sqrt{2} \end{vmatrix} = -\frac{1}{2} - \frac{1}{2} = -1$$

Therefore,

$$S^{-1} = -1 \begin{pmatrix} 1/\sqrt{2} & 0 & -1/\sqrt{2} \\ 0 & -1 & 0 \\ -1/\sqrt{2} & 0 & -1/\sqrt{2} \end{pmatrix} = \begin{pmatrix} -1/\sqrt{2} & 0 & 1/\sqrt{2} \\ 0 & 1 & 0 \\ 1/\sqrt{2} & 0 & 1/\sqrt{2} \end{pmatrix} = S^T$$

The matrix S is indeed orthogonal.

H–8. Diagonalize the matrix in Problem H–1.

The columns of the matrix S are given by the normalized eigenvectors of A:

$$\begin{pmatrix} 1/\sqrt{2} & 1/\sqrt{2} \\ 1/\sqrt{2} & -1/\sqrt{2} \end{pmatrix}$$

Since S is orthogonal, $S^{-1} = S^T$. To diagonalize A:

$$D = S^{-1}AS = S^TAS$$

$$= \begin{pmatrix} 1/\sqrt{2} & 1/\sqrt{2} \\ 1/\sqrt{2} & -1/\sqrt{2} \end{pmatrix} \begin{pmatrix} 1 & 1 \\ 1 & 1 \end{pmatrix} \begin{pmatrix} 1/\sqrt{2} & 1/\sqrt{2} \\ 1/\sqrt{2} & -1/\sqrt{2} \end{pmatrix}$$

$$= \begin{pmatrix} 1/\sqrt{2} & 1/\sqrt{2} \\ 1/\sqrt{2} & -1/\sqrt{2} \end{pmatrix} \begin{pmatrix} \sqrt{2} & 0 \\ \sqrt{2} & 0 \end{pmatrix}$$

$$= \begin{pmatrix} 2 & 0 \\ 0 & 0 \end{pmatrix}$$

The diagonal elements are the eigenvalues of A.

H–9. Diagonalize the matrix in Problem H–2.

The columns of the matrix S are given by the normalized eigenvectors of A:

$$\begin{pmatrix} 1/\sqrt{2} & 1/\sqrt{2} \\ -1/\sqrt{2} & 1/\sqrt{2} \end{pmatrix}$$

Since S is orthogonal, $S^{-1} = S^T$. To diagonalize A:

$$D = S^{-1}AS = S^TAS$$

$$= \begin{pmatrix} 1/\sqrt{2} & -1/\sqrt{2} \\ 1/\sqrt{2} & 1/\sqrt{2} \end{pmatrix} \begin{pmatrix} 1 & -2 \\ -2 & 1 \end{pmatrix} \begin{pmatrix} 1/\sqrt{2} & 1/\sqrt{2} \\ -1/\sqrt{2} & 1/\sqrt{2} \end{pmatrix}$$

$$= \begin{pmatrix} 1/\sqrt{2} & 1/\sqrt{2} \\ 1/\sqrt{2} & -1/\sqrt{2} \end{pmatrix} \begin{pmatrix} 1/\sqrt{2} + \sqrt{2} & 1/\sqrt{2} - \sqrt{2} \\ -\sqrt{2} - 1/\sqrt{2} & -\sqrt{2} + 1/\sqrt{2} \end{pmatrix}$$

$$= \begin{pmatrix} 3 & 0 \\ 0 & -1 \end{pmatrix}$$

The diagonal elements are the eigenvalues of A.

H–10. Diagonalize the matrix in Problem H–3.

The columns of the matrix S are given by the normalized eigenvectors of A:

$$\begin{pmatrix} (1+\sqrt{5})/[2(5+\sqrt{5})]^{1/2} & 0 & (1-\sqrt{5})/[2(5-\sqrt{5})]^{1/2} \\ 0 & 1 & 0 \\ \left[2/(5+\sqrt{5})\right]^{1/2} & 0 & \left[2/(5-\sqrt{5})\right]^{1/2} \end{pmatrix}$$

Since S is orthogonal, $S^{-1} = S^T$. To diagonalize A:

$$D = S^{-1}AS = S^TAS$$

$$= \begin{pmatrix} (1+\sqrt{5})/[2(5+\sqrt{5})]^{1/2} & 0 & \left[2/(5+\sqrt{5})\right]^{1/2} \\ 0 & 1 & 0 \\ (1-\sqrt{5})/[2(5-\sqrt{5})]^{1/2} & 0 & \left[2/(5-\sqrt{5})\right]^{1/2} \end{pmatrix}$$

$$\times \begin{pmatrix} 1 & 0 & 1 \\ 0 & 1 & 0 \\ 1 & 0 & 0 \end{pmatrix} \begin{pmatrix} (1+\sqrt{5})/[2(5+\sqrt{5})]^{1/2} & 0 & (1-\sqrt{5})/[2(5-\sqrt{5})]^{1/2} \\ 0 & 1 & 0 \\ \left[2/(5+\sqrt{5})\right]^{1/2} & 0 & \left[2/(5-\sqrt{5})\right]^{1/2} \end{pmatrix}$$

$$= \begin{pmatrix} (1+\sqrt{5})/[2(5+\sqrt{5})]^{1/2} & 0 & \left[2/(5+\sqrt{5})\right]^{1/2} \\ 0 & 1 & 0 \\ (1-\sqrt{5})/[2(5-\sqrt{5})]^{1/2} & 0 & \left[2/(5-\sqrt{5})\right]^{1/2} \end{pmatrix}$$

$$\times \begin{pmatrix} (3+\sqrt{5})/[2(5+\sqrt{5})]^{1/2} & 0 & (3-\sqrt{5})/[2(5-\sqrt{5})]^{1/2} \\ 0 & 1 & 0 \\ (1+\sqrt{5})/\left[2(5+\sqrt{5})\right]^{1/2} & 0 & (1+\sqrt{5})/\left[2(5-\sqrt{5})\right]^{1/2} \end{pmatrix}$$

$$= \begin{pmatrix} (1+\sqrt{5})/2 & 0 & 0 \\ 0 & 1 & 0 \\ 0 & 0 & (1-\sqrt{5})/2 \end{pmatrix}$$

The diagonal elements are the eigenvalues of A.

H–11. Diagonalize the matrix in Problem H–4.

The matrix S was determined in Problem H–7. Note that the first column of this matrix differs from the first eigenvector determined in Problem H–4 by an overall sign, which does not affect the outcome of the diagonalization. The matrix S is:

$$\begin{pmatrix} -1/\sqrt{2} & 0 & 1/\sqrt{2} \\ 0 & 1 & 0 \\ 1/\sqrt{2} & 0 & 1/\sqrt{2} \end{pmatrix}$$

Since S is orthogonal, $S^{-1} = S^T$. To diagonalize A:

$$D = S^{-1}AS = S^TAS$$

$$= \begin{pmatrix} -1/\sqrt{2} & 0 & 1/\sqrt{2} \\ 0 & 1 & 0 \\ 1/\sqrt{2} & 0 & 1/\sqrt{2} \end{pmatrix} \begin{pmatrix} 1 & 0 & -1 \\ 0 & 1 & 0 \\ -1 & 0 & 1 \end{pmatrix} \begin{pmatrix} -1/\sqrt{2} & 0 & 1/\sqrt{2} \\ 0 & 1 & 0 \\ 1/\sqrt{2} & 0 & 1/\sqrt{2} \end{pmatrix}$$

$$= \begin{pmatrix} -1/\sqrt{2} & 0 & 1/\sqrt{2} \\ 0 & 1 & 0 \\ 1/\sqrt{2} & 0 & 1/\sqrt{2} \end{pmatrix} \begin{pmatrix} -\sqrt{2} & 0 & 0 \\ 0 & 1 & 0 \\ \sqrt{2} & 0 & 0 \end{pmatrix}$$

$$= \begin{pmatrix} 2 & 0 & 0 \\ 0 & 1 & 0 \\ 0 & 0 & 0 \end{pmatrix}$$

The diagonal elements are the eigenvalues of **A**.

H–12.   If $\mathbf{D} = \mathbf{S}^{-1}\mathbf{AS}$, where **S** is an orthogonal matrix, show that $\text{Tr }\mathbf{D} = \text{Tr }\mathbf{A}$.

Since

$$D_{il} = \sum_j \sum_k S_{ij}^{-1} A_{jk} S_{kl}$$

The trace of **D** is

$$\sum_i D_{ii} = \sum_i \sum_j \sum_k S_{ij}^{-1} A_{jk} S_{ki}$$

$$= \sum_j \sum_k A_{jk} \left(\sum_i S_{ij}^{-1} S_{ki} \right)$$

$$= \sum_j \sum_k A_{jk} \delta_{jk}$$

$$= \sum_j A_{jj}$$

The orthogonal nature of **S**, $\sum_i S_{ij}^{-1} S_{ki} = \delta_{jk}$, has been applied.

H–13.   Programs such as *MathCad* and *Mathematica* can find the eigenvalues and corresponding eigenvectors of large matrices in seconds. Use one of these programs to find the eigenvalues and corresponding eigenvectors of

$$\mathbf{A} = \begin{pmatrix} a & 1 & 0 & 0 & 0 & 1 \\ 1 & a & 1 & 0 & 0 & 0 \\ 0 & 1 & a & 1 & 0 & 0 \\ 0 & 0 & 1 & a & 1 & 0 \\ 0 & 0 & 0 & 1 & a & 1 \\ 1 & 0 & 0 & 0 & 1 & a \end{pmatrix}$$

The eigenvalues are $-2 + a$, $-1 + a$, $-1 + a$, $1 + a$. $1 + a$, $2 + a$. The corresponding eigenvectors are respectively

$$\begin{pmatrix} -1 \\ 1 \\ -1 \\ 1 \\ -1 \\ 1 \end{pmatrix} \quad \begin{pmatrix} -1 \\ 0 \\ 1 \\ -1 \\ 0 \\ 1 \end{pmatrix} \quad \begin{pmatrix} -1 \\ 1 \\ 0 \\ -1 \\ 1 \\ 0 \end{pmatrix} \quad \begin{pmatrix} 1 \\ 0 \\ -1 \\ -1 \\ 0 \\ 1 \end{pmatrix} \quad \begin{pmatrix} -1 \\ -1 \\ 0 \\ 1 \\ 1 \\ 0 \end{pmatrix} \quad \begin{pmatrix} 1 \\ 1 \\ 1 \\ 1 \\ 1 \\ 1 \end{pmatrix}$$

Many-Electron Atoms

PROBLEMS AND SOLUTIONS

9–1. Show that the atomic unit of energy can be written as

$$E_h = \frac{\hbar^2}{m_e a_0^2} = \frac{e^2}{4\pi\epsilon_0 a_0} = \frac{m_e e^4}{16\pi^2\epsilon_0^2\hbar^2}$$

The atomic unit for energy is defined in Table 9.1 as

$$E_h = \frac{m_e e^4}{16\pi^2\epsilon_0^2\hbar^2}$$

Using $a_0 = 4\pi\epsilon_0\hbar^2/m_e e^2$, this becomes,

$$E_h = \frac{e^2}{4\pi\epsilon_0}\left(\frac{m_e e^2}{4\pi\epsilon_0\hbar^2}\right) = \frac{e^2}{4\pi\epsilon_0 a_0} = \frac{e^2}{4\pi\epsilon_0 a_0}\left(\frac{1}{a_0}\right)\left(\frac{4\pi\epsilon_0\hbar^2}{m_e e^2}\right) = \frac{\hbar^2}{m_e a_0^2}$$

9–2. Show that the energy of a helium ion in atomic units is $-2\,E_h$.

From Problem 7–29 with $Z = 2$, the energy of a helium atom is

$$E = -\frac{Z^2 m_e e^4}{8\epsilon_0^2 h^2} = -\frac{m_e e^4}{8\pi^2\epsilon_0^2\hbar^2} = -2E_h$$

9–3. Show that the speed of an electron in the first Bohr orbit is $e^2/4\pi\epsilon_0\hbar = 2.188 \times 10^6\ \text{m}\cdot\text{s}^{-1}$. This speed is the unit of speed in atomic units (see Table 9.1).

In Problem 1–35, the speed of an electron in the nth Bohr orbit was found to be

$$v = \frac{e^2}{2\epsilon_0 n h}$$

which for $n = 1$ becomes

$$v = \frac{e^2}{2\epsilon_0 h} = \frac{e^2}{4\pi\epsilon_0 \hbar} = 2.188 \times 10^6 \text{ m}\cdot\text{s}^{-1}$$

9–4. Show that the speed of light is equal to 137 in atomic units. (*Hint*: Use the result of the previous problem.)

From the previous problem, the atomic unit of speed is 2.188×10^6 m·s^{-1}. Thus, in atomic units,

$$c = \frac{2.997\,924\,58 \times 10^8 \text{ m}\cdot\text{s}^{-1}}{2.188 \times 10^6 \text{ m}\cdot\text{s}^{-1}} = 137.02$$

9–5. Another way to introduce atomic units is to express mass as multiples of m_e, the mass of an electron (instead of kg); charge as multiples of e, the protonic charge (instead of C); angular momentum as multiples of $\hbar$ (instead of in J·s = kg·m^2·s^{-1}); and permittivity as multiples of $4\pi\epsilon_0$ (instead of in C^2·s^2·kg^{-1}·m^{-3}). This conversion can be achieved in all of our equations by letting $m_e = e = \hbar = 4\pi\epsilon_0 = 1$. Show that this procedure is consistent with the definition of atomic units used in the chapter.

The atomic units of mass, charge, angular momentum, and permittivity given in Table 9.1 are trivially equal to one when the substitutions $m_e = e = \hbar = 4\pi\epsilon_0 = 1$ are made. This leaves only the atomic units for length and energy. For these two units, use the definitions from Table 9.1 and make the substitutions, $m_e = e = \hbar = 4\pi\epsilon_0 = 1$ to find

$$E_h = \frac{m_e e^4}{16\pi^2 \epsilon_0^2 \hbar^2} = \frac{1 \cdot 1^4}{1^2 \cdot 1^2} = 1$$

$$a_0 = \frac{4\pi\epsilon_0 \hbar^2}{m_e e^2} = \frac{1 \cdot 1^2}{1 \cdot 1^2} = 1$$

9–6. Derive Equation 9.5 from Equation 9.4. Be sure to remember that ∇^2 has units of (distance)$^{-2}$.

A quick way to solve this problem is to take the approach of the previous one and simply make the substitutions $m_e = e = \hbar = 4\pi\varepsilon_0 = 1$ into Equation 9.4 to give Equation 9.5:

$$\hat{H} = -\frac{\hbar^2}{2m_e}\nabla_1^2 - \frac{\hbar^2}{2m_e}\nabla_2^2 - \frac{2e^2}{4\pi\epsilon_0 r_1} - \frac{2e^2}{4\pi\epsilon_0 r_2} + \frac{e^2}{4\pi\epsilon_0 r_{12}}$$

$$= -\frac{1}{2}\nabla_1^2 - \frac{1}{2}\nabla_2^2 - \frac{2}{r_1} - \frac{2}{r_2} + \frac{1}{r_{12}}$$

Another method is to note that if x is a distance in meters, then the same distance in atomic units is $x' = x/a_0$. Thus, $1/x = 1/(a_0 x')$, and

$$\frac{e^2}{4\pi\epsilon_0 x} = \frac{e^2}{4\pi\epsilon_0 a_0 x'} = \frac{E_h}{x'}$$

This substitution can be used to express the three potential energy terms of the Hamiltonian operator in atomic units. Furthermore, application of the chain rule twice gives

$$\frac{d^2}{dx^2} = \left(\frac{dx'}{dx}\right)^2 \frac{d^2}{dx'^2} = \frac{1}{a_0^2}\frac{d^2}{dx'^2}$$

This means that transforming to atomic units turns ∇^2 into ∇^2/a_0^2, and the kinetic energy terms in atomic units are,

$$-\frac{\hbar^2}{2m_e a_0^2}\nabla^2 = -\frac{E_h}{2}\nabla^2$$

Thus,

$$\hat{H} = -\frac{\hbar^2}{2m_e}\nabla_1^2 - \frac{\hbar^2}{2m_e}\nabla_2^2 - \frac{2e^2}{4\pi\epsilon_0 r_1} - \frac{2e^2}{4\pi\epsilon_0 r_2} + \frac{e^2}{4\pi\epsilon_0 r_{12}}$$

$$= \left(-\frac{1}{2}\nabla_1^2 - \frac{1}{2}\nabla_2^2 - \frac{2}{r_1} - \frac{2}{r_2} + \frac{1}{r_{12}}\right)E_h$$

which upon dividing by E_h, that is, converting to atomic units for energy, gives the desired result.

9–7. In this problem, we shall show that Equation 9.10 has the form of a sum of a quadratic term in ζ and a linear term in ζ because of the form of the Hamiltonian operator of an atom. Start with $\hat{H}$ for an N-electron atom with nuclear charge Z:

$$\hat{H} = -\frac{1}{2}\sum_{j=1}^{N}\nabla_j^2 - \sum_{j=1}^{N}\frac{Z}{r_j} + \sum_{j=1}^{N}\sum_{i<j}^{N}\frac{1}{r_{ij}} = \hat{T} + \hat{V}$$

where $\hat{T}$ is the kinetic energy operator and $\hat{V}$ is the potential energy operator. Write out the expressions for $T = \langle\psi \mid \hat{T} \mid \psi\rangle/\langle\psi \mid \psi\rangle$ and $V = \langle\psi \mid \hat{V} \mid \psi\rangle/\langle\psi \mid \psi\rangle$ using a wave function of the form $\psi(x_1, y_1, \ldots, y_N, z_N)$. Now write out the expressions for $T(\zeta) = \langle\psi(\zeta) \mid \hat{T} \mid \psi(\zeta)\rangle/\langle\psi(\zeta) \mid \psi(\zeta)\rangle$ using a wave function of the form $\psi(\zeta x_1, \zeta y_2, \ldots, \zeta y_N, \zeta z_N)$, where all the coordinates are scaled by a factor of ζ. Let $x_1' = \zeta x_1$, $y_1' = \zeta y_1$, and so on to show that

$$T(\zeta) = \zeta^2 T(\zeta = 1)$$

Similarly, write out the expression for $V(\zeta) = \langle\psi(\zeta) \mid \hat{V} \mid \psi(\zeta)\rangle/\langle\psi(\zeta) \mid \psi(\zeta)\rangle$ and then show that

$$V(\zeta) = \zeta V(\zeta = 1)$$

Two results from elementary calculus are helpful in working this problem. First if $x' = \zeta x$, then

$$\frac{d}{dx} = \left(\frac{dx'}{dx}\right)\frac{d}{dx'} = \zeta\frac{d}{dx'}$$

$$\frac{d^2}{dx^2} = \zeta\frac{d}{dx}\left(\frac{d}{dx'}\right) = \zeta\left(\frac{dx'}{dx}\right)\frac{d^2}{dx'^2} = \zeta^2\frac{d^2}{dx'^2}$$

and second, if $I = \int f(x)\,dx$, then

$$\int f(\zeta x)\,dx = \frac{1}{\zeta}\int f(x')\,dx' = \frac{1}{\zeta}I$$

since $dx = dx'/\zeta$. For multiple integrations, each integration over a scaled coordinate gives a factor of ζ^{-1}. Finally, it will be useful to transform a distance r between two points $\mathbf{r_1}$ and $\mathbf{r_2}$ in the unscaled coordinates to the scaled system:

$$r = \left[(x_2 - x_1)^2 + (y_2 - y_1)^2 + (z_2 - z_1)^2\right]^{1/2}$$

$$= \left[\left(\frac{x_2'}{\zeta} - \frac{x_1'}{\zeta}\right)^2 + \left(\frac{y_2'}{\zeta} - \frac{y_1'}{\zeta}\right)^2 + \left(\frac{z_2'}{\zeta} - \frac{z_1'}{\zeta}\right)^2\right]^{1/2}$$

$$= \frac{1}{\zeta}r'$$

Therefore, if the normalization integral, $\langle\psi|\psi\rangle = I$ for $\psi(x_1, y_1, \ldots, y_N, z_N)$, then using a wave function of the form $\psi(\zeta) = \psi(\zeta x_1, \zeta y_2, \ldots, \zeta y_N, \zeta z_N)$ gives

$$\langle\psi(\zeta)|\psi(\zeta)\rangle = \zeta^{-3N}\langle\psi|\psi\rangle = \zeta^{-3N}I$$

because there are $3N$ integrations to perform. Similarly,

$$\left\langle\psi(\zeta)\left|\nabla_j^2\right|\psi(\zeta)\right\rangle = \zeta^{2-3N}\left\langle\psi\left|\nabla_j^2\right|\psi\right\rangle$$

since the second derivatives in ∇^2 give a factor of ζ^2 and the $3N$ integrations give a factor of ζ^{-3N}, and

$$\left\langle\psi(\zeta)\left|\frac{1}{r}\right|\psi(\zeta)\right\rangle = \zeta^{1-3N}\left\langle\psi\left|\frac{1}{r}\right|\psi\right\rangle$$

Starting, then, with the kinetic energy for an N-electron atom with nuclear charge Z:

$$\hat{T} = -\frac{1}{2}\sum_{j=1}^{N}\nabla_j^2$$

The kinetic energy in the scaled system is

$$T(\zeta) = \frac{\langle \psi(\zeta) | \hat{T} | \psi(\zeta) \rangle}{\langle \psi(\zeta) | \psi(\zeta) \rangle}$$

$$= \frac{-\frac{1}{2} \sum_{j=1}^{N} \langle \psi(\zeta) | \nabla_j^2 | \psi(\zeta) \rangle}{\langle \psi(\zeta) | \psi(\zeta) \rangle}$$

$$= \frac{-\frac{\zeta^{2-3N}}{2} \sum_{j=1}^{N} \langle \psi | \nabla_j^2 | \psi \rangle}{\zeta^{-3N} \langle \psi | \psi \rangle}$$

$$= \zeta^2 \frac{\langle \psi | \hat{T} | \psi \rangle}{\langle \psi | \psi \rangle}$$

$$= \zeta^2 T(\zeta = 1)$$

The potential energy operator is

$$\hat{V} = -\sum_{j=1}^{N} \frac{Z}{r_j} + \sum_{j=1}^{N} \sum_{i<j}^{N} \frac{1}{r_{ij}}$$

so that

$$V(\zeta) = \frac{\langle \psi(\zeta) | \hat{V} | \psi(\zeta) \rangle}{\langle \psi(\zeta) | \psi(\zeta) \rangle}$$

$$= \frac{-\sum_{j=1}^{N} \langle \psi(\zeta) | \frac{Z}{r_j} | \psi(\zeta) \rangle + \sum_{j=1}^{N} \sum_{i<j}^{N} \langle \psi(\zeta) | \frac{1}{r_{ij}} | \psi(\zeta) \rangle}{\langle \psi(\zeta) | \psi(\zeta) \rangle}$$

$$= \frac{-\zeta^{1-3N} \sum_{j=1}^{N} \langle \psi | \frac{Z}{r_j} | \psi \rangle + \zeta^{1-3N} \sum_{j=1}^{N} \sum_{i<j}^{N} \langle \psi | \frac{1}{r_{ij}} | \psi \rangle}{\zeta^{-3N} \langle \psi | \psi \rangle}$$

$$= \zeta \frac{\langle \psi | \hat{V} | \psi \rangle}{\langle \psi | \psi \rangle}$$

$$= \zeta V(\zeta = 1)$$

9–8. Show that when we use Equation 9.8 in Equation 9.9, the average kinetic energy and average potential energy come out to be ζ^2 and $-27\zeta/8$, respectively. (Use the results of either Problem 8–39 or 8–40.)

For this problem concerning the helium atom, $N = 2$, and

$$\hat{H} = -\frac{1}{2}\nabla_1^2 - \frac{1}{2}\nabla_2^2 - \frac{Z}{r_1} - \frac{Z}{r_2} + \frac{1}{r_{12}}$$

where the substitution $Z = 2$ will be made later. Since the wave function of Equation 9.8 is normalized, $\langle \psi | \psi \rangle = 1$, and

$$T = \left\langle \psi \left| \hat{T} \right| \psi \right\rangle = \int \int d\mathbf{r}_1 \, d\mathbf{r}_2 \, \psi(\mathbf{r}_1, \mathbf{r}_2) \left(-\frac{1}{2}\nabla_1^2 - \frac{1}{2}\nabla_2^2 \right) \psi(\mathbf{r}_1, \mathbf{r}_2)$$

The wave function $\psi(\mathbf{r}_1, \mathbf{r}_2)$ has no angular dependence. Consequently, derivatives over angular coordinates vanish, and the integrations over the angular coordinates give a factor of $(4\pi)^2$. Thus,

$$
\begin{aligned}
T = -8\zeta^6 &\left[\int_0^\infty dr_1 \, r_1^2 e^{-\zeta r_1} \left(\frac{1}{r_1^2}\frac{d}{dr_1} r_1^2 \frac{d}{dr_1} \right) e^{-\zeta r_1} \int_0^\infty dr_2 \, r_2^2 e^{-2\zeta r_2} \right. \\
&\left. + \int_0^\infty dr_1 \, r_1^2 e^{-2\zeta r_1} \int_0^\infty dr_2 \, r_2^2 e^{-\zeta r_2} \left(\frac{1}{r_2^2}\frac{d}{dr_2} r_2^2 \frac{d}{dr_2} \right) e^{-\zeta r_2} \right] \\
= -2\zeta^3 &\left[\int_0^\infty dr_1 \left(\zeta^2 r_1^2 - 2\zeta r_1 \right) e^{-2\zeta r_1} + \int_0^\infty dr_2 \left(\zeta^2 r_2^2 - 2\zeta r_2 \right) e^{-2\zeta r_2} \right] \\
= -2\zeta^3 &\left(-\frac{1}{4\zeta} - \frac{1}{4\zeta} \right) \\
= \zeta^2 &
\end{aligned}
$$

Similarly,

$$V = \left\langle \psi \left| \hat{V} \right| \psi \right\rangle = \int \int d\mathbf{r}_1 \, d\mathbf{r}_2 \, \psi(\mathbf{r}_1, \mathbf{r}_2) \left(-\frac{Z}{r_1} - \frac{Z}{r_2} + \frac{1}{r_{12}} \right) \psi(\mathbf{r}_1, \mathbf{r}_2)$$

For the first two terms in the potential energy, there is no angular dependence, and the angular integration is once again straightforward. The integral over the electron-electron repulsion term was evaluated in Problem 8–39 or 8–40, and making the correspondence $Z \rightarrow \zeta$ and converting to atomic units gives

$$
\begin{aligned}
V = 16Z\zeta^6 &\left[-\int_0^\infty dr_1 \, r_1 e^{-2\zeta r_1} \int_0^\infty dr_2 \, r_2^2 e^{-2\zeta r_2} - \int_0^\infty dr_1 \, r_1^2 e^{-2\zeta r_1} \int_0^\infty dr_2 \, r_2 e^{-2\zeta r_2} \right] \\
&+ \int \int d\mathbf{r}_1 \, d\mathbf{r}_2 \, \psi(\mathbf{r}_1, \mathbf{r}_2) \left(\frac{1}{r_{12}} \right) \psi(\mathbf{r}_1, \mathbf{r}_2) \\
= 16Z\zeta^6 &\left[-\left(\frac{1}{4\zeta^2} \right)\left(\frac{1}{4\zeta^3} \right) - \left(\frac{1}{4\zeta^3} \right)\left(\frac{1}{4\zeta^2} \right) \right] + \frac{5}{8}\zeta \\
= -2Z\zeta &+ \frac{5}{8}\zeta
\end{aligned}
$$

For helium, $Z = 2$, and $V = -2(2)\zeta + 5\zeta/8 = -27\zeta/8$.

9–9. Show that Equation 9.10 can be written as

$$E(\zeta) = T(\zeta) + V(\zeta) = \zeta^2 T(\zeta = 1) + \zeta V(\zeta = 1)$$

where $T(\zeta = 1)$ and $V(\zeta = 1)$ are the average kinetic energy and the average potential energy, respectively, calculated with Equation 9.8 with $\zeta = 1$ (cf. Problem 9–7).

From the previous problem, when $\zeta = 1$, then $T = 1$ and $V = -27/8$. Therefore,

$$T(\zeta) = \zeta^2 = \zeta^2(1) = \zeta^2 T(\zeta = 1)$$

$$V(\zeta) = -\frac{27}{8}\zeta = \zeta V(\zeta = 1)$$

$$E(\zeta) = T(\zeta) + V(\zeta) = \zeta^2 T(\zeta = 1) + \zeta V(\zeta = 1)$$

9–10. Use the results of the previous two problems to show that $T(\zeta = 1)/V(\zeta = 1) = -8/27$ when you use Equation 9.8. What should this ratio equal according to the virial theorem? Now determine the value of ζ such that the ratio $T(\zeta)/V(\zeta)$ satisfies the virial theorem and compare your result to the value of ζ obtained variationally.

Since $T(\zeta = 1) = 1$ and $V(\zeta = 1) = -27/8$ (see previous two problems), then

$$\frac{T(\zeta = 1)}{V(\zeta = 1)} = \frac{1}{-27/8} = -\frac{8}{27}$$

According to the virial theorem, this ratio should be equal to $-1/2$ (see Problem 7–31). Allowing ζ to vary gives

$$\frac{T(\zeta)}{V(\zeta)} = \frac{\zeta^2}{-27\zeta/8} = -\frac{8}{27}\zeta$$

Satisfying the virial theorem requires that

$$\frac{T(\zeta)}{V(\zeta)} = -\frac{8}{27}\zeta = -\frac{1}{2}$$

$$\zeta = \frac{27}{16}$$

which is the same value of ζ as is obtained variationally.

9–11. Show that the generalization of Equation 9.10 for a two-electron atom or two-electron ion of nuclear charge Z is

$$E(\zeta) = -\zeta^2 + 2\zeta(\zeta - Z) + \frac{5}{8}\zeta$$

Show that this equation reduces to Equation 9.10 when $Z = 2$.

This result was found in the solution to Problem 9–8, since for the two-electron atom, $T(\zeta) = \zeta^2$ independent of Z, and the general result, $V(\zeta) = (5/8 - 2Z)\zeta$, was determined. Thus,

$$E(\zeta) = T(\zeta) + V(\zeta)$$

$$= \zeta^2 - 2Z\zeta + \frac{5}{8}\zeta$$

$$= -\zeta^2 + 2\zeta^2 - 2Z\zeta + \frac{5}{8}\zeta$$

$$= -\zeta^2 + 2\zeta\ (\zeta - Z) + \frac{5}{8}\zeta$$

When $Z = 2$,

$$E(\zeta) = -\zeta^2 + 2\zeta\ (\zeta - Z) + \frac{5}{8}\zeta$$

$$= -\zeta^2 + 2\zeta\ (\zeta - 2) + \frac{5}{8}\zeta$$

$$= -\zeta^2 + 2\zeta^2 - 4\zeta + \frac{5}{8}\zeta$$

$$= \zeta^2 - \frac{27}{8}\zeta$$

which is Equation 9.10

9–12. Use a program such as *MathCad* or *Mathematica* to evaluate $E(\zeta_1, \zeta_2)$ using Equation 9.13 and then minimize the result with respect to both ζ_1 and ζ_2 to obtain $E = -2.875\,66\ E_\text{h}$.

The wave function of Equation 9.13 is

$$\psi(\mathbf{r}_1, \mathbf{r}_2) = N \left(e^{-\zeta_1 r_1} e^{-\zeta_2 r_2} + e^{-\zeta_2 r_1} e^{-\zeta_1 r_2} \right)$$

and the energy calculated for a helium atom using this trial function will depend on the values chosen for the two variational parameters:

$$E(\zeta_1, \zeta_2) = \int \int d\mathbf{r}_1\, d\mathbf{r}_2\, \psi(\mathbf{r}_1, \mathbf{r}_2) \hat{H} \psi(\mathbf{r}_1, \mathbf{r}_2)$$

with

$$\hat{H} = -\frac{1}{2}\nabla_1^2 - \frac{2}{r_1} - \frac{1}{2}\nabla_2^2 - \frac{2}{r_2} + \frac{1}{r_{12}}$$

Because the trial function has no angular dependence, the derivatives with respect to angular coordinates vanish in the kinetic energy operators, and the following two one-electron integrals can be defined:

$$h_1(\zeta_1, \zeta_2) = \frac{1}{N^2(\zeta_1, \zeta_2)} \int \int d\mathbf{r}_1 \, d\mathbf{r}_2 \, \psi(\mathbf{r}_1, \mathbf{r}_2) \left(-\frac{1}{2}\nabla_1^2 - \frac{2}{r_1}\right) \psi(\mathbf{r}_1, \mathbf{r}_2)$$

$$= \int_0^{2\pi} d\phi_1 \int_0^{2\pi} d\phi_2 \int_0^{\pi} d\theta_1 \, \sin\theta_1 \int_0^{\pi} d\theta_2 \, \sin\theta_2$$

$$\times \int_0^{\infty} dr_2 \, r_2^2 \int_0^{\infty} dr_1 \, r_1^2 \left(e^{-\zeta_1 r_1}e^{-\zeta_2 r_2} + e^{-\zeta_2 r_1}e^{-\zeta_1 r_2}\right)$$

$$\times \left[-\frac{1}{2r_1^2}\frac{d}{dr_1}r_1^2\frac{d}{dr_1} - \frac{2}{r_1}\right]\left(e^{-\zeta_1 r_1}e^{-\zeta_2 r_2} + e^{-\zeta_2 r_1}e^{-\zeta_1 r_2}\right)$$

$$= 16\pi^2 \int_0^{\infty} dr_2 \, r_2^2 \left[-\frac{1}{2}\int_0^{\infty} dr_1 \left(e^{-\zeta_1 r_1}e^{-\zeta_2 r_2} + e^{-\zeta_2 r_1}e^{-\zeta_1 r_2}\right)\right.$$

$$\times \left(\frac{d}{dr_1}r_1^2\frac{d}{dr_1}\right)\left(e^{-\zeta_1 r_1}e^{-\zeta_2 r_2} + e^{-\zeta_2 r_1}e^{-\zeta_1 r_2}\right)$$

$$\left.-2\int_0^{\infty} dr_1 \, r_1 \left(e^{-\zeta_1 r_1}e^{-\zeta_2 r_2} + e^{-\zeta_2 r_1}e^{-\zeta_1 r_2}\right)^2\right]$$

and

$$h_2(\zeta_1, \zeta_2) = \frac{1}{N^2(\zeta_1, \zeta_2)} \int \int d\mathbf{r}_1 \, d\mathbf{r}_2 \, \psi(\mathbf{r}_1, \mathbf{r}_2) \left(-\frac{1}{2}\nabla_2^2 - \frac{2}{r_2}\right) \psi(\mathbf{r}_1, \mathbf{r}_2)$$

$$= h_1(\zeta_1, \zeta_2)$$

Using the approach of Problem 8–39 or 9–57 the two-electron integral becomes, after performing the integrals over θ_1, ϕ_1, and ϕ_2:

$$g(\zeta_1, \zeta_2) = \frac{1}{N^2(\zeta_1, \zeta_2)} \int \int d\mathbf{r}_1 \, d\mathbf{r}_2 \, \psi(\mathbf{r}_1, \mathbf{r}_2) \left(\frac{1}{r_{12}}\right) \psi(\mathbf{r}_1, \mathbf{r}_2)$$

$$= 8\pi^2 \int_0^{\infty} dr_1 \, r_1^2 \int_0^{\infty} dr_2 \, r_2^2 \left(e^{-\zeta_1 r_1}e^{-\zeta_2 r_2} + e^{-\zeta_2 r_1}e^{-\zeta_1 r_2}\right)^2 \int_0^{\pi} \frac{d\theta_2 \, \sin\theta_2}{\left(r_1^2 + r_2^2 - 2r_1 r_2 \cos\theta_2\right)^{1/2}}$$

In all three of these integrals, the normalization constant is given by

$$N^2(\zeta_1, \zeta_2) = \left[16\pi^2 \int_0^{\infty} dr_1 \, r_1^2 \int_0^{\infty} dr_2 \, r_2^2 \left(e^{-\zeta_1 r_1}e^{-\zeta_2 r_2} + e^{-\zeta_2 r_1}e^{-\zeta_1 r_2}\right)^2\right]^{-1}$$

Thus,

$$E(\zeta_1, \zeta_2) = \frac{2h_1(\zeta_1, \zeta_2) + g(\zeta_1, \zeta_2)}{16\pi^2 \int_0^{\infty} dr_1 \, r_1^2 \int_0^{\infty} dr_2 \, r_2^2 \left(e^{-\zeta_1 r_1}e^{-\zeta_2 r_2} + e^{-\zeta_2 r_1}e^{-\zeta_1 r_2}\right)^2}$$

These expressions can all be evaluated by *MathCad* or *Mathematica* or other similar programs after entering (according to the formatting requirements of the particular program):

$$\psi(\zeta_1, \zeta_2, r_1, r_2) = e^{-\zeta_1 r_1}e^{-\zeta_2 r_2} + e^{-\zeta_2 r_1}e^{-\zeta_1 r_2}$$

$$D\psi(\zeta_1, \zeta_2, r_1, r_2) = \frac{d}{dr_1}\psi(\zeta_1, \zeta_2, r_1, r_2)$$

$$h_1(\zeta_1, \zeta_2) = -8\pi^2 \int_0^\infty \int_0^\infty r_2^2\psi(\zeta_1, \zeta_2, r_1, r_2)\frac{d}{dr_1}\left[r_1^2 D\psi(\zeta_1, \zeta_2, r_1, r_2)\right] dr_1\, dr_2$$

$$- 32\pi^2 \int_0^\infty \int_0^\infty r_1 r_2^2 \psi^2(\zeta_1, \zeta_2, r_1, r_2)\, dr_1\, dr_2$$

$$g(\zeta_1, \zeta_2) = 8\pi^2 \int_0^\infty \int_0^\infty r_1^2 r_2^2 \psi^2(\zeta_1, \zeta_2, r_1, r_2) \int_0^\pi \frac{\sin\theta_2}{\sqrt{r_1^2 + r_2^2 - 2r_1 r_2 \cos\theta_2}}\, d\theta_2\, dr_1\, dr_2$$

$$NN(\zeta_1, \zeta_2) = 16\pi^2 \int_0^\infty \int_0^\infty r_1^2 r_2^2 \psi^2(\zeta_1, \zeta_2, r_1, r_2)\, dr_1\, dr_2$$

$$E(\zeta_1, \zeta_2) = \frac{2h_1(\zeta_1, \zeta_2) + g(\zeta_1, \zeta_2)}{NN(\zeta_1, \zeta_2)}$$

The program is then requested to perform the minimization of $E(\zeta_1, \zeta_2)$ with respect to (ζ_1, ζ_2). Doing so gives the results,

$$\zeta_1 = 1.1885$$

$$\zeta_2 = 2.1832$$

$$E(1.1885, 2.1832) = -2.875\,66\ E_h$$

9–13. Show that the normalization constant for the radial part of Slater orbitals is

$$(2\zeta)^{n+\frac{1}{2}}/[(2n)!]^{1/2}$$

The angular portions of the Slater orbitals, $Y_l^{m_l}(\theta, \phi)$, are already normalized, so only the radial portion given in Equation 9.15, $r^{n-1}e^{-\zeta r}$, needs further consideration. Normalization of the radial portion requires that

$$1 = N_{nl}^2 \int_0^\infty r^2 r^{2(n-1)} e^{-2\zeta r}\, dr$$

$$= N_{nl}^2 \int_0^\infty r^{2n} e^{-2\zeta r}\, dr$$

$$= N_{nl}^2 \frac{(2n)!}{(2\zeta)^{2n+1}}$$

$$N_{nl} = \sqrt{\frac{(2\zeta)^{2n+1}}{(2n)!}} = \frac{(2\zeta)^{n+1/2}}{[(2n)!]^{1/2}}$$

9–14.   Compare the Slater orbital $S_{200}(r)$ to the hydrogen atomic orbital $\psi_{200}(r)$ (Table 7.3) by plotting them together. Do the same for $S_{300}(r)$ and $\psi_{300}(r)$.

The plots are shown below.

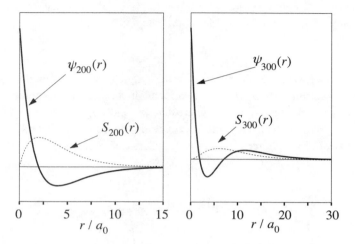

Unlike the corresponding hydrogen atomic s orbitals, these Slater orbitals go to zero as $r \to 0$ and do not possess radial nodes.

9–15.   Show that Equation 9.16 is normalized.

Equation 9.16 is a linear combination of two Slater orbitals.

$$\phi_{1s}(r) = 0.843\,785\,S_{1s}(\zeta = 1.453\,63) + 0.180\,687\,S_{1s}(\zeta = 2.910\,93)$$

Normalization requires that $\langle \phi_{1s} | \phi_{1s} \rangle = 1$.

$$\langle \phi_{1s} | \phi_{1s} \rangle = (0.843\,785)^2 \langle S_{1s}(\zeta = 1.453\,63) | S_{1s}(\zeta = 1.453\,63) \rangle$$

$$+ (0.180\,687)^2 \langle S_{1s}(\zeta = 2.910\,93) | S_{1s}(\zeta = 2.910\,93) \rangle$$

$$+ 2 (0.843\,785)(0.180\,687) \langle S_{1s}(\zeta = 1.453\,63) | S_{1s}(\zeta = 2.910\,93) \rangle$$

The two Slater orbitals are individually normalized so that $\langle S_{1s}(\zeta) | S_{1s}(\zeta) \rangle = 1$, but the overlap integral $\langle S_{1s}(\zeta_1) | S_{1s}(\zeta_2) \rangle$ must be determined. In general for two Slater orbitals with the same values for n and l, but different values of ζ,

$$\langle S_{nl}(\zeta_1) | S_{nl}(\zeta_2) \rangle = \frac{(4\zeta_1\zeta_2)^{n+1/2}}{(2n)!} \int_0^\infty r^{2n} e^{-(\zeta_1+\zeta_2)r} \, dr = \frac{(4\zeta_1\zeta_2)^{n+1/2}}{(\zeta_1 + \zeta_2)^{2n+1}}$$

Thus,

$$\langle S_{1s}(\zeta = 1.453\,63) | S_{1s}(\zeta = 2.910\,93) \rangle = 0.837\,524$$

and

$$\langle\phi_{1s}|\phi_{1s}\rangle = (0.843\ 785)^2 + (0.180\ 687)^2 + 2\ (0.843\ 785)\ (0.180\ 687)\ (0.837\ 524) = 1$$

so that $\phi_{1s}(r)$ is indeed normalized.

9–16. Compare our simple variational trial function $(\zeta^3/\pi)^{1/2}e^{-\zeta r}$ with $\zeta = 27/16$ with the SCF orbital given by Equation 9.16 by plotting them together (cf. Figure 9.1).

Plotted below are two views of the functions: the simple variational function, $\psi_{1s}(r)$, and the SCF orbital from Equation 9.16, $\phi_{1s}(r)$. The two wave functions are very similar, although as is more clearly seen in the right hand graph, where $r\psi_{1s}(r)$ and $r\phi_{1s}(r)$ are shown (squaring these last two functions gives the respective radial probability distributions), the SCF orbital has slightly increased probability at larger values of r and decreased probability closer to the nucleus.

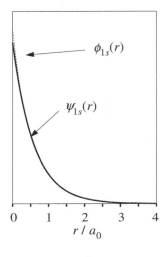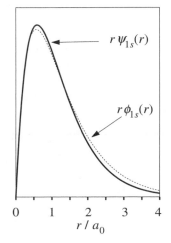

9–17. Substitute Equation 9.5 for $\hat{H}$ into

$$E = \iint d\mathbf{r}_1 d\mathbf{r}_2 \psi^*(\mathbf{r}_1)\psi^*(\mathbf{r}_2)\hat{H}\psi(\mathbf{r}_1)\psi(\mathbf{r}_2)$$

and show that

$$E = I_1 + I_2 + J_{11}$$

where

$$I_j = \int d\mathbf{r}_j \psi^*(\mathbf{r}_j)\left[-\frac{1}{2}\nabla_j^2 - \frac{Z}{r_j}\right]\psi(\mathbf{r}_j)$$

and

$$J_{11} = \iint d\mathbf{r}_1 d\mathbf{r}_2 \psi^*(\mathbf{r}_1)\psi^*(\mathbf{r}_2)\frac{1}{r_{12}}\psi(\mathbf{r}_1)\psi(\mathbf{r}_2)$$

Equation 9.5 gives

$$\hat{H} = -\frac{1}{2}\nabla_1^2 - \frac{1}{2}\nabla_2^2 - \frac{2}{r_1} - \frac{2}{r_2} + \frac{1}{r_{12}}$$

so that

$$
\begin{aligned}
E &= \iint d\mathbf{r}_1 \, d\mathbf{r}_2 \, \psi^*(\mathbf{r}_1)\psi^*(\mathbf{r}_2)\hat{H}\psi(\mathbf{r}_1)\psi(\mathbf{r}_2) \\
&= \iint d\mathbf{r}_1 \, d\mathbf{r}_2 \, \psi^*(\mathbf{r}_1)\psi^*(\mathbf{r}_2)\left(-\frac{1}{2}\nabla_1^2 - \frac{1}{2}\nabla_2^2 - \frac{2}{r_1} - \frac{2}{r_2} + \frac{1}{r_{12}}\right)\psi(\mathbf{r}_1)\psi(\mathbf{r}_2) \\
&= \iint d\mathbf{r}_1 \, d\mathbf{r}_2 \, \psi^*(\mathbf{r}_1)\psi^*(\mathbf{r}_2)\left(-\frac{1}{2}\nabla_1^2 - \frac{2}{r_1}\right)\psi(\mathbf{r}_1)\psi(\mathbf{r}_2) \\
&\quad + \iint d\mathbf{r}_1 \, d\mathbf{r}_2 \, \psi^*(\mathbf{r}_1)\psi^*(\mathbf{r}_2)\left(-\frac{1}{2}\nabla_2^2 - \frac{2}{r_2}\right)\psi(\mathbf{r}_1)\psi(\mathbf{r}_2) \\
&\quad + \iint d\mathbf{r}_1 \, d\mathbf{r}_2 \, \psi^*(\mathbf{r}_1)\psi^*(\mathbf{r}_2)\left(\frac{1}{r_{12}}\right)\psi(\mathbf{r}_1)\psi(\mathbf{r}_2) \\
&= I_1 + I_2 + J_{11}
\end{aligned}
$$

9–18. Why do you think that J_{11} in Equation 9.25 is called a Coulomb integral?

Classically, the integral

$$J = \iint d\mathbf{r}_1 \, d\mathbf{r}_2 \, \rho_1(\mathbf{r}_1)\left(\frac{1}{r_{12}}\right)\rho_2(\mathbf{r}_2)$$

is the Coulomb interaction energy (in atomic units) between two charge distributions, where $\rho_1(\mathbf{r}_1)$ and $\rho_2(\mathbf{r}_2)$ are the charge densities associated with the two distributions. In quantum mechanics, the squares of the wave functions, $\psi^*(\mathbf{r}_1)\psi(\mathbf{r}_1)$ and $\psi^*(\mathbf{r}_2)\psi(\mathbf{r}_2)$, give the probability densities, respectively, for the two electrons, and so the integral

$$J_{11} = \iint d\mathbf{r}_1 \, d\mathbf{r}_2 \, \psi^*(\mathbf{r}_1)\psi(\mathbf{r}_1)\frac{1}{r_{12}}\psi^*(\mathbf{r}_2)\psi(\mathbf{r}_2)$$

can be interpreted as representing the interaction between two charge distributions.

9–19. The normalized variational helium orbital we determined in Chapter 7 is

$$\psi_{1s}(r) = 1.2368 \, e^{-27r/16}$$

A two-term Hartree–Fock orbital is given in Equation 9.16,

$$\psi_{1s}(r) = 0.843\,785 \, S_{1s}(\zeta = 1.453\,63) + 0.280\,687 \, S_{1s}(\zeta = 2.910\,93)$$

and a four-term orbital given by Equation 9.72 is

$$\psi_{1s} = 1.347\,900\,S_{1s}(\zeta = 1.4595) - 0.001\,613\,S_{3s}(\zeta = 5.3244)$$

$$- 0.100\,506\,S_{2s}(\zeta = 2.6298) - 0.270\,779\,S_{2s}(\zeta = 1.7504)$$

Compare these orbitals by plotting them on the same graph.

The three orbtials are plotted below. The first two orbitals were compared in Problem 9–16. On the scale of these graphs, the four-term orbital given by Equation 9.72 is indistinguishable from the two-term orbital of Equation 9.16.

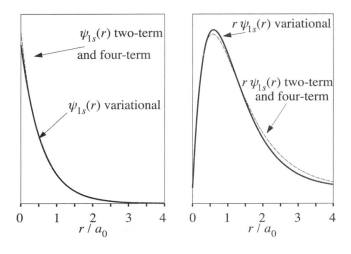

9–20.   Use the SCF orbital given in Equation 9.16 to calculate $V_1^{\text{eff}}(r_1)$ given in Equation 9.19 and compare your result to Figure 9.2.

The SCF orbital of Equation 9.16 is a linear combination of two Slater orbitals:

$$\phi_{1s}(r) = 0.843\,785\,S_{1s}(\zeta = 1.453\,63) + 0.180\,687\,S_{1s}(\zeta = 2.910\,93)$$

The calculation of V_1^{eff} will require the evaluation of the following general integral involving Slater $1s$ orbitals:

$$I_{\zeta_1\zeta_2}(r_1) = \int d\mathbf{r_2}\,S_{1s}(\zeta_1)\frac{1}{r_{12}}S_{1s}(\zeta_2)$$

$$= \frac{(4\zeta_1\zeta_2)^{3/2}}{8\pi}\int_0^{2\pi}d\phi_2\int_0^\infty dr_2\,r_2^2 e^{-(\zeta_1+\zeta_2)r_2}\int_0^\pi \frac{d\theta_2\,\sin\theta_2}{r_1^2 + r_2^2 - 2r_1r_2\cos\theta_2}$$

The integral over θ_2 was evaluated in Problem 8–39, where it was found

$$\int_0^\pi \frac{d\theta_2\,\sin\theta_2}{r_1^2 + r_2^2 - 2r_1r_2\cos\theta_2} = \begin{cases} \dfrac{2}{r_1} & r_1 > r_2 \\[2mm] \dfrac{2}{r_2} & r_2 > r_1 \end{cases}$$

Use this result in the equation for $I_{\zeta_1\zeta_2}$, and perform the simple integral over ϕ_2 to give

$$I_{\zeta_1\zeta_2}(r_1) = \frac{(4\zeta_1\zeta_2)^{3/2}}{2}\left[\frac{1}{r_1}\int_0^{r_1} dr_2\, r_2^2 e^{-(\zeta_1+\zeta_2)r_2} + \int_{r_1}^{\infty} dr_2\, r_2 e^{-(\zeta_1+\zeta_2)r_2}\right]$$

These last two integrals were also evaluated in Problem 8–39 (with $2Z$ taking the place of $\zeta_1 + \zeta_2$). After collecting similar terms, the final result for $I_{\zeta_1\zeta_2}$ is

$$I_{\zeta_1\zeta_2}(r_1) = \frac{(4\zeta_1\zeta_2)^{3/2}}{2(\zeta_1+\zeta_2)^2}\left\{\frac{2}{(\zeta_1+\zeta_2)\,r_1}\left[1-e^{-(\zeta_1+\zeta_2)r_1}\right] - e^{-(\zeta_1+\zeta_2)r_1}\right\}$$

As $r_1 \to 0$, the series expansion for the exponential terms may be used to show that $I_{\zeta_1\zeta_2}$ has a finite limiting value at $r_1 = 0$. In fact,

$$\lim_{r_1\to 0} I_{\zeta_1\zeta_2}(r_1) = \frac{(4\zeta_1\zeta_2)^{3/2}}{2(\zeta_1+\zeta_2)^2}$$

The effective Hartree-Fock potential for the SCF orbital of Equation 9.16 can be written in terms of these integrals.

$$V_1^{\text{eff}}(r_1) = \int d\mathbf{r_2}\, \phi_{1s}(\mathbf{r_2})\frac{1}{r_{12}}\phi_{1s}(\mathbf{r_2})$$

$$= (0.843\,785)^2\, I_{1.453\,63,1.453\,63}(r_1) + (0.180\,687)^2\, I_{2.910\,93,2.910\,93}(r_1)$$

$$+ 2\,(0.843\,785)\,(0.180\,687)\,(0.837\,524)\, I_{1.453\,63,2.910\,93}(r_1)$$

At this point, the best approach is to use a program such as *MathCad* or *Mathematica* to evaluate $V_1^{\text{eff}}(r_1)$ using the explicit formula developed for the $I_{\zeta_1\zeta_2}$. Alternatively, these same programs could have been used to evaluate $V_1^{\text{eff}}(r_1) = \int d\mathbf{r_2}\,\phi_{1s}(\mathbf{r_2})\frac{1}{r_{12}}\phi_{1s}(\mathbf{r_2})$ directly without the need to find an expression for $I_{\zeta_1\zeta_2}$. Either way, calculation of $V_1^{\text{eff}}(r_1)$ for a range of values of r_1 gives the solid line in the graph below. A coulombic $1/r$ potential is presented as a dashed line for comparison. The graph is seen to be identical to Figure 9.2.

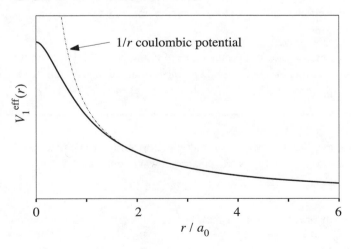

9–21. Given that $\Psi(1, 2) = 1s\alpha(1)\,1s\beta(2) - 1s\alpha(2)\,1s\beta(1)$, prove that

$$\int d\tau_1 d\tau_2 \Psi^*(1, 2)\Psi(1, 2) = 2$$

if the spatial part is normalized.

$$\int d\tau_1\, d\tau_2\, \Psi^*(1, 2)\Psi(1, 2) = \int d\tau_1\, d\tau_2\, \left[1s\alpha^*(1)1s\beta^*(2) - 1s\alpha^*(2)1s\beta^*(1)\right]$$

$$\times \left[1s\alpha(1)1s\beta(2) - 1s\alpha(2)1s\beta(1)\right]$$

$$= \int d\tau_1\, d\tau_2\, 1s\alpha^*(1)1s\beta^*(2)1s\alpha(1)1s\beta(2)$$

$$- \int d\tau_1\, d\tau_2\, 1s\alpha^*(2)1s\beta^*(1)1s\alpha(1)1s\beta(2)$$

$$- \int d\tau_1\, d\tau_2\, 1s\alpha^*(1)1s\beta^*(2)1s\alpha(2)1s\beta(1)$$

$$+ \int d\tau_1\, d\tau_2\, 1s\alpha^*(2)1s\beta^*(1)1s\alpha(2)1s\beta(1)$$

$$= 1 - 0 - 0 + 1 = 2$$

9–22. Why is distinguishing the two electrons in a helium atom impossible, but distinguishing the two electrons in separated hydrogen atoms possible? Do you think the electrons are distinguishable in the diatomic H_2 molecule? Explain your reasoning.

For two electrons in separated hydrogen atoms, the electrons are associated with two different nuclei and there are no potential energy terms that couple the two electrons. Consequently, the Hamiltonian operator of the system is separable, and the two electrons are independent of one another. The two electrons in a helium atom cannot be distinguished because the Hamiltonian operator cannot be separated into a sum of two terms that are functions of only one electron each, and the wave function of this two-electron system must be antisymmetric under the interchange of the two electrons.

Because of an electron-electron repulsion term in the Hamiltonian operator of the diatomic H_2 molecule, the electrons in that molecule are also indistinguishable. Indeed, the Pauli exclusion principle also applies to molecular orbitals.

9–23. Why is the angular dependence of multielectron atomic wave functions in the Hartree–Fock approximation the same as for hydrogen atomic wave functions?

The Hamiltonian operator used in the Hartree-Fock approximation depends only on r_1 (see Equation 9.20), and so the angular dependence of the wave functions of a multielectron atom in the Hartree-Fock approximation is the same as in the hydrogen atom.

9–24. Why is the radial dependence of many-electron atomic wave functions in the Hartree–Fock approximation different from the radial dependence of hydrogen atomic wave functions?

The radial dependence of the effective Hamiltonian operator differs from the Hamiltonian operator of a hydrogen atom due to the inclusion of $V_j^{\text{eff}}(\mathbf{r}_j)$ (see Figure 9.2 and also Problem 9–20), which causes the potential energy to deviate from a pure coulombic interaction.

9–25. Show that the atomic determinantal wave function

$$\psi = \frac{1}{\sqrt{2}} \begin{vmatrix} 1s\alpha(1) & 1s\beta(1) \\ 1s\alpha(2) & 1s\beta(2) \end{vmatrix}$$

is normalized if the $1s$ orbitals are normalized.

Expand the determinant to get

$$\psi = \frac{1}{\sqrt{2}} [1s\alpha(1)\,1s\beta(2) - 1s\alpha(2)\,1s\beta(1)]$$

Then

$$\int d\tau\, \psi^*\psi = \frac{1}{2} \int d\tau\, [1s\alpha(1)\,1s\beta(2) - 1s\alpha(2)\,1s\beta(1)]^2$$

$$= \frac{1}{2} (1 + 0 + 0 + 1) = \frac{1}{2}(2) = 1$$

where the solution of Problem 9–21 is used to evaluate the integral.

9–26. Show that the two-electron determinantal wave function in Problem 9–25 factors into a spatial part and a spin part.

$$\psi = \frac{1}{\sqrt{2}} [1s(1)\alpha(1)\,1s(2)\beta(2) - 1s(2)\alpha(2)\,1s(1)\beta(1)]$$

$$= 1s(1)\,1s(2) \left\{ \frac{1}{\sqrt{2}} [\alpha(1)\beta(2) - \alpha(2)\beta(1)] \right\}$$

9–27. Argue that the normalization constant of an $N \times N$ Slater determinant of orthonormal spin orbitals is $1/\sqrt{N!}$.

The expansion of an $N \times N$ determinant yields $N!$ terms

$$\sigma_1 + \sigma_2 + \ldots + \sigma_{N!}$$

and $(N!)^2$ terms when the wave function is squared

$$\sigma_1^2 + \sigma_2^2 + \ldots \sigma_{N!}^2 + 2\sigma_1\sigma_2 + \ldots$$

However, only the first $N!$ terms in the square of the wave function yield non-zero integrals, each of which equals one when integrated over all the electron coordinates, due to the normalization of the individual Slater spin orbitals. All the cross products yield a zero integral because of the orthogonality of at least one spin orbital. Thus, the integration of the square of the $N \times N$ Slater determinant of spin orbitals over all space results in a value of $N!$. Therefore, the normalization constant for the Slater determinant is $1/\sqrt{N!}$.

9–28. Show that the energy associated with the wave function in Equation 9.51 is

$$E = 2 \sum_{j=1}^{N} I_j + \sum_{i=1}^{N} \sum_{j=1}^{N} (2J_{ij} - K_{ij})$$

This problem is more difficult than most of the others in this chapter. Because of its determinantal nature, $\Psi(1, 2, \ldots, 2N)$ can be written as

$$\Psi(1, 2, \ldots, 2N) = \frac{1}{\sqrt{(2N)!}} \sum_{P} \epsilon_P u_1(P1) u_2(P2) \ldots u_{2N}(P2N) \tag{1}$$

where the u_i are the spin orbitals, $u_1 = \psi_1 \alpha(1)$, $u_2 = \psi_1 \beta(1)$, $u_3 = \psi_2 \alpha(2)$, etc., and where P represents a permutation of the numbers $(1, 2, \ldots, 2N)$, $\epsilon_P = +1$ if the permutation $(P1, P2, \ldots, P2N)$ differs from $(1, 2, \ldots, 2N)$ by an even number of interchanges of pairs of numbers, and $\epsilon_P = -1$ if the permutation $(P1, P2, \ldots, P2N)$ differs from $(1, 2, \ldots, 2N)$ by an odd number of interchanges. An equivalent way of writing Equation 1 is

$$\Psi(1, 2, \ldots, 2N) = \frac{1}{\sqrt{(2N)!}} \sum_{P} \epsilon_P u_{P1}(1) u_{P2}(2) \ldots u_{P2N}(2N) \tag{2}$$

In other words, the labels of the spin-orbitals are interchanged, rather than those of the electrons. Both expressions for Ψ will be used in this problem. First show that Ψ is normalized. Using Equation 1 gives

$$\int \cdots \int d\tau_1 \, d\tau_2 \ldots d\tau_{2N} \, \Psi^* \Psi = \frac{1}{(2N)!} \sum_{P} \epsilon_P \sum_{Q} \epsilon_Q \int \cdots \int d\tau_1 \, d\tau_2 \ldots d\tau_{2N} \, u_1^*(Q1)$$

$$\times u_2^*(Q2) \ldots u_{2N}^*(Q2N) u_1(P1) u_2(P2) \ldots u_{2N}(P2N)$$

These integrals will vanish unless $Q1 = P1$, $Q2 = P2$, and, in general, $Qx = Px$, where x can have any integer value between 1 and $2N$. This means that the only permutation Q that will lead to a non-zero result is when $Q = P$. This leaves

$$\int \cdots \int d\tau_1 \, d\tau_2 \ldots d\tau_{2N} \, \Psi^* \Psi = \frac{1}{(2N)!} \sum_{P} \epsilon_P^2$$

$$\times \int \cdots \int d\tau_1 \, d\tau_2 \ldots d\tau_{2N} \, u_1^*(P1) u_1(P1) u_2^*(P2) u_2(P2) \ldots u_{2N}^*(P2N) u_{2N}(P2N)$$

where $\epsilon_P^2 = 1$. Because the integration variables can be relabeled, all these integrals are equal, and equal to unity because the individual spin-orbitals are normalized. There are $(2N)!$ permutations, and so

$$\int \cdots \int d\tau_1 \, d\tau_2 \ldots d\tau_{2N} \, \Psi^*\Psi = \frac{1}{(2N)!}(2N)! = 1$$

Now consider integrals of the form

$$A_1 = \int \cdots \int d\tau_1 \, d\tau_2 \ldots d\tau_{2N} \, \Psi^* \left[\sum_{j=1}^{2N} \hat{h}_j \right] \Psi$$

where $\hat{h}_j$ from Equation 9.50 is a one-electron operator, meaning that it operates only on the coordinates of electron j. This operator can be expressed as

$$\sum_{j=1}^{2N} \hat{h}_j = \sum_{j=1}^{2N} \left[-\frac{1}{2}\nabla_j^2 - \frac{Z}{r_j} \right]$$

Using Equation 1 for Ψ, this becomes

$$A_1 = \frac{1}{(2N)!} \sum_P \epsilon_P \sum_Q \epsilon_Q \int \cdots \int d\tau_1 \ldots d\tau_{2N} \, u_1^*(Q1)u_2^*(Q2)$$

$$\times \ldots u_{2N}^*(Q2N) \left[\sum_{j=1}^{2N} \hat{h}_j \right] u_1(P1)u_2(P2) \ldots u_{2N}(P2N)$$

As before, all $(2N)!$ permutations lead to the same integral, and so

$$A_1 = \sum_{j=1}^{2N} \int u_j^*(j)\hat{h}_j u_j(j) \, d\tau_j$$

Finally, consider integrals of the type

$$A_2 = \int \cdots \int d\tau_1 \ldots d\tau_{2N} \, \Psi^* \left[\sum_{i=1}^{2N} \sum_{j>i}^{2N} \frac{1}{r_{ij}} \right] \Psi$$

Using Equation 2 for Ψ, write

$$A_2 = \frac{1}{(2N)!} \sum_P \epsilon_P \sum_Q \epsilon_Q \int \cdots \int d\tau_1 \ldots d\tau_{2N} \, u_{Q1}^*(1)u_{Q2}^*(2) \ldots u_{Q2N}^*(2N)$$

$$\times \left[\sum_{i=1}^{2N} \sum_{j>i}^{2N} \frac{1}{r_{ij}} \right] u_{P1}(1)u_{P2}(2) \ldots u_{P2N}(2N)$$

These integrals will vanish unless the subscripts of all the spin-orbitals *except* those for electrons i and j match. There will be two non-vanishing terms in the summation of all Q permutations:

$$A_2 = \frac{1}{(2N)!} \sum_P \left\{ \epsilon_P \sum_{i=1}^{2N} \sum_{j>i}^{2N} \iint d\tau_i \, d\tau_j \, u_{P1}^*(i) u_{P2}^*(j) \frac{1}{r_{ij}} u_{P1}(i) u_{P2}(j) \right.$$

$$\left. -\epsilon_P \sum_{i=1}^{2N} \sum_{j>i}^{2N} \int \int d\tau_i \, d\tau_j \, u_{P2}^*(i) u_{P1}^*(j) \frac{1}{r_{ij}} u_{P1}(i) u_{P2}(j) \right\}$$

Note that $Q = P$ for the term with the positive sign and that Q and P differ by the interchange of one pair of subscripts for the term with the negative sign. As before, all $(2N)!$ permutations lead to the same quantities, and so

$$A_2 = \sum_{i=1}^{2N} \sum_{j>i}^{2N} \iint d\tau_1 \, d\tau_2 \, u_i^*(1) u_j^*(2) \frac{1}{r_{12}} u_i(1) u_j(2)$$

$$- \sum_{i=1}^{2N} \sum_{j>i}^{2N} \iint d\tau_1 d\tau_2 u_i^*(1) u_j^*(2) \frac{1}{r_{12}} u_i(2) u_j(1)$$

$$= \sum_{i=1}^{2N} \sum_{j>i}^{2N} J_{ij}' - \sum_{i=1}^{2N} \sum_{j>i}^{2N} K_{ij}'$$

The integrals here involve spin orbitals, while the integrals expressed by Equations 9.54 through 9.56 involve only spatial orbitals. Combining the results for A_1 and A_2 and substituting into Equation 9.52 gives

$$E = \int \cdots \int d\tau_1 \ldots d\tau_{2N} \, \Psi^* \hat{H} \Psi$$

$$= \int \cdots \int d\tau_1 \, d\tau_2 \ldots d\tau_{2N} \Psi^* \left[\sum_j^{2N} \hat{h}_j \right] \Psi$$

$$+ \int \cdots \int d\tau_1 \ldots d\tau_{2N} \, \Psi^* \left[\sum_{i=1}^{2N} \sum_{j>i}^{2N} \frac{1}{r_{ij}} \right] \Psi$$

$$= A_1 + A_2$$

$$= \sum_{j=1}^{2N} \int u_j^*(j) \hat{h}_j u_j(j) d\tau_j + \sum_{i=1}^{2N} \sum_{j>i}^{2N} J_{ij}' - \sum_{i=1}^{2N} \sum_{j>i}^{2N} K_{ij}'$$

$$= \sum_{j=1}^{2N} I_j' + \sum_{i=1}^{2N} \sum_{j>i}^{2N} J_{ij}' - \sum_{i=1}^{2N} \sum_{i<j}^{2N} K_{ij}'$$

Since the overall wave function has N doubly occupied orbitals, with the spatial parts of u_1 and u_2 identical and so on, then $I_1' = I_2'$, $I_3' = I_4'$, and, in general, $I_{2n-1}' = I_{2n}' = I_n$, where n goes from 1 to N. Furthermore,

$$\begin{array}{llll} J_{12}' &= J_{11} & J_{13}' &= J_{12} & J_{14}' &= J_{12} \\ J_{23}' &= J_{12} & J_{24}' &= J_{22} \end{array}$$

and so forth. Also,

$$K_{12}' = 0 \qquad K_{13}' = K_{12} \qquad K_{14}' = 0$$

and so forth, giving

$$E = 2 \sum_{j=1}^{N} I_j + \sum_{i=1}^{N} \sum_{j=1}^{N} \left(2J_{ij} - K_{ij} \right)$$

where the summations are over spatial orbitals. If the relationships between the J'_{ij} and the J_{ij}, and those between the K'_{ij} and the K_{ij}, are unclear, take the beryllium atom as a concrete example.

9–29. Show that Equation 9.52 can be expressed through Equations 9.53 to 9.56.

This was done in answering the previous problem.

9–30. The total z component of the spin angular-momentum operator for an N-electron system is

$$\hat{S}_{z,\text{total}} = \sum_{j=1}^{N} \hat{S}_{zj}$$

Show that both

$$\psi = \frac{1}{\sqrt{2}} \begin{vmatrix} 1s\alpha(1) & 1s\beta(1) \\ 1s\alpha(2) & 1s\beta(2) \end{vmatrix}$$

and

$$\psi = \frac{1}{\sqrt{3!}} \begin{vmatrix} 1s\alpha(1) & 1s\beta(1) & 2s\alpha(1) \\ 1s\alpha(2) & 1s\beta(2) & 2s\alpha(2) \\ 1s\alpha(3) & 1s\beta(3) & 2s\alpha(3) \end{vmatrix}$$

are eigenfunctions of $\hat{S}_{z,\text{total}}$. What are the eigenvalues in each case?

For the first wave function given,

$$\hat{S}_{z,\text{total}}\psi = \frac{1}{\sqrt{2}} \left(\hat{S}_{z1} + \hat{S}_{z2} \right) [1s\alpha(1)1s\beta(2) - 1s\beta(1)1s\alpha(2)]$$

$$= \frac{1}{\sqrt{2}} \left[\left(\frac{1}{2} - \frac{1}{2} \right) 1s\alpha(1)1s\beta(2) - \left(-\frac{1}{2} + \frac{1}{2} \right) 1s\beta(1)1s\alpha(2) \right]$$

$$= 0 \left[1s\alpha(1)1s\beta(2) - 1s\beta(1)1s\alpha(2) \right]$$

Thus, ψ is an eigenfunction of $\hat{S}_{z,\text{total}}$ with eigenvalue zero.

The expansion of the second determinant given in the problem involves six terms. A typical term is the diagonal product,

$$\left(\hat{S}_{z1} + \hat{S}_{z2} + \hat{S}_{z3} \right) 1s\alpha(1)1s\beta(2)2s\alpha(3) = \left(\frac{1}{2} - \frac{1}{2} + \frac{1}{2} \right) 1s\alpha(1)1s\beta(2)2s\alpha(3)$$

There are six terms in the expansion of the 3×3 determinant. Each one of these terms has two electrons described by an α spin function and one electron with a β spin function. Consequently, the net effect of the $\left(\hat{S}_{z1} + \hat{S}_{z2} + \hat{S}_{z3}\right)$ operator is to multiply each term by 1/2. Thus, ψ turns out to be an eigenfunction of $\hat{S}_{z,\text{total}}$ with eigenvalue 1/2.

9–31.   Consider the determinantal atomic wave function

$$\Psi(1, 2) = \frac{1}{\sqrt{2}} \begin{vmatrix} \psi_{211}\alpha(1) & \psi_{21-1}\beta(1) \\ \psi_{211}\alpha(2) & \psi_{21-1}\beta(2) \end{vmatrix}$$

where $\psi_{21\pm1}$ is a hydrogen-like wave function. Show that $\Psi(1, 2)$ is an eigenfunction of

$$\hat{L}_{z,\text{total}} = \hat{L}_{z1} + \hat{L}_{z2}$$

and

$$S_{z,\text{total}} = \hat{S}_{z1} + \hat{S}_{z2}$$

What are the eigenvalues?

Use the identities for hydrogenlike wave functions (Equations 7.49 and 7.50 in atomic units)

$$\hat{L}_z\psi_{nlm_l} = m\psi_{nlm_l} \quad \hat{S}_z\alpha = \frac{1}{2}\alpha \quad \hat{S}_z\beta = -\frac{1}{2}\beta$$

Now

$$\hat{L}_{z,\text{total}}\Psi = \frac{1}{\sqrt{2}} \left(\hat{L}_{z1} + \hat{L}_{z2}\right) \left[\psi_{211}\alpha(1)\psi_{21-1}\beta(2) - \psi_{211}\alpha(2)\psi_{21-1}\beta(1)\right]$$

$$= \frac{1}{\sqrt{2}} \left[(1 - 1)\psi_{211}\alpha(1)\psi_{21-1}\beta(2) - (-1 + 1)\psi_{211}\alpha(2)\psi_{21-1}\beta(1)\right]$$

$$= 0$$

Therefore, Ψ is an eigenfunction of $\hat{L}_{z,\text{total}}$ with eigenvalue zero. Similarly,

$$\hat{S}_{z,\text{total}}\Psi = \frac{1}{\sqrt{2}} \left(\hat{S}_{z1} + \hat{S}_{z2}\right) \left[\psi_{211}\alpha(1)\psi_{21-1}\beta(2) - \psi_{211}\alpha(2)\psi_{21-1}\beta(1)\right]$$

$$= \frac{1}{\sqrt{2}} \left[\left(\frac{1}{2} - \frac{1}{2}\right) \psi_{211}\alpha(1)\psi_{21-1}\beta(2) \right.$$

$$\left. - \left(-\frac{1}{2} + \frac{1}{2}\right) \psi_{211}\alpha(2)\psi_{21-1}\beta(1)\right] = 0$$

Ψ is also an eigenfunction of $\hat{S}_{z,\text{total}}$ with eigenvalue zero.

9–32. Use a program such as *MathCad* or *Mathematica* to show that Equation 9.72 is normalized.

The ns Slater orbitals have no angular dependence, and the integration over angular coordinates gives a factor of 4π, so that normalization requires

$$4\pi \int_0^\infty r^2 \psi_{1s}^2(r)\, dr = 1$$

where $\psi_{1s}(r)$ is given by Equations 9.72 and 9.73. Evaluation of this integral using one of these programs does indeed give a result of one, showing that Equation 9.72 is normalized.

9–33. Use a program such as *MathCad* or *Mathematica* to show that the two lithium atomic orbitals ψ_{1s} and ψ_{2s} from *http://www.ccl.net/cca/data/atomic-RHF-wavefunctions/tables* are orthogonal.

From the website, the two lithium atomic orbitals are found to be linear combinations of seven Slater orbitals:

$$\psi_i = \sum_{j=1}^{7} c_{ij} S_{n_j s}(\zeta_j)$$

with the parameters given in the following table:

j	n_j	ζ_j	$c_{1s,j}$	$c_{2s,j}$
1	1	4.3069	0.141 279	−0.022 416
2	1	2.4573	0.874 231	−0.135 791
3	3	6.7850	−0.005 201	0.000 389
4	2	7.4527	−0.002 307	−0.000 068
5	2	1.8504	0.006 985	−0.076 544
6	2	0.7667	−0.000 305	0.340 542
7	2	0.6364	0.000 760	0.715 708

When the normalization integrals (see previous problem) are evaluated using one of the programs, the following results are obtained

$$4\pi \int_0^\infty r^2 \psi_{1s}^2(r)\, dr = 1.000\,000$$

$$4\pi \int_0^\infty r^2 \psi_{2s}^2(r)\, dr = 1.000\,000$$

Thus, the two orbitals are normalized.

9–34. We're going to prove Equation 9.74 in this problem. The wave function of an N-electron system is a function of $3N$ spatial coordinates and N spin coordinates, $\psi(x_1, y_1, z_1, \sigma_1, \ldots, x_N, y_N, z_N, \sigma_N)$. The normalization condition for this wave functon is

$$\langle \psi \mid \psi \rangle = \int \cdots \int \psi^*(x_1, \ldots, \sigma_N)\psi(x_1, \ldots, \sigma_N)\, d\tau_1 d\tau_2 \ldots d\tau_N = 1$$

where the integral over $d\tau_i = dx_i dy_i dz_i d\sigma_i$ symbolically signifies an integration over the spatial coordinates x_i, y_i, z_i, and a summation over the spin coordinate σ_i. Now, the probability that electron i will be found in the volume element $dx_i dy_i dz_i$ at the point x_i, y_i, z_i is given by the integral over all the coordinates (spatial and spin) except for x_i, y_i, z_i:

$$\int \cdots \int d\tau_1 \ldots d\tau_{i-1} d\sigma_i d\tau_{i+1} \ldots d\tau_N \, \psi^*(x_1, \ldots, \sigma_N) \psi(x_1, \ldots, \sigma_N)$$

The variables x_i, y_i, and z_i are not integrated over here, so the result of the integrations over all the other (spatial and spin) variables is a function of x_i, y_i, and z_i. We'll denote this function by $\rho(x_i, y_i, z_i)$ and write

$$\rho(x_i, y_i, z_i) = \int \cdots \int d\tau_1 \ldots d\tau_{i-1} d\sigma_i d\tau_{i+1} \ldots d\tau_N \, \psi^*(x_1, \ldots, \sigma_N) \psi(x_1, \ldots, \sigma_N)$$

Show that because the N electrons are indistinguishable, the probability density associated with *any* electron is just N times $\rho(x_i, y_i, z_i)$ or, equivalently, that

$$\rho(x, y, z) = N \int \cdots \int d\tau_1 \ldots d\tau_{i-1} d\sigma_i d\tau_{i+1} \ldots d\tau_N \, \psi^*(x_1, \ldots, \sigma_N) \psi(x_1, \ldots, \sigma_N)$$

This final expression for $\rho(x, y, z)$ is the probability density of finding an electron in the volume $dx\,dy\,dz$ surrounding the point (x, y, z).

Consider the SCF wave function of a helium atom, given by Equation 9.47:

$$\Psi(1, 2) = \frac{1}{\sqrt{2}} [1s\alpha(1) 1s\beta(2) - 1s\alpha(2) 1s\beta(1)]$$

Now square $\Psi(1, 2)$ and integrate over the spin coordinate of electron 1 and both the spatial and spin coordinates of electron 2 to obtain $1s^2(1)$. Electrons 1 and 2 are indistinguishable, so the probability density of a ground-state helium atom is given by $2\,1s^2(r)$.

Now let's look at the Slater determinant of a lithium atom (Problem 9–30). Show that if we square $\Psi(1, 2, 3)$ and integrate over the spin coordinate of electron 1 and all the coordinates of electrons 2 and 3, then we obtain

$$\rho(r) = 2\psi_{1s}^2(r) + \psi_{2s}^2(r)$$

The general result is

$$\rho(\mathbf{r}) = \sum n_j \, | \, \psi_j(\mathbf{r})|^2$$

where the summation runs over all the orbitals and $n_j = 0$, 1, or 2 is the number of electrons in that orbital.

The starting point for this problem is the expression for the probability that electron i will be found in the volume element $dx_i dy_i dz_i$ at the point x_i, y_i, z_i:

$$\rho(x_i, y_i, z_i) = \int \cdots \int d\tau_1 \ldots d\tau_{i-1} d\sigma_i \, d\tau_{i+1} \ldots d\tau_N \, \psi^*(x_1, \ldots, \sigma_N) \psi(x_1, \ldots, \sigma_N)$$

Now the probability that any one of the N electrons will be found in the volume element $dx_i dy_i dz_i$ at the point x_i, y_i, z_i is just the sum of probability densities for each of the electrons, or

$$\rho(x, y, z) = \sum_{i=1}^{N} \rho(x_i, y_i, z_i)$$

Because the N electrons are indistinguishable, $\rho(x_i, y_i, z_i)$ is the same for each electron, so

$$\rho(x, y, z) = \sum_{i=1}^{N} \rho(x_i, y_i, z_i)$$

$$= \rho(x_i, y_i, z_i) \sum_{i=1}^{N}(1)$$

$$= N\rho(x_i, y_i, z_i)$$

$$= N \int \cdots \int d\tau_1 \dots d\tau_{i-1} d\sigma_i d\tau_{i+1} \dots d\tau_N \, \psi^*(x_1, \dots, \sigma_N)\psi(x_1, \dots, \sigma_N)$$

The key to the equality of all the $\rho(x_i, y_i, z_i)$ terms is the determinantal nature of the wave function, which in turn reflects that the electrons are indistinguishable. This is illustrated in the two specific examples that follow.

For helium,

$$\Psi(1, 2) = \frac{1}{\sqrt{2}}[1s\alpha(1)\,1s\beta(2) - 1s\alpha(2)\,1s\beta(1)]$$

and

$$\int d\mathbf{r}_2 \, d\sigma_1 \, d\sigma_2 \, \Psi^2(1, 2) = \frac{1}{2} \int d\mathbf{r}_2 \, d\sigma_1 \, d\sigma_2 \, [1s\alpha(1)\,1s\beta(2) - 1s\alpha(2)\,1s\beta(1)]^2$$

$$= \frac{1}{2} \left[\int d\mathbf{r}_2 \, d\sigma_1 \, d\sigma_2 \, 1s\alpha(1)\,1s\beta(2)\,1s\alpha(1)\,1s\beta(2) \right.$$

$$- \int d\mathbf{r}_2 \, d\sigma_1 \, d\sigma_2 \, 1s\alpha(1)\,1s\beta(2)\,1s\alpha(2)\,1s\beta(1)$$

$$- \int d\mathbf{r}_2 \, d\sigma_1 \, d\sigma_2 \, 1s\alpha(2)\,1s\beta(1)\,1s\alpha(1)\,1s\beta(2)$$

$$\left. + \int d\mathbf{r}_2 \, d\sigma_1 \, d\sigma_2 \, 1s\alpha(2)\,1s\beta(1)\,1s\alpha(2)\,1s\beta(1) \right]$$

$$= \frac{1}{2} \left[1s^2(1) - 0 - 0 + 1s^2(1) \right] = 1s^2(1)$$

It is clear the integration over the spatial coordinates of electron 1 instead of electron 2 would have given $1s^2(2)$, which is indeed indistinguishable from $1s^2(1)$. Thus, the total probability density is $2\left[1s^2(r)\right]$.

For the lithium atom, the Slater determinant is (see Problem 9–30)

$$\Psi(1, 2, 3) = \frac{1}{\sqrt{3!}} \begin{vmatrix} 1s\alpha(1) & 1s\beta(1) & 2s\alpha(1) \\ 1s\alpha(2) & 1s\beta(2) & 2s\alpha(2) \\ 1s\alpha(3) & 1s\beta(3) & 2s\alpha(3) \end{vmatrix}$$

$$= \frac{1}{\sqrt{6}} [1s\alpha(1)1s\beta(2)2s\alpha(3) - 1s\alpha(1)2s\alpha(2)1s\beta(3)$$

$$- 1s\beta(1)1s\alpha(2)2s\alpha(3) + 1s\beta(1)2s\alpha(2)1s\alpha(3)$$

$$+ 2s\alpha(1)1s\alpha(2)1s\beta(3) - 2s\alpha(1)1s\beta(2)1s\alpha(3)]$$

Squaring this wave function would give 36 terms, but notice that as in the case of the helium atom (see also Problem 9–27), all of the integrations over the cross terms vanish leaving

$$\int d\mathbf{r}_2 \, d\mathbf{r}_3 \, d\sigma_1 \, d\sigma_2 \, d\sigma_3 \, \Psi^2(1, 2, 3) = \frac{1}{6} \int d\mathbf{r}_2 \, d\mathbf{r}_3 \, d\sigma_1 \, d\sigma_2 \, d\sigma_3 \left\{ [1s\alpha(1)1s\beta(2)2s\alpha(3)]^2 \right.$$

$$+ [1s\alpha(1)2s\alpha(2)1s\beta(3)]^2 + [1s\beta(1)1s\alpha(2)2s\alpha(3)]^2$$

$$+ [1s\beta(1)2s\alpha(2)1s\alpha(3)]^2 + [2s\alpha(1)1s\alpha(2)1s\beta(3)]^2$$

$$\left. + [2s\alpha(1)1s\beta(2)1s\alpha(3)]^2 \right\}$$

$$= \frac{1}{6} \left[1s^2(1) + 1s^2(1) + 1s^2(1) + 1s^2(1) + 2s^2(1) + 2s^2(1) \right]$$

$$= \frac{2}{3} \left[1s^2(1) \right] + \frac{1}{3} \left[2s^2(1) \right]$$

Precisely the same result would be obtained for electron 2 or electron 3, so the total probability density is

$$\rho(r) = 3 \left\{ \frac{2}{3} \left[1s^2(r) \right] + \frac{1}{3} \left[2s^2(r) \right] \right\} = 2\psi_{1s}^2(r) + \psi_{2s}^2(r)$$

9–35. Go to the website *http://www.ccl.net/cca/data/atomic-RHF-wavefunctions/tables* and write down the ψ_{1s} orbital for a beryllium atom. Use a program such as *MathCad* or *Mathematica* to show that the orbital is normalized and orthogonal to ψ_{2s} (see also Example 9–7). Using

$$\rho(r) = 2\psi_{1s}^2(r) + 2\psi_{2s}^2(r)$$

for the electron probability density of a beryllium atom, plot $r^2\rho(r)$ against r.

From the website, the two beryllium atom atomic orbitals are found to be linear combinations of seven Slater orbitals:

$$\psi_i = \sum_{j=1}^{7} c_{ij} S_{n_j s}(\zeta_j)$$

with the parameters given in the following table:

j	n_j	ζ_j	$c_{1s,j}$	$c_{2s,j}$
1	1	5.7531	0.285 107	−0.016 378
2	1	3.7156	0.474 813	−0.155 066
3	3	9.9670	−0.001 620	0.000 426
4	3	3.7128	0.052 852	−0.059 234
5	2	4.4661	0.243 499	−0.031 925
6	2	1.2919	0.000 106	0.387 968
7	2	0.8555	−0.000 032	0.685 674

It is easy to use one of the math programs to check the normalization integrals (see Problem 9–32) and obtain the following results:

$$4\pi \int_0^\infty r^2 \psi_{1s}^2(r)\, dr = 1.000\,008$$

$$4\pi \int_0^\infty r^2 \psi_{2s}^2(r)\, dr = 1.000\,003$$

Thus, within the precision of numerical data given, the two orbitals are normalized. Similarly,

$$4\pi \int_0^\infty r^2 \psi_{1s}(r)\psi_{2s}(r)\, dr = -9.015\,80 \times 10^{-6}$$

which is equal to zero within the expected precision, and the two orbitals are indeed orthogonal. Finally, the probability density is calculated using the expression given in the problem, and $r^2\rho(r)$ is plotted against r in the graph below.

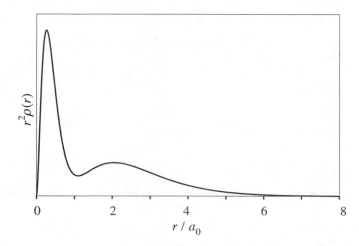

9–36. Go to the website *http://www.ccl.net/cca/data/atomic-RHF-wavefunctions/tables* to verify Koopmans's ionization energies of a neon atom and an argon atom and compare your results to those in Table 9.3.

The Koopman approximation to the ionization energy is the negative of corresponding orbital energy. The orbital energies are found in the tables at the website and converted from atomic units to $MJ \cdot mol^{-1}$.

Electron	Ionization Energy / a.u.	Ionization Energy / $MJ \cdot mol^{-1}$
Neon 1s	32.772 442	86.044 046
Neon 2s	1.930 391	5.068 242
Neon 2p	0.850 410	2.232 751
Argon 1s	118.610 349	311.411 471
Argon 2s	12.322 152	32.351 810
Argon 2p	9.571 464	25.129 879
Argon 3s	1.277 352	3.353 688
Argon 3p	0.591 016	1.551 713

After rounding to the same number of figures, these values agree with those in Table 9.3.

9–37. Go to the website *http://www.ccl.net/cca/data/atomic-RHF-wavefunctions/tables* to reproduce Figure 9.7.

The website contains the data necessary to reproduce the crosses in Figure 9.7, which are ionization energies calculated according to Koopmans's approximation. In this approximation the ionization energy for an orbital is taken as being equal to the negative of the corresponding orbital energy. Thus, the first ionization energy for a neutral atom in Koopmans's approximation is the negative of the *highest* energy orbital for an atom. Going to the website, collecting the highest orbital energies for hydrogen through xenon, and plotting their negatives against atomic number will give Figure 9.7. Looking at Figures 9.3 and 9.5 and Example 9–7, where excerpts from the website are shown, the first four points for the graph can be found. (Of course, the point for hydrogen is just 0.5 E_h.)

Element	Atomic Number	Ionization Energy / E_h
H	1	0.500 000
He	2	0.917 955
Li	3	0.196 323
Be	4	0.309 269

If experimental data are desired for comparison, these can be found at the website *http://www .physics.nist.gov/PhysRefData/ASD/levels_form.html* and added to the graph.

9–38. Use the entries in Table 9.2 to calculate the percentage of the correlation energy of the Eckart and Hylleraas wave functions for a helium atom.

The correlation energy is defined by Equation 9.31 or 9.75 in the chapter, and for the helium atom is

$$E_{corr} = E_{exact} - E_{HF} = -2.903\,724\,375\,E_h - \left(-2.861\,68\,E_h\right) = -0.042\,04\,E_h$$

The percentage of the correlation energy captured by a given calculation is then

$$\% \text{ correlation energy} = \frac{E_{calc} - E_{HF}}{E_{corr}} \times 100$$

For the Eckart wave function,

$$\frac{E_{calc} - E_{HF}}{E_{corr}} \times 100 = \frac{-2.8757\,E_h - \left(-2.861\,68\,E_h\right)}{-0.042\,04\,E_h} \times 100 = 33.35\,\%$$

For the 10-parameter Hylleraas wave function in Table 9.2,

$$\frac{E_{calc} - E_{HF}}{E_{corr}} \times 100 = \frac{-2.903\,63\,E_h - \left(-2.861\,68\,E_h\right)}{-0.042\,04\,E_h} \times 100 = 99.79\,\%$$

This wave function captures nearly all of the correlation energy. This can be compared with the less flexible 3-parameter wave function of Equation 9.17, which does better than the Eckart wave function, but not as well as the 10-parameter wave function.

$$\frac{E_{calc} - E_{HF}}{E_{corr}} \times 100 = \frac{-2.8913\,E_h - \left(-2.861\,68\,E_h\right)}{-0.042\,04\,E_h} \times 100 = 70.46\,\%$$

9–39. Verify that the virtual orbital obtained in the appendix is

$$\psi_2(r) = 1.624\,04\,S_{1s}(\zeta = 1.453\,63) - 1.821\,22\,S_{1s}(\zeta = 2.910\,93)$$

The Fock matrix from the final iteration of the iterative procedure can be found in Table 9.8:

$$F = \begin{pmatrix} -0.888\,940 & -0.904\,199 \\ -0.904\,199 & -0.285\,607 \end{pmatrix}$$

and leads to the secular equation

$$\begin{vmatrix} -0.888\,940 - \varepsilon & -0.904\,199 - 0.837\,524\,\varepsilon \\ -0.904\,199 - 0.837\,524\,\varepsilon & -0.285\,607 - \varepsilon \end{vmatrix} = 0$$

or

$$0.298\,554\,\varepsilon^2 - 0.340\,030\,\varepsilon - 0.563\,688 = 0$$

The value of the lower root is $-0.917\,935$ as indicated in the table, but it is the upper root, $\varepsilon = 2.056\,86$, that leads to the virtual orbital. Using this value in Equations 14 of the Appendix gives the system of equations

$$-2.945\,80\,c_1 - 2.626\,87\,c_2 = 0$$
$$-2.626\,87\,c_1 - 2.342\,47\,c_2 = 0$$

or, $c_1/c_2 = -0.891\,734$. The normalization condition for $\psi_2(r)$ is

$$c_1^2 + 2c_1c_2S_{12} + c_2^2 = 1$$

$$(-0.891\,734)^2\, c_2^2 + 2\,(-0.891\,734)\, c_2^2 S_{12} + c_2^2 = 1$$

$$0.301\,492 c_2^2 = 1$$

$$c_2 = \pm 1.821\,22$$

$$c_1 = \mp 1.624\,04$$

with $S_{12} = 0.837\,524$ from Table 9.7 (see also Problem 9–40). Choosing the lower set of signs for c_1 and c_2 gives

$$\psi_2(r) = 1.624\,04\, S_{1s}(\zeta = 1.453\,63) - 1.821\,22\, S_{1s}(\zeta = 2.910\,93)$$

as the virtual orbital.

9–40.   Show that

$$\langle S_{1s}(\zeta = 1.453\,63) \mid S_{1s}(\zeta = 2.910\,93)\rangle = 0.837\,524$$

In Problem 9–15 it was found that

$$\langle S_{nl}(\zeta_1)|S_{nl}(\zeta_2)\rangle = \frac{(4\zeta_1\zeta_2)^{n+1/2}}{(2n)!}\int_0^\infty r^{2n} e^{-(\zeta_1+\zeta_2)r}\, dr = \frac{(4\zeta_1\zeta_2)^{n+1/2}}{(\zeta_1+\zeta_2)^{2n+1}}$$

Thus,

$$\langle S_{1s}(\zeta = 1.453\,63)|S_{1s}(\zeta = 2.910\,93)\rangle = 0.837\,524$$

9–41.   Verify that the Ψ_{CI} trial function given by Equation 9.76 leads to an energy of $-2.875\,41\, E_h$.

From Equation 9.76,

$$\Psi_{CI}(1,\,2) = c_1\psi_1(1)\psi_1(2) + c_2\psi_2(1)\psi_2(2)$$

where c_1 and c_2 are to be found variationally. From the expressions for $\psi_1(r)$ and $\psi_2(r)$ given just before Equation 9.76 (see also Table 9.8 and Problem 9–39),

$$\begin{pmatrix} \psi_1 \\ \psi_2 \end{pmatrix} = A \begin{pmatrix} \phi_1 \\ \phi_2 \end{pmatrix}$$

$$A = \begin{pmatrix} a_{11} & a_{12} \\ a_{21} & a_{22} \end{pmatrix} = \begin{pmatrix} 0.843\,785 & 0.180\,687 \\ 1.624\,04 & -1.821\,22 \end{pmatrix}$$

$$\phi_1 = S_{1s}(\zeta = 1.453\,63)$$

$$\phi_2 = S_{1s}(\zeta = 2.910\,93)$$

Notice that ϕ_1 and ϕ_2 are the Slater orbitals used in the Appendix to Chapter 9.

The variational calculation requires the three matrix elements specified in the secular determinant just below Equation 9.76. Using the definitions above, these become, with $i, j, k, l = 1, 2$,

$$
\begin{aligned}
H_{11} &= \left\langle \psi_1(1)\psi_1(2) \left| \hat{H} \right| \psi_1(1)\psi_1(2) \right\rangle \\
&= \left\langle \psi_1(1) \left| \left(-\frac{1}{2}\nabla_1^2 - \frac{2}{r_1}\right) \right| \psi_1(1) \right\rangle \left\langle \psi_1(2) | \psi_1(2) \right\rangle \\
&\quad + \left\langle \psi_1(1) | \psi_1(1) \right\rangle \left\langle \psi_1(2) \left| \left(-\frac{1}{2}\nabla_2^2 - \frac{2}{r_2}\right) \right| \psi_1(2) \right\rangle \\
&\quad + \left\langle \psi_1(1)\psi_1(2) \left| \frac{1}{r_{12}} \right| \psi_1(1)\psi_1(2) \right\rangle \\
&= 2\sum_{ij} a_{1i}a_{1j}h_{ij} + \sum_{ijkl} a_{1i}a_{1j}a_{1k}a_{1l} \left\langle \phi_i(1)\phi_j(2) \left| \frac{1}{r_{12}} \right| \phi_k(1)\phi_l(2) \right\rangle
\end{aligned}
$$

since the ψ's are normalized and the two one-electron integrals are equivalent. The integrals remaining, h_{ij} and the two-electron integrals, are in Table 9.7, although the symmetry properties discussed there will be needed to find them all. Similarly,

$$
\begin{aligned}
H_{22} &= \left\langle \psi_2(1)\psi_2(2) \left| \hat{H} \right| \psi_2(1)\psi_2(2) \right\rangle \\
&= \left\langle \psi_2(1) \left| \left(-\frac{1}{2}\nabla_1^2 - \frac{2}{r_1}\right) \right| \psi_2(1) \right\rangle \left\langle \psi_2(2) | \psi_2(2) \right\rangle \\
&\quad + \left\langle \psi_2(1) | \psi_2(1) \right\rangle \left\langle \psi_2(2) \left| \left(-\frac{1}{2}\nabla_2^2 - \frac{2}{r_2}\right) \right| \psi_2(2) \right\rangle \\
&\quad + \left\langle \psi_2(1)\psi_2(2) \left| \frac{1}{r_{12}} \right| \psi_2(1)\psi_2(2) \right\rangle \\
&= 2\sum_{ij} a_{2i}a_{2j}h_{ij} + \sum_{ijkl} a_{2i}a_{2j}a_{2k}a_{2l} \left\langle \phi_i(1)\phi_j(2) \left| \frac{1}{r_{12}} \right| \phi_k(1)\phi_l(2) \right\rangle
\end{aligned}
$$

The off-diagonal element is slightly different, since ψ_1 and ψ_2, like ϕ_1 and ϕ_2, are not orthogonal.

$$
\begin{aligned}
H_{12} &= \left\langle \psi_1(1)\psi_1(2) \left| \hat{H} \right| \psi_2(1)\psi_2(2) \right\rangle \\[2mm]
&= \left\langle \psi_1(1) \left| \left(-\frac{1}{2}\nabla_1^2 - \frac{2}{r_1} \right) \right| \psi_2(1) \right\rangle \langle \psi_1(2)|\psi_2(2) \rangle \\[2mm]
&\quad + \langle \psi_1(1)|\psi_2(1) \rangle \left\langle \psi_1(2) \left| \left(-\frac{1}{2}\nabla_2^2 - \frac{2}{r_2} \right) \right| \psi_2(2) \right\rangle \\[2mm]
&\quad + \left\langle \psi_1(1)\psi_1(2) \left| \frac{1}{r_{12}} \right| \psi_2(1)\psi_2(2) \right\rangle \\[2mm]
&= 2 \left(\sum_{ij} a_{1i}a_{2j}h_{ij} \right) \left(\sum_{kl} a_{1k}a_{2l}S_{kl} \right) + \sum_{ijkl} a_{1i}a_{1j}a_{2k}a_{2l} \left\langle \phi_i(1)\phi_j(2) \left| \frac{1}{r_{12}} \right| \phi_k(1)\phi_l(2) \right\rangle
\end{aligned}
$$

When these expressions for the matrix elements are evaluated using the values of the coefficients given above and the values of the integrals from Table 9.7, the secular determinant following Equation 9.76 is indeed obtained. Solving for the lower root gives $E = -2.875\,41\ E_{\rm h}$.

9–42. Determine the term symbols for a $2s^1 3s^1$ electron configuration.

There are four possible microstates for this configuration because the electrons are in nonequivalent orbitals. The maximum value for M_L is 0, and M_S can be -1, 0, or 1. The four microstates are given in the following table:

M_L	M_S		
	1	0	−1
0	$0^+, 0^+$	$0^+, 0^-; 0^-, 0^+$	$0^-, 0^-$

All the states have $M_L = 0$, so they must correspond to $L = 0$. The largest value of M_S is one, so there must be a state with $S = 1$, corresponding to a ^{3}S state. This ^{3}S state accounts for one microstate from each column, leaving one microstate unaccounted for in the middle column. This microstate has $M_L = 0$ and $M_S = 0$, and so must have $L = 0$ and $S = 0$, which means a ^{1}S state. Appending the allowed values for J gives the two term symbols ^{3}S$_1$ and ^{1}S$_0$ for this electron configuration.

9–43. Determine the term symbols associated with an np^1 electron configuration. Show that these term symbols are the same as for an np^5 electron configuration.

For an np^1 electron configuration, there are six entries in a table of possible sets of m_l and m_s.

m_l	m_s	M_L	M_S	M_J
1	$+\frac{1}{2}$	1	$+\frac{1}{2}$	$+\frac{3}{2}$
1	$-\frac{1}{2}$	1	$-\frac{1}{2}$	$+\frac{1}{2}$
0	$+\frac{1}{2}$	0	$+\frac{1}{2}$	$+\frac{1}{2}$
0	$-\frac{1}{2}$	0	$-\frac{1}{2}$	$-\frac{1}{2}$
-1	$+\frac{1}{2}$	-1	$+\frac{1}{2}$	$-\frac{1}{2}$
-1	$-\frac{1}{2}$	-1	$-\frac{1}{2}$	$-\frac{3}{2}$

The M_L and M_S values given here correspond to a ^{2}P state, and the values of M_J correspond to a value of J of either 1/2 or 3/2. Thus, the term symbols associated with an np^1 electron configuration are $^2P_{3/2}$ and $^2P_{1/2}$. The ground state is determined by using Hund's rules; by Rule 3, the most stable state (and therefore the ground state) is $^2P_{1/2}$.

An np^5 configuration can be thought of as an np^1 configuration because two of the np orbitals are filled and so M_S and M_L are determined by the remaining half-filled p-orbital. Therefore, the term symbols associated with the np^5 configuration will be the same as those for an np^1 configuration. Notice that Equation 9.85 gives the same number of microstates for an np^5 configuration as for the np^1 configuration. One difference, however, is that Hund's Rule 3 gives the $^2P_{3/2}$ state as the ground state, since the subshell is more than half-filled.

9–44. Show that the number of sets of magnetic quantum numbers (m_l) and spin quantum numbers (m_s) associated with any term symbol is equal to $(2L + 1)(2S + 1)$. Apply this result to the np^2 case discussed in Section 9.10, and show that the term symbols ^{1}S, ^{3}P, and ^{1}D account for all the possible sets of magnetic quantum numbers and spin quantum numbers.

Each value of M_L associated with a value of L gives $2S + 1$ entries, and there are $2L + 1$ values of M_L for each value of L. The total number of entries for each term symbol, then, excluding the J subscript, is $(2L + 1)(2S + 1)$. Equation 9.85 with $N = 2$ and $G = 6$ shows there are 15 microstates for the np^2 configuration, so accounting for all fifteen will account for all possible sets of quantum numbers. Applying the result of this problem to the np^2 configuration gives

$$
\begin{array}{ccccccc}
^1\text{S} & & ^3\text{P} & & ^1\text{D} & & \\
(1 \times 1) & + & (3 \times 3) & + & (1 \times 5) & = & 15
\end{array}
$$

These three term symbols account for all possible sets of m_l and m_s.

9–45. Calculate the number of sets of magnetic quantum numbers (m_l) and spin quantum numbers (m_s) for an nd^8 electron configuration. Show that the term symbols ^{1}S, ^{1}D, ^{3}P, ^{3}F, and ^{1}G account for all possible term symbols.

From Equation 9.85, there are a total of

$$
\frac{G!}{N!(G - N)!} = \frac{10!}{8!2!} = 45 \text{ microstates}
$$

in a table of possible values of m_l and m_s. Proceeding as in Problem 9–44,

$$^1\text{S} \qquad\quad ^1\text{D} \qquad\quad ^3\text{P} \qquad\quad ^3\text{F} \qquad\quad ^1\text{G}$$

$$(1 \times 1) \;+\; (1 \times 5) \;+\; (3 \times 3) \;+\; (3 \times 7) \;+\; (1 \times 9) \;=\; 45$$

so these five term symbols account for all possible sets of m_l and m_s.

9–46. Determine the term symbols for the electron configuration $nsnp$. Which term symbol corresponds to the lowest energy?

There are 2 possible sets of m_l and m_s values for the ns electron and 6 possible sets of m_l and m_s values for the np electron, so there are $2 \times 6 = 12$ possible microstates for this configuration because the electrons are in nonequivalent orbitals. The maximum value for M_L is 1, and M_S can be $-1, 0$, or 1. The 12 microstates are given in the following table:

M_L	M_S = 1	M_S = 0	M_S = −1
1	$0^+, 1^+$	$0^-, 1^+; 0^+, 1^-$	$0^-, 1^-$
0	$0^+, 0^+$	$0^-, 0^+; 0^+, 0^-$	$0^-, 0^-$
−1	$0^+, -1^+$	$0^-, -1^+; 0^+, -1^-$	$0^-, -1^-$

The largest value of M_L is 1, and it occurs with $M_S = 1, 0, -1$. Therefore, there must be a state with $L = 1$ and $S = 1$, or ^3P. This state accounts for 9 microstates with $M_L = 1, 0, -1$ and $M_S = 1, 0, -1$ and leaves 3 microstates in the middle column of the table that have a maximum $M_L = 1$ ($L = 1$) and maximum $M_S = 0$ ($S = 0$), or a ^1P state. The values of J can be derived using Equation 9.86. The final result gives the term symbols

$$^3\text{P}_2 \qquad\qquad ^3\text{P}_1 \qquad\qquad ^3\text{P}_0 \qquad\qquad ^1\text{P}_1$$
$$(L+S) \qquad (L+S-1) \qquad (|L-S|) \qquad (L+S)$$

According to Hund's rules, the ground state is $^3\text{P}_0$.

9–47. How many sets of magnetic quantum numbers (m_l) and spin quantum numbers (m_s) are there for an $nsnd$ electron configuration? What are the term symbols? Which term symbol corresponds to the lowest energy?

There are 2 possible sets of m_l and m_s values for the ns electron and 10 possible sets of m_l and m_s values for the nd electron, so there are $2 \times 10 = 20$ possible sets of m_l and m_s values for this configuration. The determination of the term symbols can be carried out as in Problem 9–46. The term symbols corresponding to the 20 microstates are ^3D and ^1D. The values of J are determined using Equation 9.86, and the final result is

$$^3\text{D}_3 \qquad\qquad ^3\text{D}_2 \qquad\qquad ^3\text{D}_1 \qquad\qquad ^1\text{D}_2$$
$$(L+S) \qquad (L+S-1) \qquad (|L-S|) \qquad (L+S)$$

According to Hund's rules, the ground state is $^3\text{D}_1$.

9–48. The term symbols for an nd^2 electron configuration are ^1S, ^1D, ^1G, ^3P, and ^3F. Calculate the values of J associated with each of these term symbols. Which term symbol represents the ground state?

Use Equation 9.86 to find the values of J for each term symbol.

Term Symbol	L	S	J	Full Term Symbol
^1S	0	0	0	$^1S_0$
^1D	2	0	2	$^1D_2$
^1G	4	0	4	$^1G_4$
^3P	1	1	2, 1, 0	$^3P_2, {}^3P_1, {}^3P_0$
^3F	3	1	4, 3, 2	$^3F_4, {}^3F_3, {}^3F_2$

By Hund's rules, the ground state is represented by the term symbol $^3F_2$.

9–49. The term symbols for an np^3 electron configuration are ^2P, ^2D, and ^4S. Calculate the values of J associated with each of these term symbols. Which term symbol represents the ground state?

Use Equation 9.86 to find the values of J for each term symbol.

Term Symbol	L	S	J	Full Term Symbol
^2P	1	$\frac{1}{2}$	$\frac{3}{2}, \frac{1}{2}$	$^2P_{3/2}, {}^2P_{1/2}$
^2D	2	$\frac{1}{2}$	$\frac{5}{2}, \frac{3}{2}$	$^2D_{5/2}, {}^2D_{3/2}$
^4S	0	$\frac{3}{2}$	$\frac{3}{2}$	$^4S_{3/2}$

By Hund's rules, the ground state is represented by the term symbol $^4S_{3/2}$.

9–50. What is the electron configuration of a magnesium atom in its ground state, and what is its ground-state term symbol?

The ground state electron configuration of a magnesium atom is (from general chemistry) $1s^2 2s^2 2p^6 3s^2$, or [Ne]$3s^2$. In the initial discussion of term symbols in Section 9.9, the ns^2 electron configuration was shown to correspond to the term symbol $^1S_0$, so the term symbol for atomic magnesium in the ground state is $^1S_0$.

9–51. Given that the electron configuration of a zirconium atom is [Kr] $4d^2 5s^2$, what is the ground-state term symbol for Zr?

The term symbol for zirconium is determined by the nd^2 electrons. The ground state of an nd^2 electron configuration is $^3F_2$ (Problem 9–48).

9–52. Given that the electron configuration of a palladium atom is $[Kr] 4d^{10}$, what is the ground-state term symbol for Pd?

Because all of the subshells of palladium are filled, the term symbol is $^1S_0$.

9–53. Consider the $1s2p$ electron configuration for helium. What are the states (term symbols) that correspond to this electron configuration? What are the degeneracies of each state if we do not consider spin-orbit coupling? What will happen if you include the effect of spin-orbit coupling?

The general case for this problem (an $nsnp$ configuration) was done in Problem 9–46. The states corresponding to this electron configuration and their degeneracies are then

$^3P_2$	$^3P_1$	$^3P_0$	$^1P_1$
5	3	1	3

According to Hund's rules, the ground state is $^3P_0$. Including the effect of spin-orbit coupling removes the degeneracy of the electronic states, and so spin-orbit coupling splits the lines in an atomic spectrum.

9–54. Use Table 9.5 to calculate the wavelength of the $4d\ ^2D_{3/2} \to 3p\ ^2P_{1/2}$ transition in atomic sodium and compare your result with that given in Figure 9.9. Be sure to use the relation $\lambda_{vac} = 1.000\ 29\ \lambda_{air}$ (see Example 9–14).

Table 9.5 gives

$$\Delta E = E(4d^2D_{3/2}) - E(3p^2P_{1/2})$$

$$= 34\ 548.766\ \text{cm}^{-1} - 16\ 956.172\ \text{cm}^{-1}$$

$$= 17\ 592.594\ \text{cm}^{-1}$$

Therefore, the wavelength of this transition in vacuum is

$$\lambda_{vac} = \frac{1}{\Delta E} = 5.684\ 21 \times 10^{-5}\ \text{cm} = 5684.21\ \text{Å}$$

Converting to the wavelength observed in air,

$$\lambda_{air} = \frac{\lambda_{vac}}{1.000\ 29} = 5682.56\ \text{Å}$$

in excellent agreement with Figure 9.9.

9–55. The orbital designations s, p, d, and f come from an analysis of the spectrum of atomic sodium. The series of lines due to $ns\,^2S \rightarrow 3p\,^2P$ transitions is called the *sharp* (*s*) series; the series due to $np\,^2P \rightarrow 3s\,^2S$ transitions is called the *principal* (*p*) series; the series due to $nd\,^2D \rightarrow 3p\,^2P$ transitions is called the *diffuse* (*d*) series; and the series due to $nf\,^2F \rightarrow 3d\,^2D$ transitions is called the *fundamental* (*f*) series. Identify each of these series in Figure 9.9, and tabulate the wavelengths of the first few lines in each series. Now go to *http://www.physics .nist.gov/PhysRefData/ASD/levels_form.html* to calculate the wavelengths of the first few lines in each series and compare your results with those in Figure 9.9. Be sure to use the relation $\lambda_{vac} = 1.000\,29\,\lambda_{air}$ (see Example 9–14).

Identification of these series in the figure is easily done. The wavelengths found from the figure are

sharp series	principal series	diffuse series	fundamental series
11 404 Å	5 895.9 Å	8 194.8 Å	18 459 Å
11 382 Å	5 889.9 Å	8 183.3 Å	
6 160.7 Å	3 302.9 Å	5 688.2 Å	12 678 Å
6 154.2 Å	3 302.3 Å	5 682.7 Å	
5 153.6 Å	2 853.0 Å	4 982.9 Å	
5 149.1 Å	2 852.8 Å	4 978.6 Å	

The website with the necessary data can be found at *http://www.physics.nist.gov/PhysRefData/ ASD/levels_form.html*, which can be used, after searching for "Na I" to give (using the same procedure as in Problem 9–54, and with all the numbers given in the table expressed in wavenumbers)

sharp series	principal series	diffuse series	fundamental series
8 766.623	16 956.172	12 199.5	5 414.1
8 783.819	16 973.368	12 216.7	
16 227.307	30 266.99	17 575.4	7 884.8
16 244.503	30 272.58	17 592.6	
19 399.252	35 040.38	20 063.4	
19 416.448	35 042.85	20 080.6	

Converting these results to λ_{air} by using the formula $1.000\,29\lambda_{air}\tilde{\nu}_{vac} = 1$ gives

sharp series	principal series	diffuse series	fundamental series
11 404 Å	5 895.8 Å	8 194.7 Å	18 465 Å
11 381 Å	5 889.9 Å	8 183.1 Å	
6 160.7 Å	3 303.0 Å	5 688.1 Å	12 679 Å
6 154.1 Å	3 302.4 Å	5 682.6 Å	
5 153.3 Å	2 853.0 Å	4 982.8 Å	
5 148.8 Å	2 852.8 Å	4 978.5 Å	

in very good agreement with the results from Figure 9.9 tabulated earlier in this problem.

9–56. Use the spectroscopic data for Li, Li$^+$, and Li^{++} in *http://www.physics.nist.gov/ PhysRefData/ASD/levels_form.html* to calculate the energy of a lithium atom and compare your result to the Hartree–Fock value in Figure 9.5.

The energy of the lithium atom is the energy released when three electrons are added to a lithium nucleus or the opposite of the sum of the three ionization energies taking Li $\rightarrow$ Li$^+$ $\rightarrow$ Li^{2+} $\rightarrow$ Li^{3+}. These three ionization energies are found at *http://www.physics.nist.gov/PhysRefData/ ASD/levels_form.html*:

$$E = -\ (43\ 487.150 + 610\ 079.0 + 987\ 661.027)\ \text{cm}^{-1}$$

$$= -1\ 641\ 227.177\ \text{cm}^{-1}$$

$$= -7.478014\ E_h$$

This can be compared with the Hartree-Fock value of $-7.432\ 726\ 924\ E_h$ in Figure 9.5.

9–57. In this problem, we will derive an explicit expression for $V^{\text{eff}}(r_1)$ given by Equation 9.19 using $\phi(\mathbf{r})$ of the form $(Z^3/\pi)^{1/2}e^{-Zr}$. (We have essentially done this problem in Problem 8–39). Start with

$$V^{\text{eff}}(\mathbf{r}_1) = \frac{Z^3}{\pi} \int d\mathbf{r}_2 \frac{e^{-2Zr_2}}{r_{12}}$$

As in Problem 8–39, we use the law of cosines to write

$$r_{12} = (r_1^2 + r_2^2 - 2r_1r_2 \cos \theta)^{1/2}$$

and so V^{eff} becomes

$$V^{\text{eff}}(r_1) = \frac{Z^3}{\pi} \int_0^\infty dr_2\, e^{-2Zr_2}r_2^2 \int_0^{2\pi} d\phi \int_0^\pi \frac{d\theta \sin \theta}{(r_1^2 + r_2^2 - 2r_1r_2 \cos \theta)^{1/2}}$$

Problem 8–39 asks you to show that the integral over θ is equal to $2/r_1$ if $r_1 > r_2$ and is equal to $2/r_2$ if $r_1 < r_2$. Thus, we have

$$V^{\text{eff}}(r_1) = 4Z^3 \left[\frac{1}{r_1} \int_0^{r_1} e^{-Zr_2}r_2^2\, dr_2 + \int_{r_1}^\infty e^{-2Zr_2}r_2\, dr_2 \right]$$

Now show that

$$V^{\text{eff}}(r_1) = \frac{1}{r_1} - e^{-2Zr_1}\left(Z + \frac{1}{r_1} \right)$$

This problem is indeed essentially the same as Problem 8–39, although atomic units are used here. It is also a special case of the integral found in Problem 9–20 with $\zeta_1 = \zeta_2 = Z$. From the solution to Problem 8–39 and using the properties of spherical coordinates,

$$V^{\text{eff}}(r_1, \theta_1, \phi_1) = \frac{Z^3}{\pi} \int_0^{2\pi} d\phi_2 \int_0^\infty dr_2\, r_2^2 e^{-2Zr_2} \int_0^\pi \frac{d\theta_2 \sin\theta_2}{(r_1^2 + r_2^2 - 2r_1 r_2 \cos\theta_2)^{1/2}}$$

$$= 2Z^3 \left(\frac{2}{r_1} \int_0^{r_1} dr_2\, r_2^2 e^{-Zr_2} + 2\int_{r_1}^\infty dr_2\, r_2 e^{-2Zr_2} \right)$$

$$= 4Z^3 \left(\frac{1}{r_1} \int_0^{r_1} dr_2\, r_2^2 e^{-Zr_2} + \int_{r_1}^\infty dr_2\, r_2 e^{-2Zr_2} \right) \tag{1}$$

These integrals were solved in Problem 8–39 where it was found that

$$\frac{1}{r_1} \int_0^{r_1} dr_2\, r_2^2 e^{-2Zr_2/a_0} + \int_{r_1}^\infty dr_2\, r_2 e^{-2Zr_2/a_0} = \frac{a_0^3}{4Z^3} \left[\frac{1}{r_1} - e^{-2Zr_1/a_0} \left(\frac{Z}{a_0} + \frac{1}{r_1} \right) \right]$$

or (in atomic units)

$$\frac{1}{r_1} \int_0^{r_1} dr_2\, r_2^2 e^{-2Zr_2} + \int_{r_1}^\infty dr_2\, r_2 e^{-2Zr_2} = \frac{1}{4Z^3} \left[\frac{1}{r_1} - e^{-2Zr_1} \left(Z + \frac{1}{r_1} \right) \right] \tag{2}$$

Substituting Equation 2 into Equation 1 gives

$$V^{\text{eff}}(r_1) = \frac{4Z^3}{4Z^3} \left[\frac{1}{r_1} - e^{-2Zr_1} \left(Z + \frac{1}{r_1} \right) \right]$$

$$= \frac{1}{r_1} - e^{-2Zr_1} \left(Z + \frac{1}{r_1} \right)$$

9–58. Repeat the previous problem using the expansion of $1/r_{12}$ given in Problem 8–40.

Again, atomic units are used in this problem. The expansion of $1/r_{12}$ needed is given in Problem 8–40. It gives,

$$V^{\text{eff}}(r_1) = \frac{Z^3}{\pi} \int_0^{2\pi} d\phi_2 \int_0^\pi d\theta_2 \, \sin\theta_2 \sum_{l=0}^\infty \sum_{m=-l}^l \left(\frac{4\pi}{2l+1} \right) Y_l^m(\theta_1, \phi_1) Y_l^{m*}(\theta_2, \phi_2) \int_0^\infty dr_2\, r_2^2 \frac{r_<^l}{r_>^{l+1}} e^{-2r_2 Z}$$

As found in Problem 8–40, the angular integration requires that $l = 0$ and $m = 0$, so that

$$V^{\text{eff}}(r_1) = \frac{Z^3(4\pi)}{\pi} \left[\frac{1}{r_1} \int_0^{r_1} dr_2\, r_2^2 e^{-2r_2 Z} + \int_{r_1}^\infty dr_2\, r_2 e^{-2r_2 Z} \right]$$

$$= 4Z^3 \left[\frac{1}{r_1} \int_0^{r_1} dr_2\, r_2^2 e^{-2r_2 Z} + \int_{r_1}^\infty dr_2\, r_2 e^{-2r_2 Z} \right]$$

$$= \frac{1}{r_1} - e^{-2Zr_1} \left(Z + \frac{1}{r_1} \right)$$

The last equality uses the results of Problem 9–57, and the evaluation of the spherical harmonics was done in Problem 8–40.

9–59. Derive equation 4 of the appendix.

To determine the energy given by the trial function

$$\Psi(\mathbf{r}_1, \mathbf{r}_2) = \psi(r_1)\psi(r_2)$$

$$\psi(r) = c_1\phi_1(r) + c_2\phi_2(r)$$

where the $\phi_i(r)$ are the two Slater orbitals given in Equation 2 of the Appendix with $\zeta_1 = 1.453\,63$, and $\zeta_1 = 2.910\,93$, respectively, Ψ must be substituted into

$$E = \frac{\left\langle \Psi \left| \hat{H} \right| \Psi \right\rangle}{\langle \Psi | \Psi \rangle}$$

To do this consider the following integrals in turn. First, from the denominator of the energy expression,

$$
\begin{aligned}
\langle \Psi | \Psi \rangle &= \left\langle \psi(r_1) | \psi(r_1) \right\rangle \left\langle \psi(r_2) | \psi(r_2) \right\rangle \\
&= \left(\sum_{i=1}^{2} \sum_{j=1}^{2} c_i c_j \left\langle \phi_i(r_1) | \phi_j(r_1) \right\rangle \right) \left(\sum_{i=1}^{2} \sum_{j=1}^{2} c_i c_j \left\langle \phi_i(r_2) | \phi_j(r_2) \right\rangle \right) \\
&= \left(\sum_{i=1}^{2} \sum_{j=1}^{2} c_i c_j S_{ij} \right)^2 \\
&= \left(c_1^2 + 2c_1 c_2 S_{12} + c_2^2 \right)^2
\end{aligned}
$$

Of course, if the coefficients are chosen such that Ψ is normalized, this integral will equal one. Next consider the "electron one" portion of the Hamiltonian operator,

$$
\begin{aligned}
\left\langle \Psi \left| -\frac{1}{2}\nabla_1^2 - \frac{Z}{r_1} \right| \Psi \right\rangle &= \left\langle \psi(r_1) \left| -\frac{1}{2}\nabla_1^2 - \frac{Z}{r_1} \right| \psi(r_1) \right\rangle \left\langle \psi(r_2) | \psi(r_2) \right\rangle \\
&= \left(\sum_{i=1}^{2} \sum_{j=1}^{2} c_i c_j \left\langle \phi_i(r_1) \left| \hat{h}(\mathbf{r}_1) \right| \phi_j(r_1) \right\rangle \right) \left(\sum_{i=1}^{2} \sum_{j=1}^{2} c_i c_j \left\langle \phi_i(r_2) | \phi_j(r_2) \right\rangle \right) \\
&= \left(\sum_{i=1}^{2} \sum_{j=1}^{2} c_i c_j h_{ij} \right) \left(\sum_{i=1}^{2} \sum_{j=1}^{2} c_i c_j S_{ij} \right) \\
&= \left(c_1^2 h_{11} + 2c_1 c_2 h_{12} + c_2^2 h_{22} \right) \left(c_1^2 + 2c_1 c_2 S_{12} + c_2^2 \right)
\end{aligned}
$$

Because the electrons are indistinguishable, the "electron two" portion of the Hamiltonian operator will give the same result. The electron-electron repulsion portion of the Hamiltonian operator is a bit more complicated,

$$\left\langle \Psi \left| \frac{1}{r_{12}} \right| \Psi \right\rangle = \left\langle \psi(r_1)\psi(r_2) \left| \frac{1}{r_{12}} \right| \psi(r_1)\psi(r_2) \right\rangle$$

$$= \sum_{i=1}^{2}\sum_{j=1}^{2}\sum_{k=1}^{2}\sum_{l=1}^{2} c_i c_k c_j c_l \left\langle \phi_i(r_1)\phi_k(r_2) \left| \frac{1}{r_{12}} \right| \phi_j(r_1)\phi_l(r_2) \right\rangle$$

$$= \sum_{i=1}^{2}\sum_{j=1}^{2} c_i c_j \left[\sum_{k=1}^{2}\sum_{l=1}^{2} c_k c_l \left\langle \phi_i(r_1)\phi_k(r_2) \left| \frac{1}{r_{12}} \right| \phi_j(r_1)\phi_l(r_2) \right\rangle \right]$$

$$= \left(c_1^2 + 2c_1 c_2 S_{12} + c_2^2 \right) \sum_{i=1}^{2}\sum_{j=1}^{2} c_i c_j g_{ij}$$

where the g_{ij} are defined by Equation 6 of the Appendix. (In the first printing of the textbook, the factor $c_1^2 + 2c_1 c_2 S_{12} + c_2^2$ in the denominator was missing in Equation 6. This factor is equal to 1 if normalized wave functions are used.)

$$g_{ij} = \frac{\displaystyle\sum_{k=1}^{2}\sum_{l=1}^{2} c_k c_l \left\langle \phi_i(r_1)\phi_k(r_2) \left| \frac{1}{r_{12}} \right| \phi_j(r_1)\phi_l(r_2) \right\rangle}{c_1^2 + 2c_1 c_2 S_{12} + c_2^2}$$

Thus,

$$\left\langle \Psi \left| \hat{H} \right| \Psi \right\rangle = \left\langle \Psi \left| \hat{h}(\mathbf{r}_1) \right| \Psi \right\rangle + \left\langle \Psi \left| \hat{h}(\mathbf{r}_2) \right| \Psi \right\rangle + \left\langle \Psi \left| \frac{1}{r_{12}} \right| \Psi \right\rangle$$

$$= \left[2\left(c_1^2 h_{11} + 2c_1 c_2 h_{12} + c_2^2 h_{22} \right) + c_1^2 g_{11} + 2c_1 c_2 g_{12} + c_2^2 g_{22} \right] \left(c_1^2 + 2c_1 c_2 S_{12} + c_2^2 \right)$$

and

$$E = \frac{\left\langle \Psi \left| \hat{H} \right| \Psi \right\rangle}{\langle \Psi | \Psi \rangle}$$

$$= \frac{\left[2\left(c_1^2 h_{11} + 2c_1 c_2 h_{12} + c_2^2 h_{22} \right) + c_1^2 g_{11} + 2c_1 c_2 g_{12} + c_2^2 g_{22} \right] \left(c_1^2 + 2c_1 c_2 S_{12} + c_2^2 \right)}{\left(c_1^2 + 2c_1 c_2 S_{12} + c_2^2 \right)^2}$$

$$= \frac{2\left(c_1^2 h_{11} + 2c_1 c_2 h_{12} + c_2^2 h_{22} \right) + c_1^2 g_{11} + 2c_1 c_2 g_{12} + c_2^2 g_{22}}{c_1^2 + 2c_1 c_2 S_{12} + c_2^2}$$

which is Equation 4 of the Appendix.

9–60. Evaluate $\left\langle \phi_i \left| \dfrac{1}{r} \right| \phi_j \right\rangle$ for a $1s$ Slater orbital.

In general for two Slater orbitals

$$\phi_i = \frac{(2\zeta_i)^{n+1/2}}{[(2n)!]^{1/2}} r^{n-1} e^{-\zeta_i r} Y_l^{m_l}(\theta, \phi)$$

with the same values for n, l, and m_l but different values of ζ,

$$\left\langle S_{nlm_l}(\zeta_i) \left| \frac{1}{r} \right| S_{nlm_l}(\zeta_j) \right\rangle = \frac{(4\zeta_i\zeta_j)^{n+1/2}}{(2n)!} \int_0^\infty dr\, r^{2n-1} e^{-(\zeta_i+\zeta_j)r}$$

$$\times \int_0^\pi d\theta\, \sin\theta \int_0^{2\pi} d\phi\, Y_l^{m_l*}(\theta, \phi) Y_l^{m_l}(\theta, \phi)$$

$$= \frac{(4\zeta_i\zeta_j)^{n+1/2}}{(2n)!} \int_0^\infty r^{2n-1} e^{-(\zeta_i+\zeta_j)r}\, dr$$

$$= \frac{(4\zeta_i\zeta_j)^{n+1/2}}{2n\,(\zeta_i + \zeta_j)^{2n}}$$

Specializing this result to $1s$ Slater orbitals just requires setting $n = 1$:

$$\left\langle \phi_i \left| \frac{1}{r} \right| \phi_j \right\rangle = \frac{(4\zeta_i\zeta_j)^{3/2}}{2\,(\zeta_i + \zeta_j)^2} = \frac{4\left(\zeta_i^3\zeta_j^3\right)^{1/2}}{(\zeta_i + \zeta_j)^2}$$

9–61. Use a program such as *MathCad* or *Mathematica* to verify the numerical values in Table 9.7.

Enter the following expressions for the one-electron integrals into the program, using the necessary formatting requirements, to obtain the results shown in Table 9.7. The factors of 4π come from the integrations over the angular coordinates.

$$S_{1s}(\zeta, r) = \frac{(2\zeta)^{3/2}}{\sqrt{8\pi}} e^{-\zeta r}$$

$$\zeta_1 = 1.453\,63$$

$$\zeta_2 = 2.910\,93$$

$$\phi_1(r) = S_{1s}(\zeta_1, r)$$

$$\phi_2(r) = S_{1s}(\zeta_2, r)$$

$$S_{11} = 4\pi \int_0^\infty r^2 \phi_1^2(r)\, dr$$

$$S_{22} = 4\pi \int_0^\infty r^2 \phi_2^2(r)\, dr$$

$$S_{12} = 4\pi \int_0^\infty r^2 \phi_1(r)\phi_2(r)\, dr$$

$$S_{21} = 4\pi \int_0^\infty r^2 \phi_2(r)\phi_1(r)\, dr$$

$$D\phi_1(r) = \frac{d}{dr}\phi_1(r)$$

$$D\phi_2(r) = \frac{d}{dr}\phi_2(r)$$

$$h_{11} = -2\pi \int_0^\infty \phi_1(r)\frac{d}{dr}\left[r^2 D\phi_1(r)\right]\,dr - 8\pi \int_0^\infty r\phi_1^2(r)\,dr$$

$$h_{22} = -2\pi \int_0^\infty \phi_2(r)\frac{d}{dr}\left[r^2 D\phi_2(r)\right]\,dr - 8\pi \int_0^\infty r\phi_2^2(r)\,dr$$

$$h_{12} = -2\pi \int_0^\infty \phi_1(r)\frac{d}{dr}\left[r^2 D\phi_2(r)\right]\,dr - 8\pi \int_0^\infty r\phi_1(r)\phi_2(r)\,dr$$

$$h_{21} = -2\pi \int_0^\infty \phi_2(r)\frac{d}{dr}\left[r^2 D\phi_1(r)\right]\,dr - 8\pi \int_0^\infty r\phi_2(r)\phi_1(r)\,dr$$

The various two electron integrals are very similar to each other, differing only in the choices for the subscripts of the four ϕ's that appear in each. The general form is

$$g_{ijkl} = \left\langle \phi_i(1)\phi_j(2) \left| \frac{1}{r_{12}} \right| \phi_k(1)\phi_l(2) \right\rangle = 8\pi^2 \int_0^\infty r_1^2 \phi_i(r_1)\phi_k(r_1) \int_0^\infty r_2^2 \phi_j(r_2)\phi_l(r_2)$$

$$\times \int_0^\pi \frac{\sin\theta}{\sqrt{r_1^2 + r_2^2 - 2r_1 r_2 \cos\theta}}\, d\theta\, dr_2\, dr_1$$

where i, j, k, l are each 1 or 2. The six distinct two-electron integrals appearing in Table 9.7 can be entered into the calculation program, having made the previous definitions appearing above, to verify the results given there. It may be necessary to set a stricter than default convergence criterion in your program to achieve agreement to the 6 figures in the table.

CHAPTER 10

The Chemical Bond: One- and Two-Electron Molecules

PROBLEMS AND SOLUTIONS

10–1. Express the Hamiltonian operator for a hydrogen molecule in atomic units.

The Hamiltonian operator $\hat{H}_{el}$ for a hydrogen molecule is

$$\hat{H}_{el} = -\frac{\hbar^2}{2m_e}\left(\nabla_1^2 + \nabla_2^2\right) - \frac{e^2}{4\pi\epsilon_0 r_{1A}} - \frac{e^2}{4\pi\epsilon_0 r_{1B}} - \frac{e^2}{4\pi\epsilon_0 r_{2A}} - \frac{e^2}{4\pi\epsilon_0 r_{2B}}$$
$$+ \frac{e^2}{4\pi\epsilon_0 r_{12}} + \frac{e^2}{4\pi\epsilon_0 R}$$

Now let mass, charge, angular momentum, and permittivity in the above expression be expressed in units of m_e, e, $\hbar$, and $4\pi\epsilon_0$, respectively, to give

$$\hat{H} = -\frac{1}{2}\left(\nabla_1^2 + \nabla_2^2\right) - \frac{1}{r_{1A}} - \frac{1}{r_{1B}} - \frac{1}{r_{2A}} - \frac{1}{r_{2B}} + \frac{1}{r_{12}} + \frac{1}{R}$$

10–2. The vibrational energy levels of H_2^+ (in cm^{-1}) are given by Chase, M. W., Jr., et al., *J. Phys. Chem. Ref. Data* 1985, vol. 14, supplement no. 1, also known as the JANAF Thermochemical Tables. These tables were updated in 1998 by M. W. Chase, Jr. as the *NIST–JANAF Thermochemical Tables*, 4th ed., monograph no. 9 (parts 1 and 2), available from the American Institute of Physics. See *www.nist.gov/srd/jpcrd_28.htm#janaf*

$$G(v) = 2323.23\left(v + \frac{1}{2}\right) - 67.39\left(v + \frac{1}{2}\right)^2 + 0.93\left(v + \frac{1}{2}\right)^3 - 0.029\left(v + \frac{1}{2}\right)^4$$

The molecule dissociates in the limit that $\Delta G(v) = G(v + 1) - G(v) \to 0$, and so v_{max}, the maximum vibrational quantum number, is given by $\Delta G(v_{max}) = 0$. Plot $\Delta G(v) = G(v + 1) - G(v)$ against v and show that $v_{max} = 18$.

Use $G(v)/cm^{-1}$ to determine an expression for $\Delta G(v)$:

$$\Delta G(v) = G(v+1) - G(v) = 2323.23\left[\left(v+\frac{3}{2}\right)-\left(v+\frac{1}{2}\right)\right] - 67.39\left[\left(v+\frac{3}{2}\right)^2-\left(v+\frac{1}{2}\right)^2\right]$$

$$+ 0.93\left[\left(v+\frac{3}{2}\right)^3-\left(v+\frac{1}{2}\right)^3\right] - 0.029\left[\left(v+\frac{3}{2}\right)^4-\left(v+\frac{1}{2}\right)^4\right]$$

$$= 2323.23 - 67.39\,(2v+2) + 0.93\left(3v^2+6v+13/4\right) - 0.029\left(4v^3+12v^2+13v+5\right)$$

$$= 2191.33 - 129.577v + 2.442v^2 - 0.116v^3$$

A plot of $\Delta G(v)$ vs v is shown below:

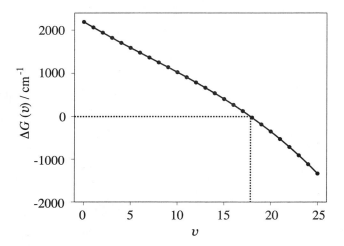

To determine v_{max}, set $\Delta G(v) = 0$.

$$2191.33 - 129.577v_{max} + 2.442v_{max}^2 - 0.116v_{max}^3 = 0$$

Of the three roots for v_{max}, two are imaginary. The real root is $v_{max} = 17.829$ (as indicated by the vertical dashed line in the figure), which, when rounded to the nearest integer, becomes $v_{max} = 18$.

10–3. The JANAF tables (see the previous problem) give a value of $D_0 = 255.76\ \text{kJ·mol}^{-1}$. Use the vibrational data in the previous problem to show that $D_e = 269.46\ \text{kJ·mol}^{-1} = 0.102\,63\,E_h$.

From the $G(v)$ expression in the last problem,

$$G(0) = \left[2323.23\left(\frac{1}{2}\right) - 67.39\left(\frac{1}{2}\right)^2 + 0.93\left(\frac{1}{2}\right)^3 - 0.029\left(\frac{1}{2}\right)^4\right]\text{cm}^{-1}$$

$$= 1144.88\ \text{cm}^{-1}$$

$$= 13.696\ \text{kJ} \cdot \text{mol}^{-1}$$

The dissociation energy measured from the bottom of the potential curve is

$$D_e = D_0 + G(0)$$

$$= 255.76 \text{ kJ} \cdot \text{mol}^{-1} + 13.696 \text{ kJ} \cdot \text{mol}^{-1}$$

$$= 269.46 \text{ kJ} \cdot \text{mol}^{-1}$$

Since $1E_h = 2625.5 \text{ kJ} \cdot \text{mol}^{-1}$, $D_e = 0.102\,63\,E_h$.

10–4. The value of D_e for H_2^+ is $0.102\,63\,E_h$, yet the minimum energy in Figure 10.2 is $-0.602\,63\,E_h$. Why is there a difference?

The value of D_e represents the energy difference between the minimum of the energy curve and the dissociation limit. The dissociation limit of the curve in Figure 10.2 is $-1/2\,E_h$, the energy of a ground-state hydrogen atom. Thus,

$$D_e = -1/2\,E_h - (-0.602\,64\,E_h) = 0.102\,64\,E_h$$

10–5. Plot the product $1s_A\,1s_B$ along the internuclear axis for several values of R.

Let the nuclei be located along the x axis at $x = -R/2$ and $x = R/2$, respectively. Then

$$1s_A = \left(\frac{1}{\pi}\right)^{1/2} e^{-|x-R/2|}$$

$$1s_B = \left(\frac{1}{\pi}\right)^{1/2} e^{-|x+R/2|}$$

and the product is

$$1s_A\,1s_B = \left(\frac{1}{\pi}\right) e^{-|x-R/2|} e^{-|x+R/2|} = \left(\frac{1}{\pi}\right) e^{-(|x-R/2|+|x+R/2|)}$$

A plot for four different values of R is shown below:

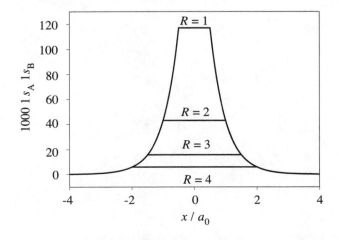

10–6. The overlap integral, Equation 10.13c, and other integrals that arise in two-center systems like H_2^+ are called *two-center integrals*. Two-center integrals are most easily evaluated by using a coordinate system called *elliptical coordinates*. In this coordinate system (Figure 10.4), there are two fixed points separated by a distance R. A point P is given by the three coordinates

$$\lambda = \frac{r_A + r_B}{R}$$

$$\mu = \frac{r_A - r_B}{R}$$

and the angle ϕ, which is the angle that the (r_A, r_B, R) triangle makes about the interfocal axis. The differential volume element in elliptical coordinates is

$$d\mathbf{r} = \frac{R^3}{8}(\lambda^2 - \mu^2)d\lambda d\mu d\phi$$

Given the above definitions of λ, μ, and ϕ, show that

$$1 \le \lambda < \infty \qquad -1 \le \mu \le 1 \qquad 0 \le \phi \le 2\pi$$

Now use elliptical coordinates to evaluate the overlap integral (Equation 10.13c):

$$S = \int d\mathbf{r} \, 1s_A \, 1s_B$$

$$= \frac{Z^3}{\pi} \int d\mathbf{r} e^{-Zr_A} e^{-Zr_B}$$

The triangle inequality shows that the sum $r_A + r_B$ can never be less than R; thus, $1 \le \lambda < \infty$. In addition, the difference $r_A - r_B$ is bounded by $\pm R$, giving $-1 \le \mu \le 1$. The angle ϕ can lie anywhere within a full revolution about the internuclear axis, or $0 \le \phi \le 2\pi$. Using these elliptical coordinates, the overlap integral is

$$S = \frac{Z^3}{\pi} \int d\mathbf{r} \, e^{-Zr_A} e^{-Zr_B}$$

$$= \frac{Z^3}{\pi} \int_0^{2\pi} d\phi \int_1^\infty d\lambda \int_{-1}^1 d\mu \, \frac{R^3}{8} \left(\lambda^2 - \mu^2\right) e^{-Z(r_A + r_B)}$$

$$= \frac{R^3 Z^3}{4} \int_1^\infty d\lambda \int_{-1}^1 d\mu \, \left(\lambda^2 - \mu^2\right) e^{-ZR\lambda}$$

$$= \frac{R^3 Z^3}{4} \int_1^\infty d\lambda \, e^{-ZR\lambda} \int_{-1}^1 d\mu \, \left(\lambda^2 - \mu^2\right)$$

$$= \frac{R^3 Z^3}{4} \int_1^\infty d\lambda \, e^{-ZR\lambda} \left(2\lambda^2 - \frac{2}{3}\right)$$

$$= \frac{R^3 Z^3}{2} \left[\left(\frac{1}{ZR} + \frac{2}{Z^2 R^2} + \frac{2}{Z^3 R^3}\right) e^{-ZR} - \frac{1}{3ZR} e^{-ZR}\right]$$

$$= e^{-ZR} \left(1 + ZR + \frac{Z^2 R^2}{3}\right)$$

10–7. In this problem, we evaluate the overlap integral (Equation 10.13c) using spherical coordinates centered on atom A. The integral to evaluate is (Problem 10–6)

$$S(R) = \frac{1}{\pi} \int d\mathbf{r}_A e^{-r_A} e^{-r_B}$$

$$= \frac{1}{\pi} \int_0^\infty dr_A e^{-r_A} r_A^2 \int_0^{2\pi} d\phi \int_0^\pi d\theta \sin\theta e^{-r_B}$$

where r_A, r_B, and θ are shown in the figure.

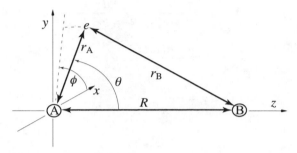

To evaluate the above integral, we must express r_B in terms of r_A, θ, and ϕ. We can do this using the law of cosines:

$$r_B = (r_A^2 + R^2 - 2r_A R \cos\theta)^{1/2}$$

So, the first integral we must consider is

$$I = \int_0^\pi e^{-(r_A^2 + R^2 - 2r_A R \cos\theta)^{1/2}} \sin\theta d\theta$$

As usual, let $\cos\theta = x$ to get

$$I = \int_{-1}^1 e^{-(r_A^2 + R^2 - 2r_A Rx)^{1/2}} dx$$

Now let $u = (r_A^2 + R^2 - 2r_A Rx)^{1/2}$ and show that

$$dx = -\frac{u\,du}{r_A R}$$

Show that the limits of the integration over u are $u = r_A + R$ when $x = -1$ and $u = |R - r_A|$ when $x = 1$. Then show that

$$I = \frac{1}{r_A R} \left[e^{-(R-r_A)}(R + 1 - r_A) - e^{-(R+r_A)}(R + 1 + r_A) \right] \qquad r_A < R$$

$$= \frac{1}{r_A R} \left[e^{-(r_A-R)}(r_A - R + 1) - e^{-(R+r_A)}(R + 1 + r_A) \right] \qquad r_A > R$$

Now substitute this result into $S(R)$ above to get

$$S(R) = e^{-R} \left(1 + R + \frac{R^2}{3} \right)$$

Compare the length of this problem to Problem 10–6.

Let $\cos\theta = x$ to rewrite the integral I:

$$I = \int_0^\pi e^{-(r_A^2+R^2-2r_A R\cos\theta)^{1/2}}\sin\theta\,d\theta$$

$$= \int_{-1}^1 e^{-(r_A^2+R^2-2r_A Rx)^{1/2}}dx$$

Let $u = (r_A^2 + R^2 - 2r_A Rx)^{1/2}$. Then

$$du = -\frac{r_A R dx}{(r_A^2 + R^2 - 2r_A Rx)^{1/2}}$$

$$= -\frac{r_A R dx}{u}$$

$$dx = -\frac{u\,du}{r_A R}$$

When $x = -1$,

$$u = (r_A^2 + R^2 + 2r_A R)^{1/2} = r_A + R$$

Only the positive square root of $(r_A^2 + R^2 + 2r_A R)^{1/2}$ is considered because r_B, and therefore, u, is a positive quantity. When $x = 1$,

$$u = (r_A^2 + R^2 - 2r_A R)^{1/2} = |R - r_A|$$

Once again, only the positive square root is considered.

The integral I becomes

$$I = \frac{1}{r_A R}\int_{|R-r_A|}^{r_A+R} ue^{-u}\,du$$

When $r_A < R$,

$$I = \frac{1}{r_A R}\int_{R-r_A}^{r_A+R} ue^{-u}\,du$$

$$= \frac{1}{r_A R}\left[-e^{-u}(1+u)\right]\Big|_{R-r_A}^{r_A+R}$$

$$= \frac{1}{r_A R}\left[e^{-(R-r_A)}(R + 1 - r_A) - e^{-(R+r_A)}(R + 1 + r_A)\right]$$

On the other hand, when $r_A > R$,

$$I = \frac{1}{r_A R}\int_{r_A-R}^{r_A+R} ue^{-u}\,du$$

$$= \frac{1}{r_A R}\left[-e^{-u}(1+u)\right]\Big|_{r_A-R}^{r_A+R}$$

$$= \frac{1}{r_A R}\left[e^{-(r_A-R)}(r_A - R + 1) - e^{-(R+r_A)}(R + 1 + r_A)\right]$$

Substituting these expressions for the I integral into the $S(R)$ expression gives

$$S(R) = \frac{1}{\pi} \int_0^\infty dr_A \, e^{-r_A} r_A^2 \int_0^{2\pi} d\phi \int_0^\pi d\theta \, \sin\theta e^{-r_B}$$

$$= 2 \int_0^R dr_A \, e^{-r_A} \frac{r_A}{R} \left[e^{-(R-r_A)}(R+1-r_A) - e^{-(R+r_A)}(R+1+r_A) \right]$$

$$+ 2 \int_R^\infty dr_A \, e^{-r_A} \frac{r_A}{R} \left[e^{-(r_A-R)}(r_A - R + 1) - e^{-(R+r_A)}(R+1+r_A) \right]$$

$$= \frac{2e^{-R}}{R} \int_0^R r_A (R+1-r_A) \, dr_A - \frac{2e^{-R}}{R} \int_0^R e^{-2r_A} r_A (R+1+r_A) \, dr_A$$

$$+ \frac{2e^R}{R} \int_R^\infty e^{-2r_A} r_A (r_A - R + 1) \, dr_A - \frac{2e^{-R}}{R} \int_R^\infty e^{-2r_A} r_A (R+1+r_A) \, dr_A$$

$$= \frac{2e^{-R}}{R} \int_0^R r_A (R+1-r_A) \, dr_A + \frac{2e^R}{R} \int_R^\infty e^{-2r_A} r_A (r_A - R + 1) \, dr_A$$

$$- \frac{2e^{-R}}{R} \int_0^\infty e^{-2r_A} r_A (R+1+r_A) \, dr_A$$

$$= \frac{2e^{-R}}{R} \left(\frac{R^3}{6} + \frac{R^2}{2} \right) + \frac{2e^R}{R} e^{-2R} \left(\frac{3R+2}{4} \right) - \frac{2e^{-R}}{R} \left(\frac{R+2}{4} \right)$$

$$= e^{-R} \left(1 + R + \frac{R^2}{3} \right)$$

Using spherical coordinates to evaluate the overlap integral in this problem is a much more lengthy process than using elliptical coordinates in Problem 10–6.

10–8. Show that the first two integrals in Equation 10.16 come out to be 1/2 and −1, respectively.

The function $1s_A$ is the $1s$ hydrogen atomic orbital centered on nucleus A:

$$1s_A = \frac{1}{\sqrt{\pi}} e^{-r_A}$$

Before evaluating the first integral in Equation 10.16, determine $\nabla^2 1s_A$:

$$\nabla^2 1s_A = \left[\frac{1}{r_A^2} \frac{\partial}{\partial r_A} \left(r_A^2 \frac{\partial}{\partial r_A} \right) + \frac{1}{r_A^2 \sin\theta_A} \frac{\partial}{\partial \theta_A} \left(\sin\theta_A \frac{\partial}{\partial \theta_A} \right) + \frac{1}{r_A^2 \sin^2\theta_A} \frac{\partial^2}{\partial \phi_A^2} \right] \left(\frac{1}{\sqrt{\pi}} e^{-r_A} \right)$$

$$= \frac{1}{\sqrt{\pi}} \frac{1}{r_A^2} \frac{\partial}{\partial r_A} \left(-r_A^2 e^{-r_A} \right)$$

$$= \frac{1}{\sqrt{\pi}} \frac{1}{r_A^2} \left(-2r_A e^{-r_A} + r_A^2 e^{-r_A} \right)$$

The first integral is then

$$\left\langle 1s_A \left| -\frac{1}{2}\nabla^2 \right| 1s_A \right\rangle = \int_0^{2\pi} d\phi_A \int_0^\pi d\theta_A \sin\theta_A \int_0^\infty dr_A \left\{ r_A^2 \left(\frac{1}{\sqrt{\pi}} e^{-r_A} \right) \left(-\frac{1}{2} \right) \right.$$

$$\left. \times \left[\frac{1}{\sqrt{\pi}} \frac{1}{r_A^2} \left(-2r_A e^{-r_A} + r_A^2 e^{-r_A} \right) \right] \right\}$$

$$= -\frac{1}{2\pi}(4\pi)\int_0^\infty dr_A \left(-2r_A e^{-2r_A} + r_A^2 e^{-2r_A} \right)$$

$$= -2\left(-\frac{1}{2} + \frac{1}{4} \right)$$

$$= \frac{1}{2}$$

The second integral in Equation 10.16 is

$$\left\langle 1s_A \left| -\frac{1}{r_A} \right| 1s_A \right\rangle = \int_0^{2\pi} d\phi_A \int_0^\pi d\theta_A \sin\theta_A \int_0^\infty dr_A \, r_A^2 \left(\frac{1}{\sqrt{\pi}} e^{-r_A} \right) \left(-\frac{1}{r_A} \right) \left(\frac{1}{\sqrt{\pi}} e^{-r_A} \right)$$

$$= -\frac{1}{\pi}(4\pi)\int_0^\infty dr_A \, r_A e^{-2r_A}$$

$$= -4\left(\frac{1}{4} \right)$$

$$= -1$$

10–9. Show that the fourth integral in Equation 10.16 is equal to $1/R$.

Since R is independent of the electronic coordinates contained in the wave function centered on nucleus A and since $1s_A$ is normalized, the fourth integral in Equation 10.16 becomes

$$\left\langle 1s_A \left| \frac{1}{R} \right| 1s_A \right\rangle = \frac{1}{R}\langle 1s_A \mid 1s_A \rangle$$

$$= \frac{1}{R}$$

10–10. Use the elliptical coordinate system of Problem 10–6 to derive analytic expressions for S, J, and K for the simple molecular orbital treatment of H_2^+.

Two of the elliptical coordinates λ and μ are related to r_A and r_B:

$$\lambda = \frac{r_A + r_B}{R}$$

$$\mu = \frac{r_A - r_B}{R}$$

Thus,

$$r_A = \frac{R(\lambda + \mu)}{2}$$

$$r_B = \frac{R(\lambda - \mu)}{2}$$

Now determine J:

$$J = \int d\mathbf{r}\, 1s_A^* \left(-\frac{1}{r_B}\right) 1s_A$$

$$= \frac{R^3}{8\pi} \int_0^{2\pi} d\phi \int_1^\infty d\lambda \int_{-1}^1 d\mu\, (\lambda^2 - \mu^2) \left(-\frac{1}{r_B}\right) e^{-2r_A}$$

$$= \frac{R^3}{8\pi} \int_0^{2\pi} d\phi \int_1^\infty d\lambda \int_{-1}^1 d\mu\, (\lambda^2 - \mu^2) \left[-\frac{2}{R(\lambda - \mu)}\right] e^{-R(\lambda+\mu)}$$

$$= -\frac{R^2}{2} \int_1^\infty d\lambda\, e^{-R\lambda} \int_{-1}^1 d\mu\, (\lambda + \mu) e^{-R\mu}$$

$$= -\frac{R^2}{2} \int_1^\infty d\lambda\, e^{-R\lambda} \left[\lambda \left(\frac{e^R - e^{-R}}{R}\right) + \frac{e^R(1-R) - e^{-R}(1+R)}{R^2} \right]$$

$$= -\frac{R^2}{2} \left\{ \left(\frac{e^R - e^{-R}}{R}\right)\left(\frac{R+1}{R^2}\right) e^{-R} + \left[\frac{e^R(1-R) - e^{-R}(R+1)}{R^2}\right] \frac{e^{-R}}{R} \right\}$$

$$= -\frac{1}{2R} \left[2 - 2e^{-2R}(1+R)\right]$$

$$= e^{-2R}\left(1 + \frac{1}{R}\right) - \frac{1}{R}$$

and K:

$$K = -\int d\mathbf{r}\, 1s_A^* \left(-\frac{1}{r_A}\right) 1s_B$$

$$= \frac{R^3}{8\pi} \int_0^{2\pi} d\phi \int_1^\infty d\lambda \int_{-1}^1 d\mu\, (\lambda^2 - \mu^2) \left(-\frac{1}{r_A}\right) e^{-(r_A + r_B)}$$

$$= -\frac{R^3}{4} \int_1^\infty d\lambda \int_{-1}^1 d\mu\, (\lambda^2 - \mu^2) \frac{2}{R(\lambda + \mu)} e^{-R\lambda}$$

$$= -\frac{R^2}{2} \int_1^\infty d\lambda\, e^{-R\lambda} \int_{-1}^1 d\mu\, (\lambda - \mu)$$

$$= -R^2 \int_1^\infty d\lambda\, \lambda e^{-R\lambda}$$

$$= -R^2 \left[\frac{e^{-R}(1+R)}{R^2}\right]$$

$$= -e^{-R}(1+R)$$

An analytic expression for S has already been derived in Problem 10–6. In the case of H_2^+, $Z = 1$, and the expression reduces to

$$S = e^{-R}\left(1 + R + \frac{R^2}{3}\right)$$

10–11. Let's use the method that we developed in Problem 10–7 to evaluate the coulomb integral, J, given by Equation 10.18. Let

$$J = -\int \frac{d\mathbf{r}\, 1s_A^*\, 1s_A}{r_B} = -\frac{1}{\pi}\int d\mathbf{r}\, \frac{e^{-2r_A}}{(r_A^2 + R^2 - 2r_A R \cos\theta)^{1/2}}$$

$$= -\frac{1}{\pi}\int_0^\infty dr_A r_A^2 e^{-2r_A} \int_0^{2\pi} d\phi \int_0^\pi \frac{d\theta\, \sin\theta}{(r_A^2 + R^2 - 2r_A R \cos\theta)^{1/2}}$$

Using the approach of Problem 10–7, let $\cos\theta = x$ and $u = (r_A^2 + R^2 - 2r_A R \cos\theta)^{1/2}$ to show that

$$J = \frac{2}{R}\int_0^\infty dr_A r_A e^{-2r_A} \int_{R+r_A}^{|R-r_A|} du = \frac{2}{R}\int_0^\infty dr_A r_A e^{-2r_A}[|R - r_A| - (R + r_A)]$$

$$= e^{-2R}\left(1 + \frac{1}{R}\right) - \frac{1}{R}$$

Hint: You need to use the integrals

$$\int x e^{ax} dx = e^{ax}\left(\frac{x}{a} - \frac{1}{a^2}\right)$$

and

$$\int x^2 e^{ax} dx = e^{ax}\left(\frac{x^2}{a} - \frac{2x}{a^2} + \frac{2}{a^3}\right)$$

Letting $\cos\theta = x$ and $u = (r_A^2 + R^2 - 2r_A R \cos\theta)^{1/2}$, the integral over θ becomes

$$\int_0^\pi \frac{d\theta\, \sin\theta}{(r_A^2 + R^2 - 2r_A R \cos\theta)^{1/2}} = \int_{-1}^1 \frac{dx}{(r_A^2 + R^2 - 2r_A Rx)^{1/2}}$$

$$= -\int_{R+r_A}^{|R-r_A|} \frac{u\,du}{r_A R u}$$

$$= -\frac{1}{r_A R}\int_{R+r_A}^{|R-r_A|} du$$

$$= -\frac{1}{r_A R}\left[|R - r_A| - (R + r_A)\right]$$

Substitute this into the J integral to give

$$J = -\frac{1}{\pi} \int_0^\infty dr_A \, r_A^2 e^{-2r_A} \int_0^{2\pi} d\phi \int_0^\pi \frac{d\theta \sin\theta}{(r_A^2 + R^2 - 2r_A R \cos\theta)^{1/2}}$$

$$= 2 \int_0^\infty dr_A \, r_A^2 e^{-2r_A} \frac{1}{r_A R} \left[|R - r_A| - (R + r_A) \right]$$

$$= \frac{2}{R} \int_0^\infty dr_A \, e^{-2r_A} r_A \left[|R - r_A| - (R + r_A) \right]$$

$$= \frac{2}{R} \left[\int_0^R dr_A \, e^{-2r_A} r_A \, (R - r_A) + \int_R^\infty dr_A \, e^{-2r_A} r_A \, (r_A - R) - \int_0^\infty dr_A \, e^{-2r_A} r_A \, (R + r_A) \right]$$

$$= 2e^{-2r_A} \left(-\frac{r_A}{2} - \frac{1}{4} \right) \Big|_0^R - \frac{2}{R} e^{-2r_A} \left(-\frac{r_A^2}{2} - \frac{r_A}{2} - \frac{1}{4} \right) \Big|_0^R$$

$$\quad + \frac{2}{R} e^{-2r_A} \left(-\frac{r_A^2}{2} - \frac{r_A}{2} - \frac{1}{4} \right) \Big|_R^\infty - 2e^{-2r_A} \left(-\frac{r_A}{2} - \frac{1}{4} \right) \Big|_R^\infty$$

$$\quad - 2e^{-2r_A} \left(-\frac{r_A}{2} - \frac{1}{4} \right) \Big|_0^\infty + \frac{2}{R} e^{-2r_A} \left(-\frac{r_A^2}{2} - \frac{r_A}{2} - \frac{1}{4} \right) \Big|_0^\infty$$

$$= e^{-2R} \left(-R - \frac{1}{2} \right) + \frac{1}{2} + e^{-2R} \left(R + 1 + \frac{1}{2R} \right) - \frac{1}{2R}$$

$$\quad + e^{-2R} \left(R + 1 + \frac{1}{2R} \right) - e^{-2R} \left(R + \frac{1}{2} \right) - \frac{1}{2} - \frac{1}{2R}$$

$$= e^{-2R} \left(1 + \frac{1}{R} \right) - \frac{1}{R}$$

10–12. Use the entries in Table 10.1 to plot ΔE_+ against R.

Using Table 10.1,

$$\Delta E_+(R) = \frac{J(R) + 1/R}{1 + S(R)} + \frac{K(R) + S(R)/R}{1 + S(R)}$$

$$= \frac{\left[e^{-2R} (1 + 1/R) - 1/R \right] + 1/R}{1 + e^{-R} (1 + R + R^2/3)} + \frac{\left[-e^{-R} (1 + R) \right] + \left(e^{-R}/R \right) (1 + R + R^2/3)}{1 + e^{-R} (1 + R + R^2/3)}$$

$$= \frac{e^{-2R} (1 + 1/R)}{1 + e^{-R} (1 + R + R^2/3)} + \frac{e^{-R} (1/R - 2R/3)}{1 + e^{-R} (1 + R + R^2/3)}$$

A plot of ΔE_+ against R is shown below:

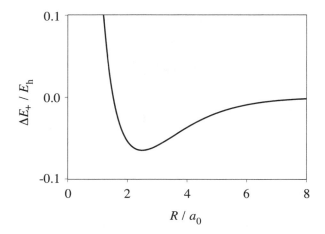

10–13.   Use the entries in Table 10.1 to plot ΔE_- against R.

Using Table 10.1,

$$\Delta E_-(R) = \frac{J(R) + 1/R}{1 - S(R)} - \frac{K(R) + S(R)/R}{1 - S(R)}$$

$$= \frac{\left[e^{-2R}\left(1 + 1/R\right) - 1/R\right] + 1/R}{1 - e^{-R}\left(1 + R + R^2/3\right)} - \frac{\left[-e^{-R}\left(1 + R\right)\right] + \left(e^{-R}/R\right)\left(1 + R + R^2/3\right)}{1 - e^{-R}\left(1 + R + R^2/3\right)}$$

$$= \frac{e^{-2R}\left(1 + 1/R\right)}{1 - e^{-R}\left(1 - R + R^2/3\right)} - \frac{e^{-R}\left(1/R - 2R/3\right)}{1 - e^{-R}\left(1 - R + R^2/3\right)}$$

A plot of ΔE_- against R is shown below:

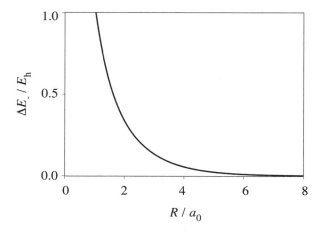

Note that this is the first excited state of H_2^+, which does not support any bound states.

10–14.   Show that

$$H_{AA} = H_{BB} = -\frac{1}{2} + J + \frac{1}{R}$$

and that

$$H_{AB} = -\frac{S}{2} + K + \frac{S}{R}$$

in the simple molecular orbital treatment of H_2^+. The quantities J and K are given by Equations 10.18 and 10.21, respectively.

The integrals for H_{AA} (Equation 10.16) are

$$H_{AA} = \left\langle 1s_A \left| -\frac{1}{2}\nabla^2 \right| 1s_A \right\rangle + \left\langle 1s_A \left| -\frac{1}{r_A} \right| 1s_A \right\rangle + \left\langle 1s_A \left| -\frac{1}{r_B} \right| 1s_A \right\rangle + \left\langle 1s_A \left| \frac{1}{R} \right| 1s_A \right\rangle$$

The first two and the last integrals have been evaluated in Problems 10–8 and 10–9. The third integral is the J integral. Therefore,

$$H_{AA} = \frac{1}{2} - 1 + J(R) - \frac{1}{R} = -\frac{1}{2} + J(R) + \frac{1}{R}$$

The integrals for H_{BB} are

$$H_{BB} = \left\langle 1s_B \left| -\frac{1}{2}\nabla^2 \right| 1s_B \right\rangle + \left\langle 1s_B \left| -\frac{1}{r_A} \right| 1s_B \right\rangle + \left\langle 1s_B \left| -\frac{1}{r_B} \right| 1s_B \right\rangle + \left\langle 1s_B \left| \frac{1}{R} \right| 1s_B \right\rangle$$

Because of the symmetry of H_2^+, the values of the integrals must remain the same if the labels of the nuclei are interchanged. Therefore, the four integrals in the H_{BB} expression become

$$H_{BB} = \left\langle 1s_A \left| -\frac{1}{2}\nabla^2 \right| 1s_A \right\rangle + \left\langle 1s_A \left| -\frac{1}{r_B} \right| 1s_A \right\rangle + \left\langle 1s_A \left| -\frac{1}{r_A} \right| 1s_A \right\rangle + \left\langle 1s_A \left| \frac{1}{R} \right| 1s_A \right\rangle$$

$$= H_{AA}$$

Using Table 10.1, the integrals for H_{AB} (Equation 10.19) are

$$H_{AB} = \left\langle 1s_A \left| -\frac{1}{2}\nabla^2 \right| 1s_B \right\rangle + \left\langle 1s_A \left| -\frac{1}{r_A} \right| 1s_B \right\rangle + \left\langle 1s_A \left| -\frac{1}{r_B} \right| 1s_B \right\rangle + \left\langle 1s_A \left| \frac{1}{R} \right| 1s_B \right\rangle$$

$$= -\left[\frac{S(R)}{2} + K(R) \right] + K(R) + K(R) + \frac{1}{R} S(R)$$

$$= -\frac{S(R)}{2} + K(R) + \frac{S(R)}{R}$$

10–15. Show that

$$\langle \hat{T} \rangle = \left\langle \psi_+ \left| -\frac{1}{2}\nabla^2 \right| \psi_+ \right\rangle = \frac{\frac{1}{2} - \frac{S}{2} - K}{1 + S}$$

$$\langle \hat{V} \rangle = \left\langle \psi_+ \left| -\frac{1}{r_A} \right| \psi_+ \right\rangle + \left\langle \psi_+ \left| -\frac{1}{r_B} \right| \psi_+ \right\rangle + \left\langle \psi_+ \left| \frac{1}{R} \right| \psi_+ \right\rangle = \frac{-1 + J + 2K}{1 + S} + \frac{1}{R}$$

The average kinetic energy and average potential energy can be calculated using Table 10.1. First calculate the average kinetic energy:

$$\left\langle \hat{T} \right\rangle = \left\langle \psi_+ \left| -\frac{1}{2}\nabla^2 \right| \psi_+ \right\rangle$$

$$= \frac{1}{2\left[1 + S(R)\right]} \left\langle 1s_A + 1s_B \left| -\frac{1}{2}\nabla^2 \right| 1s_A + 1s_B \right\rangle$$

$$= \frac{1}{2\left[1 + S(R)\right]} \left(\left\langle 1s_A \left| -\frac{1}{2}\nabla^2 \right| 1s_A \right\rangle + \left\langle 1s_A \left| -\frac{1}{2}\nabla^2 \right| 1s_B \right\rangle \right.$$

$$\left. + \left\langle 1s_B \left| -\frac{1}{2}\nabla^2 \right| 1s_A \right\rangle + \left\langle 1s_B \left| -\frac{1}{2}\nabla^2 \right| 1s_B \right\rangle \right)$$

$$= \frac{1}{2\left[1 + S(R)\right]} \left[\frac{1}{2} - \frac{S(R)}{2} - K(R) - \frac{S(R)}{2} - K(R) + \frac{1}{2} \right]$$

$$= \frac{1/2 - S(R)/2 - K(R)}{1 + S(R)}$$

The average potential energy is

$$\left\langle \hat{V} \right\rangle = \left\langle \psi_+ \left| -\frac{1}{r_A} \right| \psi_+ \right\rangle + \left\langle \psi_+ \left| -\frac{1}{r_B} \right| \psi_+ \right\rangle + \left\langle \psi_+ \left| \frac{1}{R} \right| \psi_+ \right\rangle$$

The first integral in the above expression is:

$$\left\langle \psi_+ \left| -\frac{1}{r_A} \right| \psi_+ \right\rangle = \frac{1}{2\left[1 + S(R)\right]} \left\langle 1s_A + 1s_B \left| -\frac{1}{r_A} \right| 1s_A + 1s_B \right\rangle$$

$$= \frac{1}{2\left[1 + S(R)\right]} \left(\left\langle 1s_A \left| -\frac{1}{r_A} \right| 1s_A \right\rangle + \left\langle 1s_A \left| -\frac{1}{r_A} \right| 1s_B \right\rangle \right.$$

$$\left. + \left\langle 1s_B \left| -\frac{1}{r_A} \right| 1s_A \right\rangle + \left\langle 1s_B \left| -\frac{1}{r_A} \right| 1s_B \right\rangle \right)$$

$$= \frac{1}{2\left[1 + S(R)\right]} \left[-1 + K(R) + K(R) + J(R) \right]$$

$$= \frac{1}{2\left[1 + S(R)\right]} \left[-1 + 2K(R) + J(R) \right]$$

Because of the symmetry of the wave function ψ_+, interchanging the labels of the two nuclei in the above integral gives an equivalent integral:

$$\left\langle \psi_+ \left| -\frac{1}{r_A} \right| \psi_+ \right\rangle = \frac{1}{2\left[1 + S(R)\right]} \left\langle 1s_A + 1s_B \left| -\frac{1}{r_A} \right| 1s_A + 1s_B \right\rangle$$

$$= \frac{1}{2\left[1 + S(R)\right]} \left\langle 1s_B + 1s_A \left| -\frac{1}{r_B} \right| 1s_B + 1s_A \right\rangle$$

$$= \left\langle \psi_+ \left| -\frac{1}{r_B} \right| \psi_+ \right\rangle$$

Thus, the first two integrals in the potential energy expression are the same. Finally, the remaining integral in that expression is

$$\left\langle \psi_+ \left| \frac{1}{R} \right| \psi_+ \right\rangle = \frac{1}{R} \left\langle \psi_+ \middle| \psi_+ \right\rangle = \frac{1}{R}$$

Putting the integrals together, the average potential energy is

$$\langle \hat{V} \rangle = 2 \left\{ \frac{1}{2\left[1+S(R)\right]} \left[-1 + 2K(R) + J(R)\right] \right\} + \frac{1}{R}$$

$$= \frac{-1 + J(R) + 2K(R)}{1 + S(R)} + \frac{1}{R}$$

10–16. Show that $\langle \hat{T} \rangle + \langle \hat{V} \rangle$ from the previous problem agrees with E_+ given by Equation 10.22.

$$\langle \hat{T} \rangle + \langle \hat{V} \rangle = \frac{1/2 - S(R)/2 - K(R)}{1 + S(R)} + \frac{-1 + J(R) + 2K(R)}{1 + S(R)} + \frac{1}{R}$$

$$= \frac{-\left[1 + S(R)\right]/2 + J(R) + K(R)}{1 + S(R)} + \frac{1/R + S(R)/R}{1 + S(R)}$$

$$= -\frac{1}{2} + \frac{J(R) + 1/R}{1 + S(R)} + \frac{K(R) + S(R)/R}{1 + S(R)}$$

which is the same as Equation 10.22.

10–17. The change in the total energy when two atoms come together to form a stable bond must be negative, so that $\Delta E < 0$ at R_{eq}. Therefore, $\Delta E = \Delta T + \Delta V < 0$. Show that the virial theorem requires that $\Delta V < 0$. In addition, show that $\Delta T > 0$ upon bond formation. Using the fact that the kinetic energy of a hydrogen atom is $1/2\, E_h$ (see the second entry in Table 10.1), that its potential energy is $-1\, E_h$ (see the fourth entry in Table 10.1), and the result of Example 10–3, show that

$$\Delta T = \langle \hat{T} \rangle_{H_2^+} - \langle \hat{T} \rangle_H = 0.3827\, E_h - 0.5000\, E_h = -0.1173\, E_h$$

and

$$\Delta V = \langle \hat{V} \rangle_{H_2^+} - \langle \hat{V} \rangle_H = -0.9475\, E_h - (-1.0000\, E_h) = +0.0525\, E_h$$

upon bond formation. The signs of these results are just the opposite from that predicted by the virial theorem, which implies that the simple molecular orbital given by Equation 10.26 does not provide a satisfactory interpretation of bond formation.

Using

$$\Delta T = \langle \hat{T} \rangle_{H_2^+} - \langle \hat{T} \rangle_H$$

$$\Delta V = \langle \hat{V} \rangle_{H_2^+} - \langle \hat{V} \rangle_H$$

at R_{eq} and the virial theorem, $\langle \hat{V} \rangle / \langle \hat{T} \rangle = -2$,

$$\Delta E = \Delta T + \Delta V$$

$$= -\frac{1}{2}\left(\langle \hat{V} \rangle_{H_2^+} - \langle \hat{V} \rangle_H\right) + \left(\langle \hat{V} \rangle_{H_2^+} - \langle \hat{V} \rangle_H\right)$$

$$= \frac{1}{2}\left(\langle \hat{V} \rangle_{H_2^+} - \langle \hat{V} \rangle_H\right)$$

$$= \frac{1}{2}\Delta V = -\Delta T$$

Upon bond formation, $\Delta E < 0$, and therefore, according to the last expression, $\Delta V < 0$ and $\Delta T > 0$.

Using the kinetic energy, $1/2\ E_h$, and potential energy, $-1\ E_h$, of a hydrogen atom and the kinetic energy, $0.3827\ E_h$ and the potential energy, $-0.9475\ E_h$, of H_2^+ (Example 10–3), the values for ΔT and ΔV upon bond formation are, respectively,

$$\Delta T = \langle \hat{T} \rangle_{H_2^+} - \langle \hat{T} \rangle_H = 0.3827\ E_h - 0.5000\ E_h = -0.1173\ E_h$$

$$\Delta V = \langle \hat{V} \rangle_{H_2^+} - \langle \hat{V} \rangle_H = -0.9475\ E_h - (-1.0000\ E_h) = +0.0525\ E_h$$

10–18. Show that the overlap integral between two Slater $1s$ orbitals of the form $\phi(\zeta, r) = (\zeta^3/\pi)^{1/2}e^{-\zeta r}$ is given by

$$S(\zeta, R) = e^{-\zeta R}\left(1 + \zeta R + \frac{1}{3}\zeta^2 R^2\right)$$

The overlap integral is simpler to evaluate if the elliptical coordinates described in Problem 10–6 are used.

$$S(\zeta, R) = \int d\mathbf{r}\ \left(\frac{\zeta^3}{\pi}\right)^{1/2} e^{-\zeta r_A} \left(\frac{\zeta^3}{\pi}\right)^{1/2} e^{-\zeta r_B}$$

$$= \frac{\zeta^3}{\pi}\left(\frac{R^3}{8}\right)\int_0^{2\pi} d\phi \int_1^\infty d\lambda \int_{-1}^1 d\mu\ \left(\lambda^2 - \mu^2\right) e^{-\zeta R\lambda}$$

$$= \frac{\zeta^3}{\pi}\left(\frac{R^3}{8}\right)\int_0^{2\pi} d\phi \int_1^\infty d\lambda\ e^{-\zeta R\lambda} \int_{-1}^1 d\mu\ \left(\lambda^2 - \mu^2\right)$$

$$= \frac{\zeta^3 R^3}{4}\int_1^\infty d\lambda\ e^{-\zeta R\lambda}\left(2\lambda^2 - \frac{2}{3}\right)$$

$$= \frac{\zeta^3 R^3}{4}\left[4e^{-\zeta R}\left(\frac{1}{R^3\zeta^3} + \frac{1}{R^2\zeta^2} + \frac{1}{3R\zeta}\right)\right]$$

$$= e^{-\zeta R}\left(1 + \zeta R + \frac{\zeta^2 R^2}{3}\right)$$

10–19. Verify Equations 10.31 and 10.32.

The wave function is

$$\psi_+(\zeta, r_A, r_B, R) = \frac{\phi(\zeta, r_A) + \phi(\zeta, r_B)}{\sqrt{2\,[1 + S(\zeta, R)]}}$$

Denoting $\phi(\zeta, r_A)$ and $\phi(\zeta, r_B)$ as ϕ_A and ϕ_B, respectively, and using Table 10.1, the kinetic energy is

$$\langle \hat{T}_+ \rangle = \frac{1}{2\,[1 + S(w)]} \left\langle \phi_A + \phi_B \left| -\frac{1}{2}\nabla^2 \right| \phi_A + \phi_B \right\rangle$$

$$= \frac{1}{2\,[1 + S(w)]} \left(\left\langle \phi_A \left| -\frac{1}{2}\nabla^2 \right| \phi_A \right\rangle + \left\langle \phi_A \left| -\frac{1}{2}\nabla^2 \right| \phi_B \right\rangle + \left\langle \phi_B \left| -\frac{1}{2}\nabla^2 \right| \phi_A \right\rangle \right.$$

$$\left. + \left\langle \phi_B \left| -\frac{1}{2}\nabla^2 \right| \phi_B \right\rangle \right)$$

$$= \frac{1}{2\,[1 + S(w)]} \left\{ \frac{\zeta^2}{2} - \zeta^2 \left[\frac{S(w)}{2} + K(w) \right] - \zeta^2 \left[\frac{S(w)}{2} + K(w) \right] + \frac{\zeta^2}{2} \right\}$$

$$= \frac{\dfrac{\zeta^2}{2} - \zeta^2 \left[\dfrac{S(w)}{2} + K(w) \right]}{1 + S(w)}$$

and the potential energy is

$$\langle \hat{V}_+ \rangle = \left\langle \psi_+ \left| -\frac{1}{r_A} - \frac{1}{r_B} + \frac{1}{R} \right| \psi_+ \right\rangle$$

$$= \frac{1}{2\,[1 + S(w)]} \left(\left\langle \phi_A + \phi_B \left| -\frac{1}{r_A} \right| \phi_A + \phi_B \right\rangle + \left\langle \phi_A + \phi_B \left| -\frac{1}{r_B} \right| \phi_A + \phi_B \right\rangle \right)$$

$$+ \frac{1}{R} \langle \psi_+ \mid \psi_+ \rangle$$

Because of the symmetry of the wave function, the two integrals in the parentheses are the same. Since ψ_+ is normalized, the last term in the expression is $1/R$, or equivalently, ζ/w. Therefore,

$$\langle \hat{V}_+ \rangle = \frac{1}{[1 + S(w)]} \left(\left\langle \phi_A + \phi_B \left| -\frac{1}{r_A} \right| \phi_A + \phi_B \right\rangle \right) + \frac{\zeta}{w}$$

$$= \frac{1}{[1 + S(w)]} \left(\left\langle \phi_A \left| -\frac{1}{r_A} \right| \phi_A \right\rangle + \left\langle \phi_A \left| -\frac{1}{r_A} \right| \phi_B \right\rangle \right.$$

$$\left. + \left\langle \phi_B \left| -\frac{1}{r_A} \right| \phi_A \right\rangle + \left\langle \phi_B \left| -\frac{1}{r_A} \right| \phi_B \right\rangle \right) + \frac{\zeta}{w}$$

$$= \frac{-\zeta + \zeta J(w) + 2\zeta K(w)}{1 + S(w)} + \frac{\zeta}{w}$$

10–20. In this problem we show that the form of Equations 10.31 and 10.32 are a direct result of the form of the Hamiltonian operator. Start with the Hamiltonian operator of a general molecule consisting of N electrons and n nuclei.

$$\hat{H} = -\frac{1}{2}\sum_{j=0}^{N}\nabla_j^2 + \hat{V}(x_1, y_1, z_1, \ldots, x_N, y_N, z_N, R_1, R_2, \ldots, R_m)$$

and

$$\psi = \psi(x_1, y_1, \ldots, z_N, R_1, \ldots, R_m)$$

where $\hat{V}$ consists of coulombic terms involving the coordinates of the electrons $(x_1, y_1, \ldots, z_N)$ and the m internuclear distances, $R_1, \ldots, R_m$ (m is not necessarily equal to n; consider a diatomic molecule). Suppose now we scale all the coordinates in ψ by a common factor of ζ to write

$$\psi = \psi(\zeta x_1, \ldots, \zeta z_N, \zeta R_1, \ldots, \zeta R_m)$$

and use this wave function to calculate the average kinetic energy according to

$$\langle \hat{T} \rangle = -\frac{1}{2}\left\langle \psi(\zeta) \left| \sum_{j=1}^{N}\nabla_j^2 \right| \psi(\zeta) \right\rangle \bigg/ \langle \psi(\zeta) | \psi(\zeta) \rangle$$

$$= -\frac{1}{2}\left\langle \psi(\zeta) \left| \sum_{j=1}^{N}\left(\frac{\partial^2}{\partial x_j^2} + \frac{\partial^2}{\partial y_j^2} + \frac{\partial^2}{\partial z_j^2}\right) \right| \psi(\zeta) \right\rangle \bigg/ \langle \psi(\zeta) | \psi(\zeta) \rangle$$

Let $x_1' = \zeta x_1, \ldots, z_N' = \zeta z_N, w_1 = \zeta R_1, \ldots, w_m = \zeta R_m$ and show that

$$\langle \hat{T} \rangle = \zeta^2 T(w_1, w_2, \ldots, w_m)$$

where

$$T(w_1, \ldots, w_m) = -\frac{1}{2}\left\langle \psi(\zeta=1) \left| \sum_{j=1}^{N}\nabla_j^2 \right| \psi(\zeta=1) \right\rangle \bigg/ \langle \psi(\zeta=1) | \psi(\zeta=1) \rangle$$

Similarly, use the fact that all the terms in $V(x_1, \ldots, z_N, R_1, \ldots, R_m)$ are coulombic to show that

$$\langle \hat{V} \rangle = \langle \psi(\zeta) | \hat{V} | \psi(\zeta) \rangle / \langle \psi(\zeta) | \psi(\zeta) \rangle$$

$$= \zeta \langle \psi(\zeta=1) | \hat{V} | \psi(\zeta=1) \rangle / \langle \psi(\zeta=1) | \psi(\zeta=1) \rangle$$

$$= \zeta V(w_1, w_2, \ldots, w_m)$$

Two results from elementary calculus are helpful in working this problem. First if $x' = \zeta x$, then

$$\frac{d}{dx} = \left(\frac{dx'}{dx}\right)\frac{d}{dx'} = \zeta\frac{d}{dx'}$$

$$\frac{d^2}{dx^2} = \zeta\frac{d}{dx}\left(\frac{d}{dx'}\right) = \zeta\left(\frac{dx'}{dx}\right)\frac{d^2}{dx'^2} = \zeta^2\frac{d^2}{dx'^2}$$

and second, if $I = \int f(x)\,dx$, then

$$\int f(\zeta x)\,dx = \frac{1}{\zeta}\int f(x')\,dx' = \frac{1}{\zeta}I$$

since $dx = dx'/\zeta$. For multiple integrations, each integration over a scaled coordinate gives a factor of ζ^{-1}. Finally, it will be useful to transform a distance r between two points $\mathbf{r_1}$ and $\mathbf{r_2}$ in the unscaled coordinates to the scaled system:

$$
\begin{aligned}
r &= \left[(x_2 - x_1)^2 + (y_2 - y_1)^2 + (z_2 - z_1)^2\right]^{1/2}\\
&= \left[\left(\frac{x_2'}{\zeta} - \frac{x_1'}{\zeta}\right)^2 + \left(\frac{y_2'}{\zeta} - \frac{y_1'}{\zeta}\right)^2 + \left(\frac{z_2'}{\zeta} - \frac{z_1'}{\zeta}\right)^2\right]^{1/2}\\
&= \frac{1}{\zeta}r'
\end{aligned}
$$

Therefore, if the normalization integral, $\langle \psi | \psi \rangle = I$ for $\psi(x_1, y_1, \ldots, z_N, R_1, \ldots, R_m)$, then using a wave function of the form $\psi(\zeta) = \psi(\zeta x_1, \ldots, \zeta y_N, \zeta z_N, \zeta R_1, \ldots, \zeta R_m)$ gives

$$\langle \psi(\zeta) | \psi(\zeta) \rangle = \zeta^{-3N}\langle \psi | \psi \rangle = \zeta^{-3N}I$$

because there are $3N$ integrations to perform. Similarly,

$$\left\langle \psi(\zeta) \left| \nabla_j^2 \right| \psi(\zeta) \right\rangle = \zeta^{2-3N}\left\langle \psi \left| \nabla_j^2 \right| \psi \right\rangle$$

since the second derivatives in ∇^2 give a factor of ζ^2 and the $3N$ integrations give a factor of ζ^{-3N}, and

$$\left\langle \psi(\zeta) \left| \frac{1}{r} \right| \psi(\zeta) \right\rangle = \zeta^{1-3N}\left\langle \psi \left| \frac{1}{r} \right| \psi \right\rangle$$

Starting, then, with the kinetic energy for a molecule with N electrons and n nuclei:

$$\hat{T} = -\frac{1}{2}\sum_{j=1}^{N}\nabla_j^2$$

The kinetic energy in the scaled system is

$$
\begin{aligned}
\left\langle \hat{T} \right\rangle &= \frac{-\dfrac{1}{2}\displaystyle\sum_{j=1}^{N}\left\langle \psi(\zeta) \left| \nabla_j^2 \right| \psi(\zeta) \right\rangle}{\langle \psi(\zeta) | \psi(\zeta) \rangle}\\[2em]
&= \frac{-\dfrac{\zeta^{2-3N}}{2}\displaystyle\sum_{j=1}^{N}\left\langle \psi \left| \nabla_j^2 \right| \psi \right\rangle}{\zeta^{-3N}\langle \psi | \psi \rangle}\\[2em]
&= \zeta^2\frac{\left\langle \psi \left| \hat{T} \right| \psi \right\rangle}{\langle \psi | \psi \rangle}\\[1em]
&= \zeta^2 T(\zeta = 1)
\end{aligned}
$$

The potential energy operator is

$$\hat{V} = \sum_{j=1}^{N}\sum_{i<j}^{N}\frac{1}{r_{ij}} - \sum_{a=1}^{n}\sum_{i=1}^{N}\frac{Z_a}{r_{ia}} + \sum_{b=1}^{n}\sum_{a<b}^{n}\frac{Z_aZ_b}{R_{ab}}$$

where r_{ij} is the distance between electrons i and j, r_{ia} is the distance between electron i and nucleus a, and R_{ab} is the distance between nuclei a and b. It follows that

$$\left\langle \hat{V} \right\rangle = \frac{\left\langle \psi(\zeta) \left| \hat{V} \right| \psi(\zeta) \right\rangle}{\langle \psi(\zeta) \mid \psi(\zeta) \rangle}$$

$$= \frac{\sum_{j=1}^{N}\sum_{i<j}^{N}\left\langle \psi(\zeta) \left| \frac{1}{r_{ij}} \right| \psi(\zeta) \right\rangle - \sum_{a=1}^{n}\sum_{i=1}^{N}\left\langle \psi(\zeta) \left| \frac{Z_a}{r_{ia}} \right| \psi(\zeta) \right\rangle + \sum_{b=1}^{n}\sum_{a<b}^{n}\left\langle \psi(\zeta) \left| \frac{Z_aZ_b}{R_{ab}} \right| \psi(\zeta) \right\rangle}{\langle \psi(\zeta) \mid \psi(\zeta) \rangle}$$

$$= \frac{\zeta^{1-3N}\sum_{j=1}^{N}\sum_{i<j}^{N}\left\langle \psi \left| \frac{1}{r_{ij}} \right| \psi \right\rangle - \zeta^{1-3N}\sum_{a=1}^{n}\sum_{i=1}^{N}\left\langle \psi \left| \frac{Z_a}{r_{ia}} \right| \psi \right\rangle + \zeta^{1-3N}\sum_{b=1}^{n}\sum_{a<b}^{n}\left\langle \psi \left| \frac{Z_aZ_b}{R_{ab}} \right| \psi \right\rangle}{\zeta^{-3N}\langle \psi \mid \psi \rangle}$$

$$= \zeta \frac{\left\langle \psi \left| \hat{V} \right| \psi \right\rangle}{\langle \psi \mid \psi \rangle}$$

$$= \zeta V(\zeta = 1)$$

10–21. Instead of just minimizing $E_+(\zeta, R)$ numerically, we can do it analytically. Use Equation 10.33 to show that the optimum value of ζ as a function of $w = \zeta R$ is given by

$$\zeta = \frac{-V_+(w) - w\dfrac{dV_+(w)}{dw}}{2T_+(w) + w\dfrac{dT_+(w)}{dw}}$$

Take the derivative of

$$E_+(\zeta, w) = \zeta^2 T_+(w) + \zeta V_+(w)$$

with respect to ζ and set it to zero to determine the optimum value of ζ:

$$\frac{\partial E_+(\zeta, w)}{\partial \zeta} = 0 = 2\zeta T_+(w) + \zeta^2\frac{dT_+(w)}{d\zeta} + V_+(w) + \zeta\frac{dV_+(w)}{d\zeta}$$

$$= 2\zeta T_+(w) + \zeta^2\frac{dw}{d\zeta}\left[\frac{dT_+(w)}{dw}\right] + V_+(w) + \zeta\frac{dw}{d\zeta}\left[\frac{dV_+(w)}{dw}\right]$$

$$= 2\zeta T_+(w) + \zeta w\frac{dT_+(w)}{dw} + V_+(w) + w\frac{dV_+(w)}{dw}$$

Now rearrange the equation to give

$$\zeta = \frac{-V_+(w) - w\dfrac{dV_+(w)}{dw}}{2T_+(w) + w\dfrac{dT_+(w)}{dw}}$$

10–22. Realize that ζ in the previous problem is given as a function of w, not R. (Remember that $w = \zeta R$.) Use a program such as *MathCad* or *Mathematica* to plot both $\zeta(w)$ and $R(w)$ parametrically as functions of w to obtain a plot of ζ versus R.

To plot ζ as a function of w, use the expression

$$\zeta = \frac{-V_+(w) - w\dfrac{dV_+(w)}{dw}}{2T_+(w) + w\dfrac{dT_+(w)}{dw}}$$

The $T_+(w)$ and $V_+(w)$ expressions are obtained from Equations 10.31 and 10.32, respectively:

$$T_+(w) = \frac{\dfrac{1}{2} - \dfrac{S(w)}{2} - K(w)}{1 + S(w)}$$

$$V_+(w) = \frac{-1 + J(w) + 2K(w)}{1 + S(w)} + \frac{1}{w}$$

and the overlap, coulomb, and exchange integrals can be found in Table 10.1:

$$S(w) = e^{-w}\left(1 + w + \frac{w^2}{3}\right)$$

$$J(w) = e^{-2w}\left(1 + \frac{1}{w}\right) - \frac{1}{w}$$

$$K(w) = e^{-w}(1 + w)$$

To plot R as a function of w, divide w by $\zeta(w)$ and plot against w. The two plots are shown below.

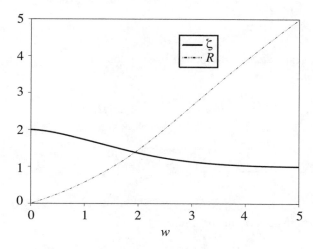

Now combine $R(w)$ and $\zeta(w)$ to arrive at the following plot:

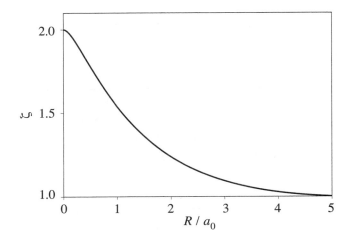

10–23. In this problem, we shall prove that $\zeta \to 2$ as $R \to 0$, as shown in Figure 10.16. First show that

$$S(w) = 1 - \frac{1}{6}w^2 + O(w^3)$$

$$J(w) = -1 + \frac{2}{3}w^2 + O(w^3)$$

$$K(w) = -1 + \frac{w^2}{2} + O(w^3)$$

as $R \to 0$. Now show that

$$T_+(w) = \frac{1}{2} - \frac{w^2}{6} + O(w^3)$$

$$V_+(w) = \frac{1}{w} - 2 + \frac{2}{3}w^2 + O(w^3)$$

Substitute these results into ζ in Problem 10–21 to obtain

$$\zeta = 2 - \frac{2w^2}{3} + O(w^3)$$

As $R \to 0$, or $w \to 0$, the series expansion $e^x = 1 + x + x^2/2! + x^3/3! + \cdots$ can be used for the e^{-w} or e^{-2w} term in each of the three integrals to give

$$S(w) = e^{-w}\left(1 + w + \frac{w^2}{3}\right)$$

$$= \left[1 - w + \frac{w^2}{2} + O(w^3)\right]\left(1 + w + \frac{w^2}{3}\right)$$

$$= 1 - \frac{w^2}{6} + O(w^3)$$

$$J(w) = e^{-2w}\left(1 + \frac{1}{w}\right) - \frac{1}{w}$$

$$= \left[1 - 2w + \frac{4w^2}{2} - \frac{8w^3}{6} + O(w^4)\right]\left(1 + \frac{1}{w}\right) - \frac{1}{w}$$

$$= -1 + \frac{2w^2}{3} + O(w^3)$$

$$K(w) = -e^{-w}(1 + w)$$

$$= -\left[1 - w + \frac{w^2}{2} + O(w^3)\right](1 + w)$$

$$= -1 + \frac{w^2}{2} + O(w^3)$$

Using these expressions, Equation 10.31, and the relation $(1 - x)^{-1} = 1 + x + x^2 + \cdots$, the kinetic energy is

$$T_+(w) = \frac{\frac{1}{2} - \frac{S(w)}{2} - K(w)}{1 + S(w)}$$

$$= \frac{\frac{1}{2} - \left(\frac{1}{2} - \frac{w^2}{12}\right) - \left(-1 + \frac{w^2}{2}\right) + O(w^3)}{1 + 1 - \frac{w^2}{6} + O(w^3)}$$

$$= \frac{12 - 5w^2 + O(w^3)}{24 - 2w^2 + O(w^3)}$$

$$= \frac{12 - 5w^2 + O(w^3)}{24\left(1 - \frac{w^2}{12}\right) + O(w^3)}$$

$$= \frac{12 - 5w^2}{24}\left(1 + \frac{w^2}{12}\right) + O(w^3)$$

$$= \frac{1}{2} - \frac{w^2}{6} + O(w^3)$$

and the potential energy (Equation 10.32) is

$$V_+(w) = \frac{-1 + J(w) + 2K(w)}{1 + S(w)} + \frac{1}{w}$$

$$= \frac{-1 + \left(-1 + \frac{2w^2}{3}\right) + 2\left(-1 + \frac{w^2}{2}\right) + O(w^3)}{1 + 1 - \frac{w^2}{6} + O(w^3)} + \frac{1}{w}$$

$$= \frac{-4 + \frac{5w^2}{3} + O(w^3)}{2\left(1 - \frac{w^2}{12}\right) + O(w^3)} + \frac{1}{w}$$

$$= \left(-2 + \frac{5w^2}{6}\right)\left(1 + \frac{w^2}{12}\right) + \frac{1}{w} + O(w^3)$$

$$= \frac{1}{w} - 2 + \frac{2w^2}{3} + O(w^3)$$

Combining the kinetic and potential energies to determine ζ:

$$\zeta = \frac{-V_+(w) - w\dfrac{dV_+(w)}{dw}}{2T_+(w) + w\dfrac{dT_+(w)}{dw}}$$

$$= \frac{-\left(\dfrac{1}{w} - 2 + \dfrac{2w^2}{3}\right) - w\left(-\dfrac{1}{w^2} + \dfrac{4w}{3}\right) + O(w^3)}{2\left(\dfrac{1}{2} - \dfrac{w^2}{6}\right) + w\left(-\dfrac{w}{3}\right) + O(w^3)}$$

$$= \frac{2 - 2w^2 + O(w^3)}{1 - \dfrac{2w^2}{3} + O(w^3)}$$

$$= \left(2 - 2w^2\right)\left(1 + \frac{2w^2}{3}\right) + O(w^3)$$

$$= 2 - \frac{2w^2}{3} + O(w^3)$$

which reduces to $\zeta = 2$ in the limit as $w \to 0$.

10–24. In this problem, we shall prove that $\zeta \to 1$ as $R \to \infty$, as shown in Figure 10.16. First show that

$$S(w) = \frac{w^2 e^{-w}}{3}$$

$$J(w) = -\frac{1}{w}$$

$$K(w) = -we^{-w}$$

as $R \to \infty$. Now show that $T(w) \to 1/2$ and $V(w) \to -1$. Substitute these results into ζ given in Problem 10–21 to show that $\zeta \to 1$ as $R \to \infty$.

Keeping only the significant terms in the overlap, coulomb and exchange integrals as $R \to \infty$, or $w \to \infty$ gives

$$S(w) = e^{-w}\left(1 + w + \frac{w^2}{3}\right)$$

$$= \frac{w^2 e^{-w}}{3}$$

$$J(w) = e^{-2w}\left(1 + \frac{1}{w}\right) - \frac{1}{w}$$

$$= -\frac{1}{w}$$

$$K(w) = -e^{-w}(1 + w)$$

$$= -we^{-w}$$

Using these expressions and Equation 10.31, the kinetic energy is

$$T_+(w) = \frac{\dfrac{1}{2} - \dfrac{S(w)}{2} - K(w)}{1 + S(w)}$$

$$= \frac{\dfrac{1}{2} - \dfrac{w^2 e^{-w}}{3} + we^{-w}}{1 + \dfrac{w^2 e^{-w}}{3}}$$

$$\to \frac{1}{2}$$

and the potential energy (Equation 10.32) is

$$V_+(w) = \frac{-1 + J(w) + 2K(w)}{1 + S(w)} + \frac{1}{w}$$

$$= \frac{-1 - \dfrac{1}{w} - 2we^{-w}}{1 + \dfrac{w^2 e^{-w}}{3}} + \frac{1}{w}$$

$$\to -1$$

The derivatives of the kinetic and potential energies are needed to determine ζ. Since the ignored terms in these energy expressions contain e^{-w}, the derivatives of these expressions, even when multiplied by w, would still approach 0 as $w \to \infty$. Thus,

$$\zeta = \frac{-V_+(w) - w\dfrac{dV_+(w)}{dw}}{2T_+(w) + w\dfrac{dT_+(w)}{dw}}$$

$$= \frac{1}{2\,(1/2)}$$

$$= 1$$

10–25.   Plot ψ_+ and ψ_- given by Equations 10.24 and 10.25 for several values of R along the internuclear axis.

Let the nuclei be located at along the x axis at $x = -R/2$ and $x = R/2$, respectively. Then

$$1s_A = \left(\frac{1}{\pi}\right)^{1/2} e^{-|x - R/2|}$$

$$1s_B = \left(\frac{1}{\pi}\right)^{1/2} e^{-|x + R/2|}$$

The wave functions are

$$\psi_+ = \frac{1}{\sqrt{2\,(1+S)}}\left(1s_A + 1s_B\right)$$

$$\psi_- = \frac{1}{\sqrt{2\,(1-S)}}\left(1s_A - 1s_B\right)$$

A plot for four different values of R (1, 2, 3, and 4) are shown below for both ψ_+ and ψ_-. For clarity, only the curves for $R = 1$ and $R = 4$ are labeled. The wave function ψ_+ is symmetric about $x = 0$, and the maxima are located at $\pm R/2$. There is a build-up of electron density between the two nuclei.

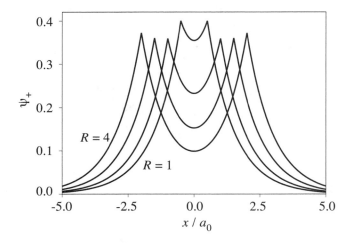

The wave function ψ_- is antisymmetric about $x = 0$; thus, there is no electron density at the midpoint of the bond. The magnitude of the wave function at the extrema, located at $\pm R/2$, is greater than that for ψ_+.

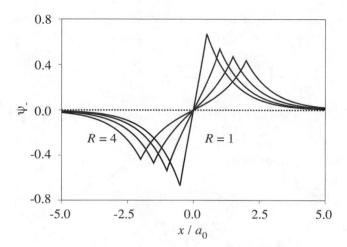

10–26. Show explicitly that an s orbital on one hydrogen atom and a p_z orbital on another have zero overlap. Use the $2s$ and $2p_z$ wave functions given in Table 7.5 to set up the overlap integral. *Hint*: You need not evaluate any integrals, but simply show that the overlap integral can be separated into two parts that exactly cancel one another.

Assume that the $2p_z$ wave function is centered on atom A and the $2s$ wave function is centered on atom B. Using atomic units, the wave functions are

$$\psi_{2p_z} = \frac{1}{4\sqrt{2\pi}} r_A e^{-r_A/2} \cos\theta$$

$$\psi_{2s} = \frac{1}{4\sqrt{2\pi}} \left(2 - r_B\right) e^{-r_B/2}$$

The overlap integral can be calculated using spherical coordinates centered on atom A depicted in the figures below. The first figure shows that the angle between R and r_A is $\theta' = \pi/2 - \theta$ for $0 \le \theta \le \pi/2$ while the second figure shows the same angle is $\theta'' = \theta - \pi/2$ for $\pi/2 \le \theta \le \pi$.

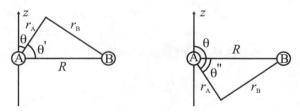

To have zero overlap, the z axis must be perpendicular to the internuclear axis. The overlap integral is

$$S = \int d\mathbf{r}_A \, \psi_{2p_z} \psi_{2s}$$

$$= \frac{1}{32\pi} \int_0^\infty dr_A \, r_A^2 r_A e^{-r_A/2} \int_0^{2\pi} d\phi \int_0^\pi d\theta \, \sin\theta \cos\theta \left(2 - r_B\right) e^{-r_B/2}$$

The integral over θ can be separated into a sum of two integrals: one integrating over $0 \leq \theta' \leq \pi/2$ and the other integrating over $0 \leq \theta'' \leq \pi/2$. In addition, r_B can be expressed in terms of r_A:

$$r_B = \left(r_A^2 + R^2 - 2r_A R \cos \theta'\right)^{1/2} \quad \text{for} \quad 0 \leq \theta \leq \pi/2$$

$$= \left(r_A^2 + R^2 - 2r_A R \cos \theta''\right)^{1/2} \quad \text{for} \quad \pi/2 \leq \theta \leq \pi$$

Thus, the integral over θ becomes

$$\int_0^\pi d\theta \, \sin \theta \cos \theta \left(2 - r_B\right) e^{-r_B/2}$$

$$= \int_0^{\pi/2} d\theta' \sin \theta \cos \theta \left[2 - \left(r_A^2 + R^2 - 2r_A R \cos \theta'\right)^{1/2}\right] e^{-\left(r_A^2 + R^2 - 2r_A R \cos \theta'\right)^{1/2}/2}$$

$$+ \int_0^{\pi/2} d\theta'' \sin \theta \cos \theta \left[2 - \left(r_A^2 + R^2 - 2r_A R \cos \theta''\right)^{1/2}\right] e^{-\left(r_A^2 + R^2 - 2r_A R \cos \theta''\right)^{1/2}/2}$$

$$= \int_0^{\pi/2} d\theta' \sin \left(\frac{\pi}{2} - \theta'\right) \cos \left(\frac{\pi}{2} - \theta'\right) \left[2 - \left(r_A^2 + R^2 - 2r_A R \cos \theta'\right)^{1/2}\right] e^{-\left(r_A^2 + R^2 - 2r_A R \cos \theta'\right)^{1/2}/2}$$

$$+ \int_0^{\pi/2} d\theta'' \sin \left(\frac{\pi}{2} + \theta''\right) \cos \left(\frac{\pi}{2} + \theta''\right) \left[2 - \left(r_A^2 + R^2 - 2r_A R \cos \theta''\right)^{1/2}\right] e^{-\left(r_A^2 + R^2 - 2r_A R \cos \theta''\right)^{1/2}/2}$$

$$= \int_0^{\pi/2} d\theta' \cos \theta' \sin \theta' \left[2 - \left(r_A^2 + R^2 - 2r_A R \cos \theta'\right)^{1/2}\right] e^{-\left(r_A^2 + R^2 - 2r_A R \cos \theta'\right)^{1/2}/2}$$

$$+ \int_0^{\pi/2} d\theta'' \cos \theta'' \left(- \sin \theta''\right) \left[2 - \left(r_A^2 + R^2 - 2r_A R \cos \theta''\right)^{1/2}\right] e^{-\left(r_A^2 + R^2 - 2r_A R \cos \theta''\right)^{1/2}/2}$$

$$= 0$$

The integral over θ' and the integral over θ'' have the same magnitude but different signs. Thus, they cancel each other, and the integral over θ is zero. As a result, the overlap integral is also zero.

10–27. Show that $\Delta E_- = (J - K)/(1 - S) + 1/R$ for the antibonding orbital ψ_- of H_2^+.

Substitute the expressions for H_{AA} and H_{AB} from Problem 10–14 into Equation 10.14 to give

$$E_- = \frac{H_{AA} - H_{AB}}{1 - S}$$

$$= \frac{\left(-\frac{1}{2} + J + \frac{1}{R}\right) - \left(-\frac{S}{2} + K + \frac{S}{R}\right)}{1 - S}$$

$$= -\frac{1}{2} + \frac{J + \frac{1}{R}}{1 - S} - \frac{K + \frac{S}{R}}{1 - S}$$

Since $-1/2$ is the ground-state energy of an isolated hydrogen atom and a bare proton,

$$\Delta E_- = E_- - E_{1s} = \frac{J + \dfrac{1}{R}}{1 - S} - \frac{K + \dfrac{S}{R}}{1 - S} = \frac{J - K}{1 - S} + \frac{1}{R}$$

10–28. Show that ψ given by Equation 10.38 is an eigenfunction of $\hat{S}_z = \hat{S}_{z1} + \hat{S}_{z2}$ with $S_z = 0$.

$$\hat{S}_z \psi = \sigma_b(1)\sigma_b(2)\left(\hat{S}_{z1} + \hat{S}_{z2}\right)\frac{1}{\sqrt{2}}\{[\alpha(1)\beta(2) - \alpha(2)\beta(1)]\}$$

$$= \frac{1}{\sqrt{2}}\sigma_b(1)\sigma_b(2)\left\{\left[\hat{S}_{z1}\alpha(1)\beta(2) - \alpha(2)\hat{S}_{z1}\beta(1)\right] + \left[\alpha(1)\hat{S}_{z2}\beta(2) - \hat{S}_{z2}\alpha(2)\beta(1)\right]\right\}$$

$$= \frac{1}{\sqrt{2}}\sigma_b(1)\sigma_b(2)\left\{\left[\frac{1}{2}\alpha(1)\beta(2) + \frac{1}{2}\alpha(2)\beta(1)\right] + \left[-\frac{1}{2}\alpha(1)\beta(2) - \frac{1}{2}\alpha(2)\beta(1)\right]\right\}$$

$$= 0\,\sigma_b(1)\sigma_b(2)\frac{1}{\sqrt{2}}\{[\alpha(1)\beta(2) - \alpha(2)\beta(1)]\}$$

Thus, ψ is an eigenfunction of $\hat{S}_z$ with eigenvalue $S_z = 0$.

10–29. Show that the determinantal wave functions in Equations 10.48 are normalized.

The six determinantal wave functions are

$$D_1 = \frac{1}{\sqrt{2}}\left[\sigma_b(1)\sigma_b(2)\alpha(1)\beta(2) - \sigma_b(1)\sigma_b(2)\beta(1)\alpha(2)\right]$$

$$D_2 = \frac{1}{\sqrt{2}}\left[\sigma_a(1)\sigma_a(2)\alpha(1)\beta(2) - \sigma_a(1)\sigma_a(2)\beta(1)\alpha(2)\right]$$

$$D_3 = \frac{1}{\sqrt{2}}\left[\sigma_b(1)\sigma_a(2)\alpha(1)\alpha(2) - \sigma_a(1)\sigma_b(2)\alpha(1)\alpha(2)\right]$$

$$D_4 = \frac{1}{\sqrt{2}}\left[\sigma_b(1)\sigma_a(2)\alpha(1)\beta(2) - \sigma_a(1)\sigma_b(2)\beta(1)\alpha(2)\right]$$

$$D_5 = \frac{1}{\sqrt{2}}\left[\sigma_b(1)\sigma_a(2)\beta(1)\alpha(2) - \sigma_a(1)\sigma_b(2)\alpha(1)\beta(2)\right]$$

$$D_6 = \frac{1}{\sqrt{2}}\left[\sigma_b(1)\sigma_a(2)\beta(1)\beta(2) - \sigma_a(1)\sigma_b(2)\beta(1)\beta(2)\right]$$

Now show that each wave function is normalized. Note that both the spatial functions and the spin functions are separately orthonormal. For the first determinantal function,

$$\int d\tau\, D_1^* D_1 = \frac{1}{2}\int d\tau\, \sigma_b(1)^*\sigma_b(2)^*\alpha(1)^*\beta(2)^*\sigma_b(1)\sigma_b(2)\alpha(1)\beta(2)$$

$$-\frac{1}{2}\int d\tau\, \sigma_b(1)^*\sigma_b(2)^*\alpha(1)^*\beta(2)^*\sigma_b(1)\sigma_b(2)\beta(1)\alpha(2)$$

$$-\frac{1}{2}\int d\tau\, \sigma_b(1)^*\sigma_b(2)^*\beta(1)^*\alpha(2)^*\sigma_b(1)\sigma_b(2)\alpha(1)\beta(2)$$

$$+\frac{1}{2}\int d\tau\, \sigma_b(1)^*\sigma_b(2)^*\beta(1)^*\alpha(2)^*\sigma_b(1)\sigma_b(2)\beta(1)\alpha(2)$$

$$=\frac{1}{2}+0+0+\frac{1}{2}$$

$$=1$$

For each of the rest of the determinantal wave functions, expansion of $D_i^* D_i$ similarly gives four terms. Two of these terms lead to integrals over orthogonal functions and are zero. The other two give integrals over normalized functions and each contributes 1/2 to $\int d\tau\, D_i^* D_i$. Therefore, each determinantal wave function is itself normalized.

10–30. The six wave functions given by Equations 10.49 are actually linear combinations of the determinantal wave functions given by Equations 10.48. Show that ψ_1 and ψ_2 are the spatial parts of D_1 and D_2, respectively; that two of ψ_3, ψ_4, or ψ_5 are the spatial parts of D_3 and D_6; that the remaining one of ψ_3, ψ_4, or ψ_5 is the spatial part of $(D_4 + D_5)/\sqrt{2}$; and that ψ_6 is the spatial part of $(D_4 - D_5)/\sqrt{2}$. The reason that we take linear combinations $(D_4 \pm D_5)/\sqrt{2}$ is because these combinations are eigenfunctions of $\hat{S}^2$. (We state this without proof.)

Factoring out the spatial parts of D_1 and D_2 gives

$$D_1 = \frac{1}{\sqrt{2}}\sigma_b(1)\sigma_b(2)\left[\alpha(1)\beta(2) - \beta(1)\alpha(2)\right]$$

$$D_2 = \frac{1}{\sqrt{2}}\sigma_a(1)\sigma_a(2)\left[\alpha(1)\beta(2) - \beta(1)\alpha(2)\right]$$

Thus, $\psi_1 = \sigma_b(1)\sigma_b(2)$ is the spatial part of D_1 and $\psi_2 = \sigma_a(1)\sigma_a(2)$ is the spatial part of D_2.

Factoring out the spin parts of D_3 and D_6 gives

$$D_3 = \frac{1}{\sqrt{2}}\alpha(1)\alpha(2)\left[\sigma_b(1)\sigma_a(2) - \sigma_a(1)\sigma_b(2)\right]$$

$$D_6 = \frac{1}{\sqrt{2}}\beta(1)\beta(2)\left[\sigma_b(1)\sigma_a(2) - \sigma_a(1)\sigma_b(2)\right]$$

Thus, two of ψ_3, ψ_4, or ψ_5 are the spatial parts of D_3 and D_6, $\sigma_b(1)\sigma_a(2) - \sigma_a(1)\sigma_b(2)$.

Combining D_4 and D_5 by adding and subtracting gives, after normalization,

$$\frac{1}{\sqrt{2}}(D_4 + D_5) = \frac{1}{2}\left\{\sigma_b(1)\sigma_a(2)\left[\alpha(1)\beta(2) + \beta(1)\alpha(2)\right] - \sigma_a(1)\sigma_b(2)\left[\beta(1)\alpha(2) + \alpha(1)\beta(2)\right]\right\}$$

$$= \frac{1}{2}\left[\alpha(1)\beta(2) + \beta(1)\alpha(2)\right]\left[\sigma_b(1)\sigma_a(2) - \sigma_a(1)\sigma_b(2)\right]$$

$$\frac{1}{\sqrt{2}}(D_4 - D_5) = \frac{1}{2}\left\{\sigma_b(1)\sigma_a(2)\left[\alpha(1)\beta(2) - \beta(1)\alpha(2)\right] - \sigma_a(1)\sigma_b(2)\left[\beta(1)\alpha(2) - \alpha(1)\beta(2)\right]\right\}$$

$$= \frac{1}{2}\left[\alpha(1)\beta(2) - \beta(1)\alpha(2)\right]\left[\sigma_b(1)\sigma_a(2) + \sigma_a(1)\sigma_b(2)\right]$$

Thus, one of ψ_3, ψ_4, or ψ_5 is the spatial part of $(D_4 + D_5)/\sqrt{2}$, $\sigma_b(1)\sigma_a(2) - \sigma_a(1)\sigma_b(2)$, and ψ_6 is the spatial part of $(D_4 - D_5)/\sqrt{2}$, $\sigma_b(1)\sigma_a(2) + \sigma_a(1)\sigma_b(2)$.

10–31. Use the result of the previous problem to show that the complete wave functions (spin functions included) given by Equations 10.49 are

$$\psi_1 = \frac{1}{\sqrt{2}}\sigma_b(1)\sigma_b(2)\left[\alpha(1)\beta(2) - \alpha(2)\beta(1)\right]$$

$$\psi_2 = \frac{1}{\sqrt{2}}\sigma_a(1)\sigma_a(2)\left[\alpha(1)\beta(2) - \alpha(2)\beta(1)\right]$$

$$\psi_3, \psi_4, \psi_5 = \frac{1}{\sqrt{2}}\left[\sigma_b(1)\sigma_a(2) - \sigma_a(1)\sigma_b(2)\right]\begin{cases}\alpha(1)\alpha(2) \\ \frac{1}{\sqrt{2}}\left[\alpha(1)\beta(2) + \alpha(2)\beta(1)\right] \\ \beta(1)\beta(2)\end{cases}$$

$$\psi_6 = \frac{1}{2}\left[\sigma_b(1)\sigma_a(2) + \sigma_b(2)\sigma_a(1)\right]\left[\alpha(1)\beta(2) - \alpha(2)\beta(1)\right]$$

Show that

$$\hat{S}_z\psi_1 = (\hat{S}_{z1} + \hat{S}_{z2})\psi_1 = 0$$

$$\hat{S}_z\psi_2 = (\hat{S}_{z1} + \hat{S}_{z2})\psi_2 = 0$$

$$\hat{S}_z\psi_3 = (\hat{S}_{z1} + \hat{S}_{z2})\psi_3 = \psi_3$$

$$\hat{S}_z\psi_4 = (\hat{S}_{z1} + \hat{S}_{z2})\psi_4 = 0$$

$$\hat{S}_z\psi_5 = (\hat{S}_{z1} + \hat{S}_{z2})\psi_5 = -\psi_5$$

$$\hat{S}_z\psi_6 = (\hat{S}_{z1} + \hat{S}_{z2})\psi_6 = 0$$

It turns out that ψ_3, ψ_4, and ψ_5 are wave functions of a triplet state and the others are wave functions of singlet states.

The following table lists the normalized spatial and normalized spin functions for the determinantal wave functions D_1, D_2, D_3, D_6, $(D_4 + D_5)/\sqrt{2}$, and $(D_4 - D_5)/\sqrt{2}$. Multiplying the corresponding spatial and spin functions together gives each of the ψ_1 through ψ_6 wave functions.

Determinantal wave function	Spatial function	Spin function	Complete wave function
D_1	$\sigma_b(1)\sigma_b(2)$	$\dfrac{1}{\sqrt{2}}[\alpha(1)\beta(2) - \beta(1)\alpha(2)]$	ψ_1
D_2	$\sigma_a(1)\sigma_a(2)$	$\dfrac{1}{\sqrt{2}}[\alpha(1)\beta(2) - \beta(1)\alpha(2)]$	ψ_2
D_3	$\dfrac{1}{\sqrt{2}}\left[\sigma_b(1)\sigma_a(2) - \sigma_a(1)\sigma_b(2)\right]$	$\alpha(1)\alpha(2)$	ψ_3
$\dfrac{1}{\sqrt{2}}\left(D_4 + D_5\right)$	$\dfrac{1}{\sqrt{2}}\left[\sigma_b(1)\sigma_a(2) - \sigma_a(1)\sigma_b(2)\right]$	$\dfrac{1}{\sqrt{2}}[\alpha(1)\beta(2) + \beta(1)\alpha(2)]$	ψ_4
D_6	$\dfrac{1}{\sqrt{2}}\left[\sigma_b(1)\sigma_a(2) - \sigma_a(1)\sigma_b(2)\right]$	$\beta(1)\beta(2)$	ψ_5
$\dfrac{1}{\sqrt{2}}\left(D_4 - D_5\right)$	$\dfrac{1}{\sqrt{2}}\left[\sigma_b(1)\sigma_a(2) + \sigma_a(1)\sigma_b(2)\right]$	$\dfrac{1}{\sqrt{2}}[\alpha(1)\beta(2) - \beta(1)\alpha(2)]$	ψ_6

Now operate on each wave function with $\hat{S}_z$:

$$\hat{S}_z\psi_1 = \frac{1}{\sqrt{2}}\sigma_b(1)\sigma_b(2)\left\{\left[\hat{S}_{z1}\alpha(1)\beta(2) - \hat{S}_{z1}\beta(1)\alpha(2)\right] + \left[\alpha(1)\hat{S}_{z2}\beta(2) - \beta(1)\hat{S}_{z2}\alpha(2)\right]\right\}$$

$$= \frac{1}{\sqrt{2}}\sigma_b(1)\sigma_b(2)\left\{\left[\frac{1}{2}\alpha(1)\beta(2) + \frac{1}{2}\beta(1)\alpha(2)\right] + \left[-\frac{1}{2}\alpha(1)\beta(2) - \frac{1}{2}\beta(1)\alpha(2)\right]\right\}$$

$$= 0$$

$$\hat{S}_z\psi_2 = \frac{1}{\sqrt{2}}\sigma_a(1)\sigma_a(2)\left\{\left[\hat{S}_{z1}\alpha(1)\beta(2) - \hat{S}_{z1}\beta(1)\alpha(2)\right] + \left[\alpha(1)\hat{S}_{z2}\beta(2) - \beta(1)\hat{S}_{z2}\alpha(2)\right]\right\}$$

$$= \frac{1}{\sqrt{2}}\sigma_a(1)\sigma_a(2)\left\{\left[\frac{1}{2}\alpha(1)\beta(2) + \frac{1}{2}\beta(1)\alpha(2)\right] + \left[-\frac{1}{2}\alpha(1)\beta(2) - \frac{1}{2}\beta(1)\alpha(2)\right]\right\}$$

$$= 0$$

$$\hat{S}_z\psi_3 = \frac{1}{\sqrt{2}}\left[\sigma_b(1)\sigma_a(2) - \sigma_a(1)\sigma_b(2)\right]\left\{\left[\hat{S}_{z1}\alpha(1)\alpha(2)\right] + \left[\alpha(1)\hat{S}_{z2}\alpha(2)\right]\right\}$$

$$= \frac{1}{\sqrt{2}}\left[\sigma_b(1)\sigma_a(2) - \sigma_a(1)\sigma_b(2)\right]\left\{\left[\frac{1}{2}\alpha(1)\alpha(2)\right] + \left[\frac{1}{2}\alpha(1)\alpha(2)\right]\right\}$$

$$= \psi_3$$

$$\hat{S}_z\psi_4 = \frac{1}{\sqrt{2}}\left[\sigma_b(1)\sigma_a(2) - \sigma_a(1)\sigma_b(2)\right]$$

$$\times \left\{\frac{1}{\sqrt{2}}\left[\hat{S}_{z1}\alpha(1)\beta(2) + \hat{S}_{z1}\beta(1)\alpha(2)\right] + \frac{1}{\sqrt{2}}\left[\alpha(1)\hat{S}_{z2}\beta(2) + \beta(1)\hat{S}_{z2}\alpha(2)\right]\right\}$$

$$= \frac{1}{\sqrt{2}}\left[\sigma_b(1)\sigma_a(2) - \sigma_a(1)\sigma_b(2)\right]$$

$$\times \left\{\frac{1}{\sqrt{2}}\left[\frac{1}{2}\alpha(1)\beta(2) - \frac{1}{2}\beta(1)\alpha(2)\right] + \frac{1}{\sqrt{2}}\left[-\frac{1}{2}\alpha(1)\beta(2) + \frac{1}{2}\beta(1)\alpha(2)\right]\right\}$$

$$= 0$$

$$\hat{S}_z\psi_5 = \frac{1}{\sqrt{2}}\left[\sigma_b(1)\sigma_a(2) - \sigma_a(1)\sigma_b(2)\right]\left\{\left[\hat{S}_{z1}\beta(1)\beta(2)\right] + \left[\beta(1)\hat{S}_{z2}\beta(2)\right]\right\}$$

$$= \frac{1}{\sqrt{2}}\left[\sigma_b(1)\sigma_a(2) - \sigma_a(1)\sigma_b(2)\right]\left\{\left[-\frac{1}{2}\beta(1)\beta(2)\right] + \left[-\frac{1}{2}\beta(1)\beta(2)\right]\right\}$$

$$= -\psi_5$$

$$\hat{S}_z\psi_6 = \frac{1}{\sqrt{2}}\left[\sigma_b(1)\sigma_a(2) + \sigma_a(1)\sigma_b(2)\right]$$

$$\times \left\{\frac{1}{\sqrt{2}}\left[\hat{S}_{z1}\alpha(1)\beta(2) - \hat{S}_{z1}\beta(1)\alpha(2)\right] + \frac{1}{\sqrt{2}}\left[\alpha(1)\hat{S}_{z2}\beta(2) - \beta(1)\hat{S}_{z2}\alpha(2)\right]\right\}$$

$$= \frac{1}{\sqrt{2}}\left[\sigma_b(1)\sigma_a(2) + \sigma_a(1)\sigma_b(2)\right]$$

$$\times \left\{\frac{1}{\sqrt{2}}\left[\frac{1}{2}\alpha(1)\beta(2) + \frac{1}{2}\beta(1)\alpha(2)\right] + \frac{1}{\sqrt{2}}\left[-\frac{1}{2}\alpha(1)\beta(2) - \frac{1}{2}\beta(1)\alpha(2)\right]\right\}$$

$$= 0$$

10–32. Show that all six wave functions in the previous problem are antisymmetric.

The wave functions ψ_1, ψ_2, and ψ_6 each has a symmetric spatial function and an antisymmetric spin function with respect to the interchange of two electrons whereas the wave functions ψ_3, ψ_4, and ψ_5 each has an antisymmetric spatial function and a symmetric spin function with respect to the interchange of two electrons. Each wave function is therefore antisymmetric overall. To show explicitly that each wave function is antisymmetric, determine $\psi_i(2, 1)$, the function that results after exchanging the two electrons in $\psi_i(1, 2)$.

$$\psi_1(2, 1) = \frac{1}{\sqrt{2}}\sigma_b(2)\sigma_b(1)\left[\alpha(2)\beta(1) - \beta(2)\alpha(1)\right] = -\psi_1(1, 2)$$

$$\psi_2(2, 1) = \frac{1}{\sqrt{2}}\sigma_a(2)\sigma_a(1)\left[\alpha(2)\beta(1) - \beta(2)\alpha(1)\right] = -\psi_2(1, 2)$$

$$\psi_3(2, 1) = \frac{1}{\sqrt{2}}\left[\sigma_b(2)\sigma_a(1) - \sigma_a(2)\sigma_b(1)\right]\alpha(2)\alpha(1) = -\psi_3(1, 2)$$

$$\psi_4(2, 1) = \frac{1}{2}\left[\sigma_b(2)\sigma_a(1) - \sigma_a(2)\sigma_b(1)\right]\left[\alpha(2)\beta(1) + \beta(2)\alpha(1)\right] = -\psi_4(1, 2)$$

$$\psi_5(2, 1) = \frac{1}{\sqrt{2}}\left[\sigma_b(2)\sigma_a(1) - \sigma_a(2)\sigma_b(1)\right]\beta(2)\beta(1) = -\psi_5(1, 2)$$

$$\psi_6(2, 1) = \frac{1}{2}\left[\sigma_b(2)\sigma_a(1) + \sigma_a(2)\sigma_b(1)\right]\left[\alpha(2)\beta(1) - \beta(2)\alpha(1)\right] = -\psi_6(1, 2)$$

10–33. Use the symmetry argument developed in Example 10–8 to show that $H_{16} = \langle\psi_1 | \hat{H} | \psi_6\rangle = 0$ for ψ_1 and ψ_6 given in Equations 10.49.

The wave functions ψ_1 and ψ_6 are

$$\psi_1 = \frac{1}{\sqrt{2}}\sigma_b(1)\sigma_b(2)\left[\alpha(1)\beta(2) - \beta(1)\alpha(2)\right]$$

$$\psi_6 = \frac{1}{2}\left[\sigma_b(1)\sigma_a(2) + \sigma_b(2)\sigma_a(1)\right]\left[\alpha(1)\beta(2) - \beta(1)\alpha(2)\right]$$

As stated in Example 10–8, interchanging A and B would have no effect on σ_b, but would change the sign of σ_a. As a result, the interchange would leave ψ_1 unchanged but change the sign of ψ_6, that is,

$$P_{AB}\psi_1 = \psi_1$$
$$P_{AB}\psi_2 = -\psi_2$$

Therefore, if $I_{AB} = \langle\psi_1|\hat{H}|\psi_6\rangle$, then

$$I_{BA} = -\langle\psi_1|\hat{H}|\psi_6\rangle = -I_{AB}$$

Since $I_{BA} = I_{AB}$, it follows that $I_{AB} = 0$, or $\langle\psi_1|\hat{H}|\psi_6\rangle = 0$.

10–34.   In this problem, we shall prove that if an operator $\hat{F}$ commutes with $\hat{H}$, then matrix elements of $\hat{H}$, $H_{ij} = \langle\psi_i|\hat{H}|\psi_j\rangle$, between states with different eigenvalues of $\hat{F}$ vanish. For simplicity, we prove this only for nondegenerate states. Let $\hat{F}$ be an operator that commutes with $\hat{H}$, and let its eigenvalues and eigenfunctions be denoted by λ and ψ_λ, respectively. Show that

$$[\hat{H}, \hat{F}]_{\lambda\lambda'} = \langle\psi_\lambda|[\hat{H}, \hat{F}]|\psi_{\lambda'}\rangle = (\lambda' - \lambda)H_{\lambda\lambda'}$$

Now argue that $H_{\lambda\lambda'} = 0$ unless $\lambda = \lambda'$. For degenerate states, it is possible to take linear combinations of the degenerate eigenfunctions and carry out a similar proof.

The matrix element $\left[\hat{H}, \hat{F}\right]_{\lambda, \lambda'}$ is

$$\left[\hat{H}, \hat{F}\right]_{\lambda,\lambda'} = \langle\psi_\lambda|\left[\hat{H}, \hat{F}\right]|\psi_{\lambda'}\rangle$$

$$= \langle\psi_\lambda|\hat{H}\hat{F}|\psi_{\lambda'}\rangle - \langle\psi_\lambda|\hat{F}\hat{H}|\psi_{\lambda'}\rangle$$

$$= \lambda'\langle\psi_\lambda|\hat{H}|\psi_{\lambda'}\rangle - \langle\hat{F}\psi_\lambda|\hat{H}|\psi_{\lambda'}\rangle$$

$$= \lambda'\langle\psi_\lambda|\hat{H}|\psi_{\lambda'}\rangle - \lambda\langle\psi_\lambda|\hat{H}|\psi_{\lambda'}\rangle$$

$$= (\lambda' - \lambda)\langle\psi_\lambda|\hat{H}|\psi_{\lambda'}\rangle$$

$$= (\lambda' - \lambda)H_{\lambda\lambda'}$$

Since $\hat{F}$ commutes with $\hat{H}$, $\left[\hat{H}, \hat{F}\right] = 0$, and therefore, $\langle\psi_\lambda|\left[\hat{H}, \hat{F}\right]|\psi_{\lambda'}\rangle = 0$. This condition is only satisfied when $\lambda = \lambda'$ or when $H_{\lambda\lambda'} = 0$. Thus, matrix elements of $\hat{H}$ between states with different eigenvalues of $\hat{F}$ vanish.

10–35. Determine the percentage of the correlation energy that the configuration-interaction calculation described in Section 10.9 gives.

The correlation energy is

$$E_{corr} = E_{exact} - E_{HF}$$
$$= -1.1744 \, E_h - \left(-1.133\,629 \, E_h\right)$$
$$= -0.0408 \, E_h$$

The correlation energy given by the configuration interaction calculation described in Section 10.9 is

$$E_{corr,CI} = E_{CI} - E_{HF}$$
$$= -1.147\,94 \, E_h - \left(-1.133\,629 \, E_h\right)$$
$$= -0.014\,31 \, E_h$$

which represents $-0.014\,31 \, E_h / (-0.0408 \, E_h) = 0.35$, or 35% of the correlation energy.

10–36. Show that Equations 10.61 through 10.63 result when Equation 10.58 is substituted into Equation 10.60.

Substitute Equation 10.58 into Equation 10.60 to give

$$E = \int d\mathbf{r}_1 \, d\mathbf{r}_2 \, \Psi^*(1,2) \left[-\frac{1}{2} \sum_{j=1}^{2} \nabla_j^2 + \sum_{j=1}^{2} \left(-\frac{1}{r_{jA}} - \frac{1}{r_{jB}} \right) + \frac{1}{r_{12}} + \frac{1}{R} \right] \Psi(1,2)$$

$$= \int d\mathbf{r}_1 \, d\mathbf{r}_2 \, \psi^*(\mathbf{r}_1)\psi^*(\mathbf{r}_2) \left[-\frac{1}{2} \sum_{j=1}^{2} \nabla_j^2 + \sum_{j=1}^{2} \left(-\frac{1}{r_{jA}} - \frac{1}{r_{jB}} \right) + \frac{1}{r_{12}} + \frac{1}{R} \right] \psi(\mathbf{r}_1)\psi(\mathbf{r}_2)$$

$$= \int d\mathbf{r}_1 \, d\mathbf{r}_2 \, \psi^*(\mathbf{r}_1)\psi^*(\mathbf{r}_2) \left(-\frac{1}{2}\nabla_1^2 - \frac{1}{r_{1A}} - \frac{1}{r_{1B}} \right) \psi(\mathbf{r}_1)\psi(\mathbf{r}_2)$$

$$+ \int d\mathbf{r}_1 \, d\mathbf{r}_2 \, \psi^*(\mathbf{r}_1)\psi^*(\mathbf{r}_2) \left(-\frac{1}{2}\nabla_2^2 - \frac{1}{r_{2A}} - \frac{1}{r_{2B}} \right) \psi(\mathbf{r}_1)\psi(\mathbf{r}_2)$$

$$+ \int d\mathbf{r}_1 \, d\mathbf{r}_2 \, \psi^*(\mathbf{r}_1)\psi^*(\mathbf{r}_2)\frac{1}{r_{12}}\psi(\mathbf{r}_1)\psi(\mathbf{r}_2)$$

$$+ \int d\mathbf{r}_1 \, d\mathbf{r}_2 \, \psi^*(\mathbf{r}_1)\psi^*(\mathbf{r}_2)\frac{1}{R}\psi(\mathbf{r}_1)\psi(\mathbf{r}_2)$$

$$= \left[\int d\mathbf{r}_2 \, \psi^*(\mathbf{r}_2)\psi(\mathbf{r}_2) \right] \left[\int d\mathbf{r}_1 \, \psi^*(\mathbf{r}_1) \left(-\frac{1}{2}\nabla_1^2 - \frac{1}{r_{1A}} - \frac{1}{r_{1B}} \right) \psi(\mathbf{r}_1) \right]$$

$$+ \left[\int d\mathbf{r}_1 \, \psi^*(\mathbf{r}_1)\psi(\mathbf{r}_1) \right] \left[\int d\mathbf{r}_2 \, \psi^*(\mathbf{r}_2) \left(-\frac{1}{2}\nabla_2^2 - \frac{1}{r_{2A}} - \frac{1}{r_{2B}} \right) \psi(\mathbf{r}_2) \right]$$

$$+ \int d\mathbf{r}_1 \, d\mathbf{r}_2 \, \psi^*(\mathbf{r}_1)\psi^*(\mathbf{r}_2) \frac{1}{r_{12}} \psi(\mathbf{r}_1)\psi(\mathbf{r}_2)$$

$$+ \frac{1}{R} \int d\mathbf{r}_1 \, d\mathbf{r}_2 \, \psi^*(\mathbf{r}_1)\psi^*(\mathbf{r}_2)\psi(\mathbf{r}_1)\psi(\mathbf{r}_2)$$

$$= \int d\mathbf{r}_1 \, \psi^*(\mathbf{r}_1) \left(-\frac{1}{2}\nabla_1^2 - \frac{1}{r_{1A}} - \frac{1}{r_{1B}} \right) \psi(\mathbf{r}_1) + \int d\mathbf{r}_2 \, \psi^*(\mathbf{r}_2) \left(-\frac{1}{2}\nabla_2^2 - \frac{1}{r_{2A}} - \frac{1}{r_{2B}} \right) \psi(\mathbf{r}_2)$$

$$+ \, J_{11} + \frac{1}{R}$$

$$= I_1 + I_2 + J_{11} + \frac{1}{R}$$

As stated in the text, since $1/R$ is a constant for a fixed geometry, it is initially ignored but added back for energy calculation later.

10–37. Show that Equation 10.61 is the same as equation 2 of the appendix when we use $\psi_1 = (1s_A + 1s_B)/\sqrt{2(1 + S)}$.

The spatial portion of the wave functions for both electrons is

$$\psi_1 = \frac{1s_A + 1s_B}{\sqrt{2(1 + S)}}$$

so that the molecular wave function is

$$\psi_1(1)\psi_2(2) = \frac{1}{2(1 + S)} \left[\phi_A(1) + \phi_B(1) \right] \left[\phi_A(2) + \phi_B(2) \right]$$

The ground-state energy of H_2 according to Equation 10.61 (with the neglected $1/R$ term added back) is,

$$E = I_1 + I_2 + J_{11} + \frac{1}{R}$$

The integral I_1 can be determined using Table 10.1 while the J_{11} integral can be determined using Table 10.8. By symmetry, $I_2 = I_1$.

The I_1 integral is

$$I_1 = \int d\mathbf{r}_1\, \psi_1^*(\mathbf{r}_1) \left(-\frac{1}{2}\nabla_1^2 - \frac{1}{r_{1A}} - \frac{1}{r_{1B}} \right) \psi_1(\mathbf{r}_1)$$

$$= \frac{1}{2\left[1+S(w)\right]} \left[\int d\mathbf{r}_1\, \phi_A(1) \left(-\frac{1}{2}\nabla_1^2 - \frac{1}{r_{1A}} - \frac{1}{r_{1B}} \right) \phi_A(1) \right.$$

$$+ 2\int d\mathbf{r}_1\, \phi_A(1) \left(-\frac{1}{2}\nabla_1^2 - \frac{1}{r_{1A}} - \frac{1}{r_{1B}} \right) \phi_B(1)$$

$$\left. + \int d\mathbf{r}_1\, \phi_B(1) \left(-\frac{1}{2}\nabla_1^2 - \frac{1}{r_{1A}} - \frac{1}{r_{1B}} \right) \phi_B(1) \right]$$

$$= \frac{1}{2\left[1+S(w)\right]} \left\{ \left[\frac{\zeta^2}{2} - \zeta + \zeta J(w) \right] + 2\left[\frac{-\zeta^2 S(w)}{2} - \zeta^2 K(w) + \zeta K(w) + \zeta K(w) \right] \right.$$

$$\left. + \left[\frac{\zeta^2}{2} + \zeta J(w) - \zeta \right] \right\}$$

$$= \zeta^2 \left\{ \frac{1 - S(w) - 2K(w)}{2\left[1+S(w)\right]} \right\} + \zeta \left\{ \frac{-2 + 2J(w) + 4K(w)}{2\left[1+S(w)\right]} \right\}$$

The J_{11} integral is

$$J_{11} = \int d\mathbf{r}_1\, d\mathbf{r}_2\, \psi_1^*(\mathbf{r}_1)\psi_1^*(\mathbf{r}_2)\frac{1}{r_{12}}\psi_1(\mathbf{r}_1)\psi_1(\mathbf{r}_2)$$

$$= \frac{1}{4\left[1+S(w)\right]^2} \left[\left\langle \phi_A(1)\phi_A(2) \left| \frac{1}{r_{12}} \right| \phi_A(1)\phi_A(2) \right\rangle + 2\left\langle \phi_A(1)\phi_A(2) \left| \frac{1}{r_{12}} \right| \phi_A(1)\phi_B(2) \right\rangle \right.$$

$$+ 2\left\langle \phi_A(1)\phi_A(2) \left| \frac{1}{r_{12}} \right| \phi_B(1)\phi_A(2) \right\rangle + 2\left\langle \phi_A(1)\phi_A(2) \left| \frac{1}{r_{12}} \right| \phi_B(1)\phi_B(2) \right\rangle$$

$$+ \left\langle \phi_A(1)\phi_B(2) \left| \frac{1}{r_{12}} \right| \phi_A(1)\phi_B(2) \right\rangle + 2\left\langle \phi_A(1)\phi_B(2) \left| \frac{1}{r_{12}} \right| \phi_B(1)\phi_A(2) \right\rangle$$

$$+ 2\left\langle \phi_A(1)\phi_B(2) \left| \frac{1}{r_{12}} \right| \phi_B(1)\phi_B(2) \right\rangle + \left\langle \phi_B(1)\phi_A(2) \left| \frac{1}{r_{12}} \right| \phi_B(1)\phi_A(2) \right\rangle$$

$$\left. + 2\left\langle \phi_B(1)\phi_A(2) \left| \frac{1}{r_{12}} \right| \phi_B(1)\phi_B(2) \right\rangle + \left\langle \phi_B(1)\phi_B(2) \left| \frac{1}{r_{12}} \right| \phi_B(1)\phi_B(2) \right\rangle \right]$$

Because of the many symmetries, the integrals in J_{11} simplify to

$$J_{11} = \frac{1}{4\left[1+S(w)\right]^2} \left[2\left\langle \phi_A(1)\phi_A(2) \left| \frac{1}{r_{12}} \right| \phi_A(1)\phi_A(2) \right\rangle + 8\left\langle \phi_A(1)\phi_A(2) \left| \frac{1}{r_{12}} \right| \phi_A(1)\phi_B(2) \right\rangle \right.$$

$$\left. + 4\left\langle \phi_A(1)\phi_A(2) \left| \frac{1}{r_{12}} \right| \phi_B(1)\phi_B(2) \right\rangle + 2\left\langle \phi_A(1)\phi_B(2) \left| \frac{1}{r_{12}} \right| \phi_A(1)\phi_B(2) \right\rangle \right]$$

$$= \frac{1}{4\left[1+S(w)\right]^2} \left[2\left(\frac{5\zeta}{8} \right) + 8\zeta L(w) + 4\zeta K'(w) + 2\zeta J'(w) \right]$$

$$= \zeta \left\{ \frac{5/16 + J'(w)/2 + K'(w) + 2L(w)}{\left[1+S(w)\right]^2} \right\}$$

Combining the I_1, I_2, and J_{11} integrals and $1/R = \zeta/w$, the ground-state energy of H_2 is

$$E = I_1 + I_2 + J_{11} + \frac{1}{R}$$

$$= 2\zeta^2 \left\{ \frac{1 - S(w) - 2K(w)}{2\left[1 + S(w)\right]} \right\} + 2\zeta \left\{ \frac{-2 + 2J(w) + 4K(w)}{2\left[1 + S(w)\right]} \right\}$$

$$+ \zeta \left\{ \frac{5/16 + J'(w)/2 + K'(w) + 2L(w)}{\left[1 + S(w)\right]^2} \right\} + \frac{\zeta}{w}$$

$$= \zeta^2 \left[\frac{1 - S(w) - 2K(w)}{1 + S(w)} \right] + \zeta \left[\frac{-2 + 2J(w) + 4K(w)}{1 + S(w)} \right]$$

$$+ \zeta \left\{ \frac{5/16 + J'(w)/2 + K'(w) + 2L(w)}{\left[1 + S(w)\right]^2} \right\} + \frac{\zeta}{w}$$

the last expression is the same as Equation 2 of the Appendix.

10–38. Substitute Equation 10.64 into Equation 10.60 to derive Equation 10.65.

To determine the energy given by the trial function

$$\Psi(\mathbf{r}_1, \mathbf{r}_2) = \psi(r_1)\psi(r_2)$$

$$\psi(r) = c_1\phi_A(r) + c_2\phi_B(r)$$

where $\phi_A(r)$ and $\phi_B(r)$ are normalized Slater orbitals with $\zeta = 1.193\,02$ centered on atoms A and B, respectively, Ψ must be substituted into

$$E = \frac{\left\langle \Psi \left| \hat{H} \right| \Psi \right\rangle}{\langle \Psi | \Psi \rangle}$$

To do this consider the following integrals in turn. First, from the denominator of the energy expression.

$$\langle \Psi | \Psi \rangle = \left\langle \psi(r_1) | \psi(r_1) \right\rangle \left\langle \psi(r_2) | \psi(r_2) \right\rangle$$

$$= \left[c_1^2 \left\langle \phi_A(1) | \phi_A(1) \right\rangle + 2c_1c_2 \left\langle \phi_A(1) | \phi_B(1) \right\rangle + c_2^2 \left\langle \phi_B(1) | \phi_B(1) \right\rangle \right]$$

$$\times \left[c_1^2 \left\langle \phi_A(2) | \phi_A(2) \right\rangle + 2c_1c_2 \left\langle \phi_A(2) | \phi_B(2) \right\rangle + c_2^2 \left\langle \phi_B(2) | \phi_B(2) \right\rangle \right]$$

$$= \left(c_1^2 S_{AA} + 2c_1c_2 S_{AB} + c_2^2 S_{BB} \right)^2$$

Of course, if the coefficients are chosen such that Ψ is normalized, this integral will equal one. Next consider the "electron one" portion of the Hamiltonian operator,

$$\left\langle \Psi \left| -\frac{1}{2}\nabla_1^2 - \frac{1}{r_{1A}} - \frac{1}{r_{1B}} \right| \Psi \right\rangle$$

$$= \left\langle \psi(r_1) \left| -\frac{1}{2}\nabla_1^2 - \frac{1}{r_{1A}} - \frac{1}{r_{1B}} \right| \psi(r_1) \right\rangle \left\langle \psi(r_2) \mid \psi(r_2) \right\rangle$$

$$= \left[c_1^2 \left\langle \phi_A(1) \left| -\frac{1}{2}\nabla_1^2 - \frac{1}{r_{1A}} - \frac{1}{r_{1B}} \right| \phi_A(1) \right\rangle + 2c_1c_2 \left\langle \phi_A(1) \left| -\frac{1}{2}\nabla_1^2 - \frac{1}{r_{1A}} - \frac{1}{r_{1B}} \right| \phi_B(1) \right\rangle \right.$$

$$\left. + c_2^2 \left\langle \phi_B(1) \left| -\frac{1}{2}\nabla_1^2 - \frac{1}{r_{1A}} - \frac{1}{r_{1B}} \right| \phi_B(1) \right\rangle \right] \left\langle \psi(r_2) \mid \psi(r_2) \right\rangle$$

$$= \left(c_1^2 H_{AA} + 2c_1c_2 H_{AB} + c_2^2 H_{BB} \right) \left(c_1^2 S_{AA} + 2c_1c_2 S_{AB} + c_2^2 S_{BB} \right)$$

Because the electrons are indistinguishable, the "electron two" portion of the Hamiltonian operator will give the same result. The electron-electron repulsion portion of the Hamiltonian operator is a bit more complicated,

$$\left\langle \Psi \left| \frac{1}{r_{12}} \right| \Psi \right\rangle$$

$$= \left\langle \left[c_1\phi_A(1) + c_2\phi_B(1) \right] \left[c_1\phi_A(2) + c_2\phi_B(2) \right] \left| \frac{1}{r_{12}} \right| \left[c_1\phi_A(1) + c_2\phi_B(1) \right] \left[c_1\phi_A(2) + c_2\phi_B(2) \right] \right\rangle$$

$$= \sum_{i=1}^{2}\sum_{j=1}^{2} c_ic_j \left[c_1^2 \left\langle \phi_i(1)\phi_A(2) \left| \frac{1}{r_{12}} \right| \phi_j(1)\phi_A(2) \right\rangle + c_1c_2 \left\langle \phi_i(1)\phi_A(2) \left| \frac{1}{r_{12}} \right| \phi_j(1)\phi_B(2) \right\rangle \right.$$

$$\left. + c_2c_1 \left\langle \phi_i(1)\phi_B(2) \left| \frac{1}{r_{12}} \right| \phi_j(1)\phi_A(2) \right\rangle + c_2^2 \left\langle \phi_i(1)\phi_B(2) \left| \frac{1}{r_{12}} \right| \phi_j(1)\phi_B(2) \right\rangle \right]$$

The indices i and j apply to both the c coefficients and the ϕ wave functions. For example, when $i = 1$, $c_i = c_1$ and $\phi_i = \phi_A$ and when $i = 2$, $c_i = c_2$ and $\phi_i = \phi_B$. Thus,

$$\left\langle \Psi \left| \frac{1}{r_{12}} \right| \Psi \right\rangle = \left(c_1^2 S_{AA} + 2c_1c_2 S_{AB} + c_2^2 S_{BB} \right) \sum_{i=1}^{2}\sum_{j=1}^{2} c_ic_j G_{ij}$$

$$= \left(c_1^2 S_{AA} + 2c_1c_2 S_{AB} + c_2^2 S_{BB} \right) \left(c_1^2 G_{AA} + c_1c_2 G_{AB} + c_2c_1 G_{BA} + c_2^2 G_{BB} \right)$$

$$= \left(c_1^2 S_{AA} + 2c_1c_2 S_{AB} + c_2^2 S_{BB} \right) \left(c_1^2 G_{AA} + 2c_1c_2 G_{AB} + c_2^2 G_{BB} \right)$$

since $G_{AB} = G_{BA}$. The G_{ij} are defined in Equation 10.67. (In the first printing of the textbook, the factor $c_1^2 S_{AA} + 2c_1c_2 S_{AB} + c_2^2 S_{BB}$ in the denominator was missing in Equation 10.67. This factor is equal to 1 if normalized wave functions are used.)

$$G_{ij} = \frac{\displaystyle\sum_{k=1}^{2}\sum_{l=1}^{2} c_kc_l \left\langle \phi_i(1)\phi_k(2) \left| \frac{1}{r_{12}} \right| \phi_j(1)\phi_l(2) \right\rangle}{c_1^2 S_{AA} + 2c_1c_2 S_{AB} + c_2^2 S_{BB}}$$

Thus,

$$\left\langle \Psi \left| \hat{H} \right| \Psi \right\rangle = \left\langle \Psi \left| -\frac{1}{2}\nabla_1^2 - \frac{1}{r_{1A}} - \frac{1}{r_{1B}} \right| \Psi \right\rangle + \left\langle \Psi \left| -\frac{1}{2}\nabla_2^2 - \frac{1}{r_{2A}} - \frac{1}{r_{2B}} \right| \Psi \right\rangle + \left\langle \Psi \left| \frac{1}{r_{12}} \right| \Psi \right\rangle$$

$$= \left[2 \left(c_1^2 H_{AA} + 2c_1 c_2 H_{AB} + c_2^2 H_{BB} \right) + c_1^2 G_{AA} + 2c_1 c_2 G_{AB} + c_2^2 G_{BB} \right]$$

$$\times \left(c_1^2 S_{AA} + 2c_1 c_2 S_{AB} + c_2^2 S_{BB} \right)$$

and

$$E = \frac{\left\langle \Psi \left| \hat{H} \right| \Psi \right\rangle}{\left\langle \Psi \mid \Psi \right\rangle}$$

$$= \frac{1}{\left(c_1^2 S_{AA} + 2c_1 c_2 S_{AB} + c_2^2 S_{BB} \right)^2} \left\{ \left[2 \left(c_1^2 H_{AA} + 2c_1 c_2 H_{AB} + c_2^2 H_{BB} \right) + c_1^2 G_{AA} + 2c_1 c_2 G_{AB} + c_2^2 G_{BB} \right] \right.$$

$$\left. \times \left(c_1^2 S_{AA} + 2c_1 c_2 S_{AB} + c_2^2 S_{BB} \right) \right\}$$

$$= \frac{2 \left(c_1^2 H_{AA} + 2c_1 c_2 H_{AB} + c_2^2 H_{BB} \right) + c_1^2 G_{AA} + 2c_1 c_2 G_{AB} + c_2^2 G_{BB}}{c_1^2 S_{AA} + 2c_1 c_2 S_{AB} + c_2^2 S_{BB}}$$

which is Equation 10.65 of the Appendix.

10–39. Verify the entries in Table 10.5.

The matrix elements in Table 10.5 are calculated for $\zeta = 1.193\,02$ at $R = 1.385\,43\ a_0$. Thus, $w = \zeta R = 1.652\,85$. Before verifying the values of these elements, it is useful to calculate the numerical values for the integrals leading to them. The definitions for the integrals can be found in Tables 10.1 and 10.8:

$$S(w) = e^{-w} \left(1 + w + \frac{w^2}{3} \right) = 0.682\,421$$

$$J(w) = e^{-2w} \left(1 + \frac{1}{w} \right) - \frac{1}{w} = -0.546\,155$$

$$K(w) = -e^{-w} (1 + w) = -0.508\,031$$

$$S'(w) = e^w \left(1 - w + \frac{w^2}{3} \right) = 1.346\,12$$

$$J'(w) = \frac{1}{w} - e^{-2w} \left(\frac{1}{w} + \frac{11}{8} + \frac{3w}{4} + \frac{w^2}{6} \right) = 0.470\,242$$

$$L(w) = e^{-w} \left(w + \frac{1}{8} + \frac{5}{16w} \right) + e^{-3w} \left(-\frac{1}{8} - \frac{5}{16w} \right) = 0.374\,466$$

$$E_1(4w) = \int_{4w}^{\infty} \frac{e^{-t}}{t} \, dt = 1.792\,53 \times 10^{-4}$$

$$E_1(2w) = \int_{2w}^{\infty} \frac{e^{-t}}{t} \, dt = 8.875\,67 \times 10^{-3}$$

$$K'(w) = \frac{1}{5} \left\{ -e^{-2w} \left(-\frac{25}{8} + \frac{23w}{4} + 3w^2 + \frac{w^3}{3} \right) \right.$$

$$\left. + \frac{6}{w} \left[S(w)^2 \left(\gamma + \ln w \right) - S'(w)^2 E_1(4w) + 2S(w)S'(w)E_1(2w) \right] \right\}$$

$$= 0.258\,721$$

The matrix elements H_{ij} (Equation 10.66) are

$$H_{ij} = H_{ji} = \left\langle \phi_i \, \left| -\frac{1}{2}\nabla_1^2 - \frac{1}{r_{1A}} - \frac{1}{r_{1B}} \right| \, \phi_j \right\rangle$$

$$= \left\langle \phi_i \, \left| -\frac{1}{2}\nabla_1^2 \right| \, \phi_j \right\rangle + \left\langle \phi_i \, \left| -\frac{1}{r_{1A}} \right| \, \phi_j \right\rangle + \left\langle \phi_i \, \left| -\frac{1}{r_{1B}} \right| \, \phi_j \right\rangle$$

and their numerical values can be determined using Table 10.1 and the numerical values of the above integrals:

$$H_{AA} = \frac{\zeta^2}{2} - \zeta + \zeta J(w)$$

$$= -1.132\,95 \, E_h$$

$$H_{AB} = H_{BA} = -\zeta^2 \left[\frac{S(w)}{2} + K(w) \right] + \zeta K(w) + \zeta K(w)$$

$$= K(w) \left(-\zeta^2 + 2\zeta \right) - \frac{\zeta^2 S(w)}{2}$$

$$= -0.974\,75 \, E_h$$

By symmetry, $H_{BB} = H_{AA} = -1.132\,95 \, E_h$.

The next set of elements are $S_{ij} = \int \phi_i \phi_j \, d\tau$. Since ϕ_A and ϕ_B are both normalized,

$$S_{AA} = S_{BB} = 1$$

and the overlap integral is

$$S_{AB} = S_{BA} = S(w) = 0.682\,42$$

Finally, the matrix elements G_{ij} (see Problem 10–38) are

$$
\begin{aligned}
G_{AA} &= \frac{1}{c_1^2 S_{AA} + 2c_1c_2 S_{AB} + c_2^2 S_{BB}} \left[c_1^2 \left\langle \phi_A(1)\phi_A(2) \left| \frac{1}{r_{12}} \right| \phi_A(1)\phi_A(2) \right\rangle \right.\\
&\quad + c_1c_2 \left\langle \phi_A(1)\phi_A(2) \left| \frac{1}{r_{12}} \right| \phi_A(1)\phi_B(2) \right\rangle + c_2c_1 \left\langle \phi_A(1)\phi_B(2) \left| \frac{1}{r_{12}} \right| \phi_A(1)\phi_A(2) \right\rangle \\
&\quad \left. + +c_2^2 \left\langle \phi_A(1)\phi_B(2) \left| \frac{1}{r_{12}} \right| \phi_A(1)\phi_B(2) \right\rangle \right] \\
&= \frac{1}{c_1^2 + 2c_1c_2 S_{AB} + c_2^2} \left[c_1^2 \left(\frac{5\zeta}{8} \right) + 2c_1c_2 \zeta L(w) + c_2^2 \zeta J'(w) \right] \\
&= \frac{0.745\,64 c_1^2 + 0.893\,49 c_1c_2 + 0.561\,01 c_2^2}{c_1^2 + 1.364\,84 c_1c_2 + c_2^2}
\end{aligned}
$$

$$
\begin{aligned}
G_{AB} = G_{BA} &= \frac{1}{c_1^2 S_{AA} + 2c_1c_2 S_{AB} + c_2^2 S_{BB}} \left[c_1^2 \left\langle \phi_A(1)\phi_A(2) \left| \frac{1}{r_{12}} \right| \phi_B(1)\phi_A(2) \right\rangle \right.\\
&\quad + c_1c_2 \left\langle \phi_A(1)\phi_A(2) \left| \frac{1}{r_{12}} \right| \phi_B(1)\phi_B(2) \right\rangle + c_2c_1 \left\langle \phi_A(1)\phi_B(2) \left| \frac{1}{r_{12}} \right| \phi_B(1)\phi_A(2) \right\rangle \\
&\quad \left. + c_2^2 \left\langle \phi_A(1)\phi_B(2) \left| \frac{1}{r_{12}} \right| \phi_B(1)\phi_B(2) \right\rangle \right] \\
&= \frac{1}{c_1^2 + 2c_1c_2 S_{AB} + c_2^2} \left[c_1^2 \zeta L(w) + 2c_1c_2 \zeta K'(w) + c_2^2 \zeta L(w) \right] \\
&= \frac{0.446\,75 c_1^2 + 0.617\,32 c_1c_2 + 0.446\,75 c_2^2}{c_1^2 + 1.364\,84 c_1c_2 + c_2^2}
\end{aligned}
$$

$$
\begin{aligned}
G_{BB} &= \frac{1}{c_1^2 S_{AA} + 2c_1c_2 S_{AB} + c_2^2 S_{BB}} \left[c_1^2 \left\langle \phi_B(1)\phi_A(2) \left| \frac{1}{r_{12}} \right| \phi_B(1)\phi_A(2) \right\rangle \right.\\
&\quad + c_1c_2 \left\langle \phi_B(1)\phi_A(2) \left| \frac{1}{r_{12}} \right| \phi_B(1)\phi_B(2) \right\rangle + c_2c_1 \left\langle \phi_B(1)\phi_B(2) \left| \frac{1}{r_{12}} \right| \phi_B(1)\phi_A(2) \right\rangle \\
&\quad \left. + c_2^2 \left\langle \phi_B(1)\phi_B(2) \left| \frac{1}{r_{12}} \right| \phi_B(1)\phi_B(2) \right\rangle \right] \\
&= \frac{1}{c_1^2 + 2c_1c_2 S_{AB} + c_2^2} \left[c_1^2 \zeta J'(w) + 2c_1c_2 \zeta L(w) + c_2^2 \left(\frac{5\zeta}{8} \right) \right] \\
&= \frac{0.561\,01 c_1^2 + 0.893\,49 c_1c_2 + 0.745\,64 c_2^2}{c_1^2 + 1.364\,84 c_1c_2 + c_2^2}
\end{aligned}
$$

10–40. Show that the normalization condition for Equation 10.64 is $c_1^2 + 2c_1c_2 S_{AB} + c_2^2 = 1$ where S_{AB} is the overlap integral involving ϕ_A and ϕ_B.

Taking c_1 and c_2 to be real, the normalization condition for Equation 10.64 is

$$\int \psi^* \psi \, d\tau = \int (c_1\phi_A + c_2\phi_B)^2 \, d\tau = 1$$

$$c_1^2 \int \phi_A^2 \, d\tau + 2c_1c_2 \int \phi_A\phi_B \, d\tau + c_2^2 \int \phi_B^2 \, d\tau = 1$$

$$c_1^2 + 2c_1c_2 S_{AB} + c_2^2 = 1$$

The two integrals, $\int \phi_A^2 \, d\tau$ and $\int \phi_B^2 \, d\tau$, equal one, since both ϕ_A and ϕ_B are normalized.

10–41. Use a program such as *MathCad* or *Mathematica* to redo the Hartree–Fock–Roothaan calculation in Section 10.10, starting with $c_1 = c_2$. Explain why the result converges so rapidly.

Since

$$c_1^2 + 2c_1c_2 S_{AB} + c_2^2 = 1$$

and $S_{AB} = 0.682\,42$ (Table 10.5), the condition $c_1 = c_2$ gives

$$c_1 = c_2 = 0.545\,15$$

Using these initial guess values of c_1 and c_2, the G_{ij} matrix elements (Problem 10–39) are

$$G_{AA} = 0.745\,64c_1^2 + 0.893\,49c_1c_2 + 0.561\,01c_2^2 = 0.653\,86$$

$$G_{AB} = G_{BA} = 0.446\,75c_1^2 + 0.617\,32c_1c_2 + 0.446\,75c_2^2 = 0.449\,00$$

$$G_{BB} = 0.561\,01c_1^2 + 0.893\,49c_1c_2 + 0.745\,64c_2^2 = 0.653\,86$$

Combining these matrix elements with others in Table 10.5, the elements of the Fock matrix are

$$F_{AA} = H_{AA} + G_{AA} = -1.132\,95 + 0.653\,86 = -0.479\,09$$

$$F_{AB} = H_{AB} + G_{AB} = -0.974\,75 + 0.449\,00 = -0.525\,75$$

$$F_{BB} = H_{BB} + G_{BB} = -1.132\,95 + 0.653\,86 = -0.479\,09$$

and the energy is

$$E = c_1^2 \left(H_{AA} + F_{AA} \right) + 2c_1c_2 \left(H_{AB} + F_{AB} \right) + c_2^2 \left(H_{BB} + F_{BB} \right) + 1/R$$

$$= -1.128\,24 \, E_h$$

Now use the F_{ij} matrix elements to solve for ε in the determinantal equation:

$$\begin{vmatrix} F_{AA} - \varepsilon & F_{AB} - \varepsilon S_{AB} \\ F_{AB} - \varepsilon S_{AB} & F_{BB} - \varepsilon \end{vmatrix} = 0$$

The smaller value of the two solutions is $\varepsilon = -0.597\,26$. This value allows the determination of the ratio for the improved c_1 and c_2:

$$\frac{c_1}{c_2} = -\frac{F_{AB} - \varepsilon S_{AB}}{F_{AA} - \varepsilon S_{AA}}$$

$$= 1.0000$$

This is the same ratio as the initial ratio used in this problem. In fact, the energy calculated is the optimized molecular orbital theory result. The calculation converges immediately because $c_1 = c_2$ is in accord with the symmetry of H_2.

10–42. Why do the energies in Table 10.6 converge after three or four iterations but the coefficients in the wave function take five iterations?

This observation is due to the fact that if a wave function is determined through first order in a perturbation, then the energy is known through third order (See Problem 8–45). The values of the energy are more stable than those of the wave function.

10–43. Use a program such as *MathCad* or *Mathematica* to plot E_{MO} given by equation 2 of the appendix against R for $\zeta = 1$, and compare your result to Figure 10.23.

The plot is made by using Equation 2 of the Appendix. The integrals in the equation can be found in Tables 10.1 and 10.8.

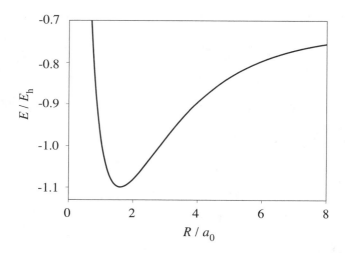

This plot is the same as the curve for $\zeta = 1$ in Figure 10.23.

10–44. Use a program such as *MathCad* or *Mathematica* and the result of Problem 10–21 to determine ζ as a function of w for E_{MO} given by equation 2 of the appendix. Plot $\zeta(w)$ and $R(w)$ parametrically to give ζ plotted against R. Compare your result to that in Figure 10.24.

To plot ζ as a function of w, use the expression

$$\zeta = \frac{-V_+(w) - w\dfrac{dV_+(w)}{dw}}{2T_+(w) + w\dfrac{dT_+(w)}{dw}}$$

The $T(w)$ and $V(w)$ expressions can be deduced from Equations 2 and 3 in the Appendix:

$$T(w) = \frac{1 - S(w) - 2K(w)}{1 + S(w)}$$

$$V(w) = \frac{-2 + 2J(w) + 4K(w)}{1 + S(w)} + \frac{5/16 + J'(w)/2 + K'(w) + 2L(w)}{[1 + S(w)]^2} + \frac{1}{w}$$

The integrals in these expressions can be found in Tables 10.1 and 10.8.

To plot R as a function of w, divide w by $\zeta(w)$ and plot against w. The two plots are shown below.

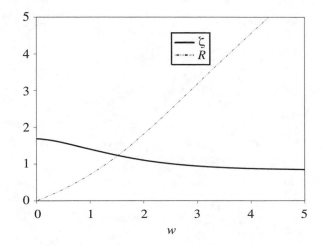

Now combine $R(w)$ and $\zeta(w)$ to arrive at the following plot:

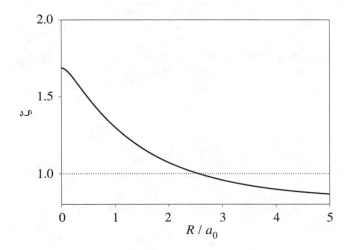

This plot is the same as that in Figure 10.24.

10–45. In this problem, we'll use the result of Problem 10–21 to show that $\zeta \to 27/16$ as $w \to 0$ for the molecular orbital calculation for H_2 in the appendix. (See also Figure 10.24.) We'll use the fact that the definition of $K'(w)$ in Table 10.8 has the following expansion:

$$E_1(x) = -\gamma - \ln x + x - x^2/4 + O(x^3)$$

(See the reference to Abramowitz and Stegun at the end of the chapter.) Using this expansion, first show that the second term in the definition of $K'(w)$ (the one multiplied by $6/w$) goes to zero as $w \to 0$. (This is a good example of one in which you should keep track of the order of the terms that you keep or ignore.) Now show that

$$K'(w) = \frac{5}{8} + O(w)$$

$$J'(w) = \frac{5}{8} + O(w)$$

$$L(w) = \frac{5}{8} + O(w)$$

In Problem 10–23, we obtained

$$S(w) = 1 - \frac{1}{6}w^2 + O(w^3)$$

$$J(w) = -1 + \frac{2}{3}w^2 + O(w^3)$$

$$K(w) = -1 + \frac{w^2}{2} + O(w^3)$$

Put this all together to show that

$$T(w) = 1 + O(w)$$

and that

$$V(w) = \frac{1}{w} - \frac{27}{8} + O(w)$$

Finally, use the result of Problem 10–21 to show that

$$\zeta \longrightarrow 27/16 \qquad \text{as } w \longrightarrow 0$$

First, consider the behavior of several terms in the definition of $K'(w)$. The two E_1 terms are

$$E_1(4w) = -\gamma - \ln(4w) + 4w - 4w^2 + O(w^3)$$

$$= -\gamma - \ln 4 - \ln w + 4w - 4w^2 + O(w^3)$$

$$E_1(2w) = -\gamma - \ln(2w) + 4w - 4w^2 + O(w^3)$$

$$= -\gamma - \ln 2 - \ln w + 4w - 4w^2 + O(w^3)$$

As $w \to 0$, all the terms involving w in $E_1(4w)$ and $E_1(2w)$, except for the $\ln w$ term, are negligible. Thus,

$$E_1(4w) = -\gamma - \ln 4 - \ln w + O(w)$$

$$E_1(2w) = -\gamma - \ln 2 - \ln w + O(w)$$

The $S(w)$ and $S'(w)$ terms as $w \to 0$ are

$$S(w) = e^{-w} \left(1 + w + \frac{w^2}{3} \right)$$

$$= \left(1 - w + \frac{w^2}{2} + \cdots \right) \left(1 + w + \frac{w^2}{3} \right)$$

$$= 1 + O(w^2)$$

$$S'(w) = e^{w} \left(1 - w + \frac{w^2}{3} \right)$$

$$= \left(1 + w + \frac{w^2}{2} + \cdots \right) \left(1 - w + \frac{w^2}{3} \right)$$

$$= 1 + O(w^2)$$

Putting these results together, the second term in the definition of $K'(w)$ becomes

$$S(w)^2 \left(\gamma + \ln w \right) - S'(w)^2 E_1(4w) + 2S(w)S'(2)E_1(2w)$$

$$= \left[1 + O(w^2) \right]^2 (\gamma + \ln w) - \left[1 + O(w^2) \right]^2 [-\gamma - \ln 4 - \ln w + O(w)]$$

$$+ 2 \left[1 + O(w^2) \right] \left[1 + O(w^2) \right] [-\gamma - \ln 2 - \ln w + O(w)]$$

$$= \left[\gamma + \ln w + O(w^2 \ln w) \right] + \left[\gamma + \ln 4 + \ln w + O(w^2 \ln w) \right]$$

$$- \left[2\gamma + 2\ln 2 + 2\ln w + O(w^2 \ln w) \right]$$

$$= 0 + O(w^2 \ln w)$$

Multiplying this term by $6/w$ as given in the definition of $K'(w)$ gives a term with $O(w \ln w)$ which goes to zero in the limit as $w \to 0$. Thus, the second term in $K'(w)$ can be ignored; only the first term needs to be considered:

$$K'(w) = -\frac{1}{5} e^{-2w} \left(-\frac{25}{8} + \frac{23w}{4} + 3w^2 + \frac{w^3}{3} \right)$$

$$= -\frac{1}{5} \left[1 - 2w + \frac{4w^2}{2} + O(w^3) \right] \left(-\frac{25}{8} + \frac{23w}{4} + 3w^2 + \frac{w^3}{3} \right)$$

$$= -\frac{1}{5} \left(-\frac{25}{8} \right) + O(w)$$

$$= \frac{5}{8} + O(w)$$

where the expansion $e^x = 1 + x + x^2/2 + \cdots$ has been applied. Using the same expansion, the $J'(w)$ and $L(w)$ integrals become

$$J'(w) = \frac{1}{w} - e^{-2w}\left(\frac{1}{w} + \frac{11}{8} + \frac{3w}{4} + \frac{w^2}{6}\right)$$

$$= \frac{1}{w} - \left[1 - 2w + \frac{4w^2}{2} + O(w^3)\right]\left(\frac{1}{w} + \frac{11}{8} + \frac{3w}{4} + \frac{w^2}{6}\right)$$

$$= \frac{1}{w} - \left[\frac{1}{w} + \frac{11}{8} - 2 + O(w)\right]$$

$$= \frac{5}{8} + O(w)$$

$$L(w) = e^{-w}\left(w + \frac{1}{8} + \frac{5}{16w}\right) + e^{-3w}\left(-\frac{1}{8} - \frac{5}{16w}\right)$$

$$= \left[1 - w + \frac{w^2}{2} + O(w^3)\right]\left(w + \frac{1}{8} + \frac{5}{16w}\right) + \left[1 - 3w + \frac{9w^2}{2} + O(w^3)\right]\left(-\frac{1}{8} - \frac{5}{16w}\right)$$

$$= \left(\frac{1}{8} - \frac{5}{16}\right) - \left(\frac{1}{8} - \frac{15}{16}\right) + O(w)$$

$$= \frac{5}{8} + O(w)$$

Combining these integrals and those for $S(w)$, $J(w)$, and $K(w)$ as $w \to 0$, the kinetic and potential energies (Equations 2 and 3 in the Appendix) are

$$T(w) = \frac{1 - S(w) - 2K(w)}{1 + S(w)}$$

$$= \frac{1 - (1 - w^2/6) - 2(-1 + w^2/2) + O(w^3)}{1 + (1 - w^2/6) + O(w^3)}$$

$$= 1 + O(w)$$

$$V(w) = \frac{-2 + 2J(w) + 4K(w)}{1 + S(w)} + \frac{5/16 + J'(w)/2 + K'(w) + 2L(w)}{[1 + S(w)]^2} + \frac{1}{w}$$

$$= \frac{-2 + 2(-1 + 2w^2/3) + 4(-1 + w^2/2) + O(w^3)}{1 + (1 - w^2/6) + O(w^3)}$$

$$\quad + \frac{5/16 + 5/16 + 5/8 + 5/4 + O(w)}{[1 + (1 - w^2/6) + O(w^3)]^2} + \frac{1}{w}$$

$$= \frac{-8}{2} + \frac{5/2}{4} + \frac{1}{w} + O(w)$$

$$= \frac{1}{w} - \frac{27}{8} + O(w)$$

To evaluate ζ, determine $w\,dT/dw$ and $w\,dV/dw$. Since the most significant term in dT/dw is a constant, $w\,dT/dw = 0$ in the limit as $w \to 0$. For $w\,dV/dw$,

$$w\frac{dV}{dw} = w\left(-\frac{1}{w^2}\right) = -\frac{1}{w}$$

The value of ζ in the limit as $w \to 0$ is

$$\zeta = \frac{-V(w) - w\dfrac{dV}{dw}}{2T(w) + w\dfrac{dT}{dw}}$$

$$= \frac{-\dfrac{1}{w} + \dfrac{27}{8} + \dfrac{1}{w}}{2}$$

$$= \frac{27}{16}$$

10–46. This problem complements the previous problem in the sense that we'll show that $\zeta \rightarrow$ 27/32 as $w \rightarrow \infty$ for the molecular orbital calculation for H_2 done in the appendix. (See also Figure 10.24 and Example 10–6.) We'll use the fact that

$$E_1(x) \longrightarrow \frac{e^{-x}}{x}\left[1 - \frac{1}{x} + O\left(\frac{1}{x^2}\right)\right] \qquad \text{as } x \longrightarrow \infty$$

(See the reference to Abramowitz and Stegun at the end of the chapter.) Show that

$$T(w) \longrightarrow 1 + O(w^2 e^{-w})$$

and that

$$V(w) \longrightarrow -\frac{27}{16} - \frac{1}{2w} + O(we^{-w})$$

Now use the result of Problem 10–21 to show that

$$\zeta \longrightarrow 27/32 \qquad \text{as } w \longrightarrow \infty$$

The behavior of the integrals $S(w)$, $J(w)$, and $K(w)$ as $w \rightarrow \infty$ has already been considered in Problem 10–24:

$$S(w) = \frac{w^2 e^{-w}}{3}$$

$$J(w) = -\frac{1}{w}$$

$$K(w) = -we^{-w}$$

Now find expressions for $S'(w)$, $J'(w)$, $L(w)$, and $K'(w)$ as $w \rightarrow \infty$, keeping only the most significant terms:

$$S'(w) = e^w \left(1 - w + \frac{w^2}{3}\right)$$

$$= \frac{w^2 e^w}{3}$$

$$J'(w) = \frac{1}{w} - e^{-2w}\left(\frac{1}{w} + \frac{11}{8} + \frac{3w}{4} + \frac{w^2}{6}\right)$$

$$= \frac{1}{w}$$

$$L(w) = e^{-w}\left(w + \frac{1}{8} + \frac{5}{16w}\right) + e^{-3w}\left(-\frac{1}{8} - \frac{5}{16w}\right)$$

$$= we^{-w}$$

$$K'(w) = \frac{1}{5}\left\{-e^{-2w}\left(-\frac{25}{8} + \frac{23w}{4} + 3w^2 + \frac{w^3}{3}\right)\right.$$

$$+ \frac{6}{w}\left[S(w)^2(\gamma + \ln w) - S'(w)^2 E_1(4w) + 2S(w)S'(w)E_1(2w)\right]\Big\}$$

$$= \frac{1}{5}\left\{-e^{-2w}\left(-\frac{25}{8} + \frac{23w}{4} + 3w^2 + \frac{w^3}{3}\right)\right.$$

$$+ \frac{6}{w}\left[\frac{w^4 e^{-2w}}{9}\ln w - \frac{w^4 e^{2w}}{9}\left(\frac{e^{-4w}}{4w}\right)\left(1 - \frac{1}{4w} + O(w^{-2})\right)\right.$$

$$\left.\left. + \frac{2w^4}{9}\left(\frac{e^{-2w}}{2w}\right)\left(1 - \frac{1}{2w} + O(w^{-2})\right)\right]\right\}$$

$$= \frac{1}{5}\left[-\frac{w^3 e^{-2w}}{3} + \frac{6}{w}\left(\frac{w^4 e^{-2w}}{9}\ln w - \frac{w^4 e^{-2w}}{36w} + \frac{w^4 e^{-2w}}{9w}\right)\right]$$

$$= \frac{1}{5}\left[-\frac{w^3 e^{-2w}}{3} + \frac{6}{w}\left(\frac{w^4 e^{-2w}}{9}\ln w\right)\right]$$

$$= \frac{2w^3 \ln w\, e^{-2w}}{15}$$

Combining these integrals and those for $S(w)$, $J(w)$, and $K(w)$ as $w \to \infty$, the kinetic and potential energies (Equations 2 and 3 in the Appendix) are

$$T(w) = \frac{1 - S(w) - 2K(w)}{1 + S(w)}$$

$$= \frac{1 - w^2 e^{-w}/3 + 2we^{-w}}{1 + w^2 e^{-w}/3}$$

$$= 1 + O(we^{-w})$$

$$V(w) = \frac{-2 + 2J(w) + 4K(w)}{1 + S(w)} + \frac{5/16 + J'(w)/2 + K'(w) + 2L(w)}{[1 + S(w)]^2} + \frac{1}{w}$$

$$= \frac{-2 - 2/w - 4we^{-w}}{1 + w^2 e^{-w}/3} + \frac{5/16 + 1/(2w) + 2w^3 \ln w\, e^{-2w}/15 + 2we^{-w}}{\left(1 + w^2 e^{-w}/3\right)^2} + \frac{1}{w}$$

$$= \left(-2 - \frac{2}{w}\right) + \left(\frac{5}{16} + \frac{1}{2w}\right) + \frac{1}{w} + O(we^{-w})$$

$$= -\frac{27}{16} - \frac{1}{2w} + O(we^{-w})$$

To evaluate ζ, determine $w\, dT/dw$ and $w\, dV/dw$. Since the most significant term in dT/dw contains e^{-w}, $w\, dT/dw = 0$ in the limit as $w \to \infty$.

The limit of ζ in the limit as $w \to \infty$ is

$$\zeta = \frac{-V(w) - w\dfrac{dV}{dw}}{2T(w) + w\dfrac{dT}{dw}}$$

$$= \frac{\dfrac{27}{16} + \dfrac{1}{2w} - w\left(\dfrac{1}{2w^2}\right)}{2}$$

$$= \frac{27}{32}$$

10–47. Show that all the integrals in Table 10.8 go to zero as $R \to \infty$ so that $H_{11} = H_{22} \to \zeta^2 - 27\zeta/16$, $H_{12} \to 5\zeta/16$, and $E_{CI} \to \zeta^2 - \zeta$.

Keeping the most significant terms as $R \to \infty$ (which is equivalent to $w \to \infty$), the integrals $S(w)$, $S'(w)$, $J'(w)$, $L(w)$, and $K'(w)$ are shown, from the last problem, to be

$$S(w) = \frac{w^2 e^{-w}}{3}$$

$$S'(w) = \frac{w^2 e^{w}}{3}$$

$$J'(w) = \frac{1}{w}$$

$$L(w) = we^{-w}$$

$$K'(w) = \frac{2w^3 \ln w\, e^{-2w}}{15}$$

which all reduce to zero in the limit as $R \to \infty$. Therefore,

$$H_{11}(\zeta, w) = \zeta^2 \left[\frac{1 - S(w) - 2K(w)}{1 + S(w)}\right] + \zeta \left\{\frac{-2 + 2J(w) + 4K(w)}{1 + S(w)}\right.$$

$$\left. + \frac{5/16 + J'(w)/2 + K'(w) + 2L(w)}{[1 + S(w)]^2} + \frac{1}{w}\right\}$$

$$= \zeta^2 + \zeta \left(-2 + \frac{5}{16}\right)$$

$$= \zeta^2 - \frac{27\zeta}{16}$$

$$H_{22}(\zeta, w) = \zeta^2 \left[\frac{1 + S(w) + 2K(w)}{1 - S(w)}\right] + \zeta \left\{\frac{-2 + 2J(w) - 4K(w)}{1 - S(w)}\right.$$

$$\left. + \frac{5/16 + J'(w)/2 + K'(w) - 2L(w)}{[1 - S(w)]^2} + \frac{1}{w}\right\}$$

$$= \zeta^2 + \zeta \left(-2 + \frac{5}{16}\right)$$

$$= \zeta^2 - \frac{27\zeta}{16}$$

$$H_{12}(\zeta, w) = H_{21}(\zeta, w) = \zeta \left[\frac{5/16 - J'(w)/2}{1 - S(w)^2}\right]$$

$$= \frac{5\zeta}{16}$$

Combining these matrix elements, the ground state energy as $R \to \infty$ is

$$E_{CI} = E_-(\zeta, w) = \frac{H_{11}(\zeta, w) + H_{22}(\zeta, w)}{2}$$

$$- \frac{1}{2}\left\{[H_{11}(\zeta, w) - H_{22}(\zeta, w)]^2 + 4H_{12}^2(\zeta, w)\right\}^{1/2}$$

$$= \zeta^2 - \frac{27\zeta}{16} - \frac{1}{2}\left[\zeta^2 - \frac{27\zeta}{16} - \left(\zeta^2 - \frac{27\zeta}{16}\right) + 4\left(\frac{5\zeta}{16}\right)^2\right]^{1/2}$$

$$= \zeta^2 - \zeta$$

10–48. In the Born–Oppenheimer approximation, we assume that because the nuclei are so much more massive than the electrons, the electrons can adjust essentially instantaneously to any nuclear motion, and hence we have a unique and well-defined energy, $E(R)$, at each internuclear separation R. Under this same approximation, $E(R)$ is the internuclear potential and so is the potential field in which the nuclei vibrate. Argue, then, that under the Born–Oppenheimer approximation, the force constant is independent of isotopic substitution. Using the above ideas, and given that the dissociation energy for H_2 is $D_0 = 430.3$ kJ·mol^{-1} and that the fundamental vibrational frequency ν is 1.32×10^{14} s^{-1}, calculate D_0 and ν for deuterium, D_2. Realize that the observed dissociation energy is given by

$$D_0 = D_e - \frac{1}{2}h\nu$$

where D_e is the value of $E(R)$ at R_{eq}.

The force constant is the curvature of the internuclear potential at its minimum,

$$k = \left(\frac{d^2E}{dR^2}\right)_{R=R_e}$$

Since the internuclear potential $E(R)$ is independent of isotopic substitution, the force constant is the same for different isotopes.

The fundamental vibrational frequency is related to the force constant and reduced mass:

$$\nu = \frac{1}{2\pi}\left(\frac{k}{\mu}\right)^{1/2}$$

Since the force constant is isotope invariant, ν_{D_2} can be determined by taking the ratio of ν_{D_2} to ν_{H_2}:

$$\frac{\nu_{D_2}}{\nu_{H_2}} = \left(\frac{\mu_{H_2}}{\mu_{D_2}}\right)^{1/2} = \left(\frac{0.5039}{1.0071}\right)^{1/2} = 0.70735$$

$$\nu_{D_2} = 0.70735\left(1.32 \times 10^{14}\text{ s}^{-1}\right) = 9.337 \times 10^{13}\text{ s}^{-1}$$

The value of D_e for H_2 is found by adding the zero point energy for H_2 to D_0 for H_2:

$$D_e = D_{0,H_2} + \frac{1}{2}h\nu_{H_2}$$

$$= 430.3\text{ kJ mol}^{-1}$$

$$+ \frac{1}{2}\left(6.626\,0755 \times 10^{-34}\text{ J·s}\right)\left(1.32 \times 10^{14}\text{ s}^{-1}\right)\left(6.022\,1367 \times 10^{23}\text{ mol}^{-1}\right)\left(\frac{1\text{ kJ}}{1000\text{ J}}\right)$$

$$= 456.636\text{ kJ mol}^{-1}$$

Since the value of D_e for D_2 is the same as that for H_2, the value of D_0 for D_2 is determined by subtracting the D_2 zero point energy from D_e:

$$D_{0,D_2} = D_e - \frac{1}{2}h\nu_{D_2}$$

$$= 456.636\text{ kJ mol}^{-1}$$

$$- \frac{1}{2}\left(6.626\,0755 \times 10^{-34}\text{ J·s}\right)\left(9.337 \times 10^{13}\text{ s}^{-1}\right)\left(6.022\,1367 \times 10^{23}\text{ mol}^{-1}\right)\left(\frac{1\text{ kJ}}{1000\text{ J}}\right)$$

$$= 438.01\text{ kJ mol}^{-1}$$

CHAPTER 11

Qualitative Theory of Chemical Bonding

PROBLEMS AND SOLUTIONS

11–1. According to the JANAF Thermochemical Tables (see References), the rotational constant of Li_2 in its ground vibrational state is 0.669 cm^{-1}. Determine the bond length of Li_2.

The moment of inertia of $^7Li_2$ is

$$I = \frac{h}{8\pi^2 c \tilde{B}}$$

$$= \frac{6.626\,0755 \times 10^{-34}\text{ J}\cdot\text{s}}{8\pi^2 \left(2.997\,924\,58 \times 10^{10}\text{ cm}\cdot\text{s}^{-1}\right) \left(0.669\text{ cm}^{-1}\right)}$$

$$= 4.184 \times 10^{-46}\text{ kg}\cdot\text{m}^2$$

The reduced mass of $^7Li_2$ is

$$\mu = \frac{(7.0160\text{ amu})\,(7.0160\text{ amu})}{7.0160\text{ amu} + 7.0160\text{ amu}} \left(1.660\,5402 \times 10^{-27}\text{ kg}\cdot\text{amu}^{-1}\right)$$

$$= 5.825 \times 10^{-27}\text{ kg}$$

Combining the moment of inertia and the reduced mass, the bond length of $^7Li_2$ is

$$r = \left(\frac{I}{\mu}\right)^{1/2}$$

$$= \left(\frac{4.184 \times 10^{-46}\text{ kg}\cdot\text{m}^2}{5.825 \times 10^{-27}\text{ kg}}\right)^{1/2}$$

$$= 2.68 \times 10^{-10}\text{ m} = 268\text{ pm}$$

which compares very well with the bond length of 267 pm listed in Table 11.2.

11–2. The vibrational energy (in cm^{-1}) of Li_2 in its ground electronic state is given in the JANAF Thermochemical Tables (see References) as

$$G(v) = 351.39 \left(v + \frac{1}{2}\right) - 2.578 \left(v + \frac{1}{2}\right)^2 - 0.00647 \left(v + \frac{1}{2}\right)^3$$

$$- 9.712 \times 10^{-5} \left(v + \frac{1}{2}\right)^4$$

Calculate the number of vibrational energy levels. (See Problem 10–2.)

As stated in Problem 10–2, the maximum vibrational quantum number is given by $\Delta G(v_{max}) = 0$. Use $G(v)/cm^{-1}$ to write an expression for $\Delta G(v)$:

$$\Delta G(v) = G(v+1) - G(v) = 351.39 \left[\left(v + \frac{3}{2}\right) - \left(v + \frac{1}{2}\right)\right] - 2.578 \left[\left(v + \frac{3}{2}\right)^2 - \left(v + \frac{1}{2}\right)^2\right]$$

$$- 0.00647 \left[\left(v + \frac{3}{2}\right)^3 - \left(v + \frac{1}{2}\right)^3\right] - 9.712 \times 10^{-5} \left[\left(v + \frac{3}{2}\right)^4 - \left(v + \frac{1}{2}\right)^4\right]$$

$$= 351.39 - 2.578 \left(2v + 2\right) - 0.00647 \left(3v^2 + 6v + 13/4\right)$$

$$- 9.712 \times 10^{-5} \left(4v^3 + 12v^2 + 13v + 5\right)$$

$$= 346.212 - 5.196v + 0.020\,58v^2 - 0.000\,3885v^3$$

To determine v_{max}, set $\Delta G(v) = 0$ and solve for v_{max}:

$$346.212 - 5.196v_{max} + 0.020\,58v_{max}^2 - 0.000\,3885v_{max}^3 = 0$$

Of the three roots for v_{max}, two are imaginary. The real root is $v_{max} = 48.65$, which, when rounded to the nearest integer, becomes $v_{max} = 48$. Since the smallest vibrational quantum number is $v = 0$, the total number of vibrational energy levels is $v_{max} + 1 = 49$.

11–3. The rotational constant of B_2 in its ground vibrational state is given in the JANAF Thermochemical Tables (see References) as 1.228 cm^{-1}. Calculate the bond length of B_2.

The moment of inertia of $^{11}B_2$ is

$$I = \frac{h}{8\pi^2 c \tilde{B}}$$

$$= \frac{6.626\,0755 \times 10^{-34}\ \text{J·s}}{8\pi^2 \left(2.997\,924\,58 \times 10^{10}\ \text{cm·s}^{-1}\right) \left(1.228\ \text{cm}^{-1}\right)}$$

$$= 2.2795 \times 10^{-46}\ \text{kg·m}^2$$

The reduced mass of $^{11}B_2$ is

$$\mu = \frac{(11.0093\ \text{amu})\,(11.0093\ \text{amu})}{11.0093\ \text{amu} + 11.0093\ \text{amu}} \left(1.660\,5402 \times 10^{-27}\ \text{kg·amu}^{-1}\right)$$

$$= 9.1407 \times 10^{-27}\ \text{kg}$$

Combining the moment of inertia and the reduced mass, the bond length of $^{11}B_2$ is

$$r = \left(\frac{I}{\mu}\right)^{1/2}$$

$$= \left(\frac{2.2795 \times 10^{-46}\,\text{kg·m}^2}{9.1407 \times 10^{-27}\,\text{kg}}\right)^{1/2}$$

$$= 1.579 \times 10^{-10}\,\text{m} = 157.9\,\text{pm}$$

which compares very well with the bond length of 159 pm listed in Table 11.2.

11–4. The rotational constant of C_2 in its ground vibrational state is given in the JANAF Thermochemical Tables (see References) as $1.811\,\text{cm}^{-1}$. Calculate the bond length of C_2.

The moment of inertia of $^{12}C_2$ is

$$I = \frac{h}{8\pi^2 c\tilde{B}}$$

$$= \frac{6.626\,0755 \times 10^{-34}\,\text{J·s}}{8\pi^2\left(2.997\,924\,58 \times 10^{10}\,\text{cm·s}^{-1}\right)\left(1.811\,\text{cm}^{-1}\right)}$$

$$= 1.5457 \times 10^{-46}\,\text{kg·m}^2$$

The reduced mass of $^{12}C_2$ is

$$\mu = \frac{(12.0000\,\text{amu})\,(12.0000\,\text{amu})}{12.0000\,\text{amu} + 12.0000\,\text{amu}}\left(1.660\,5402 \times 10^{-27}\,\text{kg·amu}^{-1}\right)$$

$$= 9.9632 \times 10^{-27}\,\text{kg}$$

Combining the moment of inertia and the reduced mass, the bond length of $^{12}C_2$ is

$$r = \left(\frac{I}{\mu}\right)^{1/2}$$

$$= \left(\frac{1.5457 \times 10^{-46}\,\text{kg·m}^2}{9.9632 \times 10^{-27}\,\text{kg}}\right)^{1/2}$$

$$= 1.246 \times 10^{-10}\,\text{m} = 124.6\,\text{pm}$$

which compares very well with the bond length of 124 pm listed in Table 11.2.

11–5. Use molecular orbital theory to explain why the dissociation energy of N_2 is greater than that of N_2^+, but the dissociation energy of O_2^+ is greater than that of O_2.

The electron configurations of N_2 and N_2^+ are

$$N_2:\quad KK\,(\sigma_g 2s)^2\,(\sigma_u 2s)^2\,(\pi_u 2p_x)^2\,(\pi_u 2p_y)^2(\sigma_g 2p_z)^2$$

$$N_2^+: \quad KK \, (\sigma_g 2s)^2 \, (\sigma_u 2s)^2 \, (\pi_u 2p_x)^2 \, (\pi_u 2p_y)^2 \, (\sigma_g 2p_z)^1$$

The bond order of N_2 is 3 and that of N_2^+ is $2\frac{1}{2}$. Thus, the dissociation energy of N_2 is greater than that of N_2^+.

The electron configurations of O_2 and O_2^+ are

$$O_2: \quad KK \, (\sigma_g 2s)^2 \, (\sigma_u 2s)^2 \, (\sigma_g 2p_z)^2 \, (\pi_u 2p_x)^2 \, (\pi_u 2p_y)^2 \, (\pi_g 2p_x)^1 \, (\pi_g 2p_y)^1$$
$$O_2^+: \quad KK \, (\sigma_g 2s)^2 \, (\sigma_u 2s)^2 \, (\sigma_g 2p_z)^2 \, (\pi_u 2p_x)^2 \, (\pi_u 2p_y)^2 \, (\pi_g 2p_x)^1$$

The bond order of O_2 is 2. Since O_2^+ has one less antibonding electron than O_2, the bond order of this species is $2\frac{1}{2}$. Thus, the dissociation energy of O_2^+ is greater than that of O_2.

11–6. Discuss the bond properties of F_2 and F_2^+ using molecular orbital theory.

The electron configurations of F_2 and F_2^+ are

$$F_2: \quad KK \, (\sigma_g 2s)^2 \, (\sigma_u 2s)^2 \, (\sigma_g 2p_z)^2 \, (\pi_u 2p_x)^2 \, (\pi_u 2p_y)^2 \, (\pi_g 2p_x)^2 \, (\pi_g 2p_y)^2$$
$$F_2^+: \quad KK \, (\sigma_g 2s)^2 \, (\sigma_u 2s)^2 \, (\sigma_g 2p_z)^2 \, (\pi_u 2p_x)^2 \, (\pi_u 2p_y)^2 \, (\pi_g 2p_x)^2 \, (\pi_g 2p_y)^1$$

The bond order of F_2 is 1. Since F_2^+ has one less antibonding electron than F_2, the bond order of this species is $1\frac{1}{2}$. Thus, F_2^+ has a greater bond strength and a shorter bond than F_2.

11–7. Predict the relative stabilities of the species N_2, N_2^+, and N_2^-.

The electron configurations of N_2, N_2^+, and N_2^- are

$$N_2: \quad KK \, (\sigma_g 2s)^2 \, (\sigma_u 2s)^2 \, (\pi_u 2p_x)^2 \, (\pi_u 2p_y)^2 \, (\sigma_g 2p_z)^2$$
$$N_2^+: \quad KK \, (\sigma_g 2s)^2 \, (\sigma_u 2s)^2 \, (\pi_u 2p_x)^2 \, (\pi_u 2p_y)^2 \, (\sigma_g 2p_z)^1$$
$$N_2^-: \quad KK \, (\sigma_g 2s)^2 \, (\sigma_u 2s)^2 \, (\pi_u 2p_x)^2 \, (\pi_u 2p_y)^2 \, (\sigma_g 2p_z)^2 \, (\pi_g 2p_x)^1$$

N_2^+ has one less bonding electron than N_2 while N_2^- has one more antibonding electron than N_2. The bond orders of N_2, N_2^+, and N_2^- are 3, $2\frac{1}{2}$, and $2\frac{1}{2}$, respectively. Thus, the relative stabilities are

$$N_2 > N_2^+ \approx N_2^-$$

11–8. Predict the relative bond strengths and bond lengths of diatomic carbon, C_2, and its negative ion, C_2^-.

The electron configurations of C_2 and C_2^- are

$$C_2: \quad KK \, (\sigma_g 2s)^2 \, (\sigma_u 2s)^2 \, (\pi_u 2p_x)^2 \, (\pi_u 2p_y)^2$$
$$C_2^-: \quad KK \, (\sigma_g 2s)^2 \, (\sigma_u 2s)^2 \, (\pi_u 2p_x)^2 \, (\pi_u 2p_y)^2 \, (\sigma_g 2p_z)^1$$

C_2^- has one more bonding electron than C_2. The bond orders of C_2 and C_2^- are 2 and $2\frac{1}{2}$, respectively. Thus, C_2^- has a greater bond strength and a shorter bond than C_2.

11–9. The force constants for the diatomic molecules B_2 through F_2 are given in the table below. Is the order what you expect? Explain.

Diatomic molecule	$k/\text{N}\cdot\text{m}^{-1}$
B_2	350
C_2	930
N_2	2260
O_2	1140
F_2	450

The force constant of a molecule is directly proportional to the bond strength of a molecule, which, in turn, is directly related to its bond order. The electron configurations of these diatomic molecules are

Molecule	Electron Configuration	Bond Order
B_2	$KK\,(\sigma_g 2s)^2\,(\sigma_u 2s)^2(\pi_u 2p_x)^1(\pi_u 2p_y)^1$	1
C_2	$KK\,(\sigma_g 2s)^2\,(\sigma_u 2s)^2\,(\pi_u 2p_x)^2\,(\pi_u 2p_y)^2$	2
N_2	$KK\,(\sigma_g 2s)^2\,(\sigma_u 2s)^2\,(\pi_u 2p_x)^2\,(\pi_u 2p_y)^2\,(\sigma_g 2p_z)^2$	3
O_2	$KK\,(\sigma_g 2s)^2\,(\sigma_u 2s)^2\,(\sigma_g 2p_z)^2\,(\pi_u 2p_x)^2\,(\pi_u 2p_y)^2\,(\pi_g 2p_x)^1\,(\pi_g 2p_y)^1$	2
F_2	$KK\,(\sigma_g 2s)^2\,(\sigma_u 2s)^2\,(\sigma_g 2p_z)^2\,(\pi_u 2p_x)^2\,(\pi_u 2p_y)^2\,(\pi_g 2p_x)^2\,(\pi_g 2p_y)^2$	1

Arranging these diatomic molecules in terms of decreasing bond order gives

$$N_2 > C_2 \approx O_2 > B_2 \approx F_2$$

This order is also reflected in the order of force constants. The bond order information, however, is qualitative, and cannot be used to determine the magnitudes of the force constants for molecules of the same bond order.

11–10. Write out the ground-state molecular orbital electron configurations for Na_2 through Ar_2. Would you predict a stable Mg_2 molecule?

Using L to represent the filled $n = 2$ shell.

$$Na_2: \quad KKLL\,(\sigma_g 3s)^2$$
$$Mg_2: \quad KKLL\,(\sigma_g 3s)^2\,(\sigma_u 3s)^2$$
$$Al_2: \quad KKLL\,(\sigma_g 3s)^2\,(\sigma_u 3s)^2\,(\pi_u 3p)^2$$
$$Si_2: \quad KKLL\,(\sigma_g 3s)^2\,(\sigma_u 3s)^2\,(\pi_u 3p)^4$$
$$P_2: \quad KKLL\,(\sigma_g 3s)^2\,(\sigma_u 3s)^2\,(\pi_u 3p)^4\,(\sigma_g 3p_z)^2$$
$$S_2: \quad KKLL\,(\sigma_g 3s)^2\,(\sigma_u 3s)^2\,(\pi_u 3p)^4\,(\sigma_g 3p_z)^2\,(\pi_g 3p)^2$$
$$Cl_2: \quad KKLL\,(\sigma_g 3s)^2\,(\sigma_u 3s)^2\,(\pi_u 3p)^4\,(\sigma_g 3p_z)^2\,(\pi_g 3p)^4$$
$$Ar_2: \quad KKLL\,(\sigma_g 3s)^2\,(\sigma_u 3s)^2\,(\pi_u 3p)^4\,(\sigma_g 3p_z)^2\,(\pi_g 3p)^4\,(\sigma_u 3p)^2$$

Since the bond order of Mg_2 is 0, it is not expected to be stable.

11–11. In Section 11.2, we constructed molecular orbitals for homonuclear diatomic molecules using the $n = 2$ atomic orbitals on each of the bonded atoms. In this problem, we will consider the molecular orbitals that can be constructed from the $n = 3$ atomic orbitals. These orbitals are important in describing diatomic molecules of the first row of the transition metals. Once again we choose the z axis to lie along the molecular bond. What are the designations for the $3s_A \pm 3s_B$ and $3p_A \pm 3p_B$ molecular orbitals? The $n = 3$ shell also contains a set of five $3d$ orbitals. (The shapes of the $3d$ atomic orbitals are shown in Figure 7.8.) Given that molecular orbitals with two nodal planes that contain the internuclear axis are called δ orbitals, show that ten $3d_A \pm 3d_B$ molecular orbitals consist of a bonding σ orbital, a pair of bonding π orbitals, a pair of bonding δ orbitals, and their corresponding antibonding orbitals.

As the case with $n = 2$, the designations for the $3s_A \pm 3s_B$ orbitals are $\sigma_g 3s$ and $\sigma_u 3s$, respectively, and the designations for the $3p_A \pm 3p_B$ orbitals are

$$3p_{z,A} \pm 3p_{z,B}: \quad \sigma_g 3p_z \quad \text{and} \quad \sigma_u 3p_z$$
$$3p_{x,A} \pm 3p_{x,B}: \quad \pi_g 3p_x \quad \text{and} \quad \pi_u 3p_x$$
$$3p_{y,A} \pm 3p_{y,B}: \quad \pi_g 3p_y \quad \text{and} \quad \pi_u 3p_y$$

The designations for the $3d_A \pm 3d_B$ orbitals are

$$3d_{x^2-y^2,A} \pm 3d_{x^2-y^2,B}: \quad \delta_g 3d_{x^2-y^2} \quad \text{and} \quad \delta_u 3d_{x^2-y^2}$$
$$3d_{z^2,A} \pm 3d_{z^2,B}: \quad \sigma_g 3d_{z^2} \quad \text{and} \quad \sigma_u 3d_{z^2}$$
$$3d_{xy,A} \pm 3d_{xy,B}: \quad \delta_g 3d_{xy} \quad \text{and} \quad \delta_u 3d_{xy}$$
$$3d_{xz,A} \pm 3d_{xz,B}: \quad \pi_g 3d_{xz} \quad \text{and} \quad \pi_u 3d_{xz}$$
$$3d_{yz,A} \pm 3d_{yz,B}: \quad \pi_g 3d_{yz} \quad \text{and} \quad \pi_u 3d_{yz}$$

11–12. Determine the largest bond order for a first-row transition-metal homonuclear diatomic molecule. (See the previous problem.)

The molecule with the most bonding electrons and fewest antibonding electrons will have the largest bond order. The first-row transition metal homonuclear diatomic molecule that fits this criterion is Cr_2, with an electronic configuration of

$$KKLLMM(\sigma_g 4s)^2(\sigma_g 3d)^2(\pi_u 3d)^4(\delta_g 3d)^4$$

All the bonding orbitals derived from the atomic $3d$ orbitals in the ground state are filled, and all the corresponding antibonding orbitals are empty. The bond order of Cr_2 is 6. It was first experimentally observed in molecular beam experiments, and the measured bond length agreed with molecular orbital calculations based on a bond order of 6.

11–13. Determine the ground-state molecular orbital electron configuration of NO^+ and NO. Compare the bond order of these two species.

The electron configurations of NO^+ and NO are

$$NO^+: \quad KK \, (\sigma_g 2s)^2 \, (\sigma_u 2s)^2 \, (\pi_u 2p_x)^2 \, (\pi_u 2p_y)^2 \, (\sigma_g 2p_z)^2$$

NO: $KK \; (\sigma_g 2s)^2 \; (\sigma_u 2s)^2 \; (\pi_u 2p_x)^2 \; (\pi_u 2p_y)^2 \; (\sigma_g 2p_z)^2 (\pi_g 2p_x)^1$

The bond orders of NO^+ and NO are 3 and $2\frac{1}{2}$, respectively.

11–14. Figure 11.10 plots a schematic representation of the energies of the molecular orbitals of HF. How will the energy-level diagram for the diatomic OH radical differ from that of HF? What is the highest occupied molecular orbital of OH?

In the energy-level diagram for the diatomic OH radical, the energies of the $1s_H$ and $2p_O$ orbitals are closer together than is the case for the $1s_H$ and $2p_F$ orbitals in HF. The highest occupied molecular orbital of OH are the nonbonding orbitals $2p_{xO}$ and $2p_{yO}$, which between them hold three electrons.

11–15. Using Figure 11.10, you found that the highest occupied molecular orbital for HF is a fluorine $2p$ atomic orbital. The measured ionization energies for an electron from this nonbonding molecular orbital of HF is 1550 kJ·mol^{-1}. However, the measured ionization energy of a $2p$ electron from a fluorine atom is 1795 kJ·mol^{-1}. Why is the ionization energy of an electron from the $2p$ atomic orbital on fluorine greater for the fluorine atom than for an HF molecule?

The bonding electrons are unequally shared between the hydrogen and the fluorine atoms in HF and are more localized at the more electronegative fluorine atom. Because of this localization, the bonding pair of electrons shields the nonbonding $2p$ electron on the fluorine atom from the nucleus more than the inner shell electrons on atomic fluorine shield the $2p$ electron. Thus, it takes more energy to ionize a $2p$ electron from the fluorine atom than an electron from the nonbonding $2p$ orbital of HF.

11–16. In this problem, we consider the heteronuclear diatomic molecule CO. The ionization energies of an electron from the valence atomic orbitals on the carbon atom and the oxygen atom are listed below.

Atom	Valence orbital	Ionization energy / MJ·mol^{-1}
O	$2s$	3.116
	$2p$	1.524
C	$2s$	1.872
	$2p$	1.023

Use these data to construct a molecular orbital energy-level diagram for CO. What are the symmetry designations of the molecular orbitals of CO? What is the electron configuration of the ground state of CO? What is the bond order of CO? Is CO paramagnetic or diamagnetic?

A molecular-orbital energy-level diagram for CO consistent with the ionization energies given for the carbon and oxygen atom is shown below:

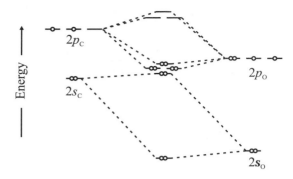

Since C and O have similar orbital energies, the bonding will be similar to that in a homonuclear diatomic molecule. The orbitals are designated by σ and π to describe those that are cylindrically symmetric about the CO axis and those with one nodal plane containing the CO axis, respectively. The energy ordering of the bonding molecular orbitals is consistent with the photoelectron spectrum in Figure 11.8. In order of increasing energy, the symmetry designations of the molecular orbitals of CO are (ignoring the $1s$ orbitals) $\sigma_b 2s$, $\sigma_a 2s$, $\pi_b 2p_x$ and $\pi_b 2p_y$, $\sigma_b 2p_z$, $\pi_a 2p_x$ and $\pi_a 2p_y$, and $\sigma_a 2p_z$, where the subscripts a and b are used to designate antibonding and bonding orbitals, respectively. The subscripts g and u are not used because CO does not have an inversion center. The electron configuration of the ground state CO is

$$K K (\sigma_b 2s)^2 (\sigma_a 2s)^2 (\pi_b 2p_x)^2 (\pi_b 2p_y)^2 (\sigma_b 2p_z)^2$$

The bond order of CO is 3. Because there are no unpaired electrons, CO is diamagnetic.

11–17. The molecule BF is isoelectronic with CO. However, the molecular orbitals for BF are different from those for CO. Unlike CO, the energy difference between the $2s$ orbitals of boron and fluorine is so large that the $2s$ orbital of boron combines with a $2p$ orbital on fluorine to make a molecular orbital. The remaining $2p$ orbitals on fluorine combine with two of the $2p$ orbitals on B to form π orbitals. The third $2p$ orbital on B is non-bonding. The energy ordering of the molecular orbitals is $\psi(2s_B + 2p_F) < \psi(2p_B - 2p_F) < \psi(2s_B - 2p_F) < \psi(2p_B + 2p_F) < \psi(2p_B)$. What are the symmetry designations of the molecular orbitals of BF? What is the electron configuration of the ground state of BF? What is the bond order of BF? Is BF diamagnetic or paramagnetic? How do the answers to these last two questions compare with those obtained for CO (Problem 11–16)?

The symmetry designations of the molecular orbitals of BF, in order of increasing energies, are $2s_F$, σ_b, $\pi_b 2p_x$ and $\pi_b 2p_y$, σ_a, $\pi_a 2p_x$ and $\pi_a 2p_y$, and $2p_B$. The electron configuration of the ground state BF is

$$K K (2s_F)^2 (\sigma_b)^2 (\pi_b 2p_x)^2 (\pi_b 2p_y)^2 (\sigma_a)^2$$

The bond order is 2 and BF is diamagnetic, while CO has a bond order of 3 and is diamagnetic.

11–18. The photoelectron spectrum of O_2 exhibits two bands of $52.398 \, \text{MJ} \cdot \text{mol}^{-1}$ and $52.311 \, \text{MJ} \cdot \text{mol}^{-1}$ that correspond to the ionization of an oxygen $1s$ electron. Explain this observation.

These two bands correspond to the ionization of a $1s$ electron with spin $+\frac{1}{2}$ and the ionization of a $1s$ electron with spin $-\frac{1}{2}$. The different energies result from spin-orbit coupling.

11–19. The experimental ionization energies for a fluorine $1s$ electron from HF and F_2 are 66.981 and 67.217 MJ·mol^{-1}, respectively. Explain why these ionization energies are different even though the $1s$ electrons of the fluorine are not involved in the chemical bond.

Although the $1s$ electrons of the fluorine atom are not involved in the chemical bond, the bonding electrons do affect the interaction of the $1s$ electrons with the nucleus. In F_2, the bonding electrons are equally distributed between the two atoms, but in HF the bonding electrons are not equally distributed and are localized at the fluorine atom. This increases the shielding of the $1s$ orbital of the fluorine atom on HF relative to that on F_2. Therefore, the ionization energy is smaller for HF than for F_2.

11–20. Go to the website *www.ccl.net/cca/data/atomic-RHF-wavefunctions/tables* and verify that the $2s$ and $2p$ orbital energies of a fluorine atom are $-1.572 \, E_h$ and $-0.730 \, E_h$, respectively.

These energies can be found on the website under the entry for fluorine.

11–21. When we built up the molecular orbitals for diatomic molecules, we combined only those orbitals with the same energy because we said that only those with similar energies mix well. This problem is meant to illustrate this idea. Consider two atomic orbitals χ_A and χ_B. Show that a linear combination of these orbitals leads to the secular determinant

$$\begin{vmatrix} \alpha_A - E & \beta - ES \\ \beta - ES & \alpha_B - E \end{vmatrix} = 0$$

where

$$\alpha_A = \int \chi_A h^{\text{eff}} \chi_A d\tau \qquad \alpha_B = \int \chi_B h^{\text{eff}} \chi_B d\tau$$

$$\beta = \int \chi_B h^{\text{eff}} \chi_A d\tau = \int \chi_A h^{\text{eff}} \chi_B d\tau \qquad S = \int \chi_A \chi_B d\tau$$

where h^{eff} is some effective one-electron Hamiltonian operator for the electron that occupies the molecular orbital ϕ. Show that

$$(1 - S^2)E^2 + (2\beta S - \alpha_A - \alpha_B)E + \alpha_A \alpha_B - \beta^2 = 0$$

It is usually a satisfactory first approximation to neglect S. Doing this, show that

$$E_\pm = \frac{\alpha_A + \alpha_B \pm [(\alpha_A - \alpha_B)^2 + 4\beta^2]^{1/2}}{2}$$

Now if χ_A and χ_B have the same energy, show that $\alpha_A = \alpha_B = \alpha$ and that

$$E_\pm = \alpha \pm \beta$$

giving one level of β units below α and one level of β units above α—that is, one level of β units more stable than the isolated orbital energy and one level of β units less stable.

Now investigate the case in which $\alpha_A \neq \alpha_B$, say, $\alpha_A < \alpha_B$. Show that

$$E_{\pm} = \frac{\alpha_A + \alpha_B}{2} \pm \frac{\alpha_A - \alpha_B}{2}\left[1 + \frac{4\beta^2}{(\alpha_A - \alpha_B)^2}\right]^{1/2}$$

$$= \frac{\alpha_A + \alpha_B}{2} \pm \frac{\alpha_A - \alpha_B}{2}\left[1 + \frac{2\beta^2}{(\alpha_A - \alpha_B)^2} - \frac{2\beta^4}{(\alpha_A - \alpha_B)^4} + \cdots\right]$$

$$= \frac{\alpha_A + \alpha_B}{2} \pm \frac{\alpha_A - \alpha_B}{2} \pm \frac{\beta^2}{\alpha_A - \alpha_B} + \cdots$$

where we have assumed that $\beta^2 < (\alpha_A - \alpha_B)^2$ and have used the expansion

$$(1+x)^{1/2} = 1 + \frac{x}{2} - \frac{x^2}{8} + \cdots$$

Show that

$$E_+ = \alpha_A \pm \frac{\beta^2}{\alpha_A - \alpha_B} + \cdots \quad \text{and} \quad E_- = \alpha_B - \frac{\beta^2}{\alpha_A - \alpha_B} + \cdots$$

Using this result, discuss the stabilization–destabilization of α_A and α_B versus the case above in which $\alpha_A = \alpha_B$. For simplicity, assume initially that $\alpha_A - \alpha_B$ is large.

If $\psi = c_1\chi_A + c_2\chi_B$ and if χ_A and χ_B are normalized, the secular determinant is

$$\begin{vmatrix} \alpha_A - E & \beta - ES \\ \beta - ES & \alpha_B - E \end{vmatrix} = 0$$

Expanding the determinant gives

$$(\alpha_A - E)(\alpha_B - E) - (\beta - ES)^2 = 0$$

$$\alpha_A\alpha_B - E\alpha_A - E\alpha_B + E^2 - \beta^2 + 2\beta ES - E^2S^2 = 0$$

$$(1 - S^2)E^2 + (2\beta S - \alpha_A - \alpha_B)E + \alpha_A\alpha_B - \beta^2 = 0$$

Neglecting S, the two roots for E are

$$E_{\pm} = \frac{\alpha_A + \alpha_B \pm [(\alpha_A + \alpha_B)^2 - 4\alpha_A\alpha_B + 4\beta^2]^{1/2}}{2}$$

$$= \frac{\alpha_A + \alpha_B \pm \left[(\alpha_A - \alpha_B)^2 + 4\beta^2\right]^{1/2}}{2}$$

If χ_A and χ_B have the same energy, then

$$\int \chi_A h^{\text{eff}} \chi_A \, d\tau = \int \chi_B h^{\text{eff}} \chi_B \, d\tau$$

or

$$\alpha_A = \alpha_B = \alpha$$

and $E_\pm$ becomes

$$E_\pm = \frac{2\alpha \pm \left(4\beta^2\right)^{1/2}}{2}$$

$$= \alpha \pm \beta$$

If $\alpha_A < \alpha_B$, then

$$E_\pm = \frac{\alpha_A + \alpha_B}{2} \pm \frac{\left[(\alpha_A - \alpha_B)^2 + 4\beta^2\right]^{1/2}}{2}$$

$$= \frac{\alpha_A + \alpha_B}{2} \pm \frac{\alpha_A - \alpha_B}{2}\left[1 + \frac{4\beta^2}{(\alpha_A - \alpha_B)^2}\right]^{1/2}$$

$$= \frac{\alpha_A + \alpha_B}{2} \pm \frac{\alpha_A - \alpha_B}{2}\left[1 + \frac{2\beta^2}{(\alpha_A - \alpha_B)^2} - \frac{2\beta^4}{(\alpha_A - \alpha_B)^4} + \cdots\right]$$

$$= \frac{\alpha_A + \alpha_B}{2} \pm \frac{\alpha_A - \alpha_B}{2} \pm \frac{\beta^2}{\alpha_A - \alpha_B} + \cdots$$

where the assumption $\beta^2 < (\alpha_A - \alpha_B)^2$ has been made and the expansion $(1 + x)^{1/2} = 1 + x/2 - x^2/8 + \cdots$ has been applied. It follows that

$$E_+ = \frac{\alpha_A + \alpha_B}{2} + \frac{\alpha_A - \alpha_B}{2} + \frac{\beta^2}{\alpha_A - \alpha_B}$$

$$= \alpha_A + \frac{\beta^2}{\alpha_A - \alpha_B}$$

$$E_- = \frac{\alpha_A + \alpha_B}{2} - \frac{\alpha_A - \alpha_B}{2} - \frac{\beta^2}{\alpha_A - \alpha_B}$$

$$= \alpha_B - \frac{\beta^2}{\alpha_A - \alpha_B}$$

The lower energy in this case is less stabilized and the higher energy is less destabilized than in the case where $\alpha_A = \alpha_B$. In fact, if the difference between α_A and α_B is very large, then the two energies obtained are simply α_A and α_B; there is no stabilization or destabilization of energy. The smaller the difference between α_A and α_B, the greater the amount of stabilization and destabilization of energy levels that occurs.

11–22. Show that filled orbitals can be ignored in the determination of molecular term symbols.

A filled σ_g or σ_u orbital has two electrons and

$$M_L = 0 + 0 = 0$$

$$M_S = \frac{1}{2} - \frac{1}{2} = 0$$

Likewise, for doubly degenerate π_u or π_g orbitals,

$$M_L = 1 + 1 - 1 - 1 = 0$$

$$M_S = \frac{1}{2} - \frac{1}{2} + \frac{1}{2} - \frac{1}{2} = 0$$

Therefore, a filled orbital does not contribute to the molecular term symbol.

11–23.   Deduce the ground-state term symbols of all the diatomic molecules given in Table 11.3.

H_2^+ $(1\sigma_g)^1$ corresponds to $M_L = 0$ and $M_S = \pm\frac{1}{2}$, or a $^2\Sigma$ term symbol. The unpaired electron occupies a molecular orbital of symmetry g and the $1\sigma_g$ wave function is unchanged upon reflection through a plane containing the nuclei. Therefore, the complete ground-state term symbol of H_2^+ is $^2\Sigma_g^+$.

H_2 $(1\sigma_g)^2$ corresponds to $M_L = 0$ and $M_S = 0$, or a $^1\Sigma$ term symbol. The symmetry is g and the $1\sigma_g$ wave function is unchanged upon reflection through a plane containing the nuclei. Therefore, the complete ground-state term symbol of H_2 is $^1\Sigma_g^+$.

He_2^+ The ground-state term symbol $(^2\Sigma_u^+)$ is derived in Examples 11–5 and 11–9.

Li_2 $(1\sigma_g)^2(1\sigma_u)^2(2\sigma_g)^2$ corresponds to $M_L = 0$ and $M_S = 0$, or a $^1\Sigma$ term symbol. The symmetry is g and the σ orbitals remain unchanged upon reflection through a plane containing the nuclei, so the complete ground-state term symbol of Li_2 is $^1\Sigma_g^+$.

B_2 Example 11–7 shows that the partial molecular term symbol is $^3\Sigma_g$. Because one of the half-filled π orbitals changes sign upon reflection through a plane containing the two nuclei, the complete ground-state term symbol of B_2 is $^3\Sigma_g^-$.

C_2 $(1\sigma_g)^2(1\sigma_u)^2(2\sigma_g)^2(2\sigma_u)^2(1\pi_u)^2(1\pi_u)^2$ corresponds to $M_L = 0$ and $M_S = 0$, or a $^1\Sigma$ term symbol. The symmetry of the molecule is g. Because the π orbital which changes sign upon reflection through a plane containing the two nuclei is filled, the molecular wave function does not change when reflected through a plane containing the two nuclei. The complete ground-state term symbol of C_2 is $^1\Sigma_g^+$.

N_2^+ $(1\sigma_g)^2(1\sigma_u)^2(2\sigma_g)^2(2\sigma_u)^2(1\pi_u)^2(1\pi_u)^2(3\sigma_g)^1$ corresponds to $M_L = 0$ and $M_S = \pm\frac{1}{2}$, or a $^2\Sigma$ term symbol. The symmetry of the molecule is g. The complete ground-state term symbol of N_2^+ is $^2\Sigma_g^+$, because the molecular wave function does not change when reflected through a plane containing the two nuclei.

N_2 $(1\sigma_g)^2(1\sigma_u)^2(2\sigma_g)^2(2\sigma_u)^2(1\pi_u)^2(1\pi_u)^2(3\sigma_g)^2$ corresponds to $M_L = 0$ and $M_S = 0$, or a $^1\Sigma$ term symbol. The symmetry of the molecule is g. The complete ground-state term symbol of N_2 is $^1\Sigma_g^+$, because the molecular wave function does not change when reflected through a plane containing the two nuclei.

O_2^+ $(1\sigma_g)^2(1\sigma_u)^2(2\sigma_g)^2(2\sigma_u)^2(3\sigma_g)^2(1\pi_u)^2(1\pi_u)^2(1\pi_g)^1$ corresponds to $M_L = \pm 1$ and $M_S = \pm\frac{1}{2}$, or a $^2\Pi$ term symbol. The symmetry of the molecule is g, since the only unfilled molecular orbital has symmetry g, so the complete ground-state term symbol of O_2^+ is $^2\Pi_g$.

O_2 The ground-state term symbol ($^3\Sigma_g^-$) has been derived in Example 11–8.

F_2 $(1\sigma_g)^2(1\sigma_u)^2(2\sigma_g)^2(2\sigma_u)^2(3\sigma_g)^2(1\pi_u)^2(1\pi_u)^2(1\pi_g)^2(1\pi_g)^2$ corresponds to $M_L = 0$ and $M_S = 0$, or a $^1\Sigma$ term symbol. The symmetry of the molecule is g and the complete ground-state term symbol of F_2 is $^1\Sigma_g^+$, because the molecular wave function does not change when reflected through a plane containing the two nuclei.

11–24. Determine the ground-state molecular term symbols of O_2, N_2, N_2^+, and O_2^+.

See Problem 11–23 for the ground-state molecular term symbols of these molecules.

11–25. Calculate the sum of the energies of one hydrogen atom in the $1s$ state and one in the $2s$ state and show that your result is consistent with Figure 11.12.

The energy of a hydrogen atom is given in Equation 7.11:

$$E_n = -\frac{m_e e^4}{32\pi^2 \epsilon_0^2 \hbar^2 n^2}$$

In terms of atomic units, where $1E_h = m_e e^4 / (16\pi^2 \epsilon_0^2 \hbar^2)$, the energy becomes

$$E_n = -\frac{1}{2n^2} E_h$$

Therefore, the sum of energies of one hydrogen atom in the $1s$ state and one in the $2s$ state is

$$E_1 + E_2 = -\frac{1}{2}E_h - \frac{1}{8}E_h = -\frac{5}{8}E_h$$

and is consistent with the value of $-0.625E_h$ shown in Figure 11.12.

11–26. The highest occupied molecular orbitals for an excited electronic configuration of the oxygen molecule are $(1\pi_g)^1(3\sigma_u)^1$. What are the molecular term symbols for oxygen with this electronic configuration?

Set up a table listing all possible values of m_l and m_s for the two electrons:

	m_{l_1}	m_{s_1}	m_{l_2}	m_{s_2}	M_L	M_S
1.	$+1$	$+\frac{1}{2}$	0	$+\frac{1}{2}$	1	1
2.	$+1$	$+\frac{1}{2}$	0	$-\frac{1}{2}$	1	0
3.	$+1$	$-\frac{1}{2}$	0	$+\frac{1}{2}$	1	0
4.	$+1$	$-\frac{1}{2}$	0	$-\frac{1}{2}$	1	-1
5.	-1	$+\frac{1}{2}$	0	$+\frac{1}{2}$	-1	1
6.	-1	$+\frac{1}{2}$	0	$-\frac{1}{2}$	-1	0
7.	-1	$-\frac{1}{2}$	0	$+\frac{1}{2}$	-1	0
8.	-1	$-\frac{1}{2}$	0	$-\frac{1}{2}$	-1	-1

Entries 1, 2, 4, 5, 6, and 8 correspond to $|M_L| = 1$ and $S = 1$, giving a $^3\Pi$ molecular term symbol. Entries 3 and 7 correspond to $|M_L| = 1$ and $S = 0$, giving a $^1\Pi$ molecular term symbol.

11–27. Show that the π molecular orbital corresponding to the energy $E = \alpha - \beta$ for ethene is $\psi_\pi = \frac{1}{\sqrt{2}}(2p_{z1} - 2p_{z2})$.

The general form of the two π molecular orbitals for ethene (Equation 11.6) is

$$\psi_\pi = c_1 \cdot 2p_{z1} + c_2 \cdot 2p_{z2}$$

and the equations for c_1 and c_2 (Example 11–11) are

$$c_1(\alpha - E) + c_2\beta = 0$$
$$c_1\beta + c_2(\alpha - E) = 0$$

Substituting $E = \alpha - \beta$ into these expressions gives the same equation:

$$c_1\beta + c_2\beta = 0$$

or $c_1 = -c_2$. The wave function corresponding to $E = \alpha - \beta$ is therefore

$$\psi_\pi = c_1\left(2p_{z1} - 2p_{z2}\right)$$

To determine c_1, we require that ψ_π be normalized, that is,

$$c_1^2(1 + 2S + 1) = 1$$

Using the Hückel assumption that $S = 0$, $c_1 = 1/\sqrt{2}$, giving

$$\psi_\pi = \frac{1}{\sqrt{2}}(2p_{z1} - 2p_{z2})$$

11–28. Generalize our Hückel molecular orbital treatment of ethene to include overlap of $2p_{z1}$ and $2p_{z2}$. Determine the energies and the orbitals in terms of the overlap integral, S.

Including the overlap of the $2p_z$ orbitals means that S_{12} and S_{21} are no longer zero. Denoting $S_{12} = S_{21} = S$, the Hückel secular determinant becomes

$$\begin{vmatrix} \alpha - E & \beta - ES \\ \beta - ES & \alpha - E \end{vmatrix} = 0$$

Expanding the determinant gives

$$(\alpha - E)^2 - (\beta - ES)^2 = 0$$

$$\alpha - E = \pm(\beta - ES)$$

$$E = \frac{\alpha \pm \beta}{1 \pm S}$$

The wave functions can be determined by evaluating c_1 and c_2. The equations for c_1 and c_2 are obtained from the algebraic equations that give rise to the secular determinant:

$$c_1(\alpha - E) + c_2(\beta - ES) = 0$$

$$c_1(\beta - ES) + c_2(\alpha - E) = 0$$

Substituting $E = (\alpha + \beta)/(1 + S)$ into the above equations gives only one independent equation:

$$c_1\left(\frac{\alpha S - \beta}{1 + S}\right) + c_2\left(\frac{\beta - \alpha S}{1 + S}\right) = 0$$

or $c_1 = c_2$. Therefore,

$$\psi = c_1\left(2p_{z1} + 2p_{z2}\right)$$

Normalization of this wave function gives

$$c_1^2\left(1 + 2S + 1\right) = 1$$

$$c_1 = \frac{1}{\sqrt{2\left(1 + S\right)}}$$

The wave function associated with $E = (\alpha + \beta)/(1 + S)$ is then

$$\psi = \frac{1}{\sqrt{2(1 + S)}}\left(2p_{z1} + 2p_{z2}\right)$$

To determine the other wave function, substitute $E = (\alpha - \beta)/(1 - S)$ into the equations for c_1 and c_2 to give:

$$c_1\left(\frac{\beta - \alpha S}{1 - S}\right) + c_2\left(\frac{\beta - \alpha S}{1 - S}\right) = 0$$

or $c_2 = -c_1$. Therefore,

$$\psi = c_1\left(2p_{z1} - 2p_{z2}\right)$$

Normalization of this wave function gives

$$c_1^2 (1 - 2S + 1) = 1$$

$$c_1 = \frac{1}{\sqrt{2(1 - S)}}$$

The wave function associated with $E = (\alpha - \beta)/(1 - S)$ is then

$$\psi = \frac{1}{\sqrt{2(1 - S)}} \left(2p_{z1} - 2p_{z2}\right)$$

11–29. Verify Equation 11.10.

There are four carbon atoms in butadiene, requiring a 4×4 secular determinant for energy determination. The general form of the determinant is

$$\begin{vmatrix} H_{11} - ES_{11} & H_{12} - ES_{12} & H_{13} - ES_{13} & H_{14} - ES_{14} \\ H_{21} - ES_{21} & H_{22} - ES_{22} & H_{23} - ES_{23} & H_{24} - ES_{24} \\ H_{31} - ES_{31} & H_{32} - ES_{32} & H_{33} - ES_{33} & H_{34} - ES_{34} \\ H_{41} - ES_{41} & H_{42} - ES_{42} & H_{43} - ES_{43} & H_{44} - ES_{44} \end{vmatrix} = 0$$

The three assertions in the Hückel molecular orbital theory are

1. the overlap integrals S_{ij} are zero unless $i = j$, where $S_{ii} = 1$;
2. all coulomb integrals are assumed to be the same for all equivalent carbons and are denoted by α;
3. the resonance integrals involving nearest-neighbor carbon atoms are assumed to be the same and denoted by β.

The first assertion eliminates all S_{ij} for $i \neq j$ and gives $S_{ii} = 1$. The second assertion gives $H_{ii} = \alpha$. The third assertion gives $H_{ij} = \beta$ if i and j are nearest neighbors. In the case of butadiene (Figure 11.17), carbon 1 and carbon 4 each has only one nearest neighbor, carbon 2 and carbon 3, respectively. Carbon 2 has two nearest neighbors: carbon 1 and carbon 3. Similarly, carbon 3 has two nearest neighbors: carbon 2 and carbon 4. Putting this together, the secular determinant becomes

$$\begin{vmatrix} \alpha - E & \beta & 0 & 0 \\ \beta & \alpha - E & \beta & 0 \\ 0 & \beta & \alpha - E & \beta \\ 0 & 0 & \beta & \alpha - E \end{vmatrix} = 0$$

which is the same as Equation 11.10.

11–30. Show that

$$\begin{vmatrix} x & 1 & 0 & 0 \\ 1 & x & 1 & 0 \\ 0 & 1 & x & 1 \\ 0 & 0 & 1 & x \end{vmatrix} = 0$$

gives the algebraic equation $x^4 - 3x^2 + 1 = 0$.

Expand the determinant about the first row of elements:

$$x \begin{vmatrix} x & 1 & 0 \\ 1 & x & 1 \\ 0 & 1 & x \end{vmatrix} - \begin{vmatrix} 1 & 1 & 0 \\ 0 & x & 1 \\ 0 & 1 & x \end{vmatrix} = 0$$

Now expand each of the two determinants about its first column.

$$x \left(x \begin{vmatrix} x & 1 \\ 1 & x \end{vmatrix} - \begin{vmatrix} 1 & 0 \\ 1 & x \end{vmatrix} \right) - \left(\begin{vmatrix} x & 1 \\ 1 & x \end{vmatrix} \right) = 0$$

$$x(x^3 - x - x) - (x^2 - 1) = 0$$

$$x^4 - 3x^2 + 1 = 0$$

11–31. Show that the four π molecular orbitals for butadiene are given by Equations 11.15.

The energies for the four π molecular orbitals of butadiene [see Equation 11.12 where $x = (\alpha - E)/\beta$] are

$$x_1 = -1.618\,04$$

$$x_2 = -0.618\,04$$

$$x_3 = 0.618\,04$$

$$x_4 = 1.618\,04$$

The general form of the four corresponding molecular orbitals is given by

$$\psi = c_1 \cdot 2p_{z1} + c_2 \cdot 2p_{z2} + c_3 \cdot 2p_{z3} + c_4 \cdot 2p_{z4}$$

The equations for c_1, c_2, c_3, and c_4 are obtained from the algebraic equations that give rise to the secular determinant:

$$c_1 x + c_2 = 0 \tag{1}$$

$$c_1 + c_2 x + c_3 = 0 \tag{2}$$

$$c_2 + c_3 x + c_4 = 0 \tag{3}$$

$$c_3 + c_4 x = 0 \tag{4}$$

Three of these equations are independent; thus, to determine each wave function, require the function to be normalized. Equations 1 and 4 give relations between c_1 and c_2 and between c_3 and c_4, respectively:

$$c_1 = -\frac{c_2}{x} \tag{5}$$

$$c_4 = -\frac{c_3}{x} \tag{6}$$

Now find a relation between c_2 and c_3. Substituting Equation 5 into Equation 2 gives

$$-\frac{c_2}{x} + c_2 x + c_3 = 0$$

$$c_3 = c_2 \left(\frac{1 - x^2}{x}\right) \tag{7}$$

For $E_1 = \alpha + 1.618\,04\beta$, that is, $x_1 = -1.618\,04$, Equation 7 gives

$$c_3 = c_2 \left[\frac{1 - (-1.618\,04)^2}{-1.618\,04}\right] = c_2$$

Since $c_2 = c_3$, Equations 5 and 6 become

$$c_1 = -\frac{c_2}{x} = 0.618\,03c_2$$

$$c_4 = -\frac{c_3}{x} = 0.618\,03c_3 = 0.618\,03c_2$$

The wave function is then

$$\psi_1 = c_2 \left(0.618\,03 \cdot 2p_{z1} + 2p_{z2} + 2p_{z3} + 0.618\,03 \cdot 2p_{z4}\right)$$

Normalization of the wave function gives

$$c_2^2(0.618\,03^2 + 1^2 + 1^2 + 0.618\,03^2) = 1$$

$$c_2 = 0.601\,50$$

Therefore, $c_3 = c_2 = 0.601\,50$ and $c_1 = c_4 = 0.618\,03c_2 = 0.371\,73$, giving the wave function

$$\psi_1 = 0.3717 \cdot 2p_{z1} + 0.6015 \cdot 2p_{z2} + 0.6015 \cdot 2p_{z3} + 0.3717 \cdot 2p_{z4}$$

The other three wave functions can be calculated using the same procedure. For $E_2 = \alpha + 0.618\,04\beta$, that is, $x_2 = -0.618\,04$,

$$c_3 = -c_2$$

$$c_1 = 1.618\,02c_2$$

$$c_4 = 1.618\,02c_3 = -1.618\,02c_2$$

Normalization gives $c_2 = 0.371\,75$ (chosen arbitrarily to be positive). Thus, the wave function is

$$\psi_2 = 0.6015 \cdot 2p_{z1} + 0.3717 \cdot 2p_{z2} - 0.3717 \cdot 2p_{z3} - 0.6015 \cdot 2p_{z4}$$

For $E_3 = \alpha - 0.618\,04\beta$, that is, $x_3 = 0.618\,04$,

$$c_3 = c_2$$

$$c_1 = -1.618\,02c_2$$

$$c_4 = -1.618\,02c_3 = -1.618\,02c_2$$

Normalization gives $c_2 = -0.371\,75$ (chosen arbitrarily to be negative). Thus, the wave function is

$$\psi_3 = 0.6015 \cdot 2p_{z1} - 0.3717 \cdot 2p_{z2} - 0.3717 \cdot 2p_{z3} + 0.6015 \cdot 2p_{z4}$$

For $E_4 = \alpha - 1.618\,04\beta$, that is, $x_4 = 1.618\,04$,

$$c_3 = -c_2$$

$$c_1 = -0.618\,03c_2$$

$$c_4 = -0.618\,03c_3 = 0.618\,03c_2$$

Normalization gives $c_2 = -0.601\,50$ (chosen arbitrarily to be negative). Thus, the wave function is

$$\psi_4 = 0.3717 \cdot 2p_{z1} - 0.6015 \cdot 2p_{z2} + 0.6015 \cdot 2p_{z3} - 0.3717 \cdot 2p_{z4}$$

11–32. Show that the four molecular orbitals for butadiene (Equations 11.15) satisfy Equation 11.25.

Equation 11.25 is

$$E_\pi = \sum_r p_{rr}^\pi \alpha_r + \sum_{\substack{r \\ r \neq s}} \sum_s p_{rs}^\pi \beta_{rs}$$

where

$$p_{rs}^\pi = \sum_i n_i c_{ri} c_{si}$$

For butadiene, $n_1 = n_2 = 2$ and $n_3 = n_4 = 0$. In addition,

$$\alpha_1 = \alpha_2 = \alpha_3 = \alpha_4 = \alpha$$

$$\beta_{12} = \beta_{23} = \beta_{34} = \beta$$

$$\beta_{13} = \beta_{14} = \beta_{24} = 0$$

$$\beta_{ij} = \beta_{ji}$$

Therefore, Equation 11.25 becomes

$$E_\pi = \left(p_{11}^\pi + p_{22}^\pi + p_{33}^\pi + p_{44}^\pi\right)\alpha + \left(p_{12}^\pi + p_{21}^\pi + p_{23}^\pi + p_{32}^\pi + p_{34}^\pi + p_{43}^\pi\right)\beta$$

Using the four π molecular orbitals for butadiene given by Equation 11.15, the expressions for p_{rr}^π are

$$p_{11}^\pi = 2c_{11}^2 + 2c_{12}^2 = 2(0.3717)^2 + 2(0.6015)^2 = 1.0000$$

$$p_{22}^\pi = 2c_{21}^2 + 2c_{22}^2 = 2(0.6015)^2 + 2(0.3717)^2 = 1.0000$$

$$p_{33}^\pi = 2c_{31}^2 + 2c_{32}^2 = 2(0.6015)^2 + 2(-0.3717)^2 = 1.0000$$

$$p_{44}^\pi = 2c_{41}^2 + 2c_{42}^2 = 2(0.3717)^2 + 2(-0.6015)^2 = 1.0000$$

and the expressions for p_{rs}^π when $r \neq s$ are

$$p_{12}^\pi = p_{21}^\pi = 2c_{11}c_{21} + 2c_{12}c_{22} = 2(0.3717)(0.6015) + 2(0.6015)(0.3717) = 0.8943$$

$$p_{23}^\pi = p_{32}^\pi = 2c_{21}c_{31} + 2c_{22}c_{32} = 2(0.6015)(0.6015) + 2(0.3717)(-0.3717) = 0.4473$$

$$p_{34}^\pi = p_{43}^\pi = 2c_{31}c_{41} + 2c_{32}c_{42} = 2(0.6015)(0.3717) + 2(-0.3717)(-0.6015) = 0.8943$$

Combining the p_{rr}^π and p_{rs}^π terms, the π-electron energy for butadiene according to Equation 11.25 is

$$E_\pi = 4\,(1.0000)\,\alpha + 2\,(0.8943 + 0.4473 + 0.8943)\,\beta$$

$$= 4.0000\alpha + 4.4718\beta$$

which agrees with Equation 11.13. Thus, the four π molecular orbitals for butadiene satisfy Equation 11.25.

11–33. Derive Equation 11.19.

Substitute Equation 11.18

$$\psi_i = \sum_r c_{ri} \cdot 2p_{zr}$$

into Equation 11.17

$$E_i = \left\langle \psi_i \,\middle|\, \hat{H}_{\text{eff}} \,\middle|\, \psi_i \right\rangle$$

to give

$$E_i = \left\langle \sum_r c_{ri} \cdot 2p_{zr} \,\middle|\, \hat{H}_{\text{eff}} \,\middle|\, \sum_s c_{si} \cdot 2p_{zs} \right\rangle$$

$$= \sum_r c_{ri}^2 \left\langle 2p_{zr} \,\middle|\, \hat{H}_{\text{eff}} \,\middle|\, 2p_{zr} \right\rangle + \sum_{\substack{r \\ r\neq s}} \sum_s c_{ri}c_{si} \left\langle 2p_{zr} \,\middle|\, \hat{H}_{\text{eff}} \,\middle|\, 2p_{zs} \right\rangle$$

$$= \sum_r c_{ri}^2 \alpha_r + \sum_{\substack{r \\ r\neq s}} \sum_s c_{ri}c_{si}\beta_{rs}$$

The last equation is Equation 11.19. Note that $\alpha_r = \left\langle 2p_{zr} \,\middle|\, \hat{H}_{\text{eff}} \,\middle|\, 2p_{zr} \right\rangle$ and $\beta_{rs} = \left\langle 2p_{zr} \,\middle|\, \hat{H}_{\text{eff}} \,\middle|\, 2p_{zs} \right\rangle$.

11–34. Show that

$$E_\pi = \sum_r p_{rr}^\pi \alpha_r + 2 \sum_r \sum_{s>r} p_{rs}^\pi \beta_{rs}$$

where α_r is the coulomb integral associated with the rth atom and β_{rs} is the exchange integral between atoms r and s. The relation serves as a good check on the calculated energy levels.

The π-electron energy is given by

$$E_\pi = \sum_i n_i E_i$$

Substitute the expression for E_i (see Equation 11.19 or Problem 11–33)

$$E_i = \sum_r c_{ri}^2 \alpha_r + \sum_{\substack{r \\ r \neq s}} \sum_s c_{ri} c_{si} \beta_{rs}$$

into the expression for E_π to give

$$E_\pi = \sum_i n_i \left(\sum_r c_{ri}^2 \alpha_r + \sum_{\substack{r \\ r \neq s}} \sum_s c_{ri} c_{si} \beta_{rs} \right)$$

$$= \sum_r \alpha_r \left(\sum_i n_i c_{ri}^2 \right) + \sum_{\substack{r \\ r \neq s}} \sum_s \beta_{rs} \left(\sum_i n_i c_{ri} c_{si} \right)$$

Defining

$$p_{rs}^\pi = \sum_i n_i c_{ri} c_{si}$$

and realizing that $p_{rs}^\pi = p_{sr}^\pi$ and $\beta_{rs} = \beta_{sr}$, the π-electron energy becomes

$$E_\pi = \sum_r p_{rr}^\pi \alpha_r + 2 \sum_r \sum_{s>r} p_{rs}^\pi \beta_{rs}$$

11–35. Derive the Hückel theory secular determinant for benzene (see Equation 11.28).

Benzene has six carbon atoms, requiring therefore a 6×6 secular determinant for energy determination. The ijth element of the secular determinant is given by $H_{ij} - E S_{ij}$. The Hückel approximation gives $H_{jj} = \alpha$, $H_{ij} = \beta$ for neighboring atoms, $H_{ij} = 0$ for distant atoms, and $S_{ij} = \delta_{ij}$. Because benzene is a cyclic molecule, carbon 6 is adjacent to carbon 1, and so $H_{16} = H_{61} = \beta$. The Hückel secular determinant is therefore

$$\begin{vmatrix} \alpha - E & \beta & 0 & 0 & 0 & \beta \\ \beta & \alpha - E & \beta & 0 & 0 & 0 \\ 0 & \beta & \alpha - E & \beta & 0 & 0 \\ 0 & 0 & \beta & \alpha - E & \beta & 0 \\ 0 & 0 & 0 & \beta & \alpha - E & \beta \\ \beta & 0 & 0 & 0 & \beta & \alpha - E \end{vmatrix} = 0$$

11–36. Show that the six roots of Equation 11.28 are $E_1 = \alpha + 2\beta$, $E_2 = E_3 = \alpha + \beta$, $E_4 = E_5 = \alpha - \beta$, and $E_6 = \alpha - 2\beta$.

Let $x = (\alpha - E)/\beta$, the secular determinant for benzene (see Equation 11.28 or Problem 11–35) after dividing by β^6 becomes

$$\begin{vmatrix} x & 1 & 0 & 0 & 0 & 1 \\ 1 & x & 1 & 0 & 0 & 0 \\ 0 & 1 & x & 1 & 0 & 0 \\ 0 & 0 & 1 & x & 1 & 0 \\ 0 & 0 & 0 & 1 & x & 1 \\ 1 & 0 & 0 & 0 & 1 & x \end{vmatrix} = 0$$

The determinantal equation is best evaluated using a program such as *MathCad* or *Mathematica*. The roots are $x = -2, -1, -1, 1, 1, 2$, which correspond to $E_1 = \alpha + 2\beta$, $E_2 = E_3 = \alpha + \beta$, $E_4 = E_5 = \alpha - \beta$, and $E_6 = \alpha - 2\beta$.

11–37. Calculate the Hückel π-electronic energies of cyclobutadiene. What do Hund's rules say about the ground state of cyclobutadiene? Compare the stability of cyclobutadiene with that of two isolated ethylene molecules.

The structure of cyclobutadiene is

Letting $x = (\alpha - E)/\beta$, the Hückel determinantal equation for cyclobutadiene is

$$\begin{vmatrix} x & 1 & 0 & 1 \\ 1 & x & 1 & 0 \\ 0 & 1 & x & 1 \\ 1 & 0 & 1 & x \end{vmatrix} = 0$$

Expanding the determinant about the first column of elements gives

$$x \begin{vmatrix} x & 1 & 0 \\ 1 & x & 1 \\ 0 & 1 & x \end{vmatrix} - \begin{vmatrix} 1 & 0 & 1 \\ 1 & x & 1 \\ 0 & 1 & x \end{vmatrix} - \begin{vmatrix} 1 & 0 & 1 \\ x & 1 & 0 \\ 1 & x & 1 \end{vmatrix} = 0$$

Expanding the first two determinants about the first column of elements and the third determinant about the third column of elements gives

$$x \left[x \left(x^2 - 1 \right) - x \right] - \left[\left(x^2 - 1 \right) - (-1) \right] - \left[\left(x^2 - 1 \right) + 1 \right] = 0$$

$$x^4 - 4x^2 = 0$$

$$x^2 \left(x^2 - 4 \right) = 0$$

$$x = -2, 0, 0, 2$$

Recalling that $x = (\alpha - E)/\beta$, the energies of the four molecular orbitals for cyclobutadiene are

$$E_1 = \alpha + 2\beta \qquad E_2 = E_3 = \alpha \qquad E_4 = \alpha - 2\beta$$

There are four π electrons in cyclobutadiene. Two electrons will occupy the orbital with $E_1 = \alpha + 2\beta$ and the other two will occupy the doubly degenerate $E_2 = E_3 = \alpha$ orbitals. According to Hund's rule, each of these degenerate orbitals will be occupied by one electron and the electrons will have the same spin. Therefore, the ground state of cyclobutadiene will be a triplet state. The π-electron energy for cyclobutadiene is

$$\begin{aligned} E_\pi &= 2E_1 + E_2 + E_3 \\ &= 2 \left(\alpha + 2\beta \right) + \alpha + \alpha \\ &= 4\alpha + 4\beta \end{aligned}$$

The energy stabilization of cyclobutadiene that arises from delocation can be determined by comparing its π-electon energy with the energy of two isolated ethene molecules (each with an energy of $2\alpha + 2\beta$):

$$\begin{aligned} E_{\text{deloc}} &= E_\pi (\text{cyclobutadiene}) - 2E_\pi (\text{ethene}) \\ &= (4\alpha + 4\beta) - 2 (2\alpha + 2\beta) \\ &= 0 \end{aligned}$$

Cyclobutadiene has the same stability as two isolated ethene molecules.

11–38. Calculate the Hückel π-electronic energy of trimethylenemethane:

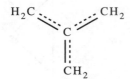

Let carbon 4 be the central carbon atom. Denoting $x = (\alpha - E)/\beta$, the Hückel secular determinant is

$$\begin{vmatrix} x & 0 & 0 & 1 \\ 0 & x & 0 & 1 \\ 0 & 0 & x & 1 \\ 1 & 1 & 1 & x \end{vmatrix} = 0$$

Expanding the determinant about the first column of elements gives

$$x \begin{vmatrix} x & 0 & 1 \\ 0 & x & 1 \\ 1 & 1 & x \end{vmatrix} - \begin{vmatrix} 0 & 0 & 1 \\ x & 0 & 1 \\ 0 & x & 1 \end{vmatrix} = 0$$

Expanding each determinant about the first column of elements gives

$$x \left[x \left(x^2 - 1 \right) - x \right] - x^2 = 0$$

$$x^4 - 3x^2 = 0$$

$$x^2 \left(x^2 - 3 \right) = 0$$

$$x = -\sqrt{3}, 0, 0, \sqrt{3}$$

Recalling that $x = (\alpha - E)/\beta$, the energies of the four molecular orbitals for trimethylenemethane are

$$E_1 = \alpha + \sqrt{3}\beta \qquad E_2 = E_3 = \alpha \qquad E_4 = \alpha - \sqrt{3}\beta$$

There are four π electrons in trimethylenemethane. Two electrons will occupy the orbital with $E_1 = \alpha + 2\beta$ and the other two will occupy the doubly degenerate $E = \alpha$ orbitals. According to Hund's rule, each of these degenerate orbitals will be occupied by one electron and the electrons will have the same spin. The π-electron energy for trimethylenemethane is

$$E_\pi = 2E_1 + E_2 + E_3$$

$$= 2 \left(\alpha + \sqrt{3}\beta \right) + \alpha + \alpha$$

$$= 4\alpha + 2\sqrt{3}\beta$$

11–39. Calculate the π-electronic energy levels and the total π-electronic energy of bicyclobutadiene:

Using the following numbering scheme for the carbon atoms

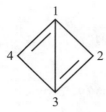

and denoting $x = (\alpha - E)/\beta$, the Hückel secular determinant is

$$\begin{vmatrix} x & 1 & 1 & 1 \\ 1 & x & 1 & 0 \\ 1 & 1 & x & 1 \\ 1 & 0 & 1 & x \end{vmatrix} = 0$$

Expanding the determinant about the second column gives

$$-\begin{vmatrix} 1 & 1 & 0 \\ 1 & x & 1 \\ 1 & 1 & x \end{vmatrix} + x\begin{vmatrix} x & 1 & 1 \\ 1 & x & 1 \\ 1 & 1 & x \end{vmatrix} - \begin{vmatrix} x & 1 & 1 \\ 1 & 1 & 0 \\ 1 & 1 & x \end{vmatrix} = 0$$

Expanding each determinant about its third column of elements gives

$$-\left[-(1-1) + x(x-1) \right] + x\left[(1-x) - (x-1) + x\left(x^2 - 1\right) \right] - \left[(1-1) + x(x-1) \right] = 0$$

$$x^4 - 5x^2 + 4x = 0$$

The four roots to this equation are $x = (-1 - \sqrt{17})/2$, 0, 1, $(-1 + \sqrt{17})/2$. Therefore, the energies of the four molecular orbitals for bicyclobutadiene are

$$E_1 = \alpha + \frac{1 + \sqrt{17}}{2}\beta = \alpha + 2.56155\beta$$

$$E_2 = \alpha$$

$$E_3 = \alpha - \beta$$

$$E_4 = \alpha + \frac{1 - \sqrt{17}}{2}\beta = \alpha - 1.56155\beta$$

There are 4 electrons, and they occupy the two lowest energy levels. The π-electron energy of bicyclobutadiene is therefore

$$E_\pi = 2E_1 + 2E_2$$

$$= 2\left(\alpha + 2.56155\beta\right) + 2\alpha$$

$$= 4\alpha + 5.1231\beta$$

11–40. Show that the Hückel molecular orbitals of benzene given in Equation 11.32 are orthonormal.

In the Hückel approximation, the overlap integral of $2p_z$ orbitals on different carbon atoms is zero. Therefore,

$$\int \psi_i^* \psi_j \, d\tau = \sum_k c_{ki} c_{kj}$$

where c_{ki} is the coefficient of the kth $2p_z$ function in the ith molecular orbital.

For $i = j$,

$$\int \psi_1^* \psi_1 \, d\tau = \left(\frac{1}{\sqrt{6}}\right)^2 \left(1^2 + 1^2 + 1^2 + 1^2 + 1^2 + 1^2\right) = 1$$

$$\int \psi_2^* \psi_2 \, d\tau = \left(\frac{1}{\sqrt{4}}\right)^2 \left[0^2 + 1^2 + 1^2 + 0^2 + (-1)^2 + (-1)^2\right] = 1$$

$$\int \psi_3^* \psi_3 \, d\tau = \left(\frac{1}{\sqrt{3}}\right)^2 \left[1^2 + \left(\frac{1}{2}\right)^2 + \left(-\frac{1}{2}\right)^2 + (-1)^2 + \left(-\frac{1}{2}\right)^2 + \left(\frac{1}{2}\right)^2\right] = 1$$

$$\int \psi_4^* \psi_4 \, d\tau = \left(\frac{1}{\sqrt{4}}\right)^2 \left[0^2 + 1^2 + (-1)^2 + 0^2 + 1^2 + (-1)^2\right] = 1$$

$$\int \psi_5^* \psi_5 \, d\tau = \left(\frac{1}{\sqrt{3}}\right)^2 \left[1^2 + \left(-\frac{1}{2}\right)^2 + \left(-\frac{1}{2}\right)^2 + 1^2 + \left(-\frac{1}{2}\right)^2 + \left(-\frac{1}{2}\right)^2\right] = 1$$

$$\int \psi_6^* \psi_6 \, d\tau = \left(\frac{1}{\sqrt{6}}\right)^2 \left[1^2 + (-1)^2 + 1^2 + (-1)^2 + 1^2 + (-1)^2\right] = 1$$

Therefore, all six wave functions are normalized.

For $i \neq j$,

$$\int \psi_1^* \psi_2 \, d\tau = \left(\frac{1}{\sqrt{24}}\right)^2 (0 + 1 + 1 + 0 - 1 - 1) = 0$$

$$\int \psi_1^* \psi_3 \, d\tau = \left(\frac{1}{\sqrt{18}}\right)^2 \left(1 + \frac{1}{2} - \frac{1}{2} - 1 - \frac{1}{2} + \frac{1}{2}\right) = 0$$

$$\int \psi_1^* \psi_4 \, d\tau = \left(\frac{1}{\sqrt{24}}\right)^2 (0 + 1 - 1 + 0 + 1 - 1) = 0$$

$$\int \psi_1^* \psi_5 \, d\tau = \left(\frac{1}{\sqrt{18}}\right)^2 \left(1 - \frac{1}{2} - \frac{1}{2} + 1 - \frac{1}{2} - \frac{1}{2}\right) = 0$$

$$\int \psi_1^* \psi_6 \, d\tau = \left(\frac{1}{\sqrt{6}}\right)^2 (1 - 1 + 1 - 1 + 1 - 1) = 0$$

The remaining pairs of orbitals can similarly be shown to be orthogonal.

11–41. Set up the Hückel molecular orbital theory determinantal equation for naphthalene.

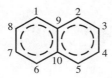

Using the numbering scheme for the carbon atoms shown in the question and denoting $x = (\alpha - E)/\beta$, the Hückel determinantal equation for naphthalene is

$$
\begin{vmatrix}
x & 0 & 0 & 0 & 0 & 0 & 0 & 1 & 1 & 0 \\
0 & x & 1 & 0 & 0 & 0 & 0 & 0 & 1 & 0 \\
0 & 1 & x & 1 & 0 & 0 & 0 & 0 & 0 & 0 \\
0 & 0 & 1 & x & 1 & 0 & 0 & 0 & 0 & 0 \\
0 & 0 & 0 & 1 & x & 0 & 0 & 0 & 0 & 1 \\
0 & 0 & 0 & 0 & 0 & x & 1 & 0 & 0 & 1 \\
0 & 0 & 0 & 0 & 0 & 1 & x & 1 & 0 & 0 \\
1 & 0 & 0 & 0 & 0 & 0 & 1 & x & 0 & 0 \\
1 & 1 & 0 & 0 & 0 & 0 & 0 & 0 & x & 1 \\
0 & 0 & 0 & 0 & 1 & 1 & 0 & 0 & 1 & x
\end{vmatrix} = 0
$$

11–42. Use a program such as *MathCad* or *Mathematica* to show that a Hückel calculation for naphthalene, $C_{10}H_8$, gives the molecular orbital energy levels $E_i = \alpha + m_i\beta$, where the 10 values of m_i are 2.3028, 1.6180, 1.3029, 1.0000, 0.6180, -0.6180, -1.0000, -1.3029, -1.6180, and -2.3028. Calculate the ground-state π-electron energy and the delocalization energy of naphthalene.

Using a program such as *MathCad* or *Mathematica*, the 10 roots for the determinantal equation in Problem 11–41 are found to be $x = -2.302\,78$, $-1.618\,03$, $-1.302\,78$, $-1.000\,00$, $-0.618\,03$, $0.618\,03$, $1.000\,00$, $1.302\,78$, $1.618\,03$, $2.302\,78$. Since $x = (\alpha - E)/\beta$, $E = \alpha - x\beta$. Comparing this equation with the energy expression given in the problem, $E = \alpha + m_i\beta$, it can be readily seen that $m_i = -x$. Thus, arranging in increasing energy, $m_i = 2.302\,78$, $1.618\,03$, $1.302\,78$, $1.000\,00$, $0.618\,03$, $-0.618\,03$, $-1.000\,00$, $-1.302\,78$, $-1.618\,03$, $-2.302\,78$. The 10 π electrons in naphthalene occupy the five lowest energy levels, giving a ground state π-electron energy of

$$
E_\pi = 2\left[(\alpha + 2.302\,78\beta) + (\alpha + 1.618\,03\beta) + (\alpha + 1.302\,78\beta) + (\alpha + 1.000\,00\beta) + (\alpha + 0.618\,03\beta)\right]
$$

$$
= 10\alpha + 13.6832\beta
$$

The delocalization energy of naphthalene is calculated by comparing the π-electron energy of naphthalene with that of five isolated ethene molecules. Each ethene molecule has a π-electron energy of $2\alpha + 2\beta$. Therefore,

$$
E_{\text{deloc}} = 10\alpha + 13.6832\beta - 5(2\alpha + 2\beta)
$$

$$
= 3.6832\beta
$$

11–43. Use a program such as *MathCad* or *Mathematica* to determine the 10 π orbitals of naptha-lene. Calculate the π-electronic charge on each carbon atom and the various bond orders.

To determine the ten π orbitals of naphthalene, use a program such as *MathCad* or *Mathematica* to find the eigenvalues and eigenvectors for the H (see Equation 11.34) matrix:

$$\begin{vmatrix} \alpha & 0 & 0 & 0 & 0 & 0 & 0 & \beta & \beta & 0 \\ 0 & \alpha & \beta & 0 & 0 & 0 & 0 & 0 & \beta & 0 \\ 0 & \beta & \alpha & \beta & 0 & 0 & 0 & 0 & 0 & 0 \\ 0 & 0 & \beta & \alpha & \beta & 0 & 0 & 0 & 0 & 0 \\ 0 & 0 & 0 & \beta & \alpha & 0 & 0 & 0 & 0 & \beta \\ 0 & 0 & 0 & 0 & 0 & \alpha & \beta & 0 & 0 & \beta \\ 0 & 0 & 0 & 0 & 0 & \beta & \alpha & \beta & 0 & 0 \\ \beta & 0 & 0 & 0 & 0 & 0 & \beta & \alpha & 0 & 0 \\ \beta & \beta & 0 & 0 & 0 & 0 & 0 & 0 & \alpha & \beta \\ 0 & 0 & 0 & 0 & \beta & \beta & 0 & 0 & \beta & \alpha \end{vmatrix} = 0$$

Writing the normalized wave functions in the form of Equation 11.18:

$$\psi_i = \sum_{r=1}^{10} c_{ri} \cdot 2p_{zr}$$

where i labels the energy levels as given in the Problem 11–42. The coefficients for each eigenfunction are listed in the following table:

i	c_{1i}	c_{2i}	c_{3i}	c_{4i}	c_{5i}	c_{6i}	c_{7i}	c_{8i}	c_{9i}	c_{10i}
1	0.3006	0.3006	0.2307	0.2307	0.3006	0.3006	0.2307	0.2307	0.4614	0.4614
2	0.2629	−0.2629	−0.4253	−0.4253	−0.2629	0.2629	0.4253	0.4253	0.0000	0.0000
3	−0.3996	−0.3996	−0.1735	0.1735	0.3996	0.3996	0.1735	−0.1735	−0.3470	0.3470
4	0.0000	0.0000	−0.4082	−0.4082	0.0000	0.0000	−0.4082	−0.4082	0.4082	0.4082
5	0.4253	−0.4253	−0.2629	0.2629	0.4253	−0.4253	−0.2629	0.2629	0.0000	0.0000
6	−0.4253	0.4253	−0.2629	−0.2629	0.4253	−0.4253	0.2629	0.2629	0.0000	0.0000
7	0.0000	0.0000	0.4082	−0.4082	0.0000	0.0000	−0.4082	0.4082	−0.4082	0.4082
8	−0.3996	−0.3996	0.1735	0.1735	−0.3996	−0.3996	0.1735	0.1735	0.3470	0.3470
9	−0.2629	0.2629	−0.4253	0.4253	−0.2629	0.2629	−0.4253	0.4253	0.0000	0.0000
10	0.3006	0.3006	−0.2307	0.2307	−0.3006	−0.3006	0.2307	−0.2307	−0.4614	0.4614

The total π-electron charge on the jth carbon atom (Equation 11.23) is

$$q_j = \sum_i n_i c_{ji}^2$$

Naphthalene has 10 π electrons, giving $n_i = 2$ for $i \leq 5$ and $n_i = 0$ for $i > 5$. Therefore, the above equation reduces to

$$q_j = 2 \sum_{i=1}^{5} c_{ji}^2$$

which gives $q_j = 1$ for each carbon atom.

The π-bond order between adjacent carbon atoms r and s can be determined using Equation 11.24:

$$p_{rs}^{\pi} = \sum_i n_i c_{ri} c_{si}$$

which, when applied to naphthalene, becomes

$$p_{rs}^{\pi} = 2 \sum_{i=1}^{5} c_{ri} c_{si}$$

Applying this equation to adjacent carbon atoms gives

$$\pi_{23}^{\pi} = \pi_{45}^{\pi} = \pi_{67}^{\pi} = \pi_{18}^{\pi} = 0.724\,56$$

$$\pi_{34}^{\pi} = \pi_{78}^{\pi} = 0.603\,17$$

$$\pi_{19}^{\pi} = \pi_{29}^{\pi} = \pi_{5,10}^{\pi} = \pi_{6,10}^{\pi} = 0.554\,70$$

$$\pi_{9,10}^{\pi} = 0.518\,23$$

11–44. Using Hückel molecular orbital theory, determine whether the linear state $(H—H—H)^+$ or the triangular state

$$\left[\begin{array}{c} H \\ \diagup \diagdown \\ H——H \end{array} \right]^+$$

of H_3^+ is the more stable state. Repeat the calculation for H_3 and H_3^-.

For the triangular state, the Hückel determinantal equation is

$$\begin{vmatrix} x & 1 & 1 \\ 1 & x & 1 \\ 1 & 1 & x \end{vmatrix} = 0$$

Expanding the determinant about the first column of elements gives

$$x\left(x^2 - 1\right) - (x - 1) + (1 - x) = 0$$

$$x^3 - 3x + 2 = 0$$

The three roots of this equation are $x = -2, 1, 1$. Since $x = (\alpha - E)/\beta$, the molecular orbital energies for the triangular state are

$$E_1 = \alpha + 2\beta \qquad E_2 = E_3 = \alpha - \beta$$

For the two-electron H_3^+, the three-electron H_3, and four-electron H_3^-, the energies are

$$E_\pi(H_3^+, \text{triangular}) = 2E_1 = 2\alpha + 4\beta$$

$$E_\pi(H_3, \text{triangular}) = 2E_1 + E_2 = 3\alpha + 3\beta$$

$$E_\pi(H_3^-, \text{triangular}) = 2E_1 + E_2 + E_3 = 4\alpha + 2\beta$$

Note that H_3^- will have a triplet ground state, since the energy level $\alpha - \beta$ is doubly degenerate.

For the linear state, the Hückel determinantal equation is

$$\begin{vmatrix} x & 1 & 0 \\ 1 & x & 1 \\ 0 & 1 & x \end{vmatrix} = 0$$

Expanding the determinant about the first column of elements gives

$$x\left(x^2 - 1\right) - x = 0$$

$$x^3 - 2x = 0$$

The three roots of this equation are $x = -\sqrt{2}, 0, \sqrt{2}$. Since $x = (\alpha - E)/\beta$, the molecular orbital energies for the linear state are

$$E_1 = \alpha + \sqrt{2}\beta \qquad E_2 = \alpha \qquad E_3 = \alpha - \sqrt{2}\beta$$

The energies of linear H_3^+, H_3, and H_3^- are

$$E_\pi(H_3^+, \text{linear}) = 2E_1 = 2\alpha + 2\sqrt{2}\beta = 2\alpha + 2.8284\beta$$

$$E_\pi(H_3, \text{linear}) = 2E_1 + E_2 = 3\alpha + 2\sqrt{2}\beta = 3\alpha + 2.8284\beta$$

$$E_\pi(H_3^-, \text{linear}) = 2E_1 + E_2 + E_3 = 4\alpha + 2\sqrt{2}\beta = 4\alpha + 2.8284\beta$$

The triangular geometry is more stable for H_3^+ and H_3 but the linear geometry is more stable for H_3^-.

11–45. Set up a Hückel theory secular determinant for pyridine.

The structure of pyridine is given below:

Note that there is a heteroatom in pyridine. The coulomb integral for a nitrogen atom is not the same as that for a carbon atom and the resonance integral involving carbon and nitrogen as nearest neighbors is not the same as that involving nearest-neighbor carbon atoms. Incorporating these considerations, and numbering the nitrogen as atom 1, the Hückel theory determinantal equation for pyridine is

$$
\begin{vmatrix}
\alpha_N - E & \beta_{CN} & 0 & 0 & 0 & \beta_{CN} \\
\beta_{CN} & \alpha_C - E & \beta_{CC} & 0 & 0 & 0 \\
0 & \beta_{CC} & \alpha_C - E & \beta_{CC} & 0 & 0 \\
0 & 0 & \beta_{CC} & \alpha_C - E & \beta_{CC} & 0 \\
0 & 0 & 0 & \beta_{CC} & \alpha_C - E & \beta_{CC} \\
\beta_{CN} & 0 & 0 & 0 & \beta_{CC} & \alpha_C - E
\end{vmatrix} = 0
$$

11–46. Calculate the delocalization energy, the charge on each carbon atom, and the bond orders for the allyl radical, cation, and anion. Sketch the molecular orbitals for the allyl system.

The structure of the allyl system is

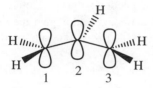

Using the above numbering scheme, the Hückel determinantal equation for the allyl system is

$$
\begin{vmatrix}
x & 1 & 0 \\
1 & x & 1 \\
0 & 1 & x
\end{vmatrix} = x^3 - 2x = 0
$$

The roots of this equation are $-\sqrt{2}, 0, \sqrt{2}$. Since $x = (\alpha - E)/\beta$, the π molecular orbital energies are

$$
E_1 = \alpha + \sqrt{2}\beta \qquad E_2 = \alpha \qquad E_3 = \alpha - \sqrt{2}\beta
$$

For the allyl radical (with 3 π electrons)

$$
E_\pi = 2E_1 + E_2 = 3\alpha + 2\sqrt{2}\beta = 3\alpha + 2.8284\beta
$$

$$
E_{\text{deloc}} = E_\pi - (3\alpha + 2\beta) = 0.8284\beta
$$

where we have subtracted $3\alpha + 2\beta$ from E_π because three localized π electrons could be thought of as the sum of one ethene molecule (of energy $2\alpha + 2\beta$) and a single carbon atom (of energy α).

For the allyl cation (with 2 π electrons)

$$
E_\pi = 2(\alpha + \sqrt{2}\beta) = 2\alpha + 2\sqrt{2}\beta = 2\alpha + 2.8284\beta
$$

$$
E_{\text{deloc}} = E_\pi - (2\alpha + 2\beta) = 0.8284\beta
$$

For the allyl anion (with 4 π electrons)

$$E_\pi = 2(\alpha + \sqrt{2}\beta) + 2\alpha = 4\alpha + 2\sqrt{2}\beta = 4\alpha + 2.8284\beta$$

$$E_{\text{deloc}} = E_\pi - (2\alpha + 2\beta) - 2\alpha = 0.8284\beta$$

To calculate the charges on each carbon atom and the bond orders, the molecular orbital associated with each value of E needs to be determined. The general form of the three molecular orbitals is given by

$$\psi = c_1 \cdot 2p_{z1} + c_2 \cdot 2p_{z2} + c_3 \cdot 2p_{z3}$$

and the equations for c_1, c_2, and c_3 are obtained from the algebraic equations that give rise to the secular determinant:

$$c_1 x + c_2 = 0 \tag{1}$$

$$c_1 + c_2 x + c_3 = 0 \tag{2}$$

$$c_2 + c_3 x = 0 \tag{3}$$

Two of these equations are independent; thus, to determine each wave function, require the function to be normalized. Equations 1 and 2 give relations between c_1 and c_2 and between c_1 and c_3, respectively:

$$c_2 = -c_1 x$$

$$c_3 = -c_1 - c_2 x = c_1 \left(x^2 - 1 \right)$$

For $E_1 = \alpha + \sqrt{2}\beta$ or $x_1 = -\sqrt{2}$,

$$c_2 = -c_1 x_1 = \sqrt{2}c_1$$

$$c_3 = c_1 \left(x_1^2 - 1 \right) = c_1$$

The wave function is then

$$\psi_1 = c_1 \left(2p_{z1} + \sqrt{2} \cdot 2p_{z2} + 2p_{z3} \right)$$

Normalization of the wave function gives

$$c_1^2 \left(1 + 2 + 1 \right) = 1$$

$$c_1 = \frac{1}{2}$$

Thus,

$$\psi_1 = \frac{1}{2} \cdot 2p_{z1} + \frac{1}{\sqrt{2}} \cdot 2p_{z2} + \frac{1}{2} \cdot 2p_{z3}$$

For $E_2 = \alpha$ or $x_2 = 0$,

$$c_2 = -c_1 x_2 = 0$$

$$c_3 = c_1 \left(x_2^2 - 1 \right) = -c_1$$

The wave function is then

$$\psi_2 = c_1 \left(2p_{z1} - 2p_{z3} \right)$$

Normalization of the wave function gives

$$c_1^2 \left(1 + 1 \right) = 1$$

$$c_1 = \frac{1}{\sqrt{2}}$$

Thus,

$$\psi_2 = \frac{1}{\sqrt{2}} \cdot 2p_{z1} - \frac{1}{\sqrt{2}} \cdot 2p_{z3}$$

For $E_3 = \alpha - \sqrt{2}\beta$ or $x_3 = \sqrt{2}$,

$$c_2 = -c_1 x_3 = -\sqrt{2}c_1$$

$$c_3 = c_1 \left(x_3^2 - 1 \right) = c_1$$

The wave function is then

$$\psi_3 = c_1 \left(2p_{z1} - \sqrt{2} \cdot 2p_{z2} + 2p_{z3} \right)$$

Normalization of the wave function gives

$$c_1^2 \left(1 + 2 + 1 \right) = 1$$

$$c_1 = \frac{1}{2}$$

Thus,

$$\psi_3 = \frac{1}{2} \cdot 2p_{z1} - \frac{1}{\sqrt{2}} \cdot 2p_{z2} + \frac{1}{2} \cdot 2p_{z3}$$

A sketch of the three molecular orbitals is shown below:

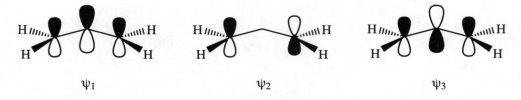

The π-electronic charge on the jth carbon atom (Equation 11.23) is

$$q_j = \sum_i n_i c_{ji}^2$$

and the π-bond order between adjacent carbon atoms r and s (Equation 11.24) is:

$$p_{rs}^\pi = \sum_i n_i c_{ri} c_{si}$$

For the allyl radical, $n_1 = 2$, $n_2 = 1$, and $n_3 = 0$,

$$q_j = 2c_{j1}^2 + c_{j2}^2$$

$$q_1 = 2\left(\frac{1}{2}\right)^2 + \left(\frac{1}{\sqrt{2}}\right)^2 = 1$$

$$q_2 = 2\left(\frac{1}{\sqrt{2}}\right)^2 + 0 = 1$$

$$q_3 = 2\left(\frac{1}{2}\right)^2 + \left(-\frac{1}{\sqrt{2}}\right)^2 = 1$$

and

$$p_{rs}^\pi = 2c_{r1}c_{s1} + c_{r2}c_{s2}$$

$$p_{12}^\pi = 2\left(\frac{1}{2}\right)\left(\frac{1}{\sqrt{2}}\right) + \left(\frac{1}{\sqrt{2}}\right)(0) = \frac{1}{\sqrt{2}} = 0.7071$$

$$p_{23}^\pi = 2\left(\frac{1}{\sqrt{2}}\right)\left(\frac{1}{2}\right) + (0)\left(-\frac{1}{\sqrt{2}}\right) = \frac{1}{\sqrt{2}} = 0.7071$$

For the allyl cation, $n_1 = 2$, and $n_2 = n_3 = 0$,

$$q_j = 2c_{j1}^2$$

$$q_1 = 2\left(\frac{1}{2}\right)^2 = \frac{1}{2}$$

$$q_2 = 2\left(\frac{1}{\sqrt{2}}\right)^2 = 1$$

$$q_3 = 2\left(\frac{1}{2}\right)^2 = \frac{1}{2}$$

and

$$p_{rs}^\pi = 2c_{r1}c_{s1}$$

$$p_{12}^\pi = 2\left(\frac{1}{2}\right)\left(\frac{1}{\sqrt{2}}\right) = \frac{1}{\sqrt{2}} = 0.7071$$

$$p_{23}^\pi = 2\left(\frac{1}{\sqrt{2}}\right)\left(\frac{1}{2}\right) = \frac{1}{\sqrt{2}} = 0.7071$$

Finally, for the allyl anion, $n_1 = n_2 = 2$, and $n_3 = 0$,

$$q_j = 2c_{j1}^2 + 2c_{j2}^2$$

$$q_1 = 2\left(\frac{1}{2}\right)^2 + 2\left(\frac{1}{\sqrt{2}}\right)^2 = \frac{3}{2}$$

$$q_2 = 2\left(\frac{1}{\sqrt{2}}\right)^2 + 2\,(0) = 1$$

$$q_3 = 2\left(\frac{1}{2}\right)^2 + 2\left(-\frac{1}{\sqrt{2}}\right)^2 = \frac{3}{2}$$

and

$$p_{rs}^\pi = 2c_{r1}c_{s1} + 2c_{r2}c_{s2}$$

$$p_{12}^\pi = 2\left(\frac{1}{2}\right)\left(\frac{1}{\sqrt{2}}\right) + 2\left(\frac{1}{\sqrt{2}}\right)(0) = \frac{1}{\sqrt{2}} = 0.7071$$

$$p_{23}^\pi = 2\left(\frac{1}{\sqrt{2}}\right)\left(\frac{1}{2}\right) + 2\,(0)\left(-\frac{1}{\sqrt{2}}\right) = \frac{1}{\sqrt{2}} = 0.7071$$

These results are summarized in the table below:

Species	E_π	E_{deloc}	q_1	q_2	q_3	p_{12}^π	p_{23}^π
allyl radical	$3\alpha + 2\sqrt{2}\beta$	0.8284β	1	1	1	0.7071	0.7071
allyl cation	$2\alpha + 2\sqrt{2}\beta$	0.8284β	1/2	1	1/2	0.7071	0.7071
allyl anion	$4\alpha + 2\sqrt{2}\beta$	0.8284β	3/2	1	3/2	0.7071	0.7071

11–47. Because of the symmetry inherent in the Hückel theory secular determinant of linear and cyclic conjugated polyenes, we can write mathematical formulas for the energy levels for an arbitrary number of carbon atoms in the system (for present purposes, we consider cyclic polyenes with only an even number of carbon atoms). These formulas are

$$E_n = \alpha + 2\beta \cos \frac{\pi n}{N+1} \qquad n = 1, 2, \ldots, N \qquad \text{linear chains}$$

and

$$E_n = \alpha + 2\beta \cos \frac{2\pi n}{N} \qquad n = 0, \pm 1, \ldots, \pm\left(\frac{N}{2} - 1\right), \frac{N}{2} \qquad \text{cyclic chains (N even)}$$

where N is the number of carbon atoms in the conjugated π system.

(a) Use these formulas to verify the results given in the chapter for butadiene and benzene.

(b) Now use these formulas to predict energy levels for linear hexatriene (C_6H_8) and octatetraene (C_8H_{10}). How does the stabilization energy of these molecules per carbon atom vary as the chains grow in length?

(c) Compare the results for hexatriene and benzene. Which molecule has a greater stabilization energy? Why?

(a) For butadiene, $n = 4$. The energies according to the formula for linear polyenes are

$$E_1 = \alpha + 2\beta \cos \frac{\pi}{5} = \alpha + 1.618\beta$$

$$E_2 = \alpha + 2\beta \cos \frac{2\pi}{5} = \alpha + 0.618\beta$$

$$E_3 = \alpha + 2\beta \cos \frac{3\pi}{5} = \alpha - 0.618\beta$$

$$E_4 = \alpha + 2\beta \cos \frac{4\pi}{5} = \alpha - 1.618\beta$$

which are in agreement with the values given in the text.

For benzene, $n = 6$. The energies according to the formula for cyclic polyenes are

$$E_0 = \alpha + 2\beta \cos \frac{(2\pi)(0)}{6} = \alpha + 2\beta$$

$$E_1 = \alpha + 2\beta \cos \frac{(2\pi)(1)}{6} = \alpha + \beta$$

$$E_{-1} = \alpha + 2\beta \cos \frac{(2\pi)(-1)}{6} = \alpha + \beta$$

$$E_2 = \alpha + 2\beta \cos \frac{(2\pi)(2)}{6} = \alpha - \beta$$

$$E_{-2} = \alpha + 2\beta \cos \frac{(2\pi)(-2)}{6} = \alpha - \beta$$

$$E_3 = \alpha + 2\beta \cos \frac{(2\pi)(3)}{6} = \alpha - 2\beta$$

which are in agreement with the values given in the text.

(b) Applying the formula for linear polyenes to linear hexatriene ($n = 6$), the energies are

$$E_1 = \alpha + 2\beta \cos \frac{\pi}{7} = \alpha + 1.8019\beta$$

$$E_2 = \alpha + 2\beta \cos \frac{2\pi}{7} = \alpha + 1.2470\beta$$

$$E_3 = \alpha + 2\beta \cos \frac{3\pi}{7} = \alpha + 0.4450\beta$$

$$E_4 = \alpha + 2\beta \cos \frac{4\pi}{7} = \alpha - 0.4450\beta$$

$$E_5 = \alpha + 2\beta \cos \frac{5\pi}{7} = \alpha - 1.2470\beta$$

$$E_6 = \alpha + 2\beta \cos \frac{6\pi}{7} = \alpha - 1.8019\beta$$

The π-electron energy and the delocalization energy of linear hexatriene are

$$E_\pi = 2E_1 + 2E_2 + 2E_3 = 6\alpha + 6.9878\beta$$

$$E_{\text{deloc}} = E_\pi - 3E_\pi(\text{ethene}) = (6\alpha + 6.9878\beta) - 3(2\alpha + 2\beta) = 0.9878\beta$$

Dividing E_{deloc} by the number of carbon atoms in the system, the delocation energy is 0.1646β per carbon atom.

For linear octatetraene ($n = 8$), the energies are

$$E_1 = \alpha + 2\beta \cos \frac{\pi}{9} = \alpha + 1.8794\beta$$

$$E_2 = \alpha + 2\beta \cos \frac{2\pi}{9} = \alpha + 1.5321\beta$$

$$E_3 = \alpha + 2\beta \cos \frac{3\pi}{9} = \alpha + 1.0000\beta$$

$$E_4 = \alpha + 2\beta \cos \frac{4\pi}{9} = \alpha + 0.3473\beta$$

$$E_5 = \alpha + 2\beta \cos \frac{5\pi}{9} = \alpha - 0.3473\beta$$

$$E_6 = \alpha + 2\beta \cos \frac{6\pi}{9} = \alpha - 1.0000\beta$$

$$E_7 = \alpha + 2\beta \cos \frac{7\pi}{9} = \alpha - 1.5321\beta$$

$$E_8 = \alpha + 2\beta \cos \frac{8\pi}{9} = \alpha - 1.8794\beta$$

The π-electron energy and the delocalization energy of linear octatetraene are

$$E_\pi = 2E_1 + 2E_2 + 2E_3 + 2E_4 = 8\alpha + 9.5176\beta$$

$$E_{\text{deloc}} = E_\pi - 4E_\pi(\text{ethene}) = (8\alpha + 9.5176\beta) - 4(2\alpha + 2\beta) = 1.5176\beta$$

Dividing E_{deloc} by the number of carbon atoms in the system, the delocation energy is 0.1897β per carbon atom.

The results from hexatriene and octatetraene show that the stabilization energy of linear polyenes per carbon atom increases as the chain grows in length.

(c) Benzene has a greater stabilization energy (2β) than hexatriene (0.9878β) because its cyclic structure allows for more delocalization than the corresponding linear structure.

11–48. The problem of a linear conjugated polyene of N carbon atoms can be solved in general. The energies E_j and the coefficients of the atomic orbitals in the jth molecular orbital are given by

$$E_j = \alpha + 2\beta \cos \frac{j\pi}{N+1} \qquad j = 1, 2, 3, \ldots, N$$

and

$$c_{jk} = \left(\frac{2}{N+1} \right)^{1/2} \sin \frac{jk\pi}{N+1} \qquad k = 1, 2, 3, \ldots, N$$

Determine the energy levels and the wave functions for butadiene using these formulas.

The equation for the energies E_j is the same as the equation for E_n in Problem 11–47 and the energy levels are given in that problem.

The general form of the four corresponding molecular orbitals for butadiene (see Equation 11.18) is given by

$$\psi_i = c_{1i} \cdot 2p_{z1} + c_{2i} \cdot 2p_{z2} + c_{3i} \cdot 2p_{z3} + c_{4i} \cdot 2p_{z4}$$

The c_{jk} equation gives, for $N = 4$,

$$c_{11} = 0.3717$$

$$c_{22} = 0.3717$$

$$c_{33} = -0.3717$$

$$c_{44} = -0.3717$$

$$c_{12} = c_{21} = 0.6015$$

$$c_{13} = c_{31} = 0.6015$$

$$c_{14} = c_{41} = 0.3717$$

$$c_{23} = c_{32} = -0.3717$$

$$c_{24} = c_{42} = -0.6015$$

$$c_{34} = c_{43} = 0.6015$$

Thus, the four wave functions are

$$\psi_1 = 0.3717 \cdot 2p_{z1} + 0.6015 \cdot 2p_{z2} + 0.6015 \cdot 2p_{z3} + 0.3717 \cdot 2p_{z4}$$

$$\psi_2 = 0.6015 \cdot 2p_{z1} + 0.3717 \cdot 2p_{z2} - 0.3717 \cdot 2p_{z3} - 0.6015 \cdot 2p_{z4}$$

$$\psi_3 = 0.6015 \cdot 2p_{z1} - 0.3717 \cdot 2p_{z2} - 0.3717 \cdot 2p_{z3} + 0.6015 \cdot 2p_{z4}$$

$$\psi_4 = 0.3717 \cdot 2p_{z1} - 0.6015 \cdot 2p_{z2} + 0.6015 \cdot 2p_{z3} - 0.3717 \cdot 2p_{z4}$$

11–49. We can calculate the electronic states of a hypothetical one-dimensional solid by modeling the solid as a one-dimensional array of atoms with one orbital per atom, and using Hückel theory to calculate the allowed energies. Use the formula for E_j in the previous problem to show that the energies will form essentially a continuous band of width 4β. *Hint:* Calculate $E_N - E_1$ and let N be very large so that you can use $\cos x \approx 1 - x^2/2 + \cdots$.

The equation for E_j in Problem 11–48 gives

$$E_1 = \alpha + 2\beta \cos \frac{\pi}{N+1}$$

$$E_2 = \alpha + 2\beta \cos \frac{2\pi}{N+1}$$

$$E_N = \alpha + 2\beta \cos \frac{N\pi}{N+1}$$

To find the width of the band, subtract E_1 from E_N:

$$E_N - E_1 = 2\beta \left(\cos \frac{N\pi}{N+1} - \cos \frac{\pi}{N+1} \right)$$

As $N \to \infty$, the above equation becomes

$$E_N - E_1 \approx 2\beta \left(\cos \pi - \cos 0 \right) = -4\beta$$

(Note that β is negative.) Thus, the width of the energy band approaches 4β for large values of N.

Now consider the spacing between the first two levels:

$$E_2 - E_1 = 2\beta \left(\cos \frac{2\pi}{N+1} - \cos \frac{\pi}{N+1} \right)$$

$$= 2\beta \left[1 - \frac{1}{2} \left(\frac{2\pi}{N+1} \right)^2 - 1 + \frac{1}{2} \left(\frac{\pi}{N+1} \right)^2 \right]$$

$$= 2\beta \left[-\frac{3}{2} \left(\frac{\pi}{N+1} \right)^2 \right]$$

As $N \to \infty$, $E_2 - E_1 \to 0$. The conclusion is true in general for the spacing between E_{j+1} and E_j. Therefore, the band becomes continuous.

11–50. The band of electronic energies that we calculated in the previous problem can accommodate N pairs of electrons of opposite spins, or a total of $2N$ electrons. If each atom contributes one electron (as in the case of a polyene), the band is occupied by a total of N electrons. Using some ideas you may have learned in general chemistry, would you expect such a system to be a conductor or an insulator?

The N energy levels are close together, and only half of them are occupied in the ground state. Therefore, when N is large, we would expect this system to be a conductor (since relatively little energy is required to excite an electron).

11–51. Verify Equation 11.39.

Let C be the matrix whose columns are formed by the N eigenvectors of H; that is, the ath column of C is $\mathbf{c}^a$. Then Equation 11.35 can be rewritten as

$$H\mathbf{c}^a = \sum_k H_{ik} C_{ka} = E_a C_{ia} = (HC)_{ia}$$

The matrix D has elements $D_{ka} = E_k \delta_{ka}$, and the product with C is given by

$$(CD)_{ia} = \sum_k C_{ik} D_{ka}$$

$$= \sum_k C_{ik} E_k \delta_{ka}$$

$$= C_{ia} E_a$$

Since $(HC)_{ia} = (CD)_{ia}$, $HC = CD$.

11–52. Write out the C matrix for butadiene and show that it is orthogonal. Recall (MathChapter G) that $C^T = C^{-1}$, or that $CC^T = C^TC = I$ for an orthogonal matrix.

Arrange the eigenvectors of butadiene (Equation 11.15) in columns to give the C matrix:

$$\begin{pmatrix} 0.3717 & 0.6015 & 0.6015 & 0.3717 \\ 0.6015 & 0.3717 & -0.3717 & -0.6015 \\ 0.6015 & -0.3717 & -0.3717 & 0.6015 \\ 0.3717 & -0.6015 & 0.6015 & -0.3717 \end{pmatrix}$$

Since the ijth element in this matrix is the same as the jith element, $C^T = C$. It follows that

$$CC^T = C^TC = \begin{pmatrix} 0.3717 & 0.6015 & 0.6015 & 0.3717 \\ 0.6015 & 0.3717 & -0.3717 & -0.6015 \\ 0.6015 & -0.3717 & -0.3717 & 0.6015 \\ 0.3717 & -0.6015 & 0.6015 & -0.3717 \end{pmatrix}$$

$$\times \begin{pmatrix} 0.3717 & 0.6015 & 0.6015 & 0.3717 \\ 0.6015 & 0.3717 & -0.3717 & -0.6015 \\ 0.6015 & -0.3717 & -0.3717 & 0.6015 \\ 0.3717 & -0.6015 & 0.6015 & -0.3717 \end{pmatrix}$$

$$= \begin{pmatrix} 1 & 0 & 0 & 0 \\ 0 & 1 & 0 & 0 \\ 0 & 0 & 1 & 0 \\ 0 & 0 & 0 & 1 \end{pmatrix}$$

Thus, the C matrix is orthogonal.

11–53. Show that Equations 11.42 and 11.43 are equivalent to Equations 11.23 and 11.24.

The kjth element of the R matrix

$$R = CnC^T$$

is

$$R_{kj} = \sum_i \sum_l c_{ki} n_{il} c_{jl}$$

Since the n matrix is diagonal, $n_{il} = n_i \delta_{il}$, where n_i is the occupation number of the ith molecular orbital. Therefore,

$$R_{kj} = \sum_i n_i c_{ki} c_{ji} \tag{1}$$

but according to Equation 11.24,

$$p_{kj}^\pi = \sum_i n_i c_{ki} c_{ji}$$

Thus,

$$p_{kj}^\pi = R_{kj}$$

which is the same as Equation 11.43.

For $k = j$, Equation 1 gives

$$R_{jj} = \sum_i n_i c_{ji}^2$$

but according to Equation 11.23,

$$q_j = \sum_i n_i c_{ji}^2$$

Thus,

$$q_j = R_{jj}$$

which is the same as Equation 11.42.

11–54. Use a program such as *MathCad* or *Mathematica* to show that $HC = CD$ for butadiene.

Using the numbering system for carbon atoms shown in Figure 11.17, the H matrix for butadiene is

$$H = \begin{pmatrix} \alpha & \beta & 0 & 0 \\ \beta & \alpha & \beta & 0 \\ 0 & \beta & \alpha & \beta \\ 0 & 0 & \beta & \alpha \end{pmatrix}$$

and the corresponding D matrix is

$$D = \begin{pmatrix} \alpha + 1.61804\beta & 0 & 0 & 0 \\ 0 & \alpha + 0.61804 & 0 & 0 \\ 0 & 0 & \alpha - 0.61804 & 0 \\ 0 & 0 & 0 & \alpha - 1.61804 \end{pmatrix}$$

The C matrix is written out in Problem 11–52. Using a program such as *MathCad* or *Mathematica*, the product HC is

$$HC = \begin{pmatrix} \alpha & \beta & 0 & 0 \\ \beta & \alpha & \beta & 0 \\ 0 & \beta & \alpha & \beta \\ 0 & 0 & \beta & \alpha \end{pmatrix} \begin{pmatrix} 0.3717 & 0.6015 & 0.6015 & 0.3717 \\ 0.6015 & 0.3717 & -0.3717 & -0.6015 \\ 0.6015 & -0.3717 & -0.3717 & 0.6015 \\ 0.3717 & -0.6015 & 0.6015 & -0.3717 \end{pmatrix}$$

$$= \begin{pmatrix} 0.3717\alpha + 0.6015\beta & 0.6015\alpha + 0.3717\beta & 0.6015\alpha - 0.3717\beta & 0.3717\alpha - 0.6015\beta \\ 0.6015\alpha + 0.9732\beta & 0.3717\alpha + 0.2298\beta & -0.3717\alpha + 0.2298\beta & -0.6015\alpha + 0.9732\beta \\ 0.6015\alpha + 0.9732\beta & -0.3717\alpha - 0.2298\beta & -0.3717\alpha + 0.2298\beta & 0.6015\alpha - 0.9732\beta \\ 0.3717\alpha + 0.6015\beta & -0.6015\alpha - 0.3717\beta & 0.6015\alpha - 0.3717\beta & -0.3717\alpha + 0.6015\beta \end{pmatrix}$$

and the product CD is

$$CD = \begin{pmatrix} 0.3717 & 0.6015 & 0.6015 & 0.3717 \\ 0.6015 & 0.3717 & -0.3717 & -0.6015 \\ 0.6015 & -0.3717 & -0.3717 & 0.6015 \\ 0.3717 & -0.6015 & 0.6015 & -0.3717 \end{pmatrix}$$

$$\times \begin{pmatrix} \alpha + 1.61804\beta & 0 & 0 & 0 \\ 0 & \alpha + 0.61804 & 0 & 0 \\ 0 & 0 & \alpha - 0.61804 & 0 \\ 0 & 0 & 0 & \alpha - 1.61804 \end{pmatrix}$$

$$= \begin{pmatrix} 0.3717\alpha + 0.6015\beta & 0.6015\alpha + 0.3717\beta & 0.6015\alpha - 0.3717\beta & 0.3717\alpha - 0.6015\beta \\ 0.6015\alpha + 0.9732\beta & 0.3717\alpha + 0.2298\beta & -0.3717\alpha + 0.2298\beta & -0.6015\alpha + 0.9732\beta \\ 0.6015\alpha + 0.9732\beta & -0.3717\alpha - 0.2298\beta & -0.3717\alpha + 0.2298\beta & 0.6015\alpha - 0.9732\beta \\ 0.3717\alpha + 0.6015\beta & -0.6015\alpha - 0.3717\beta & 0.6015\alpha - 0.3717\beta & -0.3717\alpha + 0.6015\beta \end{pmatrix}$$

Thus, HC = CD for butadiene.

11–55. Use a program such as *MathCad* or *Mathematica* to show that the R matrix (Equation 11.41) for butadiene is

$$\begin{pmatrix} 1.0000 & 0.8943 & 0 & -0.4473 \\ 0.8943 & 1.0000 & 0.4473 & 0 \\ 0 & 0.4473 & 1.0000 & 0.8943 \\ -0.4473 & 0 & 0.8943 & 1.0000 \end{pmatrix}$$

Interpret this result.

For butadiene, $n_1 = n_2 = 2$ and $n_3 = n_4 = 0$, the n matrix is

$$n = \begin{pmatrix} 2 & 0 & 0 & 0 \\ 0 & 2 & 0 & 0 \\ 0 & 0 & 0 & 0 \\ 0 & 0 & 0 & 0 \end{pmatrix}$$

The C matrix is written out in Problem 11–52. The R matrix is then

$$R = CnC^T$$

$$= \begin{pmatrix} 0.3717 & 0.6015 & 0.6015 & 0.3717 \\ 0.6015 & 0.3717 & -0.3717 & -0.6015 \\ 0.6015 & -0.3717 & -0.3717 & 0.6015 \\ 0.3717 & -0.6015 & 0.6015 & -0.3717 \end{pmatrix} \begin{pmatrix} 2 & 0 & 0 & 0 \\ 0 & 2 & 0 & 0 \\ 0 & 0 & 0 & 0 \\ 0 & 0 & 0 & 0 \end{pmatrix} \begin{pmatrix} 0.3717 & 0.6015 & 0.6015 & 0.3717 \\ 0.6015 & 0.3717 & -0.3717 & -0.6015 \\ 0.6015 & -0.3717 & -0.3717 & 0.6015 \\ 0.3717 & -0.6015 & 0.6015 & -0.3717 \end{pmatrix}$$

$$= \begin{pmatrix} 1.0000 & 0.8943 & 0 & -0.4473 \\ 0.8943 & 1.0000 & 0.4473 & 0 \\ 0 & 0.4473 & 1.0000 & 0.8943 \\ -0.4473 & 0 & 0.8943 & 1.0000 \end{pmatrix}$$

The diagonal elements of R are the fractional π electron charges on the carbon atoms. In the case of butadiene, they are all one. The off-diagonal elements of R corresponding to neighboring carbon atoms are π-bond orders. Thus, $p_{12}^\pi = 0.8943$, $p_{23}^\pi = 0.4473$, and $p_{34}^\pi = 0.8943$. Note that a bond order has physical meaning only between atoms for which there is a nonzero value of β. Thus, the values of the elements R_{13}, R_{14}, and R_{24} have no physical meaning.

11–56. Use a program such as *MathCad* or *Mathematica* to show that the R matrix (Equation 11.41) for cyclobutadiene is

$$\begin{pmatrix} 1.0 & 0.5 & 0 & 0.5 \\ 0.5 & 1.0 & 0.5 & 0 \\ 0 & 0.5 & 1.0 & 0.5 \\ 0.5 & 0 & 0.5 & 1.0 \end{pmatrix}$$

Interpret this result.

Using the numbering system for carbon atoms shown in Problem 11–37, which is reproduced in the following figure,

the H matrix for butadiene is

$$H = \begin{pmatrix} \alpha & \beta & 0 & \beta \\ \beta & \alpha & \beta & 0 \\ 0 & \beta & \alpha & \beta \\ \beta & 0 & \beta & \alpha \end{pmatrix}$$

The eigenvalues for this matrix are derived in Problem 11–37, and the corresponding eigenfunctions can be determined using either the equations for c_1, c_2, c_3, and c_4 obtained from the algebraic equations that give rise to the secular determinant, or simply using a program such as *MathCad* or *Mathematica*. The normalized eigenvectors are arranged in columns to give the C matrix:

$$C = \begin{pmatrix} \frac{1}{2} & 0 & \frac{1}{\sqrt{2}} & \frac{1}{2} \\ \frac{1}{2} & \frac{1}{\sqrt{2}} & 0 & -\frac{1}{2} \\ \frac{1}{2} & 0 & -\frac{1}{\sqrt{2}} & \frac{1}{2} \\ \frac{1}{2} & -\frac{1}{\sqrt{2}} & 0 & -\frac{1}{2} \end{pmatrix}$$

Since $n_1 = 2$, $n_2 = n_3 = 1$, and $n_4 = 0$ (see Problem 11–37), the n matrix is

$$n = \begin{pmatrix} 2 & 0 & 0 & 0 \\ 0 & 1 & 0 & 0 \\ 0 & 0 & 1 & 0 \\ 0 & 0 & 0 & 0 \end{pmatrix}$$

The R matrix is

$$R = CnC^T$$

$$= \begin{pmatrix} \frac{1}{2} & 0 & \frac{1}{\sqrt{2}} & \frac{1}{2} \\ \frac{1}{2} & \frac{1}{\sqrt{2}} & 0 & -\frac{1}{2} \\ \frac{1}{2} & 0 & -\frac{1}{\sqrt{2}} & \frac{1}{2} \\ \frac{1}{2} & -\frac{1}{\sqrt{2}} & 0 & -\frac{1}{2} \end{pmatrix} \begin{pmatrix} 2 & 0 & 0 & 0 \\ 0 & 1 & 0 & 0 \\ 0 & 0 & 1 & 0 \\ 0 & 0 & 0 & 0 \end{pmatrix} \begin{pmatrix} \frac{1}{2} & \frac{1}{2} & \frac{1}{2} & \frac{1}{2} \\ 0 & \frac{1}{\sqrt{2}} & 0 & -\frac{1}{\sqrt{2}} \\ \frac{1}{\sqrt{2}} & 0 & -\frac{1}{\sqrt{2}} & 0 \\ \frac{1}{2} & -\frac{1}{2} & \frac{1}{2} & -\frac{1}{2} \end{pmatrix}$$

$$= \begin{pmatrix} 1.0 & 0.5 & 0 & 0.5 \\ 0.5 & 1.0 & 0.5 & 0 \\ 0 & 0.5 & 1.0 & 0.5 \\ 0.5 & 0 & 0.5 & 1.0 \end{pmatrix}$$

The diagonal elements of R are the fractional π electron charges on the carbon atoms. In the case of cyclobutadiene, they are all one. The off-diagonal elements of R corresponding to neighboring carbon atoms are π-bond orders. Thus, $p_{12}^{\pi} = p_{23}^{\pi} = p_{34}^{\pi} = p_{41}^{\pi} = 0.5$. Note that a bond order has physical meaning only between atoms for which there is a nonzero value of β. Thus, the values of the elements R_{13} and R_{24} have no physical meaning (they happen to be zero here).

11–57. Use a program such as *MathCad* or *Mathematica* to show that the R matrix (Equation 11.41) for bicyclobutadiene is

$$\begin{pmatrix} 0.621268 & 0.485071 & 0.485071 & 0.621268 \\ 0.485071 & 1.37873 & -0.621268 & 0.485071 \\ 0.485071 & -0.621268 & 1.37873 & 0.485071 \\ 0.621268 & 0.485071 & 0.485071 & 0.621268 \end{pmatrix}$$

Interpret this result.

Using the numbering system for carbon atoms shown in Problem 11–39, which is reproduced in the following figure,

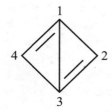

the H matrix for bicyclobutadiene is

$$H = \begin{pmatrix} \alpha & \beta & \beta & \beta \\ \beta & \alpha & \beta & 0 \\ \beta & \beta & \alpha & \beta \\ \beta & 0 & \beta & \alpha \end{pmatrix}$$

The eigenvalues for this matrix are derived in Problem 11–39, and the corresponding eigen-functions can be determined using either the equations for c_1, c_2, c_3, and c_4 obtained from the algebraic equations that give rise to the secular determinant, or simply using a program such as *MathCad* or *Mathematica*. The normalized eigenvectors are arranged in columns to give the C matrix:

$$C = \begin{pmatrix} 0.557\,345 & 0 & -0.707\,107 & -0.435\,162 \\ 0.435\,162 & -0.707\,107 & 0 & 0.557\,345 \\ 0.557\,345 & 0 & 0.707\,107 & -0.435\,162 \\ 0.435\,162 & 0.707\,107 & 0 & 0.557\,345 \end{pmatrix}$$

Since $n_1 = n_2 = 2$, $n_3 = n_4 = 0$ (see Problem 11–39), the n matrix is

$$n = \begin{pmatrix} 2 & 0 & 0 & 0 \\ 0 & 2 & 0 & 0 \\ 0 & 0 & 0 & 0 \\ 0 & 0 & 0 & 0 \end{pmatrix}$$

The R matrix is

$$R = CnC^T$$

$$= \begin{pmatrix} 0.557\,345 & 0 & -0.707\,107 & -0.435\,162 \\ 0.435\,162 & -0.707\,107 & 0 & 0.557\,345 \\ 0.557\,345 & 0 & 0.707\,107 & -0.435\,162 \\ 0.435\,162 & 0.707\,107 & 0 & 0.557\,345 \end{pmatrix} \begin{pmatrix} 2 & 0 & 0 & 0 \\ 0 & 2 & 0 & 0 \\ 0 & 0 & 0 & 0 \\ 0 & 0 & 0 & 0 \end{pmatrix}$$

$$\times \begin{pmatrix} 0.557\,345 & 0.435\,162 & 0.557\,345 & 0.435\,162 \\ 0 & -0.707\,107 & 0 & 0.707\,107 \\ -0.707\,107 & 0 & 0.707\,107 & 0 \\ -0.435\,162 & 0.557\,345 & -0.435\,162 & 0.557\,345 \end{pmatrix}$$

$$= \begin{pmatrix} 0.621\,268 & 0.485\,071 & 0.621\,268 & 0.485\,071 \\ 0.485\,071 & 1.378\,730 & 0.485\,071 & -0.621\,268 \\ 0.621\,268 & 0.485\,071 & 0.621\,268 & 0.485\,071 \\ 0.485\,071 & -0.621\,268 & 0.485\,071 & 1.378\,730 \end{pmatrix}$$

The diagonal elements of R are the fractional π electron charges on the carbon atoms. In the case of bicyclobutadiene, the π electron charges for carbons 1 and 3 are 0.621 268 while those for carbons 2 and 4 are 1.378 730. The off-diagonal elements of R corresponding to neighboring carbon atoms are π-bond orders. Thus, $p_{12}^{\pi} = p_{23}^{\pi} = p_{34}^{\pi} = p_{41}^{\pi} = 0.485\,071$ and $p_{13}^{\pi} = 0.621\,268$. Note that a bond order has physical meaning only between atoms for which there is a nonzero value of β. Thus, the value of the element R_{24} has no physical meaning.

The R matrix derived here is the same as given in the problem, except the numbering systems are different: there is a switch of carbons 3 and 4 between the two.

CHAPTER 12

The Hartree–Fock–Roothaan Method

PROBLEMS AND SOLUTIONS

12–1. Derive Equations 12.3 through 12.6.

This problem is very similar to that done for the atomic case in Problem 9–28. Because of its determinantal nature, $\Psi(1, 2, \ldots, 2N)$ can be written as

$$\Psi(1, 2, \ldots, 2N) = \frac{1}{\sqrt{(2N)!}} \sum_P \epsilon_P u_1(P1)u_2(P2) \ldots u_{2N}(P2N) \tag{1}$$

where the u_i are the spin orbitals, $u_1 = \psi_1\alpha(1)$, $u_2 = \psi_1\beta(1)$, $u_3 = \psi_2\alpha(2)$, etc., and where P represents a permutation of the numbers $(1, 2, \ldots, 2N)$, $\epsilon_P = +1$ if the permutation $(P1, P2, \ldots, P2N)$ differs from $(1, 2, \ldots, 2N)$ by an even number of interchanges of pairs of numbers, and $\epsilon_P = -1$ if the permutation $(P1, P2, \ldots, P2N)$ differs from $(1, 2, \ldots, 2N)$ by an odd number of interchanges. An equivalent way of writing Equation 1 is

$$\Psi(1, 2, \ldots, 2N) = \frac{1}{\sqrt{(2N)!}} \sum_P \epsilon_P u_{P1}(1)u_{P2}(2) \ldots u_{P2N}(2N) \tag{2}$$

In other words, the labels of the spin-orbitals are interchanged, rather than those of the electrons. Both expressions for Ψ will be used in this problem. First show that Ψ is normalized. Using Equation 1 gives

$$\int \cdots \int d\tau_1 d\tau_2 \ldots d\tau_{2N} \, \Psi^*\Psi = \frac{1}{(2N)!} \sum_P \epsilon_P \sum_Q \epsilon_Q \int \cdots \int d\tau_1 d\tau_2 \ldots d\tau_{2N} \, u_1^*(Q1)$$

$$\times u_2^*(Q2) \ldots u_{2N}^*(Q2N)u_1(P1)u_2(P2) \ldots u_{2N}(P2N)$$

These integrals will vanish unless $Q1 = P1$, $Q2 = P2$, and, in general, $Qx = Px$, where x can have any integer value between 1 and $2N$. This means that the only permutation Q that will lead to a non-zero result is when $Q = P$. This leaves

$$\int \cdots \int d\tau_1 d\tau_2 \ldots d\tau_{2N} \, \Psi^*\Psi = \frac{1}{(2N)!} \sum_P \epsilon_P^2 \int \cdots \int d\tau_1 d\tau_2 \ldots d\tau_{2N}$$

$$\times u_1^*(P1)u_1(P1)u_2^*(P2)u_2(P2) \ldots u_{2N}^*(P2N)u_{2N}(P2N)$$

where $\epsilon_P^2 = 1$. Because the integration variables can be relabeled, all these integrals are equal, and equal to unity because the individual spin-orbitals are normalized. There are $(2N)!$ permutations,

613

and so

$$\int \cdots \int d\tau_1 \, d\tau_2 \ldots d\tau_{2N} \, \Psi^* \Psi = \frac{1}{(2N)!}(2N)! = 1$$

Now consider integrals of the form

$$A_1 = \int \cdots \int d\tau_1 \, d\tau_2 \ldots d\tau_{2N} \, \Psi^* \left[\sum_{j=1}^{2N} \hat{h}_j \right] \Psi$$

where $\hat{h}_j$ from Equation 12.4 is a one-electron operator, meaning that it operates only on the coordinates of electron j. This operator can be expressed as

$$\sum_{j=1}^{2N} \hat{h}_j = \sum_{j=1}^{2N} \left[-\frac{1}{2}\nabla_j^2 - \sum_{A=1}^{M} \frac{Z_A}{r_{jA}} \right]$$

Using Equation 1 for Ψ, this becomes

$$A_1 = \frac{1}{(2N)!} \sum_P \epsilon_P \sum_Q \epsilon_Q \int \cdots \int d\tau_1 \ldots d\tau_{2N} \, u_1^*(Q1)u_2^*(Q2)$$

$$\times \ldots u_{2N}^*(Q2N) \left[\sum_{j=1}^{2N} \hat{h}_j \right] u_1(P1)u_2(P2) \ldots u_{2N}(P2N)$$

As before, all $(2N)!$ permutations lead to the same integral, and so

$$A_1 = \sum_{j=1}^{2N} \int u_j^*(j)\hat{h}_j u_j(j) \, d\tau_j$$

Finally, consider integrals of the type

$$A_2 = \int \cdots \int d\tau_1 \ldots d\tau_{2N} \, \Psi^* \left[\sum_{i=1}^{2N} \sum_{j>i}^{2N} \frac{1}{r_{ij}} \right] \Psi$$

Using Equation 2 for Ψ, write

$$A_2 = \frac{1}{(2N)!} \sum_P \epsilon_P \sum_Q \epsilon_Q \int \cdots \int d\tau_1 \ldots d\tau_{2N} \, u_{Q1}^*(1)u_{Q2}^*(2) \ldots u_{Q2N}^*(2N)$$

$$\times \left[\sum_{i=1}^{2N} \sum_{j>i}^{2N} \frac{1}{r_{ij}} \right] u_{P1}(1)u_{P2}(2) \ldots u_{P2N}(2N)$$

These integrals will vanish unless the subscripts of all the spin-orbitals *except* those for electrons i and j match. There will be two non-vanishing terms in the summation of all Q permutations:

$$A_2 = \frac{1}{(2N)!} \sum_P \left\{ \epsilon_P \sum_{i=1}^{2N} \sum_{j>i}^{2N} \iint d\tau_i\, d\tau_j\, u_{P1}^*(i) u_{P2}^*(j) \frac{1}{r_{ij}} u_{P1}(i) u_{P2}(j) \right.$$

$$\left. -\epsilon_P \sum_{i=1}^{2N} \sum_{j>i}^{2N} \iint d\tau_i\, d\tau_j\, u_{P2}^*(i) u_{P1}^*(j) \frac{1}{r_{ij}} u_{P1}(i) u_{P2}(j) \right\}$$

Note that $Q = P$ for the term with the positive sign and that Q and P differ by the interchange of one pair of subscripts for the term with the negative sign. As before, all $(2N)!$ permutations lead to the same quantities, and so

$$A_2 = \sum_{i=1}^{2N} \sum_{j>i}^{2N} \iint d\tau_1\, d\tau_2\, u_i^*(1) u_j^*(2) \frac{1}{r_{12}} u_i(1) u_j(2)$$

$$- \sum_{i=1}^{2N} \sum_{j>i}^{2N} \iint d\tau_1\, d\tau_2 u_i^*(1) u_j^*(2) \frac{1}{r_{12}} u_i(2) u_j(1)$$

$$= \sum_{i=1}^{2N} \sum_{j>i}^{2N} J_{ij}' - \sum_{i=1}^{2N} \sum_{j>i}^{2N} K_{ij}'$$

The integrals here involve spin orbitals, while the integrals expressed by Equations 12.4 through 12.6 involve only spatial orbitals. Combining the results for A_1 and A_2 and substituting into Equation 12.2 gives

$$E = \int \cdots \int d\tau_1 \ldots d\tau_{2N}\, \Psi^* \hat{H} \Psi$$

$$= \int \cdots \int d\tau_1\, d\tau_2\, \ldots d\tau_{2N} \Psi^* \left[\sum_j^{2N} \hat{h}_j \right] \Psi$$

$$+ \int \cdots \int d\tau_1 \ldots d\tau_{2N}\, \Psi^* \left[\sum_{i=1}^{2N} \sum_{j>i}^{2N} \frac{1}{r_{ij}} \right] \Psi$$

$$= A_1 + A_2$$

$$= \sum_{j=1}^{2N} \int u_j^*(j) \hat{h}_j u_j(j) d\tau_j + \sum_{i=1}^{2N} \sum_{j>i}^{2N} J_{ij}' - \sum_{i=1}^{2N} \sum_{j>i}^{2N} K_{ij}'$$

$$= \sum_{j=1}^{2N} I_j' + \sum_{i=1}^{2N} \sum_{j>i}^{2N} J_{ij}' - \sum_{i=1}^{2N} \sum_{j>i}^{2N} K_{ij}'$$

Since the overall wave function has N doubly occupied orbitals, with the spatial parts of u_1 and u_2 identical and so on, then $I_1' = I_2'$, $I_3' = I_4'$, and, in general, $I_{2n-1}' = I_{2n}' = I_n$, where n goes from 1 to N. Furthermore,

$$
\begin{array}{cccccc}
J_{12}' &=& J_{11} & J_{13}' &=& J_{12} & J_{14}' &=& J_{12} \\
J_{23}' &=& J_{12} & J_{24}' &=& J_{22} &&&
\end{array}
$$

and so forth. Also,

$$K_{12}' = 0 \quad K_{13}' = K_{12} \quad K_{14}' = 0$$

and so forth, giving

$$E = 2\sum_{j=1}^{N} I_j + \sum_{i=1}^{N}\sum_{j=1}^{N}\left(2J_{ij} - K_{ij}\right)$$

where the summations are over spatial orbitals.

12–2. Show that the coulomb integral J_{ii} is equal to the exchange integral K_{ii}.

From Equation 12.5 with $i = j$,

$$J_{ii} = \iint d\mathbf{r}_1\, d\mathbf{r}_2\; \psi_i^*(\mathbf{r}_1)\psi_i^*(\mathbf{r}_2)\frac{1}{r_{12}}\psi_i(\mathbf{r}_1)\psi_i(\mathbf{r}_2)$$

and from Equation 12.6 with $i = j$,

$$K_{ii} = \iint d\mathbf{r}_1\, d\mathbf{r}_2\; \psi_i^*(\mathbf{r}_1)\psi_i^*(\mathbf{r}_2)\frac{1}{r_{12}}\psi_i(\mathbf{r}_2)\psi_i(\mathbf{r}_1)$$

$$= \iint d\mathbf{r}_1\, d\mathbf{r}_2\; \psi_i^*(\mathbf{r}_1)\psi_i^*(\mathbf{r}_2)\frac{1}{r_{12}}\psi_i(\mathbf{r}_1)\psi_i(\mathbf{r}_2)$$

$$= J_{ii}$$

12–3. Explain why the subscript i in Equation 12.7 runs from 1 to N even though it is describing a $2N$-electron molecule.

Equation 12.7 gives the N spatial orbitals of the determinantal wave function, each of which is filled by two electrons of opposite spin.

12–4. Show that the Hartree–Fock energy is not equal to the sum of the Hartree–Fock orbital energies.

The Hartree-Fock energy is, from Equation 12.3

$$E = 2\sum_{j=1}^{N} I_j + \sum_{i=1}^{N}\sum_{j=1}^{N}\left(2J_{ij} - K_{ij}\right)$$

The Hartree-Fock orbital energies are

$$\varepsilon_i = I_i + \sum_{j=1}^{N}\left(2J_{ij} - K_{ij}\right)$$

The sum of the energies of the N Hartree-Fock orbitals is

$$\sum_{i=1}^{N} \varepsilon_i = \sum_{i=1}^{N} I_i + \sum_{i=1}^{N} \sum_{j=1}^{N} \left(2J_{ij} - K_{ij}\right) \neq E$$

Since the N Hartree-Fock orbitals are doubly occupied, one might think that twice the sum of the orbital energies would give E, but this, too, is not the case,

$$\sum_{i=1}^{N} 2\varepsilon_i = 2\sum_{i=1}^{N} I_i + 2\sum_{i=1}^{N} \sum_{j=1}^{N} \left(2J_{ij} - K_{ij}\right) \neq E$$

Rather, as given in Equation 12.14,

$$E = \sum_{i=1}^{N} \left(I_i + \varepsilon_i\right) = \sum_{i=1}^{N} \varepsilon_i + \sum_{i=1}^{N} I_i$$

12–5. Write out the basis set orbital $\sigma_u 3s\,(7.291\,69)$ in the first column of Table 12.1.

The basis set orbital $\sigma_u 3s\,(7.291\,69)$ is the difference of two Slater $3s$ orbitals, each with orbital exponent $\zeta = 7.291\,69$. From Equation 9.15 with $Y_0^0(\theta, \phi) = 1/\sqrt{4\pi}$,

$$S_{3s}(r) = \left(\frac{2\zeta^7}{45\pi}\right)^{1/2} r^2 e^{-\zeta r}$$

then

$$\sigma_u 3s\,(7.291\,69) = N\left[S_{3s}(r_A, 7.291\,69) - S_{3s}(r_B, 7.291\,69)\right]$$

$$= N\left(\frac{2\zeta^7}{45\pi}\right)^{1/2} \left(r_A^2 e^{-\zeta r_A} - r_B^2 e^{-\zeta r_B}\right)$$

where N is a normalization constant and $\zeta = 7.291\,69$.

12–6. Write out the basis set orbital $\sigma_u 3d\,(1.690\,03)$ in the first column of Table 12.1.

The basis set orbital $\sigma_u 3d\,(1.690\,03)$ is the difference of two Slater $3d$ orbitals, each with orbital exponent $\zeta = 1.690\,03$. Since σ_{nd} orbitals result from the sums and differences of d_{z^2} atomic orbitals, Equation 9.15 gives

$$S_{3d_{z^2}}(r) = \left(\frac{8\zeta^7}{45}\right)^{1/2} r^2 e^{-\zeta r} Y_2^0(\theta, \phi)$$

then

$$\sigma_u 3d\,(1.690\,03) = N\left[S_{3d_{z^2}}(r_A, 1.690\,03) - S_{3d_{z^2}}(r_B, 1.690\,03)\right]$$

$$= N\left(\frac{8\zeta^7}{45}\right)^{1/2} \left[r_A^2 e^{-\zeta r_A} Y_2^0(\theta_A, \phi_A) - r_B^2 e^{-\zeta r_B} Y_2^0(\theta_B, \phi_B)\right]$$

where N is a normalization constant and $\zeta = 1.690\,03$.

12–7. Use the information in Table 12.1 to write out the $2\sigma_u$ molecular orbital in terms of the atomic orbitals.

Use the coefficients from the fourth column of Table 12.1 to give

$$
\begin{aligned}
2\sigma_u = &-0.243\,70\left[S_{1s}(r_A, 5.955\,34) - S_{1s}(r_B, 5.955\,34)\right] \\
&-0.000\,00\left[S_{1s}(r_A, 10.658\,79) - S_{1s}(r_B, 10.658\,79)\right] \\
&+0.364\,37\left[S_{2s}(r_A, 1.570\,44) - S_{2s}(r_B, 1.570\,44)\right] \\
&+0.547\,02\left[S_{2s}(r_A, 2.489\,65) - S_{2s}(r_B, 2.489\,65)\right] \\
&-0.030\,54\left[S_{3s}(r_A, 7.291\,69) - S_{3s}(r_B, 7.291\,69)\right] \\
&-0.413\,55\left[S_{2p_z}(\mathbf{r}_A, 1.485\,49) + S_{2p_z}(\mathbf{r}_B, 1.485\,49)\right] \\
&-0.109\,45\left[S_{2p_z}(\mathbf{r}_A, 3.499\,90) + S_{2p_z}(\mathbf{r}_B, 3.499\,90)\right] \\
&-0.035\,53\left[S_{3d_{z^2}}(\mathbf{r}_A, 1.690\,03) - S_{3d_{z^2}}(\mathbf{r}_B, 1.690\,03)\right]
\end{aligned}
$$

In this equation, $\mathbf{r}_A$ and $\mathbf{r}_B$ denote the three spherical coordinates of the electron in the nonspherical S_{2p_z} and $S_{3d_{z^2}}$ orbitals. Don't forget that σ_u symmetry is achieved by the difference of s or d_{z^2} atomic orbitals, but by the sum of p_z orbitals.

12–8. Show that the matrix elements of the Fock operator consist of integrals of the form

$$
(\mu\nu \mid \sigma\lambda) = \int d\mathbf{r}_1\, d\mathbf{r}_2\, \phi_\mu(\mathbf{r}_1)\phi_\nu(\mathbf{r}_1)\frac{1}{r_{12}}\phi_\sigma(\mathbf{r}_2)\phi_\lambda(\mathbf{r}_2)
$$

The matrix elements of the Fock operator are

$$
F_{\mu\nu} = \left\langle \phi_\mu \left| \hat{F} \right| \phi_\nu \right\rangle = \int d\mathbf{r}_1\, \phi_\mu(\mathbf{r}_1)\hat{F}(\mathbf{r}_1)\phi_\nu(\mathbf{r}_1)
$$

where the basis functions are assumed real. The Fock operator comprises one-electron operators (Equation 12.9), coulomb operators (Equation 12.10), and exchange operators (Equation 12.11). The matrix elements of the coulomb and exchange operators are responsible for the multicenter integrals. Consider the matrix elements of the coulomb operator,

$$\left\langle \phi_\mu \left| \hat{J}_j \right| \phi_\nu \right\rangle = \int d\mathbf{r}_1 \, \phi_\mu(\mathbf{r}_1) \hat{J}_j(\mathbf{r}_1) \phi_\nu(\mathbf{r}_1)$$

$$= \int d\mathbf{r}_1 \, \phi_\mu(\mathbf{r}_1) \phi_\nu(\mathbf{r}_1) \int d\mathbf{r}_2 \, \psi_j^*(\mathbf{r}_2) \frac{1}{r_{12}} \psi_j(\mathbf{r}_2)$$

$$= \sum_{\sigma=1}^{K} \sum_{\lambda=1}^{K} c_\sigma c_\lambda \int \int d\mathbf{r}_1 \, d\mathbf{r}_2 \, \phi_\mu(\mathbf{r}_1) \phi_\nu(\mathbf{r}_1) \frac{1}{r_{12}} \phi_\sigma(\mathbf{r}_2) \phi_\lambda(\mathbf{r}_2)$$

$$= \sum_{\sigma=1}^{K} \sum_{\lambda=1}^{K} c_\sigma c_\lambda \, (\mu\nu \,|\, \sigma\lambda)$$

The matrix elements of the exchange operator are quite similar.

$$\left\langle \phi_\mu \left| \hat{K}_j \right| \phi_\nu \right\rangle = \int d\mathbf{r}_1 \, \phi_\mu(\mathbf{r}_1) \hat{K}_j(\mathbf{r}_1) \phi_\nu(\mathbf{r}_1)$$

$$= \int d\mathbf{r}_1 \, \phi_\mu(\mathbf{r}_1) \psi_j(\mathbf{r}_1) \int d\mathbf{r}_2 \, \psi_j^*(\mathbf{r}_2) \frac{1}{r_{12}} \phi_\nu(\mathbf{r}_2)$$

$$= \sum_{\sigma=1}^{K} \sum_{\lambda=1}^{K} c_\sigma c_\lambda \int \int d\mathbf{r}_1 \, d\mathbf{r}_2 \, \phi_\mu(\mathbf{r}_1) \phi_\sigma(\mathbf{r}_1) \frac{1}{r_{12}} \phi_\lambda(\mathbf{r}_2) \phi_\nu(\mathbf{r}_2)$$

$$= \sum_{\sigma=1}^{K} \sum_{\lambda=1}^{K} c_\sigma c_\lambda \, (\mu\sigma \,|\, \lambda\nu)$$

Thus, both of these two-electron operators lead to multicenter integrals of the given form.

12–9. Show that the Gaussian functions given by Equation 12.21 are normalized.

Using $r^2 = x^2 + y^2 + z^2$ and the two integrals (from a table of integrals, or see Problems 12–12 and 12–13)

$$\int_{-\infty}^{\infty} e^{-au^2} \, du = \left(\frac{\pi}{a} \right)^{1/2}$$

$$\int_{-\infty}^{\infty} u^2 e^{-au^2} \, du = \frac{1}{2a} \left(\frac{\pi}{a} \right)^{1/2}$$

the normalization integrals for the Gaussian functions are straightforwardly evaluated in Cartesian coordinates. For $g_s(r, \alpha)$,

$$\int d\mathbf{r} \, g_s^2(r, \alpha) = \left(\frac{2\alpha}{\pi} \right)^{3/2} \int_{-\infty}^{\infty} dx \int_{-\infty}^{\infty} dy \int_{-\infty}^{\infty} dz \, e^{-2\alpha r^2}$$

$$= \left(\frac{2\alpha}{\pi} \right)^{3/2} \int_{-\infty}^{\infty} dx \, e^{-2\alpha x^2} \int_{-\infty}^{\infty} dy \, e^{-2\alpha y^2} \int_{-\infty}^{\infty} dz \, e^{-2\alpha z^2}$$

$$= \left(\frac{2\alpha}{\pi} \right)^{3/2} \left[\left(\frac{\pi}{2\alpha} \right)^{1/2} \right]^3$$

$$= 1$$

Consequently, $g_s(r, \alpha)$ is normalized. For $g_x(r, x, \alpha)$,

$$\int d\mathbf{r}\, g_x^2(r, x, \alpha) = \left(\frac{128\alpha^5}{\pi^3}\right)^{1/2} \int_{-\infty}^{\infty} dx \int_{-\infty}^{\infty} dy \int_{-\infty}^{\infty} dz\, x^2 e^{-2\alpha r^2}$$

$$= \left(\frac{128\alpha^5}{\pi^3}\right)^{1/2} \int_{-\infty}^{\infty} dx\, x^2 e^{-2\alpha x^2} \int_{-\infty}^{\infty} dy\, e^{-2\alpha y^2} \int_{-\infty}^{\infty} dz\, e^{-2\alpha z^2}$$

$$= \left(\frac{128\alpha^5}{\pi^3}\right)^{1/2} \left(\frac{1}{4\alpha}\right) \left[\left(\frac{\pi}{2\alpha}\right)^{1/2}\right]^3$$

$$= 1$$

and this function, too, is normalized. The integrals for $g_y(r, y, \alpha)$ and $g_z(r, z, \alpha)$ are completely analogous to that for $g_x(r, x, \alpha)$, and these two functions are also normalized.

12–10. Show that a three-dimensional Gaussian function centered at $\mathbf{r}_0 = x_0\mathbf{i} + y_0\mathbf{j} + z_0\mathbf{k}$ is a product of three one-dimensional Gaussian functions centered on x_0, y_0, and z_0.

A three-dimensional Gaussian function centered at $\mathbf{r}_0$ is $e^{-a(\mathbf{r}-\mathbf{r}_0)^2}$. In squaring the vector quantity in the exponent, remember that the dot product of a unit vector with itself is one, and that the dot product of two different, orthogonal unit vectors is zero.

$$e^{-a(\mathbf{r}-\mathbf{r}_0)^2} = e^{-a[(x-x_0)\mathbf{i}+(y-y_0)\mathbf{j}+(z-z_0)\mathbf{k}]^2}$$

$$= e^{-a[(x-x_0)^2+(y-y_0)^2+(z-z_0)^2]}$$

$$= e^{-a(x-x_0)^2} e^{-a(y-y_0)^2} e^{-a(z-z_0)^2}$$

which is the product of three one-dimensional Gaussian functions centered on x_0, y_0, and z_0.

12–11. Show that

$$\int_{-\infty}^{\infty} e^{-(x-x_0)^2} dx = \int_{-\infty}^{\infty} e^{-x^2} dx = 2\int_0^{\infty} e^{-x^2} dx = \pi^{1/2}$$

For the first equality, let $x' = x - x_0$, then $dx' = d(x - x_0) = dx$, and

$$\int_{-\infty}^{\infty} e^{-(x-x_0)^2} dx = \int_{-\infty}^{\infty} e^{-x'^2} dx' = \int_{-\infty}^{\infty} e^{-x^2} dx$$

The second equality is true because a Gaussian function is even. Thus,

$$2\int_0^{\infty} e^{-x^2} dx = 2\left(\frac{\pi}{4}\right)^{1/2} = \pi^{1/2}$$

12–12. The Gaussian integral

$$I_0 = \int_0^\infty e^{-ax^2} dx$$

can be evaluated by a trick. First write

$$I_0^2 = \int_0^\infty dx\, e^{-ax^2} \int_0^\infty dy\, e^{-ay^2} = \int_0^\infty \int_0^\infty dx\, dy\, e^{-a(x^2+y^2)}$$

Now convert the integration variables from cartesian coordinates to polar coordinates and show that

$$I_0 = \frac{1}{2}\left(\frac{\pi}{a}\right)^{1/2}$$

First, write out the square of I_0 using the two different variables of integration suggested.

$$I_0^2 = \left(\int_0^\infty e^{-ax^2} dx\right)\left(\int_0^\infty e^{-ay^2} dy\right) = \int_0^\infty \int_0^\infty dx\, dy\, e^{-a(x^2+y^2)}$$

The integration is over the entire first quadrant, $0 \le x \le \infty$ and $0 \le y \le \infty$ so that in polar coordinates, the limits of integration are $0 \le r < \infty$ and $0 \le \theta \le \pi/2$. Using the substitutions $r^2 = x^2 + y^2$ and $dx\, dy = r\, dr\, d\theta$, the integral becomes

$$I_0^2 = \int_0^{\pi/2} d\theta \int_0^\infty dr\, r e^{-ar^2}$$

$$= \left(\frac{\pi}{2}\right)\left(\frac{1}{2a}\right) = \frac{\pi}{4a}$$

Taking the square root of both sides of the equation leads directly to the desired result,

$$I_0 = \frac{1}{2}\left(\frac{\pi}{a}\right)^{1/2}$$

12–13. Show that the integral

$$I_{2n} = \int_0^\infty x^{2n} e^{-ax^2} dx$$

can be obtained from I_0 in the previous problem by differentiating n times with respect to a. Show that

$$I_{2n} = \frac{1 \cdot 3 \cdot 5 \cdots (2n-1)}{2(2a)^n}\left(\frac{\pi}{a}\right)^{1/2}$$

Differentiate I_0 from the previous problem with respect to a:

$$\frac{dI_0}{da} = -\int_{-\infty}^{\infty} x^2 e^{-ax^2}\, dx = -I_2$$

$$\frac{d^2 I_0}{da^2} = \int_{-\infty}^{\infty} x^4 e^{-ax^2}\, dx = I_4$$

This can be repeated, and the pattern is obvious, although the general result could be proved via induction if desired.

$$\frac{d^n I_0}{da^n} = (-1)^n \int_{-\infty}^{\infty} x^{2n} e^{-ax^2}\, dx = (-1)^n I_{2n}$$

Starting with

$$I_0 = \frac{1}{2}\left(\frac{\pi}{a}\right)^{1/2}$$

differentiation with respect to a and application of the formula gives

$$I_2 = \frac{1}{2(2a)}\left(\frac{\pi}{a}\right)^{1/2}$$

$$I_4 = \frac{3}{2(2a)^2}\left(\frac{\pi}{a}\right)^{1/2}$$

Each additional differentiation results in a factor of $(2n-1)/2$ and an increment in the power of $1/a$, so that the general result is

$$I_{2n} = \frac{1\cdot 3\cdot 5\cdots(2n-1)}{2(2a)^n}\left(\frac{\pi}{a}\right)^{1/2}$$

12–14. Show that the product of a (not normalized) Gaussian function centered at $\mathbf{R}_A$ and one centered at $\mathbf{R}_B$—that is,

$$\phi_1 = e^{-\alpha|\mathbf{r}-\mathbf{R}_A|^2} \qquad \text{and} \qquad \phi_2 = e^{-\beta|\mathbf{r}-\mathbf{R}_B|^2}$$

is a Gaussian function centered at

$$\mathbf{R}_p = \frac{\alpha\mathbf{R}_A + \beta\mathbf{R}_B}{\alpha + \beta}$$

For simplicity, work in one dimension and appeal to Problem 12–10 for the three-dimensional proof.

In one dimension, these equations become

$$\phi_1 = e^{-\alpha(x-x_A)^2} \qquad \text{and} \qquad \phi_2 = e^{-\beta(x-x_B)^2}$$

and the Gaussian function is centered at

$$x_p = \frac{\alpha x_A + \beta x_B}{\alpha + \beta}$$

The easiest way to do this problem is to show that $\phi_1 \phi_2$ can be written as a Gaussian function centered at x_p. The product $\phi_1 \phi_2$ is

$$\phi_1 \phi_2 = \exp\left[-\alpha(x - x_A)^2 - \beta(x - x_B)^2\right]$$

$$= \exp\left(-\alpha x^2 + 2\alpha x x_A - \alpha x_A^2 - \beta x^2 + 2\beta x x_B - \beta x_B^2\right)$$

$$= \exp\left[-(\alpha x_A^2 + \beta x_B^2)\right] \exp\left[-(\alpha + \beta)x^2 + 2x(\alpha x_A + \beta x_B)\right]$$

$$= \exp\left[-(\alpha x_A^2 + \beta x_B^2)\right] \exp\left[(\alpha + \beta)\left(\frac{\alpha x_A + \beta x_B}{\alpha + \beta}\right)^2\right]$$

$$\times \exp\left\{-(\alpha + \beta)\left[x^2 - 2x\frac{\alpha x_A + \beta x_B}{\alpha + \beta} + \left(\frac{\alpha x_A + \beta x_B}{\alpha + \beta}\right)^2\right]\right\}$$

$$= \exp\left[-\frac{\alpha\beta(x_A - x_B)^2}{\alpha + \beta}\right] \exp\left[-(\alpha + \beta)(x - x_p)^2\right]$$

This is a Gaussian function centered at x_p. This could be repeated to find two additional, similar one-dimensional functions in y and z centered at y_p and z_p; the product of these three one-dimensional functions would be the three-dimensional function sought (as shown in Problem 12–10).

12–15. Show explicitly that if

$$\phi_s(\alpha, \mathbf{r} - \mathbf{R}_A) = \left(\frac{2\alpha}{\pi}\right)^{3/4} e^{-\alpha|\mathbf{r} - \mathbf{R}_A|^2}$$

and

$$\phi_s(\beta, \mathbf{r} - \mathbf{R}_B) = \left(\frac{2\beta}{\pi}\right)^{3/4} e^{-\beta|\mathbf{r} - \mathbf{R}_B|^2}$$

are normalized Gaussian s functions, then

$$\phi_s(\alpha, \mathbf{r} - \mathbf{R}_A)\phi_s(\beta, \mathbf{r} - \mathbf{R}_B) = K_{AB}\phi_{1s}(p, \mathbf{r} - \mathbf{R}_p)$$

where $p = \alpha + \beta$, $\mathbf{R}_p = (\alpha\mathbf{R}_A + \beta\mathbf{R}_B)/(\alpha + \beta)$ (see the previous problem), and

$$K_{AB} = \left[\frac{2\alpha\beta}{(\alpha + \beta)\pi}\right]^{3/4} e^{-\frac{\alpha\beta}{\alpha+\beta}|\mathbf{R}_A - \mathbf{R}_B|^2}$$

Using the result of Problem 12–14,

$$\phi_1(\alpha, \mathbf{r} - \mathbf{R}_A)\phi_2(\beta, \mathbf{r} - \mathbf{R}_B) = \left(\frac{4\alpha\beta}{\pi^2}\right)^{3/4} e^{-\alpha\beta|\mathbf{R}_A-\mathbf{R}_B|^2/(\alpha+\beta)} e^{-(\alpha+\beta)(\mathbf{r}-\mathbf{R}_p)^2} \qquad (1)$$

The normalization constant of $e^{-(\alpha+\beta)(\mathbf{r}-\mathbf{R}_p)^2}$ is A, where

$$1 = A^2 \int_{-\infty}^{\infty} dx \int_{-\infty}^{\infty} dy \int_{-\infty}^{\infty} dz \left\{e^{-(\alpha+\beta)[(x-x_p)^2+(y-y_p)^2+(z-z_p)^2]}\right\}^2$$

$$= A^2 \left[2\int_0^{\infty} e^{-2(\alpha+\beta)u^2} du\right]^3 = A^2 \left[\frac{\pi}{2(\alpha+\beta)}\right]^{3/2}$$

$$A = \left[\frac{2(\alpha+\beta)}{\pi}\right]^{3/4}$$

Therefore, the normalized Gaussian function centered at $\mathbf{R}_p = (\alpha\mathbf{R}_A + \beta\mathbf{R}_B)/(\alpha + \beta)$ is

$$\phi_{1s}(\alpha + \beta, \mathbf{r} - \mathbf{R}_p) = \left[\frac{2(\alpha+\beta)}{\pi}\right]^{3/4} e^{-(\alpha+\beta)(\mathbf{r}-\mathbf{R}_p)^2}$$

Substituting into Equation 1 gives

$$\phi_{1s}(\alpha, \mathbf{r} - \mathbf{R}_A)\phi_{1s}(\beta, \mathbf{r} - \mathbf{R}_B) = \left(\frac{4\alpha\beta}{\pi^2}\right)^{3/4} \left[\frac{\pi}{2(\alpha+\beta)}\right]^{3/4} e^{-\alpha\beta|\mathbf{R}_A-\mathbf{R}_B|^2/(\alpha+\beta)} \phi_{1s}(p, \mathbf{r} - \mathbf{R}_p)$$

$$= \left[\frac{2\alpha\beta}{(\alpha+\beta)\pi}\right]^{3/4} e^{-\alpha\beta|\mathbf{R}_A-\mathbf{R}_B|^2/(\alpha+\beta)} \phi_{1s}(p, \mathbf{r} - \mathbf{R}_p)$$

$$= K_{AB}\phi_{1s}(p, \mathbf{r} - \mathbf{R}_p)$$

where

$$K_{AB} = \left[\frac{2\alpha\beta}{(\alpha+\beta)\pi}\right]^{3/4} e^{-\frac{\alpha\beta}{\alpha+\beta}|\mathbf{R}_A-\mathbf{R}_B|^2}$$

12–16. Plot the product of the two Gaussian functions $\phi_1 = e^{-2(x-1)^2}$ and $\phi_2 = e^{-3(x-2)^2}$. Interpret the result.

The two functions and their product (shown by the solid line in the figure) are shown below.

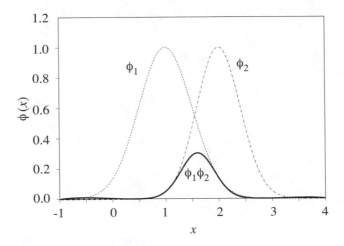

The product is a Gaussian function centered at $x = (2 \cdot 1 + 3 \cdot 2)/(2 + 3) = 8/5$.

12–17. Using the result of Problem 12–15, show that the overlap integral of the two normalized Gaussian functions

$$\phi_s = \left(\frac{2\alpha}{\pi}\right)^{3/4} e^{-\alpha|\mathbf{r}-\mathbf{R}_A|^2} \qquad \text{and} \qquad \phi_s = \left(\frac{2\beta}{\pi}\right)^{3/4} e^{-\beta|\mathbf{r}-\mathbf{R}_B|^2}$$

is

$$S(|\mathbf{R}_A - \mathbf{R}_B|) = \left(\frac{4\alpha\beta}{\pi^2}\right)^{3/4} \left(\frac{\pi}{\alpha+\beta}\right)^{3/2} e^{-\frac{\alpha\beta|\mathbf{R}_A-\mathbf{R}_B|^2}{\alpha+\beta}}$$

Plot this result as a function of $|\mathbf{R}_A - \mathbf{R}_B|$.

Using the result of Problem 12–15, the product of the two Gaussian functions given in this problem is written as a single Gaussian function.

$$\phi_s(\alpha, \mathbf{r} - \mathbf{R}_A)\phi_s(\beta, \mathbf{r} - \mathbf{R}_B) = \left(\frac{2\alpha}{\pi}\right)^{3/4} e^{-\alpha|\mathbf{r}-\mathbf{R}_A|^2} \left(\frac{2\beta}{\pi}\right)^{3/4} e^{-\beta|\mathbf{r}-\mathbf{R}_B|^2}$$

$$= K_{AB}\phi_s(\alpha + \beta, \mathbf{r} - \mathbf{R}_p)$$

$$= \left[\frac{2\alpha\beta}{(\alpha+\beta)\,\pi}\right]^{3/4} e^{-\frac{\alpha\beta}{\alpha+\beta}|\mathbf{R}_A-\mathbf{R}_B|^2} \left[\frac{2(\alpha+\beta)}{\pi}\right]^{3/4} e^{-(\alpha+\beta)(\mathbf{r}-\mathbf{R}_p)^2}$$

$$= \left(\frac{4\alpha\beta}{\pi^2}\right)^{3/4} e^{-\frac{\alpha\beta}{\alpha+\beta}|\mathbf{R}_A-\mathbf{R}_B|^2} e^{-(\alpha+\beta)(\mathbf{r}-\mathbf{R}_p)^2}$$

where $\mathbf{R}_p$ is as defined in Problem 12–15. Thus,

$$S(|\mathbf{R}_A - \mathbf{R}_B|^2) = \int d\mathbf{r}\, \phi_{1s}(\alpha, \mathbf{r} - \mathbf{R}_A)\phi_{1s}(\beta, \mathbf{r} - \mathbf{R}_B)$$

$$= \left(\frac{4\alpha\beta}{\pi^2}\right)^{3/4} e^{-\frac{\alpha\beta}{\alpha+\beta}|\mathbf{R}_A-\mathbf{R}_B|^2} \int d\mathbf{r}\, e^{-(\alpha+\beta)(\mathbf{r}-\mathbf{R}_p)^2}$$

The integral is given by (see Problems 12–10 and 12–11)

$$\int_{-\infty}^{\infty}\int_{-\infty}^{\infty}\int_{-\infty}^{\infty} dx\,dy\,dz\, e^{-(\alpha+\beta)(x^2+y^2+z^2)} = \left[2\int_{0}^{\infty} e^{-(\alpha+\beta)u^2}du\right]^3 = \left(\frac{\pi}{\alpha+\beta}\right)^{3/2}$$

so

$$S = \left(\frac{4\alpha\beta}{\pi^2}\right)^{3/4}\left(\frac{\pi}{\alpha+\beta}\right)^{3/2} e^{-\frac{\alpha\beta}{\alpha+\beta}|\mathbf{R}_A-\mathbf{R}_B|^2}$$

$$= \left[\frac{4\alpha\beta}{(\alpha+\beta)^2}\right]^{3/4} e^{-\frac{\alpha\beta}{\alpha+\beta}|\mathbf{R}_A-\mathbf{R}_B|^2}$$

The overlap function $S(|\mathbf{R}_A - \mathbf{R}_B|)$ decays as a Gaussian function of the distance between the two centers.

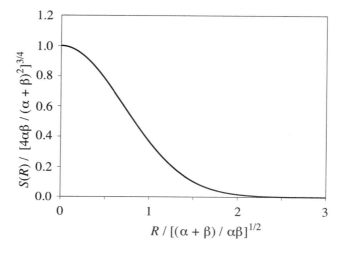

12–18. One criterion for determining the best possible "fit" of a Gaussian function to a Slater orbital is to minimize the integral of the square of their difference. For example, we can find the optimal value of α in $g_s(r, \alpha)$ by minimizing

$$I = \int d\mathbf{r}\,[S_{1s}(r,\,1.00) - g_s(r,\,\alpha)]^2$$

with respect to α. If the two functions $S_{1s}(r,\,1.00)$ and $g_s(r,\,\alpha)$ are normalized, show that minimizing I is equivalent to maximizing the overlap integral of $S_{1s}(r,\,1.00)$ and $g_s(r,\,\alpha)$:

$$S(\alpha) = \int d\mathbf{r}\,S_{1s}(r,\,1.00)g_s(r,\,\alpha)$$

Start by expanding the square in the integrand, and then use the fact that both S_{1s} and g_s are normalized.

$$I = \int d\mathbf{r} \left[S_{1s}(r, 1.00) - g_s(r, \alpha) \right]^2$$

$$= \int d\mathbf{r} \left[S_{1s}(r, 1.00) \right]^2 - 2 \int d\mathbf{r} \, S_{1s}(r, 1.00) g_s(r, \alpha) + \int d\mathbf{r} \left[g_s(r, \alpha) \right]^2$$

$$= 2 - 2 \int d\mathbf{r} \, S_{1s}(r, 1.00) g_s(r, \alpha) = 2 - 2S(\alpha)$$

Minimizing I is then equivalent to maximizing $S(\alpha)$.

12–19. Show that $S(\alpha)$ in the previous problem is given by

$$S(\alpha) = 4\pi^{1/2} \left(\frac{2\alpha}{\pi} \right)^{3/4} \int_0^\infty r^2 e^{-r} e^{-\alpha r^2} dr$$

Using a program such as *Mathematica* or *MathCad*, show that the maximum occurs at $\alpha = 0.270\ 95$.

Substitute Equations 12.22 and 12.23 into the expression for $S(\alpha)$:

$$S(\alpha) = \int d\mathbf{r} \, S_{1s}(r, 1.00) g_s(r, \alpha)$$

$$= \int_0^{2\pi} d\phi \int_0^\pi d\theta \, \sin\theta \int_0^\infty dr \, r^2 \left(\frac{1}{\pi} \right)^{1/2} e^{-r} \left(\frac{2\alpha}{\pi} \right)^{3/4} e^{-\alpha r^2}$$

$$= 4\pi^{1/2} \left(\frac{2\alpha}{\pi} \right)^{3/4} \int_0^\infty r^2 e^{-r} e^{-\alpha r^2} dr$$

This integral can be entered into a program such as *Mathematica* or *MathCad*, which is then able to find the maximum with respect to α (in most cases with a single command). When this is done, the maximum is found at $\alpha = 0.270\ 95$. A graph of $S(\alpha)$ versus α that shows the behavior of $S(\alpha)$ near the maximum is presented below.

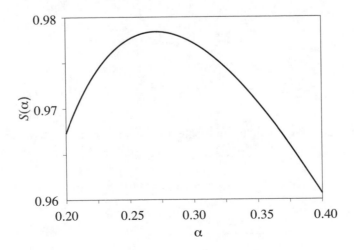

12–20.   Compare $S_{1s}(r, 1.00)$ and $g_s(r, 0.270\,95)$ graphically by plotting them on the same graph.

From Equations 12.22 and 12.23,

$$S_{1s}(r,\ 1.00) = \left(\frac{1}{\pi}\right)^{1/2} e^{-r}$$

$$g_s(r,\ 0.270\,95) = \left[\frac{2\,(0.270\,95)}{\pi}\right]^{3/4} e^{-0.270\,95r^2}$$

The two functions are shown in the graph below.

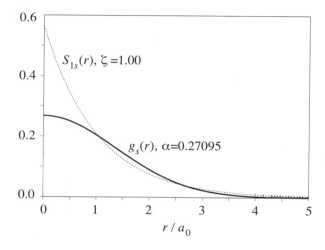

Notice the significant differences at small values of r, as well as the fact that the Slater orbital is larger than the Gaussian orbital for very large values of r.

12–21.   To show how to determine a fit of Gaussian functions to a Slater orbital $S_{1s}(\zeta \neq 1)$ when we know the fit to $S_{1s}(\zeta = 1)$, start with the overlap integral of $S_{1s}(r, \zeta)$ and $g_s(r, \beta)$:

$$S = 4\pi^{1/2}\left(\frac{2\beta}{\pi}\right)^{3/4} \zeta^{3/2} \int_0^\infty r^2 e^{-\zeta r} e^{-\beta r^2} dr$$

Now let $u = \zeta r$ to get

$$S = 4\pi^{1/2}\left(\frac{2\beta/\zeta^2}{\pi}\right)^{3/4} \int_0^\infty u^2 e^{-u} e^{-(\beta/\zeta^2)u^2} du$$

Compare this result for S with that in Problem 12–19 to show that $\beta = \alpha\zeta^2$ or, in more detailed notation,

$$\alpha(\zeta) = \alpha(\zeta = 1.00) \times \zeta^2$$

If $u = \zeta r$, then $r = u/\zeta$ and $dr = du/\zeta$. Thus,

$$S = 4\pi^{1/2} \left(\frac{2\beta}{\pi}\right)^{3/4} \zeta^{3/2} \int_0^\infty r^2 e^{-\zeta r} e^{-\beta r^2} dr$$

$$= 4\pi^{1/2} \left(\frac{2\beta}{\pi}\right)^{3/4} \zeta^{3/2} \int_0^\infty \frac{u^2}{\zeta^2} e^{-u} e^{-\beta u^2/\zeta^2} \frac{du}{\zeta}$$

$$= 4\pi^{1/2} \left(\frac{2\beta/\zeta^2}{\pi}\right)^{3/4} \int_0^\infty u^2 e^{-u} e^{-(\beta/\zeta^2)u^2} du$$

This result is equivalent to that of Problem 12–19 if

$$\frac{\beta}{\zeta^2} = \alpha$$

$$\beta = \alpha\zeta^2$$

Therefore,

$$\alpha(\zeta) = \alpha(\zeta = 1.00) \times \zeta^2$$

12–22. What is the difference between a double-zeta (DZ) basis set and a double split-valence basis set?

In a double-zeta (DZ) basis set, all atomic orbitals are represented by sums of two Slater orbitals with different values of ζ; in a double split-valence basis set, only the valence atomic orbitals are represented by sums of two Slater orbitals with different values of ζ.

12–23. What is meant by a triple-zeta basis set?

A triple zeta basis set is one in which each atomic orbital is expressed as a sum of three Slater-type orbitals:

$$\phi_{nl}(r) = S_{nl}(r, \zeta_1) + d_1 S_{nl}(r, \zeta_2) + d_2 S_{nl}(r, \zeta_3)$$

12–24. How many basis functions are used in a 6-31G basis set calculation of an ethane molecule? How many for a propane molecule?

The 6-31G basis set is a split-valence basis set in which each core orbital is represented by one contracted Gaussian function and each valence orbital is represented by two contracted Gaussian functions. Thus, each hydrogen atom is assigned two $1s$ orbitals, a compact one and an extended one; each carbon atom is assigned one $1s$ orbital, two $2s$ orbitals, and two of each of the $2p$ orbitals. This gives a total of 2 basis functions for each H atom and 9 basis functions for each C atom.

For ethane, C_2H_6, there are $6 \times 2 + 2 \times 9 = 30$ basis functions.

For propane, C_3H_8, there are $8 \times 2 + 3 \times 9 = 43$ basis functions.

12–25. How many basis functions are used in a 6-31G basis set calculation of a benzene molecule? How many for a toluene molecule?

As in the previous problem, there are 2 basis functions for each H atom and 9 basis functions for each C atom in the 6-31G basis set.

For benzene, C_6H_6, there are $6 \times 2 + 6 \times 9 = 66$ basis functions.

For toluene, C_7H_8, there are $8 \times 2 + 7 \times 9 = 79$ basis functions.

12–26. How many primitive Gaussian functions and how many contracted Gaussian functions are used in a 6-31G basis set calculation of an ethane molecule? How many for a propane molecule?

The number of contracted Gaussian functions in the 6-31G basis set for each molecule is determined in Problem 12–24, namely 30 for ethane and 43 for propane.

In the 6-31G basis set, each core orbital is a contraction of six primitive Gaussian functions, each compact valence orbital is a contraction of three primitives, and each extended valence orbital is a single Gaussian primitive function. Thus, for each H atom there are $3 + 1 = 4$ primitive functions used; for each C atom, there are $6 + 3 + 1 + 3 \times 3 + 3 \times 1 = 22$ primitive functions in the basis.

For ethane, C_2H_6, there are $6 \times 4 + 2 \times 22 = 68$ primitive Gaussian functions.

For propane, C_3H_8, there are $8 \times 4 + 3 \times 22 = 98$ primitive Gaussian functions.

12–27. How many primitive Gaussian functions and how many contracted Gaussian functions are used in a 6-31G basis set calculation of a benzene molecule? How many for a toluene molecule?

The number of contracted Gaussian functions in the 6-31G basis set for each molecule is determined in Problem 12–25, namely 66 for benzene and 79 for toluene.

Problem 12–26 gives the number of primitive Gaussian functions assigned to each atom.

For benzene, C_6H_6, there are $6 \times 4 + 6 \times 22 = 156$ primitive Gaussian functions.

For toluene, C_7H_8, there are $8 \times 4 + 7 \times 22 = 186$ primitive Gaussian functions.

12–28. Even though there are only five d orbitals, it is customary in Gaussian basis sets to use six "d" orbitals:

$$g_{xx} = \left(\frac{2048\alpha^7}{9\pi^3}\right)^{1/4} x^2 e^{-\alpha r^2} \qquad g_{yy} = \left(\frac{2048\alpha^7}{9\pi^3}\right)^{1/4} y^2 e^{-\alpha r^2}$$

$$g_{zz} = \left(\frac{2048\alpha^7}{9\pi^3}\right)^{1/4} z^2 e^{-\alpha r^2} \qquad g_{xy} = \left(\frac{2048\alpha^7}{9\pi^3}\right)^{1/4} xy e^{-\alpha r^2}$$

$$g_{xz} = \left(\frac{2048\alpha^7}{9\pi^3}\right)^{1/4} xz e^{-\alpha r^2} \qquad g_{yz} = \left(\frac{2048\alpha^7}{9\pi^3}\right)^{1/4} yz e^{-\alpha r^2}$$

Not all those functions have the same angular symmetry as d orbitals. However, show that g_{xy}, g_{xz}, g_{yz}, $(2g_{zz} - g_{xx} - g_{yy})/2$, and $(3/4)^{1/2}(g_{xx} - g_{yy})$ do have the correct symmetry. Show also that a sixth linear combination, $(g_{xx} + g_{yy} + g_{zz})/5^{1/2}$, has the symmetry of an s orbital.

The angular symmetry of g_{xy} is determined by the product

$$xy = (r \sin\theta \cos\phi)(r \sin\theta \sin\phi) = \frac{r^2 \sin^2\theta \sin 2\phi}{2}$$

which upon comparison with Equation 7.32 has the same angular dependence as a d_{xy} orbital. Similarly, for g_{xz} and g_{yz}:

$$xz = (r \sin\theta \cos\phi)(r \cos\theta) = r^2 \sin\theta \cos\theta \cos\phi$$

$$yz = (r \sin\theta \sin\phi)(r \cos\theta) = r^2 \sin\theta \cos\theta \sin\phi$$

which have the same symmetry as a d_{xz} orbital and a d_{yz} orbital, respectively.

All six functions given have the same radial dependence and normalization constant, so only the angular portions need to be considered for $2g_{zz} - g_{xx} - g_{yy}$; the factor of 1/2 is also irrelevant to the angular properties.

$$2z^2 - x^2 - y^2 = 2(r \cos\theta)^2 - (r \sin\theta \cos\phi)^2 - (r \sin\theta \sin\phi)^2$$

$$= r^2 \left[2\cos^2\theta - \sin^2\theta\left(\cos^2\phi + \sin^2\phi\right)\right]$$

$$= r^2 \left(2\cos^2\theta - \sin^2\theta\right)$$

$$= r^2 \left(3\cos^2\theta - 1\right)$$

Thus, as seen in Equation 7.32, this linear combination has the same angular symmetry of the d_{z^2} orbital.

Similarly for $g_{xx} - g_{yy}$,

$$x^2 - y^2 = (r \sin\theta \cos\phi)^2 - (r \sin\theta \sin\phi)^2$$

$$= r^2 \sin^2\theta \left(\cos^2\phi - \sin^2\phi\right)$$

$$= r^2 \sin^2\theta \cos 2\phi$$

and this linear combination has the same angular symmetry as the $d_{x^2-y^2}$ orbital.

Finally, for $g_{xx} + g_{yy} + g_{zz}$, consider

$$x^2 + y^2 + z^2 = (r \sin\theta \cos\phi)^2 + (r \sin\theta \sin\phi)^2 + (r \cos\theta)^2$$

$$= r^2 \left[\cos^2\theta + \sin^2\theta \left(\cos^2\phi + \sin^2\phi \right) \right]$$

$$= r^2 \left(\cos^2\theta + \sin^2\theta \right)$$

$$= r^2$$

This linear combination has no angular dependence, which is the symmetry of an s orbital.

12–29. Problem 12–28 describes the Gaussian functions that correspond to d orbitals. Rather than use the five combinations that have the correct symmetry for d orbitals, it is easier computationally to use all six Gaussian "d" orbitals in forming basis sets. Given that, show that a 6-31G* basis set calculation of an ethane molecule uses 42 basis functions. How many are used in a similar calculation for a benzene molecule? How many in a 6-31G** basis set calculation for a benzene molecule?

For each carbon atom, the 6-31G* basis set uses the six Gaussian "d" orbitals in addition to those of the 6-31G basis set considered in Problems 12–24 to 12–27. Thus, there are 15 basis functions for each carbon atom in the 6-31G* basis, while 2 basis functions continue to be used for each hydrogen atom.

For ethane, C_2H_6, there are $6 \times 2 + 2 \times 15 = 42$ basis functions used.

For benzene, C_6H_6, there are $6 \times 2 + 6 \times 15 = 102$ basis functions used.

Compared to the 6-31G* basis, the 6-31G** basis set adds an additional three Gaussian p orbitals to each hydrogen atom. This results in 5 basis functions used for each hydrogen atom and 15 for each carbon atom. In this basis set for benzene, C_6H_6, there are $6 \times 5 + 6 \times 15 = 120$ basis functions used.

12–30. Go to the website *http://gnode2.pnl.gov/bse/portal* to obtain the information to write out the contracted Gaussian functions for the 6-31G basis set of a nitrogen atom. Show that the two $2s$ orbitals are normalized. Plot them and show that the one represented by a single Gaussian function is larger in extent than the other.

Using the information from the website, the core $1s$ orbital is given by

$$\phi_{1s} = +0.001\,8348\, g_s(r, 4173.511\,0000) + 0.013\,9950\, g_s(r, 627.457\,9000)$$

$$+ 0.068\,5870\, g_s(r, 142.902\,1000) + 0.232\,2410\, g_s(r, 40.234\,3300)$$

$$+ 0.469\,0700\, g_s(r, 12.820\,2100) + 0.360\,4500\, g_s(r, 4.390\,4370)$$

The compact $2s$ orbital is given by

$$\phi'_{2s} = -0.114\,9610\, g_s(r, 11.626\,3580) - 0.169\,1180\, g_s(r, 2.716\,2800)$$

$$+ 1.145\,8520\, g_s(r, 0.772\,2180)$$

and the extended $2s$ orbital by

$$\phi''_{2s} = g_s(r, 0.212\,0313)$$

where $g_s(r, \alpha)$ is given by Equation 12.21.

The compact $2p_x$ orbital is given by

$$\phi'_{2p_x} = +0.067\,5800\, g_x(r, x, 11.626\,3580) + 0.323\,9070\, g_x(r, x, 2.716\,2800)$$

$$+ 0.740\,8950\, g_x(r, x, 0.772\,2180)$$

and the extended $2p_x$ orbital by

$$\phi''_{2p_x} = g_x(r, x, 0.212\,0313)$$

where $g_x(r, x, \alpha)$ is given by Equation 12.21. There are similar results for the $2p_y$ and $2p_z$ orbitals. Notice that the $2s$ and $2p$ orbitals share the same orbital exponents.

The expressions for the two $2s$ orbitals can be entered into a program such as *Mathematica* or *MathCad*, along with a definition of $g_s(r, \alpha)$, and the normalization integrals evaluated. Doing so gives the results,

$$\int_0^{2\pi} d\phi \int_0^{\pi} d\theta\, \sin\theta \int_0^{\infty} dr\, r^2 \phi'^2_{2s} = 4\pi \int_0^{\infty} dr\, r^2 \phi'^2_{2s} = 1$$

$$\int_0^{2\pi} d\phi \int_0^{\pi} d\theta\, \sin\theta \int_0^{\infty} dr\, r^2 \phi''^2_{2s} = 4\pi \int_0^{\infty} dr\, r^2 \phi''^2_{2s} = 1$$

Thus, both orbitals are normalized. With some effort for the compact orbital, this result could also be obtained by hand.

The two $2s$ orbitals are plotted below, and the extended orbital is seen to extend to larger values of r than the compact orbital.

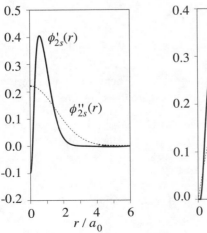 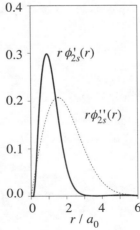

12–31. Go to the website *http://gnode2.pnl.gov/bse/portal* to obtain the information to write out the contracted Gaussian functions for the 6-311G basis set of a carbon atom. Show that the three $2s$ orbitals are normalized. Plot them and show that they are progressively larger in extent.

Using the information from the website, the core $1s$ orbital is given by

$$\phi_{1s} = +0.001\,96665\,g_s(r, 4563.240\,0000) + 0.015\,2306\,g_s(r, 682.024\,0000)$$
$$+ 0.076\,1269\,g_s(r, 154.973\,0000) + 0.260\,8010\,g_s(r, 44.455\,3000)$$
$$+ 0.616\,4620\,g_s(r, 13.029\,0000) + 0.221\,0060\,g_s(r, 1.827\,7300)$$

The compact $2s$ orbital is given by

$$\phi'_{2s} = +0.114\,6600\,g_s(r, 20.964\,2000) + 0.919\,9990\,g_s(r, 4.803\,3100)$$
$$- 0.003\,03068\,g_s(r, 1.459\,3300)$$

the first extended $2s$ orbital by

$$\phi''_{2s} = g_s(r, 0.483\,4560)$$

and the second extended $2s$ orbital by

$$\phi'''_{2s} = g_s(r, 0.145\,5850)$$

where $g_s(r, \alpha)$ is given by Equation 12.21.

The compact $2p_x$ orbital is given by

$$\phi'_{2p_x} = +0.040\,2487\,g_x(r, x, 20.964\,2000) + 0.237\,5940\,g_x(r, x, 4.803\,3100)$$
$$+ 0.815\,8540\,g_x(r, x, 1.459\,3300)$$

the first extended $2p_x$ orbital by

$$\phi''_{2p_x} = g_x(r, x, 0.483\,4560)$$

and the second extended $2p_x$ orbital by

$$\phi'''_{2p_x} = g_x(r, x, 0.145\,5850)$$

where $g_x(r, x, \alpha)$ is given by Equation 12.21. There are similar results for the $2p_y$ and $2p_z$ orbitals. Notice that the $2s$ and $2p$ orbitals share the same orbital exponents.

The expressions for the three $2s$ orbitals can be entered into a program such as *Mathematica* or *MathCad*, along with a definition of $g_s(r, \alpha)$, and the normalization integrals evaluated. Doing so gives the results,

$$\int_0^{2\pi} d\phi \int_0^\pi d\theta\, \sin\theta \int_0^\infty dr\, r^2 \phi'^2_{2s} = 4\pi \int_0^\infty dr\, r^2 \phi'^2_{2s} = 1$$

$$\int_0^{2\pi} d\phi \int_0^\pi d\theta\, \sin\theta \int_0^\infty dr\, r^2 \phi''^2_{2s} = 4\pi \int_0^\infty dr\, r^2 \phi''^2_{2s} = 1$$

$$\int_0^{2\pi} d\phi \int_0^\pi d\theta\, \sin\theta \int_0^\infty dr\, r^2 \phi'''^2_{2s} = 4\pi \int_0^\infty dr\, r^2 \phi'''^2_{2s} = 1$$

Thus, all three orbitals are normalized. With some effort for the compact orbital, this result could also be obtained by hand.

The three $2s$ orbitals are plotted below, and they are seen to extend to progressively larger values of r.

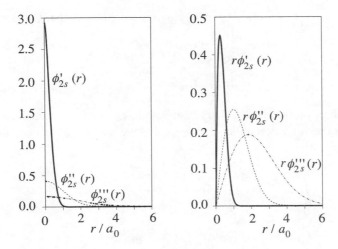

12–32. Go to the website *http://gnode2.pnl.gov/bse/portal* to obtain the information to write out the contracted Gaussian functions for the 6-31G* basis set of an oxygen atom. Show that the two $2s$ orbitals are normalized. Plot them and show that the one represented by a single Gaussian function is larger in extent than the other.

Using the information from the website, the core $1s$ orbital is given by

$$\phi_{1s} = + 0.001\,8311\, g_s(r, 5484.671\,7000) + 0.013\,9501\, g_s(r, 825.234\,9500)$$
$$+ 0.068\,4451\, g_s(r, 188.046\,9600) + 0.232\,7143\, g_s(r, 52.964\,5000)$$
$$+ 0.470\,1930\, g_s(r, 16.897\,5700) + 0.358\,5209\, g_s(r, 5.799\,6353)$$

The compact $2s$ orbital is given by

$$\phi'_{2s} = - 0.110\,7775\, g_s(r, 15.539\,6160) - 0.148\,0263\, g_s(r, 3.599\,9336)$$
$$+ 1.130\,7670\, g_s(r, 1.013\,7618)$$

and the extended $2s$ orbital by

$$\phi''_{2s} = g_s(r, 0.270\,0058)$$

where $g_s(r, \alpha)$ is given by Equation 12.21.

The compact $2p_x$ orbital is given by

$$\phi'_{2p_x} = + 0.070\,8743\, g_x(r, x, 15.539\,6160) + 0.339\,7528\, g_x(r, x, 3.599\,9336)$$
$$+ 0.727\,1586\, g_x(r, x, 1.013\,7618)$$

and the extended $2p_x$ orbital by

$$\phi''_{2p_x} = g_x(r, x, 0.270\,0058)$$

where $g_x(r, x, \alpha)$ is given by Equation 12.21. There are similar results for the $2p_y$ and $2p_z$ orbitals. Notice that the $2s$ and $2p$ orbitals share the same orbital exponents.

Finally, this basis set includes a set of $2d$ polarization orbitals. The $2d_{xx}$ orbital is given by

$$\phi_{2d_{xx}} = g_{xx}(r, x, 0.800\,0000)$$

with similar results for the other five Gaussian "d" orbitals given in Problem 12–28.

The expressions for the two $2s$ orbitals can be entered into a program such as *Mathematica* or *MathCad*, along with a definition of $g_s(r, \alpha)$, and the normalization integrals evaluated. Doing so gives the results,

$$\int_0^{2\pi} d\phi \int_0^{\pi} d\theta \, \sin\theta \int_0^{\infty} dr \, r^2 \phi_{2s}'^2 = 4\pi \int_0^{\infty} dr \, r^2 \phi_{2s}'^2 = 1$$

$$\int_0^{2\pi} d\phi \int_0^{\pi} d\theta \, \sin\theta \int_0^{\infty} dr \, r^2 \phi_{2s}''^2 = 4\pi \int_0^{\infty} dr \, r^2 \phi_{2s}''^2 = 1$$

Thus, both orbitals are normalized. With some effort for the compact orbital, this result could also be obtained by hand.

The two $2s$ orbitals are plotted below, and the extended orbital is seen to extend to larger values of r than the compact orbital.

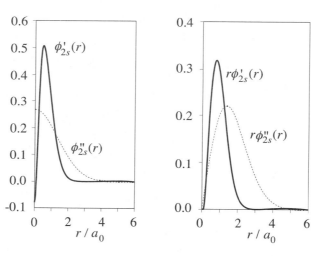

12–33. Go to the website *http://gnode2.pnl.gov/bse/portal* to obtain the STO-3G $1s$ and $2s$ orbitals of a carbon atom and show that they are normalized.

The STO-3G $1s$ and $2s$ orbitals for carbon from the website are

$$\phi_{1s} = +0.154\,328\,97\, g_s(r, 71.616\,8370) + 0.535\,328\,14\, g_s(r, 13.045\,0960)$$
$$+ 0.444\,634\,54\, g_s(r, 3.530\,5122)$$

$$\phi_{2s} = -0.099\,967\,23\, g_s(r, 2.941\,2494) + 0.399\,512\,83\, g_s(r, 0.683\,4831)$$
$$+ 0.700\,115\,47\, g_s(r, 0.222\,2899)$$

The expressions for these two orbitals can be entered into a program such as *Mathematica* or *MathCad*, along with a definition of $g_s(r, \alpha)$, and the normalization integrals evaluated. Doing so gives the results,

$$\int_0^{2\pi} d\phi \int_0^\pi d\theta \, \sin\theta \int_0^\infty dr \, r^2 \phi_{1s}^2 = 4\pi \int_0^\infty dr \, r^2 \phi_{1s}^2 = 1$$

$$\int_0^{2\pi} d\phi \int_0^\pi d\theta \, \sin\theta \int_0^\infty dr \, r^2 \phi_{2s}^2 = 4\pi \int_0^\infty dr \, r^2 \phi_{2s}^2 = 1$$

Thus, both orbitals are normalized. With some effort, this result could also be obtained by hand.

12–34. Show that the $1s, 2s', 2p'_x, 2p'_y,$ and $2p'_z$ orbitals in the 6-31G** basis set given in Figure 12.7 are normalized.

Although the normalization integrals could be evaluated by hand, it is much easier to make use of a program such as *Mathematica* or *MathCad*. Additionally, learning to use such programs through the practice provided by this and other problems in the text will pay dividends in future work.

The s orbitals can be defined in these programs as

$$g_s(r, \alpha) = \left(\frac{2\alpha}{\pi}\right)^{3/4} e^{-\alpha r^2}$$

$$\phi_{1s}(r) = \sum_{i=1}^6 d_{1si} g(r, \alpha_{1i})$$

$$\phi'_{2s}(r) = \sum_{i=1}^3 d_{2si} g(r, \alpha_{2i})$$

where the α's taken from the first column of numbers in Figure 12.7 and the d's taken from the second column are entered into the program as indexed vectors. Next, the following two integrals can be entered and evaluated in spherical polar coordinates,

$$\int_0^{2\pi} d\phi \int_0^\pi d\theta \, \sin\theta \int_0^\infty dr \, r^2 \phi_{1s}^2(r) = 4\pi \int_0^\infty dr \, r^2 \phi_{1s}^2(r) = 1$$

$$\int_0^{2\pi} d\phi \int_0^\pi d\theta \, \sin\theta \int_0^\infty dr \, r^2 \phi_{2s}'^2(r) = 4\pi \int_0^\infty dr \, r^2 \phi_{2s}'^2(r) = 1$$

Whether one has the program evaluate the integrals over the angular coordinates or just enters the factor of 4π leaving only the radial integral for the program is a matter of personal preference. It is also possible to express $r^2 = x^2 + y^2 + z^2$ and evaluate the normalization integrals in Cartesian coordinates. Just remember that the volume element $r^2 \sin\theta \, dr \, d\theta \, d\phi$ becomes $dx \, dy \, dz$.

The normalization integrals for the p functions are handled most straightforwardly in the programs using spherical polar coordinates. Enter the definitions, adapted from Equation 12.21,

$$g_x(r, \theta, \phi, \alpha) = \left(\frac{128\alpha^5}{\pi^3}\right)^{1/4} r \sin\theta \cos\phi e^{-\alpha r^2}$$

$$g_y(r, \theta, \phi, \alpha) = \left(\frac{128\alpha^5}{\pi^3}\right)^{1/4} r \sin\theta \sin\phi e^{-\alpha r^2}$$

$$g_z(r, \theta, \phi, \alpha) = \left(\frac{128\alpha^5}{\pi^3}\right)^{1/4} r \cos\theta e^{-\alpha r^2}$$

$$\phi'_{2p_x}(r, \theta, \phi) = \sum_{i=1}^{3} d_{2pi} g_x(r, \theta, \phi, \alpha_{2i})$$

$$\phi'_{2p_y}(r, \theta, \phi) = \sum_{i=1}^{3} d_{2pi} g_y(r, \theta, \phi, \alpha_{2i})$$

$$\phi'_{2p_z}(r, \theta, \phi) = \sum_{i=1}^{3} d_{2pi} g_z(r, \theta, \phi, \alpha_{2i})$$

where the α's continue to be taken from the first column of numbers in Figure 12.7 (recall that the s and p orbitals share orbital exponents), but the d's are now taken from the third column. Both are entered into the program as indexed vectors. Then enter the following integrals, which are all equal to 1 when evaluated.

$$\int_0^{2\pi} d\phi \int_0^{\pi} d\theta \, \sin\theta \int_0^{\infty} dr \, r^2 \phi'^2_{2p_x}(r, \theta, \phi) = 1$$

$$\int_0^{2\pi} d\phi \int_0^{\pi} d\theta \, \sin\theta \int_0^{\infty} dr \, r^2 \phi'^2_{2p_y}(r, \theta, \phi) = 1$$

$$\int_0^{2\pi} d\phi \int_0^{\pi} d\theta \, \sin\theta \int_0^{\infty} dr \, r^2 \phi'^2_{2p_z}(r, \theta, \phi) = 1$$

Since all the normalization integrals are equal to 1, the orbitals are all normalized.

12–35. If you have access to Gaussian, verify that a 6-311G(d,p) calculation on acetaldehyde uses 78 basis functions and 128 primitive Gaussian functions.

The necessary information is included in the Gaussian output file. If the keyword guess=only is included in the Gaussian job file, then Gaussian only sets up the basis orbitals and does no other calculation. This can be accomplished by checking the Only do guess (no SCF) option in the Guess tab in GaussView. After specifying the 6-311G(d,p) basis set, or equivalently, 6-311G**, constructing the acetaldehyde molecule, and running the job, search the output file for the third occurrence of the word "basis." You will find the following line:

```
78 basis functions, 128 primitive gaussians, 81 cartesian basis functions
```

This verifies the number of basis functions and primitive Gaussian functions used.

For Problems 12–36 to 12–40: Even if you don't have unlimited access to Gaussview or WebMO, you can log into WebMO as a guest and receive one minute of CPU time to run calculations. One minute is quite enough time to run a number of Hartree-Fock calculations on the molecules that we have been considering. Just go to the website www.webmo/net/demo/, then log into the Demo Server and then follow the instructions.

12–36. Use either GaussView or WebMO to confirm the HF/6-31G* entry for a methane molecule in Table 12.6.

After using either program to build the CH_4 molecule, specify the method (or theory) as Hartree-Fock (HF), the basis set as 6-31G(d), which is equivalent to 6-31G*, and the job type as optimization. Submit and run the job. When it is complete, open the result file and view the energy in the calculation summary. One such calculation gives $E = -40.195\,1719\,E_h$, which verifies the result in Table 12.6. There may be slight differences depending on the starting geometry or computer system used.

12–37. Use either GaussView or WebMO to confirm the HF/6-311G* entry for NH_3 in Table 12.7.

After using either program to build the NH_3 molecule, specify the method (or theory) as Hartree-Fock (HF), the basis set as 6-311G(d), which is equivalent to 6-311G*, and the job type as optimization. Submit and run the job. When it is complete, open the result file in the viewer and measure, using the viewer tools, the N–H bond length and HNH angle. One such calculation gives $r_{NH} = 0.998\,892\,55$ Å, and $\theta_{HNH} = 107.420\,448\,84°$. (The quoted results are extracted from the Gaussian output file itself. The viewer may give fewer figures.) The figures agree satisfactorily with the values in Table 12.7.

12–38. Use either GaussView or WebMO to confirm the HF/6-311G* entry for CH_4 in Table 12.7.

After using either program to build the CH_4 molecule, specify the method (or theory) as Hartree-Fock (HF), the basis set as 6-311G(d), which is equivalent to 6-311G*, and the job type as optimization. Submit and run the job. When it is complete, open the result file in the viewer and measure, using the viewer tools, the C–H bond length and HCH angle. One such calculation gives $r_{CH} = 1.083\,005\,43$ Å, and $\theta_{HCH} = 109.471\,220\,63°$. (The quoted results are extracted from the Gaussian output file itself. The viewer may give fewer figures.) The figures agree satisfactorily with the values in Table 12.7.

12–39. Use either GaussView or WebMO to confirm the HF/6-31G* entry for HF in Table 12.8.

After using either program to build the HF molecule, specify the method (or theory) as Hartree-Fock (HF), the basis set as 6-31G(d), which is equivalent to 6-31G*, and the job type as optimization. Submit and run the job. When it is complete, open the result file and view the dipole moment in the calculation summary. One such calculation gives $\mu = 1.9716$ D, which verifies the result in Table 12.8. There may be slight differences depending on the starting geometry or computer system used. Although not necessary for Hartree-Fock calculations, it is important to include the density=current keyword for calculations of the dipole moment and other electronic properties when using post-Hartree-Fock methods.

12–40. Use either GaussView or WebMO to confirm the HF/6-31G* entry for a formaldehyde molecule in Table 12.10.

After using either program to build the H_2CO molecule, specify the method (or theory) as Hartree-Fock (HF), the basis set as 6-31G(d), which is equivalent to 6-31G*, and the job type as optimization. Submit and run the job. When it is complete, open the result file in the viewer and measure, using the viewer tools, the two bond angles. One such calculation gives $\theta_{OCH} = 122.164\,845\,32°$, and $\theta_{HCH} = 115.673\,812\,43°$. (The quoted results are extracted from the Gaussian output file itself. The viewer may give fewer figures.) The figures agree satisfactorily with the values in Table 12.10.

12–41. Use the NIST website *http://srdata.nist.gov/cccbdb* to verify the entries in Table 12.6. (The Hartree–Fock limits are from other sources.)

Use a browser to access the NIST website. Once there, navigate to Calculated data. Scroll down to D.1.a, and click on Energies (equilibrium, no zero-point energy). Enter the formulas of one of the molecules from Table 12.6 in the box and click Submit. The desired information is found in the various columns of the first row of the table labeled Methods with standard basis sets. Repeat for the other three molecules. The values for water differ from those seen in Figure 12.19 because the screen shot is taken from the Energies (equilibrium, no zero-point energy) with heat correction to 298.15 K page of the website, while those in Table 12.6 are from the page referenced in this problem.

12–42. Show that the number of singly excited determinants that occur in a configuration-interaction calculation is given by $2N(2K - 2N)$ for a molecule with $2N$ electrons and $2K$ spin orbitals. Show that the number of n-tuple excitations is given by

$$\binom{2N}{n}\binom{2K - 2N}{n}$$

If there are $2K$ spin orbitals in a molecule with $2N$ electrons, then $2N$ spin orbitals are occupied, leaving $2K - 2N$ spin orbitals empty. A singly excited determinant in a configuration interaction calculation is formed by taking one of the $2N$ electrons from an occupied spin orbital and placing it in one of the $2K - 2N$ vacant spin orbitals. There are $2N$ choices of which electron to take, and $2N - 2K$ choices of where to put it, giving a total of $2N\,(2K - 2N)$ different singly excited determinants.

Similarly in an n-tuple excitation, n electrons are chosen from the $2N$ occupied spin orbitals and n of the $2K - 2N$ vacant spin orbitals are chosen to receive them. The number of ways of choosing n objects out of a total is given by the binomial coefficient. Taking the product of the two independent choices being made gives the desired result,

$$\binom{2N}{n}\binom{2K-2N}{n}$$

12–43. Use the result of the previous problem to show that the number of single and double excitations for a CI calculation on N_2 using a basis set consisting of 100 basis functions is over a million.

The N_2 molecule has 14 electrons, and 100 basis functions gives 200 spin orbitals, so use $N = 7$ and $K = 100$ in the results from the previous problem. The number of single excitations is

$$14(200 - 14) = 2604$$

and the number of double excitations is

$$\binom{14}{2}\binom{200-14}{2} = \frac{14!}{12!\,2!}\frac{186!}{184!\,2!} = 1\,565\,655$$

Thus, the total number of single and double excitations is $2604 + 1\,565\,655 = 1\,568\,259$, which is well over a million.

12–44. Why does Ψ_{CI} in Equation 12.31 truncate at the $2N$th term?

The $2N$th term in Equation 12.31 represents determinants in which all $2N$ electrons are promoted into virtual orbitals. There can be no higher-order excitations than those where all the electrons in the molecule occupy virtual orbitals.

12–45. Why are many more excited terms used in a CCD calculation than a CID calculation?

As Equation 12.37 shows, by using the exponential of the double excitation operator, $\hat{T}_2$, in a CCD calculation, the CCD wave function includes not only all double excitations, but also the quadruple, sextuple, octuple, and further excitations that are in a sense products of double excitations. A CID calculation includes only the double excitations.

12–46. What do you think a CISDQ calculation means?

CISDQ would specify a configuration interaction (CI) calculation in which single (S), double (D), and quadruple (Q) excitations are used.

12–47. In this problem, we'll give a simple proof that CCD is size-consistent but that CID is not. Consider n widely separated molecules, A. Argue that the wave function of these n molecules is given by $\psi(n\mathrm{A}) = \psi^n(\mathrm{A})$ and the energy is $E(n\mathrm{A}) = nE(\mathrm{A})$. For one molecule, CCD gives $\psi(\mathrm{A}) = \exp(\hat{T}_2)\psi_0(\mathrm{A})$, where $\psi_0(\mathrm{A})$ is the reference wave function of an A molecule. If we let the reference wave function of the n widely separated A molecules be $\Psi_0(n\mathrm{A}) = \psi_0(1)\psi_0(2)\cdots\psi_0(n)$, then show that

$$\Psi_{\mathrm{CCD}} = \exp[\hat{T}_2(1) + \hat{T}_2(2) + \cdots + \hat{T}_2(n)]\,\Psi_0(n\mathrm{A})$$

$$= [\exp(\hat{T}_2)\psi_0(\mathrm{A})]^n = \psi^n(\mathrm{A})$$

and that the corresponding energy is $nE(\mathrm{A})$. Now show that in contrast, if $\psi(\mathrm{A}) = (1 + \hat{C}_2)\psi_0(\mathrm{A})$ as in CID, then

$$\Psi_{\mathrm{CID}} = [1 + \hat{C}_2(1) + \cdots + \hat{C}_2(n)]\Psi_0(n\mathrm{A})$$

$$\neq [(1 + \hat{C}_2)\psi_0(\mathrm{A})]^n$$

For n widely separated molecules, A, the total Hamiltonian operator for the system is the sum of the n identical, individual Hamiltonian operators for the isolated molecules,

$$\hat{H}_{n\mathrm{A}} = \sum_{k=1}^{n} \hat{H}_{\mathrm{A}_k}$$

There are no interaction terms in the total Hamiltonian operator between the individual molecules. As discussed in Section 3.9, if a Hamiltonian operator is separable in this manner, then the eigenfunctions of $\hat{H}_{n\mathrm{A}}$ are given by the products of the eigenfunctions of the $\hat{H}_{\mathrm{A}_k}$. That is,

$$\psi(n\mathrm{A}) = \prod_{k=1}^{n} \psi(\mathrm{A}_k) = \psi^n(\mathrm{A})$$

where the last equality follows because the individual $\hat{H}_{\mathrm{A}_k}$ are identical and, consequently, so are their eigenfunctions. Furthermore, the eigenvalues of $\hat{H}_{n\mathrm{A}}$ are the sums of the eigenvalues of the $\hat{H}_{\mathrm{A}_k}$ or,

$$E(n\mathrm{A}) = \sum_{k=1}^{n} E(\mathrm{A}_k) = nE(\mathrm{A})$$

again because the eigenvalues are the same for the identical molecules.

In the system of n widely separated molecules, the effect of the coupled-cluster $\hat{T}_2(n\mathrm{A})$ operator on the reference wave function, $\Psi_0(n\mathrm{A})$, is to generate a weighted sum of all double excitations. These excitations can be considered to be grouped by molecule, or

$$\hat{T}_2(n\mathrm{A}) = \hat{T}_2(1) + \hat{T}_2(2) + \cdots + \hat{T}_2(n)$$

Then, in the CCD approximation (see Equation 12.36)

$$\Psi_{\text{CCD}} = \exp\left[\hat{T}_2(n\text{A})\right]\Psi_0(n\text{A})$$

$$= \exp\left[\hat{T}_2(1) + \hat{T}_2(2) + \cdots + \hat{T}_2(n)\right]\psi_0(1)\psi_0(2)\cdots\psi_0(n)$$

$$= \left\{\exp\left[\hat{T}_2(1)\right]\psi_0(1)\right\}\left\{\exp\left[\hat{T}_2(2)\right]\psi_0(2)\right\}\cdots\left\{\exp\left[\hat{T}_2(n)\right]\psi_0(n)\right\}$$

$$= \left[\exp\left(\hat{T}_2\right)\psi_0(\text{A})\right]^n$$

$$= \psi^n(\text{A})$$

once again because the n widely separated molecules are identical. Thus, the CCD wave function for the n widely separated molecules is the nth power of the CCD wave function for a single, isolated molecule. It follows that the CCD energy of the n-molecule system will be n times the energy of a single molecule, and CCD is size-consistent.

For CID, on the other hand, although it is still possible to write the double excitation operator for the n-molecule system as a sum over the individual molecules,

$$\hat{C}_2(n\text{A}) = \hat{C}_2(1) + \cdots + \hat{C}_2(n)$$

the effect on the reference wave function is different.

$$\Psi_{\text{CID}} = \left[1 + \hat{C}_2(1) + \cdots + \hat{C}_2(n)\right]\Psi_0(n\text{A})$$

$$= \Psi_0(n\text{A}) + \left[\hat{C}_2(1)\psi_0(1)\right]\psi_0(2)\cdots\psi_0(n)$$

$$+ \psi_0(1)\left[\hat{C}_2(2)\psi_0(2)\right]\cdots\psi_0(n) + \cdots + \psi_0(1)\psi_0(2)\cdots\left[\hat{C}_2(n)\psi_0(n)\right]$$

$$= \psi_0^n(\text{A}) + \sum_{k=1}^{n}\psi_0^{n-1}(\text{A})\hat{C}_2\psi_0(\text{A})$$

$$= \psi_0^{n-1}(\text{A})\left[\left(1 + n\hat{C}_2\right)\psi_0(\text{A})\right]$$

$$\neq \left[\left(1 + \hat{C}_2\right)\psi_0(\text{A})\right]^n$$

Thus, CID is not size consistent.

12–48. Correct the values of ΔH_r° at 298 K that we obtained for the dissociation reaction $\text{HCl}(g) \rightarrow \text{H}(g) + \text{Cl}(g)$ in Section 12.7 for the effect of the zero-point vibrational energy of $\text{HCl}(g)$ and the spin-orbit interaction energy of $\text{Cl}(g)$. (Go to the NIST cccbdb website for the values of these quantities.)

The zero-point vibrational energy of $\text{HCl}(g)$ and the spin-orbit interaction energy of $\text{Cl}(g)$ for the three model chemistries considered are found at the cccbdb website and collected in the following table. Since the zero-point vibrational energy of $\text{HCl}(g)$ is "extra" energy already present in the reactant, its effect is to reduce the calculated enthalpy of reaction. The spin-orbit interaction energy of $\text{Cl}(g)$ lowers the energy of a product, and it, too, reduces the calculated value. The total correction, converted to kJ·mol^{-1} is given in the table along with the ΔH_r° value from the text, and the corrected value. (In the first printing, the ΔH_r° values calculated on page 649 in the text for the two post-Hartree-Fock methods had the products and reactants switched, and they should be multiplied by -1. Additionally, on page 648, the experimental values given for CISD/cc-p-VTZ and CCSD/cc-pVTZ were also switched.)

	HF/cc-pVTZ	CISD/cc-pVTZ	CCSD/cc-pVTZ
Zero-point energy HCl(g)/ E_h	0.006 508	0.006 412	0.006 490
Spin-orbit energy Cl(g)/ E_h	$-0.001\,338$	$-0.001\,338$	$-0.001\,338$
Total correction / kJ·mol^{-1}	-20.60	-20.35	-20.55
ΔH_r°(calc)/ kJ·mol^{-1}	323.1	422.5	433.0
ΔH_r°(corr)/ kJ·mol^{-1}	302.5	402.2	412.4

12–49. Go to the NIST cccbdb website and determine the value of ΔH_r° at 298 K for the dissociation reaction HF(g) $\rightarrow$ H(g) $+$ F(g) from the Reaction Data section. Now calculate ΔH_r° from the values of ΔH_f° of the individual species and compare your results. Why do they differ?

The values can be determined from the website by following the directions just after Equation 12.39 in the text, substituting F(g) for Cl(g). On the resulting page, the experimental value of $\Delta H_\text{r}^\circ = 570.68$ kJ·mol^{-1} is found, as well as the errors associated with calculations done using various model chemistries. The ΔH_f° values for the individual species are found by clicking on Experimental data and then Enthalpy of formation to see a page asking for the species desired. The following table contains the data so obtained.

Species	ΔH_f°/ kJ·mol^{-1}
HF(g)	-273.30
H(g)	$+218.00$
F(g)	$+79.38$

The value of ΔH_r° calculated from these data is

$$\Delta H_\text{r}^\circ = \Delta H_\text{f}^\circ \,[\text{H}(g)] + \Delta H_\text{f}^\circ \,[\text{F}(g)] - \Delta H_\text{f}^\circ \,[\text{HF}(g)]$$

$$= +218.00 \text{ kJ·mol}^{-1} + 79.38 \text{ kJ·mol}^{-1} - \left(-273.30 \text{ kJ·mol}^{-1}\right)$$

$$= +570.68 \text{ kJ·mol}^{-1}$$

There is no difference between the value calculated from the values of ΔH_f° for the individual species and the value determined from the website. This is, in fact, expected for experimental data. Had theoretical results been used instead, then it would be necessary to include corrections for zero-point energy and spin-orbit interaction energy, as appropriate for the various species, to achieve agreement.

12–50. Use the NIST cccbdb website to calculate the standard molar enthalpy of combustion of propane at 298 K with a CCSD/cc-pVTZ model chemistry and compare your result to the experimental value.

The combustion reaction for propane is

$$C_3H_8(g) + 5O_2 \rightarrow 3CO_2(g) + 4H_2O(g)$$

Follow the directions just after Equation 12.39 and then enter the reactants and products for this problem. Proceed to the next page to enter the stoichiometric coefficients that balance the chemical equation, check the balancing, and then continue to the result page. The experimental value is $\Delta H^\circ_{comb} = -2043.1 \pm 0.6 \text{ kJ} \cdot \text{mol}^{-1}$. The page does not contain results for the combustion of propane using the CCSD/cc-pVTZ model chemistry. The only CCSD calculation listed uses the 6-31G* basis and gives a value of $\Delta H^\circ_{comb} = -1750.62 \text{ kJ} \cdot \text{mol}^{-1}$. Of the several calculation listed using the cc-pVTZ basis set, the frozen core MP2 calculation provides a value within $40 \text{ kJ} \cdot \text{mol}^{-1}$ of the experimental one. Other MP2 calculations with Pople basis sets do even better.

12–51. Derive Equation 12.41.

Since the N electrons are indistinguishable, as reflected in the determinantal nature of the molecular wave function, the expectation value of any single one-electron operator is identical to that of any other electron, so that the sum is over N identical terms. Furthermore, the operators affect only the spatial portion of the wave function. Thus,

$$\left\langle \psi \left| \hat{A} \right| \psi \right\rangle = \sum_{i=1}^{N} \int \cdots \int \psi^*(x_1, \ldots, \sigma_N) \hat{A}(x_i, y_i, z_i) \psi(x_1, \ldots, \sigma_N) \, d\tau_1 \, d\tau_2 \cdots d\tau_N$$

$$= \sum_{i=1}^{N} \int \int \int dx_i \, dy_i \, dz_i \, \hat{A}(x_i, y_i, z_i)$$

$$\times \int \cdots \int \psi^*(x_1, \ldots, \sigma_N) \psi(x_1, \ldots, \sigma_N) \, d\tau_1 \, d\tau_2 \cdots d\tau_{i-1} \, d\sigma_i \, d\tau_{i+1} \cdots d\tau_N$$

$$= N \int \int \int dx_1 \, dy_1 \, dz_1 \, \hat{A}(x_1, y_1, z_1)$$

$$\times \int \cdots \int \psi^*(x_1, \ldots, \sigma_N) \psi(x_1, \ldots, \sigma_N) \, d\sigma_1 \, d\tau_2 \cdots d\tau_N$$

$$= \int \int \int dx \, dy \, dz \, \hat{A}(x, y, z) \rho(x, y, z)$$

where

$$\rho(x, y, z) = N \int \cdots \int d\sigma_1 \, d\tau_2 \cdots d\tau_N \, \psi^*(x, y, z, \sigma_1, x_2, \ldots, \sigma_N) \psi(x, y, z, \sigma_1, x_2, \ldots, \sigma_N)$$

12–52. The classic example of a functional is a definite integral. Give two other examples of functionals.

The number of maxima or zeros of a function, $f(x)$, in an interval are two other examples of a mapping from a function to a number; that is, a functional.

12–53. Why might you not expect that the energy can be expressed as a functional of the electron density?

_____

Because the interelectronic interaction terms in the Hamiltonian operator are two-electron operators ($1/r_{ij}$), it is not obvious that the energy of a molecule can be expressed as a functional of the electron density.

12–54. The units of dipole moment given by Gaussian are called debyes (D, after the Dutch-American chemist, Peter Debye, who was awarded the Nobel Prize in Chemistry in 1936 for his work on dipole moments). One debye is equal to 10^{-18} esu·cm, where esu (electrostatic units) is the non-SI unit for electric charge. Given the protonic charge is 4.803×10^{-10} esu, show that the conversion factor between debyes and C·m (coulomb·meters) is $1\,D = 3.33 \times 10^{-30}$ C·m.

_____

The protonic charge, 4.803×10^{-10} esu $= 1.602\,177\,33 \times 10^{-19}$ C. Thus,

$$1\,D = 10^{-18} \text{ esu·cm} \left(\frac{1.602\,177\,33 \times 10^{-19} \text{ C}}{4.803 \times 10^{-10} \text{ esu}} \right) \left(\frac{1\,\text{m}}{100\,\text{cm}} \right)$$

$$= 3.335\,785 \times 10^{-30} \text{ C·m}$$

$$= 3.336 \times 10^{-30} \text{ C·m}$$

12–55. Show that a dipole of one debye (1 D) is equivalent to 0.39345 au. (See the previous problem for the definition of a debye.)

_____

The atomic unit for length is the Bohr radius, and that for charge is the protonic charge. Thus,

$$1\,D = 3.335\,785 \times 10^{-30} \text{ C·m} \left(\frac{1}{1.602\,177\,33 \times 10^{-19} \text{ C}} \right) \left(\frac{1}{5.291\,772\,49 \times 10^{-11} \text{ m}} \right)$$

$$= 0.393\,45 \text{ au}$$

12–56. The Gaussian output for a water molecule calculated with an HF/6-31G* model chemistry gives $R(\text{OH}) = 0.947$ Å, $A(\text{HOH}) = 105.5°$, and the partial charges on the hydrogen atoms as $+0.41$ au and on the oxygen atom as -0.82 au. Use these values to calculate the dipole moment of a water molecule and compare your result to the experimental value.

_____

The calculated dipole moment for the water molecule is given by,

$$\mu = \sum_i q_i \mathbf{r}_i$$

where q_i is the partial charge on the ith atom, $\mathbf{r}_i$ is its position, and the sum is taken over all atoms. For a neutral species, the dipole moment is independent of the choice of origin, and it is convenient to place the oxygen atom at the origin. There is only one non-vanishing component of the dipole moment vector for water, which can be taken as μ_x. In atomic units, $R_{OH} = 1.79a_0$, and

$$\mu_x = \sum_i q_i x_i$$

$$= -0.82\, e(0\, a_0) + 0.41\, e\left(1.79\, a_0 \cos \frac{105.5°}{2}\right) + 0.41\, e\left(1.79\, a_0 \cos \frac{105.5°}{2}\right)$$

$$= 0.888\, e\, a_0 = 2.26\, \mathrm{D}$$

where the result of Problem 12–55 is used to convert the calculated dipole moment from atomic units to Debye. This result can be compared with the experimental value of 1.85 D. As expected, the Hartree-Fock calculation does not give a result that is in good agreement with experiment.

12–57. The Gaussian output for an ammonia molecule calculated with an HF/6-31G* model chemistry gives $R(NH) = 1.000$ Å, $A(HNH) = 107.1°$, and the partial charges on the hydrogen atoms as +0.37 au and that on the nitrogen atom as −1.11 au. Use these values to calculate the dipole moment of an ammonia molecule and compare your result to the experimental value.

The calculated dipole moment for the ammonia molecule is given by,

$$\mu = \sum_i q_i \mathbf{r}_i$$

where q_i is the partial charge on the ith atom, $\mathbf{r}_i$ is its position, and the sum is taken over all atoms. For a neutral species, the dipole moment is independent of the choice of origin, and it is convenient to place the nitrogen atom at the origin. There is only one non-vanishing component of the dipole moment vector for ammonia, which is along the three-fold symmetry axis, and can be taken as μ_z. In atomic units, $R_{NH} = 1.890a_0$. Finding the angle, θ, between the N–H bonds and the symmetry axis of the molecule in terms of the HNH angle, α, that is given requires a little solid geometry. The desired relation is,

$$\cos\theta = \sqrt{\frac{2\cos\alpha + 1}{3}} = 0.370\,55$$

Use this in the expression for the dipole moment to give,

$$\mu_z = \sum_i q_i z_i$$

$$= -1.11\, e(0\, a_0) + 0.37\, e\left(1.890\, a_0 \cos\theta\right) + 0.37\, e\left(1.890\, a_0 \cos\theta\right) + 0.37\, e\left(1.890\, a_0 \cos\theta\right)$$

$$= 0.777\, e\, a_0 = 1.98\, \mathrm{D}$$

where the result of Problem 12–55 is used to convert the calculated dipole moment from atomic units to Debye. This result can be compared with the experimental value of 1.47 D. As expected, the Hartree-Fock calculation does not give a result that is in good agreement with experiment.

SOME MATHEMATICAL FORMULAS

$\sin(x \pm y) = \sin x \cos y \pm \cos x \sin y$

$\cos(x \pm y) = \cos x \cos y \mp \sin x \sin y$

$\sin x \sin y = \dfrac{1}{2} \cos(x - y) - \dfrac{1}{2} \cos(x + y)$

$\cos x \cos y = \dfrac{1}{2} \cos(x - y) + \dfrac{1}{2} \cos(x + y)$

$\sin x \cos y = \dfrac{1}{2} \sin(x + y) + \dfrac{1}{2} \sin(x - y)$

$e^{\pm ix} = \cos x \pm i \sin x$

$\cos x = \dfrac{e^{ix} + e^{-ix}}{2} \qquad \sin x = \dfrac{e^{ix} - e^{-ix}}{2i}$

$\cosh x = \dfrac{e^{x} + e^{-x}}{2} \qquad \sinh x = \dfrac{e^{x} - e^{-x}}{2}$

$f(x) = f(a) + f'(a)(x - a) + \dfrac{1}{2!} f''(a)(x - a)^2 + \dfrac{1}{3!} f'''(a)(x - a)^3 + \cdots$

$e^x = 1 + x + \dfrac{x^2}{2!} + \dfrac{x^3}{3!} + \dfrac{x^4}{4!} + \cdots$

$\cos x = 1 - \dfrac{x^2}{2!} + \dfrac{x^4}{4!} - \dfrac{x^6}{6!} + \cdots$

$\sin x = x - \dfrac{x^3}{3!} + \dfrac{x^5}{5!} - \dfrac{x^7}{7!} + \cdots$

$\ln(1 + x) = x - \dfrac{x^2}{2} + \dfrac{x^3}{3} - \dfrac{x^4}{4} + \cdots \qquad -1 < x \le 1$

$\dfrac{1}{1 - x} = 1 + x + x^2 + x^3 + x^4 + \cdots \qquad x^2 < 1$

$(1 \pm x)^n = 1 \pm nx + \dfrac{n(n - 1)}{2!} x^2 \pm \dfrac{n(n - 1)(n - 2)}{3!} x^3 + \cdots \qquad x^2 < 1$

$\displaystyle\int_0^{\infty} x^n e^{-ax} dx = \dfrac{n!}{a^{n+1}} \qquad (n \text{ positive integer})$

$\displaystyle\int_0^{\infty} e^{-ax^2} dx = \left(\dfrac{\pi}{4a} \right)^{1/2}$

$\displaystyle\int_0^{\infty} x^{2n} e^{-ax^2} dx = \dfrac{1 \cdot 3 \cdot 5 \cdots (2n - 1)}{2^{n+1} a^n} \left(\dfrac{\pi}{a} \right)^{1/2} \qquad (n \text{ positive integer})$

$\displaystyle\int_0^{\infty} x^{2n+1} e^{-ax^2} dx = \dfrac{n!}{2a^{n+1}} \qquad (n \text{ positive integer})$